Biotechnology – the Making of a Globe

Biotechnology is one of the fastest growing areas of scientific, technical and industrial innovation; it is also one of the most widely publicised new technologies. Innovations as varied as the development of genetic testing and therapies, genetically modified food crops, animal and stem-cell cloning have given rise to increasingly prominent public debates. Genetically modified soya and Dolly the sheep – to mention the two most prominent events – led to world-wide controversies. While the biotechnology industry initially assumed that regulatory processes were the sole hurdle, it is now apparent that a second hurdle, national and international public opinion, must be reckoned with. *Biotechnology – the Making of a Global Controversy* brings together key findings from a comparative study of public perceptions, media coverage and regulatory frameworks of a newly emergent technology. With chapters from leading international experts in the field, the book contributes important empirical and conceptual analyses to this crucial public debate.

Martin W. Bauer is Senior Lecturer in Social Psychology at the London School of Economics and Research Fellow at the Science Museum London. He co-coordinated the research presented in this volume. His research interests are the functions of resistance in social processes; longitudinal indicators of public understanding of science through media and survey research; the changing representations of science and technology in texts and images. His publications include *Resistance to New Technology: Nuclear Power, Information Technology, Biotechnology* (Cambridge, 1995).

George Gaskell is Professor of Social Psychology at the London School of Economics and Director of the School's Methodology Institute. His research interests include risk as a representation; expert and lay understanding of risk and uncertainty; the structure of public perception of biotechnology; how public opinion is shaped by and shapes policy and regulation; public opinion and regulation; and the democratisation of science and technology policy. He is currently coordinator of the research project 'Life Sciences in European Society', funded under the European Commission's 5th Framework Programme.

The coordinators' co-publications include *Qualitative Researching with Text, Image and Sound* (2000), *Biotechnology 1996–2000: the Years of Controversy* (2001) and *Biotechnology in the Public Sphere: a European Sourcebook* (1998).

Biotechnology – the Making of a Global Controversy

edited by

M. W. Bauer and G. Gaskell

CAMBRIDGE
UNIVERSITY PRESS

Published in association with the Science Museum, London

PUBLISHED BY THE PRESS SYNDICATE OF THE UNIVERSITY OF CAMBRIDGE
The Pitt Building, Trumpington Street, Cambridge, United Kingdom

CAMBRIDGE UNIVERSITY PRESS
The Edinburgh Building, Cambridge CB2 2RU, UK
40 West 20th Street, New York, NY 10011–4211, USA
477 Williamstown Road, Port Melbourne, VIC 3207, Australia
Ruiz de Alarcón 13, 28014 Madrid, Spain
Dock House, The Waterfront, Cape Town 8001, South Africa

http://www.cambridge.org

First published 2002

Printed in the United Kingdom at the University Press, Cambridge

Typeface Plantin 10/12 pt   *System* LaTeX 2_ε [TB]

A catalogue record for this book is available from the British Library

Library of Congress Cataloging-in-Publication data
Biotechnology : the making of a global controversy / M. W. Bauer &
G. Gaskell, eds.
    p.   cm.
Includes bibliographical references and index.
ISBN 0-521-77317-2 – ISBN 0-521-77439-X (pb.)
1. Biotechnology – Europe – Public opinion.   2. Mass media – Europe.
3. Biotechnology – Social aspects.   I. Bauer, Martin W.   II. Gaskell, George.
TP248.23.B563   2002
660.6 – dc21   2001052636

ISBN 0 521 77317 2 hardback
ISBN 0 521 77439 X paperback

Contents

Contents vii

MARTINA LEONARZ, FEDERICA MANZOLI, ANNA
OLOFSSON, ANDRZEJ PRZESTALSKI, TIMO RUSANEN,
FRANZ SEIFERT, ANGELIKI STATHOPOULOU AND
WOLFGANG WAGNER

Plates

Figures

Tables

List of contributors

AGNES ALLANSDOTTIR – Department of Communication Science, University of Siena, Italy

NICK ALLUM – Methodology Institute, London School of Economics and Political Science, UK

*MARTIN W. BAUER – Department of Social Psychology, London School of Economics and Political Science, UK

MARIE-LOUISE VON BERGMANN-WINBERG – Faculty of Interdisciplinary Research, Mid Sweden University, Oestersund, Sweden

ANNE BERTHOMIER – Laboratoire Communication et Politique, Centre National de la Recherche Scientifique (CNRS), Paris, France

HEINZ BONFADELLI – Institute of Mass Communication and Media Research, University of Zürich, Switzerland

DANIEL BOY – Maison des Sciences de l'Homme, Centre d'Études de la Vie Politique Française (Cevipof), Paris, France

ELEANOR BRIDGMAN – Public Understanding of Science Group, Science Museum, London

AIGLI CHATJOULI – Department of Biological Research and Biotechnology, Office of Science Communication and Bioethics, National Hellenic Research Foundation, Athens, Greece

SUZANNE DE CHEVEIGNÉ – Laboratoire Communication et Politique, Centre National de la Recherche Scientifique (CNRS), Paris, France

CARMEN DIEGO – Department of Sociology, ISCTE, University of Lisbon, Portugal

ROBIN DOWNEY – Faculty of General Studies, University of Calgary, Canada

* Co-ordinators of the project

JOHN DURANT – Director, At-Bristol, Bristol, UK

EDNA F. EINSIEDEL – Faculty of General Studies, University of Calgary, Canada

BJÖRN FJÆSTAD – Mid Sweden University, Stockholm, Sweden

HELLE FREDERIKSEN – Department of Communication, Education and Computer Science, Roskilde University, Denmark

JEAN-CHRISTOPHE GALLOUX – University of Versailles Saint-Quentin and Director of the Centre de Recherche en Droit de Technologies de l'Information et du Vivant (CRDTIV)

*GEORGE GASKELL – Methodology Institute, London School of Economics and Political Science, UK

ALEXANDER GOERKE – Faculty of Social and Behaviour Sciences, Friedrich Schiller University, Jena, Germany

PETRA GRABNER – Department of Political Science, University of Salzburg, Austria

JAN M. GUTTELING – Department of Communication Studies, University of Twente, Enschede, The Netherlands

JÜRGEN HAMPEL – Centre of Technology Assessment in Baden-Württemberg, Stuttgart, Germany

MARCUS HEINßEN – Centre of Technology Assessment in Baden-Württemberg, Stuttgart, Germany

PETRA HIEBER – Department of Social and Economic Psychology, University of Linz, Austria

ERLING JELSØE – Department of Environment, Technology and Social Studies, Roskilde University, Denmark

MERCI WAMBUI KAMARA – Department of Environment, Technology and Social Studies, Roskilde University, Denmark

MATTHIAS KOHRING – Department of Media Science, Friedrich Schiller University, Jena, Germany

NICOLE KRONBERGER – Department of Social and Economic Psychology, University of Linz, Austria

JESPER LASSEN – Department of Development and Planning, Aalborg University, Denmark

MARTINA LEONARZ – Institute of Mass Communication and Media Research, University of Zürich, Switzerland

MILTOS LIAKOPOULOS – Europäische Akademie zur Erforschung von Folgen wissenschaftlich-technischer Entwicklungen, Bad Neuwahr-Ahrwiler, Germany

FREDERICA MANZOLI – Department of Communication Science, University of Siena, Italy

ATHENA MAROUDA-CHATJOULIS – Institute of Applied Psychology, Department of Communication and Mass Media, University of Athens, Greece

CEES J. H. MIDDEN – Department of Technology Management, Eindhoven University of Technology, The Netherlands

JON D. MILLER – Centre for Biomedical Communications, Northwestern University, Chicago, USA

ARNE THING MORTENSEN – Department of Communications, Roskilde University, Denmark

TORBEN HVIID NIELSEN – Centre for Technology and Culture, University of Oslo, Norway

MARIANNE ODEGAARD – Centre for Technology and Culture, University of Oslo, Norway

SUSANNA ÖHMAN – Mid Sweden University, Oestersund, Sweden

ANNA OLOFSSON – Mid Sweden University, Oestersund, Sweden

ANDRZEJ PRZESTALSKI – Institute of Sociology, Adam Mickiewicz University, Poznan, Poland

BIANCA RIZZO – Department of Communication, University of Siena, Italy

GEORG RUHRMANN – Department of Media Science, Friedrich Schiller University, Jena, Germany

MARIA RUSANEN – Department of Social Sciences, University of Kuopio, Finland

TIMO RUSANEN – Department of Social Sciences, University of Kuopio, Finland

GEORGE SAKELLARIS – Institute of Biological Research, National Hellenic Research Foundation, Athens, Greece

MICHAEL SCHANNE – AGK Communication Consulting, Zürich, Switzerland

FRANZ SEIFERT – Institute of Advanced Studies, University of Vienna, Austria

G. CARLA J. SMINK – SWOKA Institute for Consumer Research, The Netherlands

ANGELIKI STATHOPOULOU – Department of Qualitative Research, Metron Analysis S.A., Athens, Greece

PAUL THOMPSON – Department of Philosophy, Purdue University, Lafayette, Indiana, USA

HELGE TORGERSEN – Institut für Technikfolgen Abschätzung, Österreichische Akademie der Wissenschaften, Vienna, Austria

TOMASZ TWARDOWSKI – Institute of Bioorganic Chemistry, Polish Academy of Sciences, Poznan, Poland

WOLFGANG WAGNER – Institut für Pädagogie und Psychologie, Johannes-Kepler-University, Linz, Austria

Acknowledgements

The research for this book was funded by the Fourth (EU) RTD Framework, Concerted Actions BIO4-CT98-0488 (DG12) and BIO4-CT95-0043 (DG12-SSMA), granted to John Durant at the Science Museum, London, UK. The book was written with the support of the European Commission, Research Directorate (QLRT-1999-00286). The authors and contributors acknowledge a debt of gratitude to John Durant, who played a vital role in the research but who has moved on to new challenges. Thanks also to Jane Gregory for her English language work on the chapters.

Additional funding was received from the following institutions:

Austria: Österreichischer Fonds zur Förderung der wissenschaftlichen Forschung, P11849-SOZ, and Institute of Technology Assessment, Austrian Academy of Sciences.

Canada: Social Sciences and Humanities Research Council (Science and Technology Policy Strategic Grants Program).

Denmark: Danish Social Science Research Council for research within the project 'Biotechnology and the Danish Public'.

France: Ministry of Agriculture (DGAL), INRA and CNRS (SHS and Programme 'Risques collectifs et situations de crise').

Germany: Deutsche Forschungsgemeinschaft (DFG), contracts 467/2-1 and 467/2-2.

Poland: Polish State Committee for Research.

Sweden: Mid Sweden University, the Freja Foundation, the Erinaceidæ Foundation, the Megn Bergvall Foundation and the Swedish Natural Science Research Council.

United Kingdom: Leverhulme Trust, contract F/4/BG, 'Biotechnology and the British Public'.

The publisher has used its best endeavours to ensure that the URLs for external websites referred to in this book are correct and active at the time of going to press. However, the publisher has no responsibility for the websites and can make no guarantee that a site will remain live or that the content is or will remain appropriate.

1 Researching the public sphere of biotechnology

Martin W. Bauer and George Gaskell

Biotechnology is one of the fastest-growing areas of scientific, technical and industrial innovation of recent times, and it is also one of the most prominent in public discussion. Following the development of recombinant DNA techniques in the early 1970s, modern biotechnology has burgeoned in diverse areas including pharmaceuticals, diagnostics and testing, cloning and xenotransplantation, genetically modified seeds and foods and environmental remediation. Such is the breadth of impacts across previously unrelated sectors that a new collective category, 'the life sciences', has been adopted within the industrial and scientific communities. Accompanying this research, development and commercial exploitation has been a widening range and growing intensity of public debates. These have featured issues such as the use of genetic information, the labelling of genetically modified foods, intellectual property rights, the privatisation of research activities and biodiversity, but these have also been paralleled by more fundamental considerations of the rights and wrongs of modern biotechnology as a whole.

While debates on these issues have appeared, and indeed disappeared, in different countries and at different times in Europe and the United States, modern biotechnology has become increasingly sensitive, socially and politically. In contemporary times, public opinion is not merely a perspective 'after the fact'; it is a crucial constraint, in the dual sense of the limitations and opportunities for governments and industries to exploit the new technology. Whereas the biotechnology industry assumed that regulatory processes were the sole hurdle prior to commercialisation, it is now apparent that a second hurdle, national and international public opinion, must be taken into account.

This book takes up themes explored at a conference at the Science Museum, London, in 1993, which was convened to explore the structures and functions of resistance in the development of new technologies. At that meeting, three base technologies of the post-war years were contrasted: nuclear power, information technology and genetic engineering. The main thesis of the conference was that resistance is not a problem residing in the public, rather it is a signal that something is going wrong with the technology; and that resistance acts as a catalyst for organisational and institutional learning (Bauer, 1995). With an exclusive focus on biotechnology, this idea is further developed and expanded in this volume.

Here we present results of a four-year international research project conducted between 1996 and 1999. The project, 'Biotechnology and the European Public', brought together social scientists from a variety of different disciplines, including science and technology studies, sociology, social philosophy and psychology, consumer behaviour, communication science and political science. All the members of the project, based in fourteen countries – Austria, Denmark, Finland, France, Germany, Greece, Italy, the Netherlands, Poland, Sweden, Switzerland and the UK, with associates from Canada and the USA – had at least one thing in common. They shared a keen interest in monitoring and understanding the reception of modern biotechnology in the public spheres of Europe and North America. In the research, the public sphere is defined as the intersections between public opinion as evidenced in public perceptions and media coverage, and regulation and policy-making. The objective was to chart the dynamics of public opinion and regulatory activity that accompanied the development of biotechnology, from its beginnings in 1973 until 1996, in a multinational and comparative framework.

In our previous book (Durant, Bauer and Gaskell, 1998), we published the basic empirical data together with descriptive commentaries on the national developments in regulation and public opinion. In the present book, we step back from the data and reflect on biotechnology in Europe and North America in the years up to 1996/97. With the benefit of hindsight, this proved to be a watershed in the development of this strategic technology. Towards the end of 1996 the annual cargo of American soya was shipped into Europe. For the first time this was a crop of soya that included genetically modified soya (GM soya) grown from engineered seeds made resistant to Roundup herbicide. The seeds for this new GM soya were developed by the American multinational company Monsanto, whose name became synonymous with GM products. In the heady days of new developments of genetically modified seeds, with their promise to introduce a second green revolution, Monsanto may have been pleased to see their name as the brand leader. But this unusual cargo, intended or unintended as it may have been, had consequences that changed the image of biotechnology among the European public, and spilled over into the other parts of the globe. A few months later, in February 1997, the Scottish veterinary research station at Edinburgh claimed to have achieved what hitherto had been thought impossible: it had transferred the genetic material from an adult sheep to a uterus cell, and raised a cloned, genetically identical, offspring. 'Dolly the sheep' turned science fiction into a reality. Both these events, though local in character, became global markers and symbols of the genetic society cultivating contrary visions of progress and awe against doom and anxiety.

At other times, these two events might have quickly evaporated into thin air following the knee-jerk reactions of sensationalist mass media. But it is significant that they followed a slow build-up of public debate and concern about biotechnology that had been rumbling since the early 1970s. This book demonstrates how the build-up of public awareness and information, and the contrasting euphoria and gloom that accompanied the early developments of biotechnology, set the context for the reception of these two events. By understanding this earlier period, we can appreciate why the two events achieved their significance, and how that significance influenced the development of a global biotechnology controversy in the latter part of the 1990s. In sum, we explore the preconditions of what historians may come to call the 'great European biotechnology debate' which unfolded in the last few years of the twentieth century concerning a novel technology with commercial applications in the fields of crops and food production and pharmaceuticals and medicines.

Throughout this study, the term biotechnology is generally used to mean 'modern biotechnology', i.e. those processes, products and services that have been developed on the basis of interventions at the level of the gene. In the literature, modern biotechnology thus defined is generally contrasted with 'traditional biotechnology', i.e. those processes, products and services that have been developed on the basis of interventions at the level of the cell, tissue or whole organism. Although these are justifiable distinctions from the point of view of the biotechnologist, it is important to note that, in any particular social situation, they may or may not accord with the representations of biotechnology in the public sphere. Since the public sphere is our principal object of interest, it is vital that we acknowledge what the public understands by biotechnology, irrespective of whether this 'lay definition' complies with scientific definitions.

The fact that the project dealt with eight different languages led to some semantic challenges. The denotation of 'biotechnology and new genetics' has to be recovered from a changing lexicon of words and phrases both across languages and over time. It became clear, for example, that as the technology has developed so too has the vocabulary that denotes it. In English, for example, the term 'recombinant DNA' (rDNA) was current in the 1970s but disappeared later on; the term 'biotechnology' was not commonly used until around -1980; and terms such as 'genetic engineering', 'genetic manipulation' and 'genetic modification' all appear and disappear later on, in what seems to be a complex game played with semantics in the public sphere. Indicators for this semantic uncertainty are the Ernst & Young bi-annual reports on the state of the European biotechnology business which over recent years used different terms from 'biotechnology' (1996) to 'life sciences' (1998) to 'Evolution' (2000).

The public sphere: conceptual framework and empirical foundations

We consider biotechnology as an emerging scientific-industrial complex – a growing activity complex of research, development, production and service provision. By this, we do not mean to imply that biotechnology is a unified field, complete with a single, hierarchical mechanism of command and control; rather, we regard it as a heterogeneous coalition of many different actors, institutions and interests engaged in a competitive game over the control of this complex for purposes of commercial advantage. The biotechnology complex evolves alongside and within established societal spheres – economic, legal, mass media, political, religious, and so on – that collectively constitute its environment. Developmental change occurs in part through 'challenges' of one societal system upon another, and responses to these challenges. For research purposes, any particular societal system may be foregrounded as the focus of attention. For our purposes, it is not biotechnology itself – its locations, business logic, manpower, capitalisation, and so on – but certain aspects of the political systems as part of its environment that are foregrounded in this way. In particular, we are interested in the way in which old and new structures of a modern public sphere (Habermas, 1989) shape the contents and trajectory of a new technology.

The economic, legal, political and media environments each give more or less attention to biotechnology at different times, frame the technology according to a particular logic, and all have 'eyes' on other issues. As each turns its gaze to biotechnology, it may construct a different representation of the 'object'. Thus, for the financial system the representation is likely to emphasise investment opportunities, risk and stock market performance; whereas for the mass media it may consist largely in the 'news value' of particular developments tied to novelty, human interest or scandal. In this sense, the symbolic environment of biotechnology is made by a set of observers with different levels of attention and different ways of seeing. But these are more than merely passive observers. Their gaze is an active process of selection and framing that may facilitate and/or constrain the development of biotechnology in particular ways. The presumptions underlying this research are that, in the course of its twenty-five-year development: first, biotechnology regularly presented challenges to observers within the public sphere; and, second, these observers at times responded with counter-challenges or resistance that contributed to shape the continued development of biotechnology itself.

By systematically observing the observers of biotechnology in the public sphere, we aim to document the presence and the potential influence of the public counter-challenges upon this emerging technology. Our research observes the public sphere as a tripartite structure of policy, media and perceptions, and through what we term 'representations' of biotechnology (see chapter 13). For present purposes, a public representation is simply conversation and writing within the public sphere referring to biotechnology, which is 'objectified' for the conduct of research (Bauer and Gaskell, 1999).

The research was organised in the following way. First, each participating country conducted a longitudinal (historical) analysis of the development of public policy for biotechnology over the period 1973 to 1996 (a similar longitudinal study of policy developments at the European level has also been undertaken). This period was chosen to embrace the entire history of modern biotechnology from the discovery of rDNA technology up to 1996, the year in which our research project commenced. Second, each participating country conducted a longitudinal analysis of media coverage of biotechnology in the opinion-leading press, also from 1973 to 1996. Third, all participating countries contributed to the development and analysis of a representative sample survey of public perceptions of biotechnology, Eurobarometer 46.1. This survey was carried out in October/November of 1996 in each member state of the European Union (EU). Similar surveys were also developed and carried out by affiliated teams in Norway, Switzerland, Canada and the USA.

Public policy: chronology and domains of regulation

Public policy is an important expression of the aspirations, attitudes and values of a country. Public policies may have various explicit or implicit aims: they may seek to promote public goods (for example, through the encouragement of innovation), or to prevent public harms (for example, through the imposition of health and safety regulations); they may seek to protect the interests of producers (for example, through the patent laws), or those of consumers (for example, through product labelling requirements); and they may seek to reconcile conflicting ideals or interests (for example, in the provision of guidelines for the acceptable conduct of research on human embryos). Public policy is the outcome of activities in political forums. In pluralistic democracies, these activities are necessarily multiple and multi-valent. In other words, at any particular time no single actor or interest group is likely to dominate the policy-making process to the exclusion of all others. Instead, different actors and interest groups vie for influence in a political process (part private, part public) involving competition, cooperation, lobbying, public relations campaigns, coalition-making and breaking, and compromise. In the European Union, policy-making for biotechnology takes place at both the national and the European levels. To the complex of actors and interests operating at the level of the individual nation-state must therefore be added a second complex of actors and interests (including the European Commission, the European Council and the European Parliament) operating at the level of the fifteen member states. In the end, what transpires as official policy may be something that no single actor or interest group originally intended.

In reviewing the history of biotechnology policy-making, we have concentrated in the main on formal policy-making processes; that is, on the institutionalised activities by which official public policy has been established. However, wherever possible we have also paid attention to informal influences (such as lobbying by business organisations, or the opposition activities of non-governmental organisations) on the formal sector. We have been interested in questions of three main types: those that concern the characterisation of 'frames' of biotechnology within the policy field; those that concern the mechanisms by which policy has been framed; and those that concern the relationships between individual nation-states and the European Union.

In the first category, we considered questions such as: which issues have been debated? how have these issues been framed by the selection of themes? which have been the principal sponsors/constituencies of particular themes? how has the policy process dealt with opportunities and risks in relation to biotechnology? have policy-makers concentrated on the control of processes or products? In the second category, we considered questions such as: what have been the distinctive 'policy cultures' for biotechnology in Europe? what have been the principal mechanisms for generating biotechnology policy in Europe? what has been the influence of public opinion upon policy-making? have policy-making processes tended towards the 'technocratic' or the 'participative' mode? and is there evidence of 'institutional learning' as policy-makers develop new instruments and forums in light of previous experience? In the third category, we considered the timing of policy processes: how early or late do particular countries become engaged with biotechnology policy-making?

how far do particular countries 'lead' or 'follow' in particular areas of policy-making? and what are the relations between national and European initiatives on biotechnology?

We developed a chronology of key policy developments in each country for the period 1973–96. This chronology was based partly on published sources and partly on interviews with key actors from different arenas, including government, industry and non-governmental organisations. The aims of the chronologies were: to document concisely the most important policy initiatives in each country; to provide a base-line of data for comparison with the chronology emerging from the mass media study; and to provide a base-line of data for purposes of international comparison. Once the chronologies had been completed, these were converted into a 'policy template', providing in a standardised form a concise summary of policy developments in each country over almost a quarter of a century. Policy events were classified into ten areas: reproductive technologies; gene therapy; genetic screening; transgenic animals; genetically modified food; releases of genetically modified organisms (GMOs); GMO contained use; health and safety; research and development policy; and intellectual property rights.

These chronologies and templates were complemented by a review of the 'policy culture' in each country. By policy culture in this context we mean the prevailing styles of policy-making at the national level. These were judged to be vitally important for the interpretation of national similarities and differences (see, for details, Durant, Bauer and Gaskell, 1998).

Media coverage: intensity and contents of coverage

The mass media constitute a major arena of the modern public sphere. There is general agreement in the literature that the mass media are influential, but much less agreement about the exact nature of this influence. It is variously argued that the mass media serve to 'frame' issues in the public domain, that they serve an 'agenda-setting' role, and that they pander to and therefore, by way of appeal, express public perceptions. For our purposes, the mass media are viewed as one of several modes of representing biotechnology in the public sphere. They function both to explain and legitimate formal policies ('top–down'), and to signal issues and themes to policy-making that arise from informal political forums ('bottom–up'). Throughout, it is the complex interrelationships between media discourses, policy discourses and public perceptions with respect to biotechnology that are the focus of our attention. In order to study this interrelationship empirically, we constructed a media database following the paradigm of 'parallel content analysis' (Neuman, 1989).

The media analysis comprised two elements: first, an indicator of intensity of coverage over time; and, second, the characterisation of this coverage in a longitudinal content analysis from 1973 to 1996. We established an indicator of the intensity of media coverage by estimating the number of all relevant articles on a year-by-year basis. The second key element was the creation of a corpus of media material in each country for purposes of comparative analysis. As we were dealing with the emergence of a new technology in the public sphere, we selected the opinion-leading press for study. By 'opinion-leading' we refer to outlets that are read by decision-makers for information and by other journalists for inspiration. We assumed that for each country it is possible to identify outlets that stand as proxies for the nature and intensity of media coverage more generally. In most participating countries national newspapers still act as opinion leaders. If this is doubtful, the criterion 'opinion leader' provides a functional reference for selection independent of other characteristics such as circulation or quality. In some contexts, this criterion may lead to the use of more than one newspaper within the longitudinal study. One newspaper alone may not cover the opinion-leading function, or the newspapers may change their function over time. Furthermore, newspapers make convenient and reliable sources for purposes of data collection.

Establishing a comparable sampling frame for media analysis across twelve countries was a challenge in itself. Our strategy was to establish functional equivalence across the countries. Some newspapers offer a historical index of articles. This constitutes a self-classification by journalists. For the early years, several of us relied on this entry point, although we were aware that this classification was not necessarily exhaustive. Such indices were checked by manual scanning, under a protocol according to which the number of issues scanned was inversely proportional to the amount of relevant material they were expected to contain (the smaller the number of expected articles, the greater the number of issues that need to be checked in order to establish a reliable intensity index). With on-line resources such as FT-Profile or CD-ROMs from certain newspapers, the complexity of sampling is reduced to the question: what are relevant keywords or search strings? To answer this question, the project defined a core set of key words translatable into all of the eight languages involved. These were: *biotech**, *genetic**, *genome*, and *DNA*.

The coding method indicated for our purpose was classical content analysis (e.g. Bauer, 2000). We chose this approach from a multitude of textual analysis techniques because: first, it allows for systematic (i.e. publicly accountable and replicable) comparisons on the basis of a common coding frame; second, it can cope with large amounts of material; and, third, it is sensitive to symbolic material, albeit through a process of coordinated local interpretation. The aim of the analysis was to deliver a systematic and comparable interpretation of the mass media traces of biotechnology since 1973.

The coding frame provided a grid of comparison of coverage in terms of framing, thematic structure and evaluation of biotechnology. Frames, themes and evaluations were further differentiated. The unit of analysis was the 'single press article', which was read by the coders and interpreted in the light of the questions posed by the coding frame. As most coders were highly educated members of the national research teams, their readings are likely to reflect the subcultural features of those who produced the articles and thus to constitute in this sense a valid, albeit not a universal, reading.

For each article, the coding frame assessed journalistic features such as the section of the newspaper in which it appeared; the size of the article, as an indicator of news importance; the format of the article; and whether the article appeared to be controversial. The news event was characterised by authorship, the actors identified with biotechnology, the themes, their location, the attributed consequences in terms of risks and benefits, and the implicit evaluation of biotechnology. A key feature of the coding was the identification of 'frames' of coverage (see Gamson and Modigliani, 1989).

Quality management of the process included careful negotiation of sampling and coding procedures, familiarisation with the procedures in the context of local constraints, revision of the coding frame to take account of local pilot work, and formal reliability checks for both within-country and cross-country consistency (see Durant, Bauer and Gaskell, 1998: 297–8 for reliability assessments).

Public perceptions: knowledge, attitudes and images

The third module of the research was concerned with measurement of public perceptions by means of random sample social survey. By 'public perceptions' in this context we mean all of the considerations, expressed in interviews, that people may have concerning biotechnology. As such, the term embraces interest and involvement in, understanding of and attitudes towards biotechnology; but it also includes the images, hopes, fears, expectations and even forebodings that people may experience when they think about biotechnology. The term 'perception' includes the processes of imagination at any moment in time. The importance of imagination lies in its capacity to go beyond the present reality by re-presenting 'things' independently of space and time: to locate events that happen at other places, to recollect or link past events, and to anticipate a negative or positive future that inspires present-day actions, even to play on fantasies without any constraints such as science fiction. In the case of biotechnology – literally, the technology of life itself – the importance of individual and collective imagination can scarcely be exaggerated. The cultural resonance of key phrases – 'test-tube baby', 'genetic engineering', 'cloning', 'the blueprint of life', 'frankenfood', 'Boys from Brazil' – has as much to do with their metaphorical and mythopoeic powers as with their scientific and technological significance.

From the outset, our research was organised around the opportunity to conduct a social survey through the Eurobarometer Office of the European Commission. After extensive qualitative research using individual and focus group interviewing in the spring of 1996, a survey instrument was designed and pilot tested. Following necessary modifications, the Eurobarometer on Biotechnology (46.1) was conducted during October and November 1996. The survey was carried out in each member state of the European Union, using a multi-stage random sampling procedure providing a statistically representative sample of national residents aged 15 and over. The total sample within the European Union was 16,246 respondents (i.e. about 1,000 per EU country). In addition, similar samples were achieved in Norway and Canada (1996) and in Switzerland and the USA (1997).

The survey drew on the questionnaire employed in 1991 and 1993 in previous Eurobarometers on Biotechnology (35.1 and 39.1). Where possible, questions were repeated for purposes of trend analysis; but changes both in biotechnology itself and in the public debate about biotechnology dictated the need for a number of new question sets in the survey instrument. The revised questionnaire included items on the following topics: optimism/pessimism about the impact of specified technologies (including biotechnology and genetic engineering); elementary scientific knowledge relating to biotechnology; beliefs about the role of nature and nurture in the development of human attributes; specific attitudes on six applications of biotechnology measured on four dimensions of usefulness, risk, moral acceptability and support; general attitudes towards the regulation of biotechnology and its agencies; confidence in different institutions to tell the truth about biotechnology; future expectations about the contributions of biotechnology to society; the importance of the issue, sources of information and attentiveness to the issue; political orientation; and, finally, socio-demographic characteristics such as age, religious orientation, sex, income and level of education.

By integrating the results from the 1996 Eurobarometer (46.1) with the systematic media and policy analyses, the project maps the main contours of the different 'national public landscapes' within which biotechnology is developing in Europe and in North America. The project was designed to study public opinion in the public sphere. This necessitates going beyond social survey data, thus avoiding the trap of equating public opinion with public opinion polls and social survey data.

The structure of the present volume

The book presents a series of empirical studies exploring the interrelationships between mass media coverage, public perceptions and public policy in the development of biotechnology to 1996/97. Together these provide an analysis of the preconditions of the recent public controversies about biotechnology across Europe. The book is divided into five parts.

The first part provides a longitudinal perspective dealing with the long-term cultivation of symbolism around biotechnology in Europe since 1973.

Chapter 2 reviews the development of biotechnology regulations across Europe and in the European Community from the 1970s onwards. In their extended *tour d'horizon*, Torgersen et al. depart from the traditional notion that societies adapt to a technology after the fact, with a cultural lag. The struggle for national and international regulation has the double character of preparing for, as well as reacting to, developments. They offer a carefully crafted periodicity of European regulatory events since the Asilomar Conference in 1975. They explain the puzzle of why the controversies that flared briefly during the 1980s, and were widely thought to be closed, re-emerged in Europe, with due national variation, during the first half of the 1990s and fully reopened during the watershed of 1996/97. They explore several explanations and weigh the evidence: the failure of risk regulation in the face of diverging expert observations and public perceptions; the specificity of biotechnology in dealing with the enigma of life; the ethical implications of the technology, which make it unsuitable for a simple technological fix; and, finally, the possibility that in many contexts biotechnology is only the pretext, while people's real sources of grievance lie elsewhere, for example in the realities of European integration. Public opinion projects onto this new technology, which acted almost as a sounding board for a range of cultural discontents at the end of the twentieth century.

In polemic mode, the media are often blamed for causing the 'difficulties' in which some sectors of the biotechnology movement find themselves: the media misinform the public and thereby stir up irrational anxieties. That this is short-sighted as well as just factually wrong is the topic of the third chapter. At an early stage a technology has mainly a symbolic existence in the form of visions, images and ideas. Most citizens depend on the media to raise their awareness and to provide information and images of things to come. The chapter analyses the logic of selection and framing that is characteristic of media reportage, using national elite press as a comparative proxy. The dynamics of the coverage of biotechnology in elite newspapers in different European countries are compared for the period between 1973 and 1996. The description of the evolution of the intensity of press reportage, and the featured actors, themes and location of actions, provide a detailed picture of the shifting frames within which biotechnology is represented in public. Gutteling et al. finally explore some correlations between the contents of press coverage and people's awareness and evaluation of biotechnology as at 1996.

Chapter 4 by Galloux et al. explores the emergence of the ethical concern in the biotechnology discussion. In a comparison of Denmark, France, Italy and Greece, they reconstruct the institutionalisation of ethics as applied to issues of biotechnology over the past twenty-five years. They identify three different frames or ethical arguments that have come to bear on issues in biotechnology: arguments of utility, democracy and veneration. They explore the trends towards, and the dangers of, intentionally or unintentionally reintroducing a technocratic logic under a different guise.

The last chapter with an explicit diachronic outlook explores the interrelations between controversy, media coverage and public knowledge about biotechnology. Bauer and Bonfadelli take up the long-standing discussions on mass media effects, in particular the knowledge gap hypothesis, and put several propositions to the test. This hypothesis argues that media coverage on issues far from everyday experience, such as emerging technologies, will increase the inequalities of knowledge and awareness in society owing to differential public attention to the mass media. However, public controversies will increase the penetration of the mass media and thereby reduce the disparities in the representation of biotechnology in public. Mass media coverage of new issues increases the educational disparities that already exist in a country, while large-scale controversies will reduce them by speeding up and deepening the circulation of information. Controversies thus raise awareness and educate the public. The authors show that knowledge and other gaps are associated with the intensity of media coverage, and that the level of controversies is associated with a narrower knowledge gap. However, the longitudinal test shows that intensified coverage over time is associated with widening and narrowing knowledge gaps, not directly related to the level of controversy; other factors come to play a role. This leads to the conclusion that the knowledge gap hypothesis is underspecified as a dynamic model.

The second part of the book explores the structure and functions of public perceptions across Europe at the end of 1996, as evidenced mainly in survey data. The chapters do this from several theoretical and analytic orientations.

Chapter 6 takes up the idea of a social movement that is forming against biotechnology. Nielsen et al. explore, with the help of survey data, the constituencies of such resistance. By analogy to trends in environmentalism, they hypothesise that two ideological positions can be typified: a 'blue' or conservative and backward-looking resistance, and a 'green' or modernist and forward-looking resistance. They are able to demonstrate a cluster of ideas consistent with this distinction in most, but not all, European countries. In profiling the socio-demographic make-up of the two positions, they identify a north–south diversity. A blue in the north is not the same social constituency as a blue in the south; the same is true for the greens. The chapter closes with the construction of the prototypical arguments for a blue and a green position, drawing on different key metaphors and coming to different conclusions over the regulation of biotechnology.

Chapter 7 considers people's attitudes towards biotechnology. Classical social psychological attitude theory considers the strength, based on knowledge, and the structure or dimensionality of attitudes. It explores schematic information-processing in making judgements. Midden et al. draw a map of favourability and expectations towards biotechnology in Europe, and show that Europeans' attitudes are structured (based on a Lisrel model) along two independent dimensions: a utility–morality judgement, and a risk assessment. The low level of knowledge involved in these judgements suggests that, as is to be expected in the early stages of a public debate, these attitudes are not-yet-attitudes – not yet crystallised into a stable mental disposition. Their character as non-attitudes makes them inherently unstable and prone to change in the future. The relationship between knowledge and attitudes shows a low overall correlation. However, knowledgeable people are likely to hold extreme positions on the attitude scales, either positive or negative. This observation contradicts the popular notion of a knowledge deficit underlying public concerns over biotechnology. Or, as common sense has it: to know is to love it; on the other hand, familiarity breeds contempt.

Chapter 8 takes a different look at the same survey data. Allum et al. engage with the discussions in public understanding of science over the attitude–knowledge complex that struggle to prove or disprove the simplistic deficit model. They explore the 'post-industrial model of public understanding of science', according to which the knowledge deficit model (to know is to love it) is neither true nor false in general, but may have or may not have validity in particular socio-economic contexts.

They consider work on the economic regionalisation of Europe and test the variation of knowledge and attitudes, and the relationship between attitudes and knowledge, in a multilevel analysis, with the individuals, the regions and the countries as sources of variation. All three levels contribute to variations in knowledge of biotechnology, and therefore the authors conclude that regions matter. They map perceptions of biotechnology into a four-fold typology of European regions and show characteristic patterns of perception along these lines; however, these differences are not in line with the expectations of the post-industrial model of public understanding of science. Here the authors meet the limitation of the hypothesis and of the attempts to test a diachronic hypothesis on synchronic evidence.

It is widely known and accepted that attitudinal judgements depend on knowledge and images of the object to be judged. The conceptualisation of this phenomenon is theoretically controversial. Chapter 9 demonstrates the representations that underpin attitudinal judgements of biotechnology. Wagner et al. investigate this with data from the same survey as the previous three chapters, but use in addition a particular type of data, free associations. Respondents were asked to answer the open question 'What comes to your mind when you think about modern biotechnology in a broad sense, that is including genetic engineering?' The authors systematically analysed several thousand answers to this question as an index of common sense in Austria, the UK, France, Germany, Norway and Sweden. The idea is that discursive frames evidence the process of symbolic coping by which the public takes up the challenge of biotechnology in different contexts. The ambiguous evaluation evident in the data is related to the images of biotechnology: curing diseases, unnatural food, diagnostic testing and monsters. The identification of general images of biotechnology is complemented by particular national images with iconic quality.

The third part of the book presents two case studies from the watershed years of 1996/97 that triggered an international and synchronous public controversy over biotechnology.

Chapter 10 systematically analyses the events of late 1996 and early 1997 when Monsanto's shipments of soya that contained admixtures of a GM variety entered European ports. Lassen et al. show that official approval of Monsanto by the European Commission was undermined by the issue of labelling, which was left unresolved and which provided an entry for public controversy. Different patterns of immediate events are analysed for Denmark, Sweden, Italy and the UK in terms of socioeconomic standing, political tradition, the strength of the actors involved and the discursive environment. The soya controversy had three effects:

it reopened the debate simultaneously in many countries; it reframed the discourses; and it left Monsanto as a corporate pariah on the European stage.

The first ten days of mass media coverage of Dolly the sheep across eleven European countries plus Canada are the focus of Chapter 11. In February 1997, eighteen months after its conception, a sheep was presented to the globally alerted mass media in Edinburgh, highlighting an achievement that had been deemed impossible. Whereas Round-up Ready soya was the trigger for the first European-wide biotechnology controversy, Dolly the sheep put the whole world into synchrony over biotechnology for the first time. Einsiedel et al. document the elaboration of a global trigger event in a cascade of mass media stories, evidencing a confluence of the operating logic of journalist practices and the logic of social representation – the metaphorical anchoring and pictorial objectification of strange, and potentially threatening, happenings. The proliferation of stories fluctuated wildly between the frames of progress and doom, and documented the ongoing struggle to represent technology in the late-modern age. The global moral panic over potentials for human cloning masked not only the national peculiarities in the imagining of cloning, but also what the Roslin research was intended to achieve: the efficient production of specific pharmaceuticals through live factories.

The fourth part of the book comprises the penultimate chapter, which takes a look across the Atlantic. A constant puzzle in our project was the apparent absence in the United States of concerns and voices analogous to those in Europe. Gaskell et al. explore this issue with a comparison of public perceptions and media coverage in Europe and the USA to 1996, supplemented by an excursion into the history of biotechnology regulations in the USA. The authors typify three logics of public reasoning: opposed, risk-consciously in favour, and in favour of biotechnology applications. Comparison of different applications reveals generally lower concerns in the USA than in Europe, although sensitivities are different across the Atlantic. The authors explore three putative explanations of the observed transatlantic differences: differences in media coverage, in public knowledge and in trust in the regulatory system. The characterisation of the history of debates and policy-making in the United States argues that, whereas the 1990s were a period of relatively low controversy in the USA, the 1970s and 1980s were marked by more lively debates. Through a mixture of normal and revolutionary policy-making the USA is found to have secured high levels of trust in the checks and balances of the regulatory system as a whole. However, there seem to be signs that Dolly and GM crops and food might reopen the public debate.

In the final part, chapter 13 offers a reflection on the ideas that developed in the course of this study, and proposes the integrative concept of a 'biotechnology movement' that carries this new technology through history. Centred on the scientific-industrial complex, the biotechnology movement is an actor network characterised by conflicting visions of the future. The biotechnology movement presents itself in material form in products and services and is symbolically re-presented in the tripartite public sphere of mass media coverage, policy and regulations, and public perceptions. Bauer and Gaskell identify a lacuna in the social studies of science and technology, namely a non-specified notion of the dynamics of the public sphere. They elaborate a 'three-arena model of the public sphere', comprising regulation, mass media and everyday conversations. Biotechnology constitutes a challenge for the public sphere, which in turn poses the counter-challenge for the biotechnology movement. The public sphere modelled as a tripartite arena allows us to characterise the changing symbolic environment of the biotechnology movement and, at the same time, the potentials for organisational and institutional learning that may result from public resistance. Two forms of collective learning – assimilation and accommodation – are introduced for this purpose. The traditional technological strategy has been to get society to adopt technological innovations by way of assimilating society to the corporate image of technology. However, resistance from the public sphere may lead to a change in the trajectory of technology, by way of accommodating corporate activities to societies.

In toto, this volume draws on empirical observations from policy-making, media reporting and public perceptions. It addresses conceptual and theoretical issues to document and explain the ways in which different European public spheres converged on controversies over biotechnology triggered by the watershed events in 1996/97. Beyond the particular focus on biotechnology, the book offers a conceptual framework and an empirical paradigm for observing new technologies as socio-technical movements within the constraints of the modern public sphere in the twenty-first century. In so doing, it is intended to contribute both to the continuous public debates over biotechnologies as well as to a growing corpus of scholarly interdisciplinary literature on science, technology and society.

REFERENCES

Bauer, M. (1995) 'Towards a Functional Analysis of Resistance', in M. Bauer (ed.) *Resistance to New Technology – Nuclear Power, Information Technology, Biotechnology*, Cambridge: Cambridge University Press; 2nd edn, 1997.
    (2000) 'Classical Content Analysis – a Review', in M.W. Bauer and G. Gaskell (eds.), *Qualitative Researching with Text, Image and Sound – a Practical Handbook*, London: Sage, pp. 131–51.

Bauer, M.W. and G. Gaskell (1999) 'Towards a Paradigm for Research on Social Representations', *Journal for the Theory of Social Behaviour* 29: 163–86.

Durant, J., M.W. Bauer and G. Gaskell (eds.) (1998) *Biotechnology in the Public Sphere – a European Sourcebook*, London: Science Museum.

Gamson, W.A. and A. Modigliani (1989) 'Media Discourse and Public Opinion on Nuclear Power: a Constructivist Approach', *American Journal of Sociology* 95: 1–37.

Habermas, J. (1989) *The Transformation of the Public Sphere*, Cambridge: Polity Press.

Neuman, W.R. (1989) 'Parallel Content Analysis: Old Paradigms and New Proposals', *Public Communication and Behavior* 2: 205–89.

Part I

The framing of a new technology:
1973–1996

2   Promise, problems and proxies: twenty-five years of debate and regulation in Europe

Helge Torgersen, Jürgen Hampel, Marie-Louise von Bergmann-Winberg, Eleanor Bridgman, John Durant, Edna Einsiedel, Björn Fjæstad, George Gaskell, Petra Grabner, Petra Hieber, Erling Jelsøe, Jesper Lassen, Athena Marouda-Chatjoulis, Torben Hviid Nielsen, Timo Rusanen, George Sakellaris, Franz Seifert, Carla Smink, Tomasz Twardowski and Merci Wambui Kamara

Introduction: biotechnology – a chequered field

By the end of the twentieth century, disputes about biotechnology and its agricultural products had reached a peak in Europe. Industry had failed to win the hearts of European consumers and, de facto, there was still no real harmonisation of product assessment within the European Union (EU). This is noteworthy since, after a quarter of a century of rapid scientific and technological development on the one hand, and contentious debates about real or possible implications on the other, biotechnology has 'come of age'. Industry has finally engaged in biotechnology on a broad base, forming large conglomerates that deal with seeds as well as with pharmaceuticals under the heading of 'life science'. Ten (or even five) years ago, most spectators expected disputes over biotechnology to be settled imminently; nevertheless, issues are again subject to conflicts to such a degree that the future of some applications of biotechnology in Europe is uncertain. Obviously, traditional ways of introducing this new technology have failed. But can we establish why?

Classically, authors understood the relationship between technology and society to involve society adapting to technological developments (see Ogburn, 1973; Ellul, 1964). They saw technology as relatively independent of society, following its own rationality, or even forcing society to adapt to it. According to this view, it is inevitable that society 'culturally lags' behind technology, because technological development is faster than social and cultural adaptation (Ogburn, 1973). Society seeks to catch up, while new technological developments constantly open new gaps.

21

Indeed, the introduction of a new technology often seemed to follow this schema. For example, communication and information technologies raised debates in the 1980s (stimulated by Orwell's famous novel *1984*); but now, after the 'cultural lag' has been narrowed, they constitute welcome elements of everyday life. However, in the case of biotechnology, this classical model of explaining technological innovation obviously falls short. Here, even an apparent calming of the debate in the early 1990s proved to be only temporary and conflicts have reappeared.

From early on, uncertainty over possible hazards nourished struggles over biotechnology. It is a truism that every technological innovation is associated with risks or even hazards – remember explosions involving steam engines, or airplane crashes in the pioneer days of aviation. The difference here was that there was uncertainty as to whether there were any hazards at all. Opponents of biotechnology strenuously advanced claims of risks, while supporters denied them with equal assiduity. It is not our objective to determine whether any of these claims have been substantiated, or whether the risks identified are only hypothetical. Rather, in this chapter we attempt to trace *opinions* about whether there are risks, and, if there are, which kind, as well as the political consequences thereof.

At the heart of many controversies about risk lies a difference between public and expert experience of what biotechnology 'really' is. For the general public, this is not a technology like any other, since specific installations and artefacts are not easily discernible.[1] An atomic power plant serves as a symbol for nuclear technology, and computers are typical products of microelectronics. Without these practical applications, the technology would not have existed. Biotechnology, in contrast, is usually invisible to laypeople. Although applied in many fields, its techniques are employed behind closed doors in laboratories indistinguishable from outside. As such, at first glance, its products are not at all specific. Biotechnology consequently remains enigmatic, at the same time as dealing with 'the secret of life'. For experts, on the other hand, its character is no different from that of other technologies. Biotechnology laboratories are, for them, definitely recognisable, containing, as they do, specific apparatuses for research and production. To the expert, their products are clearly distinguishable by their properties, and no other technology could have produced them. Moreover, for the skilled, biotechnology did not pose intrinsic problems that could not be understood, and controlling the mechanisms of heredity, in principle, does not present insurmountable problems.

In order to deal with uncertainty, and to make risks in many fields calculable and manageable, engineers and insurers have developed (technical) risk assessments. Although twenty-five years have elapsed without a single major accident caused by biotechnology,[2] expectations among scientists that they would be able comprehensively and credibly to identify, control, compare and evaluate the technical risks of biotechnology have, nonetheless, failed. One major reason for this is that, owing to the subject of biotechnology being 'life' itself, ethical considerations played a significant role from the very beginning. Hence, debates went beyond the discussion of technical aspects, and especially those involving risks that could be technically determined. Indeed, in its status as a contested issue, biotechnology served as a sounding board, or a projection screen, for deep differences in interests and world-views. This resulted in a wide opening of the debate and the forging of links with other issues considered to be at stake.

A similar broadening of the debate could be observed in the case of nuclear energy and, to a much lesser degree, with microelectronics. Both technologies had acquired the position of strategic technologies long before biotechnology, although the latter was assigned this status soon after its birth (Bauer, 1995). Differences arose in the perceived need to regulate these technologies: with nuclear energy, the demand for regulation by the government was obvious; all but a few, however, agreed that the regulation of microelectronics was hardly feasible. In contrast, the struggle over biotechnology led to an ambiguous result, since different actors failed to agree on the issue of regulation, and different groups sought to delay or to support the use of biotechnological applications. Debates went beyond the question of whether or not a law should be passed, to the issue of the appropriate contents of regulatory policies. The latter became an especially contested topic not only within Europe, but also between the USA – as a promoter of a regulatory approach based on scientific evidence alone – and the EU, which was opting for a broader basis of argument.

Responding to competing interests, governments adopted different permutations of a two-sided policy: on the one hand, they provided regulation as a means of containing risks; on the other hand, they backed research and application in order to stay, or become, economically competitive (Jasanoff, 1995). This strategy seemed appropriate in order to 'make biotechnology happen'.[3] Obviously, governments thought that biotechnology was something worth developing and they supported it with alacrity. Yet they also styled themselves as impartial regulators of what many perceived to be a risky endeavour. This ambiguity later proved to be one of the sources of public distrust.

Despite this general pattern, the style and background of the debates and their political outcomes differed markedly from country to country, and subjects, actors and forums changed considerably over time. Even among EU member states,[4] we encounter an astonishing multitude of reactions to the challenges that biotechnology presented, in terms both of public debate as well as of regulation. Even differences in regulatory style influenced how biotechnology was perceived and which of the problems raised by different actors were deemed legitimate areas of state involvement. In order to gain an understanding of what may appear, at first glance, an odd and almost impenetrable jungle of rhetoric and facts, policies and interests, opportunities and risks, wishes and fears, attitudes and world-views, we have to examine this varied picture and explore its determining factors.

Much has been written about this issue over time.[5] As such, this chapter draws upon a wealth of data, resulting from four years of concerted research by an international group of scientists from various disciplines. This research covers policy initiatives, media coverage and public attitudes towards biotechnology in thirteen European countries, as well as in the USA and Canada (compiled in Durant et al., 1998).[6]

Concepts and methodology

In the following, we will trace the tangled history of biotechnology debate and policy-making during the last quarter of a century. We will keep in mind the four general issues summarised in table 2.1: first, the transgression of the boundaries of technical risk issues in the debates, which indicate that biotechnology acquired the role of a sounding board for the articulation of deeper concerns; secondly, the challenging of the technocratic paradigm of an autonomous chain of innovation; thirdly, the prevalence of regulatory responses intended to make biotechnology happen in the context of a two-sided policy of regulation and support; and, finally, the difficulties for the EU common market presented by the existence of profound differences among the nations of the EU in their attitudes towards biotechnology. These themes are, of course, interrelated; for example, the transgression of risk issues in the debates contributed significantly to those delays and detours (even dead-ends) encountered in attempts to implement biotechnology that served to challenge the

Table 2.1 *General issues in the biotechnology debate*

Sounding board: the transgression of risk issues
Challenging the traditional innovation chain paradigm
Dual strategy in order to make biotechnology happen
National variations across the EU

Table 2.2 *Definitions of terms*

Terms	Definition
Biotechnology	Recombinant DNA techniques as well as methods for their practical application in various fields
Recombinant DNA techniques, genetic engineering	Combination of methods for splitting, sequencing, splicing and constructing genes
Genetically modified organisms	Organisms that carry genes modified by these techniques
Products of biotechnology	Results of production processes applying such methods
(General) public	Societal entity that carries public perceptions
Regulation	Means to put an aspect of biotechnology under rules

innovation chain paradigm. And many governments' dual strategies for encouraging biotechnology were inevitably embedded in national political styles. Different emphases in the various countries on the perceived problems concerning biotechnology made the sounding-board play different tunes, and facilitated or hindered governments falling back into their perception of the innovation chain.

Over time, we observe developments not only in biotechnology per se, but also in the meaning of the word. So, before looking at the history of this technology we should provide some provisional definitions of terms, since at different times stakeholder groups have used different terminology (see table 2.2).[7] For the sake of simplicity, we will stick to the term 'biotechnology' even if we refer to developments that scientists would consider to fall outside this categorisation. In general, however, we will focus on applications made possible by research on nucleic acids, which emerged during the 1970s as 'genetic engineering'. However, because biotechnology, even in this sense, is a very broad field, we have to restrict the scope of this investigation somewhat. We will therefore concentrate on those issues that are commonly covered by national biotechnology regulation, and give particular emphasis to the problems surrounding the release and marketing of genetically modified organisms (GMOs), since it is such products that have triggered the most intense public debate in recent years (BEP, 1997; Durant et al., 1998).

Debates arose not only among regulators and those involved in research and development (R&D), but also among the general public. It is not easy to define this 'public': there are many conceptualisations of the 'public' among social scientists, and there are probably also different 'layers' of the 'public'. Thus, our pragmatic approach is to adopt an understanding of the public as an entity that carries 'public perceptions' about biotechnology (see Gaskell et al., 1998: 9–10).[8]

The political problems associated with biotechnology changed over time as the technology matured from a set of methods for basic research to techniques for applications and products in various fields. Consequently, 'biotechnology policy' meant something different in each of these fields. Debates arose within many political arenas, such as parliaments, political parties and professional societies, but also in newspapers and other media. Actors, such as scientists, politicians, industry representatives, activists from non-governmental organisations (NGOs), and well-respected intellectuals, demanded different degrees of regulation.

Since the political reasons for regulation varied among countries, a multitude of ways of regulating biotechnology emerged, including special laws, adaptations of existing regulation, statutory guidelines and self-regulation. Clearly, regulation could take on very different forms, according to the type of actor each intervention was aimed at, and the functions it was intended to fulfil. As with most new technologies, from the very beginning the overall aim of attempts to control aspects of biotechnology was to provide a reliable frame within which it could be pursued safely while it was being implemented and promoted. Thus, keeping in mind the varieties of approaches involved, when we speak of regulation we refer to any means of applying rules to an aspect of biotechnology; if this is by law, we call it legal regulation.

Phases

In this chapter we are dealing with the social and political reactions to, and conflicts over, technological, social and political developments with respect to biotechnology. This task is complicated not only because of the complexity and variability of the technology itself, but also because the conflicts, debates and political reactions occurred in a bewildering array of temporal and spatial contexts. If we compare the history of debates and policy-making across Europe (Durant et al., 1998), we see numerous periods in which public attention and policy initiatives were negligible, alternating with more active periods. Likewise, we see activity in different countries at different points in time.

There are many reasons for this variegated picture. In Europe in this period, a loose economic community was changing its character to become an increasingly integrated system of member states. Yet, in spite of this integration, the political cultures and legal systems in European countries remain very different. Some legal systems are oriented towards Germanic law, whereas others follow Roman law. Countries with a first-past-the-post electoral system sit alongside countries with proportional representation, which lowers the hurdles for the formation of new political parties. Some countries saw the development of green parties in this period; others did not. The way in which the political systems reacted to public unease varied considerably; and the responsiveness of governments may have changed over time, since, during the twenty-five years covered, each European country experienced at least one change of government.

Initially, the asynchrony across Europe was most noticeable between the north and the south. Countries in the vanguard of technological development, political debates and regulation from the north and west of Europe had been dealing with the issue since the 1970s. Some of them witnessed widespread debate and others tried out various styles of regulation. These nations set the agenda for other countries, but some of the latter followed only after a delay of up to twenty years. Regarding the form of regulation, almost every country adopted a different position. The variety of approaches and responses in the European countries, and the fact that the EU institutions added another layer of regulation in an attempt to impose some harmonisation, make generalising extremely difficult. Thus, at first glance, producing a comparative analysis of the political debates and regulation of biotechnology in Europe might seem to be a hopeless enterprise.

However, regardless of all the intra- and international differences, there are some common characteristics. Some similarities and general patterns emerge with the help of comparative media analysis (Durant et al., 1998), which reveals common themes, or 'frames', representing the main interpretations of the opportunities and challenges for biotechnology. These frames were characterised by a shared set of problems, reference to similar stages in the development of the technology, and the expression of specific views that predominated among important actors. From our analysis of the main issues in public debates and policy-making over the past twenty-five years, we have discerned four phases, exemplified by distinct and overarching frames (table 2.3). Thus, it seems that, after a short period of time, the media emphasis shifted from laboratory science to industrial and economic aspects, subsequently moving on to the question

Table 2.3 *Phases in the debate about biotechnology*

Phase	Main issues
Phase 1 (from 1973)	Scientific research
Phase 2 (from 1978)	Competitiveness, resistance and regulatory responses
Phase 3 (from 1990)	European integration
Phase 4 (from 1996)	Renewed opposition: consumers

of the EU's harmonisation of regulation, and, finally to the issue of consumer market products.

Biotechnology as a *scientific endeavour* was the first frame. Early 1970s
coverage was, almost universally, characterised by a conceptualisation of
biotechnology as an area of promising scientific research. Although the
issue of risk was stressed, it was discussed by scientists themselves, who
soon concluded that the risks were manageable. When, in some countries, public opposition emerged, scientists made attempts to reassure the
public.

Later, this view of biotechnology gave way to a more practically oriented
perception. It came to be recognised that biotechnology might gain enormous *economic* importance in an increasingly competitive world. Hence, it
had to be defended against *resistance* arising out of safety fears. Scientists
from other fields had developed counter-arguments, and these were taken
up by critics. Opponents also asserted links with other contested issues,
and the moral status of biotechnology – rendering life itself malleable –
gave rise to concerns. It therefore became clear to many governments
that self-regulation by scientists could not provide a satisfactory degree
of protection against the perceived risks. In an attempt to confine the issues at stake, some governments began to *regulate* the field on a national
basis.

The divergence of views about the appropriate way of regulating biotechnology eventually led the EU's institutions to take responsibility. It
was hoped that this would engender some degree of *harmonisation* in
the common market. This proved to be problematic, since regulatory
styles and the history of the debate in the member countries varied so
profoundly. Eventually, however, resistance faded and it appeared as if
biotechnology had become a well-regulated and *accepted technology*.

Yet, when *consumer* products materialised, old conflicts reopened over
new issues and the *opposition* was renewed. As consumers, citizens had
a powerful voice, and it became obvious that existing modes of regulation appeared inadequate to many, and that national governments and
EU institutions would have to find new solutions. Conflicts reached an
unprecedented level even in countries where people had previously been
fairly positive towards biotechnology.

The phases of the debate outlined above not only reflect the progression of the technology, as conventional innovation theory would predict,
but also represent changes in the political, social, cultural and economic
environment of biotechnology. This is illustrated by certain prevailing
frames in the debates and by the character of particular policy initiatives
at different stages. These frames were certainly not synchronous everywhere; rather, they overlapped differently for each country and field of
application. Indeed, challenges did not occur at the same time in different

parts of Europe and problems had not necessarily been solved when the new phase began. The subjects of earlier phases retained relevance in later stages. As such, the phases we stipulate indicate the addition of further layers of complexity rather than fundamental changes in the issues at stake (see table 2.3).

The debates expanded and contracted during each period. When biotechnology was seen as a scientific endeavour, debates could mostly be settled by scientists voicing safety reassurances. When biotechnology entered society at large, and commercial interests gained momentum, debates expanded once again. Controversy then began to cover applications and to emphasise world views, which prompted regulatory responses that often appeased resistance and led to a contraction of debates. The EU took up the issue at this juncture and attempted to harmonise regulatory frameworks. This became a new subject for debate and inflamed existing controversies; although, when biotechnology was formally endorsed, debate contracted once more. However, the commercial release of GMOs again triggered widespread opposition, and today we are in another phase of widespread debate.

This 'phase' structure, underpinned by concrete events, allows us to develop a framework for analysing the development of biotechnology debates and regulations in Europe. It also permits us to trace the four general issues mentioned above as characterising the four phases. In the following we will examine the four phases in turn, elaborating those aspects that are important for understanding this rich and complex field. In the conclusion we will then revisit the general issues and, finally, speculate on what the future might hold.

Phase 1 (1973–1978): scientific research

When biotechnology first appeared as a public issue, debate was clearly shaped by existing notions of scientific progress and by demands for scientific accountability. From the beginning, the issue of risk was emphasised and diffused from scientific discussions into public debate. The objective was to promote scientific research, while minimising technical risks and coping effectively with an emerging public debate. We will see how some countries, with a strong tradition in biotechnology, major industry interests and/or an alerted public, met the new challenges.

A call for scientific accountability

Methods of splitting, sequencing and splicing genes were developed in the USA. The first recombinant DNA (rDNA) result, in 1973, involved the successful *in vitro* transfer of a gene between species. With the advent of these new techniques, concerns arose in the scientific community about possible safety hazards. This resulted in a moratorium being agreed in order to allow time for scientists to think about the consequences of further research. Hence, biotechnology was accompanied, from birth, by the attribution of risk. This perception continues to this day.

Details published in a letter, entitled 'Potential Biohazards of Recombinant DNA Molecules', to the magazine *Science* (185, 1974: 303) outlined the scientists' concerns. The authors, Paul Berg and colleagues, called for a cautious approach until the hazards were better understood and an international conference could agree on how those risks should be contained. The scientific community felt that controls were necessary in order for work to be continued safely, and they accepted a voluntary moratorium. After eight months, at the Asilomar Conference, it was agreed that the moratorium could be lifted, and it was left to national governments to act upon the recommendations discussed there. In 1976, the US National Institutes of Health (NIH) issued technical guidelines for recombinant DNA research in line with the proposals, which were generally accepted world-wide as a standard for ensuring laboratory safety. Covering all work that was funded by the federal government, these famous NIH guidelines specified norms for risk evaluation in genetic engineering, for the safe handling of GMOs and for the adequate design of laboratories. With the help of these guidelines, funding agencies in the USA and other countries, as well as national governments, could easily adopt an internationally compatible frame of what to consider risky or safe. However voluntary it was, this was the first instance of a regulation being introduced in order to ensure that biotechnology could be developed.

From 1974 on, researchers and regulators responded promptly to the concerns raised in the Berg letter in an international, collaborative approach to regulating certain aspects of biotechnology. They perceived the need for rules long before the politicians. Scientists appreciated that research in biotechnology had huge potential but that this potential would be realised only if controls were put in place. Thus, the new NIH standard was welcomed. Yet it raised the political question of how to implement regulation on a national level, since countries differ hugely in their scientific landscapes. Some countries, such as Denmark, France, Germany, the Netherlands, Sweden, Switzerland and the UK, had a strong tradition in molecular and cellular biological research and major industrial interests, and some of these 'forerunners' also had an alerted public that would react to the new challenge. Other countries lacked such traditions, interests or a similar level of public alertness.

The solution was to leave it to the scientists to decide. The US debate that led to the moratorium had taken place mostly among the scientific community, and this group constituted the major actor in Europe as well. Since many researchers had for long been committed to international collaboration, they reacted in a fairly uniform manner in defining the problem in terms of science-based risk assessments. Most of the 'forerunner' countries quickly opted for a voluntary approach based on the NIH guidelines, and advisory committees were set up in these countries in similar ways. Since the emerging public policy issues focused on the perceived health and safety risks, it was no wonder that the different regulatory solutions were strikingly uniform from the practical point of view. The European countries had a common interest in safeguarding the future scientific development of biotechnology, so (voluntary) regulation aimed to provide the conditions in which the new and exciting field could flourish. Scientists themselves, it was felt, should narrow the gap between technology and society by virtue of their technical expertise.

Despite a general impression of uniformity, however, there were already some differences in these early days. Some of the 'forerunner' countries tried to keep the problem strictly within the scientific frame of technology and left the solution to technical experts alone, while others adopted a more inclusive approach, soliciting views from other sections of society. Indeed, it was soon clear that biotechnology implicated issues beyond the purely technical aspects of risk and safety.

From science to politics

Reactions among European 'forerunners' Among the first countries to react to scientific concerns was the UK, where the Department of Education and Science had implemented its own moratorium. Research Councils were keen to ensure that appropriate guidelines were introduced in order to allow researchers to return to work quickly. These guidelines were based on the existing regulatory framework for hazards from such sources as chemicals, and they even preceded the NIH guidelines. Although the UK did not see any significant public debate in this period, and discussions remained firmly within the scientific arena, the members of the Genetic Manipulation Advisory Group founded in 1976 (the term 'manipulation' was changed later to 'modification' owing to its perceived contentiousness) were recruited in a very British mode. Although they only advised on technical matters, such as risk assessment and physical containment levels, not all of them were scientific or medical experts – some were drawn from trade unions, management, industry and public interest groups. This composition provided for some information flow into other sectors of society (Cantley, 1995: 516), but the scientific frame was not questioned. Much as in the USA, the media focused on scientific (particularly medical) progress (see Gutteling et al. in chapter 3 in this book). Nevertheless, the UK was one of the first countries to introduce a regulation (in 1978) seeking to ensure a safe approach to genetic modification. Under existing legislation, laboratories had to notify the Health and Safety Executive of experiments. Scientists were advised on a 'voluntary' case-by-case basis, but they were not required by law to follow the advice; they only had to submit the information.

The Swedish Scientific Committee, too, set up in 1975 by the Academy of Sciences, was composed not only of scientists but also of representatives from government agencies, industry and NGOs. Their tasks were slightly different from those of the British: besides the review of research proposals, they had a mandate to keep the general public informed and to advise the Cabinet. This indicates that the issue was held to be of general concern and not confined to the scientific community. This perception proved to be a realistic assessment, although, according to another interpretation, it helped to trigger the first public debate on biotechnology in Europe when the announcement to build a special laboratory attracted media attention in 1978. For some time, as in the USA and other European countries, biotechnology had been a story of a promising new technology that focused on the USA. Now surveys showed strong public opposition to the plans, as biotechnology appeared dangerous and morally questionable. The Swedish government's answer to public concerns was regulation. In 1980, biotechnology was brought under existing laws, still following the NIH guidelines. The director of the powerful Committee for Industrial Safety and Occupational Health chaired the new governmental Recombinant DNA Advisory Agency, indicating that the new issue had found its place in the existing framework of risk management. A general feeling that trusted public institutions had taken care of the matter allowed public debate to wane; however, it never ceased altogether.

Comparing Britain and Sweden, we may identify similarities in that both were 'forerunners' in science; both saw the need for public and government involvement in the issue; both acknowledged, implicitly or explicitly, legitimate interests beyond the purely technical; and they both set up advisory committees. In the UK this resulted in rapid regulation, while in Sweden it resulted in the first public debate on biotechnology in Europe (regulation came later). The difference obviously lay in the fact that in Sweden a special incident (the new laboratory) elicited public attention. In the UK there was no such event in phase 1 or, at least, similar events were not covered in the same way, and public debate arose only much later. Such 'key events' of various kinds played a significant role in many countries because they raised public interest and eventually kicked off public debates, in turn leading to regulatory initiatives. However, this happened at different points in time, which contributed to the striking European diversity in the response to biotechnology.

Elsewhere, politicians (as well as the public) saw any problem related to biotechnology as an issue to be resolved exclusively by the scientific expert community. This also resulted in the idea that matters with which scientific expertise was unable to deal were to be considered irrelevant. France's regulatory approach came closest to a system of scientific self-regulation. Institutions performing rDNA research were to sign a contract with a research agency promising that they would follow French and international guidelines (Gottweis, 1995: 150). Although ethical considerations were emphasised early on (Cantley, 1995: 587 ff.), they were to be dealt with only in special committees. This approach was complemented by government's state-centred modernisation efforts. The institutionalisation of molecular biology was guided by a deliberate science-push policy, in line with a traditional understanding of the path of technical innovation. The French regulatory system, with its tradition of long-term planning and strong top–down orientation, succeeded for a long time in keeping the issue within the realms of the scientific and regulatory expert community. In order not to jeopardise this official policy, the French government was reluctant to respond to any concerns on issues other than physical risk.

Switzerland and Germany, which were among the technological 'forerunners' with strong pharmaceutical industries, also followed the path of scientific self-regulation, but initially with less emphasis placed on coordinating state action than in France. As early as 1975, the Swiss scientific community adopted the NIH guidelines and the Academy of Sciences established an Advisory Committee. Also in 1975, the German Research Association founded a Commission for Recombinant DNA. Later, the chemical engineers' association extended their concept of biotechnology to include the 'new genetics'. In 1978, at the same time as in the UK, the German government enacted safety guidelines for biotechnological research. Attempts by the German government to be the first in formulating a draft law for 'the protection from risks associated with modern biotechnology' in the same year triggered resistance by scientific organisations and industry (Gill, 1991). Scientists perceived it as an insult to their freedom of research when government tried to impose regulations on a scientific field. Otherwise, during phase 1, there was little indication of the serious conflicts that would later emerge.

'Latecomer' countries did not exhibit particular strategies with respect to biotechnology, and consequently adopted regulation only later (this group included the Mediterranean countries, Ireland, Finland, Austria,

Belgium, Luxembourg and Norway[9]). They had either a less important research base or an industry that had no genuine interest in lobbying for support. Alternatively, they simply followed, after a short delay, the path of the 'forerunners'. As long as there was no public debate or reaction to perceived risks, 'latecomer' countries avoided actively addressing this issue, adopting an attitude of 'wait and see'. Eventually, some of the 'latecomers' adopted strategies resembling those of phase 1 in the 'forerunner' countries, at a point in time when the latter had already reached subsequent phases. As we shall see, others went their own way.

Europe and the USA parting   The debate on genetic manipulation had culminated in the USA in 1977 when an increasing number of scientists concluded that the risks had been exaggerated and the moratorium was a mistake. What had begun as a small group of scientists attempting to act responsibly towards the public was seen to have spiralled out of control because it resulted in the public perception of biotechnology as a field replete with risks. Consequently, and pointing to the fact that no major accidents had yet been reported, the US scientists started publicly to deny that there were any novel 'biohazards', i.e. any risks that would justify special legislation, let alone a moratorium. Having opened the debate on biotechnology, over the next couple of years the scientific community also managed to close it. At least for the time being, genetic technology ceased to be a hot topic for public debate in the USA (if it ever had been one).

At this juncture, the European and the American scenes diverged. At the end of phase 1, when public debate was beginning to wane in the USA, events attracting media attention led to new debates in Sweden, and at a later point in time in Denmark, Germany, the Netherlands and other countries. European countries also took up various issues beyond those of technical risk, thus widening the scope of the debate. It is important to note that until the mid-1990s these debates were diverse and strictly national, and they were often confined to local contexts. At the EU level, however, the Directorate General XII responsible for R&D (DG XII), accepted the dominant framing of biotechnology as an entirely scientific endeavour, much in line with the US approach (Cantley, 1995: 518 ff.).[10]

Summary

An initial wave of concern originating from the scientific community in the USA led to a self-imposed moratorium that subsequently gave rise to the NIH guidelines. Doubts arose about these concerns – and about whether the price to be paid (the self-imposed restrictions) was too high. In the USA, public debate waned after scientists had reassured the public that risks were both marginal and manageable if voluntary guidelines were followed. In certain European countries, public opposition was mobilised when a particular event or project attracted enough media attention. The conflicts often resulted in demands for the regulation of biotechnology under special laws. However, the US example prompted many regulators to abandon any such plans. By 1980, self-regulation by the scientific community in line with the US guidelines, and made statutory because of links to the existing regulation of other fields (for example, chemicals), seemed to provide a flexible and sufficient means of ensuring an acceptable safety level for the handling of rDNA. Concerns other than those of technical risk were dismissed as unscientific or were left to be dealt with separately. This arrangement served effectively to assuage public debates in most countries. For the regulators, the problem appeared to have been settled. However, contrary to the situation in the USA, it was only a matter of time before new conflicts arose.

Phase 2 (1978–1990): competitiveness, resistance and regulatory responses

Phase 2 saw the general entry of biotechnology into the public sphere. New frames arose and new debates emerged as it became clear that biotechnology had great economic potential. But, at the same time, the issue began to attract the interest of new actors. Starting from controversies over risk, the debates opened up, borrowing arguments from other contested fields in society. In particular, when GMOs were grown in large quantities or were to be released into the environment, biotechnology was discussed from a variety of perspectives and no longer within the scientific frame alone. Political responses themselves were different according to the regulatory styles and policy traditions of each nation. Over time, regulatory events and arguments raised in the various debates influenced each other across borders and boundaries, and biotechnology became a European issue.

Biotechnology meets society at large: the new economic frame

In phase 1, it was mainly scientists who were concerned about the risks of their daily work in the laboratory. In phase 2, biotechnology began to branch out from basic research into applied science. In pharmaceutical, chemical and, later, also plant-breeding companies, biotechnology improved both R&D as well as production processes. Its promise was to create products that had never before been available, or at least not in such quantities or at such high quality.[11] This produced an enlargement in the range of legitimate points for consideration in discussions of regulation. Economic expectations became more important, and in order to speed up the handling of applications, and thus to enhance national competitiveness, there was pressure for regulation to be simplified.

Another factor supporting this development was the general political climate. The 1980s witnessed a 'paradigm shift' according to Peter Hall (Hall, 1993), manifested in the rise of conservative governments and the expansion of supply-side economic thinking. International competitiveness became important, and the early commercialisation of biotechnology fits neatly into this paradigm. Unemployment increased in many countries and started to become a serious political problem. Competing in the high-technology innovation race appeared to be the only means of maintaining established levels of welfare in the face of rising unemployment and strained social security systems. In this context, biotechnology gained the reputation of being a 'key technology' and an 'industrially strategic area' (Sharp, 1985).

Starting in the USA in the early 1980s, public attention also shifted to the industrial potential of the technology (Pline, 1991). The press emphasised the importance of biotechnology in a new era of the scientific-technological race. Biotechnology was framed as a lead technology, along with information technology, which at that time had already started to have an immense impact on industrial production processes.[12] Accordingly, it was believed that those who missed the chance to be at the forefront of technological innovation would put the wealth of their nation at peril: biotechnology had to be made to happen. Thus, supporters of biotechnology warned national governments not to delay its application by 'unnecessary' regulation. Especially in the USA, where commercial biotechnology was most advanced, regulation was seen as a threat to economic competitiveness. Research had been going on without incident for several years, and the NIH had continuously modified the guidelines with a view to facilitating R&D and, implicitly, encouraging the successful application of its results. Small start-up firms, founded as spin-offs from academic research, struggled to develop their techniques while keeping up the promise for new products in order to raise money on the stock markets (Oakey et al., 1990).

Following this imperative, by 1978 the UK government had recategorised much rDNA work as low risk and advised that it should take place under 'good microbiological practice'.[13] In general, whereas in the 1970s regulation in the form of guidelines for laboratory safety had been brought to the fore, now regulatory efforts were reduced. However, an encouraging regulatory environment was only one element in securing a successful future for biotechnology. Promoters demanded more: namely an explicit and dedicated support strategy for translating scientific insights into commercial products, in other words, to accelerate the speed of the innovation chain.[14]

So, support for R&D gained momentum with the complementary strategy of governments' two-sided approach to fostering biotechnology. In most European 'forerunner' countries, governments funded huge programmes to promote research.[15] However, paramount in realising the promised potential of biotechnology was support for new firms. Some European governments invested in new biotechnology start-up firms or forged relations between academic researchers and industry. Companies such as Celltech in Britain and Kabi in Sweden, but also Novo in Denmark, Hoechst in Germany and Gist Brocades in the Netherlands, embraced biotechnology as a field of basic innovation with enormous economic opportunities.[16]

Emerging debates: commercial applications of genetically modified organisms

New safety problems arose with the possibility of commercial application. 'Large scale' and 'deliberate release' were terms that came to symbolise concerns that went beyond the conditions of basic scientific research (Cantley, 1995: 528).[17] In particular, the regulatory challenges presented by large-scale production had to be reconciled with the desire to support industrial biotechnology.[18] The growth of modified micro-organisms in large fermenters, as opposed to tiny articles of laboratory apparatus, posed a considerable risk. If an accident occurred, a huge number of GMOs, usually bacteria, would be released into the environment, with unknown consequences. However, there were tools at hand to deal with such risks. The NIH guidelines had already defined containment measures for various classes of organisms according to their potential hazard. The first products, available from the mid-1980s, were biologically active substances such as human growth hormone, insulin, interferon and vaccines against an increasing variety of diseases. Other types of products entering the market stage were enzymes for washing purposes and chymosin for cheese production.

The second area of commercial application was agriculture, which implied the release of GMOs into the environment without any containment. There were no blueprints for guidelines at hand, in contrast to the situation with large-scale production. Consequently, uncertainty was much greater, and scientists too had some doubts about their ability to predict risks. The opposition of anti-biotechnology NGOs was more far-reaching (see Beck, 1986). They considered the risks posed by the genetic modification of living organisms to be 'new', because such organisms could not be encountered in nature. The release of such an organism was seen as an irreversible experiment with unforeseeable consequences. In the light of the precautionary principle,[19] opponents claimed that such uncertainty demanded a ban. Proof of the absence of risk was argued to be a proper precondition for the use of GMOs. And, since large-scale production posed the risk of unintentional release, the same argument was advanced for other applications.

Proponents of biotechnology considered the NGOs' demands to be illogical nonsense, since the absence of something cannot be proven. They pointed to the fact that every innovation to date had entailed some uncertainty, otherwise it would not have been innovative. According to proponents, in the case of biotechnology in general, and even releases of GMOs, a risk/benefit analysis would reveal huge benefits combined with minimal risks. This opened up a second path of argumentation. Yet opponents also claimed that the new biotechnological products had disadvantages compared with traditional ones, which meant that the risk/benefit relation would always be negative in their view.

The new discussions about the deliberate releases of GMOs emerged across the Atlantic, when in 1982 researchers in the USA for the first time sought permission for a release, which was granted after years of litigation and conflict (Krimsky, 1991: 113–32).[20] The photograph of a worker wearing a protective astronaut-like suit while spraying bacteria on a field became one of the governing images of what the release of GMOs could mean. While the debate in the USA waned,[21] or became confined to scientific circles, interest was intensified in Europe in a political climate shaped by several other conflicts over technologies. In the early 1980s, some European countries, such as Sweden, Denmark, Germany and the Netherlands, saw frequent public debates about technology, and the struggle over nuclear power was at its height. In a climate in which NGOs and sections of the public were geared to engage in debates, new developments in the commercial application of biotechnology swiftly evoked opposition. For example, when the Hoechst Company announced its plan to construct a plant for recombinant insulin production, this became the German 'key event' and triggered widespread public debate, NGO action and ligitation.

However, we have to keep in mind that reactions varied from country to country. For example, the question of whether or not GMOs should be released elicited virtually no reaction in France but led to a heated controversy in Germany.

Biotechnology as a sounding board

It was not only the risk issue involved in the move from basic research to applied science that had huge consequences for debates on biotechnology. Once elicited, debates rarely stuck simply to the technology itself: they came to inhere in much broader societal discourses. Together with interests in commercial application, those interests focused on biotechnology that ran counter to the kind of innovations the new technology would bring about.

An example was the conflict over recombinant Bovine Somatotropin (rBST).[22] This case sheds light on the fact that, from early on, there was acute sensitivity about applications of biotechnology in agriculture, and revealed that more was at stake than the health of people or animals. Here, socio-economic considerations about the structure of agricultural production were clearly at issue. Opponents considered biotechnology to be spearheading an agricultural rationalisation and industrialisation oriented solely towards economic criteria.[23] Although such arguments were dismissed as irrelevant when it came to the approval of single veterinary drugs, these concerns were nevertheless considered implicitly and played a role in the decision-making process (Tichy, 1995).[24]

Even more fundamental was the debate on medical diagnostics, because this clearly touched on issues of medical ethics. As the German sociologist Ulrich Beck (1988) speculated, such applications would not only permit, but reinforce, eugenic sentiments.[25] In some (Romanic) countries, ethical considerations had played a role for some time in the discourse among experts and churches. But Beck's concerns were derived from both ethical and societal considerations, since improved medical diagnostics might have major consequences for health insurance systems and for social solidarity.

Characteristic of this phase was also the fact that arguments were not focused on the different forms of application, but often transgressed the boundaries and conceived of biotechnology as a general entity. Critics tried to construe, in all applications, the underlying rationale of biotechnology as a technological and reductionist perception of life itself, one that seeks to instrumentalise life for the sake of profit. For their part, promoters tried to demonstrate the contributions to the common good that would emerge from the new methods. They stressed enhanced precision, widening possibilities for medicine, environmental remedies and industrial production. Ethical problems were sometimes acknowledged, but were usually left to expert advice.

Opponents often sought advice from scientists opposed to the analytical/synthetical approach of molecular biology, and often found them in the ecological sciences. Hence, in the struggles over biotechnology, scientific expertise of two different kinds was deployed, either denying the possibility of hazards or emphasising uncertainty and risks.[26] This contributed to the bewildering array of laypeople and politicians who demanded 'impartial' expertise from the sciences, since for most of them it was inconceivable that there was more than one version of 'objective' truth. Ultimately, this would contribute to the erosion of trust in scientific advice.

As the debate grew more intense, the argumentation patterns became even more concentrated on social and environmental aspects and the implications of the technology than on the technological methods themselves. In dealing with social, economic, ethical and environmental subjects, the frame became enlarged. As well as debating new technological opportunities and risks, the basic relation between man and nature was questioned. And, by looking at the performance of the actors involved, the relation between scientific and political elites and the general public came into focus (Kliment et al., 1994).

The media reported two different stories about biotechnology. Industry emphasised commercial applications including large-scale fermenting and deliberate release. Exaggerations of future promises were commonplace as companies tried to attract venture capital, despite the fact that very few of them had an actual product. On the other hand, environmentalists put forward not only risk arguments – whether or not they were substantiated – and abstract ethical considerations, but also concrete and serious doubts about the track record of the chemical and agro-industry and the financial entanglement of senior researchers with start-up companies (Cantley, 1995: 547).

Thus, the image of biotechnology as a potentially controversial technique became established in part of the elite's mind. In phase 1, the critical arguments originated from within the scientific community; in phase 2, all sorts of organisations, including churches and NGOs, contributed to the debate. Biotechnology became a widely discussed political topic, where underlying interests and world-views played a significant role. Also the meaning of 'risk' changed: in phase 1, a technical risk definition was still adequate, whereas in phase 2 critics linked the issue to other contested areas, such as reproductive technology, feminism, animal rights and even the desired path of future development. The argumentation patterns reflected the technology in its social and environmental context. Along with new technological opportunities came new questions: who should have the profit? And what should the relationship be between the scientific and political elites and ordinary citizens? By going beyond the narrow range of issues that most scientific experts deemed legitimate for discussion, the seemingly 'natural' way of implementing the new technology and entering the next stage in the innovation chain was seriously challenged.

In order to understand the evolving controversies, we have to look at how the political systems in the European countries reacted to this challenge. Not surprisingly, we meet a stunning variety of responses. Clearly, there is no such thing as a standard political response; rather, each country developed its own way of dealing with the emerging challenge posed by different stakeholders' demands.

Political responses

Proponents wanted to continue the expert-oriented self-regulatory practice established in phase 1. This was perceived to be sufficient for all legitimate concerns. After all, according to scientists, genetic engineering methods were inherently safe; and they discounted the potential for intentional misuse or remotely possible unintentional hazards. Critics tried to abolish industry's privilege of self-regulation, which, they argued, inevitably entailed a conflict between commercial interest and safety. They wanted to broaden the scope of regulation to cover all applications, or even to ban the technology altogether.

There was major variation in these lines of thought across Europe, owing to differences in the development of the biotechnology industry and research bases, but also because of differing political situations. Some countries had green parties in their parliaments, or NGOs integrated into the national political system, and others had not. The role of churches had been different, and governments were confronted with public awareness of the potential risks of genetic engineering to differing degrees. Some countries experienced severe economic crises during the 1980s, and the period witnessed the rise of conservative governments and an increased emphasis on economic issues.

Accordingly, variations in national regulation at the beginning of phase 2 continued or even increased over time.[27] The task for each government was to realise the promises of biotechnology as a 'key technology', but the strategies employed differed. Governments did not heed public demands for a ban on biotechnology; rather, they looked for possibilities of integrating concerns in a way that did not threaten the economic status of national industries or the competitiveness of the research base. They also sought to maintain efforts to support R&D activities. These political strategies, designed to avoid widespread conflicts, were more or less successful in most countries. This does not mean that, in reality, a national 'consensus' was achieved. Rather, the opposing forces – with some notable exceptions – did not intend, or were unable, to provoke an open conflict fought out in a public arena.[28]

Countries and policy styles

Within the range of European diversity, some countries provide exemplars of the different types of political response. We can distinguish four (idealised) ways of dealing with the task of regulation and with public concerns:

1. exclusive or elite decision-making (as in France and the UK);
2. co-option (as in the Netherlands and Sweden);
3. public participation (as in Denmark);
4. delegation to the European level (as in most southern countries, among others).

Model I: exclusive or elite decision-making   France had experienced a relatively intensive and controversial debate on nuclear energy, which had even been 'significantly more aggressive' than that in Germany (Rucht, 1995: 282). In comparison, public awareness of biotechnology remained negligible. France was the only 'forerunner' country that succeeded in maintaining an entirely expert-based regulatory style without the incorporation of public concerns. The partnership between public authorities, industry and research institutes was supported by the recruitment system dominated by the 'Grands Corps', which permitted common orientations and personal ties to develop among the different societal elites. Until 1996, other institutions such as churches, consumer or environmental organisations never succeeded in making (or never intended to make) biotechnology a controversial issue. In phase 2, this closed regulatory process was not confronted by any public mobilisation as far as biotechnology is concerned.

Like France, the UK had a closed political culture, supported by the electoral system. But British science and technology policy-making, although oriented towards expert knowledge, favoured a pragmatic case-by-case approach, in which the main actors cooperated in the introduction of a series of more-or-less ad hoc regulatory mechanisms through a complex network. 'The British style of policy-making...tends to be informal, co-operative, and closed to all but a selected inner circle of participants. Disputes are resolved as far as possible through negotiation within this socially bounded space' (Jasanoff, 1995: 325f.). The major difference from France was that this closure was handled pragmatically, with channels provided for the integration of NGOs. Advisory committees to government (a key part of UK technology policy-making) could take account of wider public concerns because of their membership (they often included public interest representatives), as well as through various forms of public consultation.[29] Thus, government could reconcile the critics' demands comparatively well without jeopardising its elitist decision-making style and without becoming truly participatory. When experiencing early difficulties with the public acceptance of biotechnology, the British authorities included environmentalists on the advisory committees. This pattern secured the co-option of moderate groups to include 'a voice of reasoned dissent that could be internalised without seriously jeopardising the evolution of technology' (Jasanoff, 1995: 319). Such committees were also flexible enough to adapt to new policies: following recognition of the industrial use of biotechnology, in 1984 the advisory body was replaced, marking the shift from a regulatory phase to that of a concrete promotion strategy.

Early in phase 2, the German public was mobilised. This was the peak of a development, which began with the student movement of the late 1960s, questioning the traditional path of technological modernisation. Different concerns 'merged together over the years and formed a highly active and politicised movement sector with considerable overlaps by the end of the 1970s without losing its political and organisational heterogeneity' (Rucht, 1991: 185). The government tried to channel the emerging public debate back to the scientific experts (Gill, 1991) by concentrating on technical details.[30] But these attempts to close the debate failed. With the Green Party, debate moved to parliament, and there a Commission of Enquiry discussed the (economic) opportunities versus the (technical) risks of biotechnology on a rather broad base. When attempts to ban the technology were dismissed, the Green Party started to mobilise the public (Gill, 1991: 172). German industry, at first strictly opposed to a special law, came to appreciate the need for a clear legal framework as a protection against litigation. The German gene law of 1990 offered, as a partial concession, participation on the government's key advisory committee and created a new public hearing process for deliberate release applications. These innovations seemed responsive to the enactment. In practice, however, the first hearings deteriorated into administrative wrangles and rhetorical stand-offs that led the government to revise the law in 1993 (Jasanoff, 1995: 323).

Model II: co-option The situation in the Netherlands was quite different from that in the UK and in France. As in Germany, radical groups had been involved in the debate on biotechnology, and in the late 1980s fields with GM crops were destroyed (van Praag, 1991). But social support in general for radical political action was low. 'All movements contain important reformist factions, primarily concerned with influencing decision-making, which therefore attach great importance to a pragmatic bond with political parties, trade-unions and various ministries' (van Praag, 1991: 312). Dutch biotechnology policy had always been explicitly two-sided: technology stimulation to harness the potential benefits was combined with attention to potential risk and public accountability.[31]

[T]he relationship between [new] social movements and the government... differs strongly from countries such as the Federal Republic of Germany. In the Netherlands, this relationship must be seen against the background of the traditional bargaining model which has always governed labour relations and the political practice... of subsidising various social groups and, if necessary, involving them in policy preparation. After the initial shock of the various protest manifestations in the sixties, the government soon decided to modify its politics with regard to the... [different] movements. Cooperation and consultation, rather than repression, were the government's first priorities.... Up till now, this policy has been rather successful; the Netherlands is one of the least [politically] violent countries in Europe.... One could say that the Dutch political system is relatively open to new demands and social needs. (van Praag, 1991: 311)

In social democratic Sweden, the relatively tough regulatory regime was seen as a barrier to competitiveness and international collaboration despite certain relaxations of the regulations (Sharp, 1985). Concerns still focused on the need to ensure the management of health and safety risks, but in an environment that would not restrict R&D, since technological development was seen as being part, or even the basis, of societal progress. As it became clear that defining parameters for risk and support required a close relationship between the academic and industrial communities, the regulators' answer was to bring stakeholders from all backgrounds together in several scientific advisory committees, thus enlarging the scope of the problem definition while keeping it under strict control.

Model III: public participation Denmark, according to the results of the Eurobarometer survey (Durant et al., 1998), had one of the most 'sceptical' publics in Europe. Biotechnology issues were hotly discussed and environmental movements had been important in creating a public debate (Gundelach, 1991: 278). Yet public debate seemed to be more differentiated, and Denmark appeared to be the showpiece for public participation. This participatory political approach also reflected the regulators' inclusive political tradition. 'In Denmark... politicians have been very efficient in taking the wind out of the movements by entering into discussion with the movements and accepting several of their demands' (Gundelach, 1991: 274). Using the consensus conference model,[32] Denmark engaged in a public debate that was deemed legitimate and was able to ensure at least public tolerance of biotechnology. In 1986, it became the first country in the world to pass a special 'horizontal' law on genetic engineering, covering various applications in research and industrial production under a single piece of legislation. This largely followed the NIH guidelines for work with recombinant organisms in the laboratory, but it also provided a means to apply biotechnology on a large scale. Concerning releases, there were certain regulatory idiosyncrasies, which met the critics' demands in a rhetorical way.[33]

Model IV: delegation to European regulators In 'latecomer' countries – mostly southern ones such as Italy, Spain and Greece, but also countries at the European periphery such as Ireland – biotechnology was not as developed as in the 'forerunners', and public awareness was more or less absent. Hence they could choose simply to proceed with the application of the NIH guidelines until such time as the EU developed regulations aimed at harmonising the common European market.[34]

During the 1980s, as at other times, most governments aimed to promote the research base, and there was no need to develop their own regulatory activities, which could have led to the public's becoming interested or suspicious. Debates about biotechnology, if they arose, remained national, and even in countries sharing a common language, such as Germany, Switzerland and Austria, the situations were entirely different. Whether or not the public 'caught fire' remained largely a matter of chance, or depended on a mobilising event such as the first release of a GMO or the construction of a particular production facility. In most 'latecomer' countries, there was no such crucial event during the 1980s.

Cross-country repercussions and EU regulation

Although debates remained confined to national settings in phase 2, any national regulation – such as the German gene law – was, of course, not totally detached from what was going on elsewhere in Europe. This piece of legislation, for example, took into account the precedent of the Danish law as well as those of the USA, the UK and the Netherlands in attempting to legislate for the entire range of biotechnological applications. The German law also built on the 'Recombinant DNA Safety Considerations' of the Organisation for Economic Cooperation and Development (OECD, 1986; see below),[35] and mirrored the discussions at the EU level of proposals for several special directives on biotechnology.[36] In fact, the EU proposals on the idea of separate legislation may have weakened those forces in Germany that still resisted a special national gene law.

Conversely, regulatory developments in Germany also had repercussions in other European countries. In France, in the late 1980s, policymakers attempted to counteract what they perceived as rising public distrust (Gottweis, 1995: 394–402).[37] What happened in Germany was considered anathema to the promoters of biotechnology. But, unlike the British and German governments, the French did not pass any legally binding regulations, in particular because a regulatory decision was soon to be expected at the EU level. Rather, they followed a 'business as usual' strategy, keeping matters within the realm of expert decision-making.[38] From 1986, the Commission Génie Biomoléculaire at the Ministry of Agriculture promoted the objective of France becoming the leading European country in the field of GMO releases. Indeed, the number of French releases did increase, and by the end of 1998 they were almost double those in Italy, the country with the second-largest number of releases in Europe. Only in Canada and the USA were there more releases.

Regulation and the economic frame

The European Commission had long been watching the emergence of regulations in European countries. It feared that this trend would jeopardise the common market in a future field of high technology, and over the 1980s it made serious efforts to define a common European legal standard (for reviews see Cantley, 1995; Galloux et al., 1998). The first such attempt, a proposal in 1979 by DG XII, had been abandoned because at that time non-regulation was the favoured option (Cantley, 1995: 526; Galloux et al., 1998: 180). Since the methods of biotechnology had not resulted in any apparent negative incidents, the intention was not to endanger the 'key technology's' promise of industrial growth and new products.

In response to changing perceptions of the need for regulation, in 1982 the Council recommended the development of oversight structures. The general feeling over the next three years was that new applications would not pose insurmountable problems as far as safety was concerned. To enhance European competitiveness,[39] the Commission put more emphasis on a coherent strategy to promote R&D through DG XII and its Biotechnology Steering Committee.[40] However, with the small budget available, this proved to be difficult for the Commission, and industry hesitated to cooperate. 'At the end of the 1980s, the Commission saw itself confronted with the fact that its aim, namely to promote cooperation between public and private research and to construct a European research network, could not be reached' (Bongert, 1997: 127, our translation).

Extension of the frame of regulation

The political efforts for the promotion of R&D changed in the mid-1980s. Paramount was 'the entry of environmental interests into the policy debates on biotechnology . . . , as the public authorities at national and Community levels interpreted their general responsibilities for the protection of the environment in relation to the challenges of the new processes and products resulting from biotechnology' (Cantley, 1995: 546). This marked a major widening of the scope of what were perceived to be legitimate problems involved with biotechnology regulation. Although the 'precautionary principle' was not explicitly adopted, the environment became a focus of attention, along with economic and agricultural interests, while the question of workers' protection was still unresolved. The views of the 'sounding boards' in the individual countries were heard in the European Commission.

This was even more the case with the European Parliament, which in 1985 came up with a view more extreme than that of the Commission: it demanded special regulation.[41] In the same year, the Commission set up a committee with members from several Directorates General (Galloux et al., 1998: 181), with the aim of reviewing existing rules and determining whether they adequately dealt with commercial applications. Its composition (representatives of the rival DGs for industry, social affairs, agriculture, environment and science, internally competing for influence) suggested a broader understanding of the issues at stake in terms both of the direction of commercial application and of environmental concerns. At the same time, an amendment to the Chemicals Directive provided a blueprint for much of the possible biotechnology regulation.[42]

The way to the EU directives

The 'hinge year' proved to be 1986 (Cantley, 1995: 549). The first genetic engineering law, passed in Denmark in order to provide an overarching legal frame for the activities of biotechnology research and industry, put the Commission under pressure to act. The German Commission of Enquiry, influential across the borders of the country, also had the implicit aim of devising a national biotechnology law. Both the Danish and German initiatives, as well as the European Parliament's report, stressed the need for special legislation.

On the other hand, both Britain and the Netherlands felt comfortable with their pragmatic approach under existing sectoral legislation. They received backing from the USA, which defined competencies among government agencies under existing legislation in its Coordinated Framework for the Regulation of Biotechnology. In the same year, the OECD issued its 'Recombinant DNA Safety Considerations' (OECD, 1986), introducing recommendations for the risk assessment of GMOs, e.g. the 'step-by-step' principle and the idea of Good Industrial Large Scale Practice (GILSP), while postponing guidelines for deliberate releases. The OECD report was in line with the US Framework and held strongly the position that 'there is no scientific base for specific legislation to regulate the use of recombinant organisms'; this was also the way sectoral DGs of the European Commission, as well as industry, preferred to see the issue. Inadvertently, however, the report helped to pave the way for viewing recombinant organisms as constituting a distinct category.

The question of whether a transgenic organism should fall under special regulation solely on the grounds of its genetic modifications, and irrespective of its application, was hotly debated. It is still a moot point whether such 'horizontal' regulation is adequate, as opposed to 'vertical' regulation for each product according to its properties. Although industry and scientists repeatedly denied the need for – and the feasibility of – a special regulation for GMOs, the Commission finally proposed 'horizontal' regulation in a 'Community Framework for the Regulation of Biotechnology'. The reasons for this step (see Cantley, 1995: 550 ff.) were threefold and may be conceptualised in terms of harmonisation, risk reduction and dealing with uncertainty. First, the Common Market was to be protected from idiosyncratic national regulation. Secondly, the intention was to apply appropriate guidelines for containing levels of production universally. Thirdly, there was still uncertainty about the criteria for the risk assessment of GMOs to be released to the environment, so a case-by-case approach was designed to systematise the accumulation of experiences.

Clearly, one additional, and openly admitted, aspect of the pressure for regulation was 'public concern, in part a spillover effect from other areas of technology where accidents of an unexpected nature or scale had occurred'.[43] In line with the European Parliament's position, and rhetorically fortified by public concerns, the Commission decided to issue a regulation on GMOs. The proposals for the Directives on Contained Use, as well as on the Deliberate Release and the Placing on the Market of Products, exclusively dealt with GMOs.[44] So, although other directives or drafts had been issued that were more sectorally oriented,[45] this approach was abandoned in favour of a more horizontal one.

Industry strongly opposed such horizontal regulation, which it feared would lead to the 'stigmatisation' of the technology of genetic modification. Consequently, the proposals prompted harsh criticism from industry and scientists. These critics were supported by the USA, which claimed that there was no scientific basis for regulating GMOs as a special category; that the directives would hinder R&D; and that they would introduce non-tariff trade barriers on future products.[46] The opposition prompted the EC to construct the Directive on Deliberate Release and the Placing on the Market of Products (which became known as Directive 90/220/EEC) in such a way as to facilitate future 'sectoral' legislation (in spite of the 'horizontal' scope of the proposal). The Commission immediately started work on formulating such legislation for 'novel food'.

Two opposing demands had to be balanced. On the one hand, the European Parliament as a representative of a fictitious 'European public' conveyed the plea for a very cautious approach following the public unease expressed in several member countries, where fears of long-term and indirect effects clearly played a role. On the other hand, science and industry demanded a 'purely science-based' approach focusing only on organismic and product properties, in order not to jeopardise innovation by regulatory demands that remained ambiguous to many. Industry wanted a narrowly defined legal framework so as to enter the race for the application of a 'key technology'. It was sympathetic to the explicitly deregulatory US approach of officially relating the degree of regulation to the amount of 'objective risk' posed by the organism.

The political solution seemed to be to stress uncertainty, and the Commission defended the need for a special regulation with the argument that the technique was novel. So the purpose of the regulation was announced to be proactive (i.e. put in place before any accidents had occurred), in accordance with (at least a certain understanding of) the precautionary principle. The directives were definitely not enacted on the basis of proven risks from the methods of genetic engineering; rather, the rationale was to use discretion until more experience was gained and regulation could be relaxed.[47] By emphasising uncertainty, this regulatory solution tried to force together two apparently incongruent approaches to risk assessment, namely one that built on scientific evidence ex post, and one that built on scenarios of hypothetical risks ex ante. It also acknowledged the dual nature of the 'biotechnology problem' – technical as well as a matter of public perception – without openly addressing issues beyond risk that could not be dealt with by scientific experts. This artistic and delicate balancing act attempted to bridge the gap between the different regulatory styles and public attitudes in various European countries, in order to provide a unifying framework for future technological innovations. Since the stakes were set so high, it was no wonder that difficulties later arose.

Summary

In phase 2 biotechnology and the debate surrounding it diversified and extended, as it became clear that the new methods could be applied in various fields and contexts. The argumentation reflected two lines of thought prominent in the 1980s. On the one hand, the economic exploitation of research results (as of anything else) was prevalent in a climate shaped by the increasing dominance of supply-side and market economic thinking. On the other hand, technology critics placed biotechnology on the list of contentious technologies, together with nuclear energy, that were considered to play a crucial role in the 'risk society'. Uncertainty about both possible or perceived risks and promises of economic opportunities led some European governments to 'balance chances and risks' via national regulation. The European Commission recognised several challenges, all of which highlighted the need for unified conditions in the application of biotechnology. First, industry demanded a defined legal framework for its activities. Secondly, national regulations threatened to jeopardise the common market for new products. Thirdly, since public opposition in mostly Scandinavian or German-speaking countries supported the demand for a special law, there was a need for compromises in order to contain criticism and to prevent its spread.

Phase 3 (1990–1996): European integration

The European Community's biotechnology directives were an attempt to arrive at consistent and homogeneous regulation across the EU. The aims were to prevent the passage of idiosyncratic national laws that would jeopardise the implementation of this key technology in a common market, and to provide universal safety standards. Aside from this, the directives had a variety of impacts in a period in which the EU was considerably enlarged, and when biotechnology regulation became a hotly debated issue across the Atlantic. Although the homogenisation of regulatory approaches proved to be difficult, public debates waned as biotechnology proceeded and became entrenched in science and the economy.

Directives

In 1990, after long deliberations, the European Commission passed a set of directives on biotechnology.[48] The Contained Use Directive imposed a categorisation scheme for micro-organisms and the split between the application of such organisms in 'small-scale' (e.g. for research purposes) and 'large-scale' devices (e.g. for production). The Deliberate Release

Directive rendered mandatory the principles of 'step-by-step' (from laboratory experiments to small research releases to releases of a large number of organisms and, finally, to commercialisation) and 'case-by-case' (which implied that each release application had to be considered separately and, although some provisions for a later exemption were made, there was no categorisation). Permissions for the commercialisation of products were granted Community-wide, whereas permissions for releases remained within national competencies. The directives did not permit socio-economic criteria to be considered during the assessments of planned releases or products to be commercialised; only risk issues were intended to be dealt with. This should have served to restrict the range of permissible arguments and to keep the regulatory process firmly within the scope of scientific risk assessment. However, it turned out that this expectation could not be entirely fulfilled.[49]

This programme was, in fact, a compromise. Although influenced by public debate, it fell short of a comprehensive regulation that took into account all the concerns that had been voiced in the various countries.[50] The directives provided a legal framework, but industry was not very enthusiastic: it perceived the scientific base of the technology-centred regulation to be flawed, and feared difficulties in implementation as well as an 'unnecessary burden' for biotechnology. Despite their perceived shortcomings, and the Commission's reluctance to see the directives as a final piece of legislation, their importance was enormous.

Now biotechnology had become a truly European matter. Until then, problem awareness among scientists and regulators had frequently been triggered by developments in the USA. Public debates, on the other hand, had arisen almost exclusively in response to certain national events triggering a reaction in the media and arousing the interest of NGOs.[51] The relative failure of the European Commission to create a Community-wide research network by the end of the 1980s had mirrored the national preoccupation with the research base as well. Hence, before 1990, matters of biotechnology had largely been either transatlantic or national issues in the 'forerunner' countries and non-issues everywhere else. The directives marked a significant shift, since the intervention of the EU in the regulatory process had consequences for all EU members as well as for other countries.

- The 'forerunners' had to adapt their national regulations to the new European standard.
- Other member states of the EU had to adopt the regulation of a hitherto unregulated field.
- New member states had to implement the European regulation.

- After the collapse of the Soviet system, East and Central European states became liberal democracies and developed market economies. As part of their drive towards membership, they adopted relevant EU regulations.
- In order to prevent future trade wars due to 'non-tariff trade barriers', when GM products were to be placed on the market ways had to be found to reconcile the directives with the US and Canadian way of (de-)regulating biotechnology.

Impact of the directives on European countries

'*Forerunners*' The 'forerunner' countries had, over the years, set up rules in a system of national interest intermediation. The delicate balance of power between the various actors arose from written and unwritten rules and structural conditions varying from country to country. These rules and conditions changed when regulations started to be negotiated in Brussels instead. The strategic balance in biotechnology conflicts changed dramatically thereafter. Interest groups lost their influence unless they joined European organisations, where there were different rules and different actors had their say.

In general, a multi-layered regulatory system like the EU tends to favour more potent actors such as industry, since they have better resources (Bandelow, 1997). Consequently, the shift of the decision-making power to Brussels had a negative impact on the influence of national NGOs, whose resource base was relatively weak. Industry radically reorganised their lobby apparatus in Brussels in the early 1990s, creating a new common organisation called Europa-Bio that served to increase their visibility for the Commission and other actors. Consumer and environmental organisations were much slower in securing representation at the European level. This might have been an additional reason why regulation in phase 3 primarily followed industry's interests, and why regulatory changes were oriented to deregulation rather than to public concerns as communicated by NGOs.

The Deliberate Release Directive mentioned the possibility of public participation in national decision-making on some issues, but this was not mandatory and was implemented in only a few countries. Jülich (1998: 56) analysed the formal possibilities for participation in such processes and distinguished two groups of countries. The Netherlands, Luxembourg and, somewhat differently, Denmark and Norway (not an EU member, see below), but also new members of the EU such as Sweden and Austria, offered guaranteed channels for public participation. In other member states such as Belgium, France, the UK, Ireland, Italy,

Portugal and Spain, opportunities for public involvement in decision-making processes were minimal. Germany was in the first group until 1993, when the revised gene law restricted participation.

The directives elicited different reactions in each member state. In some countries, implementation entailed stricter rules or even initiated regulation; in others the existing regulation was relaxed. Some countries felt the necessity of emphasising areas that were not formally regulated by the directives, for example, ethical considerations, but did so by means of committees and not by law. Thus, though EU regulation was confined to scientific assessments of risks to health and the environment, this did not prevent the broadening of risk issues in public debate and even as a subject to be dealt with by official institutions.

The UK's answer to the directives, beyond some legal changes, was to establish an Advisory Committee on Releases to the Environment (ACRE). The UK's flexible political mode also allowed for adaptation to new questions that were not dealt with by the directives: in 1991 the launch of the Nuffield Council for Bioethics signalled the increasing importance of ethical issues. Since then, a plethora of new Advisory Committees to Government has been created in order to deal with ethical and social issues arising out of particular medical applications.[52]

In Denmark, the most important change in the Genetic Engineering Act brought about by the implementation of the directives was the endorsement of the EU approval system for marketing GMOs, which replaced the general prohibition on releases. This demanded free marketing of all products approved in the EU, balanced (rhetorically) by environmental guarantees. After the revision of the Danish regulation, political and public activities around biotechnology abated. It appeared as if Denmark had found a level of regulation that was acceptable to most actors; though some frustration on the part of the opponents may have persisted. Later on, this temporary frustration manifested itself in renewed opposition.

In France, the directives forced stricter regulation than previously and ended the particular French method of scientists' self-regulation. Genetic engineering as a topic found its way into political debate, but the French system was also confronted by new questions. As in Britain, problems of bioethics officially entered the political arena, although they were not formally regulated. Ethical issues had been debated for a considerable time, and French contributions to this field had been influential both at the EU level and within the United Nations Educational, Scientific, and Cultural Organisation (UNESCO).

In the Netherlands, some adaptations to existing laws were necessary, but did not arouse much public interest compared with the advent of the first 'transgenic' (i.e. genetically modified) animal, 'Herman' the bull.[53] In 1992, shortly after the directives had been implemented, this event revealed a shortcoming in the Deliberate Release Directive: its inadequate treatment of transgenic animals. This case elicited a heated public debate: there were things at stake other than risk–benefit calculations, such as animals' rights and their intrinsic dignity, which many considered were being violated by genetic manipulation.[54] The political reactions within the pragmatic and open tradition in the Netherlands consisted of parliamentary interventions, technology assessments,[55] and the establishment of a new ethics committee.

In Germany, the European directives had a considerable impact on national regulation. Soon after it had been passed, the German gene law was criticised as being too strict and a threat to competitiveness. Promoters of biotechnology saw the directives as a chance quickly to get rid of some of the most unwelcome paragraphs. The 1993 revision of the gene law, to comply with the directives, led to considerable deregulation.[56] In the early 1990s, debates saw a shift in focus from environmental problems to economic perspectives (Gill et al., 1998: 22), and the government eventually succeeded in linking biotechnology and economic perspectives within the frame of competitiveness. Consequently, over the 1990s, Germany became the most prominent promoter of deregulation at the European level.

'*Latecomers*' Before 1990, regulation was perceived as a problem mainly in the 'forerunner' countries. After the directives had been passed, the situation changed, especially in the Mediterranean countries, where the public had hitherto taken little notice of biotechnology. However, developments were not at all uniform. Although public debates comparable to those in 'forerunner' countries could still not be observed after the implementation of the directives, over the 1990s GMOs slowly became an issue in Greece, and involved international NGOs at the beginning of the debate. The situation in Italy was different: the issue was hardly on the agenda, although the Catholic Church had argued publicly against biotechnology. The impact of the different economic situation in each country was more decisive for the position of biotechnology than was public debate. In Greece, EU programmes and cooperation increased the funding and strengthened the position of biotechnology, whereas the crisis of the pharmaceutical industry in Italy during the same period slowed down its development.

'*Newcomers*' The directives also became binding for the new member countries Sweden, Finland and Austria, which joined the EU in

January 1995. In Finland, the gene law passed in 1995 closely reflected the EU regulation. This law, however, did not involve a simple adoption of EU policy, but resulted from extensive discussions between governmental bodies and a plethora of different institutions such as universities, trade unions and NGOs. The openness of the regulative process reflected the intentions of official bodies rather than of public pressure, since biotechnology had never been a subject of public or political debate in Finland.

This contrasted with neighbouring Sweden, where the first public debates about biotechnology had taken place in the 1970s. As in Finland, NGOs and a wide variety of interest groups were vocal in 1994. The Swedish gene law exhibited a substantial difference from other European regulations in that it also considered ethical aspects. Other attempts to establish stricter regulation failed. Nevertheless, in spite of the generally critical attitude of the Swedish public towards biotechnology, the debate receded once more after the enactment of the law in 1995.

In Austria, the European directives had already been integrated into the national regulations before Austria joined the EU, solving the conflict over whether or not a specific law was necessary. A parliamentary Commission of Enquiry was set up, based on the German example, with the task of anticipating public concerns. Eventually, however, industry interests became more influential, as reflected in the law of 1994. The Commission's work had hardly any impact beyond the integration of a paragraph on the 'social sustainability' of products. This raised doubts about its compatibility with Directive 90/220, since it also employed a socio-economic criterion (Seifert and Torgersen, 1997). Such criteria were perceived by the EU as raising non-tariff trade barriers, which were contrary to the overall aim of the Common Market. Since the Austrian paragraph on social sustainability was in practice irrelevant, it remained in the law.

Although the Norwegians decided in a referendum not to join the EU, Norway remained a member of the European Economic Area, which required that it comply with the Community's biotechnology regulations. Part of the Norwegian law was in even more striking contrast to the Release Directive's focus on scientifically assessable risk issues than the Austrian social sustainability criterion. The Norwegian law demanded that GMOs be used in an ethically and socially justifiable way, and that special emphasis must be given to the benefit to society and the contribution to sustainable development.[57] Being a 'latecomer' compared with its neighbours Sweden and Denmark, Norway thus adopted the most restrictive gene law in Europe. Obviously, industry interests were unable to counter such moral considerations and national peculiarities (Nielsen, 1999).

The new democracies Phase 3 also saw the decline of the Soviet system and a new political order in Europe. The general situation switched from post-war to post-Cold War, and circumstances in East-Central and Eastern Europe changed dramatically. As a consequence, former Soviet 'satellite' states adopted the Western system and tried to become members of the EU. Whereas most of them had followed a US style of regulation, applying guidelines inspired by the NIH, it now became necessary to adopt the EU directives. Thus, in states associated with the EU (the Czech Republic, Poland and Hungary), the directives led to a reorientation and served as a model for national regulations.

The Czech Republic had no law on biotechnology until 1998, but when joining the OECD it agreed to introduce regulation within two years based on the EU Directives 90/219 and 90/220. In Hungary, the competent authority secured EU conformity by explicitly following the OECD guidelines. In Poland, regulation started earlier with a Patent Law (1993, reflecting the influence of US views) and the Biodiversity Convention (1995). In 1997, Poland submitted the first draft of a new biotechnology law on the basis of the directives, to be implemented with the assistance of international organisations (the OECD, the United Nations Environment Programme and the United Nations Industrial Development Organisation). In all three states, there were no signs of substantial public or political debate on biotechnology. Regulation was left to experts, whose role was questioned neither by domestic NGOs nor by a concerned public.[58]

Political implications

US point of view: product versus process From the American point of view, European regulation was perceived to be process oriented, since biotechnology as such was regulated in the directives, irrespective of what kind of product would result from the process (i.e., the technology) involved. The American position was that it is only the product that can be regulated – irrespective of the process applied in its production. This, they argued, was essential to obviate discrimination against novel technologies. The complicated US system of diverse regulatory rulings and various authorities with competence in this area closely mirrored this attitude. Yet, the EU insisted that its approach was not actually process based. It claimed that, because there was not enough experience with the new technology, as a temporary measure 'horizontal' regulation had been put in place. The EU argued that it expected the horizontal regulation to be replaced, in due time, by sectoral or 'vertical' laws according to the type of product.[59]

The USA was constantly suspicious that the EU would introduce forms of regulation based not upon 'sound science' (i.e. on evidence of harm), but rather on a doubtful precautionary principle open to voluntary and unpredictable interpretations. By and large, the US position was that the products of modern biotechnology were not substantially different from traditional products. And, by invoking this claim, under the heading of deregulation the USA had started, early on, to exempt certain crop plants from the requirement of obtaining permission prior to release. This was hardly acceptable to the EU with its case-by-case principle. Marketing authorisations that proved virtually unobtainable in the EU were more readily granted in the USA. Consequently, the EU was accused of protectionism, because the products in question could not be proven, by 'sound science', to pose any significant risk. Hence the USA perceived any objections posed by the EU as attempts to protect its internal market from US imports. The manifestation of these different regulatory styles in the EU and the USA led eventually to a renewed public engagement in phase 4.

Problems integrating European diversity However, the European stance on GMO releases and marketing conditions was by no means as uniform as it appeared to the USA. There was profound variation in how the obligatory risk assessment was performed; what factors were considered; what 'familiarity' meant; how the step-by-step principle was interpreted; and how to proceed on a case-by-case basis in the face of an increasing number of release applications (Levidow et al., 1996).[60] National regulators implementing the EU directives had to establish their own normative standards and in doing so they made implicit value statements (von Schomberg, 1998). Hence, strong national differences persisted in spite of the existence of common EU regulations. Disharmony was aggravated by different regulatory cultures in the EU member countries, for example with respect to the disclosure of data from the application files.[61] The Commission, aware of the implementation problems, created a high-level inter-DG Biotechnology Coordination Committee in 1991 in order to oversee the implementation of the directives. This powerful committee proved to be open to requests for a flexible revision of the regulation,[62] 'according to technical progress, easing administrative requirements' (Galloux et al., 1998: 181).

Debates waning: towards acceptance

Despite the continuing national differences, the EU regulation contributed to a calming of the debate on biotechnology in most of the European countries. The economic side and the frame of competitiveness seemed to have suborned public concerns.

Progress in medical applications, which had become generally accepted, was one of the reasons the public opposition cooled in many countries where biotechnology had been vigorously contested. It was clear that such medical progress was intimately linked to various areas of biotechnological research. A range of highly welcome new drugs and vaccines had emerged over time. Forensic identification of suspects had made huge progress and had been accepted by courts of law; and in several spectacular cases such methods had led to the identification and sentencing of criminals. The Human Genome Project made great progress during the 1990s in developing methods of establishing the base sequence of various organisms and – ultimately – of the human genome. Although in the USA voices warned of the societal implications for medical insurance of the newly developed predictive medicine, this triggered less concern in Europe with its public health systems providing universal coverage, so that debate relating to this problem was confined to academic circles.

Aside from medicine, a handful of biotechnological consumer products had entered the market in some countries.[63] In general, applications that appeared to be of advantage for the consumer seemed to have gained general acceptance by 1995. When a legal basis had been created that officially satisfied the formal regulative needs within a scientific frame of risk assessment,[64] the first products were sold on the market. Hence, existing regulation (at least in the UK) succeeded in ensuring that innovative products could enter the market.

As early as 1995 the first 'genetically engineered' food product, a canned tomato paste made from transgenic fruit, was sold in the UK by two supermarket chains. For the first time, consumers were directly confronted with such products, albeit in a very small market segment. Even when labelling was not statutory, the companies selling the tomato paste labelled voluntarily, arguing that consumers have the right to choose. These market releases, for the first time allowing consumers to choose between GM and traditional products, did not elicit any public debate in the UK or elsewhere, and the products sold well. Although survey results indicated persistent public unease towards biotechnology (INRA, 1993; Macer, 1994), it seemed that the conflict had lost its ferocity.

However, the picture was more complicated and could not simply be explained as an adaptation process, 'narrowing the gap' between society and technology. In countries such as Germany, 'green' issues had lost some of their appeal in favour of efforts to create jobs, but this was a superficial development. The themes remained in the vocabulary and, as further events showed, also in people's minds. What was actually observed (as, for example, shown by the 1993 Eurobarometer survey; INRA, 1993) was a bifurcation in the public's attitude towards biotechnology: on the one hand, applications that led to new treatments, drugs and vaccines were now rarely disputed; on the other, agricultural applications continued to arouse suspicion.[65] Although, by 1995, the progress of biotechnology in agriculture appeared inexorable,[66] general diffusion conditions for companies engaged in agricultural biotechnology became more difficult.[67]

The biotechnology industry

Not until the early 1990s did biotechnology emerge as a 'key technology' of economic significance. New developments were now being fed into the innovation chain without reference to public debate, especially in the USA, which had taken a substantial lead.

These years were characterised in many European countries by the aftermath of the late 1980s recession; the ever-increasing dynamics of globalisation; and a strong focus on technological innovation. This also led to the emergence of a 'biotechnology industry'. Up to then, pharmaceutical companies and producers of agrochemicals had strong links with organic chemistry. Now, links were being established between the pharmaceutical industry and seed producers, since both depended on biotechnological methods and the patenting of genes (this was of increasing importance), and both were facing rapidly increasing costs for R&D. It resulted in mergers of companies globally engaged in the pharmaceutical and seed sectors. These companies' engagement in particular countries was said to show the latter's competitiveness. In official language, anything that jeopardised efforts to sustain or attain this degree of national competitiveness was to be assiduously avoided. The state's role was perceived to be restricted to providing a congenial environment for industrial performance, and it was no longer considered appropriate for the state to promote other societal goals when regulating biotechnology (Bongert, 1997).

Meanwhile, the biotechnology industry underwent massive expansion. From 1993 to 1996, the numbers employed by that industry more than doubled, although it was still small in comparison with more traditional sectors. The number of patent applications increased sharply between 1990 and 1996, and the sector expected a growth rate of 20 per cent. Nevertheless, European industry still lagged heavily behind that of the USA. This was largely because of structural weaknesses: from the industrial and economic point of view there was 'insufficient collaboration between academia and industry, lack of coordination of research between EU member states, [lack of] shared access to resources and infrastructure, and inadequate venture capital' (Galloux et al., 1998: 177). The aim of the EU's policy was to obviate these shortcomings and to narrow the gap between the EU and the USA.

This disparity was especially pronounced in agriculture. In the USA and Canada, but also in China, GM crop plants, such as maize, tobacco, oilseed rape and cotton, began to cover ever-increasing percentages of the total agricultural area.[68] Aside from the more favourable regulatory climate in such countries, the EU also suffered because the sorts of plants that had been developed were most suited to large-scale agricultural systems like that of the USA, and were less suitable for the small-scale production typified by many European countries. Only France and neighbouring Spain were expected to become significant producers of transgenic crops, with other countries lagging behind.

Summary

In phase 3, all EU member states had implemented national regulations. The political elites were able to handle biotechnology within their scientific and economic frames, and these became more and more dominant. The EU made attempts to homogenise the rules for the application of biotechnology in order to create a single common market for the products to come. This proved to be more difficult than originally expected, especially since the Release Directive did not resolve, once and for all, national differences in thinking about how to proceed with risk assessment and so on. Nevertheless, the directives created a compulsory frame that all member countries had to adopt, and that non-members also took as a reference point (especially because some of them subsequently sought to become full EU members). The shift of decision-making competence to Brussels led to a new power balance in favour of industry. Public debate waned in most countries owing to other problems that had emerged after the collapse of the former socialist empire and in the light of increasing pressure from globalisation. Public resistance against biotechnology, as such, more or less waned. However, this was not due to overwhelming acceptance; rather, the applications were now being differentiated in the public mind. Thus, medical applications were more or less welcomed, whereas agricultural biotechnology was still not so well received. This development was intensified after the marketing of the first commercial food products made from GM commodity crop plants.

Phase 4 (1996–2000): renewed opposition and consumers' distrust

The backlash

In phase 3, economic competitiveness and European harmonisation had been the dominating perspectives and the division in perceptions between medical and agricultural biotechnology, along with the shift to other concerns, had led to a calming of the debate. In phase 4, the issues at stake diversified again, and debates that had been considered closed re-emerged. The importation of the first GM food crops from the USA in 1996 marked the division between phase 3 and phase 4, as consumer products of biotechnology reached the market stage. This trigger event led to renewed public and NGO protests. Later, when new cloning techniques impinged upon ethically sensitive issues, a wave of concern about the disparity between what was technically possible and what was ethically defensible spread across the globe.

So phase 4 brought a double challenge: on the one hand, renewed conflicts about health risks combined with a struggle for consumers' rights, and, on the other hand, deep ethical concerns about the overstepping of boundaries affected the image of experimental biology and also biotechnology. Essentially, the issues upon which these conflicts were based were by no means new, and their outline was already visible during phase 2. But the vigour and rapid spread of the newly arising debate had serious implications for individual countries' policies as well as for those of the EU, and especially for the relationship with the USA.

Notwithstanding the European debates, biotechnology had become a global issue. International trade as well as the struggle to secure intellectual property rights became the most important fields of conflict between the EU and the USA on the one hand, and between the industrialised world and the Third World on the other. Biotechnology had finally, and undeniably, acquired the status of a 'key technology'. Ironically, as soon as biotechnology began to take off, the term itself began to be replaced by other less contested but more comprehensive descriptions.

Prelude: BSE and the loss of trust in experts and regulators

One reason the evolving conflicts over food issues became so ferocious was the earlier scandal in the EU over BSE.[69] This was totally unrelated to the issue of GM crop plants, but it set the scene in terms of consumers' concerns about food safety.

A few months before the first arrival of GM soya in a European harbour in 1996, the British government conceded, after years of rumour, that there might be a link between human Creutzfeld–Jakob disease and Bovine Spongiform Encephalitis. This led to a collapse in the beef market and a ban by the EU on British beef. The main impact of the ensuing 'BSE crisis' was a growing distrust in scientific experts and political regulators in matters of food safety, since they had been reassuring the public for a long time that there was no evidence for such a link. Consequently, regulators had to change their way of dealing with risks: they acknowledged that regulation could not be left to industry, contrary to the approach preferred by the UK government, and they were forced to accept the fallibility of expert committees.

The UK government and the European Commission were harshly reproved for their handling of this scientifically controversial issue. In 1997 the European Parliament criticised the secretive decision-making processes and the complex and undemocratic system of scientific committees. Threatened with a vote of no confidence, the Commission conceded a strengthening of consumer representation in DG XXIV (Consumer Affairs) and more transparency, indicating increasing parliamentary influence vis-à-vis the Commission (Baggott, 1998: 70ff.).[70]

Critics in some countries quickly asserted parallels between the BSE scandal and the way in which GM food products were entering the market. The issues were obviously scientifically unrelated, but opponents considered them both to be consequences of industrialised agriculture. At the same time, organic farming began to leave the small circles of eco-sectarianism in some Central and North European countries, and emerged as a counterpoint to industrialised agriculture and biotechnology. Products from organic farming could be bought in ordinary supermarkets and served a new consumer interest. This paved the way for later cooperation between NGOs and retailers fearing consumer boycotts.

Soya and maize: the European labelling debate

The soya case changed the frames of the regulatory debate away from an emphasis on scientific risk assessment towards finding ways to deal with consumer interests. When the importation of GM soya and maize from the USA to Europe began some months after the outbreak of the BSE crisis, trust in experts who had consistently reassured the public about food safety had already been shaken. The assumption that consumers would generally accept GM food (were it available) proved to be wrong in spite of previously successful market introductions into the UK.

After the EU had granted an approval for the US seed company Monsanto's GM soya, in the winter of 1995/6, the first actual imports from the USA arrived in November 1996. Obviously in an attempt to create a *fait accompli*, the US-grown modified soya had been mixed with conventional soya. Although legal from a US regulatory point of view, this precluded meaningful labelling. However, consumer and environmental organisations had demanded the 'right to choose' between GM and traditional products. The fact that consumers would not be given such a choice was successfully used by a combination of consumer and environmental NGOs to mobilise the public in most EU member countries. As a result, in many countries retailing chains that were concerned about consumer confidence joined consumer organisations in pushing for clear labelling, and renewed public opposition led to a heated debate about the appropriate labelling of such products, which industry was unwilling to guarantee.

This event was widely reported in the media and resulted in a change of priorities for regulatory efforts at both national and European levels. The most important effect was that the EU was now pushed to finalise the Novel Food Regulation. Plans for such a regulation had existed since the Release directive was debated, and the Commission had issued the first proposal as early as 1992 (see Lassen et al., chapter 10 in this volume). From the very beginning, the Novel Food Regulation was meant to cover the introduction of all new food products – including products of biotechnology that were in principle regulated by the Directive on Deliberate Release. Since the directive did not specify labelling, however, there were major differences between the approvals already granted for maize and soya and those applicable to future products.

The policies of the member countries, the Commission and the European Parliament differed significantly on the question of labelling (Behrens et al., 1997).[71] The Commission and the Council had consistently been more sympathetic to industry's arguments than Parliament was, and they had tried to regulate the issue within the economic and scientific frame of safety. However, as in the conflict over BSE, influence began to shift from the Commission to the Parliament in late 1995. The Commission accepted a proposal for the Novel Food Regulation that allowed for compulsory labelling under certain conditions[72] – if the new product was recognisably different from existing equivalents; if the product might give rise to ethical concerns or had health implications; or if the actual GMO was present in the product. When marketing approvals for GM soya and maize were pending, the labelling question became more and more pressing;[73] and, when the Novel Food Regulation was passed in January 1997 (shortly *after* the approval for maize), it took most of the Parliament's suggestions into account. This marks a turning point in that the Parliament's views now exerted the greatest influence. However, with the passing of the new regulation, problems lingered on.[74] Owing to delayed implementation, by 1999, on shelves throughout Europe, there were still unlabelled products that contained modified soya products, if only trace amounts.

Voluntary labelling had been devised as a means to ensure the public acceptance of biotechnology food products. The reluctance of the food industry to label, and the political pressure required in compelling it to do so, resulted in a conflict that created obstacles to market entry. Public opposition had reached such a level in some countries that any comprehensive labelling appeared to be a hazard for the producer. Consequently, industry became very reluctant to issue new GM products in the EU. Lacking consumer acceptance for its food products, agricultural biotechnology was in danger of never becoming established in Europe.

In line with a new emphasis on ethics in all fields, ethical questions also began to play a role with respect to the marketing of GMOs. Critics referred to the old problems of how to deal with consumer risk and protection. The growing uncertainty concerning the merits, dangers or moral ambivalence of biotechnology fostered and energised new actor constellations, such as organisations claiming to address consumer interests and retailer chains. These actors differed in importance, means of influence and power among the European countries but, particularly after 1996, they commanded considerable trust in the public sphere. Their success was facilitated by the provocative way in which multinational companies had tried to push their GM products in the European market. They made it easier for critics to portray GMO producers as villains. Additionally, references to agriculture stimulated particular national sensitivities.[75]

However, it was not only plant biotechnology that generated fierce dispute. As the following short excursus shows, other areas of biological research elicited heated responses; and in this case the moral dimension was writ large.

Dolly and the international moral accord

When, in late February 1997, the first cloning of a mammal from a somatic cell made the headlines in all Western countries, the reactions to 'Dolly the cloned sheep' as a media sensation were of a different quality from the public anxieties about soya and maize. Controversy in this case was not related to consumer concerns, fears of environmental disaster or outrage at regulatory misconduct. Instead, this event evoked the 'moral danger' of human cloning. The common feeling was that the transgression of moral boundaries was imminent and had to be prevented. In an interplay between international political actors and the media, within days a consensus developed across the whole industrialised world that human cloning should be prohibited (see Einsiedel et al., chapter 11 in this volume). In Europe, the media event triggered a more or less synchronous mobilisation of the various national publics, but the issue was still discussed at a national level. As with the GM soya case, there was still no common 'European' public, even if debates were now emerging in countries in which the public had hitherto been virtually silent on the issue of biotechnology (for instance, in Greece, Italy, Poland and Ireland).[76]

As an affirmation of this moral consensus, instant policy responses emanated from the Pope, the American President, the British government and the European Commission. The case of Dolly gave authorities an opportunity to demonstrate their responsiveness to public moral sentiment by announcing prohibitions in the form of various pieces of legislation on reproductive medicine and patenting. Their swift action was surely designed to renew trust in regulators and prevent a spillover effect on the image of biological research. However, in the public eye, Dolly became linked to other aspects of biotechnology, again going beyond the borders of a purely scientific understanding. Thus, by evoking human presumptuousness in interfering with (arguably) natural or sacred orders, biotechnology assumed clearly negative moral connotations, in spite of the international moral accord.

National responses to public opposition

Fading consumer trust prompted national publics to exert pressure on their governments. Additionally, in 1996 and 1997 some 'latecomer' countries experienced the trigger events of the first experimental releases of GMOs, and these prompted public debates of a hitherto unknown character. Another factor militating in favour of opposition was that in 1997 and 1998 elections transferred political power to centre–left governments in France, the UK and Germany. The new governments increased the attempts begun under their predecessors to rebuild public accountability in matters of food safety. With governmental changes, critics' arguments became more influential than before, and the balance of power in the EU Council of Ministers also changed.

'Latecomers' catching up

The first country to witness a broad public mobilisation against biotechnology during phase 4 was the new EU member Austria. Although there was no noticeable public or media interest until 1996, the first release of GMOs caused considerable turmoil for the Austrian authorities and the biotechnology industry.[77] A 'people's initiative' (an official petition) calling for a moratorium on agricultural applications and patents received high levels of consent and placed the government and administration under considerable political pressure. In early 1997, the authorities imposed a ban on the import and agricultural use of Bt-maize, despite the Commission's market approval, thus deliberately violating EU regulations. Austria upheld the ban, while the Commission, mainly for procedural reasons, failed to enforce its regulation.

In other small member states, such as Greece and Ireland, with no previous debates on biotechnology, the first GMO releases triggered public opposition in an atmosphere affected by the Dolly story and by controversies over food safety. In Greece the first release took place in spring 1997. This encountered opposition from NGOs, which occupied test fields and succeeded in gaining media attention. As in parts of Austria, Greek agriculture is small in scale and there is an increasing market for organic products. In November 1998, Greece followed the Austrian path in banning GM oilseed rape, which had previously been approved EU-wide.

Because of its close trade relations with the UK, Ireland was particularly affected by the BSE crisis. As in Greece, the first releases took place in spring 1997, meeting NGO resistance that manifested itself in the destruction of fields. In the 1997 election campaign, agricultural biotechnology was a contentious topic among the critical Irish public. During a temporary moratorium, the Irish government organised (public) consultations over the summer of 1998. Subsequently it announced a moderate policy that undertook to consider consumer protection to be a priority, but would aim to secure economic and employment opportunities, revoking its initial promise to prohibit agricultural biotechnology.

'Forerunners' decelerating

Changes in public and governmental attitudes towards biotechnology were not confined to small countries. The most astonishing reversal took place in France, where the conservative government discovered the merits of involving the public in decision-making before the election in spring 1997 (which brought a coalition of the socialist and green parties to power). The scientific-elitist style of biotechnology policy of the previous phases would have rendered such a step almost absurd. Yet, even though it was the French administration that had filed the request for the EU-wide marketing of the US soya/maize imports, it nevertheless decided to suspend distribution of GM maize. Under the new government, environmentalist NGOs, agricultural syndicates and the green Ministry of the Environment took a tough stance against agricultural biotechnology, which in turn had strong vested interests in agriculture, industry and research on its side. In autumn 1997, a 'citizens' conference', designed after the Danish model of consensus conferences, was scheduled for June 1998. Future policies were to be linked with the outcome of this consultation.[78] The French government followed the recommendations that were formulated, and imposed a two-year moratorium on marketing authorisations for certain plants.[79] Remarkably, after the conference the government's position was similar to that in Ireland subsequent to its consultation process: although consumer protection and environmental safety gained high priorities, biotechnology was still officially viewed as a key technology with huge economic potential that had to be supported.

Unexpectedly, the UK too experienced a reversal in its earlier policy. Compared with many other countries, British consumers had never found GM food particularly unacceptable, and biotechnology in agriculture had not been a contentious issue for environmental NGOs. The British system of selectively involving potential dissenters in decision-making had worked well in containing conflicts. The BSE crisis, however, had created a new situation.[80] Before the election, the Conservative government had faced a severe loss of public trust for various reasons. After its victory, the new Labour government was more reluctant than its predecessor to see biotechnology as a predominantly economic issue. Shortly after the election, Britain experienced an intensification in public debate.[81] The British government even considered a temporary moratorium, but instead promised a series of 'public consultations'.[82] In 1999, links with BSE were asserted after allegations were made that experimental data showing negative health effects linked to the consumption of transgenic potatoes by rats had been suppressed. These accusations resulted in a public uproar over regulatory misconduct.

In the early 1990s, Germany had abolished the cautious policy it had adopted during the late 1980s, when it had accepted a number of green caveats. Now it moved in the direction of other European countries and assumed a more positive attitude towards biotechnology. The conservative–liberal government had decided to embrace new technologies, even if the public did not particularly approve of them, in order to regain German competitiveness. Because other issues were at stake, for instance reunification and the rise in unemployment, public opposition had been hardly visible when regions had been applying for special funding to set up new biotechnology facilities. However, surveys showed continuing public unease about agricultural biotechnology. When the Social

Democrat/Green coalition took over in 1998, they acted more cautiously with respect to these hidden public anxieties, although they did not relinquish their industrial modernisation commitments. Agricultural and food biotechnology received less support, while the backing of medical research and the production of drugs and vaccines remained stable.

The Scandinavian countries had always harboured populations that were very critical of biotechnology.[83] The BSE crisis had surprisingly little impact, but the soya/maize episode strengthened the bargaining positions of NGOs opposed to biotechnology, thus affirming already sceptical attitudes. Food retailers and NGOs together lobbied for extensive labelling and a cautiously gradual introduction of GMOs. In Denmark, opposition groups and lobbyists tried to influence regulation policy, or at least its implementation, and they attempted to prohibit the marketing and selling of products that were not properly labelled. Norway continued to hold the most critical position and tried to maintain its commitment to criteria of 'sustainability' and 'societal benefit' in evaluating particular GMOs.

In the Netherlands, the government's obligation to inform the public of both positive and potentially negative effects of the new products had already been a constituent of official policy. This was intended to permit consumers freedom of choice and to provide a basis for public acceptance. The government was, likewise, to guarantee the safety of products and to be responsive to uncertainty and concerns among citizens (van Vugt and Nap, 1997: 31 f.). Thus, in 1996, the Netherlands was very active in informing the public and the media through workshops about market introductions of GM food, and the government promoted participation in EU research programmes for stimulating consumer acceptance of biotechnological products. Differences between consumers, retailers, producers and NGOs could have been divisive (Smink and Hamstra, 1996), but negotiations in line with the Dutch regulatory tradition succeeded in reconciling most of the disagreements. As a result, the Dutch government continued to be in a position to demand significant deregulation, even after the soya conflict.

Exceptions

Two countries did not follow the general European trend. Although a Scandinavian country, Finland's positive and pragmatic attitude towards biotechnology seemed to have been unaffected by general European developments. As in previous phases, biotechnology was still seen as a means to modernise the country and to gain economic advantages, although there were still only a few small biotechnology companies.

In early summer 1998, a referendum on biotechnology was held in Switzerland in order finally to resolve a political conflict that had been ongoing since the late 1980s. At that time, the developments described in other European countries were already under way. The outcome was open, but the referendum represented a clear defeat for the critics of biotechnology. The reasons for this development, which appears to run counter to the new 'critical' European trend, are two-fold. First, though the debate about biotechnology was old, the Swiss political system had reacted very slowly to this issue because of its traditionally time-consuming public participation procedures. Hence, the wording of the referendum did not mirror the recent preoccupation with GM food, but emphasised transgenic animals and, therefore, medical research. Secondly, an estimated 200 pharmaceutical companies and specialised research institutes constituted a major asset to the country's economic well-being. Aware of the importance of biotechnology in maintaining national competitiveness, Swiss industry, researchers, students and government formed a broad alliance and engaged in a coordinated campaign. One reason for their success was that, in the Swiss debate, they concentrated on medical research and pharmaceutical applications rather than on food or agricultural products, in clear contrast to the rest of Europe. Another factor may have been that Swiss pharmaceutical companies are technological world leaders, and are consequently linked to national pride as well as to economic success.

The European consumer

After imported GM crops had reached most European countries, reactions to biotechnology became pan-European, and debates arose in countries in which the public had hitherto barely been aware of biotechnology. In addition, some NGOs became players at the European level. The increased consumer anxiety felt in many European countries indicated a certain harmonisation of beliefs, and the European institutions' actions could be interpreted as reflecting them. This raises the question of whether something like a 'European public', in contrast to phases 1, 2 and 3, had emerged. The preceding periods had seen public controversies remaining strictly confined to national contexts. There had been neither a common European debate nor a European public, since the basic prerequisites of a common language, European mass media or opinion leaders, let alone a common identity, had not existed.

Yet, though the new conflict was no longer confined to single countries, debates were not truly 'European'. The above prerequisites still were lacking, and the publics in the respective countries remained more or less isolated. Instead, from 1996 onwards the GM food controversy triggered parallel, but separate, reactions from the European national publics. The media concentrated on national events, and only occasionally covered controversies in other countries.

The pressure on the national governments and, in consequence, on the European regulatory system led to a redistribution of influence. At the EU level, the European Parliament succeeded in presenting itself as the representative of a fictitious European public opposed to the European government represented by the Commission. This stance was based on an understanding of public interest in terms of consumers' right to choose, notwithstanding issues of risk, but clearly taking account of the recent experiences with BSE.

On the other hand, this consumer orientation conflicted with international agreements, made during phase 3, to liberalise trade. These agreements ruled against non-tariff trade barriers that might prohibit imports. Promoters of free technology flow intended such agreements to make the handling of biotechnology more equal across countries.[84] But these legal requirements left little room for political manoeuvre in response to public opinion or even statutory processes of public participation. Everything relied on scientific risk assessment, and social value judgements were explicitly forbidden. Nor could the precautionary principle override the requirements of the agreements, rendering it more or less toothless (Vogel, 1996).

Consequently, the Commission was more inclined to seek expert advice and to strengthen the role of scientific committees in defining risks in accordance with scientifically defensible, 'hard facts'.[85] The role of scientific experts had become statutorily paramount. Rational as this appeared, it was problematic from a public policy point of view since public trust in experts appeared to be declining. Thus, situations such as that in Austria could occur, where a people's initiative demanded a ban on GM food, whereas EU regulation – in line with international treaties – enforced importation. Such a divisive juxtaposition of the 'people's will' and international trade agreements triggered further public opposition.

Global issues

The USA: a different agenda

The declining public faith in experts and regulators in Europe again highlighted the differences between the US and the European regulatory approaches. Comparative studies of the development of regulation and market introduction of consumer products had already revealed an important division. In general, EU regulation was more paternalistic, involving reliance on experts who were considered to be proactively protecting the citizenry (McKelvey, 1997). This was partly a response to the variety of regulatory styles among the member countries. When, as a reaction to consumer concerns, the frames of the debate on biotechnology in Europe broadened, the debate on protectionism versus the free market intensified and market relations between the USA and the EU came in for serious scrutiny. From the perspective of the USA and Canada, EU policy was protectionist and represented an illegitimate attempt to gain economic advantage (McKelvey, 1997: 135). Considering the virtual absence of public opposition to GM food or releases of GMOs in North America, it was hardly conceivable to the Americans that this represented a substantive political problem in the EU. But, for democratic reasons, the EU could not ignore public opposition. As with previous transatlantic quarrels over food issues, this dispute was fed by differences in public attitudes as well as market interests.[86]

In contrast, in the USA during the mid-1990s the results and implications of the Human Genome Project were causing the furore, rather than GM releases and products. In particular, the public was anxious about the possible use of knowledge about individual risk factors for diseases or behavioural traits by health and life insurance companies, as well as in the labour market. In European countries, these themes attracted little attention, presumably because, in European social security systems, individual risk factors play lesser roles in terms of eligibility and premiums.

Patents: a global asset

Another contested area was that of intellectual property rights. In 1996, the OECD issued a report in which it described different approaches to patenting (OECD, 1996), building on an older report from 1985. Patenting had been a recurrent issue at least since the first patent of an animal was granted in the USA in 1988 and in the EU in 1992. Most disputes, however, arose over two issues. First, the patenting of genetic sequences triggered controversy because it was not clear whether such sequences were indeed true 'inventions', or whether they were just 'found' in nature. Secondly, sequence data were considered 'raw material' for basic research and future development that should not be withdrawn from the public domain. As a result, researchers themselves were involved in a conflict of interests. Some scientists pointed to industry's growing role as a promoter of basic research, and stressed the necessity of securing intellectual property rights as a reward for the companies' spending of research money. Others saw the quick and free flow of scientific data, a prerequisite for basic research, to be in jeopardy. The issue highlighted the differences in understanding between the USA and European countries about what an invention is, and about the proper relations between private enterprise and the public domain.[87]

NGO representatives saw this area as the real challenge for the future. Might genetic information be privatised? Should companies be allowed to acquire the right to do what they wished with such information once a patent had been granted? And should groups such as farmers (especially in the Third World) be denied the right to produce seed? In their view, the issue touched on the more general problem of the ever-increasing power of international capital to command the resources of life.

Yet, though NGOs perceived the patenting question to be crucial, this issue did not have the same power as GM food in mobilising the public. This is because it had no immediate impact on the consumer, but instead concerned such elusive themes as equity, international relations and future development. Obviously, such issues were less contentious to the public than domestic affairs and, especially, risks to human health and the immediate environment. Furthermore, the intricacies of the patenting debates were far too involved and complex to be presented as the catchy and simplistic stories required for effective public communication. Nevertheless, patenting remained an issue of dispute for some time in the European Parliament and elsewhere.[88] It was eventually resolved in 1998, when the Parliament accepted a Commission proposal for a Patent directive largely following industry interests, after long and complicated political negotiations.

In the reservations on patenting, one can trace the impact of the Dolly story. In the aftermath of this media event, the public, governments and EU institutions were unanimous that human cloning had to be rejected. Dolly rendered it easy to insert a prohibition into the directive because, although the economic significance of human cloning remains doubtful, the matter touched upon a common moral code. The prohibition on the patenting of human cloning techniques may be regarded as a high gain, in terms of political publicity, at low cost, since no substantial (industry) interests had to be violated. In contrast, the debate about GM food and the accompanying regulatory turmoil had little influence on the final debate about the Patenting Directive. This was an indication that the general debate on biotechnology had split into diverse conflicts over deliberate release and food products, over cloning and xenotransplantation, over intellectual property rights, and over the 'ownership' of genes.

Life sciences

Intellectual property rights as a basis for the control of genetic material and biotechnological methods had profound implications in two key, previously almost unrelated, areas. The mergers of gigantic companies engaged in the production of pharmaceuticals as well as seeds indicated that the combination of these two areas would indeed secure future gains.[89] It became clear that biotechnology had, finally, achieved a significant status. Even if there were problems concerning the public acceptance of GM food products in Europe, the progress of biotechnology now seemed inexorable. The industry's promises, which had won it backing on the stock market, finally began to come to fruition in the form of multinational companies whose activities centred on biotechnology.

Ironically, this marked the end of the use of the term 'biotechnology' itself. Companies now stopped using the word since it had acquired negative connotations. Instead, they defined the area as 'life sciences', a term with suggestions of more welcome medical applications. This term had been in use for decades to identify the fields of biomolecular and basic medical research, and was employed in the titles of university faculties and scientific journals. It gained a new meaning when industry adopted it to identify companies that apply particular methods of biotechnology, irrespective of the field, to developing new products and acquire patents on genetic sequences as a basic resource. The 'life industry' evolved from the chemical industry through the adoption of biotechnological methods and the development of pertinent products, finally shedding traditional chemical activities in favour of modern biotechnology. This indicated that the use of life (Bud, 1993) was indeed its core field of interest.

Summary

In phase 4, when the first GM crop plants entered the Common Market, concerns about agricultural biotechnology brought about a shift in public debate and policy both in the member countries and at the EU level. The dominating economic frames of previous phases became weaker as public opposition increased, while the directives failed in their objective of harmonising the member countries' interpretations of product assessment. The signals from consumer groups, retailers and NGOs were now explicit: they did not want their food to be produced using biotechnology, and they demanded the right to choose, regardless of safety issues. As a consequence, labelling became paramount. The debate coincided with the advent of new and contentious cloning techniques, re-establishing a link to reproduction that had largely vanished. It seemed as if the debate on biotechnology had returned to a starting point, notwithstanding its preliminary fading in phase 3, although debates diversified and became centred around certain application and problem fields.

In response to public opposition, and supported by government shifts in important member countries, EU policy turned towards a more cautious implementation of GM food products. This, however, caused problems with international and transatlantic trade agreements. Even more far-reaching policy issues centred around intellectual property rights, especially the patenting of genetic sequences. While companies merged to form conglomerates engaged in both medical and agro-biotechnology, the term 'biotechnology' began to disappear from corporate language, being replaced by the even more inclusive, but less contentious, 'life sciences'.

Summary

Regulation to make biotechnology happen

From the very beginning, most national governments, as well as the EU Commission, tried to 'make biotechnology happen'. This motive became dominant in the 1980s, when the economic point of view was advanced, and it has remained paramount ever since. Despite various obstacles, governments adopted a double strategy. On the one hand, they fostered industrial and scientific research even to the extent that they accepted responsibility for the establishment of commercial firms. They succeeded to differing degrees, depending on the national industry's capabilities for engaging in the new technology, and on the willingness of domestic publics to accept its products. They also tried to reassure a critical public – by implementing credible regulation – that risks could be managed. The major questions were: who should be in the position to regulate, and what would such a credible regulation involve? Was self-regulation by the scientific community sufficient, or were there issues that could not be dealt with in such a system? A major problem was to decide what biotechnology could be compared with: is it something entirely new or merely an extension of older techniques? is it like nuclear power or like bread baking? If a solution was found that was generally held to be trustworthy, debates often calmed down.

Over time, however, this strategy became less effective, reflecting a general loss of trust in government and in scientific institutions and experts. Attempts to solve the regulatory problems within an expert system, as in phase 1, failed later on, since governments had to react to the social changes of the 1980s. While some still tried to contain the debate within a strictly scientific frame, others integrated critical actors in the decision-making process as a sign of openness towards the public. It was hoped that this would prevent the spread of the critiques advanced by NGOs without jeopardising the economic potential of biotechnology. However, another reason safety regulation failed to close the debates was that it could no longer cover the issues at stake.

The failure of risk assessment

When biotechnology was introduced, supporters within science, policy and industry expected that it was only a question of time until it became generally accepted by the public. From their points of view, possible risk, to be assessed in a scientifically rational way, was the only obstacle to popular acceptance. Consequently, innovators intended to restrict the debate to classical risk assessment, but they eventually failed to do so because of the lack of universally shared criteria. Studies on risk perception indicate that lay people perceive risks differently from scientific experts (Slovic, 1987; see also Douglas and Wildavsky, 1982: 49–66; Jungermann and Slovic, 1993). Experts tend to keep the problem within what they perceive to be a purely technical frame, whereas lay people implicitly emphasise behavioural, cultural, social and economic aspects. This was hard for scientists to comprehend. Because they followed a rationality that demanded that value considerations be put to one side, they could not understand why risk debates were going beyond the issue of (physical) hazards and involving questions of equity and accountability. Such value choices, however, are the most decisive factor shaping public perceptions (BEP, 1997). When experts kept affirming that risks were negligible or manageable and need not hinder progress, they were reproached by the public for downplaying the potential risks. Because scientists were in a position to make moral judgements, and were stakeholders themselves, they could be accused of being partisan and hiding conflicts of interest. This undermined their credibility.[90]

As a consequence, the question of whether there were significant risks – notwithstanding the apparent lack of accidents – could never be adequately answered. This led to the adoption of the cautious principles of case-by-case and step-by-step assessment in EU regulation, which were themselves predicated on the precautionary principle. Uncertainty was the reason regulation was put in place until 'more experience' was acquired; again, this was subject to different interpretations and hence to regulatory uncertainty, leading to delays. Thus, the dispute over risks slowed down the pace of innovation; it even temporarily stopped it or forced it into sidings. Yet, the slow speed of implementation may have contributed to the safety of the new technology. We may speculate that, had it not been for scientific, regulatory and public scrutiny, we might indeed have experienced accidents.

Biotechnology as a sounding board

As biotechnology left the laboratories and entered society, its image varied considerably depending on what was considered technically feasible and what was deemed desirable. The focus on applications meant that the term 'biotechnology' became a symbol with a dual, and contradictory, meaning, depending on whether the user of the term had promoting or preventing interests.[91] Supporters and critics competed in establishing the 'meaning' and definition of biotechnology, and this resulted in terminological shifts. Applications themselves shaped the varying definitions of demands and risks with respect to fields, interests and world-views. Projections of future demand built on promises, sometimes vague, of new products to come that would outperform conventional ones or be entirely novel. According to the particular application, claims of risk were extended beyond human health to a wide variety of issues, from environmental protection to socio-economic factors. Opposition was often (strategically) based on the logical impossibility of proving the absence of any negative long-term outcomes.

Regulators had to find a way through the complexity of the arguments, while uncertainty about risks prevailed. Uncertainty controversies featured two types of argumentation. On the one hand, physical hazards (to health and the environment) were emphasised; on the other, societal and moral issues came to the fore, involving consumer rights, trust in experts and regulation, and the role of agriculture in post-industrial economies.[92] Both types of argument influenced attitudes but, in general, societal and moral arguments turned out to have the greatest impact on consumer behaviour (BEP, 1997; Durant et al., 1998). However, risk issues were the more commonly addressed, since they were (seemingly) independent of individual world-views. Any biotechnology regulation (except the Norwegian and, to some extent, the Swedish and Austrian gene laws) permitted the authorities to address only physical risks, so they had to cover other concerns under risk arguments.[93] In contrast, parliaments were able to take up societal and moral issues as well, even if they had little influence on actual decision-making.[94]

A similar division could be observed at the EU level. While the Commission statutorily focused on risk only, such arguments helped the previously less important European Parliament to realise its potential as the representative of a common but invisible European public. It acquired, temporarily, a role as consumers' advocate.[95] So, during the BSE crisis, the Parliament took up risk issues, but eventually went beyond these and addressed the societal/moral arguments that the Commission could not adequately deal with. Later, in the same vein, the labelling of GM products became an issue not only of paternalistic 'risk prevention', but also of 'consumer democracy' in the Parliament and elsewhere.[96]

National diversity and European integration

In Europe, the new technology encountered a different 'climate' in each country. Its reception varied according to each state's academic tradition, its industry research base and the players that had tended to influence government policy. Obstacles to broad and smooth implementation also often emerged from cultures critical of technology, in the same countries that, during the 1970s and 1980s, had seen social and environmental conflicts over nuclear power, environmental issues, disarmament, women's rights or Third World support. Later, such conflicts found their expression in the presence of green parties in several parliaments. Another factor involved the existence of religious traditions that objected to the 'manipulation of Creation'. This diversity in response to biotechnology made it more likely that national debates would be confined to national boundaries. Thus, in general, early debates throughout Europe focused upon national events. Only when the first transgenic crop plant came on the market was simultaneous media attention provoked in almost all European countries. Similarly, the BSE crisis had a huge transnational impact, and later the advent of Dolly the sheep became an international media event. However, even in these cases, the actual debates remained mostly national.

There was no general rule about how governments dealt with biotechnology conflicts. Their response depended, among other factors, on the way political problems were generally handled and whether there was a generally adversarial or consensual style of resolving them. Government reaction was also affected by whether or not other conflicts allowed biotechnology to appear on the agenda, and by the particular understanding of what biotechnology 'really' means. If regulators understood biotechnology to be similar to other technologies, regulation took into account the varied forms of application, and a strong emphasis was placed upon scientific expertise. In contrast, if biotechnology was seen as something new, entailing unknown risks that had to be broadly debated, regulators tended to enact universal laws affecting the technology itself. Some governments tried to keep issues within the realm of scientific expertise;

others emphasised public accountability. All this contributed to the striking levels of national diversity.

When, in the late 1980s, biotechnology appeared as an economically promising area of technological development, the USA had already taken the lead and Europe was lagging behind. The EU decided to create a common market for biotechnological products in Europe in order to exploit the industrial opportunities. A precondition for this was the implementation of unified regulation, involving a compromise that recognised national differences in regulatory culture and public opinion. After all, Germany appeared as a menacing example of what could happen if public opinion was allowed a free rein. The EU guidelines on contained use and deliberate release met German demands in that the biotechnology itself was regulated, and a kind of precautionary approach was applied, not least in response to public demands.[97]

The directives were intended to serve multiple purposes beyond that of ensuring harmonisation by defining risk assessment requirements. Following international trade agreements, the directives had to be compatible with US regulations, though the USA believed EU regulations to fall short of this.[98] Additionally, the directives turned out to be the master copy for countries outside the EU that lacked regulation but that had close trade relations with Europe, and eventually became EU members. Alas, these aims were set too high. As with any standard resulting from a compromise, the directives suffered from ambiguities. In particular, the wording of Release Directive 90/220 left room for interpretation, in the clear expectation that a common European understanding would develop over time. However, despite intensive negotiations over the years, binding rules on crucial issues such as risk assessment criteria for products could not be achieved. Divergence and convergence did not produce an equilibrium.[99]

Although industry and large NGOs had become European actors, national idiosyncrasies were even more influential when biotechnology entered the agricultural sphere. Since the USA demanded free access to the European market for American GM crops, a veritable trade war was looming. Attempts to 'verticalise' the regulation via directives on Novel Food, Pesticides, Feed, and so on turned out to threaten various other, previously less contentious, areas, rather than to solve the problem. National governments, as well as the European Commission, realised that, after ten years of harmonisation attempts, disharmony among EU member states over biotechnology policy had not been eliminated and a comprehensive regulation would need to be very complicated. Finally, the century ended with a temporary halt to new products in Europe.

Conclusion: no end to conflicts

After a quarter of a century, the future of biotechnology is open. Fields such as basic research and development, the production of pharmaceuticals, food additives and enzymes, forensics and diagnostics have become unthinkable without biotechnological methods. Yet in other fields, such as predictive medicine, gene therapy or environmental remediation, practical implementation has barely begun. Certain areas lag behind for various reasons: seeds commercialised so far have been an agricultural success in the USA but remain contested in Europe; conversely, genetic testing for disease preconditions[100] may entail problems with insurance coverage, and is more debated in the USA than in Europe owing to different health care systems. It has become clear that biotechnology, like other modern technologies, is embedded in society – or, rather, in the societies of various countries – and is subject to their differences. Innovation will continue to accelerate, but one may not take it for granted that a product will be welcomed just because it is new or advantageous for the producer. Public concern brought to the fore (or triggered) by NGOs and attitudes towards the public accountability of regulators and industry will contribute significantly to the success of new products.

Diversity, as it exists at the European level, is also apparent on the global scale. Multinational companies merge and increase their market power in already monopolised fields, such as seeds and pharmaceuticals, and international agreements will enforce homogeneous assessments on a seemingly scientific base. Nevertheless, differences in culture and interests over time and among countries and continents will remain. From a global perspective, the strategies of actors engaged in the promotion or rejection of technology will change, but they will always be influential political factors.

Ten years from now, we may be able to write the history of the debate on biotechnology. At present, however, the controversy is still ongoing, and there are few indications that it will soon come to an end. Whether the rapid cultural and technological changes of our times are too difficult to cope with or the technology is too complex to understand, it has become obvious that the current regulatory scheme in Europe does not fulfil what Sheila Jasanoff (1995: 311) postulated to be the essence of functioning regulation: 'it is a kind of social contract that specifies the terms under which state and society agree to accept the costs, risks and benefits of a given technological enterprise.' To set up such a contract is no less of a challenge today than it was twenty-five years ago.

Acknowledgements

Helge Torgersen is supported by the Österreichischer Fonds zur Förderung der wissenschaftlichen Forschung (project no. P11849), and the Institute of Technology Assessment, Austrian Academy of Sciences. The authors are grateful to Andrea Lenschow and Michael Nentwich for their comments on previous drafts of this chapter.

NOTES

1. According to Rammert, technology comprises 'the inventory of instruments and installations, as well as the repertoire of skills and knowledge, to achieve desired conditions, and to avoid unwanted conditions, in handling the physical, biological and symbolic world' (Rammert, 1993: 10, our translation).
2. Apart from the Showa-Denko case of tryptophane contamination, which was mainly due to a failure of the purification procedure.
3. One could argue that (statutory) regulation is always put in place in order to make possible what it regulates (Majone, 1989).
4. The European Community (EC) changed its name in 1992 to the European Union after the Maastricht Treaty. For the sake of simplicity and consistency, we will refer to both the EC and the EU as the 'EU', irrespective of the actual name used at the time referred to.
5. See, for example, the extensive reviews published earlier by Bud (1993) or Cantley (1995).
6. If no reference is given for a particular claim made in this chapter, then the claim is based on results from the above-mentioned research project. Detailed data may be found in Durant et al. (1998).
7. If we define biotechnology in a literal sense, as an activity putting living organisms to work for humankind, everything from agriculture to fermentation would be included. However, the term is mostly referred to with respect to activities that exceed well-known techniques such as baking or brewing, more in the sense of what has frequently been called modern biotechnology. We may link its onset to the beginning of broader debates arising over recombinant DNA techniques or genetic engineering. Frequently, the term biotechnology was used as a proxy for the combination of the methods for splitting, sequencing, splicing and constructing genes, including methods for their application in various fields such as basic research, industrial production, medicine and agriculture. Genetically modified organisms (GMOs) carry genes modified by these techniques. Consequently, products of biotechnology are the results of production processes applying such methods. In the following, we will stick to this understanding because it seems to be a common denominator, and not because it provides a scientifically exact definition. It should, however, be noted that the term 'biotechnology' has not been restricted exclusively to the process of recombining DNA. Its broader definition includes processes such as protoplast fusion and cell and tissue culture. This is why the notion of, for example, cloning can be considered a biotechnological activity. Although not particularly related from a scientific point of view, this example highlights the links to other fields of modern biological R&D, which gave rise to another set of innovations.

Even if cloning as a technique has nothing to do with genetic engineering, it has a lot to do with how the public perceives modern biotechnology, as the case of the famous sheep Dolly clearly shows (see below). In public debates, these innovations were at times difficult to separate owing to popular representations (Bauer et al., 1998). Therefore we will also have a look at debates and policy responses in such related fields.

8. Obviously different actors conceptualise the public differently but, interestingly, the same actors do too, depending on the context. Hill and Michael, for example, describe what they identified as the decision-maker's concept of 'the ordinary member' of that public, being constructed from ideas of the citizen and of the consumer, respectively. An idealised layperson is thus 'an admixture of (at least) autonomous, thoughtful citizen and concerned, rational decision-making consumers' (Hill and Michael, 1998: 213). This interpretation is in striking contrast to the notion also frequently encountered among scientific and regulatory elites of the 'uninformed public' being basically uninterested and dependent on mass media that keep highlighting certain facts and hiding others.

9. Some of the then Warsaw Pact states oriented themselves towards the US policy.

10. In 1978, DG XII proposed a directive against 'conjectural risks' associated with rDNA work, hence the proposal oriented itself towards the technology. However, various reports from European scientific organisations stressed that the risks had been exaggerated and relaxation was necessary and possible. This argumentation, following the evolution of policy opinion in the USA, prompted DG XII to withdraw the proposal and to opt for non-regulation in the form of a 'recommendation' only.

11. For example, the US Congress's Office of Technology Assessment issued a series of reports emphasising the prospects for medical research, pharmaceutical production, agriculture and environmental remediation (e.g. OTA, 1984).

12. Later, the OECD also assigned to biotechnology the status of a crucial cross-sectional technology, like electricity or microelectronics (OECD, 1988).

13. An annual licence for low-risk experiments replaced notification in advance, and scientific committees checked the safety on their own.

14. In 1980, a British report conveyed the message to government that commercialisation of biotechnology was both possible and desirable; however, competitiveness relied on better technology transfer, which would require more effective support from both government and industry.

15. In France, the early 1980s appeared retrospectively as the 'golden age' of promotional biotechnology policy-making (Gottweis, 1995: 225), since strong government backing led to considerable achievements and a catching-up with leading countries such as the UK.

16. However, firms, especially in the chemical sector, were reluctant to take up this entirely new approach. One reason was that, among chemists only a decade before, the 'old' biotechnology on which the new techniques were based had been deemed outdated compared with the more 'scientific' field of organic chemistry (see note 40).

17. The OECD also issued advice about large-scale industrial production (OECD, 1986), but it took them six more years to come up with recommendations

for deliberate releases for research purposes, which indicated the associated problems (OECD, 1992).

18. Consequently, in 1980 a new British risk assessment scheme served to facilitate large-scale fermentation.

19. There are many interpretations of the precautionary principle, from the rigorous driver's advice for overtaking: 'If in doubt, don't'; to the more cautious environmental protection version: 'If there are serious doubts about the outcome, then the one that is less able to defend itself should be given the benefit of the doubt' (in this case, the environment).

20. Engineered bacteria (the so-called ice-minus strains) would protect strawberry plants from freezing. Opponents made claims of (hypothetical) risks associated with the release. These even included the risk of climatic changes. After extensive review by the NIH, permission was granted to carry out the experiment in 1984. However, NGOs took up the issue and succeeded in placing 'genetic engineering' on the agenda again. Activist Jeremy Rifkin brought a law suit challenging NIH's competence as the first in a series of litigations that significantly contributed to raising awareness or, as others put it, to exaggerating the risks of biotechnology and causing widespread fear. Rifkin's litigation series triggered mostly local opposition against single projects. One of the results was that the US Environmental Protection Agency (EPA) took up biotechnology as an issue and the releases of GMOs became regulated under existing law (Krimsky, 1991).

21. 'As a public controversy, ice minus was history just as soon as the field was sprayed by the moon-suited scientists. A precedent had been set insofar as the first major barrier to the environmental application of GEMs [genetically engineered organisms] was overcome' (Krimsky, 1991: 132).

22. rBST is a bovine hormone for stimulating milk production in cows that is produced in GM bacteria. Its approval for use in dairying triggered debates, legal actions and a Commission of Enquiry of the German parliament in 1989 (Deutscher Bundestag, 1989). After protracted disputes, and contrary to the US government, the EU Commission finally denied its approval for use in milk production, which eventually gave rise to a transatlantic trade conflict.

23. For example, German opponents to biotechnology conceptualised the conflict as one between two different paths of development: 'One path will increase the industrialisation, the technical control and the re-shaping of nature to allow better exploitation. It is feared that this functionalisation . . . of life will not be limited to plants and animals. The other, preferred path of development is described as a path where technological and non-technological solutions to problems are developed that guarantee the protection and sustainability of nature' (Tappeser, 1990: 10 ff.).

24. Later, the ban on rBST was interpreted by the USA as the erection of an unfair trade barrier by the EU.

25. Genetic screening allows the selection of embryos according to their characteristics. Parents obviously want healthy children, but what 'healthy' means depends on cultural definitions. The birth of ill or handicapped children thus becomes the result of a voluntary decision by the parents.

26. For example, the Ökologie-Institut in Freiburg provided counter-expertise, focused on ecological criteria, issuing from an institutionalised base. This was intended to match the expertise of molecular biologists and industry in the biotechnology struggles and litigations during the 1980s in Germany.

27. This had serious practical consequences, because differences in regulatory styles were the basis of frequent quarrels within the EU over the focus of risk assessment (Levidow et al., 1996) and the definition of 'sound science' in their assessment of products.

28. There were different possible reasons for this: they could not find enough support among the public or from the media; their arguments were not heard by the opinion-leaders; they found other means to achieve partial successes; or, given the fact that most NGOs relied heavily on individual activists, they simply compromised on personal grounds.

29. 'The British state either integrates social movements quite well, either directly, in the policy-making process, or through political parties, and gives them some limited influence in exchange for co-operation, or it shuts them out completely, denying them any opportunity to influence policy-making' (Rüdig et al., 1991: 137).

30. 'Through the early 1980s, the strategy of containing regulatory debate within carefully structured expert committees ensured a relatively narrow focus on the physical risks of rDNA research and correspondingly muted attention to the social and political consequences of the new technology' (Jasanoff, 1995: 322).

31. Around 1989, the government tried to counteract the lack of public acceptance by supporting initiatives for multi-actor workshops and setting up an ethics committee on environmental safety. A foundation to support knowledge-based opinion formation about biotechnology by consumer NGOs is still functioning.

32. Its 'inventor', the Danish Board of Technology (founded by the Danish government in the 1980s), is a body created to assess technology, stimulate public debate and advise parliament.

33. The law prohibited releases of GMOs unless the Minister of the Environment found that there were 'special circumstances' for granting an exemption. Eventually, all applications received were exempted from the ban, even without mention of 'special circumstances'.

34. This was also the case in non-member countries joining the EU at a later point in time, for example in Austria and Finland. Finland saw biotechnology as the means to 're-industrialise' R&D in some limited areas. Countries on the other side of the Iron Curtain under Soviet domination, such as Poland and Hungary, slowly tried to establish a research base while orienting themselves towards the US policies, again in the absence of any significant public debate.

35. The OECD is based in Paris; there are twenty-nine members (1999), including the most important industrialised nations and the EU.

36. This was necessary because, if EC regulation was pending, national laws had to be compatible with what was to come.

37. Especially after an unauthorised deliberate release of GMOs in 1987.

38. The ethical debate, which had always been pronounced in France, centred on questions of human medicine and had little influence on the government's position concerning other applications of biotechnology.
39. In particular, the big chemical companies in Germany (with the exception of Hoechst) were reluctant to engage in biotechnology. This was not because of fundamental disadvantages in competitiveness. Ulrich Dolata (Dolata, 1995: 463) located the reason for German reluctance in the different culture of the chemical engineers leading these companies, and their devotion to classic organic chemistry, as opposed to that of the biologists, with their emphasis on biological processes. Meanwhile, US start-up companies had flourished and had partially been incorporated into big established pharmaceutical concerns (Oakey et al., 1990). This had led to a technology transfer, and often entailed their reorientation towards biotechnology.
40. This was in line with an OECD report entitled 'Biotechnology and the Changing role of Government' (OECD, 1988), which emphasised its main role of coordinating R&D in this field and promoting its commercial exploitation.
41. In its report of 1985, the Parliament 'summed up the political situation for biotechnology at the European Community level. It showed a broad awareness for the potential of biotechnology, and the need for a coherent strategy, responding to the need for international competitiveness. But with respect to regulation it saw "special risks with genetic engineering methods" . . . , demanding a complete ban on field releases "until binding Community safety directives have been drawn up" . . . , and with similar restrictive views on gene therapy . . . and animal transgenesis' (Cantley, 1995: 543).
42. The Sixth Amendment (79/831/EEC), shaped by DG XI (Environment) after OECD proposals and the American Toxic Substances Control Act.
43. Cantley (1995: 551), quoting from the summary of a meeting between the Commission and member state representatives.
44. The first Directive proposal on workers' safety covered not only GMOs but 'biological agents' in general.
45. Directives 87/22/EEC for the Production of Pharmaceuticals by New Technologies, 90/679/EEC on Workers' Protection from Risks Related to Biological Agents, or the proposal for a Directive on Intellectual Property Rights in Biotechnology and for a Plant Variety Rights System.
46. On the other hand, the European Parliament demanded more restrictive amendments, put forward by the German rapporteur who favoured an approach oriented to possible risks.
47. Another field of debate was the inclusion of the so-called 'fourth hurdle'. NGOs had demanded that applications of biotechnology should be linked to the demonstration of need, additional to the traditional criteria of purity, safety and efficacy in conventional drug assessment. The Commission strongly opposed such demands since it considered them to be an invitation to raise non-tariff trade barriers. Later, only the Austrian and Norwegian laws took up socio-economic criteria, a move with few practical consequences (see below).
48. Council Directive 90/219/EEC of 23 April 1990 on the Contained Use of Genetically Modified Micro-organisms, *Official Journal*, L117, 08/05/1990, and Council Directive 90/220/EEC of 23 April 1990 on the Deliberate Release

into the Environment of Genetically Modified Organisms, *Official Journal*, L117, 08/05/1990, to be implemented in national regulation within October 1991 (which was not the case in all member countries). The directives were issued by the European Economic Community (after 1993 the European Community) within the European Union, hence it is actually inappropriate to speak about 'EU directives', although they were mandatory throughout the European Union. Nevertheless we will stick to the term for the sake of simplicity.

49. 'According to Technical Progress', an amendment to the Release Directive, was issued in 1994 (Directive 94/15/EC), specifying categories of plants intended for deliberate release and clarifying details of the risk assessment.

50. The EU deliberately did not regulate genetic testing, gene therapy or any other possible applications, fields in which national regulations were implemented in some European countries, but not in all. In particular, there was no reference to ethical considerations other than in relation to risks to health and safety and the environment. Any links to reproduction techniques, cloning and the like were omitted. It even remained questionable how the Release Directive would cover the handling of transgenic animals. Another issue that was not touched upon was the dispute about intellectual property rights. In 1988, a US patent had been granted on the Harvard 'Onco-Mouse', which had raised discussions on the 'patenting of life'. In the aftermath, the issue had also created controversy among scientists. The EU prepared a draft directive that was heavily debated, and it took another ten years before a directive on Intellectual Property Rights could finally be passed (see below).

51. For example, the question of how to deal with deliberate releases of GMOs showed up on European countries' agendas when the release of ice-minus bacteria was debated in the USA. A truly public debate, however, arose only when applications for the release of transgenic plants were made within particular countries.

52. For example, gene therapy, xenotransplantation and genetic testing.

53. Made for the procreation of cows that could produce an anti-microbial protein from human milk in their own milk.

54. The case of Herman the Bull anticipated some of the debates that were to arise later over GM animals as organ donors (xenotransplantation), although medical risks were not at stake.

55. In order to broaden the basis for decision-making, a consensus conference was held on transgenic animals.

56. This step was made possible *inter alia* because the Green Party was no longer in parliament after reunification.

57. A way to reconcile the differences was to assign 'societal benefit' to every release for research purposes, since a gain in knowledge was automatically deemed socially beneficial and capable of promoting sustainable development. This was different for certain product applications, however.

58. Occasional and sudden outbursts of distrust, as in Poland in 1997, were mostly triggered by NGO actions from abroad.

59. The OECD recommendations of 'Good Developmental Principles' (OECD, 1992, part 2) for small-scale field releases did not succeed in playing a similarly unifying role for the US and the EU positions as the 'Recombinant DNA

Safety Considerations' (OECD, 1986) had done six years before. Attempts to link plant biotechnology closely to traditional breeding as a 'baseline for assessing modern biotechnology' (OECD, 1993b) were officially welcomed by both the EU and the USA, but they did not resolve the question of how in practice to assess newly introduced genes conferring traits that had not been seen before in crop plants.

60. There were even more far-reaching differences, since it was by no means clear what an acceptable outcome of the risk assessment was, or what exactly should be prevented when GMOs were released.

61. Although disclosure of such files was prohibited, more data could be obtained in Denmark and the Netherlands than, for example, in France or Germany (Jülich, 1998).

62. Mostly from industry, as well as from the governments of Germany, the UK and the Netherlands.

63. For example, a British company sold 'vegetarian' cheese, which was made with genetically engineered chymosin as an alternative to calf rennin. Most modern washing powders contained recombinant enzymes.

64. In 1994, the Directive on Deliberate Release was updated (Directive 94/15/EC).

65. Environmental remediation with the help of biotechnology was not an issue during this period. On the one hand, there seemed to be too little substantial progress, and on the other hand the release of GMOs in order to eliminate pollution appeared, to many Europeans engaged in environmental protection, as an idea derived from the most cynical technocratic thinking.

66. Sheila Jasanoff (1995: 311) described the study of the political regulation of biotechnology as the 'study of the process by which technological advances overcome public resistance and are incorporated into a receptive social context'. This diagnosis, taken from US experiences, appeared also to hold true for Europe until the mid-1990s.

67. This was acknowledged by an OECD report on 'Biotechnology, Agriculture and Food' in 1993: 'High levels of uncertainty surround the innovation process.' 'Biotechnology innovation involves new forms of co-operation between economic actors situated at different points in the agro-food system.' And: 'Successful innovation demands greater responsiveness to end-users whether they be other firms or the final consumer' (OECD, 1993a: 143).

68. The world-wide acreage covered with transgenic crops exploded between 1996 and 1998, e.g. for soya, from 0.5 to 17.0 million hectares (according to the Austrian Press Agency, 3 December 1998).

69. BSE (Bovine Spongiform Encephalitis) is a nervous tissue disease transmitted by prions (agents consisting of only protein); as it is known as Scrapie disease in sheep. Mostly British cattle acquired the disease probably from being fed food additives derived from sheep carcasses, although experts had deemed any transmission across species borders impossible. There is a very rare and similar condition in humans called Creutzfeld–Jakob disease (CJD). When an unusual number of new cases of this disease occurred in the UK, experts again reassured the public that there were no indications of a trans-species transmission between cattle and humans. In 1995, however, the possibility of a link was officially conceded after the leaking of information from a secret report.

70. The biggest problem for the EU Commission had been the collapse of the common beef market. Apparently giving in to British pressure, the Commission lifted the ban on some beef products in 1996 against heavy opposition. '... a simple appraisal of costs and benefits confirms that in terms of both socio-economic indicators and political legitimacy the policy adopted had disastrous consequences.... the costs of the crisis ... far outweigh the short-term benefits of the approach pursued by the UK government and the European Union institutions' (Baggott, 1998: 64).

71. Although the labelled tomato paste had not evoked any public debates, industry and the EU Commission still tried to prevent mandatory labelling, and food additives were excluded from the regulation's scope. On the other hand, the European Parliament's Economic and Social Select Committee had emphasised the consumer's right to choose and demanded that labelling become mandatory.

72. The labelling debate revealed two fundamentally different approaches. Conventional risk prevention demands a guarantee from the authorities that there are no significant risks for human health with any product. The way the product was generated is considered to be of no interest to the consumer, since only the (physical) properties count. Such an understanding allows labelling only when there is a more or less established risk for the public or for certain persons, e.g. those suffering from allergies. This was basically the position of the US Federal Drug Administration. On the other hand, the consumer choice approach demands labelling in order to indicate the production process. It provides a choice between products that may be substantially identical but produced differently (i.e. a GM versus a non-modified tomato). This is also in line with 'negative' labelling if it can be proven that the product does not contain anything that was genetically modified. Clearly, this approach is much more 'political' and implicitly contradicts the philosophy of international trade agreements (see below). They built on the concept of 'substantial equivalence' proposed by the OECD (OECD, 1993a), which was based on comparison between the modified and unmodified food product. It was acknowledged that uncertainty might play a role, but material differences that should 'matter' had to be established.

73. Although the Commission approved the GM soybeans, it rejected further demands by Parliament for a modification of the Novel Food Regulation. In spring 1996 the EU institutions' stances differed to such an extent that a time-consuming mediation procedure set in.

74. Although the regulation was binding from April 1997 in all member states, labelling requirements were still unclear. The Commission had to specify them for soya and maize that had been permitted before the regulation was enacted. Now labelling was also required if modified DNA or protein were present.

75. For example, the French had always had difficulties in accepting an American dictate in matters of agricultural trade, whereas Austria as a full EU member experienced a new regulatory impotence in the face of binding EU regulations, and the Danes rediscovered an unease over agricultural biotechnology that had been buried under a layer of consensus-oriented deliberations.

76. See the country reports in Durant et al. (1998).

77. The fact that this release was conducted without official permission triggered an immediate response by NGOs and the mass media.

78. During the one-year run up, French political actors kept making reference to this planned event, at the EU level, in international political arenas and even in summit talks of the World Trade Organization (WTO). The conference itself raised consumers' worries about health risks and scepticism about the independence of experts advising the government.

79. In autumn 1998 it decided to keep the harvest of modified maize completely off the market.

80. Like the French, the previous Conservative UK government had serious problems during the BSE crisis, which also had repercussions in the debates over the soya and maize imports pending.

81. During 1998, test fields were destroyed by activists. A court battle took place over contamination by genetically modified plants of an organic farm, among others, and biotechnology companies were 'named and shamed' by government advisers for flouting field trial regulations. The media emphasised consumer concerns about food and choice. The Prince of Wales (amongst others) spoke out against the use of genetically modified foods in favour of 'natural' products from organic farming.

82. A number of other publicly sensitive issues were addressed by the government, including cloning and issues of genetic testing and insurance.

83. Except Finland; see below.

84. The WTO Agreement on Sanitary and Phytosanitary Measures (SPS) demanded transparent, solely science-based risk assessment as the only legitimate basis for even the shallowest trade restrictions. However, 'in the context of the WTO, it is unclear what criteria might justify an import restriction under SPS, short of evidence of significant negative impact following release of a GMO' (Wyndham and Evans, 1998: 2 ff.)

85. Within DG XXIV, an expert committee was set up to assess genetically modified products in order to circumvent differences between evaluations by national governments. The evaluations had proven to be hard to reconcile with each other.

86. For example, the American stance towards artificially processed products that were 'nutritionally enhanced' (e.g. de-cholesteroled or sugarless) usually was more relaxed than European perceptions of an 'adulteration' of such food (Hoban, 1997). On the other hand, the perceived European preoccupation with 'naturalness' as well as the frequently rather lukewarm attitude, compared with American enthusiasm, to competition and economies of scale appeared irrational on the other side of the Atlantic.

87. It forced some revisions also in traditional plant variety protection, including such issues as the right to propagate an organism, e.g. in order to produce seed for own purposes. A different aspect of this issue was the exploitation of genetic data from developing countries' indigenous species or crop varieties by companies from the industrialised world. Issues of intellectual property rights were linked to attempts to secure biological diversity through environmental protection. When this was negotiated at the 1992 UN Conference on Environment and Development in Rio de Janeiro, the issue of sustainable

development was emphasised. In the end, 171 parties signed the Convention on Biological Diversity, including the EU, but the USA did not agree to the negotiated results. This too had consequences, especially in EU countries, for public opinion of the role of US companies and their attempts to commercialise GMOs.

88. The neglect of ethical concerns eventually led to a turning down of the EU Commission's proposal for a Patenting Directive in the European Parliament in 1995, which in turn flagged up the issue for industry (Galloux et al., 1998: 182).

89. For example, the mergers between Hoechst and Rhône-Poulenc and between Zeneca and Astra in late 1998 resulted in conglomerates that covered exactly these two areas, after they had reduced most other activities they had previously been engaged in.

90. Another problem arose from the time perspective of risk assessment. Classical technical risk assessment is in general retrospective: the risk of future hazards can be determined by applying statistical methods to past experiences. Many opponents of biotechnology held a different concept of risk, which was oriented not to past experiences but to possible futures (see Krohn and Krücken, 1993). This change in time perspective had severe consequences for the debates on risk. Because the future cannot be controlled, it is logically impossible to exclude future hazards and debates on risk cannot be closed.

91. For example, when Dolly the sheep appeared, the question of whether this was biotechnology or not split even the experts' community.

92. This distinction can also be made for example in the case of BSE: arguments about the hazard of acquiring Creutzfeld–Jacob disease concern a physical risk; those about the behaviour of governments (e.g. whether the British government was right in not disclosing ambiguous data) are societal or moral arguments.

93. A good example is the different reasons why labelling was demanded. On the one hand, labelling should serve to indicate health risks for those consumers who are, for example, allergic to certain ingredients. On the other hand, labelling of a product with respect to the technology that was used to produce it was seen as a genuine consumers' right. It was perceived as the only means by which a consumer decision on the market could give signals to producers about the acceptability of their products, rather than legal prohibition.

94. Both the German and the Austrian Commissions of Enquiry were considered to have had little impact on actual biotechnology policy (Grabner and Torgersen, 1998).

95. This role shed some light on the presently prevailing understanding of democracy in the economy-centred common market: impotent citizens (or, for paternalising state authorities, subjects) should possibly turn into powerful consumers in order to pursue a moral matter. See also note 9 above.

96. This was also reflected in new alliances between NGOs, consumer organisations and retailers.

97. The EU did not follow the European Biotechnology Council's recommendation to adopt the British model, which built much more on scientific evidence.

98. This led to the conflict over 'product versus process'.

99. Among the forces towards convergence were the pressure exerted by international competition and the globalisation of trade and industry; the necessity to adhere to guidelines issued by international organisations such as the OECD or the World Trade Organisation; and concerns about Europe as an area with high wages and standard of living that demands the production of goods with high added value. Among the forces towards divergence were the increasing importance of NGOs, acting mostly nationally, which could command high public trust, as compared with the erosion of trust in expert knowledge after food scandals; developments triggered by the election of governments formed by more leftist or green parties; a strengthening of the importance of national parliaments by governments built on weak majorities; and the acknowledgement of the importance of dealing with public concerns by state institutions and companies.

100. Apart from testing in cases of severe inherited diseases, which is generally welcomed.

REFERENCES

Baggott, R. (1998) 'The BSE Crisis. Public Health and the "Risk Society"', in P. Gray and P. 'tHart (eds.), *Public Policy Disasters in Western Europe*, London and New York: Routledge, pp. 61–78.

Bandelow, N. (1997) 'Ausweitung politischer Strategien im Mehrebenensystem, Schutz vor Risiken der Gentechnologie als Aushandelsmaterie zwischen Bundsländern, Bund und EU', in R. Martinsen (ed.), *Politik und Biotechnologie. Die Zumutung der Zukunft*, Baden-Baden: Nomos, pp. 153–68.

Bauer, M. (1995) 'Resistance to New Technology and Its Effects on Nuclear Power, Information Technology and Biotechnology', in M. Bauer (ed.), *Resistance to New Technology: Nuclear Power, Information Technology and Biotechnology*, Cambridge: Cambridge University Press, pp. 1–41.

Bauer, M., J. Durant and G. Gaskell (1998) 'Biology in the Public Sphere: a Comparative Review', in J. Durant, M. Bauer and G. Gaskell (eds.), *Biotechnology in the Public Sphere*, London: Science Museum, pp. 217–27.

Beck, U. (1986) *Die Risikogesellschaft. Auf dem Weg in eine andere Moderne*, Frankfurt am Main: Suhrkamp.

(1988) *Gegengifte. Die organisierte Unverantwortlichkeit*, Frankfurt am Main: Suhrkamp.

Behrens, M., S. Meyer-Stumborg and G. Simonis (1997) *Genfood. Einführung und Verbreitung, Konflikte und Gestaltungsmöglichkeiten*, Berlin: Sigma.

BEP (Biotechnology and the European Public Concerted Action Group) (1997) 'Europe Ambivalent on Biotechnology', *Nature* 387: 845–7.

Bongert, E. (1997) 'Towards a "European Bio-Society"? Zur Europäisierung der neuen Biotechnologie', in R. Martinsen (ed.), *Politik und Biotechnologie. Die Zumutung der Zukunft*, Baden-Baden: Nomos, pp. 117–34.

Bud, R. (1993) *The Uses of Life. A History of Biotechnology*, Cambridge: Cambridge University Press.

Cantley, M. (1995) 'The Regulation of Modern Biotechnology: A Historical and European Perspective', in D. Brauer (ed.), *Biotechnology*, vol. 12, New York: VCH, pp. 505–681.

Deutscher Bundestag (1989) *Zum gentechnologisch hergestellten Rinderwachstum-shormon*. Enquete-Kommission 'Gestaltung der technischen Entwicklung; Technikfolgen–Abschätzung und-Bewertung' des 11. Deutschen Bundestages, BT-Drucksache 11/4607, Bonn.

Dolata, U. (1995) 'Nachholende Modernisierung und internationales Innovationsmanagement. Strategien der deutschen Chemie- und Pharmakonzerne', in T. von Schell and H. Mohr (eds.), *Biotechnologie–Gentechnik. Eine Chance für neue Industrien*, Berlin: Springer, pp. 456–80.

Douglas, M. and A. Wildavsky (1982) *Risk and Culture. An Essay on the Selection of Technological and Environmental Danger*, Berkeley: University of California Press.

Durant, J., M. Bauer and G. Gaskell (eds.) (1998) *Biotechnology in the Public Sphere*, London: Science Museum.

Ellul, J. (1964) *The Technological Society*, New York: Alfred A. Knopf.

European Council (1990), Directive 90/219/EEC of 23 April 1990 on the Contained Use of Genetically Modified Micro-organisms; and Directive 90/220/EEC of 23 April 1990 on the Deliberate Release into the Environment of Genetically Modified Organisms, *Official Journal of the European Communities* L117/1–27, Brussels.

(1994), Directive 94/15/EC of 15 April 1994 for the First Adaptation of the Directive 90/220/EWG on the Deliberate Release of GM Organisms Into the Environment to Technical Progress. Abl./L 103/20, Brussels.

Galloux, J.-C., H. Prat Gaumont and E. Stevers (1998) 'Europe', in J. Durant, M. Bauer and G. Gaskell (eds.), *Biotechnology in the Public Sphere*, London: Science Museum, pp. 177–85.

Gaskell, G., M. Bauer and J. Durant (1998) 'The Representation of Biotechnology: Policy, Media and Public Perception', in J. Durant, M. Bauer and G. Gaskell (eds.), *Biotechnology in the Public Sphere*, London: Science Museum, pp. 3–12.

Gill, B. (1991) *Gentechnik ohne Politik. Wie die Brisanz der Synthethischen Biologie von wissenschaftlichen Institutionen, Ethik- und anderen Kommissionen systematisch verdrängt wird*, New York and Frankfurt am Main: Campus.

Gill, B., J. Bizer and G. Roller (1998) *Riskante Forschung. Zum Umgang mit Ungewißheit am Beispiel der Genforschung in Deutschland. Eine sozial- und rechtswissenschaftliche Untersuchung*, Berlin: Edition Sigma.

Gottweis, H. (1995) 'Governing Molecules. The Politics of Genetic Engineering in Britain, France, Germany, and in the European Union', habilitation paper, Faculty of Humanities, University of Salzburg.

Grabner, P. and H. Torgersen (1998) 'Österreichs Gentechnikpolitik – Technikkritische Vorreiterrolle oder Modernisierungsverweigerung?' *Österreichische Zeitschrift für Politikwissenschaft* 1: 5–27.

Gundelach, P. (1991) 'Research on Social Movements in Denmark', in D. Rucht (ed.), *Research on Social Movements. The State of the Art in Western Europe and the USA*, Frankfurt am Main: Campus, pp. 262–94.

Hall, P.A. (1993) 'Policy Paradigms, Social Learning and the State. The Case of Economic Policymaking in Britain', *Comparative Politics*, 25(3): 275–96.

Hill, A. and M. Michael (1998) 'Engineering Acceptance: Representations of "The Public" in Debates on Biotechnology', in P. Wheale, R. von Schomberg and P. Glasner (eds.), *The Social Management of Genetic Engineering*, Aldershot: Ashgate, pp. 201–18.

Hoban, T.J. (1997) 'Consumer Acceptance of Biotechnology: an International Perspective', *Nature Biotechnology* 15: 232–4.

INRA (1993) 'Biotechnology and Genetic Engineering. What Europeans Think about It in 1993', survey conducted in the context of Eurobarometer 39.1.

Jasanoff, S. (1995) 'Product, Process or Programme: Three Cultures and the Regulation of Biotechnology', in M. Bauer (ed.), *Resistance to New Technology. Nuclear Power, Information Technology and Biotechnology*, Cambridge: Cambridge University Press, pp. 311–31.

Jülich, R. (1998) 'Öffentlichkeitsbeteiligung im Geltungsbereich der EG-Richtlinien 90/219 und 90/220 im internationalen Vergleich. Die Ausgestaltung von Informations- und Partizipationsrechten in den EU-Mitgliedstaaten, der Schweiz und Norwegen', Öko-Institut, Freiburg, Darmstadt, Berlin.

Jungermann, H. and P. Slovic (1993) 'Characteristics of Individual Risk Perception', in Rück Bayerische (ed.), *Risk Is a Construct*, Munich: Knesebeck, pp. 85–101.

Kliment, T., O. Renn and J. Hampel (1994) 'Die Wahrnehmung von Chancen und Risiken der Gentechnik aus der Sicht der Öffentlichkeit', in T. von Schell and H. Mohr (eds.), *Biotechnologie–Gentechnik. Eine Chance für neue Industrien*, Berlin/Heidelberg: Springer, pp. 558–83.

Krimsky, S. (1991) *Biotechnics and Society. The Rise of Industrial Genetics*, New York: Praeger.

Krohn, W. and G. Krücken (1993) 'Risiko als Konstruktion und Wirklichkeit. Eine Einführung in die sozialwissenschaftliche Risikoforschung', in W. Krohn and G. Krücken (eds.), *Riskante Technologien: Reflexion und Regulation. Eine Einführung in die sozialwissenschaftliche Risikoforschung*, Frankfurt am Main: Suhrkamp, pp. 9–44.

Levidow, L., S. Carr, R. von Schomberg and D. Wield (1996) 'Regulating Agricultural Biotechnology in Europe: Harmonisation Difficulties, Opportunities, Dilemmas', *Science and Public Policy* 23: 135–7.

Macer, D. (1994) 'Bioethics for the People by the People', Christchurch, New Zealand/Tsukuba, Japan: Eubios Ethics Institute.

McKelvey, M. (1997) 'Moving to Commercialisation: Invited Response', in *Transgenic Animals and Food Production, Kungelik Skogs- och Lantbruksakademiens Tidskrift* 20: 133–8.

Majone, D. (1989) *Evidence, Argument and Persuasion in the Policy Process*, New Haven, CT: Yale University Press.

Nielsen, T.H. (1999) 'Bioteknologi og biopolitik, 1976–1999', manuscript, Oslo.

Oakey, R., W. Faulkner, S. Cooper and V. Walsh (1990) *New Firms in the Biotechnology Industry*, London: Pinter.

OECD (1986) 'Recombinant DNA Safety Considerations', Paris.

 (1988) 'Biotechnology and the Changing Role of Government', Paris.

 (1992) 'Safety Considerations for Biotechnology', Paris.

 (1993a) 'Biotechnology, Agriculture and Food', Paris.

(1993b) 'Traditional Crop Breeding Practices: an Historical Review to Serve as a Baseline for Assessing the Role of Modern Biotechnology', Paris.

(1996) 'Intellectual Property. Technology Transfer and Genetic Resources. An OECD Survey of Current Practices and Policies', Paris.

Ogburn, W.F. (1973) 'The Hypothesis of Cultural Lag', in Eva Etzioni-Halevy and Amitai Etzioni (eds.), *Social Change: Sources, Patterns, and Consequences*, 2nd edn, New York: Basic, pp. 477–80.

OTA (Office of Technology Assessment, US Congress) (1984) *New Developments in Biotechnology. 4: US Investment in Biotechnology*, OTA-BA-360, Washington D.C.: US Government Printing Office.

Pline, C. (1991) 'Popularizing Biotechnology: the Influence of Issue Definition', *Science, Technology and Human Values* 16(4): 474.

Praag, P. van (1991) 'The Netherlands: Action and Protest in a Depillarized Society', in D. Rucht (ed.), *Research on Social Movements. The State of the Art in Western Europe and the USA*, Frankfurt am Main: Campus, pp. 295–320.

Rammert, W. (1993) *Technik aus soziologischer Perspektive*, Opladen: Westdeutscher Verlag.

Rucht, D. (1991) 'The Study of Social Movements in Western Germany: between Activism and Social Science', in D. Rucht (ed.), *Research on Social Movements. The State of the Art in Western Europe and the USA*, Frankfurt am Main: Campus, pp. 175–202.

(1995) 'Impact of Anti-Nuclear Power Movements', in M. Bauer (ed.), *Resistance to New Technology. Nuclear Power, Information Technology and Biotechnology*, Cambridge: Cambridge University Press, pp. 277–91.

Rüdig, W., J. Mitchell, J. Chapman, P.D. Lowe (1991) 'Social Movements and Social Sciences in Britain', in D. Rucht (ed.), *Research on Social Movements. The State of the Art in Western Europe and the USA*, Frankfurt am Main: Campus, pp. 121–48.

Schomberg, R. von (1998) 'An Appraisal of the Working in Practice of the Directive 90/220/EEC on the Deliberate Release of GM Organisms', STOA, European Parliament, Luxembourg.

Seifert, F. and H. Torgersen (1997) 'How to Keep out What We Don't Want. An Assessment of "Sozialverträglichkeit" under the Austrian Genetic Engineering Act', *Public Understanding of Science* 6: 301–27.

Sharp, M. (1985) 'The New Biotechnology – European Governments in Search of a Strategy', *Industrial Adjustment and Policy IV Series*, Sussex European Papers No. 15.

Slovic, P. (1987) 'Perceptions of Risk', *Science* 236: 280–5.

Smink, G.C.J. and A.M. Hamstra (1996) *Informing Consumers about Foodstuffs Made with Genetic Engineering*, Leiden: SWOKA.

Tappeser, B. (1990) in 'Kurzcommentar für die Arbeitsgemeinschaft ökologischer Forschungsinstitut', K. Grosch, P. Hampe and J. Schmidt (eds.), *Herstellung der Natur? Stellungnahmen zum Bericht der Enquete-Kommission 'Chancen und Risiken der Gentechnologie' des 10. Dentschen Bundestags*, Frankfurt/New York: Campus.

Thielemann, H. (1998) 'Kommunikation im Konflikt um die gentechnische Insulinherstellung bei der Hoechst AG', in O. Renn and J. Hampel (eds.), *Kommunikation und Konflikt. Fallbeispiele aus der Chemie*, Würzburg: Königshausen & Neumann, pp. 153–81.

Tichy, G. (1995) 'Sozialverträglichkeit – ein neuer Standard zur Technikbewertung?', *Österreichische Zeitschrift für Soziologie* 19(4): 50–61.

Vogel, D. (1996) *Trading up. Consumer and Environmental Regulation in a Global Economy*, Cambridge, MA: Harvard University Press.

Vugt, F. van and A.M.P. Nap (1997) 'Regulatory and Policy Issues as Viewed within a Cultural Framework: a European Perspective', in *Transgenic Animals and Food Production, Kungl. Skogs- och Lantbruksakademiens Tidskrift* 20: 31–6.

Wyndham, A. and G. Evans (1998) 'National Biosafety Legislation and Trade in Agricultural Commodities', *BINAS News* 4(2 & 3): 2–6.

3 Media coverage 1973–1996: trends and dynamics

*Jan M. Gutteling, Anna Olofsson, Björn Fjæstad,
Matthias Kohring, Alexander Goerke, Martin W. Bauer
and Timo Rusanen. With the further cooperation of:
Agnes Allansdottir, Anne Berthomier, Suzanne de
Cheveigné, Helle Frederiksen, George Gaskell, Martina
Leonarz, Miltos Liakopoulos, Arne Thing Mortensen,
Andrzej Przestalski, Georg Ruhrmann, Maria Rusanen,
Michael Schanne, Franz Seifert, Angeliki Stathopoulou
and Wolfgang Wagner*

Introduction

Owing to the complexity and vulnerability of their technical and social systems, modern societies are sometimes characterised as 'risk societies' (Beck, 1986). A dynamic and transnational exchange of potentially conflicting information occurs between actors participating in the discourse about the consequences of risk-related decisions (Hilgartner and Bosk, 1988). In these complex societies, in which more and more immediate information is required in coping with everyday life, individuals have become highly dependent on the mass media. In most European countries, the media have assumed additional functions to informing the public: for instance, serving as society's watchdog, signalling injustice or unwanted developments, as well as acting as the main source of entertainment. As the developments and applications of modern biotechnology become more apparent, the mass media will also begin to provide these services for this particular domain of technological innovation. Furthermore, the level of individual dependency on the media will be especially high in the case of modern biotechnology because it is virtually impossible to gather information on this subject through direct personal experiences. Genetically modified soya, gene therapy, the square tomato, Dolly the sheep and Herman the bull have become well-known public 'icons' of modern biotechnology, not through personal acquaintance but through the information provided by the electronic and mass print media. This also implies that the public can construct an image of the 'biotechnology reality' only on the basis of what the media themselves decide to convey.

95

Thus, it is reasonable to identify the media as important modifiers in the processes of shaping popular perceptions and influencing the public's decision to either accept or reject new developments such as biotechnology.

Many scholars of mass communication have formulated theories of the media's role in society, in which realistic and constructionist approaches can be distinguished. The realist notion is based on the assumption of an existing set of events, which can be reported in an objective and balanced manner by competent, fair and unbiased journalists. According to this realistic approach, journalism is like a mirror on reality (see, for example, Kepplinger, Ehmig and Ahlheim, 1991). In keeping with this perspective, many (biotechnology) scientists will view the task of the press as involving the accurate and unbiased reporting of technical facts. They will argue that the complexity of the technology lends authority to the expert point of view and compels the press to highlight precisely this aspect (see, for example, Hagedorn and Allender-Hagedorn, 1997).

In contrast to this perspective, other scholars have stated that the media construct meaning by presenting a mediated world, rather than mirroring a more or less 'objective' reality (see, for example, DeFleur and Ball-Rokeach, 1989). In many conceptions of the construction of mediated reality, processes of news selection play a crucial role, particularly in the sense that a variety of social factors determine the press's notions of what is newsworthy and how issues ought to be framed. Clearly, on any day, many more events occur than the media are able to report, owing to limitations of time, space and technology. Therefore, at diverse stages in the process of news production, information is being processed: it is traced, collected, translated, edited, shortened or expanded, and transferred. After the final editorial process the public receives information about news events with little conception of these processes. In their theory of news value, Galtung and Ruge (1965) postulated that news value is mainly determined by the actuality of the event, in a temporal, geographical or psychological sense, and by its significance. A radical event, with long-lasting consequences for a large group of people, has a high news value. However, many studies in this area indicate that initially establishing a particular event as newsworthy depends on the journalist's highly personal and subjective assessments of its news value (see, for an overview, Servaes and Tonnaer, 1992).

The framing of news can be understood as the process through which complex issues are reduced to journalistically manageable dimensions, resulting in a particular focus on a certain issue. This reasoning implies that journalistic framing may also lead to journalistic selection, for example by placing heavy reliance on the information from very particular sources. With regard to modern biotechnology, US newspaper coverage is characterised by an inordinate dependence upon industrial and scientific sources, most of which emphasise economic considerations and potential benefits (Hornig, 1994). Nonetheless, if journalists are aware of differences in perception among experts and laypeople, whether owing to selective outlooks or to conflicting value orientations, they may be more inclined to emphasise their role as watchdogs. This is especially likely in the context of discussions of the potential risks and benefits of modern biotechnology, in which journalists may define their role as that of seeking out and framing the issue in terms of danger, controversy, and so on (Hornig, 1992; Frewer, Howard and Shepherd, 1997). These dynamic processes may encourage the media to highlight some aspects of biotechnological innovation at particular times, but they may also encourage the media to entirely neglect other issues related to biotechnology (Kitzinger and Reilly, 1997).

Research questions

In this chapter, we will describe the dynamics of the coverage of modern biotechnology in the opinion-leading press across Europe, from the perspective of journalistic selection and framing processes. Our study of media coverage spans a twenty-four-year period, starting in 1973 with the invention of recombinant DNA (rDNA) technology, a year considered to represent the inception of a new era of modern biotechnology. Our study ends in December 1996, a few months before the breaking of the Dolly story, which may be seen as another landmark in the development of modern biotechnology. (In chapter 11 of this book more will be said about Dolly and the media.) It is clear that the public's perception of modern biotechnology may be greatly influenced by the press's earlier coverage of modern biotechnology in its broadest sense. So the present study sheds light on the media reactions, and to some extent also the public reactions, to modern biotechnology.

As we have seen, many studies of the functioning of the mass media focus on journalistic selection and framing processes. The selection of events to be reported and the way events are portrayed may have a profound impact on the public's perceptions of these events, and may even influence policy formulation. Consequently, by studying journalistic products we may be able to gain insight into these influential processes. Thus, the basic questions we ask are: 'What is reported?' and 'How is it reported?', referring to selection and framing processes respectively. Our study of the development of the coverage of modern biotechnology proceeds from these fundamental questions. We will present data from the opinion-leading press of twelve European countries, enabling us to look at differences and similarities across Europe. Furthermore, we will look at developments across time.

First, we will describe the media coverage of modern biotechnology in general terms, analysing the main themes, actors and locations that recur in the coverage across Europe. This analysis will provide the basis for our exploration of journalistic selection processes. Next, we will focus on framing: we will analyse the negative or positive tone of the coverage, the framing of modern biotechnology in terms of hope or doom, and the distribution of risk and benefit arguments in the coverage. Indeed, the focus throughout our study will be on the press's treatment of issues related to the potential risks and benefits of modern biotechnology. The journalistic framing of modern biotechnology as either a salutary development or one in which risks are believed to outweigh all foreseeable benefits is likely to have a major impact on readers (see also Hornig, 1992; Frewer, Howard and Shepherd, 1997). In other words, we will analyse the contents of articles in which risks or benefits, or both, are described. Finally, we will look at the relationship between the media coverage of modern biotechnology and public perceptions across Europe.

Method

Content analysis may be described as a systematic preparatory and data-reducing method that seeks to achieve the objectivity of other social scientific research methodologies (Holsti, 1969; Krippendorff, 1980). In this study, the unit of analysis comprises any article containing references to modern biotechnology found in any part of the opinion-leading newspapers of the twelve European countries (with the exclusion of advertisements). These articles were further subdivided and coded according to whether they were describing conditions for, or the procedures and consequences of, research, analysis, practical handling and intervention into the genome of humans, animals and/or plants. Different examples include genetic engineering, gene therapy, modern biotechnology as a business enterprise, the genetic modification of organisms, novel foods, genetic fingerprinting, prenatal genetic diagnostics, the ethics of modern biotechnology, the statutory and non-statutory regulation of genetic technology, and initiatives supporting or opposing genetic technology. Articles addressing these diverse themes could be found in virtually all sections of the newspapers: national and international news, business pages, science pages, editorials and debate, and so on. Each relevant article was coded and the information incorporated in a transnational database.

We adopted this broad conception of modern biotechnology for several reasons. First, we wanted to ensure a thorough coverage that did not occlude the variation among different nations. Secondly, we wanted our unit of analysis to be concordant with the two other parts of the international research project, namely the analysis of the development of policy processes in the field of modern biotechnology and the analysis of public perceptions of modern biotechnology across Europe (as measured by Eurobarometer Biotechnology 1996, see 'Biotechnology and the European Public', 1997, and chapters 6, 7, 8 and 12 of this book). Finally, modern biotechnology is, in itself, a very broad issue with applications in many different fields, each of which demands consideration.

The assumption underlying our definition of what constitutes the 'opinion-leading press' is that they may be considered to be exponents of the 'media arena' in each country. Across Europe, certain newspapers and news magazines are identifiable as opinion-leading sources of information, and have assumed this status in relation to other media, to important decision-makers (such as politicians, civil servants, experts and industrialists), as well as to the general public. So, by analysing the opinion-leading press, we can realistically expect to gain an accurate impression of the social dynamics of information processing relating to modern biotechnology. In addition, we will gain insight into the development of these information flows over time. This method of choosing the opinion-leading press of the participating countries automatically leads to the inclusion of a very diverse set of media: in some countries large-circulation dailies, in others small-circulation elite newspapers, and, in still others, news weeklies. Other news media (for example, television) may also have an important opinion-leading function. However, we were able to analyse only print material because, unlike television broadcasts, these are systematically preserved – in paper, microfiche or electronic form – for the entire twenty-four-year period covered by our study. Appendix table 3A.1 contains some key data on the media in our study. This table also incorporates information on the manual or on-line selection procedures that were employed in the various countries. It should be noted that on-line searching with national equivalents for each keyword allows one to locate every article on the subject, whereas manual searching would probably lead to an underestimation of the actual number of articles in that particular period.

Coding frame (variables)

The three types of variables we studied were incorporated in a common coding frame, identical for all countries. Native speakers performed all of the coding tasks. In order to improve the comparability of the coding across countries, the coding frame was discussed thoroughly and revised after a trial coding of substantial numbers of articles in the various countries. Inter-coder reliability was also achieved so as to ensure consistent coding across and within countries. However, this process could not resolve all transnational coding issues. The more substantial differences in the characteristics of the research material in the various countries could not be obviated. All comparisons between countries should be interpreted with this caveat in mind.

The coding frame consisted of registration variables, variables relating to journalistic processes of selection and framing and, finally, variables reflecting an evaluation or judgement of the content of the article. We registered each article that qualified as a unit of analysis according to country and year of publication. Year of publication was then recoded and allotted to one of three equal-size eight-year periods, which we call octades: octade I – 1973–80; octade II – 1981–8; and octade III – 1989–96. These time periods were used to study developments, or trends, in the coverage of modern biotechnology. Because of a lack of clear internal criteria that could be used over all countries simultaneously, this periodisation was based on a criterion external to developments in modern biotechnology (see the twelve country profiles of media coverage of modern biotechnology in Durant, Bauer and Gaskell, 1998).

With respect to the variables relating to journalistic processes of selection and framing, we identified the appearance of manifest descriptions of the risks and benefits of modern biotechnology. These articles then underwent five coding stages. First, their contents were coded according to their inclusion, or otherwise, of thirty-five categories or sub-themes. Then, the coding of article content was further reduced to nine themes: 'Medical', 'Basic research', 'Animal and agricultural', 'Economic', 'Regulatory and policy', 'Ethical', 'Identification', 'Safety and risk' and 'Other themes'. In each article three different thematic codings were permitted. A second coding stage involved classifying articles according to the sorts of actors mentioned. Thirty-five classifications were identified, which were then consolidated to form nine actor groups: 'Scientists', 'NGOs', 'Politicians', 'Industry', 'Ethical committees', 'Media and public voice', 'International actors', 'EU' and 'Other actors'. The category 'Media and public voice' comprises references to modern biotechnology in other media or accounts of public opinion. In each article two actor codings were allowed. The third criterion for coding was the geographical location of the biotechnological activities identified in the articles. Locations (two codes allowed per article) were categorised according to a coding frame that comprised over fifty countries, including all participating European countries and the USA. Next these locations were collapsed into three location categories: 'Own country', 'Other European countries' and 'USA'. Further, the framing of modern biotechnology was coded in terms of the presentation of either progress/utility scenarios (the belief that biotechnology will have positive benefits) or doom scenarios (a pessimistic world-view in which biotechnology is conceived in terms of runaway technology or likened to Pandora's box). The final coding stage was termed 'coder impression of modern biotechnology'; the coder's impressions were rated on two separate five-point scales relating to positive and negative assessments, respectively. These scales were combined and recoded to provide a single score.

Results

Our database of the opinion-leading press in the twelve European countries contains 5,404 articles. In the following sections we will describe these articles in terms of our basic research questions. First, we briefly examine the database (table 3.1). Then we address the 'What?' and 'How?' questions. In doing so, we will describe the major themes relating to modern biotechnology covered in Europe's opinion-leading press (figure 3.1 and appendix table 3A.2), followed by a consideration of the main actors (figure 3.2 and appendix table 3A.3) and the main locations of activity in the European coverage of modern biotechnology (figure 3.3 and appendix table 3A.4). Then we switch to the analysis of the overall negative or positive characteristics of the coverage (table 3.2). Finally, the data from this media study are compared with data on the public perception and public knowledge of modern biotechnology from one of the other modules of this international project (table 3.3).

A first glance at the data

Table 3.1 presents, per country, the number of articles in the database and the relative distribution of these articles over the three octades. Looking at the table, we notice that, for most countries, more articles were found in octade II (1981–8) than in octade I (1973–80), and more in octade III (1989–96) than in octade II. This may suggest an increase in intensity in the coverage of the subject of biotechnology across Europe in the twenty-four-year period under study. Of course, we should bear in mind that a degree of sampling bias may have distorted these figures (see also appendix table 3A.1). However, if we take into account the different sampling strategies and then calculate – for each country and each octade – the average number of relevant articles in a random issue of each national opinion-leading daily or weekly, indications of an increase in coverage over the octades is reinforced. The exception here is Greece, where the probability of finding an article on biotechnology remains very small throughout the period studied.

Table 3.1 *Number of articles and distribution across Europe by octade, 1973–1996*

Country	Number of articles and type of medium	Octade I (1973–80) Selected	Random issue	Octade II (1981–88) Selected	Random issue	Octade III (1989–96) Selected	Random issue
Austria	191 (daily)	36	0.05	66	0.14	89	0.32
	111 (weekly)	17	0.04	31	0.09	63	0.30
Denmark[a]	300 (daily)	13	0.01	127	0.05	160	–
Finland	375 (daily)	26	0.01	97	0.04	252	0.10
France[b]	623 (daily)	17	0.05	216	0.13	390	0.63
Germany	418 (daily)	42	0.00	132	0.16	244	0.29
	170 (weekly)	13	0.00	44	0.21	113	0.54
Greece[a,c]	65 (daily)	11	0.00	13	0.01	41	0.00
Italy	340 (daily)	51	0.00	128	0.07	161	–
Netherlands[c]	1,119 (daily)	5	0.01	184	0.13	930	0.37
Poland	132 (daily)	14	0.03	53	0.04	65	0.16
	76 (weekly)	31	0.07	16	0.04	29	0.07
Sweden	734 (daily)	99	0.04	254	0.10	381	0.15
Switzerland	211 (daily)	17	0.05	61	0.18	133	0.39
United Kingdom	539 (daily)	107	0.67	156	1.01	276	2.21
Total	5,404						

Notes: (–) cannot be computed, or insufficient data.
[a] Two dailies; no correction applied.
[b] In octade 1973–80 only 1975 studied.
[c] In octade 1973–80 only four years.

Main themes in the press coverage of modern biotechnology across Europe

Figure 3.1 and appendix table 3A.2 present the distribution of the main themes in the coverage of modern biotechnology across Europe. Figure 3.1 shows, for each country, the frequency with which the eight basic themes arise ('Medical', 'Basic research', 'Economics', 'Animal and agricultural', 'Ethics', 'Regulatory and policy', 'Identification', and 'Safety and risk'). Appendix table 3A.2 contains information on the six most commonly mentioned themes, and has been subdivided into 'risk-only', 'benefit-only' and 'risk and benefit' articles (articles in which neither risks nor benefits are mentioned are ignored in this table).

As we can see from figure 3.1, in most countries 'Medical' and 'Basic research' issues are the most frequently cited themes in the press coverage. 'Regulatory' and 'Animal and agricultural' issues follow, and then 'Economic' and 'Ethical' issues. 'Animal and agricultural' issues are relatively important in Finland, the Netherlands and Poland, and are accorded considerable attention in Denmark, but they were rarely mentioned in Greece. 'Regulatory' issues are discussed relatively frequently

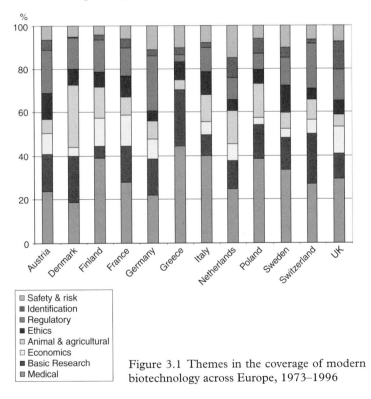

Figure 3.1 Themes in the coverage of modern biotechnology across Europe, 1973–1996

in Switzerland and Germany, but, again, the Greek press seldom broaches the issue. 'Economic' issues are comparatively important in the opinion-leading newspapers of France and Finland, but this issue is barely addressed in Poland and Denmark and is entirely neglected in Greece. 'Ethical' issues receive relatively higher coverage in the Austrian, Swedish, French, Greek and British opinion-leading press.

Table 3A.2 identifies the six most frequently encountered themes (in order of frequency, these are 'Medical', 'Basic research', 'Regulatory and policy', 'Animal and agricultural', 'Economics' and 'Ethics'), which collectively comprise more than 87 per cent of the total. Overall, these six themes are the most frequently observed in all octades. (One exception to this general rule is that the theme of 'Safety and risk' ranks third in the first octade, with 14 per cent, but only seventh in the cumulative ranking, with 7 per cent). In addition, this table shows the relative emphasis of press coverage on the issue of risks versus benefits (coded into three categories: 'risk only', 'benefit only' or 'mixed risk and benefit').

As one might expect, another general rule is that 'Medical' issues and 'Basic research' are presented with greatest frequency in 'benefit-only' articles. The relationship between risk–benefit assessment and the theme of 'Animal and agricultural' issues is more complex. In some countries (Finland, France and Italy), 'benefit-only' articles predominate in this context; in others, 'Animal and agricultural' issues are found almost as much in other types of article as well (Sweden, the Netherlands and, most clearly, Denmark). In eight of the countries 'Regulatory and policy' issues are reported most frequently as 'risk only', but in France, the UK, Italy and Poland these issues appear most frequently in 'mixed risk and benefit' articles. In general, 'Ethical' issues are found most commonly in 'risk-only' articles. Throughout Europe, 'Economic' issues appear most frequently in 'benefit-only' articles, though in Denmark they can also appear in 'mixed risk articles'.

Dynamics in the coverage of themes   Looking at the dynamics in the reporting of the six main biotechnological themes across Europe, a fairly consistent picture emerges. The theme of 'Basic research' declined in importance over the entire twenty-four-year period. The only – partial – exceptions are Poland, which accorded it most attention in octade II, and Greece, where 'Basic research' was ascribed most significance in octade III. The situation for the treatment of 'Medical' issues is a little more complex. In several countries (Austria, Finland, Germany, Switzerland and the UK) there was a decrease in the relative treatment of 'Medical' issues over the whole twenty-four-year period. In other countries, however, there was an overall increase (Denmark, France, Poland and Sweden). In Greece and the Netherlands, 'Medical' issues received a similar proportion of the biotechnology coverage, but with strong increases and decreases, respectively, in octade II. In most countries, there was a relative increase in the attention accorded to 'Animal and agricultural' issues between octade I and octade II. In Poland and the UK this was followed by a decrease during octade III. Changes in the treatment of 'Regulatory' issues may be separated into three clusters. In some countries, the proportion of coverage was relatively constant over the three octades (Finland, Germany, Italy, Poland, Sweden and the UK); whereas in others the attention devoted to these issues either diminished (France, Greece and the Netherlands) or increased (Austria, Denmark and Switzerland). The same can be said for the coverage of 'Ethical' issues. In some countries interest in this aspect of the biotechnology debate did not appreciably change during the three octades (Austria, Germany, Greece, Italy and Switzerland); in other countries the attention devoted to these issues either diminished (Denmark and Poland) or increased (Finland, France, the Netherlands, Sweden and the UK). 'Economic' issues received relatively more attention during our research period in Finland, France, Germany, the Netherlands and the UK, but less so in Austria.

For the third octade we studied the intercorrelations among the various themes across the participating countries. The results of this analysis indicate that there is no significant correlation between the occurrence of the themes 'Basic research', 'Economic' issues, 'Ethical' issues or 'Animal and agricultural' issues. There is, however, a negative correlation between 'Medical' and 'Regulatory' issues (Spearman rank order correlation $-.69$, $p < .005$). This indicates that, in countries with a relatively high coverage of 'Medical' issues, relatively few articles were found on 'Regulatory' issues.

Main actors

Figure 3.2 and appendix table 3A.3 show which actors or spokespersons played a role in the coverage of modern biotechnology across Europe. Figure 3.2 presents for each country the percentages of the observations made by the nine actor groups: 'Scientists', 'Industry', 'Politicians', 'NGOs', 'Media and public voice', 'Ethical committees', 'International actors', 'EU' and 'Others'. Table 3A.3 contains the data for only the four most frequently mentioned actor groups, subdivided into 'risk-only', 'benefit-only' and 'mixed risk and benefit' articles (articles in which neither risks nor benefits are mentioned were not included in this table).

In most countries, 'Scientists' and 'Industry', groups that focus primarily on the beneficial aspects of modern biotechnology, constitute more than half the references to actors. In Poland, Finland, and Greece, they make up at least 75 per cent of the references. Yet, in Denmark, Switzerland, Austria, the Netherlands and Sweden, these two actor groups enjoy considerably less prominence. In Denmark and Switzerland, the

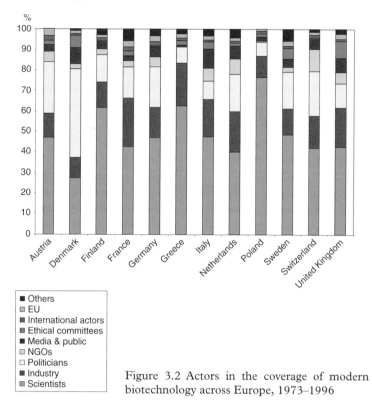

Figure 3.2 Actors in the coverage of modern biotechnology across Europe, 1973–1996

views of 'Politicians' and 'NGOs' rank highest, whereas in Finland, Greece and Poland these two actor groups receive considerably less attention. 'Media and public voice' is relatively important in Italy (10 per cent), Denmark (8 per cent) and the UK (7 per cent), but is hardly discernible in Poland, France and Greece (2 per cent or less). 'Ethical committees' gained a relatively high degree of coverage in the UK (9 per cent), Denmark (6 per cent) and Sweden (5 per cent); but in other countries, coverage of the views of such groups is insubstantial (3 per cent or less). Similarly, 'International actors' and the 'EU' receive little attention as actor groups, with the views of the EU achieving most attention in Austria (3 per cent).

Table 3A.3 identifies four frequently encountered actor groups ('Scientists', 'Politicians', 'Industry' and 'NGOs'), which together comprised more than 85 per cent of the total coverage. 'Scientists' take up more than 46 per cent of all codings, followed by 'Politicians' (18 per cent), 'Industry' (17 per cent) and 'NGOs' (about 5 per cent). In general, these four actor groups are commonly observed in all octades. The remaining, less prominent, actor groups have not been included in this table. In most countries, 'Scientists' are the most frequently cited actor group in 'benefit-only' and 'mixed risk and benefit' articles. The views of industry spokespersons and 'Politicians' are also, though less commonly, found in this context. In Poland, more than three-quarters of all identified actors are 'Scientists', and in many other countries 'Scientists' represent by far the largest single group of newspaper commentators. In contrast, in Denmark the 'Politicians' have assumed this status irrespective of the type of article.

In 'risk-only' articles the relative importance of the different actor groups is slightly different. In a number of countries, 'Scientists' again comprise the main actor group (Finland, France, the UK, Italy and Poland), but in other countries 'Politicians' are the largest group (Sweden, Switzerland and Denmark) or share this rank with 'Scientists' (Greece, Austria, the Netherlands and Germany). The journalistic selection and framing processes regarding 'NGOs' are also quite complex. In some countries, 'NGOs' are important actors in 'risk-only' and 'mixed risk and benefit' articles (most clearly so in France, Austria, the Netherlands, Germany, the UK and, perhaps, Italy), whereas in others they are most strongly associated with 'risk-only' stories (Sweden and Switzerland). In other countries (Finland, Greece, Poland and, perhaps surprisingly, Denmark), NGOs do not seem to play a very significant role (in terms of numbers involved).

Dynamics in references to actor groups Looking at the dynamics of actor-group coverage across Europe, the following picture emerges. The relative importance of 'Scientists' diminished in almost all countries (their status was stable in Poland, and there was an increase in Sweden). Over the entire twenty-four-year period, the relative prominence of 'Politicians' increased in many countries (Austria, Denmark, Finland, France, Italy, Switzerland and the UK), but stayed reasonably stable in Germany, the Netherlands, Poland and Sweden. The role of 'Industry' spokespersons in the coverage of modern biotechnology increased in almost all countries from the first to the second octade (Denmark, Finland, France, Germany, Greece, Italy, Poland, Switzerland and the UK); in many countries this was followed by a decrease during octade III. In Sweden, however, an incremental decrease in the attention accorded to 'Industry' spokespersons is discernible throughout the twenty-four-year period.

Relations between main actors and themes To look for correlations between the coverage of the main actors and the main themes in the coverage of modern biotechnology across Europe we calculated rank order correlations between these variables. We confined ourselves here to octade III, which, for most countries, contained sufficient numbers of observations of each variable to permit analysis.

A negative correlation between the occurrence of 'Politicians' and 'Industry' actor groups (Spearman rank order correlation $-.79$, $p < .001$) indicates that, in countries in which 'Politicians' were regular commentators in the coverage of modern biotechnology, fewer articles were to be found articulating the views of 'Industry' spokespersons. A similar relation pertains between 'Scientists' and 'NGOs' $(-.77$, $p < .005)$, and there is a positive correlation between 'Politicians' and 'NGOs' $(.61, p < .05)$. This indicates that, in countries in which 'NGOs' enjoyed little coverage, 'Politicians' also received fewer mentions, but references to 'Scientists' were proportionately greater. Regarding themes and actors, significant relationships were found between 'Medical' issues and 'Scientists' $(.64, p < .05)$, 'Politicians' $(-.72, p < .005)$ and 'NGOs' $(-.69, p < .01)$. In countries characterised by many articles on 'Medical' issues, more 'Scientists' tended to serve as actors, but relatively few 'Politicians' or 'NGOs'. There are also significant correlations between 'Regulatory' issues and particular actor groups: 'Industry' $(-.50$, $p < .05)$, 'Politicians' $(.71, p < .005)$ and 'NGOs' $(.54, p < .05)$. These correlations indicate that, in countries that accorded a relatively high degree of attention to 'Regulatory' issues, 'Industry' spokespersons received relatively little attention and, conversely, 'Politicians' and 'NGOs' enjoyed a higher proportion of the coverage.

Main locations of activity of modern biotechnology across Europe

Figure 3.3 and appendix table 3A.4 display the results of our assessment of the 'locations of activity' for modern biotechnology identified in the European press (the categories were 'Own country', 'Rest of Europe', 'USA', 'Other' and 'None mentioned'). Overall, 94 per cent of the articles contained at least one reference to a location of activity for modern biotechnological activity, and many articles also included a second reference. Of these, 46 per cent were made to 'Own country', which is by far the largest category. 'USA' was cited in 22 per cent of all observations and 'Rest of Europe' in 18 per cent; 'other locations' comprised 8 per cent of references. This pattern is observable throughout Europe, with the exception of Greece, in which only 13 per cent of references concerned 'Own country', with 'Rest of Europe' (40 per cent) and 'USA' (34 per cent) exceeding this proportion. Italy follows the general trend, but less conspicuously: only 32 per cent of references were made to 'Own country' whereas slightly more than 30 per cent of references were made to 'USA'.

Table 3A.4 identifies the three most important 'locations of activity' in terms of the appearance of 'risk-only', 'benefit-only' and 'mixed risk and benefit' articles. In many countries, reference to 'Own country' tended to be made in the context of 'risk-only' and 'mixed risk and benefit' articles. The exception here is again Greece. Concerning 'benefit-only' articles,

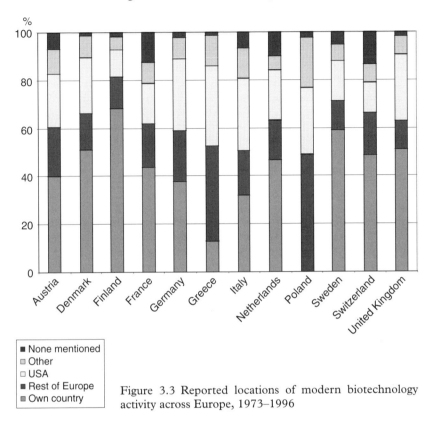

Figure 3.3 Reported locations of modern biotechnology activity across Europe, 1973–1996

'Own country' is, again, the most frequently cited 'location of activity'; however, in several countries 'USA' is the most frequently mentioned location (Germany, Greece and Italy).

Dynamics in references to locations of activity Over the twenty-four years studied, references to 'Own country' increased in many countries (Austria, Denmark, Finland, Germany, the Netherlands, Switzerland and the UK) and decreased in a few others (Greece and Sweden, the latter of which had a relatively high proportion of references to 'Own country' in octade I). In the first set of countries, during this period one can also observe a decrease in the proportion of citations of 'USA' as 'location of activity' for modern biotechnological research (with the exception of Finland). Relatively more attention was accorded to 'USA' in France, Italy, Poland and Sweden. The proportion of references to 'Rest of Europe' varies, with more references being made to this location in Denmark, Greece, the Netherlands and Sweden than in Austria, France, Italy, Poland, Switzerland and the UK.

Relations between locations of activity, actors and themes In looking for relationships among the coverage of 'locations of activity', the main actor groups and biotechnological themes, we calculated rank order correlations. Again, we confined our analysis to the third octade.

This analysis indicates that references to 'Own country' as the 'location of activity' correlate negatively with references to 'Rest of Europe' ($-.98$, $p < .001$) and 'USA' ($-.74$, $p < .005$). References to 'Rest of Europe', however, correlate positively with mentions of 'USA' ($.67$, $p < .05$). This indicates that, in countries in which references to 'Own country' were frequently made, few references to other countries were found. There is also a significant relationship between the mention of 'Economic issues' and references to 'Own country' ($.65$, $p < .05$). Among the remaining themes and actor groups, no significant correlations with 'location of activity' were found.

The overall negative or positive characteristics of modern biotechnology

Table 3.2 contains the data on journalistic framing in terms of: progress versus doom scenarios; the coder's overall impression of the coverage; and the proportions of 'risk-only' and 'benefit-only' articles per country. These data give an impression of the framing of modern biotechnology in the European opinion-leading press.

Overall, the framing of modern biotechnology in terms of 'progress/ utility' is much more commonplace than framing in terms of doom scenarios (runaway technology, Pandora's box, etc.). In 42 per cent of articles, 'progress/utility' was emphasised (reversed ranking in table 3.2), whereas in only 5 per cent of articles was doom predicted. In Italy, Greece and Poland the 'progress/utility' scenario was especially dominant; though in Denmark, Austria, Sweden and (somewhat surprisingly) Greece, we observed a relatively high proportion of articles anticipating disaster.

Table 3.2 also displays the combined measure of the coder's impression of the positive versus negative treatment of modern biotechnology. In most countries this combined measure results in a positive overall impression, with Finland, Greece, Italy and Poland having the most positive overall evaluations. In Denmark and Sweden, however, the general assessment is negative.

With respect to the proportions of 'risk-only' and 'benefit-only' articles (reversed ranking in table 3.2) a fairly consistent picture emerges. Across Europe, 11 per cent of the articles are coded as 'risk only' and 42 per cent

Table 3.2 Journalistic framing of modern biotechnology in the opinion-leading press across Europe, 1973–1996

Country	Doom scenario		Progress/utility scenario		Coder impression		Risk-only articles		Benefit-only articles		Overall rank
	%	Rank	%	Rank (reversed)	Mean	Rank	%	Rank	%	Rank (reversed)	
Austria	8	3[a]	46	6	0.8	6	13	5	51	7	6
Denmark	12	1	25	2	−0.1	2	19	1[a]	21	1	1
Finland	4	9	51	8[a]	1.3	9[a]	5	11	53	8	10
France	6	6	37	4	0.6	5	11	6	43	6	5
Germany	5	7[a]	49	7	0.9	7	6	9[a]	56	9[a]	8
Greece	9	2	72	11	1.3	9[a]	6	9[a]	63	11[a]	11
Italy	7	5	64	10	1.9	11	18	3	42	5	7
Netherlands	2	10[a]	22	1	0.3	4	9	7	29	3	4
Poland	5	7[a]	74	12	2.7	12	2	12	63	11[a]	12
Sweden	8	3[a]	42	5	−0.2	1	16	4	36	4	3
Switzerland	2	10[a]	29	3	0.0	3	19	1[a]	27	2	2
United Kingdom	2	10[a]	51	8[a]	1.2	8	6	9[a]	56	9[a]	9
Overall	5		42		0.7		11		42		

[a] Tied ranks.

as 'benefit only', indicating a generally positive attitude towards modern biotechnology. Denmark and Switzerland fail to conform to this trend, both countries exhibiting a relatively high proportion of 'risk-only' articles and a relatively low proportion of 'benefit-only' articles. In Finland and Poland this situation is reversed; here we observe a relatively low proportion of 'risk-only' articles and a relatively high proportion of 'benefit-only' articles. The proportion of 'risk-only' articles varies from 2 per cent (in Poland) to 19 per cent (in Switzerland and Denmark); the incidence of 'benefit-only' articles varies from 21 per cent (in Denmark) to 63 per cent (in Poland and Greece).

If the national data from table 3.2 are assigned to different groups according to their overall level of negative/positive coverage of modern biotechnology, one finds a clear affinity among four of the five indicators: 'progress/utility scenarios', 'coder impression', 'risk-only' articles and 'benefit-only' articles. The Spearman rank order correlations among these variables are all significant and are between .62 ($p < 0.05$) and .89 ($p < .001$). None of the correlations between the four variables and 'framing of doom' is, however, significant. Most of the twelve countries display a similar pattern with respect to these four indicators (only the Italian ranking on coder impression is inconsistent with the other rankings). Looking at the overall ranking based on these indicators, we conclude that in Denmark the coverage was coded as most negative, followed by Switzerland and Sweden. Overall, the most positive coverage was found in Finland, Greece and Poland.

Dynamics in negative or positive coverage of modern biotechnology
Looking at the dynamics in the framing of risk and benefit information, we observe that, for the transition from octade I to octade II, five countries show an increase in 'risk-only' articles and four countries a stationary situation (not shown in table). In three countries a decline in the proportion of 'risk-only' articles can be observed over these years. Regarding the 'benefit-only' articles, an increase in ten countries can be observed, but only in six of these countries (Sweden, France, Austria, Germany, the UK and Poland) was this trend accompanied by a decrease, or no change, in the number of 'risk-only' articles over the same period. This result is in accordance with the coder impression data, which suggest that, in the transition from octade I to octade II, more positive coverage appeared in the UK, France, Germany and Sweden, but it remained more or less the same as before in Austria and Poland. Combined, these results suggest that, at least in these six countries, the overall presentation of modern biotechnology became more positive between 1973 and 1988.

However, looking at the framing of risk and benefit in the transition from octade II to octade III, a different picture emerges. At this stage, many countries show either an increase in the proportion of 'risk-only' articles accompanied by a relative decrease in 'benefit-only' articles (Switzerland, Austria, Germany, the UK and Denmark), or no change in the proportion of 'risk-only' articles accompanied by a decrease in the proportion of 'benefit-only' articles (France and Italy). In six of these seven countries, this picture is reinforced by evidence of a change to a more negative coder impression of modern biotechnology from octade II to octade III (the exception being Germany, where a change to a more positive evaluation is observed). Overall, then, these results suggest the emergence of a more negative framing of modern biotechnology in newspaper reports over this period. In only three countries do we observe the reverse process: a decrease or absence of change in the proportion of 'risk-only' articles and an increase in 'benefit-only' articles (Finland, Greece and Sweden). For Greece and Sweden this is supported by a more positive coder impression. In these two countries alone, it would seem that a more positive general impression of modern biotechnology appeared between 1981 and 1996 in the opinion-leading press.

Media coverage and public perception and knowledge of modern biotechnology

The connections between the media coverage of modern biotechnology and the study of public perceptions of this technology can, of course, be conceived in a variety of ways. It is reasonable to expect the way biotechnology is portrayed in the media to have an effect on public perceptions, especially since few other sources of information are available. But it is also clear that the media tend to adapt their coverage to the perceived attitudes of their readership. Indeed, one may argue that causality flows both ways. So far in this chapter, we have examined the media coverage alone, but in the project as a whole we have also collected detailed information on the public perception and knowledge of modern biotechnology in the various countries covered in this media study (see chapters 4, 5, 11, 12 and 13).

For octade III, taking our data on the relationships among themes, actor groups, 'locations of activity' and evaluations of risks versus benefits (see above and table 3.2), we performed an additional analysis. For all these variables, rank order correlations were computed with data relating to the public perception of modern biotechnology across Europe (combined attitude score for encouragement of six applications of biotechnology)

Table 3.3 *Spearman rank order correlations across countries between public perception and knowledge, with positive or negative overall characteristics of the coverage of modern biotechnology, and themes, actor groups and location of activity*

	Public perception	Public knowledge	Characteristics of media coverage
Public perception	–	n.s.	.53*
Public knowledge	n.s.	–	−.63*
Main themes			
Basic research	n.s.	n.s.	n.s.
Medical issues	.69*	n.s.	.51*
Economic issues	n.s.	n.s.	n.s.
Ethical issues	n.s.	n.s.	n.s.
Regulatory & policy issues	−.78**	n.s.	n.s.
Animal & agricultural issues	n.s.	n.s.	n.s.
Main actor groups			
Scientists	n.s.	n.s.	.61*
Industry	.72**	n.s.	n.s.
Politicians	−.66**	n.s.	−.71**
NGOs	n.s.	n.s.	n.s.
Main location of activity			
Own country	n.s.	.53*	n.s.
Rest of Europe	n.s.	−.56*	n.s.
USA	n.s.	n.s.	n.s.

* $p < .05$; ** $p < .01$
n.s. = not significant.

and public knowledge (based on a nine-item knowledge scale from Euro-barometer 1996). Table 3.3 summarises these correlations. The first observation from this analysis is that a positive correlation exists between the public's attitude towards modern biotechnology and the overall positive or negative characteristics of the media coverage (Spearman rank order .53, $p < .05$). This suggests that countries in which the public is positive about biotechnology also tend to have a positive media coverage, and vice versa. In addition, there is a relationship between the public's knowledge of biotechnology and the media's relative emphasis upon risks and benefits: in countries with the more negatively framed coverage, the public has more knowledge about modern biotechnology ($−.63$, $p < .05$); and in countries with a positive media framing, the public has less knowledge (see chapters 7 and 8 for more detailed information). Figure 3.4 summarises the relationships between attitude and knowledge across Europe and positive/negative coverage (countries in figure 3.4 are ordered on the

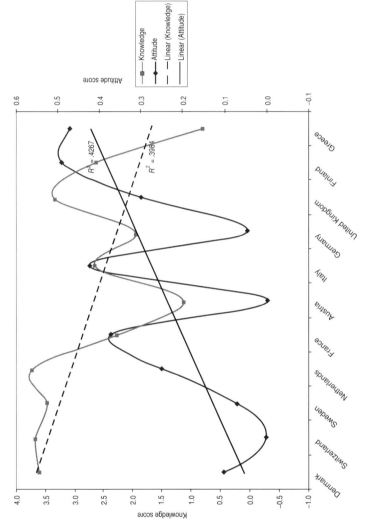

Figure 3.4 Relation between negative or positive coverage of modern biotechnology and public attitude and knowledge across Europe

Note: The countries are ordered from left to right according to their overall negative or positive coverage (Denmark has the most negative coverage; Greece has the most positive coverage).

x-axis from left to right according to their overall ranking of negative versus positive coverage; see table 3.2). Figure 3.4 also contains linear regressions of these relations indicating moderately strong relations ($R^2 = .43$ for the relation between attitude and negative/positive coverage, and $R^2 = .40$ for knowledge).

Table 3.3 indicates that the public perception of modern biotechnology in a country is more positive when the opinion-leading media pay a considerable amount of attention to 'Medical' issues, and relatively little to 'Regulatory' issues. In addition, these data suggest that public attitudes towards biotechnology are more positive when media coverage accords the views of 'Industry' spokespersons a high profile, and when it is rather less inclined to quote the views of 'Politicians'. Yet there are no significant correlations between the frequency of references to particular actor groups and public knowledge about modern biotechnology; and there is no obvious relationship between public attitudes and the cited 'location of activity'. The overall positive or negative characteristics of the coverage of modern biotechnology appear not to correlate systematically with the occurrence of the various major themes or the location of activity. Only attention to 'Medical' issues is an exception here. There is also a significant positive correlation with references to 'Scientists' and a negative correlation with 'Politicians', indicating that, in countries where the overall characteristic of the coverage of modern biotechnology is more positive, relatively more references to 'Medical' issues or 'Scientists' appear in the opinion-leading press, and relatively fewer references to 'Politicians'.

Discussion and conclusions

This study started with several goals in mind. In the first place we wanted to analyse systematically the dynamics involved in the European opinion-leading press coverage of modern biotechnology over the past twenty-four years. A second goal of our media study was to attune our analyses with other modules of the European Biotechnology project, namely the analysis of developments in modern biotechnology policy across Europe and studies of European public opinion towards modern biotechnology as assessed by the 1996 Eurobarometer Biotechnology survey. For two main reasons modern biotechnology has been conceptualised rather broadly in this media study: the technology domain itself has potential applications in many different fields; and we wanted national issues and particulars to be included within the study. Our final database consists of more than 5,400 articles from twelve European countries, which were selected from the original print material or more sophisticated storage media, and then coded and assessed according to a common coding frame. The participating countries provide an excellent coverage of the entirety of Europe, from the north to the south, and from the east to the west.

This procedure allowed us to apprehend the changing content and tone of the coverage of modern biotechnology within the various European countries, while being sensitive to the differences and similarities among these countries. As far as we are aware, this is the most substantial study of the media coverage of modern biotechnology so far undertaken. But, having said this, some additional remarks about the research, and its limitations, are due. Electing to study only the opinion-leading press in the various European countries means that our data sets contain partial incompatibilities, because in some countries major daily newspapers were analysed, whereas in others newspapers as well as news magazines were included, and in some countries only elite newspapers were selected. It should also be repeated that, for practical reasons, no attempt was made to incorporate material from other opinion-leading media, such as television.

Examining the results of this study, it is immediately obvious that the entire twenty-four-year period saw a clear increase in the attention devoted to modern biotechnology in most European countries (most of the analyses were performed at a country level). A second important conclusion is that, in general, the coverage of modern biotechnology is fairly positive. On the other hand, the countries studied vary – sometimes sharply – in terms of negative versus positive journalistic framing. In this investigation, we used several indicators to evaluate journalistic framing: whether articles invoke 'doom' or 'progress/utility' scenarios; their relative emphasis upon risks and benefits; and a subjective measure of coder impression. With the exception of the invocation of 'doom scenarios', each of these indicators has a high rank order correlation, indicating that these countries can be placed on a single scale from the positive (with Poland, Greece and Finland as exemplars) to the more negative framing of modern biotechnology (typified by Denmark, Switzerland and Sweden). Because 'doom scenarios' did not correlate with the other indicators, they were not used in ranking the various countries.

Another conclusion is that the European opinion-leading press pays considerable attention to 'Medical' issues related to modern biotechnology as well as to 'Basic research'. Indeed, the relatively positive framing of the coverage in some countries is correlated with a high proportion of press articles dealing with 'Medical' issues. This indicates that a major part of the coverage this technology receives emphasises the potential benefits for human health and progress that its further development may provide. In another study, it was found that US newspapers also tend to give much attention to the salutary aspects of modern biotechnology (Hornig, 1994). So, in this respect, the overall European and US media approaches exhibit clear similarities.

Over the whole twenty-four-year period, the attention devoted to 'Basic research' diminished. However, the dynamics involved in the coverage of 'Medical' issues are more complicated, with some countries displaying an increase in the press coverage of this aspect, and others a decrease. The European opinion-leading press also pays attention to 'Regulatory and policy' issues, but these were accorded less attention than either 'Medical' issues or 'Basic research'. 'Regulation and policy' is a more prominent issue in those countries in which the press coverage of biotechnology is mostly negatively framed; these questions were usually addressed in the context of elucidating the potentially disadvantageous implications of modern biotechnology, and of conveying to the public that biotechnology may encompass unwanted developments and that prudence is imperative. Over the whole research period we also observed an increase in the attention devoted to 'Animal and agricultural' issues in many countries, presumably owing to the fact that this area of biotechnology has experienced significant growth in the twenty-four years under study. Considerably less attention was given to either 'Ethical' or 'Economic' issues, and the dynamics of the coverage of these issues do not present a very clear picture.

So, in conclusion, the hypothesis put forward by some biotechnology proponents – that the European opinion-leading press has painted a dire picture of this industry – is at odds with the facts. Overall, coverage has been positive and varied in its themes. On the other hand, the press seems to have failed to communicate the importance that policy-makers attach to economic issues, and has neglected the ethical issues that are very important to the general public throughout Europe.

In the European opinion-leading press, 'Scientists' and 'Industry' spokespersons are the most frequent sources of information and opinions, followed at a certain distance by actors from the political or societal domain, such as 'Politicians' or 'NGOs'. Again, here there is a parallel between the European situation and the results from an earlier study carried out in the USA (Hornig, 1994), as well as from the more recent study by Kohring, Goerke and Ruhrmann (1999), which incorporated both US and European data. Among the European countries, however, differences in the relative prominence of the various actors' groups can be found. Particularly noteworthy is the finding that, in countries in which the coverage of modern biotechnology is framed in a positive manner, 'Scientists' enjoyed more coverage than 'Politicians'. Conversely, in countries with a more negative framing, we observed more references to 'Politicians' than to 'Scientists'. In addition, over the entire twenty-four-year period most countries witnessed a decrease in the input of 'Scientists' as opinion-leading press sources, in favour of 'Industry' and 'Politicians'. Most references to the locations of activity of modern biotechnology were to 'Own country', followed by 'USA' and other European countries.

The 'Biotechnology and the European Public' project offers the opportunity of analysing the relationship between, on the one hand, public perception and public knowledge about modern biotechnology and, on the other, the results of our media analysis. These analyses are informed by the notion that media coverage of a particular issue, through processes of selection or framing, may have an impact upon what the public thinks or knows about that issue. In this study we did observe a clear relation between the positive/negative framing in the various countries and the public perception of modern biotechnology, as assessed by the Eurobarometer Biotechnology 1996 (see 'Biotechnology and the European Public', 1997). In countries with a relatively negative media framing of biotechnology, the public perception of this new technology was also more negative; whereas in countries with a more positively framed coverage the public's attitude was more positive. Negative or positive framing also correlates with the level of public knowledge about modern biotechnology. However, in this case, greater public knowledge was found in those countries in which the opinion-leading press negatively framed the coverage of modern biotechnology; and lower levels of knowledge were found in countries in which a more positive framing occurred. Further studies will have to establish the determinants of these relationships.

A final conclusion of our study is that journalistic selection and framing processes play an important role in determining how the European opinion-leading press covers the many facets of modern biotechnology, and that these processes may reflect varying national sensibilities. Studying the media coverage of biotechnology, we frequently encountered differences in the national treatment of this subject. The level of the coverage itself, the emphasis upon the different themes, the reliance placed upon various actors and stakeholders, and the positive/negative framing of this technology all exhibit considerable inter-national heterogeneity. Moreover, in this respect, our finding that there is a correlation between the nature of framing and the inclusion of certain issues in discussions of biotechnology, on the one hand, and the public's perception of, and knowledge about, modern biotechnology, on the other, is especially significant. This finding emphasises the importance of the media coverage of this technology. Although the existence, or otherwise, of a causal relationship between the media coverage of biotechnology and the public's knowledge and opinions of it cannot be deduced from this correlation analysis, finding evidence for causality will be an interesting task for future studies.

Appendix

Table 3A.1 *Main characteristics of the research material*

Country	Newspaper name (and type)	Circulation: population	Estimation of total coverage	Size of sample in study	Sampling procedures	Selection procedures
Austria	*Die Presse* (daily)	1:100 *(Presse)*	1,396	302 *(Presse* 191; *Profil* 111)	*Presse:* 1973–9, 28% *Presse:* 1980–5, 20% *Presse:* 1986–96, 10% *Profil:* 1973–85, 100% *Profil:* 1986–96, 50%	1973–93 manual 1994–6 on-line
	Profil (weekly)					
Denmark	*Information* (daily)	1:34 *(Politiken)*	–	300 *(Information* 168; *Politiken* 132)	1973–89, 1996, 100% 1990–5 (only large articles)	1973–89 manual, index 1990–6 on-line
	Politiken (daily)					
Finland	*Savon sanomat* (daily)	1:55	375	375	1973–96, 100%	Manual
France	*Le Monde* (daily)	1:179	1,483	549	1975, 1982–6, 100% 1987–96, 25%	1973–86 manual, index 1987–96 on-line
Germany	*Frankfurter Allgemeine Zeitung* (daily)	1:162 *(FAZ)*	1,594 *(FAZ* 1,254; *Spiegel* 340)	588 *(FAZ* 418; *Spiegel* 170)	*FAZ*, 33% *Spiegel*, 50%	Manual
	Der Spiegel (weekly)					
Greece	*Kathemerini* (daily)	1:283	88	65 *(Kathemerini* 16; *Eleftherotypia* 49)	1973–85, every 2nd yr, 100 days/year 1986–96, 100%	Manual
	Eleftherotypia (daily)	1:89				

Country	Newspaper					
Italy	*Corriere della sera* (daily)	936	340	—	1973–86 (est.), 100% 1987–96, 20 days/year	1973–84 manual 1984–91 on-line
Netherlands	*Volkskrant* (daily)	1,185	1,119	1:41	1973–85, every 2nd yr, 50% 1986–96, 100%	1973–92 manual 1993–6 on-line
Poland	*Polityka* (weekly) *Rzeozpospolita* (daily) *Trybuna ludu* (daily)	113 (*Polityka*)	254 (*Polityka* 76; *Rzeozpospolita* 160; *Trybuna ludu* 18)	1:200	*Polityka*, 100% Others, 1 day/week	Manual *Trybuna ludu* 1973–82 *Rzeozpospolita* 1983–96
Sweden	*Dagens Nyheter* (daily)	734	734	1:21	1973–96, 100%	1973–92 manual 1992–6 on-line
Switzerland	*Neue Zürcher Zeitung* (daily)	1,537	211	1:45	1973–96, 14%	Manual
United Kingdom	*Times* (daily) *Independent* (daily)	5,471	539 (*Times* 256; *Independent* 283)	1:83 1:226	1973–87, 1996, 20 day/year 1988–95, 15 day/year	*Times* 1973–87, *Independent* 1988–96 1973–80 manual, index, 1981–96 on-line

Table 3A.2 *Main themes in the coverage of modern biotechnology across Europe: trends, 1973–1996*

Country	Type of article	Basic research %	Basic research I–II	Basic research II–III	Medical %	Medical I–II	Medical II–III	Animal/Agricultural %	Animal/Agricultural I–II	Animal/Agricultural II–III	Regulatory %	Regulatory I–II	Regulatory II–III	Ethical %	Ethical I–II	Ethical II–III	Economic %	Economic I–II	Economic II–III
Austria	Risk only	3	→	↔	17	—	→	4	⌐	↔	32	←	←	30	↔	→	0	↔	↔
	Benefit only	28	—	⌐	32	↔	↔	5	↔	↔	13	↔	→	2	↔	↔	14	↔	→
	Risk & benefit	6	→	↔	14	←	→	9	⌐	↔	17	⌐	←	25	—	→	9	↔	—
	Observations	*98*	*21%*		*110*	*24%*		*30*	*7%*		*90*	*20%*		*56*	*12%*		*43*	*9%*	
Denmark	Risk only	14	←	↔	25	→	↔	28	→	↔	17	←	↔	9	⌐	↔	0	↔	↔
	Benefit only	30	←	↔	20	←	↔	30	←	↔	10	←	↔	1	↔	↔	5	⌐	↔
	Risk & benefit	20	→	—	16	↔	←	29	⌐	↔	15	←	↔	9	—	↔	5	↔	↔
	Observations	*173*	*21%*		*155*	*19%*		*238*	*29%*		*118*	*14%*		*59*	*7%*		*33*	*4%*	
Finland	Risk only	7	⌐	↔	31	→	⌐	7	⌐	↔	22	→	←	13	←	→	2	↔	↔
	Benefit only	4	⌐	—	42	→	—	17	—	↔	10	←	↔	2	↔	↔	17	↔	←
	Risk & benefit	6	—	—	38	→	↔	13	←	↔	18	⌐	↔	13	⌐	↔	9	↔	↔
	Observations	*44*	*5%*		*320*	*39%*		*117*	*14%*		*122*	*15%*		*56*	*7%*		*105*	*13%*	
France	Risk only	14	→	↔	20	←	⌐	3	⌐	↔	12	→	↔	24	←	—	5	⌐	↔
	Benefit only	21	↔	—	28	↔	⌐	10	←	↔	11	→	↔	2	↔	↔	24	←	—
	Risk & benefit	9	→	↔	33	←	—	5	⌐	↔	16	←	↔	18	←	↔	10	⌐	↔
	Observations	*145*	*17%*		*248*	*29%*		*63*	*7%*		*115*	*13%*		*87*	*10%*		*134*	*15%*	
Germany	Risk only	7	—	↔	13	↔	↔	3	↔	↔	33	←	↔	11	↔	←	2	⌐	⌐
	Benefit only	28	⌐	→	28	↔	↔	10	⌐	↔	15	↔	←	1	↔	↔	11	⌐	↔
	Risk & benefit	6	—	↔	18	↔	→	9	↔	⌐	31	↔	↔	8	⌐	↔	10	↔	↔
	Observations	*206*	*17%*		*267*	*22%*		*102*	*8%*		*308*	*25%*		*58*	*5%*		*113*	*9%*	
Greece	Risk only	0	↔	↔	33	←	←	0	↔	↔	17	←	→	33	→	→	0	↔	↔
	Benefit only	33	—	←	52	←	→	3	→	↔	0	↔	↔	2	↔	↔	0	↔	↔
	Risk & benefit	16	↔	←	32	←	→	8	↔	←	8	→	↔	20	←	→	0	↔	↔
	Observations	*25*	*26%*		*42*	*44%*		*4*	*4%*		*3*	*3%*		*8*	*8%*		*0*	*0%*	

		I	I–II	II		III	II–III				
Italy	Risk only	7	⌐	40	↔	11	↔	13	↑	20	4
	Benefit only	14	—	45	↔	19	↔	5	↔	1	7
	Risk & benefit	7	—	36	↓	8	⌐	14	⌐	15	6
	Observations	76	10%	307	40%	96	12%	80	10%	103	1%	48	6%
Netherlands	Risk only	7	⌐	15	→	20	↑	19	→	19	3
	Benefit only	15	→	36	↔	19	↑	7	—	1	12
	Risk & benefit	8	↑	18	↑	25	↑	14	↑	10	7
	Observations	260	14%	498	27%	307	17%	204	11%	103	6%	163	9%
Poland	Risk only	2	↑	6	↑	0	↔	0	↔	3	0
	Benefit only	43	↔	60	↑	42	—	9	⌐	4	16
	Risk & benefit	21	⌐	48	↔	30	↑	6	↑	18	2
	Observations	82	17%	185	39%	74	15%	15	3%	25	5%	18	4%
Sweden	Risk only	8	→	18	↑	6	⌐	20	↑	27	0
	Benefit only	17	→	55	↑	8	—	4	—	1	8
	Risk & benefit	16	→	28	↑	8	⌐	9	↔	18	2
	Observations	178	15%	403	34%	90	8%	154	13%	150	13%	45	4%
Switzerland	Risk only	3	⌐	17	↑	10	↔	41	↑	14	1
	Benefit only	32	—	29	↑	14	↔	8	↔	1	12
	Risk & benefit	16	⌐	25	↑	9	↑	22	↑	12	8
	Observations	111	23%	130	27%	44	9%	130	27%	25	5%	31	6%
United Kingdom	Risk only	6	—	21	↑	2	⌐	22	↑	18	5
	Benefit only	15	—	33	↑	6	⌐	9	↔	1	13
	Risk & benefit	5	↔	20	—	8	↔	23	↔	13	12
	Observations	106	12%	267	29%	53	6%	127	14%	58	6%	109	12%

Notes: 1. Trends in the coverage of modern biotechnology are indicated in the columns headed I–II and II–III (representing the three octades I:1973–80, II:1981–8, and III:1989–96). The symbols in the columns indicate relative increases (⌐ = 5–10%; ↑ > 10%), relative decreases (⌐ = 5–10%; ↓ > 10%), or an absence of change (↔, < 5%).

2. We present articles containing information only about the risks of modern biotechnology, only about the benefits, or about both risks and benefits. Articles containing neither risk nor benefit information are not included in this table. The percentages indicate, for each country, the proportion of observations of different themes of modern biotechnology for each type of article (three observations of a theme per article allowed). The six themes mentioned most frequently in the overall coverage are presented in this table; the other themes are not included.

3. *Observations* rows indicate the number of observations per theme and their proportion compared with all themes.

Table 3A.3 *Main actors in the coverage of modern biotechnology across Europe: trends, 1973–1996*

Country	Actors	Risk only %	I–II	II–III	Benefit only %	I–II	II–III	Risk & benefit %	I–II	II–III
Austria	Scientists	32	↓	↓	60	|	↑	40	|	⟩
	Politicians	32	↑	↑	20	|	↔	23	↑	↔
	Industry	4	↔	⟩	14	⟩	↓	15	↔	↔
	NGOs	13	⟩	⟩	1	↔	↔	10	↔	↔
	Observations	*54*			*188*			*83*		
Denmark	Scientists	21	↓	|	32	↑	↔	27	↓	↔
	Politicians	43	↑	↔	35	↑	↓	47	↑	↓
	Industry	7	↑	↓	16	↑	↔	9	↑	|
	NGOs	3	↔	⟩	3	↔	↔	2	↔	↔
	Observations	*103*			*116*			*335*		
Finland	Scientists	46	↑	↓	66	⟩	|	59	↓	↓
	Politicians	29	↑	↔	11	↔	⟩	14	↑	↔
	Industry	9	⟩	↔	15	↔	⟩	10	↔	↔
	NGOs	3	↓	↔	2	↔	↔	5	⟩	↔
	Observations	*35*			*332*			*222*		
France	Scientists	39	↑	↓	46	↓	↔	43	|	↓
	Politicians	13	↔	|	14	↑	↔	14	|	↑
	Industry	13	↑	↔	29	⟩	↓	18	↑	⟩
	NGOs	8	↓	↔	2	↔	↔	6	⟩	↔
	Observations	*72*			*326*			*201*		
Germany	Scientists	31	|	↓	64	↔	↓	30	↓	|
	Politicians	31	↔	↔	12	|	⟩	27	↑	↓
	Industry	7	⟩	↔	16	↔	⟩	19	|	↑
	NGOs	8	↑	↔	1	↔	↔	9	⟩	↑
	Observations	*86*			*469*			*341*		
Greece	Scientists	33	↑	↑	64	⟩	|	68	↑	↓
	Politicians	33	↑	↓	5	↓	⟩	5	↓	|
	Industry	0	↔	↔	24	↔	↑	18	↔	↑
	NGOs	0	↔	↔	0	↔	↔	0	↔	↔
	Observations	*6*			*59*			*22*		
Italy	Scientists	39	|	↓	54	↓	⟩	45	|	↓
	Politicians	15	↑	↔	2	↑	↓	13	↔	↔
	Industry	11	↑	↔	26	↓	↑	14	⟩	|
	NGOs	9	↑	↔	7	↓	⟩	5	⟩	↑
	Observations	*54*			*199*			*255*		

Table 3A.3 (*cont.*)

Country	Actors	Risk only %	Risk only I–II	Risk only II–III	Benefit only %	Benefit only I–II	Benefit only II–III	Risk & benefit %	Risk & benefit I–II	Risk & benefit II–III
Netherlands	Scientists	22	↑	↑	51	↓	↔	29	↑	↓
	Politicians	22	↑	↓	11	|	↔	24	↑	↔
	Industry	12	↑	|	25	↘	↔	27	↑	↑
	NGOs	17	↑	↓	5	↔	↔	12	↘	↘
	Observations	*161*			*506*			*304*		
Poland	Scientists	50	↓	↑	33	↔	↔	72	|	↔
	Politicians	13	↑	↓	7	↓	↘	3	↘	|
	Industry	0	↔	↔	11	↑	↓	12	↑	↔
	NGOs	0	↔	↔	0	↔	↔	2	↔	↔
	Observations	*8*			*232*			*117*		
Sweden	Scientists	26	↔	↓	67	↑	↑	51	|	↔
	Politicians	33	↓	↑	10	↓	↔	17	↔	↔
	Industry	7	|	↔	16	↔	↓	11	|	↔
	NGOs	8	↔	↔	0	↔	↔	2	↔	↔
	Observations	*151*			*362*			*252*		
Switzerland	Scientists	10	↑	↓	53	↓	↓	31	↓	↔
	Politicians	38	↑	↔	13	↘	↑	29	↑	↓
	Industry	8	↘	↔	24	↑	↔	21	↓	↔
	NGOs	34	↔	↑	3	↔	↔	6	↔	↑
	Observations	*77*			*95*			*48*		
United Kingdom	Scientists	40	↓	↔	56	↓	↓	32	↓	↔
	Politicians	16	↑	↓	10	↔	↑	18	↘	↔
	Industry	9	↔	↑	25	↑	↔	21	↑	|
	NGOs	7	↑	↓	4	↔	↘	10	↘	↔
	Observations	*45*			*391*			*196*		

Notes: 1. See note 1 to table 3A.2.

2. We present articles containing information only about the risks of modern biotechnology, only about the benefits, or about both risks and benefits. Articles containing neither risk nor benefit information are not included in this table. The percentages indicate the proportion of observations of different actor groups per type of article (two observations of actors per article allowed). The four actor groups mentioned most frequently in the overall coverage are presented in this table; the other actor groups are not included.

3. *Observations* rows indicate the number of observations of actors in risk-only, benefit-only and risk and benefit articles, respectively.

Table 3A.4 *Main location of reported activity in the coverage of modern biotechnology across Europe: trends, 1973–1996*

Country	Main location of activity	Risk only %	I–II	II–III	Benefit only %	I–II	II–III	Risk & benefit %	I–II	II–III
Austria	Own country	45	↑	)	33	↔	↓	36	↑	)
	Other Europe	23	)	|	24	↓	)	19	↓	↔
	USA	15	↓	↔	25	↔	↔	29	↓	↔
	Observations (N)	*53*			*219*			*80*		
Denmark	Own country	61	↑	↓	55	↑	)	47	↑	↓
	Other Europe	15	)	↑	9	)	)	18	)	↑
	USA	16	)	↔	18	↑	↔	28	↓	↔
	Observations (N)	*86*			*99*			*284*		
Finland	Own country	51	↑	↓	72	|	↓	65	↔	|
	Other Europe	17	)	↑	13	↓	↑	13	|	↔
	USA	14	↑	|	11	↑	↑	14	|	)
	Observations (N)	*35*			*308*			*215*		
France	Own country	28	↓	↑	46	↔	↔	48	)	|
	Other Europe	21	↑	↓	18	↓	↔	19	↓	)
	USA	12	↑	|	19	↑	↔	19	↑	|
	Observations (N)	*67*			*301*			*168*		
Germany	Own country	52	↑	↓	31	)	↔	41	↑	↓
	Other Europe	11	|	↑	24	↓	)	19	↓	)
	USA	22	↔	↔	35	↔	|	28	↓	↔
	Observations (N)	*64*			*450*			*277*		
Greece	Own country	20	↓	↔	11	↓	↑	14	↓	↑
	Other Europe	20	↔	↑	38	)	↔	55	↑	|
	USA	20	↑	↓	40	)	↔	23	↓	↑
	Observations (N)	*5*			*55*			*22*		
Italy	Own country	41	↑	|	29	↑	)	31	)	↓
	Other Europe	19	↓	↑	18	↓	)	21	↓	)
	USA	16	)	↔	35	↓	↔	31	↑	)
	Observations (N)	*37*			*153*			*156*		
Netherlands	Own country	61	↓	↑	45	↑	)	54	↑	↑
	Other Europe	9	↑	↔	17	↑	↔	9	)	↔
	USA	6	↑	↓	28	↓	↔	23	↑	↓
	Observations (N)	*111*			*382*			*220*		
Poland	Own country	?	?	?	?	?	?	?	?	?
	Other Europe	16	↔	↑	41	↓	↔	49	↔	↑
	USA	33	↓	0	36	↑	↑	20	|	|
	Observations (N)	*6*			*160*			*75*		
Sweden	Own country	62	↓	↓	53	↑	|	69	|	↓
	Other Europe	14	↔	↑	13	|	↔	11	↔	↑
	USA	9	↔	|	23	)	↔	13	↔	↔
	Observations (N)	*139*			*337*			*205*		

Table 3A.4 (*cont.*)

Country	Main location of activity	Type of coverage								
		Risk only			Benefit only			Risk & benefit		
		%	I–II	II–III	%	I–II	II–III	%	I–II	II–III
Switzerland	Own country	88	↑	↔	37	↘	↑	63	↑	↓
	Other Europe	5	↑	|	24	↔	↔	6	↔	↑
	USA	0	↔	↔	16	↑	↘	9	↓	↑
	Observations (*N*)	42			70			32		
United Kingdom	Own country	62	↑	↓	50	↘	↑	55	↓	↑
	Other Europe	10	↔	↑	13	↔	|	11	↘	↔
	USA	21	↓	↑	29	|	↔	26	↘	↓
	Observations (*N*)	39			377			149		

Notes: 1. See note 1 to table 3A.2.
2. We present articles containing information only about the risks of modern biotechnology, only about the benefits, or about both risks and benefits. Articles containing neither risk nor benefit information are not included in this table. The percentages indicate, for each country, the proportion of observations of different locations of reported activity per type of article (two observations of location of activity per article allowed). The locations are 'Own country', 'Other European countries', and 'USA', respectively. Other locations are not included in this table.
3. *Observations* rows indicate the number of observations of location of activity in risk-only, benefit-only and risk and benefit articles, respectively.

REFERENCES

Beck, U. (1986) *Die Risikogesellschaft*, Frankfurt am Main: Suhrkamp.
Biotechnology and the European Public (1997) 'Europe Ambivalent on Biotechnology', *Nature* 387: 845–7.
DeFleur, M. and S. Ball-Rokeach (1989) *Theories of Mass Communication*, White Plains, NY: Longman.
Durant, J., M.W. Bauer and G. Gaskell (1998) *Biotechnology in the Public Sphere: a European Sourcebook*, London: Science Museum.
Frewer, L., C. Howard and R. Shepherd (1997) 'Public Concerns in the United Kingdom about General and Specific Applications of Genetic Engineering: Risk, Benefit, and Ethics', *Science, Technology and Human Values* 22(1): 98–124.
Galtung, J. and Ruge, M.H. (1965). 'The Structure of Foreign News: The Presentation of the Congo, Cuba and Cyprus Crises in Four Norwegian Newspapers'. *Journal of Peace Research*, 2, 64–91.
Hagedorn, C. and S. Allender-Hagedorn (1997) 'Issues in Agricultural and Environmental Biotechnology: Identifying and Comparing Biotechnology Issues from Public Opinion Surveys, the Popular Press and Technical/Regulatory Sources', *Public Understanding of Science* 6: 233–45.
Hilgartner, S. and C.L. Bosk (1988) 'The Rise and Fall of Social Problems: a Public Arenas Model', *American Journal of Sociology* 94(1): 53–78.

Holsti, O.R. (1969) *Content Analysis for the Social Sciences and Humanities*, Reading, MA: Addison Wesley.

Hornig, S. (1992) 'Framing Risk: Audience and Reader Factors', *Journalism Quarterly* 69(3): 679–90.

—— (1994) 'Structuring Public Debate on Biotechnology: Media Frames and Public Response', *Science Communication* 16(2): 166–79.

Kepplinger, H.M., S. Ehmig and C. Ahlheim (1991) *Gentechnik im Widerstreit. Zum Verhältnis von Wissenschaft und Journalismus*, Frankfurt am Main: Campus.

Kitzinger, J. and Reilly, J. (1997). 'The Rise and Fall of Risk Reporting: Media Coverage of Human Genetics Research, "False memory syndrome" and "Mad cow disease".' *European Journal of Communication*.

Kohring, M., A. Goerke and G. Ruhrmann (1999) 'Das Bild der Gentechnologie in den internationalen Medien. Eine Inhaltsanalyse meinungsführender Zeitschriften', in O. Renn and J. Haempel (eds.), *Gentechnik aus der Sicht der Öffentlichkeit*, Frankfurt am Main: Campus.

Krippendorff, K. (1980) *Content Analysis: an Introduction to Its Methodology*, Newbury Park, CA: Sage Publications.

Servaes, J. and C. Tonnaer (1992) *De nieuwsmarkt: vorm en inhoud van de internationale berichtgeving*, Groningen: Wolters-Noordhoff.

4   The institutions of bioethics

Jean-Christophe Galloux, Arne Thing Mortensen,
Suzanne de Cheveigné, Agnes Allansdottir, Aigli
Chatjouli and George Sakellaris

The discovery of genetic recombination techniques, at the beginning of
the 1970s, opened up extraordinary new possibilities for the biological
modification of organisms; and the potential for these interventions immediately prompted profound ethical concerns. As the call for a moratorium before the Asilomar Conference (held in California in February
1975) showed, microbiologists were well aware of these issues. The risks
associated with unintended genetic combinations, particularly the danger of disturbing ecological balances and, ultimately, threatening human
welfare, were seriously contemplated. Such concerns resulted in a voluntary moratorium during which the acceptability of certain experiments
could be discussed. Since then, public references to 'bioethics' (the term
was coined in the USA in 1971 by the oncologist Van R. Potter, 1971)
have grown consistently. Roughly speaking, the term 'bioethics' relates
to a body of reflections, involving multidisciplinary approaches, that deal
with all aspects of the responsible management of human life in the face of
complex and rapid developments in biomedicine. In the present chapter,
we will provide a sketch of the history of the development of bioethics,
including the media background, we will clarify the philosophical notions
behind the term and, finally, we will trace the process of institutionalisation that bioethics has undergone over the past twenty to thirty years.
In doing so, we will concentrate on four European countries: Denmark,
France, Italy and Greece. We will also consider these issues at the level
of Europe as a whole.

A short history of biotechnology and ethics

The development of genetic recombination techniques inspired speculation about the possibility of designing living organisms, and provoked
ethical discussions concerning the proper boundaries for the application
of microbiological techniques. However, during the 1970s, these dangers did not seem sufficiently realistic, or imminent, to demand political

initiatives. The discussions were, therefore, largely confined to professional circles, with the first ethical research committees appearing during the mid-1960s in the USA, following the 1964 Helsinki declaration on medical ethics. The birth of bioethics was, in fact, the result of diverse and sometimes paradoxical anxieties within a scientific community trying to come to terms with the enormity of the power it was acquiring; and it represented a search for an international consensus on values in the life sciences domain. But it also symbolised a desire to evade formal state regulation. Largely inspired by liberal Anglo-Saxon ideologies, life scientists were striving to achieve a – preferably institutionalised – professional self-regulation.

In this period, however, the public debate developed slowly. References were regularly made to ethical issues in the European elite press when addressing the subject of biotechnology (see Gutteling et al., chapter 3 in this volume). Yet, it is illuminating to compare the four countries examined in this study in terms of the prominence of ethical concerns in discussions of biotechnology.[1] In Italy, ethics represent, on average, 10.4 per cent of all themes considered in the context of biotechnology, in France 9.0 per cent, in Greece 7.8 per cent and in Denmark only 6.7 per cent. The European average is 7.1 per cent of biotechnology discussions.

We can regard this variation in the prominence of the ethics theme as an indicator of the relative attention given by the national elite press to ethical issues in the context of biotechnology. Figure 4.1 illustrates the development of the ethics theme in the newspaper coverage of biotechnology in the four countries under study. There are clearly considerable variations over time in the relative salience of the ethics theme in the media debate. However, a general trend is discernible in which the media's interest in this issue peaked in the 1970s, declined in importance during the more optimistic 1980s, only to resurface during the 1990s. There are also strong local variations that reflect national differences in the form of this debate. The data from both the Danish and the Italian media show that a substantial amount of attention was accorded to ethical issues in the years immediately before the institution of formal bioethics bodies. At the moment of institutionalisation (in 1988 in Denmark and 1990 in Italy), however, the salience of the ethics theme diminished somewhat. A similar phenomenon is observable in France at the time of the creation of the national ethics committee in 1983 and coinciding with the passage of the Bioethics Law in 1994. This would tend to support the sociological hypothesis associated with 'public arena' models, which stipulate that, once an issue has been taken up by a formal policy body, public interest in the issue is diminished or diverted towards other concerns (Hilgartner and Bock, 1988).

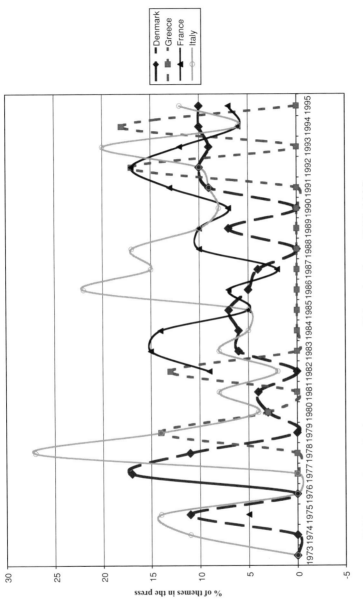

Figure 4.1 Ethics as a theme in the press coverage of biotechnology, 1973–1995

In the early days of biotechnological development, scientists were generally considered trustworthy enough to be able to handle the new risks and ethical problems. This changed during the 1980s, primarily as a result of industrial pressure. As a research field, modern biotechnology was then growing rapidly, and it was producing techniques and ideas that were obviously commercially promising. As a result, the medical industry – as well as the agrochemical and other industries – entered the public scene as powerful agents. In most European countries, perceptions regarding modern biotechnology in the early 1980s were distinctly positive. Generally speaking, biotechnology was treated in the media as another striking example of scientific progress, and even in Denmark (of the four countries compared in this study, Denmark has the most influential tradition of criticising new technologies) the attitude was positive (Durant et al., 1998). Indicative of this, very few people rejected the claim that modern biotechnology is one of the most important fields for future scientific development. Nor did they criticise the argument that industry and agriculture should exploit the opportunities biotechnology presents in the interests of economic growth and for the sake of progress itself.

In Denmark, more than in the other three countries, 'counter-experts' and public movements were able to influence the political agenda; although the critics insisted on the implementation of safety measures and the imperative for political control, they did not demand absolute prohibitions. As a result of public discussions in the early 1980s, the political majority in Denmark became convinced that the common goal – that of achieving a congenial climate for biotechnological research and development – could best be accomplished by creating a legal and administrative framework that would prevent the risks of the new technology getting out of control. The Gene Technology Act, passed in 1986, answered this need. Yet bioethics did not become an essential political issue before the establishment, in 1987, of the Danish Council of Ethics. Similar councils on bioethics were created in France (1983), in Italy (1990) and in Greece (1992). Before we look at them in detail, a comment on the primary arguments invoked in bioethical discussion is necessary.

Ethical discourse

As a philosophical realm, 'ethics' covers all efforts to found normative positions upon reasoned argument rather than upon intuitions or emotive reactions. Appealing to ethics is to demand reflection and a rational approach to the study of values. In popular usage, however, the concept of ethics is more narrowly applied. Here, reasoning based on almost universally accepted normative principles is not classified as ethical; instead, the concept tends to define only normative reasoning based on controversial principles. Another difference between the philosophical and the popular use of the word is that 'ethics' in everyday language tends to be equivalent to what philosophers call 'normative ethics' or 'meta-ethics'. For this study, these terminological variations may cause problems in both interpreting the raw data and discussions thereof. For instance, in the media study carried out in the Biotechnology and the European Public project, newspaper articles were coded as treating biotechnology in a 'progress' frame or in an 'ethical' frame, 'ethical' here denoting that biotechnology is discussed as an issue of debatable moral acceptability, while notions of 'progress' seem to be inherent in a range of unquestioned values (Durant et al., 1988). Yet, in fact, the notion of 'progress' is in itself essentially ethical, since it implies something along the lines of 'a change for the better'. Moreover, what constitutes 'better' is certainly open to dispute. Thus, the borderline between these two frames is very hard to delineate. Nevertheless, it seems obvious that there is an important everyday distinction to be made here. It is for this reason that we seek below to clarify the different ethical frameworks that are called upon in bioethics discussions, as well as the types of argument that are employed.

Ethical frames

Seeing 'ethics' as a rational enterprise – an effort to clarify arguments rather than just to sustain attitudes – we need to examine current ethical theories and determine whether the main differences between the philosophers' and non-philosophers' (scientists, professionals, politicians, laypeople, etc.) conceptions of ethics reflect important differences in the way ethical questions are conceived by these two groups.[2] In this sense, looking for logically elaborated theories is less important than elucidating arguments that belong to the same family. Such similarities can, indeed, be found by looking for evidence of shared axioms. Our hypothesis is that three principles are essential and relevant to characterising the different ways in which bioethics became a political issue during the 1980s and 1990s: the principle of utility, the principle of democracy and the principle of veneration.

The principle of utility According to the principle of utility, moral questions are questions about the optimal balance between happiness or well-being and suffering or discomfort. According to utilitarians, our actions should be judged by the consequences they have for the 'sum of happiness' in the world. Happiness is measured in terms of individuals' feelings or preferences, which should in themselves be rational since a moral agent must be able to evaluate his/her actions in relation to rational criteria. For instance, to argue that soya plants should be genetically modified because this is a way to increase crop yields while reducing expenditure on herbicides is to apply the principle of utility as an underlying premise. We call this kind of reasoning Utility-argumentation or, for short, U-argumentation.

For three main reasons, U-argumentation is closely associated with science. First, it sees human activity in terms of finding means of reaching predetermined ends, and this is exactly what science seeks to accomplish. Secondly, in U-argumentation, human ends are considered to be 'natural', in the sense that they are related to biological, social and psychological needs. Moral action would clearly need to be based upon a good understanding of these human needs and science offers the most efficacious means of acquiring this understanding. Third, utilitarians presuppose that choices between alternative courses of action would, ideally, be based upon some kind of methodical evaluation, and this might well entail the development of scientific procedures.

In view of the above, the 'welfare state' can be seen as a product of utilitarian thinking. According to utilitarian thinking, the most important relation between one person and another is one of mutual care and it is everybody's duty to maximise the sum of human happiness. Therefore, the stronger are responsible for the welfare of the weaker, with respect to both humans and animals. As for the relationship between humans and Nature, the general position is that Nature exists to be tamed and used by humans. Thus all values in Nature – aesthetic, religious or otherwise – derive from humans' usage and needs.

The principle of democracy According to the principle of democracy, every person has the right and a duty to make personal judgements about what constitutes right and wrong, so long as they accept the equal right of other members of society to do the same. As a rational moral agent, each person is considered autonomous, and in principle one is obliged (and entitled) to consider oneself as having exactly the same status as everybody else within a given society. For instance, to argue that food containing genetically modified ingredients should be labelled because we have the right to know what we eat is to apply the principle of democracy as a presupposed premise. We call this approach D-argumentation.

D-argumentation is a way of thinking that has developed in European philosophy from the seventeenth- and eighteenth-century contract theories of society up until, among others, Habermas's studies of the communicative conditions for democracy. Central to this development is Kant's definition of ethics as a study of the forms of ethical reasoning rather than of their content. Decisions about right and wrong, good or bad, must in principle be left to the autonomous person. This autonomous individual is considered to be his or her own – universally obliged – master and law-maker. According to this approach, in politics ethical positions should be institutionalised as rules of procedure rather than as specific prohibitions and allowances, leaving it up to the agents themselves to make their own concrete judgements.

The principle of veneration According to the principle of venera-tion, moral questions are questions about the relation between the moral agent (an individual or some collective 'we', such as 'humankind') and an 'other', possessing value, integrity and power of its own and, as such, demanding respect or even deference. Duties and values are, essentially, founded in this relation to something or somebody in possession of value and power independent of the agent. For instance, from this perspective one might argue that experiments involving the genetic modification of human reproductive cells should not be allowed because this constitutes interfering with Nature. Here the presupposed premise would be the principle of veneration. We term this kind of approach V-argumentation.

The most obvious example of ethical veneration concerns ethical think-ing based on religious assumptions: God is the universal source of values and the human *raison d'être* is, essentially, to worship divine perfection. But it should be remembered that religious ethics is not the only exam-ple of V-argumentation. Philosophers and non-philosophers have created refined positions based on veneration for 'the other' person involving all living beings, or Nature defined as something more than an entity existing for human exploitation. In this conception, Nature is deemed infinite in its complexity; therefore, because humans have little understanding of it, the idea of their 'taming' Nature is fundamentally absurd. This position demands the establishment of ethical principles that are in accordance with Nature, not opposed to it.

Ethical argumentation

These are three general categories of ethical thinking, each with its dis-tinct profile, and although they compete with each other they are not necessarily contradictory. In current social and political practice, 'ethics' is a battleground of disagreements, but conflict is magnified in areas, such as modern biotechnology, that raise new problems and dilemmas. To define these disagreements as 'ethical' – and not as merely 'political' or 'economic' – is to stress the presuppositions of rationality upon which the debate is conducted. However, before we examine the contents of bioethical discourse and, in particular, the activities of the different bioethics bodies, some preliminary remarks are required.

- The themes covered by ethical committees (especially the central ones) are similar among different countries.
- The 'solutions' adopted, on the other hand, are not always the same. To give an example, the Group of Advisers on the Ethical Implications of Biotechnology (GAEIB) did not reject requests for patenting human genes in its opinion of 1996, whereas the French national committee did so in December 1991. The reasoning and the arguments involved are not always employed in the same way. However, on the main 'bioethical' questions, such as human cloning, there is an obvious consensus. The differences that exist are indicative of the cultural diversity within Europe.
- The 'values' referred to, in the opinions expressed by bioethical committees, are often drawn from traditional legal frameworks. This is particularly the case at the level of the GAEIB/European Group of Ethics and of the French bioethics committee. The GAEIB, for instance, refers formally, in its opinions, to legal texts (directives or international conventions). Though some legal principles are of moral origin, this is certainly not always the case. The borders between ethical norms and legal norms are obvious and must be preserved. In some cases, however, this border has been crossed by bioethical committees. There is a danger of confusion here that is not always perceived by the actors concerned.

This raises more general questions: what are the sources of bioethics? if ethical norms have to be established, who is entitled to do so? if ethical norms have to be found among pre-existing values in a given society, is a new discipline such as bioethics required? These questions remain unanswered at the theoretical level. Yet, in practice, they were resolved in most European countries by the establishment of official bioethical committees and by the institutionalisation of bioethics.

The process of institutionalisation of bioethics

The purpose of this section is to describe how ethics has become an explicit issue in politics, and to point out some of the similarities and differences between countries in handling this issue. In most European countries, bioethics has become institutionalised by the creation of political bodies with consultative and administrative functions. In principle, this common development seems to indicate a need for, and a will to respect, fairly high standards of rationality in making ethical decisions.

Nonetheless, the ways in which these decisions are arrived at differ considerably.

Science has been a dominant factor in European culture for over a century. In all countries, the support scientific culture now confers to U-argumentation allows it to displace V- and D-argumentation. One would, however, expect some degree of compromise to be reached. In addition, it is reasonable to expect variation in the nature of conflicts depending upon the way in which they are handled. Thus, in different countries, the debate may be fought around, variously, Utilitarian versus Democratic positions, or around Utilitarian versus Veneration positions. In our comparisons, this sort of variation is visible at a national level, with Denmark differing from the other three countries and, to a lesser degree, France diverging from the other countries.

The institutionalisation of bioethics indicates that ethics has become an explicit issue in politics. The discussion of ethics at an institutional level is now considered an acceptable means of resolving conflicts concerning biotechnology and an appropriate way to deal with more basic issues related to the nature of ethics. The establishment of special bodies dedicated to bioethics has been a pivotal event in this field. But, as we have noted, the dangers of institutionalisation are two-fold: first, it can have the effect of stifling public debate; second, committees can become part of the intellectual Establishment.

The establishment of the ethics committees

The 1980s in Europe witnessed the inception of central bioethics committees and the beginning of the regulation of biotechnological activities. In all four countries examined here, and at the European level, bioethics committees were established in the nine years between 1983 and 1992:

- 21 February 1983, French National Ethics Committee (CCNE: Comité Consultatif National d'Ethique pour la santé et les sciences de la vie);
- 3 June 1987, Danish Council of Ethics (DER: Det Etiske Raat);
- 28 March 1990, Italian National Committee for Bioethics (CNB: Comitato Nazionale di Bioetica);
- 20 November 1991, Group of Advisers on the Ethical Implications of Biotechnology to the European Commission (GAEIB);
- 1992, the creation of the Greek Bioethics Committee was approved, although the committee was only actually set up in 1997.

At this stage, state interventions were largely directed at encouraging self-regulation, and, as such, few administrative constraints were imposed. But at the European level, this emphasis upon 'soft' modes of regulation began to change into 'hard' regulation, prescriptive and prohibitive, from the beginning of the 1990s. This shift followed the lead of Denmark, which had passed a Gene Technology Act in 1986, and Germany, which had passed a Genetic Engineering Law in 1990.[3]

Considering the fact that reproductive technologies had been under development in most European countries since the mid-1970s, there was something of a time-lag before governments took serious action. Indeed, the debates this technology almost immediately raised show that the life sciences were already having an impact upon society. However, bioethics forums were already in existence some time before the emergence of the central bodies mentioned above; and it is clear that institutionalisation is a protracted process. The local committees, hospital committees or research committees established in the USA towards the end of the 1960s appeared in Europe only at the beginning of the 1970s. In this period there is also evidence of formalisation: the activities of local bodies were coordinated and new bodies were created at the national level.

Thus, in 1974 the main state medical research institution in France, INSERM (Institut National de la Santé et de la Recherche Médicale), set up its own bioethics committee (Comité Consultatif d'Ethique Médicale). In Denmark, a research ethics committee was created in 1973 in the Council of Medical Research. But this movement did not spread to Italy until the 1980s (Spagnolo and Sgreccia 1987–8). In Greece, a ministerial order of 1978 demanded the creation of ethics committees in every hospital, yet the plan was rarely implemented (Dalla-Vorgia et al., 1993).

Institutions   The process of institutionalisation is characterised by the centralisation of existing local and specialised bodies. But, in addition to this, the sanction of the state must be obtained, involving the legitimisation of existing bodies by the legal authorities. This stage generally involves an administrative or legal act and the provision of central funding.

With regard to bioethics, this first took place in Denmark, where the Danish Council of Ethics was created by law (3 June 1987) and the Central Scientific Ethical Committee of Denmark, established in 1978, was confirmed by another law in 1992. The Council is now attached to the parliament and to the Ministry of the Interior, and the Committee is attached to the Ministry of Health. In France, the National Ethics Committee (CCNE) was introduced by a decree of 21 February 1983 issued by the President of the French Republic himself. More recently, the existence of the Committee was confirmed by a law of 29 July 1994. The CCNE is attached to two ministries: Public Health and Research. The Greek Bioethics Committee was also instituted by law (No. 2071/92), and is attached to the General Secretary of Research and Technology in the Ministry of Development. In Italy, the National Committee for Bioethics was established by a decree issued by the President of the Council on 28 March 1990; this committee is attached directly to the Presidency of the Council of Ministries. We shall consider each of these cases in more detail in the next section, but first let us examine the institutionalisation of bioethics at the European level.

At the European institutional level, interest in bioethics emerged in the mid-1980s. From the point of view of the Commission, this was stimulated by its participation at several international meetings, including that in Hakone, Japan, in 1985, and the 1988 Rome meeting on the ethical implications of human genome sequencing. In 1989, the Commission hosted its own meeting on environmental ethics. In the wake of these, at the end of the 1980s the Commission took the initiative of adding bioethics to its existing concerns. Then, in 1989, it launched its first European Bioethics Conference, concerning the human embryo in medical biological research. At the beginning of the 1990s, it went on to establish ad hoc groups dealing with bioethics. The Commission's involvement in the field deepened during the following years. In November 1991, it set up the Group of Advisers on the Ethical Implications of Biotechnology (GAEIB) and imposed ethical standards on all research funded within the Fourth Framework Programme. It could be argued that these initiatives were attempts to address the concerns expressed to the Commission by the Parliament and other bodies. Yet this is not entirely the case. These initiatives (which included the Eurobarometer) were also involved with the broader questions of public acceptance of biotechnology, which the Commission was addressing at the same time.

A new body replaced the GAEIB in February 1998 – the European Group on Ethics in Sciences and New Technologies (EGE). Its existence was recognised by article 7 of Directive 98/44/EC on the legal protection of biotechnological inventions, dated 6 July 1998. The EGE is attached to the Secretariat General of the Commission. It should be noted that, before any debate on the moral aspects of patenting in biotechnology developed, the Commission was already aware of the ethical dimensions of this question.[4] With the creation of the EGE, the Commission once again took the initiative by widening the scope of ethics; now it deals not only with biotechnology but also with science and new technologies in general.

The Council of Europe followed, in parallel, the same path and held its first meeting on bioethics in December 1989. At the opening conference, the Secretary General, Catherine Lalumière, asked for the creation of a European bioethics committee. A preliminary body, the CAHBI (Ad Hoc Committee on Bioethics), had already been established in 1985, and was renamed in 1989. Gradually, from the 1980s onwards, bioethics became a part of the administrative process of regulation at the European level.

Who pushed for institutionalisation and for what purpose? The European Group on Ethics was clearly established for political reasons: it was the Commission's answer to the demands made by the European Parliament. It also enables the Commission to legitimate its policies in fields in which ethical issues are at stake. But at the time of its inception, discussion of a directive had made the problem of the patentability of biological products a subject for vigorous debate in the Parliament. The GAEIB twice issued an opinion on this question (30 September 1993 and 25 September 1996), and on each occasion GAEIB's opinion was invoked by the Commission in attempting to buttress its position during negotiations with the Parliament. In its Proposal on the competitiveness of European industry, the Commission stated:

It is desirable that the Community have an advisory structure on ethics and biotechnology which is capable of dealing with ethical issues where they arise in the course of Community activities. Such a structure should permit dialogue to take place where ethical issues which Member States or other interested parties consider to require resolution could be openly discussed.... The Commission considers that this would be a positive step towards increasing the acceptance of biotechnology and towards ensuring the achievement of a single market for these products. (European Commission, 1996)

In Italy, the CNB was created in the context of national debates concerning reproductive technologies, and its launch saw confrontations between religious and secular ideas on bioethics. The creation of the CNB had followed a Christian Democrat deputy's proposal, which had been advanced in order to settle a heated parliamentary debate. A multi-party coalition had proposed a bill that would stimulate discussions about abortion and the genetic manipulation of life. At the same time, communist groups proposed the launch of a parliamentary commission to study the ethical problems arising from genetic engineering and reproductive technologies. From the outset, this was an attempt to reconcile religious and secular bioethics, and sought to alleviate the tensions by means of institutionalisation. Bioethics is still a contested domain in Italy; but this is unsurprising in a country with a strong Catholic tradition, relative political instability and scientific lobbies able to exert considerable pressure.

France has a long tradition of 'administrative consultancy'. Since the absolutist monarchies of the mid-seventeenth century, the French state has been replete with advisory councils in every sector of its competence. The activity of these consultative bodies tends to be proportional to the activity of the administration, and they often constitute an important part of the decision-making process. The CCNE belongs to this tradition. President Mitterrand involved himself personally in the creation of the CCNE; as such, its inception can be seen as a political act. But this involved only the formalisation, at a national level, of the pre-existing INSERM bioethics committee. It retained the same function, the same president (Professor J. Bernard) and a similar composition. The establishment of the CCNE appears to have satisfied two principal requirements: first, the French scientific community's need for clear guidelines for the conduct of its research in the developing field of biotechnology; secondly, the state's need for moral legitimacy in formulating legislative initiatives in regulating this new field. Neither in France nor in Italy were bioethics bodies set up in response to the demands of public, practitioners or lawyers. The chief source of appeals was the scientific community.

In Denmark, however, the system differs from those of France and Italy. There are two central bodies: the first, the Central Scientific Ethical Committee (CSEC), deals with research in the field of biomedicine; the second, the Danish Council of Ethics (DER), tackles general questions raised by the development of the life sciences. The CSEC presides over seven regional committees; it is in charge of the harmonisation of the solutions these committees suggest, and also functions as a sort of appeal court (Solbakk, 1991). The CSEC system of committees was organised, in accordance with the Second Helsinki Declaration, solely for the benefit of the research sector, and not the public. The DER, in contrast, was established in response to the political will of the Danish parliament, and was designed as an instrument by which public debate on the controversial aspects of biotechnology could be fostered.

The Greek Bioethics Committee styles itself as an intermediary between the scientific community and the public, disseminating information and facilitating the expression of opinions from a variety of different parties. For increasing the level of public dialogue, it also presents itself as a 'catalyst' for bioethics discussions. This body was initiated by a law, passed by the Greek parliament in 1992 (No. 2071/92), which provided a general mandate for the creation of advisory committees; but the Bioethics Committee was formed only on 11 April 1997 by an act of the Minister of Development. The Committee is attached to the General Secretary of Research and Development, which in turn belongs to the Ministry of Development. On 12 September 1998, the Committee underwent a transition within the ministry, and its focus changed from medical ethics to bioethics in general.

To conclude this section, several general observations about these bodies may be made. Most are consultative and multidisciplinary. They are not formally dependent upon governments, as is the case with other administrative bodies; however, this legal autonomy does not preclude constraints of a less formal political or sociological nature. All of these bodies participate at a national level in helping governments address the various ethical issues arising out of progress in science and technology. Nevertheless, it must be recognised that 'this ethical dimension is gradually being institutionalised' (Lenoir, 1998: 2). Central ethics committees have become the tools of central policy-makers; they are established by state authority and come to function as part of the administrative system. With some reservations vis-à-vis the Danish Council of Ethics, their role can be seen as that of providing governments with technical expertise. Furthermore, although the legitimacy of these bodies derives from the state, in many instances they in turn provide the state with scientific legitimacy.

Ethics committee for the establishment?

This generalised movement towards the institutionalisation of bioethics raises important questions. For whose benefit were ethics committees created? For instance, does the institutionalisation of bioethics in governmental committees mean that the bioethical discourse has been appropriated by a small minority? Are these bodies really efficient in influencing state policy related to the field of bioethics? With respect to the first question (excepting Denmark's DER), it is clear that the demand for official bioethics committees came, in the first instance, from the scientific community and, secondarily, from politicians.

The composition of the committees provides an immediate indication of the purpose for which they were established. The Danish Council of Ethics comprises seventeen members, the majority of whom are from a non-scientific background, with men and women being equally represented. This body works for the administration of the Ministry of the Interior and the parliament and enjoys considerable fiscal support. Eight members are designated by the Parliament and nine by the Ministry.

The French National Ethics Committee is a typical administrative and expert body, and it contrasts sharply with the more public-oriented Danish model. It is composed of forty-one members who are designated by a wide range of authorities: the President of the Republic, the ministries of Health, Research, Communication, Education, Industry and Justice, as well as some of the 'grands corps d'Etat' such as the Conseil d'Etat and the Cour de Cassation. The most prestigious state research bodies or academies can also make appointments, such as INSERM, Institut National de la Recherche Agronomique, Centre National de la Recherche Scientifique, the Academy of Sciences, the Collège de France, the Institut

Pasteur and the Conference of University Presidents, as can the presidents of the National Assembly and the Senate. The chief objective of this complex system of designation is to achieve an accurate and balanced representation of the scientific community. But, as a consequence, between 1996 and 1998 members of the CCNE were predominantly male (more than two-thirds) and university professors involved in the life sciences (twenty-seven out of forty-one). There were only four lawyers, none of whom were academics, and, even more surprisingly, there were only four theologians, philosophers or ethicists. In addition, more than two-thirds of the committee is Parisian. The appointment procedures are rather obscure and the turnover is low, suggested by the fact that eight out of the forty-one members have been in place for the full fifteen years.

The European Group on Ethics has twelve members appointed for their technical or scientific competence or on the more ambiguous basis of 'personal qualities'. These members are appointed by the Commission. Presently they are five women and seven men; seven of the twelve are university professors; and five of the twelve are there purely for their experience in, and knowledge of, the life sciences. In contrast to the French national bioethical body, life sciences and social sciences are fairly evenly represented (five for the first, seven for the second). However, analogous to the French CCNE, the nomination process is largely uncontrolled; further, like the CCNE, the EGE has clearly been conceived as an administrative expert body in bioethics. In view of the limited number of its members, not all EU countries are represented in the EGE; however, there is only one member per member state. North/south origin is carefully balanced.

The Italian CNB consists of forty members, who are nominated on a multidisciplinary basis. This body is comparable to the French CCNE in the sense that it is designed to act as an administrative body of experts. This committee basically comprises individuals from a scientific or political background, and their opinions reflect the divergence in views characteristic of the Italian public. Thus, discussions on major issues, such as the status of the embryo, generate no greater consensus inside the CNB than they do in the public sphere.

The Greek Bioethics Committee (GBC) is composed of seven members. Appointed by the General Secretary of Research and Technology, they are all men and all university professors, with the notable exception of a journalist. In terms of background, there are roughly equal numbers of social scientists and biologists. All of them are specialists in the field of biomedicine and bioethics. Therefore, the GBC can also be seen as an administrative expert body.

In conclusion, with the exception of the Danish Council of Ethics, the central committees are designed as expert bodies, whose members come predominantly from the scientific community or have had some involvement with scientific research. Overall, it can be seen that the national bioethics committees mostly represent what might be called a 'scientific establishment', even if, in the legal texts, the desirability of members being drawn from a plurality of backgrounds is generally endorsed. Does this mean that this minority has managed to appropriate the ethical discourse on biotechnology? In other words, has the institutionalisation of bioethics led to an impoverishment of the debate on bioethics? Two additional factors may help in addressing this issue: first, the existence and role of ethical bodies in 'competition' with the aforementioned committees; second, the power of the central bioethics bodies in relation to the media.

The European Group on Ethics has – at its level in the category of European central bodies – only one competitor: the Advisory Group on Ethics (AGE) set up by Europabio, the European Association of Bioindustries, at the beginning of 1998. AGE was created to advise Europabio and its members on ethical questions raised by their research into, and the exploitation of, biotechnology. To date, it has not yet released an opinion and, therefore, it is extremely difficult to evaluate its activities. But the EGE also has connections with other national, central bodies, with which it meets, and to which some of the EGE's members belong. But some of the 'younger' national central bodies, those that have not yet clearly defined their positions, may be susceptible to influence by the prominent and well-established EGE. To a certain extent this does indeed seem to have been the case for the Greek Bioethics Committee. This is certainly the impression given by the title and contents of the first volume of a booklet edited by the GBC: 'Views of the European Bioethics Committee about the Ethical Consequences of Biotechnology'. However, the other groups do tend to act independently of the EGE. This situation helps to maintain a minimum of diversity among bioethical forums. Nevertheless, it must be recognised that the EGE is the only expert body that is consulted and listened to by the Commission. This factor obviously considerably strengthens its authority.

In Denmark, the danger of limiting the participation in, and diversity of, bioethics debates is largely obviated by the coexistence of two separate bodies: the Central Scientific Ethical Committee, which comprises technical experts and is devoted to research, and the Danish Council of Ethics, a body open to the public and geared to stimulating public debate. In a similar vein, Italy hosts numerous unofficial institutions dealing with bioethics, such as the Fondazione Centro San Raffaele del Monte

Tabor, the Centro di Bioetica dell' Università Cattolica del Sacro Cuore in Rome, the Centro di Bioetica della Fondazione Instituto Gramsci of Rome, and the Instituto Giano di Bioetica di Genova. Their number and their vitality help to maintain a genuine public debate on bioethics, and they do not hesitate to criticise the advice proffered by the National Committee for Bioethics. In France, the CCNE retains a leading position that is unchallenged by any other body. Unlike the CCNE, the local research committees set up since 1988 are unable to deal with general questions, and there are no private bodies that can do so. In Greece, the Orthodox Church established its own Bioethics Committee. And, although it does not have a formal relationship with the state, its opinion is taken seriously by state authorities. From 1997 onwards, a number of other small-scale advisory committees have also been set up to deal with bioethical issues, and these have assumed advisory roles within the different ministries.

Relationships among these central bodies, and between their members and the media, also have a profound impact upon the public debate on bioethics. Some institutions, and members, have established very close and fruitful relationships with the mass media. The cases of Noëlle Lenoir, head of the EGE, and Axel Kahn who were both members of the French CCNE, provided good examples of this. Their close ties with the media maximised the prominence and influence enjoyed by their organisations. Nonetheless, generally speaking, the central ethics committees are extremely careful with regard to communicating their positions to the media and publicly providing advice. This is in accordance with the principles on which most of these bodies were founded: in most cases, they received a mandate from the state to educate the public (this is the case in Denmark, involving the DER, and in France and Greece; in Italy this task is not specified for the national committee). As a result, close ties between the media and bioethics central committees are deemed acceptable only when they can be seen to be fostering public debate on questions of bioethics. On the other hand, if the public debate does not really exist, media coverage may have the effect of enhancing the institutional authority of state bodies, and conferring disproportionate influence on them. This is, by and large, the case in France. In Italy, on the other hand, the number of publications on the subject of bioethics prepared by 'competitors' (for example, *Politeia*, *Problemi di Bioetica*, *Quaderni di Etica e Medicina*) ensures a far greater plurality in public debates.

We may now turn to the second main question: What is the real impact of these bodies on the public policies adopted in the field of biotechnology? If these bodies, because of their institutionalisation and their role in the regulatory process, are asked for advice by governments, they will almost certainly exert some influence. However, this does not mean that their advice is always adhered to by governments and legislators. Several reasons for this may be identified. First, the problems the committees deal with do not always need a legal solution. Second, governments and legislators have to take into account political as well as ethical considerations when formulating policy. Political factors may stymie the adoption of new laws (a good example of this is Italy concerning the problem of reproductive technologies), or lead to the acceptance of alternative legislative solutions. Third, advice offered may conflict with existing laws. A recent example is provided by the Commission's request for the EGE to decide whether or not it is ethical to fund research on embryos. The EGE's advice was released in November 1998 and, despite being fairly balanced, it was not accepted by Jacques Santer (president of the European Commission at the time) and Edith Cresson (head of DG XII). For the first time, advice requested by the Commission had been subordinated to political considerations.

Conclusions

To conclude, the institutionalisation of bioethics raises serious problems concerning the development of public debate in the field of bioethics. In particular, there is the serious danger of suppressing the diversity of ethical opinions traditionally expressed within our societies, and, instead, imposing upon society the 'ethics of the scientific establishment'. The different ethical positions we considered in the first section are not always adequately represented. Moreover, with the establishment of a central European bioethics committee, a related risk is that of losing the diversity of ethical approaches among the different European countries.

These outcomes are not especially surprising. Most of the national bodies are designed to advise governments, in their capacities as expert bodies, and to educate the public as teachers. In other words, their mission is not to host public discussions or to stimulate public debate. At any rate, these bodies have imposed their views less than their institutional presence in the regulatory process or in the decision-making process regarding biotechnology. These central bioethical committees have become a tool to legitimate the regulatory process in the field of biotechnology. Yet, in seeking to enshrine values within the law, there is always the risk of creating confusion between legal and ethical norms, as well as between morals and the law.

A final danger is that of limiting the public debate on bioethics. The demand for discussions of bioethics is part of a general phenomenon in modern Western societies, reflecting a public desire for the establishment of clearly stated, fundamental norms. One can also discern this prevalent trend in the multiplication of declarations of rights and an increasing emphasis on basic human rights. As such, it is appropriate for debates on bioethics to encourage reflection upon values and the public's discussion of moral choices. But these values should not be treated as concepts that can be adopted as and when required in order to help legitimise political or scientific interests. Avoiding this sort of scenario requires public participation in bioethical bodies to be substantially increased. Bioethics bodies must follow the example of Denmark, and come down, as it were, from their ivory towers. The lack of consensus conferences illustrates just how insufficient public involvement has been, and points to an obvious means by which it could be enhanced.

NOTES

1. In this paper we analyse only the data sets from the four countries under study. In Denmark, the newspapers analysed were *Information* and *Politiken*, and the data set is a sample from total press coverage ($n = 300$). In France, the newspaper analysed was *Le Monde*; the data set is a sample from total press coverage ($n = 622$). The data set from the French press does not include coverage before 1982 apart from a single year, 1975. In Greece, the newspapers analysed were *Kathimerini* and *Eleftheropia*, and the data set was constructed by a fixed sampling ratio ($n = 65$). In Italy, the newspaper analysed was *Il Corriere della Sera* and the data set was constructed through a variable sampling ratio from the total press coverage ($n = 340$). Each article was coded for up to three themes (see chapter 3 in this volume). This analysis makes no distinction among the order of themes coded.
2. Readers unfamiliar with philosophical ethics may find good introductions and discussions in, for instance, Singer (1991), Rachels (1993), or Dyson and Harris (1994).
3. For details, see Durant et al. (1998).
4. *Directive Proposal on Patentability of Biotechnological Inventions*, Doc COM (88) 496 final/SYN 169 (21/10/88).

REFERENCES

Dalla-Vorgia, P., V. Kalapothaki and A. Kalandidi (1993) 'Bioethics in Greece', *Quaderni di Bioetica e Cultura* January: 101.
Durant, J., M.W. Bauer and G. Gaskell (eds.) (1998) *Biotechnology in the Public Sphere. A European Sourcebook*, London: Science Museum.
Dyson, A. and J. Harris (eds.) (1994) *Ethics and Biotechnology*, London: Routledge.
European Commission (1996) 'Proposal for a Council Decision Implementing a Programme of Community Action to Promote the Competitiveness of European Industry', *Bulletin* EU 1/2-1996, point 1.3: 77.

Hilgartner, S. and C.L. Bock (1988) 'The Rise and Fall of Social Problems: a Public Arenas Model', *American Journal of Sociology* 94: 53–78.

Lenoir, N. (1998) 'Introduction to the EGE' at http://europa.eu.int/comm/european_group_ethics/gee1_en.htm#introd.

Potter, Van R. (1971) *Bioethics: Bridge to the Future*, Englewood Cliffs, NJ: Prentice Hall.

Rachels, J. (1993) *The Elements of Moral Philosophy*, 2nd edn, New York: McGraw-Hill.

Singer, P. (ed.) (1991) *A Companion to Ethics*, Cambridge, MA: Blackwell Publishers.

Solbakk, H. (1991) 'Ethics Review Committees in Biomedical Research in the Nordic Countries', in *Arsmelding (1990)*, Oslo: NAVF, p. 19.

Spagnolo, A. and E. Sgreccia (1987–8) 'I comitati e commissioni di bioética in Italia e nel mondo', *Vita e Pensiero*, 500–14.

5 Controversy, media coverage and public knowledge

Martin W. Bauer and Heinz Bonfadelli

During their life cycles, some new inventions leave the inventor's laboratory and enter the public sphere. However, before the new technology has materialised into new services and products, it first assumes a largely symbolic existence. Thus, its 'real' appearance is preceded by requests for venture capital, from the state, banks or the stock market, and by attempts to persuade investors, producers and consumers to endorse the new technology in imagination. In this way, new technologies enter the public sphere to confront existing expectations, concerns and value. The further development of a technology is thus determined by many variables, and the contingency of the trajectory it assumes will often obviate the possibility of predicting the outcomes.

In this chapter we explore the emerging representation of modern biotechnology in Western Europe during the early 1990s. Our focus is on the intensity of representation rather than on its quality. We assume that public attention and awareness are a precondition for the formation of clear images, opinions and attitudes about an emergent technology. We also recognise that biotechnology is represented in a variety of different modes: it exists as knowledge in the individual 'mind'; as an informal topic of conversation at work or over coffee, beer or dinner; and as a theme in formalised mass media coverage. The relations among these different modes of representation are of key theoretical interest (Bauer and Gaskell, 1999). It is crucial to ask how important an issue biotechnology is in the various modes of individual cognition and informal/formal communication. Further, how do these modes of representation interact to create a public image of biotechnology that has the power to influence its future development? And do these modes of representation interact with each other to increase or decrease the salience of biotechnology and, in this way, affect its development?

The 'knowledge gap hypothesis' is a mid-range theory of mass media effects that explores the intensity of representations of non-local issues by examining the relations among the intensity of media coverage, the level of controversy and the level of pluralism characterising a particular public sphere. Representation is, therefore, defined in terms of the distribution of knowledge on a given topic in the public sphere. The theory predicts that, over a number of years of media coverage of an issue, this distribution will change. First, more information will not necessarily lead to more equal distribution of knowledge; on the contrary, the differential between the knowledge of the elite and that of the remainder of the public will widen (the knowledge gap). This gap is likely to close only under specific conditions. Secondly, with time, the elite section of society will be saturated with knowledge (the ceiling effect), and the other public strata will have an opportunity to catch up. Thirdly, public controversy on the issue will fuel an information flow and reduce the knowledge gap, *ceteris paribus*. Finally, the more plural the public sphere, the less rapidly this gap will close even during a controversy. In other words, pre-existing pluralism in the public sphere neutralises the effect of controversy on the distribution of knowledge.

The production of knowledge in modern society has assumed historically unprecedented proportions. The emergent knowledge–ignorance paradox highlights the fact that, as the supply of new knowledge increases, levels of relative ignorance grow in parallel. Given the finite capacities of the individual mind, an individual's relative share in the stock of available public knowledge will inevitably decrease. Any one person becomes less and less well informed on any specific topic. In view of this, a premium is placed upon making intelligent selections of the available knowledge and achieving a meaningful integration of subjects and perspectives. However, in the context of societies in which ignorance is unavoidable, it is legitimate to ask: Why should one be at all concerned by the unequal distribution of knowledge relating to biotechnology? After all, considering that knowledge is now very far from static, an equality of knowledge would seem to be a far-fetched aspiration. Yet we believe that this is simply not the case. In evincing this claim, we will advance two arguments, one in the context of mass media studies, the other in the context of research into the public understanding of science. Both arguments support our contention that a well-informed public is vital to carry political decisions, and therefore imperative for the effective functioning of modern democracies.

Do the mass media influence or inform the public?

The concept of the knowledge gap is situated within the tradition of media effect studies. Media effects research was dominated, for a long time, by a focus upon purposeful short-term persuasive effects that were measured by studying how an individual's opinions, attitudes and stereotypes were influenced by the media. The key question was: how can modern mass media be used effectively to influence political or social behaviour, in general, and the acceptance of new technologies such as biotechnology, in particular? This focus upon media persuasion did not change until the early 1970s, when longer term cognitive phenomena, for instance, agenda-setting, knowledge gain and the cultivation of images of the social world, became the new focuses of media effects research (McComb, 1994: 3).

One important reason for this paradigm shift was a recognition that, before the media can begin to influence specific attitudes, the topic (or attitude object) has first to be generated by personal experience, informal conversations or formalised mass media coverage. As a result, the academic focus shifted from the question of 'How do people feel about' an issue, to 'How many people think about it at all?' This new question assumes that the public will learn about a topic, such as biotechnology, from the news media. In this respect, the most important effect of the mass media is their ability to confer salience on a topic in the minds of their audience. Thus, journalists employ news value and gate-keeping practices to select and to frame information in particular ways. People may be in a position to develop opinions and attitudes only on the basis of information that has been mediated in this way.

For this reason, the study of learning by mass media is of special interest in the domain of new and complex technologies. The technicalities of genetic engineering and modern biotechnology are complex, abstract and far beyond the majority of people's personal experience. Thus, by necessity, knowledge concerning science and technology is to a great extent cultivated and conveyed by way of media representations. Indeed, several studies of the media have demonstrated their importance as sources of information related to technology (Wade and Schramm, 1969; Robinson, 1972; Griffin, 1990), and especially where danger may be implicated (Mazur, 1981; Dunwoody and Peters, 1992; Coleman, 1993).

In 1970, Tichenor, Donohue and Olien, researchers at Minnesota University, formulated their knowledge gap hypothesis (KGH) in a programmatic paper entitled 'Mass Media and Differential Growth in Knowledge'. In the initial version, the hypothesis stated: 'As the infusion of mass media information into a social system increases, segments of the population with higher socio-economic status tend to acquire this information at a faster rate than the lower status segments, so that the gap in knowledge between these segments tends to increase rather than decrease' (Tichenor et al., 1970: 159–60). Since then, more than twenty-five years of research and more than a hundred empirical studies have dealt with knowledge gap phenomena, and this has stimulated further theoretical elaboration and refinement of the model (Bonfadelli, 1994; Gaziano and

Gaziano, 1996; Viswanath and Finnegan, 1996). The knowledge gap perspective questions normative assumptions about the role of the mass media in democratic societies by adducing empirical evidence for the existence of a chronically uninformed public. This model allows one to connect the timing of information diffusion in society to social structures. The underlying assumption is that a mere increase in information will not automatically result in a better and more equally informed public. Instead, the well educated segments of society are able to use the media more efficiently than the less well educated, and are better placed to take advantage of the abundance of mass media information. As a result, the knowledge gaps between the different social segments will increase rather than decrease. Furthermore, information dissemination conforms to the 'Matthew principle': those who have it will be given more.

Observations of a widening of knowledge gaps come from a variety of empirical studies conducted in the USA and Europe. Numerous cross-sectional studies have reported strong correlations between education and social status and between education and knowledge levels relating to such issues as politics, but especially science, technology and health. Furthermore, several studies have shown that the well educated tend to be reached more effectively by mass media information campaigns, and that mass media news items travel more rapidly in the higher social strata. But there are inconsistencies in this general picture. Several panel studies have reported both a stabilisation in the knowledge gaps and, in some cases, a narrowing.

These deviations from the original hypothesis have been explained in recent years in terms of contingent factors mediating between mass media coverage on the one hand, and audience reception on the other (Bonfadelli, 1994). In 1975, Donohue, Tichenor and Olien refined their KGH at the macro level by studying two mediating factors. They argued that gaps tend to be higher, and to increase, in more pluralistic public spheres and in the absence of controversy; whereas gaps are smaller, and tend to decrease, in more homogeneous settings and when controversy and conflict are involved. Additional factors operate at a micro level (Ettema and Kline, 1977). Audience factors that condition the knowledge gap include interest in a topic and the related motivation to seek information, communication skills, habitualised access to information-rich print media and frequent informal conversations on a topic. Factors related to media content and format that make a difference to the knowledge gap are the knowledge topic (whether it is practical or abstract), the type of knowledge (agenda, factual or structural knowledge), media channels (newspaper versus television) and the duration and intensity of media coverage.

Public controversy as a lever of public understanding of science

The KGH contributes to the discussion of controversies over new technologies in two key respects. First, in its elaborated version KGH emphasises the functionality of controversy for the development of a new technology. It identifies social controversy as a mediating factor that militates against the tendency for normal media to amplify the knowledge gap between the educated and the less educated sections of the public.

Secondly, controversy and conflict play a role in stimulating and maintaining the participation of citizens in the polity. The greatest enemies of democracy are not conflict and controversy, but apathy and indifference towards public affairs. Controversies, including those concerning new technologies, revitalise old values, redefine them through their applications to new situations, strengthen the cohesion of those engaged in conflict on all sides, and accelerate the circulation of information and knowledge. Newspapers, radio, television, magazines and other mass media play a vital role in such processes. These contributions of conflict to the sustainable development of a new technology are easily forgotten by short-sighted actors who envisage a quiet world of expert control in which time and money are saved through exclusive decision making and the consequent avoidance of public controversy. By way of example, the Swiss referendum on gene technology of June 1998 required a concerted effort by all parties involved, over a period of several years, first to launch the referendum and then to persuade the public of the rightness of their positions. Myopic, technocratic voices abound who consider these efforts to be a waste of time and money.

Controversies involving new technologies, such as biotechnology, complement the normal educational channels for the diffusion of knowledge. They provoke informal conversations on the topic in the home, at work or at leisure, and direct attention to media coverage that might otherwise be overlooked, in turn prompting further media coverage of the topic. This intensifies the controversy and mobilises an ever wider range of people to participate in the representation of the issue in the public sphere. Clearly, controversy fuels a virtuous circle of attention, awareness and knowledge. And although awareness and knowledge of a new technology cannot guarantee its future in modern industrial society, we can assume them to be necessary conditions.

In a democracy, public familiarity with an issue is a precondition for realistic and sustainable decision-taking pertaining to it. Democracy feeds on the culture of checks and control that high levels of public knowledge permit. Decisions that are unintelligible to the public are likely to raise suspicions and quickly become unpopular. An informed public is, therefore, necessary for efficient democratic government in the face of complex, non-local issues; and in a deregulated economy only informed consumers make for efficient market operations. Furthermore, a new technology requires a prominent public profile if it is to attract venture capital and to recruit enthusiastic young people for its activities.

The Anglo-Saxon debate on the public understanding of science has, to date, considered knowledge mainly as an independent variable. From the educationalist perspective, knowledge is mainly of interest as a predictor of attitudes (Miller, 1983), a common-sense expectation inspiring funding agencies and policy-makers ('to know it is to love it'). Many people therefore expect that further education on modern biotechnology will produce a public that is more supportive of its projects and products. This is also, from a polemical point of view, a convenient claim. By logical reversion it discredits the sceptics: if knowledge breeds love, ignorance breeds contempt, therefore people who challenge science and technology must be ignorant. Yet this claim is at variance with the empirical evidence.

The hopeful expectation of finding a link between knowledge of and support for science and technology (justifying, among other things, increased educational endeavours) has encouraged almost two decades of survey research into the public's understanding of science. However, the empirical evidence for a positive correlation between these factors is slim. Instead, models need to be highly context sensitive. Thus, some of us have developed a model of the public understanding of science in which this very relationship is variable and shifts as societies move from an industrial to a more post-industrial mode of production (Bauer, 1995; Durant et al., 2000).

The debate on the function of knowledge in public understanding of science research has meanwhile diversified into three areas. First, researchers are seeking to define the form of 'knowledge' that requires explication. Most of the time we are measuring the text-book enculturation of 'facts-out-of-context' rather than the local appropriation of knowledge for practical purposes. Unfortunately, in the past, conceptual confusion on this issue has led to unproductive polemics on methodology (Irwin and Wynne, 1996). Secondly, researchers analyse the context of different knowledge–attitudes relationships (Bauer et al., 1994; Durant et al., 2000). Thirdly, researchers investigate knowledge and its antithesis, ignorance, as dependent variables, and model their relations with various individual or structural determinants (Bauer, 1996). Our present concern with the KGH extends and elaborates the third concern by focusing on the distribution of knowledge in relation to structural variables, such as the intensity of media coverage, the level of public controversy and the pluralism of the public sphere.

The distribution of knowledge and its constraints: a two-level model

Based on these considerations, several research hypotheses can be formulated. The diffusion of information concerning biotechnology is constrained by the knowledge level of the single individual, and by structural factors at the level of the public sphere. In our case, the boundaries of the public spheres correspond to the political boundaries of European countries. The Eurobarometer survey of 1996 gathered data on knowledge levels, education and social status, and was coordinated and conducted in seventeen European countries. Our individual-level model stipulates a chain of influence as schematized in model I:

Model I: education > relevance > media discrimination/conversation
> knowledge

General education provides the citizen with a sense that biotechnology is a relevant issue, even if remote from immediate local concerns. This sense of relevance will lead to their selecting articles in the press or programmes on television or radio for immediate attention, or it may prompt informal conversations with friends or colleagues, or in the family, related to biotechnology. Citizens with higher levels of education, who have a sense of the relevance of biotechnology, who are inclined to pay attention to the media coverage and who discuss it, will eventually arrive at an awareness and knowledge of the specific issues involved. In other words, *differences* in terms of education, estimations of the relevance of biotechnology, the selection of media topics and the frequency of conversations on this subject will, to a large extent, explain the distribution of biotechnological knowledge in a society.

We have formulated three basic hypotheses on the distribution of knowledge about biotechnology in European countries.

1. Better-educated groups will be more knowledgeable about biotechnology and will be more interested in the topic. Since the level of formal education differs in the various countries of Europe, we also expect that the level of knowledge about biotechnology will vary according to the country's overall educational attainment.
2. Knowledge levels will be related to message discrimination: the more people in a country who attend to biotechnology news items, the higher the level of knowledge in that country.

3. The more people who converse about biotechnology in a country, the higher the level of knowledge it will exhibit.

At the national level, model II stipulates structural factors that mediate the flow of information:

Model II: K-gap $= f$(media coverage/level of controversy/

pluralism of public sphere)

Testing this model suggests three further hypotheses:

4. Knowledge gaps will increase as media coverage intensifies.
5. Knowledge gaps will decrease as media coverage intensifies *if* there is public controversy over biotechnology, as this will lead to an expansion of the information flow. Controversy may encourage media channels to cover the topic more frequently, and channels that would otherwise have neglected it, may begin coverage and thus bring the topic to a wider audience.
6. Knowledge gaps will increase as media coverage intensifies as a function of the plurality of the public sphere in the country. The mass media in a more pluralistic public sphere are likely to cater for a variety of news values and concerns. In a pluralistic context a non-local topic, such as biotechnology, may be 'ghettoised' to a few channels with specific concerns.

Testing model I requires survey data; model II requires, in addition, measures of the intensity of media coverage, of the level of controversy and of the pluralism of the public spheres. The basic fact we aim to establish concerns the existence of knowledge gaps relating to biotechnology in various European countries. This constitutes the explanandum of our models.

Inequalities of knowledge of biotechnology

The Eurobarometer study of 1996 contained ten knowledge questions concerning biotechnology (see table 5A.1 in the appendix to this chapter). The percentage of correct answers varies, from item to item, between 21 per cent and 84 per cent. Correct answers were summated to provide an indicator of biotechnological knowledge for each individual and an aggregate for each country, with an overall mean score of 5 out of 10. Better-educated people gave more correct answers on every knowledge item. Table 5.1 shows that the knowledge gaps, measured as the differences in knowledge between respondents with low and high education, vary from country to country (from 1.5 in Austria, Denmark and

Table 5.1 *Knowledge levels and knowledge gaps in different countries*

Country	OECD education[a]	Knowledge level[b] Knowledge-10	Individual education level[c] Low	Medium	High	Knowledge gap Diff.[d]	Corr.[e]
Germany	1	4.7	4.0	4.6	5.7	+1.7	.28
Switzerland	2	5.9	4.7	5.5	6.6	+1.9	.31
Norway	3	5.0	3.7	4.6	5.7	+2.0	.35
UK	4	5.7	4.9	5.6	6.8	+1.9	.31
Sweden	5	5.9	5.0	5.9	6.6	+1.6	.32
Austria	6	4.0	3.2	4.3	4.7	+1.5	.27
France	7	4.9	3.8	4.8	6.0	+2.2	.39
Finland	8	5.5	4.1	5.2	6.3	+2.2	.40
Denmark	9	6.0	4.8	5.7	6.3	+1.5	.28
Netherlands	10	6.3	5.1	6.0	7.1	+2.0	.37
Belgium	11	4.6	3.7	4.4	5.6	+1.9	.32
Ireland	12	4.2	3.7	4.1	5.4	+1.7	.24
Greece	13	3.8	3.0	4.0	4.8	+1.8	.36
Italy	14	5.1	4.2	5.2	6.0	+1.8	.34
Luxembourg	15	4.9	4.2	4.6	5.7	+1.5	.26
Spain	16	4.2	3.3	4.6	5.5	+2.2	.38
Portugal	17	3.8	3.1	4.9	5.5	+2.4	.44
Total		5.0	3.8	4.9	6.1	+2.3	.38

[a] Ranking according to OECD 1995 secondary-level educational attainment of 25–64 year olds.
[b] Score on ten-item Eurobarometer 1996 questionnaire.
[c] Based on school-leaving age; $N = 16{,}500$.
[d] Difference between low and high individual education.
[e] Correlation of knowledge-10 and individual level of education.

Luxembourg to 2.4 in Portugal). This picture of a direct relationship between education and knowledge holds at the country level too: the higher the country's educational attainment, the higher is the average level of knowledge of biotechnology ($r = .57$, $n = 17$).

The KGH postulates an unequal distribution of biotechnological knowledge between the most and least educated segments in the different countries of Europe. The knowledge gap, measured as the correlation between the individual's level of education and knowledge score (knowledge-10), is $r = .38$ ($n = 16{,}500$) and is discernible in every country, varying between $r = .24$ for Ireland and $r = .44$ for Portugal. Comparisons among the different countries also demonstrate knowledge gaps between countries: knowledge levels are particularly high in countries with high levels of education measured by secondary school attainment, such as the Netherlands, Denmark and Switzerland, whereas knowledge levels are low in countries with educational deficits, such as Greece, Spain and Portugal. However, the knowledge gaps in the different countries do not vary significantly with the level of educational attainment of these countries ($r = .26$, $n = 17$).

The personal relevance of biotechnology

Education functions as a motivational factor for interest in topics that are abstract and not of existential personal relevance. Therefore, one would expect that the higher the level of education individuals have received, the more they will perceive the personal relevance of biotechnology; and, at the macro level, issues related to modern biotechnology are deemed more important in countries with higher levels of education.

Clearly, knowledge is one thing, but the perceived relevance of biotechnology is quite another. People may know about it but consider it to be remote from their daily concerns. A person acquires further knowledge only on those issues that are perceived to be relevant – topics that trigger a pre-existing interest or express a particular concern. Perceived relevance can therefore be seen as the first in a chain of factors that incline a citizen to look out for more information on this novel issue.

Personal relevance was measured with question 15 in Eurobarometer 46.1 (see Durant, Bauer and Gaskell, 1998): 'We've been discussing several issues to do with modern biotechnology. Some people think these issues are very important whilst others do not. How important are these issues to you personally?' The response was given on a scale from 1, not at all important, to 10, very important.

There is only a weak correlation between self-assessments of the personal relevance of modern biotechnology and individuals' levels of education ($r = .16$, $n = 16,500$). Individual education does not seem to affect the interest that European citizens display in modern biotechnology. The average level of interest in the low education group registers 5.9 points on a ten-point scale and increases to 7.0 points in the high education group. This pattern at the individual level is mediated by factors at the country level, as shown in table 5.2. Switzerland, Austria, Finland and Sweden are countries with high levels of education in which biotechnology is perceived to be of high personal relevance; and in Ireland, Belgium, Spain and Portugal both factors are at low levels. However, Norway, Germany and the UK are countries with high levels of education but in which biotechnology registers low assessments of perceived personal relevance (note that this study was performed at

Table 5.2 *Relevance of biotechnology and level of education by country*

Constellation		Country	Education Per cent[a]	Education Rank	Relevance Per cent[b]	Relevance Rank
Education	High	Switzerland	82	2	43	6
Relevance	High	Austria	69	6	47	4
		Finland	65	8	43	6
		Sweden	75	5	48	1
Education	High	Norway	81	3	28	17
Relevance	Low	UK	76	4	36	11
		Germany	84	1	33	13
						10
Education	Lower	Denmark	62	9	48	1
Relevance	High	Netherlands	61	10	42	7
		Greece	43	13	48	1
						10
Education	Low	Ireland	47	12	30	15
Relevance	Low	Belgium	53	11	31	14
		Italy	35	14	37	10
		Spain	28	16	29	16
		Portugal	20	17	34	13

[a] OECD 1995 secondary-level educational attainment of 25–64 year olds; median = 63 per cent.
[b] Eurobarometer 1996 percentages of 'high' perceived relevance (8–10 points); median = 37 per cent; $N = 16,500$.

the end of 1996, before the controversies over GM food and Dolly the sheep). In contrast, Denmark, the Netherlands, Greece, Luxembourg and Italy have lower levels of secondary educational attainment but interest in biotechnology is relatively high.

Attention to biotechnology: message discrimination in different media

People may buy a newspaper or magazine, listen to the radio or watch television news. Yet this does not mean that they will pay attention to what they see or hear. For this reason, our survey asked whether respondents remembered having encountered anything related to biotechnology in the previous three months in newspapers or magazines or on the radio or television. This item was used as a measure of message discrimination, i.e. as an index of the amount of interest the public pays to biotechnology.

Table 5.3 *Message discrimination by level of education and relevance (%)*

Media channels[a]	Total	Education			Corr.[b] K_{tau}	Relevance			Corr.[c] K_{tau}
		Low	Medium	High		Low	Medium	High	
Television	35	24	36	44	+.20	27	35	46	.17
Newspapers	25	12	24	37	+.25	18	23	34	.15
Magazines	12	6	12	19	+.13	8	12	19	.13
Radio	11	5	10	16	+.11	7	9	15	.10
Print and AV channels	20	9	20	29	+.25	14	18	28	.20
Print only	11	6	11	16		9	11	14	
AV only	18	17	18	19		16	19	21	
Channel not known	6	6	6	7		6	6	7	
No discrimination	45	62	45	29		56	45	31	
Number of channels:					+.24				.21
2+	23	12	23	34		16	21	32	
1	32	26	32	38		27	34	37	
0	45	62	45	29		57	45	31	

Source: Eurobarometer 46.1, 1996.
Notes: $N = 16,500$.
[a] AV = audio-visual (radio or television); print = newspaper or magazines.
[b] Correlation between education and message discrimination.
[c] Correlation between relevance and message discrimination.

Education is a prerequisite for habitual media use and message discrimination. As such, the habitual use of media as sources of information about biotechnology is influenced by an individual's educational level. This is apparent from the fact that more people with higher education have heard about biotechnology than have those with lower education. Education correlates with the number of channels on which individuals discriminate biotechnological messages. The relationship between education and media sources is stronger for print media than for television and radio. Access to media sources is influenced not only by education but also by personal relevance of biotechnology.

Table 5.3 shows the use of different mass media channels, their combination and their number in relation to levels of education and perceptions of the personal relevance of biotechnology, expressed as percentages. Education clearly correlates with discrimination of biotechnology messages in the mass media: $K_{tau} = .25$. Whereas 62 per cent in the low education group could not remember having heard any mention of biotechnology in the mass media over the previous three months, fewer than 30 per cent of the higher educated respondents fell into this category. Furthermore, education correlates with the number of different forms of media in which respondents could recall reference to biotechnology having being made. In addition, the more highly educated consulted the print media significantly more frequently than people with a low level of education. The correlation between education and reading of newspapers is stronger than that between education and reliance upon the audio-visual media. These communication gaps are also apparent at the national level. Altogether, our data support the notion that attention to and discrimination of biotechnology media messages are unequal phenomena.

However, biotechnology message discrimination is not only influenced by education and perceived personal relevance. The combination of education and motivation shows that both factors interact in an additive way. Only 31 per cent of people with lower levels of education and a low interest in biotechnology had access to information about biotechnology in the mass media. In contrast, 71 per cent of the well-educated and highly interested group had been informed by the mass media.

As can be seen from table 5.3, 35 per cent of respondents had heard, over the preceding three months, about issues concerning modern biotechnology through television, 25 per cent from newspapers, 12 per cent from magazines and 11 per cent by way of radio. Television seems to be the most important single media channel for disseminating information about biotechnology (this is far from surprising, considering that the same was true for all non-local news items in most countries by the 1990s). Taken together, 55 per cent acquired their information on biotechnology via mass media channels. In comparison, 52 per cent derived their understanding from discussion of the topic with other people on at least one occasion. Indeed, since biotechnology is perceived, by many, to be of real importance, informal conversations have achieved a significance not far short of that of the mass media channels.

Biotechnology as a topic of conversation

The mass media are the most important sources of information in modern societies. This is especially true with respect to questions of science and technology. But interpersonal forms of communication can also be important, particularly where a topic is of high personal relevance. Furthermore, television displaced newspapers as the main information source, as many media studies in the USA and in Europe show. Our hypotheses in this section are that, first, television, rather than the print media or interpersonal communication, is the most widely utilised information source both for general information and for data concerning biotechnology; and, secondly, the higher the perceived personal relevance of biotechnology, the more people will use the mass media and interpersonal channels of communication.

Our data show that only a minority of people relied exclusively upon either the media (16 per cent) or conversation (11 per cent) in procuring information on biotechnology. However, 39 per cent used both channels and 34 per cent used neither media nor interpersonal channels. Furthermore, there is a significant correlation between the perceived relevance of biotechnology and use of media or interpersonal channels. So, the higher the personal importance of biotechnology, the more likely it is that the mass media, as well as interpersonal channels, will be recognised as important sources of information about modern biotechnology.

Slightly more than half of respondents had talked about biotechnology with other people, and the intensity of interpersonal communication varies strongly between countries. The countries associated with high-intensity discussions on biotechnology are Switzerland, Denmark, Austria, Finland, Germany and Norway, in which between 75 and almost 90 per cent have engaged in such discussions. Conversely, in Greece, Portugal, Spain and Ireland biotechnology does not seem to be a salient topic of day-to-day discussion among citizens by the end of 1996, as shown in figure 5.1.

Communication gaps exist not only in the use of mass media but also at the level of interpersonal communication. Thus, people with higher education talk about biotechnology with the greatest frequency; as a corollary of this, the amount of interpersonal communication about biotechnology is higher in those countries with higher levels of formal education. Clearly, there are significant communication gaps at the level of interpersonal communication about biotechnology similar to the gaps in media access. Table 5.4 shows that about half of the respondents with a high level of education talk occasionally – or even frequently – with other people about biotechnology. In contrast, almost three-quarters of the less well educated have not discussed biotechnology. These communication gaps also exist among countries. In those with a high educational attainment – most notably Switzerland, Denmark and Norway – informal discussions on biotechnology are frequent, with more than 70 per cent of respondents having participated in them. In countries with lower standards of education – such as Greece, Spain, Portugal and Italy – discussions of biotechnology among citizens are rare, with only about 30 per cent ever having conversed on the subject.

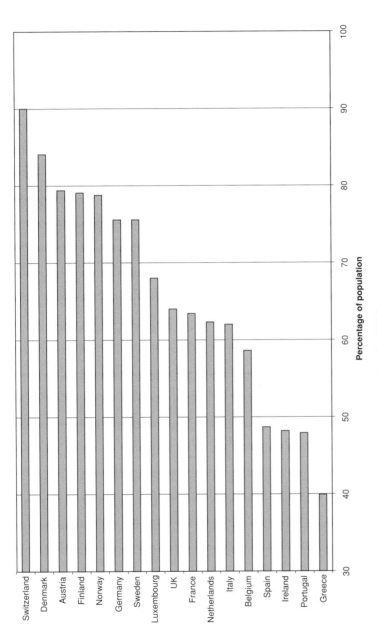

Figure 5.1 Biotechnology as a topic of conversation by 1996

Note: Percentage of yes responses to either of these questions: 'Over the last three months, have you heard anything about issues involving modern biotechnology?' or 'Before today, had you ever talked about modern biotechnology to anyone?'

Table 5.4 *Interpersonal communication by level of education and relevance (%)*

Amount of interpersonal communication	Total	Education			Relevance		
		Low	Medium	High	Low	Medium	High
Frequent	7	3	6	14	4	5	12
Occasional	27	14	26	41	18	27	39
Only once or twice	16	11	18	18	14	18	16
Never	50	73	50	30	64	50	33
Total	100	100	100	100	100	100	100
Correlation (K_{tau})			+.30			+.25	

Source: Eurobarometer 46.1, 1996.
Note: $N = 16,500$.

Knowledge, education and attention paid to biotechnology

Levels of message discrimination from different media provide an indication of an individual's knowledge of a given subject. In this case, people who remember having heard something about biotechnology in the media are likely to know more than people who have no recollection of having access to biotechnological information via the media. Media discrimination and level of education seem to have an additive effect upon individual knowledge. Moreover, table 5.5 shows that the mass media serve to narrow the knowledge gap somewhat, since levels of knowledge acquisition from access to newspaper or to television news are slightly more pronounced among the less well-educated segments of the data set. This notwithstanding, overall the mass media have less impact upon knowledge levels than does educational background, and therefore had little effect on the distribution of knowledge about biotechnology at the end of 1996.

Controversy and the polarisation of the public sphere

An important effect of social controversy is to direct the flow of information into regions that it would not conventionally reach. As such, in public controversies, the knowledge gap between the elite and lower strata is reduced faster than one might otherwise expect. This theoretical claim can be tested empirically using our data.

Table 5.5 *Knowledge levels by education and message discrimination*

	Knowledge level[a]				Knowledge gap	
		Education				
	Total	Low	Medium	High	Diff.[b]	Corr.[c]
Overall mean score	5.0	3.8	4.9	6.1	2.3	.38
Newspaper						
Yes	5.9	4.8	5.6	6.5	1.7	.29
No	4.7	3.7	4.7	5.8	2.1	.37
Media effect gap	+1.2	+1.1	+0.8	+0.7		
Television						
Yes	5.6	4.6	5.5	6.4	1.8	.32
No	4.6	3.6	4.6	5.8	2.2	.38
Media effect gap	+1.0	+1.0	+0.9	+0.6		
Channels						
More	6.0	4.8	5.7	6.6	1.8	.30
Print only	5.6	4.6	5.3	6.2	1.6	.29
Television/radio	5.3	4.4	5.2	6.1	1.7	.31
Source not remembered	4.8	3.8	4.6	5.6	1.8	.34
No media	4.2	3.4	4.6	5.5	2.1	.35
Media effect gap	+1.8	+1.4	+1.1	+1.1		
Combination of channels						
Media + interpersonal	5.9	4.8	5.6	6.5	1.7	.29
Interpersonal only	5.4	4.3	5.2	6.2	1.9	.33
Media only	4.8	4.1	4.8	5.7	1.6	.29
No communication	3.9	3.3	4.2	5.0	1.7	.28
Communication effect gap	+2.0	+1.5	+1.4	+1.5		

Source: Eurobarometer 46.1, 1996.
Notes: N = 16,500.
[a] Mean score on ten-item Eurobarometer 1996 questionnaire.
[b] Difference between low and high education.
[c] Correlation of knowledge score and level of education.

The reference point for a functional analysis is the national 'polity' in which citizens make daily decisions. We shop around for food, clothing and entertainment, take decisions on medication, invest trust in authorities and experts, and engage – periodically – in political activity by signing petitions, joining protests or casting votes on substantive issues. It is a basic feature of modern democracy that higher levels of knowledge on controversial issues stimulate greater political participation and improve the quality of decision-making in terms of both substance and legitimacy.

Knowledge allows one to operate with distinctions that would otherwise not make sense.

Conflicts and controversy will foster greater information flow with two measurable effects: knowledge gaps will be lowered in countries with high conflict and accentuated in countries that lack public debate on biotechnology; knowledge gaps will be higher in countries that have a homogeneous public sphere, and will be lower in countries with a segmented or pluralistic public sphere. For the purposes of this study, we have used three indicators of the level of controversy involving biotechnology in European countries: the scale of activities of the polity, the intensity of media coverage and the percentage of controversial media coverage focusing on the controversial aspects of biotechnology. We have also employed two indicators of pluralism: left–right political polarisation and religious polarisation (see the appendix for details).

Figure 5.2 shows the relationship between polity activity in any one country between 1985 and 1998 and the knowledge gap in 1996. The correlation is negative: $r = -.75$ ($n = 12$). As predicted by the hypothesis, public controversy measured in terms of polity activity is associated

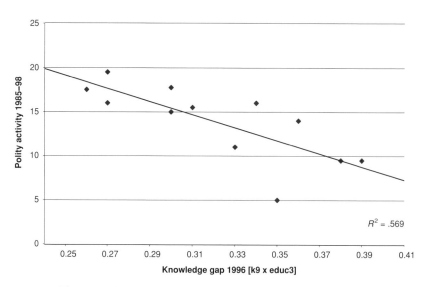

Figure 5.2 Knowledge gap, 1996, and level of polity activity, 1985–1998
Note: The knowledge gap is measured as the correlation between knowledge (k9 = nine knowledge items) and level of education (educ3 = three levels of education). For the measurement of 'political activity' see the appendix.

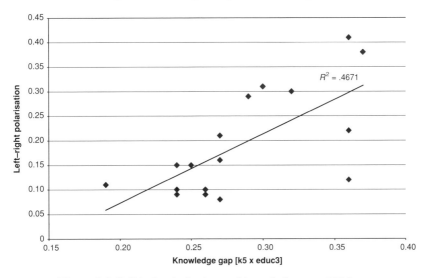

Figure 5.3 Political polarisation and knowledge gap, 1996
Note: The knowledge gap is measured as the correlation between knowledge (k9 = nine knowledge items) and level of education (educ3 = three levels of education). For the measurement of 'political polarisation' see the appendix.

with smaller knowledge gaps across twelve European countries. In countries where biotechnology and genetic engineering have been the subjects of policy-making, parliamentary (or other) debates, public hearings, demonstrations, petitions or even referenda, a higher proportion of the public will be aware of, and know about, this new technology. In this sense, public controversy surrounding biotechnology is an important factor in educating the public on this topic.

Figure 5.3 shows the relationship between the polarisation of the public sphere in seventeen European countries and the knowledge gap on biotechnology (both relating to 1996). As predicted, this correlation is positive: $r = .68$ ($n = 17$). The more plural the public sphere in terms of left–right political polarisation, the larger were the 1996 knowledge differentials. More pluralistic publics are likely to have more varied mass media, each with a different agenda and therefore promoting a more unequal distribution of knowledge on such non-local issues as the emergence of new technologies. Additionally, ethical concerns over biotechnologies would suggest that religious feelings and orientations play a part in stimulating public controversy. However, the degree of religious pluralism in a country does not affect the distribution of knowledge ($r = .16$).

Table 5.6 *Knowledge gaps, changes in knowledge gaps, intensity of media coverage, and trends in coverage*

Country	Knowledge gap[a]	Change in k-gap, 1993 to 1996[b]	Media coverage, 1995 + 1996[c]	Trend in coverage[d]
UK	.31	1.50	1,567	44
France	.39	1.43	541	46
Germany	.28	1.47	285	52
Austria	.27	–	440	57
Netherlands	.37	0.93	378	47
Switzerland	.31	–	357	47
Sweden	.32	–	130	48
Italy	.34	1.07	113	24
Finland	.40	–	82	43
Denmark	.28	1.20	28	43
Greece	.36	1.03	10	26
Spain	.38	1.16	–	–
Belgium	.32	1.13	–	–
Portugal	.44	1.44	–	–
Luxembourg	.26	1.00	–	–
Ireland	.24	0.70	–	–
Norway	.35	0.96	–	–
Median	.33	1.10	370	46.5

Source: Durant et al. (1998).

[a] The correlation between knowledge measured on nine items and the level of individual educational attainment.

[b] Ratio of k-gap for 1996 over k-gap for 1993 measured by the correlation of five-item knowledge and educational attainment. A figure >1 indicates increasing gap, a figure <1 decreasing gap.

[c] Estimates of the total number of articles in the opinion-leading newspaper for 1995 and 1996.

[d] The number of articles in 1995–6 divided by the number of articles in 1991–6. A figure >33 indicates increasing coverage, and a figure <33 decreasing coverage, by the mid-1990s.

Knowledge gaps may be either high or low in both homogeneous or more plural religious climates.

All previous results are based on cross-sectional comparisons within and among countries. Table 5.6 summarises the dynamics of the knowledge gaps within several countries. Thanks to an earlier comparable survey (Eurobarometer 37.1 of 1993), we are in a position to compare these knowledge gaps longitudinally between 1993 and 1996. This reveals that in six countries – Ireland, the Netherlands, Norway, Luxembourg, Greece and Italy – knowledge gaps remained stable; whereas in Belgium,

Spain, Denmark, France, Germany, Portugal and the United Kingdom they increased. The greatest expansions in the knowledge gap – close to 50 per cent – were recorded for the UK, France, Germany and Portugal.

For the Netherlands, Greece, Italy, Denmark, France, Germany and the United Kingdom, we are able to relate changes in the size of the knowledge gap to the level of media attention, and thereby assess the effects of media coverage and controversy on the gap. The intensity of the media's coverage of biotechnology in 1995 and 1996 relates to a widening of knowledge gaps ($r = .60$, $n = 7$). But, though intensified media coverage of biotechnology in the year before the survey accentuated the knowledge gap, it may do so with an effect that diminishes with the level of previous coverage. The knowledge gap effect in relation to already established levels of coverage is clearly a problem that requires further investigation.

In addition to examining cross-country variations in the intensity of media coverage, we were able to quantify within-country shifts in the media coverage. Countries that witnessed large increases in the media debate about biotechnology during the mid-1990s – such as Germany, France and the United Kingdom – experienced an enlarging of the knowledge gap; whereas Greece and Italy, countries with declining press coverage by the mid-1990s, saw the emergence of more stable knowledge gaps. The Netherlands presents a more ambiguous picture: this is a country with the highest average public knowledge about biotechnology, a large knowledge gap, relatively low levels of polity activity, and relatively intense (and increasing) media coverage, yet it displays a stable – or even narrowing – knowledge gap.

Conclusion: KGH is underspecified as a dynamic model

We have explored the representation of biotechnology in seventeen European countries during the mid-1990s. Proceeding from the 'knowledge gap' perspective, we compared the different representations of biotechnology: individual cognition, informal communication and mass media coverage. The hypotheses we generated stipulated the existence of knowledge gaps in the public's understanding of biotechnology, and the mediation of these gaps by various structural factors, in particular the intensity of media coverage and the degree of controversy over biotechnology in, and the pluralism of, the national public sphere.

In summary, our results present a rather ambiguous picture. Cross-sectional data clearly support the KGH. Knowledge gaps concerning biotechnology are present in all European countries, albeit to differing degrees. Individuals with a more general education are likely to know more about biotechnology than those who are less well educated. In all countries analysed, controversies on this subject in the polity, between 1985 and 1998, had the effect of decreasing these knowledge gaps, while greater left–right political polarisation worked in the opposite direction. However, longitudinal data for seven European countries show that knowledge gaps involving biotechnology increased between 1993 and 1996 as a function of both the high intensity of media coverage in the mid-1990s and an upward trend in media coverage by 1996. The Netherlands is the exception to these general rules: its already large knowledge gap remains stable despite intensified media coverage and relatively low levels of public controversy.

An assumption underlying our models concerns the nature of systemic relationships. Our claims imply linear relationships among intensity of media coverage, the knowledge gap, level of controversy and pluralism in the public sphere. However, most systemic relationships are linear only within circumscribed limits, and beyond this range it would be more realistic to assume non-linearity. Thus, it is conceivable that, once coverage reaches a certain level, the knowledge gap ceases to increase and stabilises, as in the Italian and Greek cases, or actually decreases, as in the Netherlands where there was reduced public controversy in the years up to the survey. Indeed, if controversy may, *ceteris paribus*, decrease the knowledge gap, sharp intensification of the public debate may result in the gap stabilising, or even widening. The expected effect of conflict may be the case only within a limited range; beyond this range conflict may freeze knowledge gaps in the debate, preventing rather than enhancing circulation of information in the system. It may therefore be possible to distinguish stages, or bands, of knowledge gap increase and decrease in the continuum of levels of conflict. The data to test dynamic relationships over time are currently not available. Nonetheless, by using cross-sectional data from two time-points (1993 to 1996), a rudimentary and limited dynamic comparison was performed for at least some of the countries.

Few studies looking at knowledge gaps have examined longitudinal dynamics, partly because longitudinal data of a comparative nature are hard to come by. However, longitudinal data may show that the KGH is an underspecified dynamic model. As it presently stands, the three main propositions of the model stipulate a dynamic system of interacting influences: first, increasing information flows on new technology lead to increased knowledge gaps – the educated have easier access to this information than the less educated; secondly, controversy reduces these knowledge gaps by widening the information flows; and, thirdly, pluralism in the public sphere increases knowledge gaps through a variety of media with different focuses of attention. This raises several fundamental questions, however. How do the opposing influences of public controversy and media pluralism interact over time? Is there a specific level of controversy that will necessarily reduce the size of the knowledge gaps in differentially pluralistic media structures? Is it the case that the more pluralistic the public, the larger the controversies will need to be in order to close the knowledge gaps? Is there an upper threshold of public controversy, a 'crisis' point, beyond which nobody can absorb further information? Is this crisis point different for different degrees of public plurality? These are questions that require further theoretical development and longitudinal data to allow empirical conclusions to be drawn. We leave these questions open for future research.

Fortunately, the robust longitudinal data needed for these questions to be investigated may be at our disposal in the coming years, as we continue to monitor the status of biotechnology among the European public, in terms of levels of public knowledge, media coverage and polity activity. The controversies over Monsanto's 'Roundup Ready' soybeans, GM foods in general since November 1996 and Dolly the sheep after February 1997 have ushered in a different phase in the controversy over biotechnology and genetic engineering in many countries. These new features need to be carefully compared with those of the previous phases outlined in this chapter. In this sense, we have provided a baseline against which we can compare and contrast future developments in the salience of biotechnology representations in the public spheres of Europe.

Appendix

Methodological considerations

The KGH is principally a dynamic model that makes statements about the development of variables over time: if coverage of biotechnology increases, *ceteris paribus* the knowledge gap will widen; if coverage of biotechnology increases and a public controversy arises, *ceteris paribus* the knowledge gap will decrease. Ideally, these relationships would be studied in one country over a longer period of time. In the absence of significant longitudinal data, these dynamic relationships are modelled in cross-section by making assumptions about a homogeneous trajectory of all units of comparison.

Another possibility is to compare individuals with high and low media use or countries with high and low media input on biotechnology at a single point in time. For example, we assume that different levels of media coverage represent the 'same system' at different stages of development.

We ignore the fact that different systems may have different 'natural' levels of coverage of biotechnology owing, for example, to the particular popular science culture. We also overlook the fact that such cross-sectional comparisons ignore the different contexts in which early developers and latecomers find themselves. As time progresses, the contexts are not the same for particular stages in the development of a technological controversy.

Measures

Knowledge gap Table 5A.1 lists the knowledge items used in the Eurobarometer 1996 survey. The items require either true (y) or false (f) as a correct answer. These measures are indicators of enculturation into the biotechnology system rather than a measure of any practical knowledge. The knowledge gap is measured by the correlation between knowledge and level of education. We use a ten-, nine- or five-item knowledge scale (k10, k9 or k5) and a three-point scale for education level (educ3) based on school-leaving age. The change in the knowledge gap is based on the correlations between k5 items and educ3 for both 1993 and 1996. The coefficient of 1996 is divided by that of 1993. A resulting index of >1 indicates an increase in the knowledge gap, an index of <1 a decrease (see figure 5.6).

Media intensity is the estimated number of articles in the elite press of the country for 1995–6. The changes in press coverage are measured by the number of articles in the elite press in 1995–6 as a percentage of the total number of articles during the period 1991–6. A figure of more than 33 per cent indicates an upsurge in coverage in the two years.

Controversy Measurement of controversy is based on measures of polity activity, media intensity and media controversy in each country. *Polity activity* is based on a Guttman index with nine dichotomous criteria, such as parliamentary debate, public hearings, protest, new legislation and petitions, for three periods: 1985–1991, 1992–4 and 1995–8. The judgement for each criterion is expert based and was collected mid-1998. The judgements for the three periods add up to an overall country score. *Media controversy* is measured by the percentage of articles reporting a controversy on biotechnology for 1991–6.

The level of polity activity and the intensity of media coverage are related: the more media coverage, the more polity activity; however, the required increase in media coverage is exponential ($r = .54$, $n = 12$). Polity activity and the level of media controversy are not related ($r = .15$):

Table 5A.1 *Knowledge indicators by level of education*

| | Correct answers (%) | | | | |
| | | | Education[c] | | K-gap |
Knowledge items[a]	Total[b]	Low	Medium	High	corr.[d]
1. There are bacteria which live from waste water (y)	84	75	85	91	.16
2. Possible to find out Down's syndrome during pregnancy (y)	80	71	80	87	.16
3. Yeast for brewing beer consists of living organisms (y)	69	55	70	80	.25
4. More than half of human genes identical to those of chimpanzees (y)	50	44	50	56	.12
5. Eating gene-modified fruits could modify one's genes (f)	49	33	49	63	.30
6. Cloning produces exactly identical offspring (y)	46	31	46	58	.27
7. Ordinary tomatoes do not contain genes (f)	36	22	35	51	.29
8. Genetically modified animals are always bigger (f)	36	20	34	54	.34
9. It is possible to transfer animal genes into plants (f)	31	26	31	36	.10
10. Viruses can be contaminated by bacteria (f)	21	11	19	34	.23
Knowledge score (0–10 points)	5.0	3.8	5.0	6.0	.38

Notes:

[a] (y) items need to be recognised to be true/yes to score a correct answer; (f) items need to be recognised to be false to score a correct answer. Item 4 is sometimes excluded from the total score.

[b] 'Total' indicates the percentage of correct answers for each item.

[c] 'Education' is the percentage of people that give a correct item for each of three levels of education measured by school-leaving age.

[d] The 'knowledge gap' for each item, and overall, is indicated by the bi-serial correlation between 'education' and giving a correct or incorrect response.

some countries have low-level polity activity and high levels of media controversy, and vice versa. Polity activity during and after 1995 and the trend in press coverage in 1995–6 are related ($r = .58$, $n = 12$); polity activities led to an increase in press coverage, or vice versa. The causality is not clear.

Pluralism We developed two indices for *pluralism* of the public sphere. Pluralism 1 reflects the left–right polarisation in a country – the number of left or right positions as a percentage of middle positions (left or right is based on a self-classification). Pluralism 2 is based on religious polarisation: the number of very religious plus non-religious as a percentage of some religiosity.

For further details on Eurobarometer survey 46.1 of 1996 and the media content analysis for 1973–96, see appendices 1–4 in Durant, Bauer and Gaskell (1998).

REFERENCES

Bauer, M. (1995) 'Industrial and Post-Industrial Public Understanding of Science', paper delivered to the Chinese Association for Science and Technology, Beijing, 15–19 October.
  (1996) 'Socio-economic Correlates of dk-Responses in Knowledge Surveys', *Social Science Information* 35(1): 39–68.
Bauer, M., J. Durant and G. Evans (1994) 'European Public Perceptions of Science', *International Journal of Public Opinion Research* 6(2): 163–86.
Bauer, M. and G. Gaskell (1999) 'Towards a Paradigm for Research on Social Representations', *Journal for the Theory of Social Behaviour* 29(2): 163–86.
Bonfadelli, H. (1994) *Die Wissenskluft-Perspektive. Massenmedien und gesellschaftliche Information.* Konstanz: UVK.
Coleman, C.-L. (1993) 'The Influence of Mass Media and Interpersonal Communication on Societal and Personal Risk Judgements', *Communication Research* 20(4): 611–28.
Donohue, G.A., P.J. Tichenor and C.N. Olien (1975) 'Mass Media and the Knowledge Gap: a Hypothesis Reconsidered', *Communication Research* 2: 3–23.
Dunwoody, S. and H.P. Peters (1992) 'Mass Media Coverage of Technological and Environmental Risks: a Survey of Research in the United States and Germany', *Public Understanding of Science* 1: 199–230.
Durant, J., M.W. Bauer and G. Gaskell (eds.) (1998) *Biotechnology in the Public Sphere: a European Sourcebook*, London: Science Museum.
Durant, J., M. Bauer, G. Gaskell, C. Midden, M. Liakopoulos and E. Scholten (2000) 'Two Cultures of Public Understanding of Science' and Technology in Europe, in M. Dierkes and C. von Grote (eds.), *Between Understanding and Trust: The Public, Science and Technology*, Amsterdam: Harwood Academic Publisher, pp. 131–56.

Ettema, James S. and Gerald F. Kline (1977) 'Deficits, Differences, and Ceilings. Contingent Conditions for Understanding the Knowledge Gap', *Communication Research* 2: 179–202.

Gaziano, C. and E. Gaziano (1996) 'The Knowledge Gap: Theories and Methods in Knowledge Gap Research Since 1970', in Michael B. Salwen and Don W. Stacks (eds.), *An Integrated Approach to Communication Theory and Research* Mahwah, NJ: Erlbaum, pp. 127–143.

Griffin, R.J. (1990) 'Energy in the Eighties: Education, Communication, and the Knowledge Gap', *Journalism Quarterly* 67(3): 554–66.

McCombs, Maxwell (1994) 'News Influence on Our Pictures of the World', in Jennings Bryant and Dolf Zillmann (eds.), *Media Effects. Advances in Theory and Research*, Hillsdale, NJ: Erlbaum, pp. 1–16.

Mazur, A. (1981) 'Media Coverage and Public Opinion on Scientific Controversies', *Journal of Communication* 31(2): 106–15.

Miller, J. (1983) 'Scientific Literacy: a Conceptual and Empirical Review', *Daedalus* 112: 29–48.

Robinson, J.P. (1972) 'Mass Communication and Information Diffusion', in Gerald F. Kline and P.J. Tichenor (eds.), *Current Perspectives in Mass Communication Research*, Beverly Hills, CA: Sage, pp. 71–93.

Tichenor, P.J., G.A. Donohue and C.N. Olien (1970) 'Mass Media and Differential Growth in Knowledge', *Public Opinion Quarterly* 34: 158–70.

Viswanath, K. and J.R. Finnegan (1996) 'The Knowledge Gap Hypothesis: Twenty-Five Years Later', in Brant R. Burleson (ed.), *Communication Yearbook*, vol. 19, New Brunswick, NJ: International Communication Association, pp. 187–227.

Wade, S. and W. Schramm (1969) 'The Mass Media as Sources of Public Affairs, Science, and Health Knowledge', *Public Opinion Quarterly* 33: 197–209.

Part II

Public representations in 1996: structures and functions

6 Traditional blue and modern green resistance

Torben Hviid Nielsen, Erling Jelsøe and Susanna Öhman

In Europe, scepticism towards modern biotechnology is widespread and persistent. However, although there are considerable differences among European countries in terms of the expression and prevalence of scepticism, the data from each country suggest some common themes. In particular, nowhere do sceptics fall into distinctive groupings. And, generally speaking, neither traditional demographic categories nor political/cultural orientations seem at first glance to be very important. In addition, on the political stage, resistance against new biotechnologies is articulated by parties and political groups that represent widely differing positions on the political spectrum. In view of this, there is a need to establish whether a more fine-grained analysis of the data will reveal whether sceptics of biotechnology segregate according to alternative criteria.

It has long been known that within the related, albeit much broader, field of environmentalism there is a division between a modernist group, politically to the left, and a conservative group, oriented towards the preservation of nature and its resources. With regard to resistance to biotechnology, however, few attempts have been made to explore whether analogous, or alternative, commitments might explain patterns of scepticism. Yet such a study promises to shed much-needed light on the nature of resistance to modern biotechnology. Recently, T. Hviid Nielsen (1997a, 1997b) has shown that opponents to modern biotechnology in Norway can be divided into two camps, a traditionalist or 'blue' segment and a modernist or 'green' segment. This analysis was made on the basis of survey data from the Norwegian version of the 1993 Eurobarometer survey on attitudes to biotechnology. Further evidence was derived from documenting the arguments used by various political parties in the Norwegian debate about modern biotechnology. When the data from the 1996 Eurobarometer survey became available, Hviid Nielsen repeated his analyses, using the new data, and once again he found a division in accordance with the traditionalist/modernist dichotomy. By means of 'argumentation analysis' it was also possible to show that the 'blue' and 'green' arguments are based on fundamentally different values and concerns (Nielsen, 1997b[1]).

Moreover, there are strong arguments in favour of the hypothesis that the traditionalist/modernist pattern of segmentation is a general phenomenon. The 1996 Eurobarometer survey on biotechnology, which provides an extensive database covering seventeen European countries (fifteen EU countries, Norway and Switzerland), offers an opportunity for testing this hypothesis. The purpose of the present chapter has been to report on all of the European countries in order to explore the prevalence of the traditionalist/modernist division in the case of biotechnology. In addition, to study the features of the various nationally specific patterns of scepticism, comparisons will be made among countries, as well as, if possible, among groups within the countries.

The gap in expectations

In recent years a large, and apparently widening, gap seems to have developed between the industry's and the general public's expectations of the status of modern biotechnology.

In some scenarios, biotechnology is depicted as the new megatechnology, destined to assume an economic importance analogous to that achieved by the electronic computer, and before that by the oil and petrochemical industries (Freeman, 1995). The European Association for Bioindustries predicts that this expanding sector will be worth US$285 billion by the year 2005 (*Time*, 1998). Ernst & Young, backed by the European Commission, expect 'Europe's entrepreneurial bioscience sector . . . to continue growing at about 20 per cent a year' (Ernst & Young, 1997). And even committed critics such as Jeremy Rifkin anticipate the twenty-first century becoming 'The Biotech Century'. 'Many of the scientific breakthroughs we predicted more than twenty years ago are now moving out of the laboratory and into widespread commercial use', he argues. 'The genetic revolution and the computer revolution are just now coming together to form a scientific, technological, and commercial phalanx' (Rifkin, 1998: xv). However, these hopes and promises have clearly not yet been fulfilled.

On the other side of the expectation gap, ethical concerns and scepticism about potential risks are accompanied by scientific scepticism as to the technological feasibility of the industry meeting its objectives. Biotechnology might, it is argued, be a 'bubble' of overinvestment and

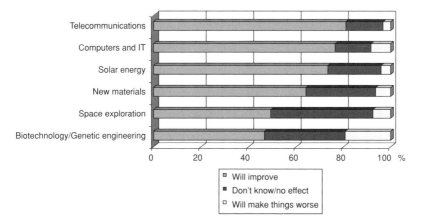

Figure 6.1 Anticipated effects of new technologies in the next twenty years: EU 15, 1996
Source: Eurobarometer 46.1 (1996).

failed investments – a dangerous hit-or-miss technology, based upon 'a crude, outmoded, reductionist view of organisms' (Mae-Wan et al., 1998). Indeed, the 1996 Eurobarometer survey (INRA, 1997) documents high and even slightly growing scepticism towards biotechnology among the European public.

In terms of the perceived ability of biotechnology to generate positive benefits, Europeans still rank modern biotechnology lowest of six new technologies: telecommunication, data and information technology, solar energy, new materials and space exploration (see figure 6.1). The largest proportion of respondents, although still less than half (47 per cent), expect biotechnology to improve our way of life in the next twenty years; but 20 per cent expect it to make things worse, 9 per cent anticipate 'no effect', and as many as 25 per cent answer 'don't know'.

Furthermore, the number of 'optimists' compared with 'pessimists' declined markedly over the five years 1991 to 1996 (see figure 6.2). The 1991 Eurobarometer survey (INRA, 1991) recorded 41 per cent more optimists (51 per cent) than pessimists (10 per cent). By 1993 (INRA, 1993) the difference had decreased to 37 per cent (48 per cent optimists and 11 per cent pessimists), and in 1996 there was a further reduction to 28 per cent (47 per cent optimists and 19 per cent pessimists).

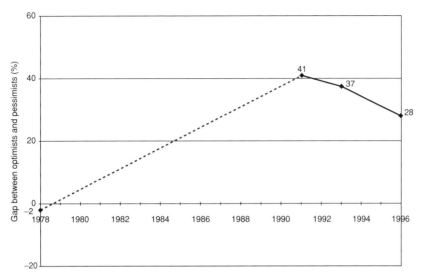

Figure 6.2 General expectations about biotechnology in the next twenty years: the balance of opinion in the EU, 1978–1996
Sources: Eurobarometer surveys 10A (1978), 35.1 (1991), 39.1 (1993), 46.1 (1996).
Note: The gap between optimists and pessimists is measured as the difference between the percentage expecting that biotechnology 'will improve way of life' and the percentage expecting that it 'will make things worse' ('worthwhile' vs. 'unacceptable risk' in 1978).

'Blue' and 'green' resistance

Expectations of biotechnology differ widely across, and within, European nations. Spain, Portugal and Italy have the highest proportion of optimists, and (apart from Greece, which has a strikingly high number of 'don't knows') Germany, Norway and Austria have the smallest proportion (see figure 6.3). This pattern clearly does not fit into the stereotype of a 'restrictive', Catholic south distinct from a 'liberal', Protestant north. Rather, expectations are generally highest in the nations where the implementation and application of the technology have been slow and on a modest scale. Conversely, expectations are lowest where the introduction of biotechnology is further advanced. An 'intuitive' public understanding of 'declining marginal utility' would seem, therefore, to be a better predictor of scepticism towards biotechnology than general religious background.

Moreover, as a group, the 'optimists' share some of the characteristics that technological-innovation theory generally expects to find among

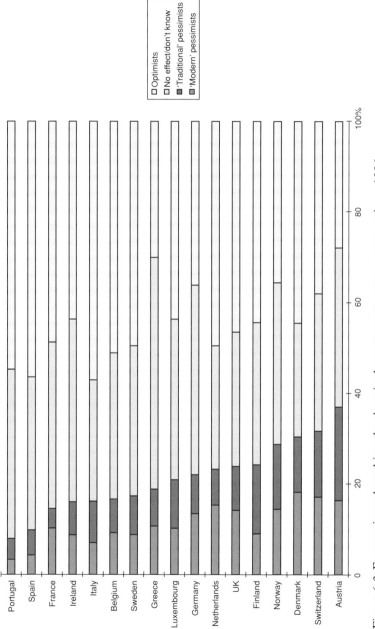

Figure 6.3 Expectations about biotechnology in the next twenty years across nations, 1996

Source: calculations based on Eurobarometer 46.1 (1996).

Note: Optimists think biotechnology will 'improve way of life'; pessimists think it will 'make things worse'.

entrepreneurs. The typical 'optimist' is a young, or younger, man with an extended education and urban residence. He combines relatively good knowledge of biotechnology with a perception of low attendant risks. And he considers himself to the right, or around the centre, of the traditional political spectrum.

The high number of 'undecided' (i.e. 'no effect' and 'don't know') respondents throughout Europe might suggest the existence of a realistic public appreciation of the fact that the future of biotechnology is contingent, and will be dependent upon the nature of regulations and the nature of its applications. The number of respondents in the 'undecided' camp is hardly affected by the ratio between optimists and pessimists, though Denmark and the Netherlands show that the number of 'don't know' optimists may be reduced in nations that cultivate high levels of public knowledge in the context of open and intense public debate. In such circumstances, a higher proportion are prepared to adopt an unambiguous point of view. It is also noticeable, however, that former 'don't knows' distribute themselves fairly evenly between the optimist and pessimist groups.

Cluster analyses on each of the national data sets for the seventeen European countries (EU countries, Norway and Switzerland) reveal that those respondents who think that modern biotechnology will make things worse can, in almost every European country, be divided into two highly distinct groups: the 'traditional' and the 'modern'. These groups can be found in fifteen of the seventeen European countries, even if the group characteristics differ somewhat among the countries (see the appendix to this chapter for details). It is only in Finland and Austria that these distinct divisions in public perception are not found.

The 'traditional' and the 'modern' groups of pessimists differ from the optimists in containing a high proportion of women, as well as high levels of rural residence and a perception of high risk concerning modern biotechnology. Aside from these three common characteristics, however, the two types of pessimists may be systematically separated. The typical representative of the 'traditionalist' group is older and will have completed his/her education after primary school rather than attending university like the typical 'modern'. The traditional group also has an inferior knowledge of biotechnology. And, where the 'traditional' inclines towards the centre and right of the political spectrum, the 'modern' is oriented towards the left. Further, the traditional tends to be strongly religious whereas the modern is inclined to be a strong non-believer. Finally, the traditional group may be described as materialists, and the modern as post-materialists.

The pattern of 'blue' and 'green' resistance in Europe

A general pattern is observable in which the number of opponents is greater in northern than in southern Europe and the number of 'green' opponents is relatively greater in the north, while 'blue' opponents are relatively more numerous in the south.

The following observations are consistent with these findings and help in explaining this pattern. Catholic nations have relatively more 'blue' and Protestant nations relatively more 'green' opponents. A greater exposure to biotechnology and/or a perception of declining marginal utility might explain why expectations are lowest in the nations where biotechnology is already established and opposition is low and expectations high in the peripheral nations (mostly in the south but also including Finland), where biotechnology is not widespread but is still seen as an integral aspect of progress and development.

Among European countries there is marked variation in the size and characteristics of the traditional and modern groups. On this basis, with the exceptions of Finland and Austria, the different states may be divided into five groups. The profiles for five representative countries – Germany, Norway, the UK, Spain and Italy – are presented in table 6.1.

In Germany, the modern 'green' group of opponents is considerably larger than the traditional 'blue' group – 14 per cent compared with 9 per cent. The traditional group is characterised by higher female than male representation; a mean age of 59 years; relatively low education; religiousness; politically right-wing sympathies; low knowledge of, and a high risk perception concerning, modern biotechnology. In contrast, the modern 'green' group is characterised by a mean age of 37 years; higher education; politically left-wing sympathies, post-materialistic values; a high level of knowledge; and a high risk perception.

In Norway, the two groups of opponents are equally large – about 14 per cent in each category. The traditional group is characterised by a mean age of 52 years; relatively low levels of education; religiousness; politically right-wing sympathies; materialistic values; and a high risk perception. Conversely, the modern 'green' group is characterised by a mean age of 35 years; higher education; a low level of religiousness; political left-wing sympathies; post-materialistic values; high levels of knowledge; and a high risk perception concerning modern biotechnology.[2]

In the UK, the modern 'green' group of opponents is also larger than the traditional 'blue' group – 14 per cent compared with 10 per cent. The traditional group is characterised by a mean age of 60 years; relatively low education; religiousness; politically right-wing sympathies; materialistic

Table 6.1 *Profiles of 'blue' and 'green' opponents in five European countries, 1996*

Country	'Blue' opponents	'Green' opponents
Germany	Size of cluster (8.6%) Female (11%) Age 55+ yrs (37%) Education <15 yrs (34%) Low knowledge (10%) Very religious (8%) Political right (7%) High risk perception (11%)	Size of cluster (13.5%) Age 15–39 yrs (22%) Higher education (18%) High knowledge (10%) Political left (14%) Post-materialistic (13%) High risk perception (10%)
Norway	Size of cluster (14.3%) Age 55+ yrs (6%) Education <15 yrs (6%) Very religious (15%) Political right (12%) Materialistic (25%) High risk perception (6%)	Size of cluster (14.5%) Age 25–39 yrs (17%) Higher education (14%) High knowledge (9%) Not religious (17%) Political left (25%) Post-materialistic (17%) High risk perception (17%)
UK	Size of cluster (9.7%) Age 55+ yrs (34%) Education <15 yrs (26%) Medium knowledge (15%) Very religious (8%) Political right (22%) Materialistic (17%) High risk perception (9%)	Size of cluster (14.2%) Female (7%) Age 25–39 yrs (11%) Higher education (18%) High knowledge (12%) Not religious (16%) Political left (12%) Post-materialistic (12%) High risk perception (11%)
Spain	Size of cluster (5.5%) Female (15%) Age 55+ yrs (14%) Education <15 yrs (25%) Low knowledge (14%) Very religious (12%) Political centre (11%)	Size of cluster (4.4%) Male (7%) Age 25–39 yrs (20%) Higher education (20%) High knowledge (23%) Not religious (23%) Political left (30%) Post-materialistic (17%) High risk perception (36%)
Italy	Size of cluster (9.1%) Age 40–55+ yrs (11%) Education <15 yrs (14%) Low knowledge (9%) Very religious (15%) Materialistic (11%)	Size of cluster (7.1%) Age 15–39 yrs (26%) Higher education (24%) High knowledge (20%) Not religious (27%) Political left (14%) Post-materialistic (17%) High risk perception (16%)

Note: The figures represent percentage overrepresentation compared with the total population, including the 'undecided'.

values; and a high risk perception concerning modern biotechnology. The modern group is characterised by greater female than male representation; a mean age of 36 years; higher education; a low level of religiousness; politically left-wing sympathies; post-materialistic values; high levels of knowledge; and a high risk perception.

If we turn to Spain, the two groups of opponents are almost equally small – about 5 per cent in the traditional group and 4 per cent in the modernist group. The former is characterised by a higher proportion of females than males; a mean age of 48 years; relatively low levels of education; religiousness; centrist political sympathies; and low relative knowledge about modern biotechnology. The modern group is characterised by a higher proportion of men; a mean age of 34 years; higher levels of education; a low level of religiousness; politically left-wing sympathies; post-materialistic values; a high level of knowledge; and a high risk perception concerning modern biotechnology.

Finally, in Italy the traditional segment comprises 9 per cent and the modern segment 7 per cent. The traditional group is characterised by a mean age of 48 years; relatively low levels of education; religiousness; materialistic values; and a low level of knowledge about modern biotechnology. The modern group is characterised by a mean age of 33 years; higher education; low levels of religiousness; left-wing political sympathies; post-materialistic values; a high level of knowledge; and a high risk perception.

These data strongly suggest that the 'traditional' and the 'modern' categories of sceptic are well defined and distinct. Though there is some national variation, owing to differing cultural contexts, this is clearly overshadowed by the striking commonalties in the nature of scepticism across Europe.

In fact, of the seventeen countries studied, Finland and Austria are the only nations for which the data do not fit this model. To some extent surprisingly, these countries are extreme cases. Finland does not cluster with the other sceptic Scandinavian nations; instead it contains an unexpected number of optimists, analogously to the pro-development southern periphery of Europe. In stark contrast, the Austrians are not only sceptical, but extremely so.

The particular national backgrounds of these two countries might explain their deviation from the general European pattern. Thus, Finland has undergone rapid urbanisation as well as having experienced a very fast transformation from an agrarian society to a post-industrial information society. Biotechnology came late on the scene, but symbolises a positive approach to new technologies in a country that generally equates new technologies with rapid improvements in the standard of living. This optimistic attitude to technological innovation, one that tends to efface notions of the potentially negative consequences of biotechnology, is likely to have blurred the division of attitudes between optimism and pessimism.[3] The Austrian data are the more surprising because in most respects Austria conforms very closely to the Swiss pattern. Austria's high level of scepticism may be the result of the fact that the public imbibed from neighbouring countries the association between scepticism and political correctness.

Jon Miller has kindly tested this model against US data collected after the genetically modified (GM) soya and Dolly the sheep events. The data are not quite compatible: for example, the US data lack any equivalent of the materialist–post-materialist index. But his analysis does indicate that the 'blue'/'green' distinction cannot be found in America, where opposition generally is small, where religious fundamentalism is more prevalent, and where a widespread pragmatic scientific discourse seems to suppress 'green' views.

Regulation and public consultation – some differences between 'blue' and 'green'

If we turn to other characteristics of the 'blue' and 'green' opponents to modern biotechnology, the pattern is consistently stable. By means of cross-tabulations with other variables in the Eurobarometer survey, we find that the 'green' modernist group trust non-governmental organisations (NGOs) much more than the 'blue' traditionalist group do, especially environmental organisations, but also consumer organisations. The traditionalists, on the other hand, have a higher proportion of respondents who express themselves uncertain about who is trustworthy. They also place a higher degree of trust in the medical profession; and in some Catholic countries they are inclined to invest confidence in religious organisations. The modernist group display a much higher level of participation in the modern biotechnology discourse. They have both heard more about modern biotechnology than the traditionalists, and discussed it with greater regularity.

The 'green' opponents have much the strongest opinions about the regulation of modern biotechnology. They are considerably less inclined to believe that current regulations are sufficient to protect people from the potential risks linked to modern biotechnology, and they tend to reject the idea that the regulation of modern biotechnology should be left mainly to industry. The modernists also believe that GM food should be labelled and that public consultations about modern biotechnology are worthwhile exercises. In contrast, the 'blue' traditionalist opponents are more likely to argue that the public ought to accept some degree of risk attendant upon the use of modern biotechnology, so long as it enhances the economic competitiveness of Europe. Further, in most Catholic countries, the 'blue' traditional group are the more willing to defend the idea that religious organisations should have a say in how modern biotechnology is regulated.

Moreover, the 'green' modernists are more sceptical than the traditionalists of the ability of biotechnology to fulfil the expectations of rapid economic growth outlined above. Nor do they believe that modern biotechnology will usher in the scale of change over the next twenty years anticipated by the traditionalist group. The modernists are, in general, more sceptical about the levels of risk *and* benefit commonly associated with biotechnology. If they are rather less convinced that biotechnology will lead to environmental pollution, the emergence of dangerous new diseases and the production of 'designer babies', they are also sceptical of the capacity of biotechnology substantially to reduce world hunger. Indeed, the modernists are especially doubtful of the validity of the more positive expectations. For their part, the traditionalists are more likely to answer 'don't know' when questioned about the expected benefits and risks of modern biotechnology.

The conclusion must be that the 'green' opponents are more in favour of regulating modern biotechnology and of giving the public a choice, through labelling but also through more direct means of public consultation. This is probably linked to their higher degree of knowledge about biotechnology and their higher level of participation in the biotechnology debate. They do not think it a waste of time, or the issues too complicated, for the public's views to be taken into consideration.

Two arguments

'Traditional' and 'modern' sceptics share the assumption that modern biotechnology will reduce our quality of life. But the general values they articulate are substantiated and underpinned by different arguments and conclusions (Toulmin, 1958; Toulmin et al., 1979). These are represented in figure 6.4.

The 'blue' argument (figure 6.4(a)) has no external references. It is predicated entirely on the conviction that technological intervention in nature is a priori unacceptable. The position can be undermined only by challenging this underlying value judgement.

The 'green' argument (figure 6.4(b)) points to the (perceived) uncertainties and risks related to biotechnology as its principal justification. Proponents of this view focus on stressing the unpredictability of biotechnology and emphasising the level of risk compared with potential benefits.

(a)

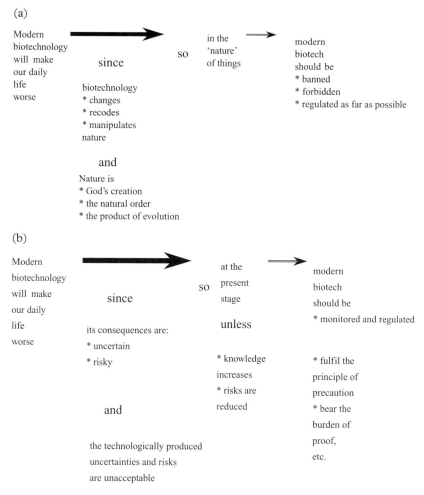

Modern biotechnology will make our daily life worse

since

biotechnology
* changes
* recodes
* manipulates
nature

and

Nature is
* God's creation
* the natural order
* the product of evolution

so

in the 'nature' of things

modern biotech should be
* banned
* forbidden
* regulated as far as possible

(b)

Modern biotechnology will make our daily life worse

since

its consequences are:
* uncertain
* risky

and

the technologically produced uncertainties and risks are unacceptable

so

at the present stage

unless

* knowledge increases
* risks are reduced

modern biotech should be
* monitored and regulated

* fulfil the principle of precaution
* bear the burden of proof, etc.

Figure 6.4 The sceptics' arguments: (a) The 'blue' critique (b) 'Green' scepticism

Its modality centres on 'the present stage' of technological development and knowledge concerning it. In this case, however, an increased knowledge of the scale of risk involved might force an alternative conclusion, one more congenial to the optimists.

Mephistopheles and Frankenstein

The underlying concerns of the 'green' and 'blue' arguments are as distinct as the different social groups that tend to express them. The 'blue' argument is supported by moral (or religious) values, the 'green' by notions of uncertainty and risk. In this sense, the 'blue' critique is 'Faustian' – the whole enterprise of biotechnology may be conceived as a modern covenant with Mephistopheles. Even though biotechnology represents technological success, insight into, and the manipulation of, Nature is considered problematic. In contrast, the 'green' scepticism is more 'Frankensteinian'. The problematic they identify concerns the insufficient knowledge of potential consequences, but intervention in nature is not held to be reprehensible per se. The danger is seen to lie in the creation of a 'monster' capable of developing in unforeseen and uncontrollable ways, and conceivably coming to usurp its creators.

Around the turn of the year 1997, two major events had the effect of mobilising scepticism, and arguments based on both ethics and risk were invoked. Shortly before Christmas 1996, the first ships carrying GM soya from fields in the USA docked in European ports. Then, in February 1997, news emerged of the cloning of the sheep 'Dolly', using a cell from a six-year-old sheep and an egg emptied of any genetic information, at the Roslin Institute in Scotland.

The soya ships on their way across the Atlantic generated fear and uncertainty about the *risks* of biotechnology. These concerns are associated with large-scale production and the consumption of GM foods. The GM soybeans were engineered to be resistant to a herbicide called 'Round-up'. The product had been approved and produced in the USA, where the soybeans had subsequently been mixed with conventional, unmodified, beans. But these shipments were to be introduced into a European market that had so far approved only field trials for, and the marketing of, a few modified products. As yet, Europe had no general rules for the regulation and labelling of GM foodstuffs.

The soya ships therefore excited fears of biotechnology as capable of unleashing a Frankensteinian monster. Could the GM soybeans spread, the public asked, and transfer their genes to other plants in the natural environment? Do we know enough about the long-term impacts of the 'Round-up' herbicide used? Does the soybean oil have the same nutritional properties? What might happen to those individuals who consume bread, chocolate, ice-cream and numerous other processed foods containing the oil? And would it be possible to reverse any unforeseen and unintended consequences?

Initially, the soybean ships were faced with direct action and symbolic protests from environmental and consumer organisations, whose campaigns were closely followed in some European countries. They subsequently managed to increase the political sensitivity of the issue at both the national and the EU level. Consumer demands for more restrictive regulation in this field could not be ignored.

The cloned sheep, Dolly, which just a few months later was pictured on the front pages of newspapers around the world (even before her creation was announced in *Nature*) rendered topical questions about the *ethical* implications of otherwise successful scientific achievements. Even though 277 trials had been performed before the researchers succeeded in their attempts to unite the young cell with the empty egg, the fact that the cloning had taken place was considered to be a scientific breakthrough. But the press soon shifted its focus away from a fascination with this narrowly scientific dimension to concerns about whether science's success portended a moral failure.

Thus, Dolly happened to mobilise the second of the classical tropes in criticising technology: the Faustian notion of an unholy bargain. This incident raised many morally pertinent questions: how old ought Dolly to be considered? who were her parents? how would a human being created in the same way form and understand its own identity? what might come next, now that humans had achieved something that a decade earlier had been considered totally unfeasible? are there now any limits? will humans now begin to create people from the genetic material of geniuses or dictators, or in the form of mannequins?

Of course, Dolly herself was only an innocent and ignorant sheep in a laboratory stable, created by techniques developed in the production of medicines for humans. But she soon acquired a symbolic importance at the highest levels of secular and religious authority. From the President of the USA to the Pope, the technology was met with condemnation, and bans were placed on the use of the same technology involving humans.

It is no coincidence that both of the classical tropes related to the dangers of new technologies – unforeseen risks and going beyond ethical limits – were conjured up as expectations of further scientific and commercial breakthroughs increased. The steps from laboratory to market, and from production to consumption, in themselves generate heightened expectations, increase levels of scepticism and serve to widen the gap between the two.

As a technological 'use of life' (Bud, 1993) and 'a cultural icon' (Nelkin and Lindee, 1995), modern biotechnology is commonly confronted by both a 'pre'-industrial critique of intervention in 'nature's order', as well as a 'post'-industrial critique of the potential risks involved with the new technology. The classic diffusion model of technological innovation is thus contradicted, even falsified, by the fact that these critical discourses have achieved such a wide and lasting appeal. And the mobilisation model demands modification in view of the two types of opposition we have elucidated: one in favour of older traditions, the other directed against the new risks ushered in by biotechnology.

The image of modern biotechnology

Do these two different segments of the public – the traditionalists and the modernists – invoke 'blue' and 'green' arguments and values, respectively, when articulating their scepticism towards modern biotechnology? In addressing this question we have analysed the Swedish data from the Eurobarometer study. In terms of people's mental associations with biotechnology and genetic engineering (in response to the open-ended question 'What comes to your mind when you think about modern biotechnology in a broad sense, that is including genetic engineering?') there are systematic differences between the two different segments of opposition. However, because the groups are not particularly large, the number of answers used in the analysis was small. In Sweden, the 'blue' group comprises 8.5 per cent and the 'green' group 8.9 per cent of respondents.

In qualitative terms, typical arguments used by the traditional group were:

'Nature should be left alone.'
'Mankind has no right to play God.'
'It is wrong, it is not moral or ethical.'
'I'm afraid that there is too much tampering with nature.'

These arguments very much mirror those encountered in the 'argumentation analysis', in which the 'blue' segment tended to raise moral and ethical issues, and used these as a basis for arguing against biotechnology.

Turning to the modernist, 'green', opponents, the typical arguments are based more on dangers and risk.

'There is a danger, the risks are not clear, not for humans, animals or plants.'
'We will have new plants and the genes will spread to the weed and then they will take over.'
'The animals suffer, their bodies are gigantic.'

These answers have also been categorised and the main differences between the two segments in the Swedish data clearly support our general claim that different arguments are employed by the different opposition groups. Among the 'green' segment, fear, risks and danger are the common themes. In fact, 42 per cent of people belonging to this group state that they are afraid of the risks and dangers involved in modern biotechnology, whereas only 34 per cent of the 'blue' group claim to be. On the

Table 6.2 *Valuation of biotechnology by the 'blue'*
and 'green' segments in Sweden (%)

Valuation	'Blue' opponents	'Green' opponents	All
Negative	69	60	45
Ambivalent	17	19	22
Neutral	13	20	25
Moral undertones	35	28	23
Fear	34	42	25

Note: The question was 'What comes to your mind when you think about modern biotechnology in a broad sense?'

other hand, the 'blue' group more frequently express associations with ethical issues: 35 per cent cite moral implications in their arguments, compared with 28 per cent for 'green' opponents. The 'blue' group are also more negative than the 'green' group (69 per cent versus 60 per cent), and the 'green' opponents are rather more likely to be neutral in their description of biotechnology (20 per cent for the 'green' group against 13 per cent for the 'blue' group). Table 6.2 summarises the results for Sweden.

One conclusion to be drawn from this analysis is that, simply on the basis of asking people about their cognitive associations with biotechnology, the results reinforce those obtained from the 'argumentation analysis'. Moreover, in general, people seem to use the same lines of argument against modern biotechnology as are employed by politicians, and/or vice versa.

The traditionalist–modernist segmentation as a general scheme for the resistance against new technology?

The pattern of resistance expressed in the 'blue'/'green' segmentation seems, at first glance, to be specific to biotechnology. This might be expected considering the strong moral commitment of the 'blue' critique – revealed through 'argumentation' analysis – prompted by fears that biotechnology will enable us to manipulate the nature of life itself. One would not expect to find such a reaction to other areas of technological development. Furthermore, the debates about the various technologies have developed differently over time, and in response to different perceptions of risk. Nevertheless, technological change in general is associated with alterations in the conditions of life and always affects various groups of the population in different ways. Indeed, the kind of modernist/traditionalist pattern of reactions to new technology identified above has been an element in most debates about new technology in both the past and the present. It would, therefore, be of interest to explore whether a similar pattern of reaction exists in relation to other areas of technology.

The Eurobarometer questionnaire included a battery of questions related to six major areas in which new technologies are currently developing. One of these is biotechnology/genetic engineering. On the basis of whether respondents believed this technology would improve or worsen the quality of life over the next twenty years, they were categorised as either optimistic or sceptical (see the appendix). Opponents to the other technologies may be defined according to the same criterion. However, the questionnaire does not permit analysis of the resistance to technologies other than modern biotechnology; and the questions asked in each case do not admit of inter-comparison.

Despite these difficulties, we attempted to produce cluster analyses on the basis of the responses obtained from opponents to two of the other technologies included in the questionnaire. We excluded those variables that are specific to biotechnology and, as a control, we performed the same analysis (i.e. using the variables non-specific to biotechnology) on modern biotechnology. The other technologies selected were 'computers and information technology' and 'space exploration'. Even though the number of opponents to both of these areas was significantly smaller than for biotechnology, the numbers were still large enough to make cluster analyses feasible (in contrast to, for instance, solar energy, to which there were very few opponents).

The results of the cluster analyses for biotechnology, in which the variables specific to biotechnology were excluded, turned out to be similar to the results for cluster analyses made when all variables were included in the analysis. This was true for all nine countries for which analyses were performed.[4] Not surprisingly, the variance in the two variables specific to biotechnology, and not included in these analyses, was somewhat smaller; but that was the most significant difference between the cluster analyses made with and without the variables specific to biotechnology. We regard this finding as confirmation of the general validity of the 'blue'/'green' segmentation. At the same time, it strongly suggests that we can make meaningful comparisons among the clusterings of opponents to different areas of technology using general variables and excluding criteria specific to any one type of technology.

Looking at the results of the cluster analyses for the two other areas of technology mentioned above, the general conclusion is clear: there is in both cases a segmentation of the opponents to these types of technology that is similar to the 'blue'/'green' segmentation for biotechnology. The only exceptions to this rule were Sweden and the UK vis-à-vis 'computers and information technology', and Sweden and Norway with respect to 'space exploration'. For seven out of the nine countries the segmentation was consistently in accordance with the modernist/traditionalist pattern within both areas of technology. Yet there are some interesting contrasts in the detailed segmentation pattern between attitudes towards these other technologies and those relating to biotechnology.

Regarding 'computers and information technology', in some countries religion loses much of the importance it has in influencing the segmentation of opponents to biotechnology. This is particularly true in the case of Italy. Conversely, political views and education become more important as criteria for segmentation. With respect to 'space exploration', the differences regarding religion are generally just as important as for biotechnology. The relative sizes of the modernist and the traditionalist groups tend to follow the same pattern as for biotechnology; thus the 'blue' group is largest in the southern European Catholic countries and the 'green' group is larger elsewhere.

Given the deficits of the model used in the analyses, the significance of the more specific segmentation patterns for technologies other than biotechnology should not be overemphasised. However, our findings do suggest that there are similar segmentation patterns for both biotechnology and the other areas of technology. This would be an interesting subject for further research.

Traditional and modern 'optimists'

It is obviously relevant to ask whether a similar segmentation to the one we have found for opposition to biotechnology has a counterpart among the *proponents* of biotechnology, i.e. the 'optimists'. To answer this question we have performed cluster analyses on the responses of the proponents of biotechnology for eleven of the fifteen countries for which a 'blue'/'green' segmentation has been found among sceptics. The results show that, almost invariably, a modernist/traditionalist segmentation also exists among the proponents. However, there are certain distinctive differences between the two groups. Among proponents, the male group is larger than the female. Further, whereas this difference holds across all countries for the modernist proponents, the picture is more complex vis-à-vis the traditionalists. Here the female group is the largest in several of the countries studied. Similarly, there is no consistent picture regarding risk perception. In almost half of the countries, risk perception is slightly lower among modernists than traditionalists.

Yet certain other observed differences between opponents and proponents do not seem to have any significance in relation to the segmentation

pattern. Thus, proponents are, on average, a few years younger than opponents, and politically they are also more to the right. In any case, the
modernists are consistently more to the left than the traditionalists and
the difference in age between modernists and traditionalists is generally
just as big among proponents as it is among opponents.

Only in the case of Norway did it prove impossible to show a meaningful segmentation for the proponents, even though there is a clear
'blue'/'green' segmentation among the opponents. We will not enter
into a detailed discussion or further analysis of the segmentation of the
proponents found by means of the cluster analysis. But it is hardly surprising to find such a segmentation among the proponents. Yet it is clear
that traditionalists are likely to experience a conflict among different considerations in formulating their attitude towards biotechnology. On the
one hand, they may tend to oppose it for the reasons we have already
discussed above. On the other hand, the imperative of economic growth
and technological innovation, which is closely connected with the market
economy, may also influence their attitudes, since they tend to be politically to the right. Consequently, a large group of traditionalists are likely
to be in favour of biotechnology even though they may have some reservations concerning the manipulation of nature, etc. In terms of the modernist group, their political orientation, which is on average slightly to the
left of centre, might be considered surprising since, in other respects, it
seems to display entrepreneurial characteristics. But the modernist group
of proponents, which is quite large in most countries, probably also contains groups of well-educated social-liberals who consider the prospect
of biotechnology important, and who believe that any environmental and
other potential problems caused will be ameliorated.

Conclusion and outlook: ethics and risk

The findings presented in this chapter are based on a 'secondary use' of
data insofar as the Eurobarometer survey has not been designed for the
kind of analysis that we have performed. Against this background it is
interesting, and significant, that a 'blue'/'green' segmentation is found in
fifteen out of the seventeen European countries included in the survey.
The assumptions characterising the two segments have been found to
be consistent across the analysis of answers to the open-ended question
('What comes to your mind when you think about modern biotechnology
in a broad sense?'), as well as in cross-tabulations with other variables in
the survey. This adds credibility to our general findings regarding segmentation. It is also noteworthy that the 'blue'/'green' pattern is found across
Europe in countries with large differences in terms of both the strength of

resistance to modern biotechnology and the history of biotechnology debates. This again points to the importance of the traditionalist/modernist segmentation in relation to attitudes to modern biotechnology.

The prevalence of the 'blue'/'green' segmentation is also unaffected by the considerable socio-cultural differences between northern and southern Europe. There are, however, differences regarding the specific features of the segmentation between the Protestant north and the Catholic south which reflect these socio-cultural differences, most notably the greater importance of the 'blue' segment in Catholic countries, and the significance religion has for the clustering pattern.

The implications of the 'blue'/'green' segmentation for political debate are clearly important. The political discourse on the question of biotechnology has concentrated on 'value-free', scientific, risk arguments, and this discourse has underpinned decisions vis-à-vis the regulation of modern biotechnology in its agricultural and industrial applications. Yet this risk discourse has, at most, an appeal to the 'green' sceptics. As a potential basis for a 'social contract' of biotechnological regulation it obviously overlooks the 'blue' opposition. This blindness to an important element in the resistance against new biotechnologies points to a major weakness in the current regulation of biotechnology.

It would probably be misleading to suggest that the scientifically oriented risk discourse is generally attractive to the 'green' sceptics. Attempts to demonstrate a stronger emphasis on moral acceptance among the 'blue' sceptics by means of statistical analysis have been unsuccessful; for 'green' sceptics, moral acceptance is also a strong predictor of their willingness to endorse applications of modern biotechnology. But there is no doubt that 'green' sceptics generally understand the far-reaching consequences of the interventions in nature that may follow from the application of modern biotechnology and, therefore, they are likely to reflect on the necessity of setting limits to these interventions. They tend to assume this position not as a sacred principle, as is the case with the 'blue' sceptics, but rather because of rational and pragmatic considerations concerning our responsibility to our descendants. The growing orientation towards ethical arguments, which is not a special characteristic of the 'blue' opponents, may also in a broader sense be linked to the emergence of reflexive modernisation, or, in the words of Giddens, 'The ethical issues which confront us today with the dissolution of nature have their origin in modernity's repression of existential questions. Such questions now return in full force and it is *these* we have to decide about in the context of manufactured uncertainty' (Giddens, 1994: 217).

Thus, an orientation towards ethical questions in relation to modern biotechnology may be a characteristic of both modernists and traditionalists, even though the specific nature of that orientation is obviously qualitatively different for the two groups. Furthermore, risk in itself has an ethical dimension that is not reflected in the scientific, expert-oriented, risk discourse, but may be apparent to laypeople when considering the far-reaching and partly unknown risks associated with the application of a new technology.

Appendix

The segmentation of the opponents to biotechnology and genetic engineering was established by means of a cluster analysis (SPSS K-means cluster analysis[5]). Through an optimisation procedure, relatively homogeneous groups of cases are identified on the basis of selected characteristics. The number of clusters must be specified. The cluster analysis begins by using the value of the first k cases in the data file as temporary estimates of the cluster means, k being the number of clusters specified by the user. The final cluster centres are found through an iterative process.

The variables used to characterise the clusters were the following:

- *Gender*
- *Age*
- *Education*: respondents are divided into three categories based on the age when they left full-time education
- *Urban/rural*: a country-specific variable indicating either degree of urbanisation or size of town
- *Religiousness*: is self-reported. In this cluster analysis the answers were divided into four categories ranging from 'very religious' to 'not religious'
- *Political orientation*: is self-reported on a ten-point left–right scale
- *Materialism/post-materialism*: as an alternative to the Inglehart index, which could not be constructed on the basis of the Eurobarometer questionnaire, *materialists* are identified by using the political importance attached by the respondent to 'criminality' and 'poverty' as an indicator, while *post-materialists* are identified by using the political importance attached by the respondent to 'environment' and 'racism' as an indicator. An index for materialism/post-materialism is then obtained by subtracting the indicators for materialism and post-materialism from each other

- *Knowledge*: an additive index of nine items designed to measure knowledge of relevant basic biology
- *Risk perception*: the sum for the perceived risk of six different applications of modern biotechnology measured on a four-point scale.

These variables represent a combination of social/demographic variables and variables based on relevant attitude measurements. It is important to note that the Eurobarometer questionnaire has not been designed for this type of cluster analysis and the choice of variables is thus a compromise between what was desirable and what was possible. The use of a self-constructed index for materialism/post-materialism as a substitute for the Inglehart index is mentioned above. Despite the ad hoc character of this variable it is a good discriminator for most countries, and the cluster means are consistently in accordance with the predicted difference between 'blue' and 'green' opponents. It would have been desirable also to include a variable for view of nature, but that has not been possible.

The scales of the variables that have been used are different. Therefore, the variables were standardised to z-scores before running the cluster analysis. This is the procedure recommended in the SPSS Applications Guide. Some statisticians warn against standardisation of variables for cluster analyses, because it may 'dilute' the differences between groups on the variables that are the best discriminators.[6] Alternatively, all variables could be transformed to variables with the same number of categories. This will inevitably lead to a loss of useful information. For this reason standardisation was used as the main procedure. As a control, the cluster analysis was also carried out for four of the countries with all variables converted to three-category variables (except of course the gender variable). The results of this approach showed good qualitative agreement with the results when using standardised variables.

For two countries, Austria and Finland, it turned out to be impossible to identify clusters with the 'blue' and 'green' characteristics. Furthermore, it is difficult to obtain stable solutions for these two countries. Small variations in input, for instance minor changes in the variable constructions, may lead to considerable changes in the clustering, with different variables being the most important discriminators. This seems to indicate that the data sets for these two countries are unstable with respect to the cluster analysis. It is quite likely that 'blue'/'green' segmentations of opponents to modern biotechnology exist in Finland and Austria, too, but that we cannot identify them. The cluster analysis was also carried out on the Austrian data set using the three-category variables as mentioned above, but still with negative results.

As a criterion for defining the group of opponents to biotechnology we used the answers to the question 'Do you think it will improve our way of life in the next 20 years?' Respondents who answered 'it will make things worse' to this question are defined as opponents. This group of respondents on average make up 20.2 per cent of the respondents for the EU as a whole, but ranges from 8.4 per cent in Portugal to 36.5 per cent in Austria. Generally, the group is only about one-fifth of the population, even though this result may be influenced by the 'split ballot' questioning. An alternative criterion might be based on the answers to questions regarding the willingness to endorse six different applications of modern biotechnology. An aggregated measure of 'willingness to encourage' the six applications leads to a different characterisation of the opponents, probably because they tend to answer differently when confronted with specific applications of biotechnology. Furthermore, the question about whether biotechnology or genetic engineering in general is likely to improve our way of life is part of a battery of similar questions about other major technologies and the context for this question is therefore different.

For all countries, the group of opponents to modern biotechnology defined by willingness to endorse the six applications is also larger than the group of opponents who expect biotechnology or genetic engineering to make things worse in the next twenty years. For this reason, too, it was considered appropriate to perform cluster analyses using 'willingness to encourage' the six applications as the criterion for defining opponents. This was done for four countries, and in every case a 'blue'/'green' pattern of clustering was obtained. Although it was quantitatively different from the solutions found by using the other criterion for defining the opponents, which is not surprising since the groups are different, this result indicates that the 'blue'/'green' segmentation is detectable despite such variations in the definition of the group of opponents. But these different possibilities of defining the group of opponents and the quantitative differences in the resulting clusters are also warnings against an overly detailed interpretation of the results.

NOTES

1. Some sections of this article have been reused in the present chapter.
2. Corresponding analyses of Norwegian data have been reported in two articles by Torben Hviid Nielsen (1997a, 1997b). The small differences compared with the results reported in this chapter are due to the use of 1993 data in the first article and the use of a different indicator for materialism/post-materialism in the second article. Hviid Nielsen used the Inglehart index for materialism/post-materialism on the Norwegian data. Since this indicator was

not available in the Eurobarometer survey for the EU countries, a substitute indicator had to be constructed, as explained in the appendix.
3. We are indebted to Timo Rusanen for comments on the Finnish situation. See also Rusanen et al. (1998).
4. The nine countries are Denmark, Norway, Sweden, Germany, the Netherlands, the UK, Italy, Spain and Greece.
5. See SPSS Base 8.0, Applications Guide, 1998.
6. See, for instance, Everitt (1993).

REFERENCES

Bud, Robert (1993) *The Uses of Life. A History of Biotechnology*, Cambridge: Cambridge University Press.
Ernst & Young (1997) 'European Biotech 97: "A New Economy"', *The Fourth Annual Ernst & Young Report on the European Biotechnology Industry* (April).
Everitt, Brian S. (1993) *Cluster Analysis*, 3rd edn, London: Edward Arnold.
Freeman, Christopher (1995) 'Technological Revolutions: Historical Analogies', in Martin Fransman, Gerd Junne and Annemieke Roobeck (eds.), *The Biotechnology Revolution?* Oxford: Basil Blackwell, pp. 7–24.
Giddens, A. (1994) *Beyond Left and Right: the Future of Radical Politics*, Cambridge: Polity Press.
INRA (1991) Eurobarometer 35.1, 'Opinions of Europeans on Biotechnology in 1991'.
    (1993) Eurobarometer 39.1, 'Biotechnology and Genetic Engineering: What Europeans Think about It in 1993', October.
    (1997) Eurobarometer 46.1, 'Les Européens et la Biotechnologie Moderne', Draft, February.
Mae-Wan Ho, Hartmut Meyer and Joe Cummins (1998) in *The Ecologist* 28(3): 146–53.
Nelkin, Dorothy and M. Susan Lindee (1995) *The DNA Mystique. The Gene as a Cultural Icon*, New York: W.H. Freeman.
Nielsen, Torben Hviid (1997a) 'Modern Biotechnology – Sustainability and Integrity', in Susanne Lundin and Malin Ideland (eds.), *Gene Technology and the Public. An Interdisciplinary Perspective*, Lund: Nordic Academic Press.
    (1997b) 'Behind the Color Code of "No"', *Nature Biotechnology* 15 December: 1320–1.
Rifkin, Jeremy (1998) *The Bioetch Century*, New York: Jeremy P. Tarcher/Putman.
Rusanen, Timo et al. (1998) 'Finland', in John Durant, Martin Bauer and George Gaskell (eds.), *Biotechnology in the Public Sphere. A European Sourcebook*, London: Science Museum, pp. 43–50.
Time (1998) 'Alien Seed?' 24 August.
Toulmin, Stephen (1958) *The Uses of Arguments*, Cambridge: Cambridge University Press.
Toulmin, Stephen, R. Reicke and A. Janik (1979) *An Introduction to Reasoning*, London: Macmillan.

7   The structure of public perceptions

Cees Midden, Daniel Boy, Edna Einsiedel, Björn Fjæstad, Miltos Liakopoulos, Jon D. Miller, Susanna Öhman and Wolfgang Wagner

Social surveys are usually conducted under the assumption that people have attitudes and that the survey taps into something called 'public opinion'. The percentages of respondents agreeing or disagreeing with a particular statement then acquire the status of a social fact. However, too much weight may be ascribed to such facts, particularly when the issue is either unfamiliar to respondents or of low salience. Since biotechnology was novel and unfamiliar to most people in 1996, it is reasonable to question the validity of survey data on this issue. Thus, in this chapter, we employ survey responses in asking the question: 'Can we talk of a public opinion on biotechnology that is based on well-formed attitudes and drawn from existing knowledge and values?'

Our analysis is based on the Eurobarometer survey of 1996, which was carried out in all European Union countries, as well as in Norway and Switzerland. The main purpose of this chapter is to improve our understanding of the attitudes expressed by examining their nature and strength and by analysing underlying factors that might explain the observed attitudinal differences. The chapter proceeds from a description of the general and specific judgements on biotechnology elicited by the survey, identifying the general level of acceptance and differences in judgements between applications and between nations. After this descriptive analysis, we will evaluate various aspects of the attitudinal data: the consistency of expectations among the different countries involved in the survey; the relations between differing attitudes and the particular applications of biotechnology; relations between such factors as 'informedness' and perceptions of this technology; the effect of more general attitudes towards technology per se; the range of attitudes towards biotechnology; and, finally, the roles of knowledge and values in determining public attitudes.

Attitudes in Europe

Biotechnology was invented and developed in the context of a Western culture that is, in general, favourably disposed towards inventions and

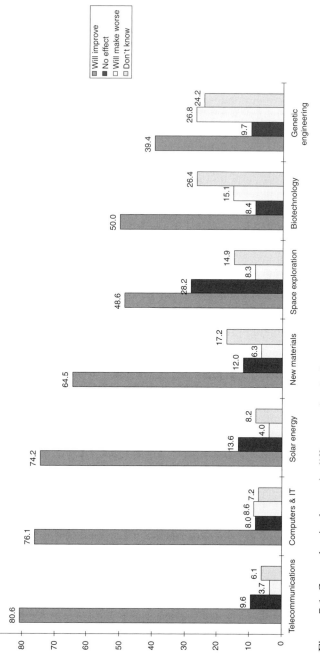

Figure 7.1 General attitudes to six different technologies

new technologies. In order to locate the position of this new technology relative to other technologies, we asked the respondents to rate six different technologies according to whether they thought they would improve our way of life in the next twenty years, have no effect, or make things worse.

As can be seen in figure 7.1, two-thirds or more of the European respondents are optimistic (in declining order) about telecommunications, computers and information technology, solar energy, and the development of new materials and substances. A mere 3 to 8 per cent are pessimistic. Half of the respondents are also optimistic about space exploration and, again, very few are pessimistic. Questions relating to our target technology, modern biotechnology, were formulated in two different ways. Half the sample were asked about biotechnology, the other half were asked about genetic engineering. The latter received the lowest score in terms of optimism (39 per cent) and the highest score in terms of pessimism (27 per cent) of the entire range of technologies cited. For both wordings, about a fifth of EU citizens expect modern biotechnology to make things worse; as many as a quarter don't know. Italy, Spain, Portugal and Belgium contain the highest percentages of optimists (all over 50 per cent), and the highest proportions of pessimists were found in Austria, Denmark, Norway, the Netherlands and the United Kingdom (all over 25 per cent). The tendency to answer 'don't know' is relatively high in some countries, especially Greece (47 per cent) but also Ireland and Portugal (33 and 31 per cent, respectively), indicating that in these countries unfamiliarity with biotechnology and genetic engineering is especially prevalent. The lowest proportions of 'don't knows' were reported from the Netherlands and Denmark (both 16 per cent).

Applications of genetic engineering

Genetic engineering comprises a number of biotechnological procedures used in a wide variety of applications. Thus, in order to refine our attitudinal measurements, we identified six different applications and asked respondents to judge them according to four criteria with relevance to the European debate: the usefulness to society of each application; how risky they are; their moral acceptability; and the extent to which their further development and use should be encouraged.

The mean results are reported in figure 7.2 on a five-point scale running from +2 to −2. The value +2 indicates maximum usefulness, minimum perception of risk, maximum moral acceptability, and maximum encouragement. Overall, EU citizens are clearly in favour of the two medical applications: the development of vaccine and drug-producing bacteria and

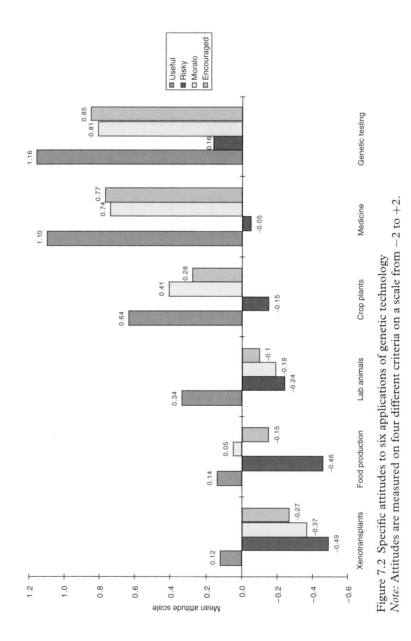

Figure 7.2 Specific attitudes to six applications of genetic technology
Note: Attitudes are measured on four different criteria on a scale from −2 to +2.

the testing for hereditary diseases. To a lesser degree, Europeans would also encourage introducing foreign genes into crop plants to protect them against pests. The remaining three applications (xenotransplantation, genetically modified (GM) food and laboratory animals) all registered close to, or below, the zero line. Least approval is accorded to the breeding of GM animals for xenotransplantation.

All six applications are judged as being useful, but to varying degrees. Again, the two medical applications and GM crops are considered to have the highest utility. These three applications are also evaluated as being the most morally acceptable. Least morally acceptable are xenotransplants, which are also seen as being most risky, together with the use of modern biotechnology in food production.

To permit comparison of the attitudes in the different EU countries, we have calculated the mean 'encouragement' scores for the six applications for each country (see figure 7.3). Most inclined to encourage the use of genetic technology is Portugal, followed by the other southern countries of the EU and the one furthest to the north, Finland. The common characteristic of these countries would seem to be their peripheral location to central Europe. The middle group comprises the western–central parts of the EU; and below them, in terms of the level of encouragement expressed, are a group of countries in the northern half of the EU. Finally, Austria registers the lowest level of encouragement. Thus, with a few notable exceptions, there is a clear south–north dimension to attitudes towards biotechnology and genetic engineering.

Expectations of genetic engineering in general

Modern biotechnology has been hailed by some as a panacea for a wide array of world problems, and by others as a threat to both humanity and nature. In order to gauge the variety of these expectations, ten possible consequences of genetic engineering were presented to the respondents, who were then asked to judge whether they were likely or unlikely to materialise within the next twenty years. The results are presented in table 7.1.

The perceived likelihood of genetic engineering bringing about both good and bad effects is striking. A majority of EU citizens believe that this technology will be able to cure most hereditary diseases, while, at the same time, two out of three hold that it will probably create dangerous new diseases. Many, but not a majority, are also optimistic about impacts on the environment and on world hunger, and are sceptical of the notion that babies will soon be made to order.

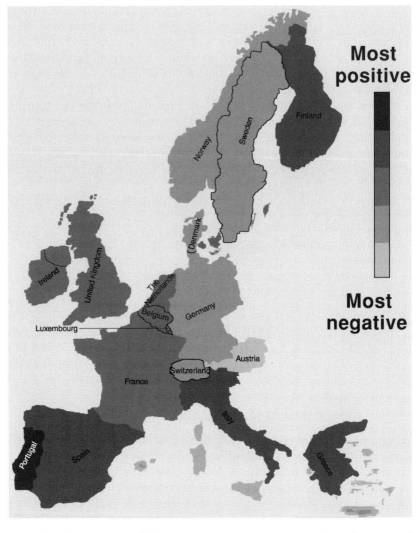

Portugal	0.61	Belgium	0.38	Denmark	0.10
---------------		France	0.37	Sweden	0.07
Spain	0.52	United Kingdom	0.30	Germany	0.05
Finland	0.49	Luxembourg	0.25	Norway	−0.09
Greece	0.47	Ireland	0.25	Switzerland	−0.16
Italy	0.42	The Netherlands	0.25	------------------	
---------------		----------------------		Austria	−0.37

Figure 7.3 Mean aggregate attitudes across Europe

Table 7.1 *Positive and negative expectations about biotechnology*

	Likely	Unlikely
Positive consequences		
Solving more crimes through genetic fingerprinting	71	19
Curing most genetic diseases	56	33
Substantially reducing environmental pollution	47	46
Substantially reducing world hunger	37	56
Negative consequences		
Creating dangerous new diseases	68	21
Allowing insurance companies to ask for genetic tests	43	42
Producing designer babies	40	50
Reducing the range of fruit and vegetables we can get	28	60
Other consequences		
Getting more out of natural resources in the Third World	54	32
Replacing most existing food products with new varieties	45	43

Table 7.2 *The positions of European countries in terms of evaluative consistency*

Evaluative consistency	High negative expectations	Low negative expectations
High positive expectations	SP, P, GR	
Low positive expectations	A	A, NL, S, D

Note: A = Austria; D = Germany; GR = Greece; NL = the Netherlands; P = Portugal; S = Sweden; SP = Spain.

It would be reasonable to assume that respondents who consider positive outcomes likely would also tend to see negative predictions as unrealistic, and vice versa. And to some extent this is borne out by our data. A quarter of Europeans belong to the positive group, and 14 per cent to the negative group – close to 40 per cent in total. Table 7.2 shows the national differences.

Congruent with the attitude results reported above, the lowest proportion of positive respondents (see table 7.2) was found in Austria (15 per cent). Yet, surprisingly, the country with the lowest share of negative respondents was also Austria (10 per cent). This apparent paradox is largely resolved if one recognises that the highest proportion of respondents with few expectations of biotechnology is again to be found in Austria, where close to 70 per cent belong to this group. This group of people with 'non-expectations' is by far the largest in the EU total. Other countries with a high proportion of respondents with low expectations are the Netherlands (64 per cent), Sweden (60 per cent), and Germany (58 per cent). In general, countries in the north of Europe tend to have lower expectations than those in the south, with west–central Europe located between the extremes. Ambivalence, i.e. the tendency to attribute both good and bad effects to biotechnology, is also correlated with the south–north axis, with the highest proportion of ambivalence found in Portugal, Spain and Greece (20–21 per cent), and the lowest found in Finland, the Netherlands and Sweden (all 5 per cent).

Attitude structures related to specific applications

As described above, respondents were asked detailed questions about six specific applications of biotechnology. They were asked to assess these six applications according to the criteria of utility, risk, morality and the proper level of encouragement. In this section we assess the coherence of the attitudes articulated. Where attitudes towards the six applications are more or less related, one can impute the presence of a meaningful general concept of biotechnology. A weak relationship, however, would suggest that specific applications tend to be judged more individually. In addition to this analysis of the structural characteristics of attitudes, we will seek to identify factors and predilections that underpin the level of support expressed for particular applications.

A confirmatory factor analysis (Lisrel8 procedure) was conducted in which twenty-four judgements were brought together (four characteristics per application). The analysis shows that the twenty-four judgements can be condensed into *two* main factors (Chi-square = 157.9, d.f. = 123; RMSEA = .017; 90 per cent CI = .007; .024). Factor 1 combined all the judgements on use, moral acceptability and encouragement. Factor 2 combined the six risk judgements alone. These findings carry two implications. The first is that the six applications are, indeed, judged in a highly consistent manner. The second is that there are two independent components to the attitudes expressed: the first comprises judgements on the proper level of encouragement integrated with judgements on moral acceptability and utility; the second concerns solely judgements as to perceived risk.

Overall, these results demonstrate that respondents employ the four criteria (use, risk, moral acceptability and encouragement) as a two-dimensional structure, which is applied across the range of biotechnological applications. Thus, when thinking about risks related to medical applications, the same basis for judgement is involved as when the respondent is judging agricultural and food applications.

One consequence of this analysis is that we are able to create an aggregate measure over the six applications. A further outcome is that we can construct two attitude indicators: a combined summary score for the judgements relating to use, moral acceptability and encouragement, and one indicator for perceived risk, all based on the respondents' evaluations of the six applications.

The core attitude structure

The correlations found among perceived risk, moral acceptability, perceived utility and desired encouragement suggest that the general evaluative reaction towards the six applications reflects a subjective assessment including each of these criteria, with perceived risk forming a separate dimension. Yet this does not necessarily imply that these factors are integrated, by respondents, in a thoughtful manner, since intuitive responses might just as well comprise these criteria. The unrelatedness of risk to the core attitude structure is a surprising finding considering the centrality of risk and safety issues in the regulation of biotechnology. A possible explanation is that the risk debate, which usually consists of technical and scientific arguments about the effects of processes and products on health and the environment, is not well understood by the average member of the public; debate on these complex and contested issues requires specialised knowledge beyond the experience of the majority of people. However, this is not to say that the public believes biotechnology to be without risk. As we have previously observed, the overall perception of risk is high with respect to most applications of biotechnology. The present analysis, however, shows that risk perception is not related to the overall evaluation of the technology.

The core attitude – composed of judgements as to utility, moral acceptability and encouragement – also requires explication. The strong correlation among these criteria shows the appeal arguments of these types have to the wider public. Moral acceptability will tend to comprise an assessment of whether an application is either 'good' or 'bad'. Combined with the utility judgement, one has a general indication of the public's sense of the rightness of biotechnological applications and their value for society at large. The high correlations suggest that this general attitudinal expression should be interpreted not as the result of a trade-off among various judgements, but as the outcome of a fundamental, irreducible impression.

This conclusion forms the entrée into the next stage of our analysis, in which we try to explain this core attitude component, as well as the perceived risk component. The objective is to develop a structural model that describes the determining factors of the attitudes expressed. This model will take into account the interconnections among the criteria identified, and their relative importance in affecting respondents' judgements as to the overall acceptability of biotechnological applications.

Modelling attitude-influencing factors

It seems unlikely that public evaluations of the six applications were usually based on a rational process of judgement based on an elaboration and integration of relevant beliefs. In view of this, we hypothesised that more general attitudes play a role in the formation of attitudes towards specific applications. In particular, we anticipated that general expectations about biotechnology would, in general, be employed by respondents where their knowledge of particular applications was ambiguous and/or incomplete. Beyond this, even more general perceptions of technology per se might be activated and applied to more specific technologies and applications.

This expectation is consistent with the notion of schematic processing (see, for example, Smith, 1998). A schema can be understood as a knowledge structure represented in the form of memory that is related to a certain class of objects. Schemas are based on earlier experiences and help a person to understand the world by linking new information to existing knowledge. In this way a person can make use of earlier experience and does not have to conceptualise new objects and situations *de novo*. Thus, schematic processing refers to a process of attitude formation in which objects are judged not according to object-specific observable characteristics, but with reference to characteristics that are connected to a general category of objects. This may occur if the specific object at hand is categorised as being connected to an existing category. Schematic processing is, therefore, an efficient means of apprehending a new concept and of obviating the need for time-consuming deliberation. Indeed, it is all but essential when a person lacks the capacity to process information thoroughly because of, for example, time constraints, distractions or a lack of required basic knowledge. Schematic processing is likely to be the customary way of forming judgements. Object-specific processing, on the other hand, requires not only cognitive capacity but also the motivation to process, and this latter factor will be contingent on the degree of personal relevance the individual ascribes to the object (see, for examples, Eagly and Chaiken, 1993). Consistent with this, we would expect the level of available knowledge and information on the topic of biotechnology to influence judgements relating to its application. Knowledge can be seen as a precondition for thoughtful object-specific judgement.

Finally, we may expect that the process of attitude formation will be different for people from different backgrounds. More specifically, we expect education to be a major determining factor, especially insofar as it may relate to levels of prior information and basic knowledge. Education might also increase an individual's sense of the importance and relevance of new technologies. Other factors are not so readily identifiable – studies in different countries report divergent results. However, the effects of age and gender will be explored.

A structural model of antecedent factors of attitudes to biotechnology

In order to test the expectations advanced in the previous section, a structural model was developed and tested applying LISREL8 (Joereskog and Soerbom, 1994). Figure 7.4 depicts a fitted LISREL solution that describes the relations among the attitude factors and the following explanatory factors: positive expectations of biotechnology in general, negative expectations of biotechnology in general, optimism or pessimism towards technology in general, level of 'informedness', and the group characteristics of education, age and gender. The model shows both direct and indirect effects upon attitudes towards biotechnology. Before discussing the relationships we must first explain the factors involved in the model.

> **Attitudes.** The attitudes are represented, first, by the core attitude component (composed of perceived utility, moral acceptability and desired level of encouragement) and, secondly, by the perceived risk component. Both components represent the aggregation of specific attitudes toward the six applications, as

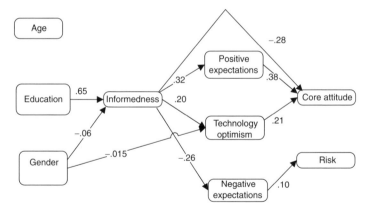

Figure 7.4 A structural model representing antecedent factors of attitudes toward biotechnology
Notes: Results of the Lisrel analysis: RMSEA = 0.26; 90 per cent CI = .014; .038; Chi-square = 38.8; d.f. 19; $R^2_{\text{core attitude}}$ = .32; R^2_{risk} = .06.

indicated by the Principal Component Analysis analysis described above.

Positive and negative expectations of biotechnology in general. As discussed above, a set of questions asked respondents to assess the likelihood of certain positive and negative outcomes of biotechnology, in general, over the next twenty years. From these items two scales were created: one for positive and one for negative attitudes (Cronbach's alpha for scale 1 = 0.68; for scale 2 = 0.66).

Optimism or pessimism about technology in general. As discussed above, respondents were asked if they were optimistic or pessimistic about six main areas of technology. These judgements (with the exclusion of biotechnology) were aggregated into an indicator representing optimism or pessimism about the future of technology in general (Cronbach's alpha = 0.66).

'Informedness'. This factor represents the level of knowledge (based on the knowledge scale as described in the section 'Knowledge and attitudes' below, and the level of awareness of biotechnology as measured by questions on having previously talked, or heard, about biotechnology). Combining these factors gave a three-level indicator of 'informedness'.

Age, education and **gender** were also included, using direct measures.

The model shows that the core attitude, reflecting perceptions of the general utility of biotechnology for society, is directly explained by three factors: the extent of positive expectations of biotechnology in general; levels of optimism or pessimism concerning the use of technology in general; and levels of 'informedness'. The pattern and direction of these effects is interesting. 'Informedness' has a direct negative effect upon general attitudes, suggesting that the more information the public has about biotechnology, the more critical its attitudes will tend to be. At the same time, however, we see that 'informedness' also exerts an influence in the opposite direction. Higher levels of knowledge and comprehension seem to foster technological optimism, and thereby this factor generates a positive effect on the core attitude. A parallel finding is that 'informedness' has a positive effect on the level of positive expectations, which also implies an indirect positive effect on the core attitude. Finally, 'informedness' is negatively correlated with public expectations of biotechnology, and thus has an indirectly negative influence on perceived levels of risk. In sum, the core attitude is influenced by positive expectations about biotechnology and by a more general technological optimism/pessimism;

'informedness' does not exert a simple effect upon attitudes to biotechnology. Its influence is mixed in the sense that knowledge of biotechnology directly promotes negative attitudes, but *in*directly it can favour a more positive core attitude and risk perception. The risk component of the attitude is directly influenced in the model by negative expectations of biotechnology.

The group characteristics (age, gender and education) have, at most, indirect effects. There is no evidence that age has any effect on attitudes towards biotechnology. Gender is weakly and negatively related to 'informedness' and to optimism or pessimism about technology in general. Education is the most relevant characteristic. It is, obviously, strongly connected with 'informedness'. Yet, the total explained variance for the risk component is very low ($R^2 = .06$). For the core attitude component the results are meaningful, albeit moderate ($R^2 = .32$).

Nonetheless, it can be concluded, first, that general ideas and beliefs about both biotechnology *and* technology were able to affect the public's attitude towards specific biotechnological applications. This suggests that respondents applied general schemas relating to biotechnology and technology to the more focused applications about which they were questioned. Secondly, we have found that 'informedness' has an important effect on attitudes. The highly informed expressed more negative views on the utility of the specific applications of biotechnology. At the same time, the model indicates that high 'informedness' is related to higher levels of optimism concerning technology *in general*, which contributed indirectly, in a positive sense, to public attitudes towards biotechnology. The highly informed also tend to have fewer negative expectations of biotechnology in general, contributing to a lower level of perceived risk.

In general, this analysis indicates that more highly informed groups are critical of specific applications. However, this critical attitude appears to be softened by positive attitudes towards technology and biotechnology in general. These findings would suggest that initially positive opinions of biotechnology, as a new technology, will recede once the public becomes better informed on the specific attributes and implications of this technology. When such information becomes available, the effect of general schemata will be reduced.

Differentiation and the strength of attitudes

The foregoing analyses have shown that attitudes among European citizens to biotechnology are unstable. This finding may be explained if we look in more detail at the strength of the attitudes expressed. New technologies are often not well known and difficult to grasp for many people; and it takes time for the public to become sufficiently well

informed to make a judgement. The meaning of attitudes differs according to the strength of an individual's convictions. Superficial attitudes are
unstable and susceptible to change in response to new information. They
are also sensitive to peripheral cues such as suggestive headlines and the
content of public relations campaigns. Furthermore, superficial attitudes
are poor predictors of people's actions. Therefore, attitudes towards new
technologies can be properly understood only if attention is paid to the
strength of the attitudes articulated.

In social psychology, numerous indicators for attitude strength have
been employed (see, for an overview, Krosnick et al., 1993), but in this
study two indicators for attitude strength have been constructed: attitude
'extremity' and attitude 'differentiation'. Extremity indicates the extent
to which respondents give answers at the extremes of the scale, on the
assumption that extreme answers reflect more convinced and certain attitudes than answers that appear in the middle of the scale. This assumption is supported by many attitude studies in social psychology (see, for
example, Eagly and Chaiken, 1993). The extremity scale is based on the
six scores on levels of encouragement deemed appropriate for the specific
biotechnological applications. A neutral answer was given a score of 0, a
score tending towards agreement or disagreement received a score of 1,
and strong agreement or disagreement was assigned a score of 2. Attitude
differentiation reflects the difference in the scores for the six applications
for each respondent. It is assumed that persons whose responses to the
six applications are differentiated have more carefully thought out attitudes. The scale is constructed by calculating the variance of the six
encouragement questions per respondent.

The frequency distribution shows that the majority of the respondents
have a moderate extremity score between 6 and 9 (range 1–12), 20 per
cent have a high extremity score and 10 per cent a low score. The differentiation index shows that about 15 per cent exhibit no variation and 66 per
cent express a variance of less than 1 (range 0.0–4.5). We can conclude
that there is a large group of respondents who have a low differentiation
in their attitudes. That people find it hard to differentiate among the applications is also suggested by the high interrelations between the specific
attitudes.

Figure 7.5 shows country differences in average attitude extremity. The
findings suggest that the countries in Western Europe whose populations seem to express opinions on biotechnology with the greatest conviction also display the most extreme attitudes. This is especially true for
Denmark, the Netherlands and Sweden. However, their scores do not
differ strongly from those of other nations. Moreover, this analysis reveals some unexpected results; in particular, Greece, with a rather low

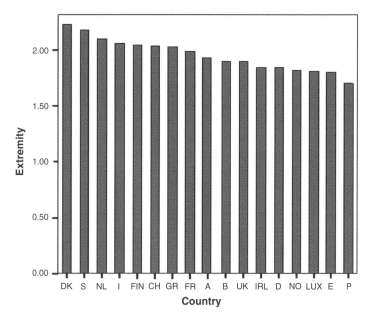

Figure 7.5 Country differences in average attitude extremity

level of public knowledge on biotechnology, also exhibits rather extreme scores. Obviously there are other factors influencing extremity. In the next section the relations between attitude extremity and knowledge will be analysed at the individual level.

Knowledge and attitudes

School education and academic knowledge are traditionally considered to be prime predictors of attitudes towards new technologies. Many scientists have the expectation that opposition to science and technology is grounded in a lack of knowledge and familiarity. Education, it is thought, may counter this alleged deficiency. From a social psychology point of view, however, one need not assume that knowledge and attitudes are related in so uncomplicated a manner. According to one alternative view, the relationship depends more on the evaluation of attributes connected to the particular technology. Knowledge may then influence the attributes that are included in the forming of an attitude, and will thus have an indirect effect upon this process (Hisschemoller and Midden, in press). In the present case we analysed the knowledge items in the 1996 Eurobarometer survey and their relationship to attitudes.

Knowledge indicators

The questionnaire comprised nine knowledge items, six of which were designed to measure 'objective' knowledge and three 'subjective knowledge'. Objective knowledge items were defined as questions whose correct answer can be found by studying a basic modern textbook on biology. These include the uses of microscopic organisms (e.g. cleansing wastewater via bacteria, and brewing beer with yeast cells), technical issues (e.g. cloning of organisms, and prenatal diagnosis of Down's Syndrome), and basic knowledge (e.g. that viruses cannot be 'contaminated' by bacteria, and that humans share a very high proportion of genes with chimpanzees).

Subjective knowledge items were defined as questions whose correct answer is either presupposed or implied by academic knowledge but cannot ordinarily be found by consulting a textbook index. Examples of such knowledge items are that all tomatoes – not only those that have been genetically altered – contain genes; that genetically modified organisms are always bigger than ordinary ones; and that genes from 'higher' species (e.g. animals) can be inserted into the genomes of 'lower' species (e.g. plants). The term 'subjective' refers to the fact that not knowing how to answer such questions makes one more prone to adopt inaccurate images of biotechnology, for example that of genes being injected into natural tomatoes to alter them genetically, a graphic illustration that was, in fact, used in some newspapers (see plate 9.1 in chapter 9).

To generate a knowledge score, the number of items answered correctly by each respondent was counted and the scores trichotomised for use in correspondence analysis. A correspondence analysis of a matrix cross-tabulating the fifteen EU-countries by the three 'objective' knowledge (OK) categories plus the three 'subjective' knowledge (SK) categories (low, medium high) reveals that both knowledge scores are highly correlated. The first dimension captures 90 per cent of variance, with low objective and subjective knowledge on one pole and high knowledge on the other (figure 7.6). The correlations between the two knowledge scores are also rather high ($r = .56$).

From this analysis, three clusters of countries appear quite clearly. At the high end of the knowledge pole are Denmark, the Netherlands and Sweden, in the middle are France, Germany, Italy and Luxembourg, and on the low end of the knowledge pole are Austria, Greece, Portugal and Spain. Finland, Ireland and the UK assume peculiar positions, outside of the principal clusters. The UK, for example, seems to be high on the scale of objective knowledge, but low in terms of subjective knowledge, whereas Finland is the reverse.

Table 7.3 *Relation between knowledge and general and specific attitudes*

	Perception of applications			
Knowledge	Useful	Risk	Morally acceptable	Encouragement
Knowledge total	.21	.20	.20	.18
Subjective knowledge	.17	.16	.17	.16
Objective knowledge	.19	.18	.18	.16

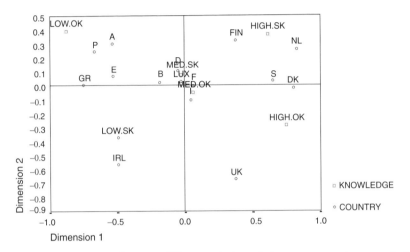

Figure 7.6 Country differences in knowledge: correspondence analysis, canonical normalisation

It should be noted that the so-called subjective knowledge items were more difficult to answer than the objective knowledge items. On average, only a third of respondents got the three subjective items correct, but the number of correct answers to the objective items ranged from 30 per cent to over 80 per cent. Neither the objective nor the subjective knowledge scores are correlated with the general attitude of the populations (optimism/pessimism) and they are only slightly (positively) correlated with perceptions of usefulness, risk, moral acceptability and appropriate encouragement. A positive correlation would indicate that the more a respondent knows, the more he or she agrees that applications are useful, risky, morally acceptable and worthy of encouragement, but this relationship never explains more than 4 per cent of the variance (table 7.3).

Knowledge and attitude extremity

It might be expected that people with high levels of knowledge would have the more definite attitudes towards biotechnology, because they will have a greater ability to process new information on this subject. Thus, we calculated the correlations between the general knowledge scale (nine items) and the extremity and differentiation indicators. Knowledge and attitude differentiation appear to be unrelated ($r = .04$). The expectation that more knowledge enhances the differentiation in judgements appears, therefore, to be invalid. Nonetheless, we found a modest, but meaningful, positive relationship between knowledge and attitude extremity ($r = .24$), which confirms that, in general, more extreme attitudes tend to be based on a higher degree of knowledge.

Values and attitudes

To what extent can attitudes to biotechnology be explained by taking the value structures of the respondents into account? Briefly put, the answer is 'not very much'. When looking at both general attitudes towards biotechnology and genetic engineering (figure 7.1 above) and a summation of the specific attitudes towards six applications (figure 7.2 above), a remarkably small percentage of the variation can be explained in terms of ordinary background variables. Entering political position (on a ten-point left–right wing scale), religiosity (on a four-point scale) and environmentalism (on a two-point scale) into a multiple regression enabled us to explain only 1 per cent and 3 per cent, respectively, of the variation in the two attitudinal measures. Some characteristics emerge as contributing significantly to this 1 per cent, mainly owing to the unusually large sample: persons with non-green values and persons to the right of the political centre tend to be more positive than those with green values and those to the left of the political centre. But, when looking at additional specific value indicators, it emerged that neither views on risk-taking nor views on the regulation of biotechnology contribute significantly to explanatory power. On the other hand, 'breeding conservatism' does. This was measured by two questions relating to the wisdom and the efficiency of relying upon traditional breeding methods. Introducing this extra variable into the regression analysis produced beta coefficients of .14 and .21 respectively, more than doubling the explained variance. However, it might be argued that this measure of 'breeding conservatism' in itself could be construed as an indicator of attitudes to modern biotechnology and genetic engineering.

Conclusions

What have we learned from our exploration of these attitudinal data? The most important and striking conclusion is that the public at large did not appear to have, as yet, well-formulated or crystallised attitudes towards biotechnology. It seems likely that the attitudes of many respondents were, at most, superficial. This chapter has underscored the crucial point that, although an issue may appear to be intensely contested, this does not necessarily mean that the public *at large* is heavily involved. Debates on political issues can be restricted (at a certain point in time) to what may be called political elites or to what have been identified as 'attentive publics' (Miller, 1983). Indeed, it would seem that only recently have clear and resilient attitudes begun to emerge, largely in response to the news coverage of new biotechnological applications and GM products entering the market.

Rather than expressing pre-existing judgements, evaluations and perceptions, many respondents will have rapidly formed attitudes when confronted with the necessity of cooperating in the survey. Although responses elicited in such a context can still be meaningful, they should be interpreted with the utmost caution. People's answers may have been influenced more strongly than is usual by circumstantial factors and by information that became available as the result of the survey process. (Yet the fact that the data show large differences between European citizens and between the various countries suggests that background factors were far from unimportant.)

How do Europeans feel about biotechnology?

Half of Europeans had few expectations about biotechnology, a trend that was stronger in southern Europe than in the north. With a few notable exceptions, there was also a south–north axis of decreasingly favourable attitudes. A quarter of Europeans expressed a basically positive attitude, and about 14 per cent a negative one; 12 per cent felt ambivalent.

At the level of content structure we find that attitudes towards biotechnology broke down into two independent components. The first, and most dominant, aspect can be understood as deriving from a recognition of the potential economic and health benefits of biotechnology, which is at least tempered by a sense of the social and moral threats attendant upon these developments. The majority of people did not seem consciously to calculate trade-offs among these various factors; instead, these aspects were conjoined at an intuitive level and produced a general reaction to the survey questions. The final response may be interpreted as an indication of how 'good' or 'bad' the general effect of biotechnology on society was considered likely to be.

The second component in our analysis constitutes the perceived level of risk. Surprisingly, this factor appears to be independent of the core attitude. Yet our modelling efforts were unable to shed light on the underpinnings or consequences of risk perceptions. Instead, we were able to find only some *effects* of risk perception upon respondents' positive and negative expectations of biotechnology in general. This suggests that people's risk perceptions were poorly formulated and were, at least partly, based on more general notions related to biotechnology.

Looking at the structure of attitudes it would seem that these are, to a large extent, uncrystallised. A majority of the respondents have a moderate extremity score and a large majority have low differentiation scores. Moreover, there was no differentiation in public attitudes among the different biotechnological applications: although different applications produced different levels of acceptance/opposition, neither the decision-making process nor the attitudinal outcome tended to be affected by the nature of the particular application.

Our structural modelling analysis shows that specific attitudes are predicted by more general evaluative expectations of biotechnology and are related to an unspecified feeling of optimism or pessimism towards technology.

We have also shown that the level of 'informedness' has interesting effects. It appears that the degree of 'informedness' has, via different paths, both positive and negative effects upon attitudes to biotechnology. People who are better informed are more sceptical about the possible benefits and moral acceptability of this technology, but at the same time they are less inclined to consider biotechnology to be high risk and they exhibit greater optimism towards technology in general.

Similarly, knowledge of biotechnology does not have a unilateral effect upon attitudes. We find that people with a higher level of knowledge express more extreme convictions, suggesting that their attitudes are more refined; but knowledge does not seem to correlate with attitude differentiation. Independently of knowledge level, we find that attitudes and the attribution of risk, moral acceptability and utility to the six specific applications are highly consistent.

Our general conclusion, that attitudes were not very well crystallised, is also reflected in the fact that attitudes towards biotechnology do not seem to have been anchored to standard value systems. We did not find strong relations between perceptions of this technology and such value orientations as political preference, religiosity and environmental consciousness.

What attitudinal changes can be expected in the near future? It seems plausible that, as biotechnology further develops, clearer attitudes will emerge. It should be noted that our data were collected before the important media events of, for example, the modified soya imports and the birth of Dolly the sheep, issues that projected biotechnology to the forefront of social debate. As it becomes apparent to more people that biotechnology is a diverse field, capable of producing a variety of very different products and services, people are likely to develop more differentiated attitudes.

The public's currently rather undifferentiated view nonetheless deserves the attention of policy-makers and companies. The introduction of new products may be hampered because these products are likely to be evaluated according to more generalised perceptions of biotechnology. Moral objections to certain applications may in this way also inhibit the development of applications that are, in themselves, highly desirable.

It seems important for us to gain a better understanding of the factors that determine perceptions of utility, moral acceptability and risk. Such an enterprise promises to be an important contribution to public policy. At the same time, our study indicates the need for well-designed and targeted information campaigns through which to inform the European public, in greater detail, about the potential pros and cons of biotechnology, and to stimulate the emergence of well-elaborated attitudes towards technology and to biotechnology in particular.

REFERENCES

Eagly, A.H. and S. Chaiken (1993) *The Psychology of Attitudes*, Fort Worth, TX: Harcourt Brace Jovanovich College Publishers.

Hisschemoller, M.H. and C.J.H. Midden (in press) 'Improving the Usability of Research on the Public Perception of Science and Technology for Policy-Making', *Public Understanding of Science* 8(1): 17–34.

Joereskog, K.G. and D. Soerbom (1994) *Lisrel8: Structural Equation Modeling with the SIMPLIS Command Language*, Hillsdale, NJ: Lawrence Erlbaum.

Krosnick, J.A., D.S. Boninger, Y.C. Chuang, M.K. Berent, and C.G. Carnot (1993) 'Attitude Strength: One Construct or Many Related Constructs?' *Journal of Personality and Social Psychology* 65(6): 1132–49.

Miller, J.D. (1983) *The Role of Public Attitudes in the Policy Process*, New York: Pergamon Press.

Smith, E. (1998) 'Mental Representations and Memory', in D.T. Gilbert, S.T. Fiske and G. Lindzey (eds.), *Handbook of Social Psychology*, Boston: McGraw-Hill.

European regions and the knowledge
deficit model

Nick Allum, Daniel Boy and Martin W. Bauer

Prototype cultures of public understanding of science

Previous chapters have explored the individual or national variation in
knowledge and attitudes to modern biotechnology. In this chapter we
will go a step further and introduce European regions as a level of comparison. Regional analysis is a speciality of geographers who map the
similarities of regions across national borders in terms of their economic
and social dynamic (e.g. Rodriguez-Pose, 1998a, 1998b).[1] The globalisation of capital and other productive factors leads to the expectation that
regional variation is levelled and geo-spatial factors in social explanation
lose their power, hence the question: do regions matter?

There are contradictory expectations about the significance of European regions in the future. Visions of European integration are tied to
a rise in the political and economic importance of regions. As a correlate to European integration, regional autonomy will be strengthened,
while the nation-state weakens. The sociological imagination concerning
the dynamics of industrial development provides some specification of
the contexts for our exploration into attitudes towards and knowledge
of modern biotechnology in the 1990s. In recent years, much has been
written about the cultural, economic, social and political changes associated with the transition from industrial to what may be termed post-
industrial or knowledge societies. In phrases such as 'post-industrialism',
'post-materialism', 'postmodernity' or 'the risk society', sociologically
minded writers have sought to evoke different axes of this transition. Post-
industrialism emphasises the dominance of knowledge-based activities
and industries (Bell, 1973; Touraine, 1969); post-materialism emphasises changing societal values (Inglehart, 1990); postmodernity emphasises a demise in the certainties about the direction of history (Lyotard,
1984); and the notion of the risk society emphasises the globalisation
of risks and their distribution (Beck, 1992; Boehme, 1993). The prefix
'post' may indicate a transitional period between the waning industrial
age and the establishment of a new social formation.

The use of the term 'post-industrial' is perhaps deliberately chosen to affirm the lack of a deterministic center by calling attention to the idea that this social formation stands in transition between more distinctive types of societies, namely the industrial society and a future social formation, not yet endowed with the same degree of specificity. (Stehr, 1994: 89)

In all of these arguments the claim is made that major structural transformations accompany the progress of industrial societies, and that this transition is likely to occur first in the most advanced industrial countries.

Macro-sociological theory concerning this transition observed as a function of economic development provides a specified context for interpreting our data on the public perception of science in general and of biotechnology in particular. The literature on public understanding of science has hitherto tried to support or reject the so-called deficit model (for example, Irwin and Wynne, 1996), which in our view expresses a common-sense idea among policy-makers rather than a comprehensive conceptualisation of the phenomenon of public perceptions. According to the deficit model, knowledge of science and technology encourages more positive attitudes. In other words, the more science you know, the more you like it. Negative attitudes towards science reflect an ignorant public, hence the notion of a knowledge 'deficit'. Empirical evidence and conceptual thinking are somewhat at odds with this simplistic idea (for example, Evans and Durant, 1995). We suggest a formulation of a model of public perception of science that focuses on varieties of perceptions of science in different contexts. For this purpose we need to characterise differences in public perception as well as differences in the contexts in which they are socially embedded. Our effort could therefore be labelled a 'specified' contextual model of public perception. The call for contextual models is not new (Wynne, 1993); what is new is the attempt to specify the contexts and to work with testable hypotheses for which evidence is available.

We reconsider a model of scientific cultures in Europe previously suggested by one of us (Bauer, 1995; Bauer et al., 1994): the post-industrialism hypothesis of public understanding of science. The model projects a two-dimensional matrix onto a set of analytic units. The first dimension represents the hypothetical transition from industrial to post-industrial societies, and the second dimension represents what may be termed the ordered deviation from the idealised path of this transition in response to local circumstances. In a sense, the first dimension attempts to capture a general process, and the second dimension attempts to take account of the particularities of place and time. We have previously tested this model with fairly good fit to national variations in knowledge, interest and attitudes to science in general (Durant et al., 2000). Here, we test

Table 8.1 *Public understanding of science in the transition from industrial to post-industrial society*

Industrial society	Post-industrial society
Scientific knowledge is confined to a small social elite	Scientific knowledge is widely distributed in society
Scientific knowledge is strongly socio-economically stratified	Scientific knowledge is weakly socio-economically stratified
Public interest in science is relatively high	Public interest in science is relatively low
Popular scientific knowledge is unified	Popular scientific knowledge is specialised
The relationship between knowledge and support is positive	The relationship between knowledge and support is increasingly chaotic
Univocality:	Ambiguity:
TO KNOW IS TO LIKE IT	TO KNOW IS TO LIKE IT; FAMILIARITY BREEDS CONTEMPT

this model on variations in attitudes and knowledge of a particular new technology and across regional variation.

The features of the hypothetical transition on the first dimension are set out in table 8.1 as a dichotomy of industrial and post-industrial society. In industrial societies, it is expected that scientific knowledge is confined to a relatively small, well-educated elite; socio-economic factors play a relatively large part in determining the distribution of scientific knowledge; interest in science (as a symbol of industrial and economic progress) is relatively high in all sections of society; a relatively unified canon of popular scientific knowledge exists; and there is a correlation between knowledge of and support for science. The image here is of a society in which science and technology, having achieved limited penetration into society, are extensively idealised as the preferred route to economic and social progress.

In post-industrial knowledge societies, by contrast, it is expected that scientific knowledge is more widely distributed; socio-demographic factors such as education, income, age or gender play a relatively small part in determining the distribution of scientific knowledge; interest in science is relatively low, because science is more generally taken for granted as part of the everyday stock of knowledge; there is a proliferation of specialist knowledges in the public domain; and the relationship between knowledge of and support for science becomes unpredictable, as different sections of the community develop different points of view on particular issues. The image is of a society in which science, having already achieved a high level of penetration into the public sphere, is no longer idealised but is critically evaluated by a public that expects to obtain continuing benefits but is also increasingly alert to the possibility of problems or disbenefits. Thus there is an apparent paradox: as the public understanding of science expands and scientific knowledge becomes more diffused, science becomes more problematic for the public.

A further refinement of the model proposed here is an extension of these hypotheses to the development of European regions. As European integration continues apace, accompanied by rising labour mobility and the development of a highly efficient communications infrastructure, it is likely that regions within nation-states will become more economically autonomous. As a corollary to this, one would expect regional political autonomy to increase, as it has done recently in the UK with the devolution of Scotland and Wales. The thesis of this chapter is that, just as we can observe differences in public understanding of science according to the stage of post-industrialisation at a national level, so we will observe the same phenomenon between European regions that are presently at different stages on this developmental continuum. This assumes that the representations of science and technology (in the present case, of biotechnology) that circulate within the public sphere are to some extent produced in relation to the local as well as the national social and economic climate. Regions that share similarities in their stage of economic development are expected to share similarities in their publics' attitudes to biotechnology, independently of the country in which they are located.

Needless to say, this model of transition of the structure of public understanding of science is a simplification. To characterise different scientific cultures adequately, more dimensions are necessary. However, radical simplification may be justified by its heuristic value of generating testable hypotheses. A second limitation of our argument will be our data. Whereas the hypotheses are longitudinal in nature, our data are cross-sectional. We are assuming a uniform transition from industrial to post-industrial mode, and any moment in time will show how countries and regions are differentially advanced in this transition.

The relationship between knowledge and attitudes

In order to measure the relationship between attitudes towards biotechnology and the level of knowledge, we use an indicator that comprises nine knowledge items, and a combined indicator across six applications of biotechnology (see Durant et al., 1998). For each application, responses regarding usefulness, moral acceptability and encouragement were coded as +2 (strong agreement) or +1 (agreement), and disagreements were coded respectively as −1 or −2; non-responses coded as 0. The assessment of risk of any of these applications was coded in the following way: perception of risk was coded −2, denial of risk was coded +2. The final index is the sum of individual responses across six applications and four criteria of judgement. Our indicator of post-industrialism is GDP per capita in purchasing power parities in 1996.

For Europe as a whole, measured at the level of the individual, the relationship between knowledge of and attitudes to biotechnology is positive but weak, although statistically significant: Pearson's r is .14 ($p < .01$). Europe-wide, knowledge explains less than 2 per cent of the variance of attitudes towards biotechnology. This would be consistent with current social psychological reasoning, according to which knowledge is not a predictor of attitudes but an indicator of their quality: attitudes based on knowledge are more elaborated and hence more resistant to change (Eagly and Chaiken, 1993).

This low correlation might be taken as an indication that, at the level of Europe as a whole, the post-industrial mode of public perception prevails; a substantial relationship between knowledge and attitudes is not present. However, we know that from north to south and east to west, European countries vary considerably, and the characteristics of a post-industrial condition are not equally distributed. How does the relationship between knowledge and attitudes vary, taking into account the different developmental contexts?

National-level analysis

Country-by-country analysis of the correlation between knowledge and attitudes shows considerable variation. How can we explain this variation? Can we verify our post-industrial hypothesis, according to which the correlation is high for countries close to an industrial mode and low for countries close to a post-industrial mode?

Our data, aggregated to mean country levels, show a negative correlation between GDP and the attitude score ($r = −.48$, $p = .05$) and a positive correlation between knowledge and GDP ($r = .43$, $p < .10$). Figure 8.1 plots the within-country correlation of knowledge and attitudes against GDP per capita. The figure clearly shows that the knowledge–attitude correlation is stronger in countries where GDP is lower, for example Greece, Spain and Portugal. This link is much weaker in the wealthier nations, for example Switzerland and Luxembourg. In between these extremes is a band of richer countries between which GDP per capita is not a good discriminator (Finland, the UK, Sweden, Italy and France).

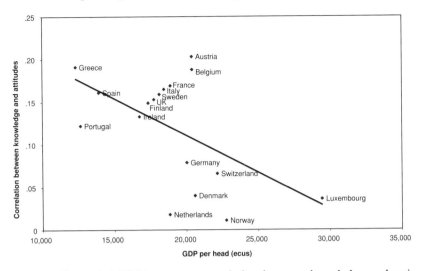

Figure 8.1 Within-country correlation between knowledge and attitudes to biotechnology in relation to GDP per head
Source: EUROSTAT 1996.

It is important, of course, that we do not make the 'ecological' error of inferring from these aggregated data that a different cognitive mechanism operates at the level of the individual in different countries (although that cannot be ruled out). However, assuming that the level of GDP (1996 in purchasing power parities) is a valid measure of the transition from an industrial to a post-industrial mode of the economy, this is exactly what our hypothesis predicts. In post-industrial mode the relationship between knowledge and attitudes becomes more complex when we make comparisons at the societal level.

Looking at figure 8.1, we observe that some countries are outliers in the linear model. Austria and Belgium have a higher correlation than their level of development might suggest, while, on the other hand, for the Netherlands, Norway and Denmark the correlation is well below expectations. In this middle range of development, the variance between countries is very large. To explain this variation we have to consider the second, as yet unspecified, dimension of our model: explaining the variations in the post-industrial model according to local circumstances.

A possible factor could be the status of the public debate on biotechnology in these countries. The Netherlands and Denmark have had intensive public controversies on biotechnology (see Jelsoe et al., 1998; Midden et al., 1998), whereas Austria and Belgium had not had significant public debates by the end of 1996 (see Wagner et al., 1998). We therefore conclude that our post-industrial model needs to include an auto-catalytic factor: public controversies. In the aftermath of public controversies over a new technology, which in themselves are likely to be a feature of post-industrial societies, the correlation between knowledge and attitudes disappears. Public controversies enhance the circulation of new knowledge in society, while not prejudicing attitudes in post-industrial mode (see chapter 5).

Multilevel regression analysis

As we have seen, there are significant differences between European countries in the relation between textbook knowledge of biology and the evaluation of a set of biotechnological applications when looking at aggregated data within the sample. It also appears that the pattern of correlations between knowledge and attitudes aggregated at the regional level is rather weak and not statistically significant in most cases. Aggregating the data tells an ambivalent story, and another form of analysis is required to disentangle the complex relationships that exist within and between different analytic levels. A suitable form of analysis for this problem is hierarchical multilevel linear regression (Goldstein, 1995; Bryk and Raudenbush, 1992). The multilevel approach treats individual cases as existing within hierarchical levels that correspond to real world situations (where individuals live in regions and those regions are situated in countries). Using this approach, no aggregation of data is necessary because the estimation procedure takes into account the hypothesised similarities between individuals within the same higher-level units and provides measures of the relative importance of factors at each level that affect the outcome variable.

A multiple regression model was set up with knowledge as the dependent variable. Attitude towards biotechnology appears alongside several new variables as predictors. The assumption here is not that this is a causal model, with attitudes driving knowledge, although this is, of course, one of the possibilities. We are viewing the set of relationships between these variables simply as associations, not causes and effects. Our primary interest is to explore the distribution of variations in these associations according to the various hierarchical levels in the analysis – individual, regional and national.

New variables

The number of people living in a household is known to correlate positively with the number of school-age children in the household. This

Table 8.2 *The variables used in the analysis of the multilevel model groups*

Level	Variables
Dependent variable (individual level)	Textbook knowledge about biotechnology
Level 1 (individual level)	Attitude towards biotechnology (negative to positive)
	Number of people in the household
	How much the respondent has talked about biotechnology in the past
Level 2 (regional level)	Long-term unemployment in the region (percentage of working population)
	Ratio of tertiary-sector employment to total employment in the region

simply reflects the fact that most households comprising more than two people are those in which there are families with children. The type of knowledge that is measured in the present survey – knowledge of a 'textbook' nature – would be more likely to circulate within a family that contains school-age children whose education will generally include basic biology (see table 8.2).

The extent to which someone has talked about biotechnology in the past can be taken as a measure of active engagement with the subject. One might consider it as a type of informal, subjective knowledge insofar as to talk about something, or simply to refer to it, involves conceptualising it in some way, anchoring it in a familiar frame of reference. Hence this variable is expected to be positively associated with the formal 'textbook' knowledge measured.

Individual, family and household relationships can be constituted in a multitude of ways. That said, there is less heterogeneity between individuals and groups of whatever kind who occupy the same social milieu. In the present case, we expect to see that the way knowledge about and attitudes towards biotechnology relate to the type of household, and the extent to which the topic is discussed, are mediated to some degree by the socioeconomic climate of the region as well as the country. The general thesis of this chapter is that the relationship between attitudes to and knowledge about science, and, in this case, biotechnology, changes according to the degree of post-industrialisation. While this is usually conceptualised at the level of the country, here the aim is to see if these differences are also present according to the socio-economic characteristics of regions, independently of their national location.

Two variables are used as economic indicators for the European regions. The proportion of long-term unemployment in the region is assumed to be a salient feature, impacting as it does on the types of economic opportunities that its residents enjoy, and most likely on attitudes of optimism and pessimism about future prospects. As far as the post-industrialisation thesis is concerned, the level of long-term unemployment in any given region may signal transition towards post-industrialisation or, alternatively, economic stagnation. Thus the direction taken by any association between long-term unemployment and knowledge is an interesting empirical question. The proportion of employment in the tertiary sector of the economy (service sector) is another of the indicators of post-industrialisation. This measure is by no means perfect. There are a few regions in the sample where there is a strongly developed service sector but very little industrialisation and a heavy reliance on agriculture. These are primarily tourist areas in Greece, Portugal and Spain and do not resemble post-industrialised economies. Nevertheless, from the data available, it is the single best proxy measure for post-industrial economic development.

Method

The economic data were taken from the European Commission's statistical service Eurostat. A reduced sample of 9,228 respondents was used for the analysis owing to missing data in the Eurostat data set. This left nine countries and fifty-one regions covered.

The models presented here are 'variance componence' models with up to three levels – individual, regional and country. In a single-level analysis, at the individual level, the assumption in a standard regression model is that each individual's score on the dependent variable is independent from all other individuals' scores. However, we hypothesise here that individuals share similarities in their knowledge about biotechnology according to the area in which they live. We would expect there to be similarities between respondents from the same region and country that remain undetected in a single-level model. One way to approach this would be to add a separate term for each country and region. This is clearly inefficient in that a huge number of coefficients would need to be estimated. Another approach is to aggregate data to the regional or country level, taking the mean score on all the variables over all individuals in the region or country. Although this is often done, and with useful results, any causal explanation at the level of the individual is not strictly warranted by discarding individual-level information, which is what aggregation does. Also, it is a well-documented statistical phenomenon that correlations tend to be exaggerated and can even change their sign when aggregated data are used (the so-called 'ecological fallacy').

The variance componence models presented here assume that the regions and countries (levels 2 and 3) are a sample of a possible population of regions and countries. The variation between countries and between regions is modelled as random variance between regression intercepts. The general model is of the form:

$$y_{ijk} = a_{jk} + b + e_{ijk}.$$

There are random disturbance terms for each level and these are conceptualised as variance at each level. As the level 2 and 3 (j and k) variance is random only for the constant, a, in the model, one can visualise the result as a set of regression lines having the same slope but different intercepts corresponding to each region and country. If no systematic differences exist between individuals in different regions and countries, we would not expect the intercepts of the regression slopes for each of these to vary significantly. The converse is to think of this as demonstrating that there are similarities between individuals within these regions and countries. If we see significant variance between regions and/or countries (undetected heterogeneity), we can then go on to add our regional (level 2) variables to the model, and to see how much, if any, of that heterogeneity can be accounted for by these economic indicators. This would be observed as a reduction in the unexplained variance at that level.

Results

Full details of each model are reported in the appendix to this chapter.

Stage 1: two-level model with individual-level predictors

All three independent variables have a significant effect on level of knowledge, all in a positive direction. Whether a respondent had talked about biotechnology to anybody prior to the survey is the single best predictor of their textbook knowledge about it. The more they had talked about it, the better is their knowledge ($r = .64$). There is a small association between the number of people in a respondent's household and their knowledge ($r = .15$). There is also a positive association between knowledge and attitude. Those with a favourable attitude towards biotechnology are likely to know more about it. The variance between regions (see table 8.3) is significant at $< .05$. Approximately 13 per cent of the variance in people's knowledge is associated with unknown factors (outside of the model) that are active within regions.

Table 8.3 *The variance explained in the multilevel models*

Model	Level 1 (individuals)		Level 2 (regions)		Level 3 (countries)	
	δ_e^2	% variance	δ_u^2	% variance	δ_v^2	% variance
Model 1	3.285	87.4	0.472	12.6	–	–
Model 2 (with level-2 variables)	3.359	92.2	0.284	7.8	–	–
Model 3	3.361	89.7	0.088	2.3	0.297	7.9
Model 4 (with level-2 variables)	3.363	91.8	0.053	1.4	0.248	6.8

Stage 2: two-level model with individual- and regional-level predictors

In this model, two variables at the regional level have been introduced. Both are significant at < .05. Higher tertiary-sector employment is associated with higher knowledge about biotechnology ($r = .22$). Higher long-term unemployment is associated with lower knowledge. The magnitude of the other predictors' coefficients is slightly reduced with the economic data added in. Overall, this model is significantly better at explaining changes in knowledge than the previous one (reduction in $-2^* \log$ likelihood: 19976; 2 d.f.).

The most interesting difference between models 1 and 2 is the difference in between-region variance. In model 1, between-region variance was approximately 13 per cent of the total variance. In model 2, it has fallen to approximately 8 per cent (.28). A significant proportion of the unexplained regional-level heterogeneity can be explained by the employment conditions in the area.

Stage 3: three-level model with individual-level predictors

To test whether these regional effects are disguised national effects, a further level was added. Model 3 shows the partitioning of variance across three levels – individual, regional and national – with only the level-1 independent variables, as in stage 1. Here we can clearly see that heterogeneity between countries accounts for most of the variance attributed to the regions in model 1. Approximately 8 per cent (.30) of the unexplained variance is between countries, while only 2 per cent (.09) is now accounted for by the regions. However, it is important to note that, even when country-level variance is modelled, regional effects are still significant.

Stage 4: three-level model with individual- and regional-level predictors

To complete the analysis, our level-2 predictors are introduced into the model. The addition of these reduces the amount of between-region and between-country variance compared with the model at stage 3. Reduction in between-country variance is approximately 16 per cent between stages 3 and 4, while about 40 per cent of between-region variance is concurrently reduced. This demonstrates that the economic conditions in regions are still correlated with inhabitants' knowledge even after controlling for country-wide similarities.[2] These results are summarised in table 8.3.

This analysis supports the hypothesis that regional economic factors are significant in predicting people's knowledge about biotechnology. It also suggests, unsurprisingly, that most of the unexplained factors that affect people's knowledge operate at the level of the individual. Which country a respondent comes from is more important than which region they live in, when considering all nine countries. We might therefore want to say that national cultures are better predictors of an individual's knowledge about biotechnology than regional cultures are. But there is still a significant variation in people's knowledge according to the region in which they live. A substantial proportion of this is associated with differences in local economic conditions.

Regional clusters of public understanding of biotechnology

We have thus far identified a statistical relationship between GDP and attitudes towards biotechnology, and between GDP and knowledge about biotechnology across European regions. We have also presented estimates for the significance of regional variation in explaining people's cognitive involvement with biotechnology. We now focus on the regional level of analysis to explore the structural features that determine people's engagement with modern biotechnology beyond the socio-political unit of the nation-state.

For this purpose we conducted a cluster analysis of European regions. The variables used were regional means on the following measures: people's general level of education measured by school-leaving age, their knowledge of, and attitude to, biotechnology, and the frequency of their previous conversations about biotechnology. The analysis yielded

Table 8.4 *Some European regions classified by economic dynamism*

Capital	Industrial declining	Intermediate	Peripheral
Berlin (D)	Wales (UK)	East/West Midlands (UK)	Sardegna (I)
Brussels (B)	Wallonie (B)	Emilia Romagna (I)	Galicia (E)
Île de France (FR)	Nord Westfalen (D)	Anatoliki (GR)	Northern Ireland (UK)
Madrid (E)	North West (I)	Mediterraneé (FR)	Ipeiros (GR)

Source: Rodriguez-Pose (1998a,b).
Note: B = Belgium; D = Germany; E = Spain; FR = France; GR = Greece; I = Italy; UK = United Kingdom.

a three-cluster solution.[3] In region type A, people have high knowledge and education, have talked only moderately about biotechnology and have moderate attitudes to biotechnology. In region type B, people have moderate levels of education and knowledge of biotechnology, have talked about it a lot and have rather negative attitudes. Finally, in region type C, people have lower levels of general education and knowledge and have not talked much about biotechnology but hold very positive attitudes towards it.

The second classification considers the socio-economic structure and dynamism of the regions as shown in table 8.4. Here we follow the regional classification developed by geographers in the context of the debate on regional convergence (Rodriguez-Pose, 1998a, 1998b, 1999). European regions are classified by level of GDP in 1996 and their rates of economic growth over the period 1980–96, standardised within each country. European regions face the secular challenge of globalisation of capital, speed of innovations and new production technologies in different ways. Rodriguez-Pose identifies four types of regional responses to these challenges. Response pattern I is characteristic of urban capital regions, world cities or megalopolises. GDP is above the European average and there is a high concentration of service industries, in particular financial sectors. Response pattern II is characteristic of declining industrial areas. Traditional heavy industries are concentrated here, with over 33 per cent of employment in the secondary sector, and major difficulties in changing the production system. Response pattern III defines intermediate areas that are neither urban centres nor old industrial or agricultural areas, but show clear economic dynamism often based on small enterprises. Finally, response pattern IV comprises the peripheral regions of Europe where the remaining agricultural employment is concentrated. Economic

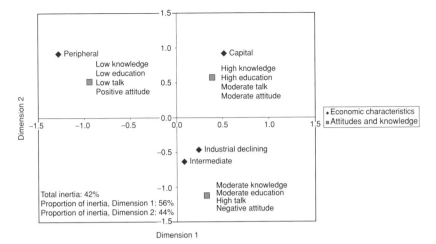

Figure 8.2 Correspondence analysis of clusters of engagement with biotechnology and patterns of regional development across Europe
Notes: The measures of engagement with biotechnology were knowledge, level of education, frequency of conversations about it and attitudes. Total inertia = 42%; proportion of inertia, dimension 1 = 56%; proportion of inertia, dimension 2 = 44%.

dynamism, if at all, arises mainly from tourism, creating a single-sector dependency.

For our problem it is now of interest to see whether these patterns of regional development over the past twenty years are systematically associated with patterns of the public's engagement with biotechnology in these regions.

Figure 8.2 shows the joint plot from a correspondence analysis (Greenacre and Blasius, 1994) of clusters of engagement with biotechnology and regional types. We can observe from this plot that the attitudinal and economic regional typologies are quite closely associated. Urban capital regions show a dominance of an engagement pattern that is characterised by high levels of general education and knowledge about biotechnology, moderate intensity of talking about it and moderate attitudes; 72 per cent of the urban population shows this response pattern. Peripheral regions are clearly associated with a pattern of engagement that is characterised by lower levels of education and knowledge, little talking about biotechnology and very positive attitudes towards biotechnology; 87 per cent of the population in peripheral regions show this response pattern. Industrial-declining and intermediate regions share a very similar pattern of engagement with biotechnology: moderate levels of education and knowledge of biotechnology and high levels of talking, but rather negative attitudes. However, the pattern is not as dominant as in other regions; only 38 per cent and 41 per cent, respectively, of the population show the characteristic pattern of response.

This is apparent from the correspondence analysis too. Looking at the first dimension of the joint plot, one can see that it is anchored by the peripheral regions at one end and the urban capital at the other. In fact the contribution of these two points to the total inertia of the dimension is 97 per cent. Intermediate and industrial-declining areas contribute to only 3 per cent of the inertia of dimension 1 and 44 per cent to dimension 2. Dimension 1 could be interpreted as an axis of post-industrialisation, but, given the small contribution of the intermediate and industrial-declining regions, it is perhaps better conceptualised more simply as marking out the contrast between peripheral and urban capital economies.

Notwithstanding the ambiguity of the intermediate and industrial-declining regions, we further hypothesised that the individual-level correlation between knowledge and attitudes would decline as one moves from the peripheral to the urban capital clusters. The evidence does not confirm such an expectation. The correlation, after pooling all respondents into the regional clusters, is approximately $r = .15$ in all four clusters. Although there is a clearly differentiated typology of attitudes and knowledge amongst the regions, this does not appear to be associated with the knowledge–attitude correlation predicted by the post-industrialisation hypothesis.

Conclusion

In this chapter a number of analyses have been presented that explore a complex of attitudes towards and knowledge about biotechnology at various levels – individual, regional and national. The post-industrialisation model of public understanding of science (PUS) receives some corroboration. The correlation of knowledge and attitudes on a country-by-country basis reveals the pattern of correlations predicted by this model. The correlation between knowledge and attitudes is lower in countries that are closer to the post-industrial ideal-type than in those where the economy is less developed. Evidence directly corroborating this particular hypothesis when extended to the level of European regions was not found. However, the multilevel analysis shows that there is a significant intra-regional correlation in the relationship between knowledge, both formal and informal, and attitudes. It has also been shown that part of this within-region homogeneity is associated with its modal type of economic production and the relative prosperity of the region. Furthermore, in using a multilevel approach we have shown that *individual-level* similarities exist between respondents in the same region that cannot be attributed to artifacts of aggregation.

The patterns of correspondence between the knowledge–attitude clusters developed here and the independently developed economic typology also lend support to the notion that regional variation in attitudes towards biotechnology across Europe is not entirely random. Although the precise basis of this regularity is as yet unclear, local social and economic conditions clearly play a part. For the extreme ends of the economic spectrum in the regions of Europe, the post-industrialisation hypothesis of PUS is also corroborated. In the peripheral regions of Europe, predominantly in the south, biotechnology is seen as progress, part of a vision of modernisation that is welcomed for its positive benefits. Attitudes are likely, though, to be rather unstable, based as they are on a low knowledge-base. The picture in the capitals of Europe is of citizens at once more knowledgeable about biotechnology but more circumspect in their expectations of what benefits it may bring.

Some implications for further research are suggested by the results of our investigations here. First, a base-line has been established that can serve as the beginning of a programme of longitudinal research on the role of the regions using future Eurobarometer surveys on biotechnology. We are already undertaking such research. Secondly, the results presented here beg a more general question: do attitudes towards science in general follow the same pattern of regional variation, or is what has been established here limited to the domain of biotechnology as a very new and unfamiliar technology?

We started this chapter with the question 'do regions matter?' The answer is 'yes, they do matter', in Europe at least, but we have only just begun to understand how, why and when. It will be the interesting task of future research to establish some firmer answers to these questions.

Appendix

Variables

know   textbook knowledge about biotechnology
cons1   constant
ztotatt attitude towards biotechnology (negative to positive)

zpeople   number of people in the household
ztalked   how much the respondent has talked about biotechnology
              in the past
zltunemp   long-term unemployment in the region (percentage of
              working population)
zemp03r   ratio of tertiary-sector employment to total employment
              in the region

All independent variables were standardised (mean = 0, s.d. = 1) in order to facilitate comparisons of the relative effect of each. Because the regression analysis can only be carried out using listwise deletion in software package MLwiN, approximately one-third of cases from the survey are treated as missing because of the gaps in the regional economic data. This left 9,228 cases in nine countries to be analysed (Belgium, France, Germany, Italy, Luxembourg, the Netherlands, Portugal, Spain and the UK).

Model 1 (only individual-level variables)

$\text{know}_{ij} \sim N(XB, \Omega)$

$\text{know}_{ij} = \beta_{0ij}\text{cons}1 + 0.236(0.016)\text{ztotatt}_{ij} + 0.149(0.016)\text{zpeople}_{ij}$
$\qquad + 0.643(0.016)\text{ztalked}_{ij}$

$\beta_{0ij} = 4.336(0.075) + u_{0j} + e_{0ij}$

$|u_{0j} \sim N(0, \Omega_u) : \Omega_u = 0.472(0.075)$

$e_{0ij} \sim N(0, \Omega_e) : \Omega_e = 3.285(0.039)$

$\qquad -2^*log(like) = 57481.840$

Model 2

$\text{know}_{ij} \sim N(XB, \Omega)$

$\text{know}_{ij} = \beta_{0ij}\text{cons}1 + 0.208(0.020)\text{ztotatt}_{ij}$
$\qquad + 0.693(0.020)\text{ztalked}_{ij} + 0.159(0.020)\text{zpeople}_{ij}$
$\qquad + -0.155(0.067)\text{zltunemp}_j + 0.218(0.077)\text{zemp03r}_j$

$\beta_{0ij} = 4.445(0.079) + u_{0j} + e_{0ij}$

$|u_{0j} \sim N(0, \Omega_u) : \Omega_u = 0.284(0.061)$

$e_{0ij} \sim N(0, \Omega_e) : \Omega_e = 3.359(0.050)$

$\qquad -2^*log(like) = 37505.350$

Model 3

$\text{know}_{ijk} \sim N(XB, \Omega)$

$\text{know}_{ijk} = \beta_{0ijk}\text{cons1} + 0.210(0.020)\text{ztotatt}_{ijk}$
$\qquad\qquad + 0.697(0.020)\text{ztalked}_{ijk} + 0.156(0.020)\text{zpeople}_{ijk}$

$\beta_{0ijk} = 4.407(0.189) + v_{0k} + u_{0jk} + e_{0ijk}$

$|v_{0k} \sim N(0, \Omega_v) : \Omega_v = 0.297(0.152)$

$u_{0jk} \sim N(0, \Omega_u) : \Omega_u = 0.088(0.024)$

$e_{0ijk} \sim N(0, \Omega_e) : \Omega_e = 3.361(0.050)$

$\qquad -2^*log(like) = 37482.070$

Model 4

$\text{know}_{ijk} \sim N(XB, \Omega)$

$\text{know}_{ijk} = \beta_{0ijk}\text{cons1} + 0.211(0.019)\text{ztotatt}_{ijk}$
$\qquad\qquad + 0.695(0.020)\text{ztalked}_{ijk} + 0.158(0.020)\text{zpeople}_{ijk}$
$\qquad\qquad + 0.177(0.054)\text{zltunemp}_{jk} + 0.150(0.049)\text{zemp03r}_{jk}$

$\beta_{0ijk} = 4.406(0.172) + v_{0k} + u_{0jk} + e_{0ijk}$

$|v_{0k} \sim N(0, \Omega_v) : \Omega_v = 0.248(0.125)$

$u_{0jk} \sim N(0, \Omega_u) : \Omega_u = 0.053(0.017)$

$e_{0ijk} \sim N(0, \Omega_e) : \Omega_e = 3.363(0.050)$

$\qquad -2^*log(like) = 37470.510$

NOTES

1. We thank Andres Rodriguez-Pose of the Department of Geography at the London School of Economics for access to the regional database he had compiled.

2. From a technical point of view it might be argued that, with only nine level-3 units, the normality assumption for these country-level effects is questionable. We therefore tried representing the nine countries as fixed effects in the two-level models (stages 1 and 2) by creating dummy variables for each. In both models, the unexplained level-2 variance and the coefficients of the economic variables took on similar magnitudes to those in stages 3 and 4. Both remained significant at $< .05$. Hence we conclude that the multilevel analysis discussed here is a suitable method for representing the relative contributions of individual, regional and national variations.

3. A hierarchical cluster analysis was carried out in SPSS using squared Euclidian distances to create the distance matrix and the 'between groups' linkage method

for the clustering. The resulting dendrogram indicated that a three-cluster solution was optimal. In order to assign each region unambiguously to one cluster, a k-means cluster analysis was carried out, requesting a three-cluster solution. In all the cluster analyses, Austrian regions were omitted owing to an extreme configuration of attitudes and knowledge, giving them the status of 'outliers' for our analysis.

REFERENCES

Bauer, M. (1995) 'Industrial and Post-Industrial Public Understanding of Science', paper presented to the meeting of the Chinese Association for Science and Technology, Beijing, October.

Bauer, M., J. Durant and G. Evans (1994) 'European Public Perceptions of Science: an Exploratory Study', *International Journal of Public Opinion Research* 6(2): 163–86.

Beck, U. (1992) *Risk Society: Towards a New Modernity*, London: Sage.

Bell, D. (1973) *The Coming of Post-Industrial Society: a Venture in Social Forecasting*, New York: Basic Books.

Boehme, G. (1993) *Am Ende des Baconschen Zeitalters*, Frankfurt am Main: Suhrkamp.

Bryk, A. and S. Raudenbush (1992) *Hierarchical Linear Model: Applications and Data Analysis Methods*, London: Sage.

Durant, J., M.W. Bauer and G. Gaskell (1998) *Biotechnology in the Public Sphere: a European Sourcebook*, London: Science Museum Publications.

Durant, J., M. Bauer, G. Gaskell, C. Midden, M. Liakopoulos and E. Scholten (2000) 'Two Cultures of Public Understanding of Science and Technology in Europe', in M. Dierkes and C. von Grote (eds.), *Between Understanding and Trust. The Public, Science and Technology*, Amsterdam: Harwood Academic Publishers, pp. 131–56.

Eagly, A.H. and S. Chaiken (1993) *The Psychology of Attitudes*, Fort Worth, TX: Harcourt Brace College Publishers.

Evans, G.A. and J.R. Durant (1995) 'The Relationship between Knowledge and Attitudes in the Public Understanding of Science', *Public Understanding of Science* 4: 57–74.

Goldstein, H. (1995) *Multilevel Statistical Models*, London: Arnold.

Greenacre, M.J. and J. Blasius (eds.) (1994) *Correspondence Analysis in the Social Sciences. Recent Developments and Applications*, London: Academic Press.

Inglehart, R. (1990) *Culture Shift in Advanced Industrial Society*, Princeton, NJ: Princeton University Press.

Irwin, A. and B. Wynne (eds.) (1996) *Misunderstanding Science? The Public Reconstruction of Science and Technology*, Cambridge: Cambridge University Press.

Jelsøe, E., J. Lassen, A.T. Mortensen, H. Frederiksen and A.W. Kamara (1998) 'Denmark', in J. Durant, M.W. Bauer and G. Gaskell (eds.), *Biotechnology in the Public Sphere: a European Sourcebook*, London: Science Museum Publications, pp. 29–42.

Lyotard, J.F. (1984) *The Post-modern Condition: a Report on Knowledge*, Manchester: Manchester University Press; Canadian French original, 1976.

Midden, C., A. Hamstra, J. Gutteling and C. Spink (1998) 'The Netherlands', in J. Durant, M.W. Bauer and G. Gaskell (eds.), *Biotechnology in the Public Sphere: a European Sourcebook*, London: Science Museum Publications, pp. 103–9.

Rodriguez-Pose, A. (1998a) 'Social Conditions and Economic Performance: the Bond between Social Structure and Regional Growth in Western Europe', *International Journal of Urban and Regional Research* 22(3): 443–59.

(1998b) *Dynamics of Regional Growth in Europe*, Oxford: Clarendon Press.

(1999) 'Convergence or Divergence? Types of Regional Responses to Socio-economic Change in Western Europe', *Tisdschrift voor Economische en Sociale Geografie* 94(4): 367—78.

Stehr, N. (1994) *Knowledge Societies*, London: Sage.

Touraine, A. (1969) *La Société post-industrielle*, Paris: Denoel.

Wagner, W., H. Torgersen, P. Grabner, F. Seifert and S. Lehner (1998) 'Austria', in J. Durant, M.W. Bauer and G. Gaskell (eds.), *Biotechnology in the Public Sphere: a European Sourcebook*, London: Science Museum Publications, pp. 15–28.

Wynne, B. (1993) 'Public Uptake of Science: a Case of Institutional Reflexivity', *Public Understanding of Science* 2: 321–37.

9   Pandora's genes – images of genes and nature

*Wolfgang Wagner, Nicole Kronberger, Nick Allum,
Suzanne de Cheveigné, Carmen Diego, George Gaskell,
Marcus Heinßen, Cees Midden, Marianne Ødegaard,
Susanna Öhman, Bianca Rizzo, Timo Rusanen and
Angeliki Stathopoulou*

Why do some Austrians, when asked to talk about biotechnology, conjure up images of children being carried to term by genetically modified pigs in place of human mothers? Why do Swedes associate biotechnology with the monster bull 'Belgian Blue', even though this breed of cattle is a product not of biotechnology but of traditional breeding methods? In this chapter, we explore how citizens of all different European countries express the concern that industrial biotechnology involves playing God by tampering with nature. Nature, the argument goes, will eventually strike back with unforeseen consequences. Be it monster bulls, playing God or other associations the public might have with modern biotechnology, these images are the fundamental bases upon which public rejection and acceptance are based.

When a new technology enters the marketplace of everyday life, it frequently creates ambivalent feelings among the general public. On the one hand, technologies are meant to make everyday chores easier and life more comfortable. On the other hand, many technologies, and particularly radically new ones, involve operations that are underpinned by complex scientific achievements. This was the case with satellite-related space technology in the 1950s and with nuclear energy technology in the 1960s, and it is the case with biotechnology's use of genetic manipulation to tailor the properties of living organisms in the 1990s. An in-depth understanding of the science involved, however, is not the prime interest of the majority of the general public, and rightly so.

However, if an understanding of the scientific basis of a technology is not necessary, people wishing to assess the value of a new technology will nevertheless have to struggle in comprehending what its effects will be. This is especially the case with modern biotechnology because it so profoundly impinges upon everyday life. It has, for instance, led to the introduction of crops and foodstuffs that are, or will be, offered on supermarket shelves to ordinary people who have only the vaguest understanding of industrial genetics. Moreover, people's own bodies might eventually be subjected to genetic tests and cures, and there are even experts who – for whatever reasons – talk of producing clones of humans, who might one day be an individual's closest relatives.

This struggle for understanding may occur at different levels. Some people, i.e. the more scientifically literate ones, might read the relevant columns in newspapers or buy and consult professional literature on the topic. This is the straightforward way. However, the proportion of people who have at their disposal the necessary educational resources or the time required for such study is unlikely to be very high. Hence, many people need to resort to other means of understanding, most commonly those governed by common sense.

The present chapter attempts to highlight some of the modes by which everyday people arrive at a common-sense understanding of modern biotechnology. This is rarely if ever an individual's achievement, but rather the result of collective processes beyond the individual's realm of influence; and it involves events at the level of policy-making and implementation, extended media discourse, and conversations in everyday life.

Such complexity requires a conceptual model that is capable of integrating social as well as individual processes, such as social representation theory (Moscovici, 1984; Jodelet, 1989; Bauer and Gaskell, 1999; Wagner et al., 1999). When a new phenomenon, such as modern biotechnology and its products, appears to threaten a social group's normal course of practice, it needs to be collectively coped with in both material and symbolic terms (Wagner, 1998). 'Material coping' involves technical and, to some extent, legalistic measures that are aimed at containing the potential risks implied by the novel technology. 'Symbolic coping' refers to the naming of the new phenomenon and attempts to understand its qualities and consequences. In other words, it involves assigning the phenomenon a place in the symbolic universe of everyday thinking and common sense. Symbolic coping results in a social representation: an ensemble of beliefs, images and feelings about a phenomenon that is shared by the members of a social unit.

A social representation of biotechnology, for instance, will rarely if ever be veridical in the sense of scientific correctness. Rather than serving a scientific understanding, the beliefs and images that constitute a representation are products of personal and media discourse, which unfold in the course of symbolic coping and serve the purpose of everyday communication. Representations straddle the interface between the individual and the collective because they are generated in collective discourse and their elements are shared to a large extent by the individuals of the group concerned.

Representations that comprise attitudes, beliefs and feelings are frequently structured in a pictorial way. Using images and metaphors is one way of symbolically coping and involves attempts to understand an unfamiliar object in the light of a more familiar one. In this view, images are not just a matter of illustrative words, but rather a way of understanding and experiencing one kind of thing in terms of another (Lakoff and Johnson, 1980). Images not only describe the object and its features, but are also interpretative and evaluative. Thus, a well-suited metaphor is 'good to think with' (Wagner et al., 1995). Such images are not a property of the individual mind but are frequently shared by a group or society when addressing an object. Their use is automatic and effortless and they are perceived not as illustrations of the unfamiliar but as the object itself. They appear as pictures, text or talk in personal and collective discourse.

Overview and method

Asked about what comes to mind when thinking about modern biotechnology, Austrians, Britons, Dutch, Finns, Germans, Greeks, Italians, Norwegians, Portuguese and Swedes show distinct patterns of response. These responses were categorised according to content and evaluative tone, then cross-tabulated and correspondence analysed. In this sample of European nations, the evaluative tone of respondents ranges from neutral (Germans) to negative and/or ambivalent (Austrians and Swedes) to distinctly positive (Italians and Portuguese) (figure 9.1).

People from countries with a more negative evaluation also raise moral concerns in their responses. Although agricultural, medical, economic and fertilisation issues are common to all countries, the more negative or ambivalent respondents are, the more they also mention the possibility of monsters being created by applying biotechnology (Sweden, Finland and Austria). Similarly, the more neutral the respondents, the more likely they are to think in terms of food issues (Norway and Germany). A belief in the potential for biotechnology to foster progress is expressed by the positively inclined Italian, Greek, Portuguese and, to a lesser extent, UK respondents. In general, it appears that the less people know about biotechnology ('don't know' responses), the more a country's population assumes a positive attitude towards this technology.

Figure 9.1 can give only a sketchy overview of the situation in Europe.[1] Results presented in the following sections will fill in this sketch with an in-depth analysis of discourses from diverse sources in six European countries.

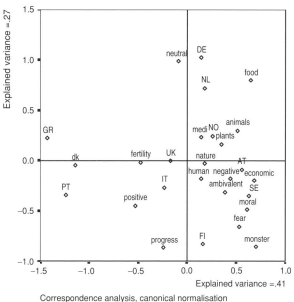

Correspondence analysis, canonical normalisation

Figure 9.1 Evaluations and topics mentioned by respondents from ten European countries when asked about biotechnology: correspondence analysis, canonical normalisation

Note: AT = Austria, DE = Germany, FI = Finland, GR = Greece, IT = Italy, NL = The Netherlands, NO = Norway, PT = Portugal, SE = Sweden, UK = United Kingdom; dk = 'don't know', 'human' = human monsters, 'medi' = medical applications, 'monster' = non-human monsters.

Method

The aim of analysing different data corpora from Austria, France, Germany, the UK, Norway and Sweden is to address the following questions: what is the public perception of biotechnology? and what attitudes, images and linguistic repertoires guide thoughts on and discussions of this topic?

The qualitative data analysed in this chapter comprise a representative survey in various European countries (Eurobarometer 46.1) in which open-ended answers to the question 'What comes to your mind when you think about modern biotechnology in a broad sense, that is including genetic engineering?' were collected. Data were also gathered *via* focus groups and in-depth interviews, which provided extensive verbal responses about different aspects of biotechnology.

Besides being categorised (see figure 9.1), the survey data for each country were subjected to separate ALCESTE analyses (Reinert, 1983,

1990, 1998; Kronberger and Wagner, 2000; see the methodological appendix). Interview and focus group data were available only for Austria, the UK and Sweden. Data from the first two countries were classified using ALCESTE, while data from Sweden were analysed according to the grounded theory approach (Glaser and Strauss, 1967).

Working with different sets of qualitative data we faced the problem of depicting and comparing complex repertoires of arguments. Thus, in order to ensure an intelligible presentation of the data, we decided to focus upon similarities and differences with regard to the contents of national discourses rather than quantitative or methodological aspects. Beginning with the classification results for the open question responses, the dominant representations in different countries are described and discussed. The answers to the open question are short and include the most salient aspects of the topic. It is not surprising, of course, that many aspects mentioned in these associations also occur in the more in-depth discourses of the interviews and focus groups. Therefore the results of the latter data sources are used to complement the results of the open question.

Discursive frames

Overview

Table 9.1 gives a very general overview of the discursive classes found in the responses to the open question from Austria, France, Germany, the UK, Norway and Sweden.[2] One has to keep in mind that the presence or absence of a discourse in a category of the table does not mean that this aspect is not mentioned in a country, but rather that it is not mentioned frequently or explicitly enough to form its own class. It is also possible that one class comprises two or even more categories of the table (if this is the case, it is listed in between the two relevant categories).

Taking a look at the table, it can be seen that, in different countries, similar aspects of biotechnology are mentioned. At the same time, however, these are addressed with differing emphasis and intensity. There is only one discourse class that is clearly country specific: the Swedish association of biotechnology with the 'monster bull', Belgian Blue.

On a general level, we find that people interpret the survey question in two different ways. The two central questions – implicitly posed and explicitly answered by the respondents – produce different frames in which the topic of biotechnology is addressed. First, there is the question 'What is biotechnology?' Answers to this question produce statements that are not primarily evaluative, but rather descriptive. This is a form of 'knowledge discourse', that is, descriptions of what respondents believe

Table 9.1 *Comparison of the analysis results of the open question responses for six European countries*

Country	What is biotechnology? (Focus on content)					Is biotechnology good or bad? (Focus on evaluation)						Lacking knowledge		
	General (rather neutral)		Specific: domains of application (evaluation involved)			Positive	Ambivalent		Negative		Country Specific	Echo[c]	Guessing[d]	Don't know
	Research (progress)	Manipulation/alteration	Food	Reproduction	Medicine	Good	Good but risky[a]	Risky/dangerous[b]	Expression of fear	Interfering with nature				
Austria		Biotechnology is a scientific activity applied to plants, animals and humans (food, reproduction, medicine) (27%)					Good but risky/Dangerous (fear) (22%)	Unknown effects/Dangerous (16%)		Interfering with Nature STOP! (36%)			*see* Interfering with Nature	*see* Unknown effects
France	Research (11%)		Food/Agriculture (15%)	Reproduction (2%)	Medicine (14%)	Improvement (10%)	Dangerous/risky although there can be good effects (also morally dangerous) (8%)		Fear/Against nature (18%)			Echo (3%)	Guessed (16%)	Don't Know (3%)
Germany		Manipulation of plants, animals, humans/Agriculture (16%)	Food (also Medicine and Reproduction) (15%)		Medicine (12%)		Good but risky/Risky/dangerous (fear) (37%)			Interfering with Nature STOP! (11%)			*see* Medicine/Good but risky	Don't Know (10%)
Norway	Research (8%)	Alteration of plants, animals, humans (21%)	Food (8%)	Food and Reproduction (16%)	Medicine (14%)		Good but frightening/Non-specific worry (22%)			Interfering with Nature (10%)			*see* Medicine/Good but frightening	
Sweden	Research (19%)	Manipulation of plants and animals (11%)	Food and Reproduction (7%)				Good if used the right way/Dangerous (15%)		Fear too fast (19%)	Interfering with Nature (21%)	Belgian Blue (9%)		*see* Research	
UK			Food (21%)	Reproduction (7%)	Medicine (21%)			Unspecific worry/Dangerous (fear) (16%)		Interfering with Nature (18%)			*see* Medicine	Don't know (17%)

Note: Numbers in parentheses indicate for each country the percentage of responses classified in a specific discourse.
[a] Good but risky: may have good effects but is risky and dangerous, therefore must be applied properly; demand for control.
[b] Risky and dangerous: biotechnology is unpredictable and therefore dangerous; fear of loss of control.
[c] Respondents repeat technologies mentioned in the preceding question ('telecommunication', 'solar energy', etc.)
[d] Associations evoked by the terms 'bio', 'gene' and 'technology' (mostly positive; e.g. ecologically beneficial or optimistic view of science).

biotechnology to involve. The second question focuses on an evaluation of biotechnology ('Is biotechnology good or bad?'). This question can be answered according to different concerns: a 'moral discourse' evaluates biotechnology according to ethical standards ('Is it morally acceptable?'), whereas a 'risk/danger discourse' considers the advantages and disadvantages as well as the potential benefits and dangers of applying genetic engineering ('Will it have salutary effects for us?'). These spontaneously emerging evaluative discourses show that, in the public view, biotechnology is not only an issue that must be understood, but also one that has to be integrated into an existing evaluative framework.

Of course, there is no clear distinction between descriptive or evaluative in the responses. Nonetheless, a certain emphasis on one or the other is discernible when people are talking about biotechnology. Some respondents evaluate biotechnology in general terms, while others refer to specific aspects and contents of this new technology (which can, nevertheless, be depicted in a more or less evaluative tone).

Different countries reveal differing emphases on these descriptive and evaluative aspects. In the Austrian and Swedish samples, for example, more than 50 per cent of the statements (for Austria, even more than 70 per cent) can be judged as primarily evaluative in nature. These are countries in which the public had been exposed to a high level of public discourse prior to the Eurobarometer survey (see figure 9.2).

In Austria, for example, a country in which media coverage started late in comparison with other countries, a people's initiative in early 1997 obliged the public to consider the 'pros' and 'cons' of biotechnology even though few had a profound knowledge of the subject (see figure 9.1). In most of the other countries, respondents highlighted different applications of biotechnology, such as food, medicine or reproduction, more frequently and in a more differentiated way than, for example, the Austrians did in 1996.[3]

In the following section, different kinds of discourse on biotechnology will be described. After taking a look at representations of what biotechnology constitutes, different ways of evaluating biotechnology are discussed. Then the most salient fields of application in public discourse (medicine, food, reproduction and xenotransplantation) will be presented by considering evaluative as well as metaphorical forms of addressing the topic.

What do Europeans associate biotechnology with?

When asked what comes to mind when thinking about biotechnology, a range of respondents in all countries described what they took this term

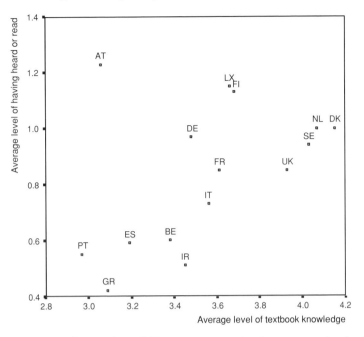

Figure 9.2 Scatterplot of European countries by average level of textbook knowledge and average level of having heard or read about biotechnology
Note: AT = Austria, BE = Belgium, DE = Germany, DK = Denmark, ES = Spain, FI = Finland, FR = France, GR = Greece, IR = Ireland, IT = Italy, LX = Luxembourg, NL = The Netherlands, PT = Portugal, SE = Sweden, UK = United Kingdom.

to mean. These statements represent a thumbnail sketch of biotechnology by referring to this topic as an action (scientific research) as well as by mentioning objects and domains of application. When biotechnology is interpreted in terms of research, statements tend to be of a general nature, related to scientific investigation and developments, and are frequently expressed in the form of optimistic views of science and new technologies. Biotechnology conceived as manipulation/alteration is taken to imply the modifying, manipulating and cloning of plants, animals and humans. The most salient objects discussed in this context are fruit and vegetables (with tomatoes being prototypical), but also animals and children. As a result of this scientific manipulation, respondents imagine the products to have become 'bigger' and 'artificial' in contrast to 'natural' products, which are free of genetic interference. The manipulation of plants, animals and humans is associated with specific domains of application, with food and medicine being the most commonly cited in all countries. The fields of reproduction and embryology are less frequent associations, but still represent salient topics in relation to biotechnology in most of the countries studied. Of course, the descriptions of biotechnological applications in the different domains also comprise evaluative (moral as well as risk-related) aspects.

On closer inspection one can see that this descriptive 'knowledge' need not correspond with scientific definition. Examples such as 'with the help of wind and solar energy, as well as with hydroelectric power plants, one can provide jobs and preserve nature' (AT, 006405[4]) or 'biotechnology also comprises winning new food stuffs out of the sea, plankton and the like' (GE, 1367) suggest that a term such as 'bio'-technology evokes certain images even in people who lack knowledge about its technical use (see the category 'Guessing' in table 9.1). To some of the respondents the word biotechnology seems to be connected with environmental responsibility and scientific progress, that is, with being in harmony with nature ('bio'); and for some the term 'bio' refers to pesticide-free agriculture. For others, biotechnology, even when poorly understood, is situated within an optimistic view of science and technology: 'things to do with the vegetable kingdom; food and cloth textiles; growing tomatoes in fluid rather than earth; trying to grow food in space or those sorts of conditions' (UK, 2219).

Another group of respondents admits to not having any idea about biotechnology at all. It is not a topic of salience or interest for them. Furthermore, a small percentage of respondents simply repeat technologies mentioned in the preceding question of the survey without referring to biotechnology in a specific sense (category 'Echo' in table 9.1).

How Europeans evaluate biotechnology

Optimism Table 9.1 shows that there is relatively little unquestioning acceptance of biotechnology in the countries covered by this analysis. Although there are optimistic evaluations concerning biotechnology in most of the countries, only in France does one find a considerable percentage of people favouring biotechnological development. This group generates its own class of statements, such as 'improvement of the life of the individual, health, better living, better food and better ageing' (FR, 542).

Ambivalence This discourse depicts biotechnology as a new technology that can, to some extent, improve our way of life, but is nonetheless risky and dangerous for mankind. 'I think that's good, there are so many people in the world that want to live and to be well-fed, but the control of biotechnology must be increased' (GE, 3240). People fear that the new possibilities offered by biotechnology could be exploited by unscrupulous and irresponsible people, and they are therefore concerned about the 'proper' application of this scientific knowledge: 'Biotechnology is positive but it should be dealt with in a responsible way. It can improve a lot but applied by the wrong hands it is dangerous' (GE, 0285). Therefore a strong demand for control, delimiting activities in relation to biotechnology by legal as well as moral standards, is expressed. 'I think man has got to conserve and preserve with technology the skills he has been given. It has to be used properly, not abused. He can just as easily destroy as create' (UK, 0227).

Rejection based on risk This discourse involves worrying about the future effects of genetic engineering in terms that suggest these to be far-reaching, incalculable and risky: 'The risk is too big because one cannot estimate the consequences' (GE, 1269); 'very dangerous subject they're getting into. It will create more problems than it will solve' (UK, 0300); or 'We will poison ourselves altogether, nobody knows what will happen, unfortunately one can't do anything about this' (AT, 016701). Although most of the time potential consequences are not explicitly identified, the consequences of biotechnology are judged to be dangerous in nature. The respondents frequently view themselves to be in the hands of science, industry and politics without having any power themselves. This discourse involves a clearly negative evaluation of biotechnology.

An image that illustrates this fear of conjuring up a scenario beyond our control is Goethe's literary figure of the Sorcerer's Apprentice. This image is most frequently invoked in France: 'It's playing the Sorcerer's Apprentice, it is necessary to elaborate some ethics before doing it practically' (FR, 260); 'Mad cows, considering animals as things, the pride of humans playing the Sorcerer's Apprentice' (FR, 937). The wish to make life easier by using only a little-known (magic or scientific) formula, it is asserted, may end up producing unexpected and uncontrollable results and, ultimately, chaos. In one scenario, this involved the 'uncontrolled spreading of artificial life' (NW, 362). According to this picture, scientists are seen as powerful on the one hand (like the apprentice, they have the necessary information at their disposal) but lacking the master's wisdom to apply their knowledge in a useful and responsible manner. On a less literary level, statements such as the following refer to the same concern: 'They just do experiments all the time and destroy earth, it would be better to spend the money in a more useful way' (AT, 006403); 'I think that there are already enough bad inventions in this world that we aren't able to control anymore like the hydro- or the atom bomb. I think this is a field in which we can't estimate the risks, we should not give these inventions, like computer-machines, too much power over us' (GE, 1100). This fear of creating far-reaching problems, without being able to handle them, is also expressed by a further reference, this time to Greek mythology's Pandora: 'It raises a whole Pandora's box of medical and ethical questions' (UK, 2048).

Ideological rejection   This discussion, one that has clearly emerged in all the countries covered by this analysis, judges biotechnology against the background of general thoughts, values and assumptions about the nature of humans and their relationship with their environment. The cultural practices of science and technology are perceived as tools that help humans to shape nature according to their own ideas. Regarding biotechnology, this discourse therefore deals with the question of whether or not humans should be permitted to interfere with the 'natural' harmony of nature. Many respondents, in all countries, think that genetic engineering constitutes an inappropriate tinkering with life and, more specifically, with the meaningful order of nature in which every species has its place and its purpose and where natural boundaries should not be transgressed by unnatural means: 'I'm against genetic engineering. If nature wanted any changes it would already have made them itself. Any interference in natural developments brings about risks that can't be estimated by anyone' (GE, 2466).

Statements such as this are predicated on a view of nature as an organism with its own laws and rhythms and, therefore, one that can never be totally predictable or controlled. An important dimension here is time; nature is associated with slow continuous development and humans are posited as interfering with this 'natural' evolution: 'It is scary, the development is too fast' (SE, 890245); 'They should go more slowly and think about the consequences' (SE, 050585). This view of biotechnology, as a development that is proceeding too fast, is stressed most commonly by Swedish respondents.

The idea of nature as some kind of 'living being' is conceived in two ways: as a sagacious, benevolent being on the one hand, and as a wild force that will take revenge on the other. It is wise, many assert, to live in 'harmony with nature', or to live 'within' nature. From this point of view, biotechnology and its applications are perceived as a harmful intervention in natural processes. Therefore, one must expect nature to 'hit back' and take revenge: 'When man interferes too much in nature it always comes back like a boomerang' (GE, 1290); 'Tampering with nature may be all right in the present, but it will hit back at us in a couple of years' (NW, 716). In countries such as Austria and Germany, respondents frequently accompany their rejection with an imperative to 'stop', prompting statements such as 'One should leave everything as it is, one should leave Nature alone' (AT, 009607), or 'The manipulating of normal things should be stopped' (GE, 3009).

This 'interfering with nature' discourse also appears in the interview and focus group data. A morally challenging point that emerges here is that biotechnology allows not only the modification of plants and animals, but also the manipulation of a very special kind of 'nature': the nature of being human. One has to note the double interpretation of 'human' used in this discourse. On the one hand, 'human as scientists' are seen as actively shaping nature and, on the other, 'humans' are considered to be the object of that manipulation. The moral concern, therefore, is humans interfering with human nature. This means not only modifying specific human characteristics such as intelligence, physical appearance and ageing, but also altering the very essence of what it is to be human. Respondents anticipate this bringing about an unpredictable change in the nature of human life. For example, with the introduction of cloning, biological fathers or mothers will become superfluous. To be born as a result of a scientist's intervention impacts upon human dignity and, according to many, contravenes any moral sense of 'good and right': 'Humans originate from humans according to a development which cannot be planned. If anybody could determine what I will become, then I would have the impression that I lack something of being human' (AT, focus group). As will be seen later in this chapter, questions of how biotechnology might undermine the essence of being human also extend to such issues as the importance of a 'natural' end to life and concerns about human uniqueness.

After all, this notion of humans 'designing' humans contradicts the claim that there are things that only a 'god' can do: in this view, humans are able to do god-like things now. This is generally judged to amount to the grossest impudence, a conviction usually expressed in the form of the persuasive image of humans 'playing God'. This also appears in the open question data: 'It is this manipulation of man, it is this alteration of nature. This is an interference in Creation, man plays God' (AT, 008002). In this view, scientists and industry are consistently viewed as playing with our lives. In accordance with the metaphor of the Sorcerer's Apprentice, they are viewed as doing something that they are not entitled to do. But, whereas the apprentice's impatience to do things before he has the requisite knowledge to do so is viewed as dangerous (because of the potential for bringing about an uncontrollable situation), the wish to 'design' and to determine the world like a god is judged as morally unacceptable: 'I'm fearful about it, it's in men's hands the use to which it's put, it could be a tool of the devil to do with altering the genetic make-up in humans' (UK, 2295).

In this context, people involved in genetic research and industry are considered not to be acting responsibly, but rather to be thoughtlessly following their – mostly egoistic – desires for power and economic gain: 'We've already come to know different technologies, modern ones like nuclear energy, the atom bomb, chemical weapons, weapons in general – did these improve human life? Only for the few who make profits out of them, totally neglecting the suffering of innocent people. With biotechnology it will be the same because again there will be some powerful people taking advantage in an unscrupulous way' (GE, 3338).

Domains of application

Fighting diseases

Medical application is one of the most common references to biotechnology appearing in the public discourse of most countries. The potential for biotechnology to help cure hitherto incurable diseases is generally judged to be an important potential improvement. Unsurprisingly, biotechnological applications in relation to medicine are evaluated in a mostly positive tone: 'curing diseases' (AT, 008806); 'fighting hereditary diseases, early diagnosis of diseases' (GE, 0190); 'development of new drugs for control of cancer, arthritis and other diseases' (UK, 0061).

In this domain, the image of 'progress' clearly emerges. Diseases such as cancer and AIDS are frequently cited by respondents expressing the hope that biotechnology will provide new remedies. 'For coping with diseases like AIDS, BSE and cancer, technologies like that are really necessary and must be encouraged' (GE, 1028). Only a few negative consequences of applying biotechnology to the domain of medicine were mentioned.

'I'm for genetic engineering because my wife suffers from cancer and maybe they will find new remedies against diseases like that' (GE, 1099). It is not surprising that personal suffering and the hope for medical advances encourage the endorsement of this aspect of biotechnology. Interestingly, this positive evaluation changes to ambivalence when different potential medical applications are distinguished. An Austrian interviewee stated: 'the term "for medical purposes" is a very vague one because being able to cure cancer is different from surgery for cosmetic purposes and is different from breeding perfect humans' (AT, focus group). This more differentiated discourse takes place in interviews and focus groups and only to a small extent in the context of answering the open question. Surprisingly, however, medical applications involving reproduction and xenotransplantation are more commonly discussed as moral than as medical questions.

Unnatural food

The food discourse occurs in the context of evaluating the advantages and disadvantages of genetically manipulated (GM) food. The validity of applying genetic manipulation techniques to food is often assessed by opposing the consumer's and producer's points of view. On the producer's side it is suggested that GM food is easier, more efficient and more profitable to grow. More frequently, however, responses are framed around the supermarket consumer shopping for the best-quality produce, and making decisions about what constitutes healthy, good food. Respondents, therefore, mainly refer to characteristics of the products, invoking words such as 'better', 'cheaper', 'bigger' or more 'beautiful', and having a longer shelf life. The disadvantages of GM food mentioned focus on the effects of genetic food manipulation for humans as well as for the ecosystems. Generally speaking, the French, British, German and Norwegian discourses on GM food are expressed in a fairly neutral, sometimes ambivalent tone, whereas evaluations in Austria and Sweden are rather negative. The Swedes stress potentially negative consequences for the biological diversity of plants and animals. Respondents from Austria and Germany more frequently assume that eating GM food is unhealthy and that it will bring about new diseases and allergies, or produce resistance to antibiotics.

An important issue from the consumer's point of view is having a choice of which products to buy and to eat: 'I don't buy it any more because now I know that it is genetically modified rape. I tried it, it had a good taste, but I won't buy it any more' (AT, focus group). This is a question of labelling, and comes down to the fear of not being able to determine whether or not a product is genetically modified. The feeling of being denied the ability to choose results in statements like those expressed by the Swedish media, in which GM food products are referred to in such terms as 'leaking uncontrollably over the borders', 'gene food sneaking into the shelves' or 'mixing modified and natural beans so that we will not know'.

As mentioned above, the objects of genetic manipulation that appear most conspicuously in the public discourse are fruit and vegetables, with tomatoes as the ideal-type. These new products are imagined to be 'bigger', 'unnatural' and 'artificial', in contrast to 'natural' foodstuffs that have not been genetically manipulated: 'giant fruit, bigger potatoes, cattle with more meat, wheat yielding more flour' (GE, 0454); 'manipulated giant tomatoes, ... artificial colours, artificial children' (GE, 0453). The paradox in this image is that products artificially produced by genetic manipulation are conceived of not in their own terms but as products whose inherent characteristics have been intensified: 'making tomatoes rounder – apples larger and greener' (UK, 0034); 'sugar-beets were genetically modified, the proportion of sugar is increased' (GE, 1194). Less frequently, people mention fruit and vegetables that have been visibly altered: 'I know they can make tomatoes of all colours, as big as pumpkins and melons' (FR, interview); 'a tomato that is a mixture of tomato and potato and that tastes like a banana' (FR, interview); 'square tomatoes' (AT, 001204).

Topics that are regularly associated with biotechnology in several countries are nuclear energy and radioactivity (the two issues seem to be perceived as similar because in both cases the technology involved is potentially dangerous and may have incalculable effects). People in different countries talk of 'irradiated' food: 'fear of BSE and irradiated vegetables' (GE, 0242); 'Nature should be left alone, otherwise we'll soon have the atom bomb in our vegetables' (AT, 015202). Understood in metaphorical terms, statements such as these express a perception of biotechnology as something dangerous and threatening. And, as was the case with GM fruit and vegetables above, irradiated food is understood as an extreme form of the original product.

In addition, biotechnology is sometimes associated with hormones. This is most typical among Norwegian respondents: 'disturbance in animals' growth hormones' (NW, 004); 'messing with the hormones in food production' (NW, 149). Hormones are most often mentioned when respondents are referring to the size of manipulated objects. This is not surprising since one of the first biotechnological applications of which the Norwegians heard was the implantation of growth hormones in salmon, an important Norwegian export product. In terms of public perception, genes and hormones are conceptualised in similar ways: both are small and invisible, and seem to provide a sound, scientific basis for understanding otherwise inscrutable aspects of organismic biology.

Although the molecular alterations are invisible to the human eye, GM food is perceived as having been changed from its natural to an unnatural and artificial state. The ubiquitous opposition of 'natural' and 'artificial' subsumes two of the themes within the discourses of health and illness as well as discussions of biotechnology: first, the association of 'artificial' products with an 'infringement of the natural state' (Herzlich, 1973: 32); and, secondly, the notion that this infringement has connotations of 'heterogeneity' or the introduction of foreign elements into nature. Moreover, if taken a little further, these heterogeneous elements may be interpreted as pollution, and this explains the use of 'unnatural' and 'unhealthy' as synonymous in some countries. The representation of the

Plate 9.1 Photo of tomato in Austrian newspaper

products of biotechnology as big and beautiful, but at the same time artificial and unnatural products (prototypically the tomato), containing heterogeneous elements, is expressed in Plate 9.1, taken from an Austrian newspaper.

Similar representations are discernible with respect to the manipulation of animals. As was the case with manipulated fruit and vegetables, GM animals are imagined to be bigger while being, at the same time, unnatural. A typical association among the Swedish public is that of biotechnology and the 'monster bull Belgian Blue',[5] a cattle breed that has no connections with genetic engineering from a scientific point of view. Nonetheless, 9 per cent of the Swedish respondents refer to this 'giant bull' when thinking of biotechnology. In the public mind, the Belgian Blue is a living embodiment of all the characteristics attributed to GM creatures: it is big and monstrous on the one hand, but weak, bizarre and unnatural on the other. This image of the Belgian Blue was also reported extensively in the Swedish mass media, attention being focused on the breed's physical characteristics, such as its giant body with large muscles, weak legs and small internal organs, together with its calving problems (see Plate 9.2). Only a few days before the survey, the Swedish newspaper *Aftonbladet* published a story with the headline 'Calf slaughtered with chain saw – inside the cow'. Similar descriptions are provided by respondents in the

Plate 9.2 Photo of 'Belgian Blue'

survey: 'I think about the Belgian Blue – monster bulls which must be delivered by sawing the cow in half' (SE, 980247).

Images such as the 'monster bull' Belgian Blue or 'mad cows' (BSE), neither of which issues is directly related to biotechnology, nevertheless occur in the public discourse and provide persuasive images informing the public's perception of the new technology. The fact that cattle are 'mad', 'degenerate' and 'bizarre' is perceived to be a result of breeders striving for better profits while paying insufficient regard to the limits set by nature. Mad cow disease is frequently quoted when respondents are referring to the unpredictable risks of applying genetic manipulation. This image recurs when respondents talk of scientists, industrialists and political authorities, which are viewed as failing in their duty to protect their country's citizens: 'I do not agree with it at all, we wouldn't have had BSE if they hadn't bloody well interfered. No, it's just a feeling of interfering, you can't alter nature without paying for it. Nature always balances' (UK, 2404). (According to the logic of the images discussed in the evaluative discourses, this could be conceived in terms of the plagues coming out of Pandora's box, as well as a first sign of Nature's revenge.)

Testing and designing humans

Upon being asked to think about biotechnology, a significant proportion of respondents mentioned aspects of reproduction and embryology. This domain is less frequently associated with biotechnology than either food or medicine, but is still of remarkable salience. Interestingly, as far as the public is concerned, this discourse is separated from, rather than connected to, the medical context. (Table 9.1 shows that in most countries the medical discourse is separated from the others; the vocabulary and associations employed in talking about reproduction are more similar to those used in discussing food.)

The most frequently mentioned reproductive associations involve 'artificial fertilisation' and 'test-tube babies'; yet neither of these is directly related to biotechnology in the current understanding. Nonetheless, these images are well suited for encapsulating a whole set of concerns relating to both technology and the creation of life. The image of a baby created in a test tube captures the 'unnatural' and 'artificial' character of biotechnology. Statements such as 'artificial children' (GE, 0453), 'the lady having a baby from her dead husband's sperm' (UK, 0213) or 'a woman who had eight babies' (UK, 0429) all depict putatively 'unnatural' aspects of human reproduction. In this way, the stress on the 'unnatural, weak or bizarre' products of the genetic manipulation of food and animals is recapitulated in the discourse surrounding the possibility of human genetic manipulation. This gives rise to such remarks as 'test-tube babies are often handicapped' (GE, 0453), 'it will lead to manipulation and the creation of degenerate species, both humans and plants' (SE, 462362) or 'it frightens me, all the horrific changes we could do with it; we could end up making little monsters' (UK, 2047).

Besides the 'test-tube babies', there are other images that contrast reproductive applications of biotechnology with 'natural births'. An Austrian image mentioned during the interviews involves a pig carrying a child (the pig is styled as a 'loan mother'). Here biotechnology is seen to offer those mothers who want to preserve their bodily youth and beauty the possibility of having children without the burden of a pregnancy. The sarcastically uttered statement 'my mother the pig' (AT, focus group) can be understood in the double sense of imagining a child being confronted with a mother-pig (What defines being a mother? Who will be the biological mother, then – pig or human?) as well as judging a mother who would do such a thing as morally unacceptable.

In the same way as depicted in the 'interfering with nature' discourse, the topic of an 'unnatural' beginning of life (whether being 'produced' in a test tube, originating from conserved sperm of an already dead person or being carried to term by a pig) is morally challenging since it questions the 'nature' of being human and therefore also the social identity of human beings. Furthermore, there is the topic of cloning, a practice viewed as a very special kind of reproduction. As mentioned earlier, cloning makes it unnecessary to have both a biological father and mother. Indeed, the possibility of asexual reproduction prompts people to wonder why there should be men and women at all: 'It must not go so far because then we wouldn't need males and females anymore, no, I'm against that. Natural reproduction, at least this should be true. I'm very much against that' (AT, focus group).

Genetic engineering in relation to reproduction is also often perceived as a matter of life and death, being seen as a tool with which to select those human beings considered worth living and to eliminate those that are not. This discourse is embedded in a general discussion about abortion and focuses on genetic engineering as enabling the early diagnosis of illnesses and handicaps: 'terminating pregnancy if foetus has illness' (UK, 2397); 'they will use their knowledge to abort babies with only small defects' (SE, 462362). In the interviews this discourse focuses on moral questions along the lines of: What makes a life worth living? What are we doing when we abort a child diagnosed as handicapped? Is the prevention of suffering a justification for overriding an organism's 'right to live'?

This moral discourse is embedded in a subjective context of personal involvement. The questions are not dealt with in a general or a neutral way; instead they tend to elicit a personal point of view: a concrete decision framed in terms of 'What would I do?' In the UK and Austria these discussions can mostly be traced back to interviews with women, and involved the issues, practical and ethical, that would inform their decisions if faced with the possibility of genetic screening and other biotechnological applications. Being a moral issue, motives and interests are of crucial importance; and because a moral conflict is often involved in deciding in favour of a single option, evaluations are generally ambivalent. In this case, the moral concern is a matter of life and death, and more specifically it is the right of humans to decide about the lives of others. Furthermore, the discussion deals with the question of what defines a 'good' life, for parents as well as for children, and which forms of interfering with the 'natural' processes of living and dying are morally legitimate.

A topic related to abortion is that of biotechnology enabling the selection of humans for certain attributes: 'makes a woman have a boy or girl if she wants it, there could be production of babies by test-tube creation, it's a lot of rubbish' (UK, 0489); 'not good: making children in tubes and sorting the children according to if they are good or bad' (NW, 174).

The images of 'perfect children', 'humans made-to-measure' or 'designer babies' suggest that parents (or scientists) can choose from a 'catalogue' of characteristics for their offspring's attributes, for example physical appearance and intelligence. This kind of discourse also links the 'production' of humans, equipped with certain features, to financial and economic factors. An English interviewee imagines a future scenario of buying children in shops: 'I mean, you just visualise, you know, the typical child, saying "where are we going to get the baby from", do you go to Marks & Spencer, eventually you're going to be able to do that' (UK, focus group). Clearly, the buying of children contradicts any sense of 'normal' birth.

Moreover, imagining some kind of industry 'producing' babies implies that people will have to pay for what they order: 'But to be able to go one step and choose your blue eyed, blonde haired, intelligent child, is going to come down to money because you're going to have to pay somebody because it's not a necessity' (UK, focus group). This impinges upon another moral issue: not everybody will be able to afford these services, and new social injustices will consequently arise.

Of course, the 'designing' of babies is not far away from creating humans for specific purposes. When discussing cloning, and emphasising the dangers of the new power relations that might emerge, science fiction scenarios are often depicted: 'This is the story of subhuman creatures, of cloning human robots, of being cloned from some physically extraordinary humans but provided with less intelligence, so they can do dangerous jobs' (AT, focus group). Huxley's 'Brave New World' is sometimes mentioned in this context. Images are commonly invoked that involve subhumans and supermen, armies bred only to fight, or living machines created to do dirty jobs. An association with eugenics is frequently cited. And it is far from surprising that in some countries the name 'Hitler' is strongly associated with cloning: 'It's frightening, in the end they will manage to do what Hitler failed to do' (UK, 2208). This raises fundamental questions: What should one be allowed to manipulate and where should the line be drawn? Who should make this sort of decision?

Monsters and matters of life and death

An application only rarely mentioned in the open question data is animal–human transplantation. This – at least in 1996 – was not a salient association in relation to biotechnology for most respondents. During the interviews, the topic was introduced by the interviewer whenever it was not mentioned by the interviewees themselves. It is interesting to note that the issue is discussed only marginally in relation to medicine. Here the motive of 'rescuing human life' is of crucial importance. But, above all, the issue evokes divergent moral questions.

The topic of xenotransplantation stimulates fantasies about combining the body parts of different species, and thereby constructing monsters.

This discourse deals with the 'natural' make-up of beings, as well as with the 'natural' borders between species: 'mice with an ear on their back, pigs with two more ribs' (GE, 0208); 'pigs with cows' heads' (GE, 0227).[6]

In the public mind, the aim of xenotransplantation to provide healthy organs is visualised in terms of 'spare parts': 'If we accept that we're going to breed a spare parts bin, do we accept that it is right to take those parts and put them into anything we so desire, be it a human being, another animal, whatever?' (UK, focus group). Animals are no longer perceived as living organisms but rather seen as sets of body parts that can be combined in different ways. But this raises the question of whether such organisms should still be considered to be living beings, or just machines: 'We produce living machines, and this must be rejected' (AT, focus group).

The idea of animals providing 'spare parts' is also extended to humans (for example, human beings without a head existing to provide organs and body parts): 'and then they will take "inferior" humans, they will say that they are a bit handicapped but they have a healthy heart, and so on, it will go on like this only for the reason that we can live eternally' (AT, focus group); 'I don't like genetic engineering, I don't want to end up with other people's bits of body' (UK, 0070). People come to think of human and animal 'monsters', and the idea of mixing different humans or combining parts of humans and animals seems to infringe 'natural' boundaries.

In the interviews, this topic is also discussed as an issue of animal welfare. Respondents question the ethics of breeding, manipulating and slaughtering animals for the purpose of curing disease and prolonging human life. Genetic engineering is here posed as a matter of life and death for humans as well as for animals. Most of the time, the motive of 'rescuing human life' is acknowledged as an important value that should be endorsed; but the problem with xenotransplantation, in the respondents' view, is that saving the life of a human costs the life of an animal. In a broader sense, the discourse also deals with the topic of organ donation in general. Respondents refer to the fact that an individual continuing to live with an implanted organ always demands the death of another (whether this is an animal or a person dying in an accident). The discourse therefore focuses on questions such as the following: Is human life worth more than that of an animal? Why not breed and slaughter animals with the purpose of obtaining organs? After all, we do so in order to produce meat. Which is more morally acceptable: waiting for a seventeen-year-old motorcyclist to die or using a pig as a 'spare parts bin'?

Xenotransplantation is viewed as a tool for prolonging human life by delivering healthy organs. A belief that emerged in the Austrian interview data is that everybody has the right to live his or her 'whole' life. That is, when the 'natural' lifespan of an individual is threatened by illness, or by other factors, then help should be available. This subsumes the motive of wanting to be immortal or 'forever young', a dream that some feel might be realised once we are capable of 'producing' young and healthy organs, and of using animals as 'spare parts bins'. However, the fulfilment of this extreme form of the wish is considered morally unacceptable. At this point, the question concerns the proper limits of xenotransplantation. Do we have the right to deny help to an ill person if this technology could be developed and employed? What is the difference between implanting an organ into a fifty-year-old person or into an eighty-year-old person? More fundamentally, when is a person old enough to die? And who should make the decision on this question?

Generally speaking, respondents' evaluations change when different motives and perspectives are considered. The motive of curing disease and enabling individuals to live 'normal' lives of 'normal' length is generally considered in a positive light. The wish to be 'forever young', however, is considered egoistic and unacceptable. Using animals to, as it were, pay for our sins (our unhealthy way of life) is also considered to be improper. But if our life is in danger, not through our own fault but because of an 'unjust' illness, then the case is different again.

Like any moral discourse, the discussion focuses upon 'limits'. What is the difference between having a new set of artificial teeth and having a new heart grown in a pig? What is the difference between receiving a donor heart from a person who died and one from a pig? What is the difference between organs bred in animals and those bred in humans? What is still acceptable and what is not?

'National' images

Figure 9.3 shows the associations between different countries and the typical images evoked by biotechnology. Generally speaking, one can see that there are few country-specific images, a major exception being the 'Belgian Blue', which is a typically Swedish association. Most of the images, though, are located around the centre of the graph and occur in most, or all, of the countries. The general (moral) images of 'interfering' or 'tampering' with nature, life and humans are central and ubiquitous. These are images that refer to biotechnology in general and do not explicitly identify specific objects of manipulation. As mentioned above, this 'interfering with nature' discourse occurs in all six countries: it depicts biotechnology as an unwise, harmful or dangerous interference in nature and life.

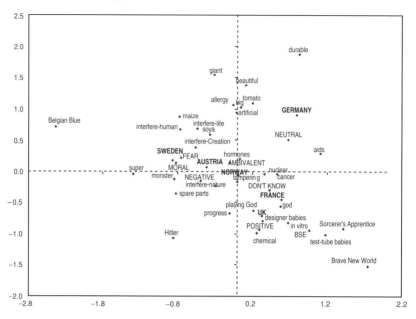

Figure 9.3 Correspondence analysis of open question for countries, evaluative tone and images.
Note: Global categorisations are in capital letters; labels in lower case letters indicate images referred to in respondents' answers to the open question.

Most of the images are mentioned in every country studied, and the national variation comprises respondents placing different emphases on the various domains of application. Images concerning food more frequently occur in Germany and Austria (tomato, big, artificial, giant, durable, etc.); Swedish respondents more often refer to animals (Belgian Blue, spare parts, monster); whereas the domain of reproduction and embryology is more frequently mentioned by British and French respondents (test-tube babies, *in vitro* fertilisation, designer babies). Comparisons to literary scenarios such as Huxley's 'Brave New World' or Goethe's 'The Sorcerer's Apprentice' are mainly drawn by French respondents. Norway, being located in the centre of the graph, does not show any special association with any of these diverse forms of application.

What do the differences and similarities in imagery among these countries mean? It must be noted that, in this context, images or metaphors not only constitute means of expressing ideas, but also function as socio-cognitive devices that help to organise the understanding of a phenomenon. Because they are metaphorically structured at a cognitive level, the manifestations of imagery in language use are also metaphorical. For example, in all six languages we find a range of respondents describing biotechnology as 'interfering', 'messing', 'tinkering' or 'tampering' with nature or life. At a more general level, all these statements express the concept that 'biotechnology is a (harmful) intervention'. This rather general concept can be understood as an image schema that structures the perception as well as the evaluation of the issues in question. The concept therefore delimits all the possible interpretations of this technology by focusing on some aspects and neglecting others. From this point of view, it is unlikely that biotechnology would be evaluated as something beneficial that will improve our quality of life. But the image schema are flexible enough to allow the expression of various evaluations: 'interfering' implies a certain degree of neutrality, whereas 'messing' and 'tampering' are clearly negative.

However, respondents in all of these countries employ very similar images of biotechnology. Therefore, it is reasonable to suppose that, because of the limited number of organising concepts, biotechnology discourses are heavily circumscribed.[7] These concepts 'establish a range of possible patterns of understanding and reasoning. They are like channels in which something can move with a certain limited, relative freedom' (Johnson, 1987: 137). Yet, although the number of interpretations of biotechnology available in the public discourse may be delimited, within the limits of any structuring concept there is a measure of freedom or variability of understanding that is heavily context dependent. The concrete words and images that are used to allude to these underlying schemas will be mainly determined by local, historical and personal contexts.

These contexts determine which images are most salient for the public in thinking about a topic such as biotechnology. Thus the images stem from different sources: first, there are images such as BSE or the Belgian Blue that reflect current discussions in political, social and everyday life in different countries; secondly, there are historical references to the past, such as to Hitler and the Nazi regime; thirdly, people may refer to mythology or literature, in this case in the form of Frankenstein, Brave New World, Pandora's box or the Sorcerer's Apprentice; and, finally, there are images that cannot be linked to any of the sources mentioned, but that grasp the essence of biotechnology with everyday words such as 'artificial food', 'interfering with nature' or 'designing babies'.

It is clear that underlying concepts and feelings can be expressed by reference to a wide range of current images. Thus, the idea that manipulated objects are bigger, but unnatural and bizarre, can be articulated in terms of mad cows or a mouse with a human ear on its back. But individuals can also draw upon mythology and literature, invoking fantastic monsters or Frankenstein. The importance of local context is especially clear in the case of Britons citing mad cows and the Swedes tending to think in terms of the monster bull Belgian Blue, even though both groups of respondents were expressing their concern that biotechnology may bring about bizarre, weak and unwell animals. Considering the media attention BSE was receiving in Britain at the time, it is far from surprising that the association was made. The Belgian Blue cattle breed was an even more localised issue (with only a few Norwegian respondents also citing it), but since the import of twenty-five Belgian Blue bulls to Sweden in 1995 this topic had attracted a lot of attention in the Swedish media, which referred to them as 'monster bulls'. Likewise, a historical reference to Hitler can express the same fear of authoritarianism as quoting Huxley's *Brave New World*. Clearly, different images can stand for similar concerns; the concrete words and terminologies used can only be understood as being largely derived from the speaker's context.

The general semantics of biotechnology discourse

There seem to be two global representations of biotechnology in which the different images are organised: one recapitulates the theme of humankind's progress in overcoming natural obstacles and the other involves humans' unwarranted interference with nature.

Fighting enemies of mankind

Discourses that express a positive evaluation of biotechnology frequently focus on a war metaphor, which depicts the modern world as confronting a range of dangers that must be ameliorated. In this view, biotechnology is a weapon that helps in the 'fight' against diseases, such as cancer or Aids, as well as world hunger. This discourse emphasises progress, the creation of a better world. Nature represents a complex set of mechanisms that can be incrementally better understood, and whose more pernicious aspects can be overcome. Accordingly, the manipulated object receives little attention, and this discourse is able to draw upon few, if any, images.

Tailoring living nature: fake life

Biotechnology seems to offer the possibility of creating and fashioning a world according to human plans and ideas ('designer babies', 'made-to-measure humans'); humans are considered able to imitate the Divine ('man plays God' or 'interferes with Creation'). But respondents believe that it would be unacceptable for humankind to do so; humans, it is claimed, are not entitled to interfere with nature in this way. For this reason, comments on biotechnology are couched in terms of irresponsible 'playing', 'tampering', 'messing' and 'tinkering' with Nature. According to this view, the expected result must be an ultimate loss of control, with humankind being exposed to chaos and the suffering implicit in references to the Sorcerer's Apprentice and Pandora's box.

Notions of human interference in 'living nature' subsume several different themes.

- Interference with respect to upsetting the natural balance and harmony of ecological systems – this presupposes an anthropomorphic image of nature: since biotechnology is 'harming nature', it will eventually 'hit back' and 'take revenge'.
- Interference that can lead to humankind itself becoming 'de-naturalised' because of the possibility of 'unnatural birth' ('test-tube babies', 'artificial fertilisation', '*in vitro* fertilisation', 'designer babies', 'pig as loan mother', 'buying babies'); these technologies are seen as allowing 'unnatural lifetimes' ('eternal living'), made possible by the use of living 'spare parts', and this is perceived to imperil the uniqueness and value of being human ('producing life', 'mass production of babies', 'armies breed to fight', 'Brave New World').
- Biotechnological interference that represents a threat to 'natural' social orders and social justice, respectively (Hitler, Brave New World, living machines doing dirty jobs, subhumans and supermen).

From this perspective, there are a variety of potential outcomes of biotechnological intervention, and these relate to different conceptions of Nature. In all cases, however, modifying nature means that the result is, by definition, no longer 'natural', it becomes 'artificial'. But what does this nature 'made' by humans look like?

Fruit and vegetables are characterised as being 'bigger', 'more beautiful' and 'more durable' and as yielding 'more profits'; manipulated humans are supposed to be 'more beautiful', 'perfect', 'super-intelligent', 'geniuses' or 'supermen'. In this way, certain desirable attributes of living beings are magnified. Where tomatoes are just 'redder' and apples 'greener', the modification is perceived in a way analogous to the fairytale notion of the most tempting apple, for example, containing the poison. The manipulated food is depicted as aesthetically pleasing but artificial, unhealthy or toxic ('radiated food', 'chemical food', 'messing with hormones of animals'). Even where there is no evidence of the modification of food, respondents consider it possible that the food could have been insidiously altered; and concerns are expressed that it is impossible to determine whether or not a product is genetically manipulated ('gene food sneaking onto the shelves'). The tempting, desirable results of biotechnology bring with them invisible dangers, and the 'natural-looking unnatural' is viewed with either suspicion or ambivalence.

Where the results of manipulation are easy to perceive, i.e. where the results are bizarre or defective, respondents think in terms of the unintended and incalculable side-effects of biotechnology. The danger of inadvertently creating organisms (plants, animals and humans) that are degenerate, bizarre, weak or ill ('monster', 'Frankenstein', 'mad cows', 'monster bull', 'handicapped children') rises to the fore. This is also the case for applications of biotechnology in which normal traits are exaggerated and the attributes of different kinds of organism are combined ('mouse with human ear', 'pig with a cow's head', 'tomato with banana taste').

Biotechnology, and especially cloning, are viewed as enabling the 'production' of life by humans. Images such as 'artificial life', 'living machines', 'human robots', 'producing life' and animals as 'spare parts bin' all depict 'unnatural life', that is, a form of life that is simultaneously real and fake. Any modification of the nature of life raises a paradox, since the very essence of natural life is that it cannot be 'made'. The images employed express a feeling of unease, embodying the notion that, although it might appear to constitute life, life manufactured by humans cannot be real.

Conclusion

The philosopher Peter Sloterdijk (1987) views the modern world since Copernicus as characterised by not only a cosmological but also an epistemic shock. He argues that it is clear that what we see and what comprises reality can no longer be considered synonymous. The dilemma is that, when we believe we are seeing the sun rising, the sun is not actually rising, and the earth can no longer be viewed as the centre of the universe. The modern corollary of this is that it is necessary to suppose a world that is not as we perceive it; there is a disjunction between the world and our perceptions of it. More and more, our knowledge depends on information produced using technical apparatus and not on our own perceptions. Nowadays, we live in a world full of invisible radiation, ozone and toxic chemicals. This is our 'real' life. Equally there is no doubt in the public mind that genetic manipulation exists and that it alters the objects to which it is applied. The same certainty is conferred on this notion as was enjoyed by witchcraft in former times; in both cases, the arguments are scientifically justified (Luhmann, 1991). In this sense, biotechnology is just one of many scientific or technological facts that cannot be perceived directly but that must be dealt with.

Since Copernicus, modern life has been defined by scientific and industrial change capable of profoundly altering the world, as well as by the emergence of world-views in which it is impossible to take anything for granted (Sloterdijk, 1987: 63). The Copernican abolition of obviousness in modern life has provoked counter-reactions that seek to restore a sense of normality to the world. Thus, common sense is concerned with making available, not 'models of reality' but rather 'models for reality' (Geertz, 1973: 93). The invisibility of genetic manipulations that might, nonetheless, bring about massive change is, therefore, countered with attempts to impose upon it rigid classifications that enable us to come to terms cognitively with this potent new technology. Visualising biotechnology with the help of images brings about a socially shared truth in which it is depicted in a way that helps to resolve the ambiguities surrounding the new technology.

This understanding involves selecting descriptive categories. The central one, in the present context, is that of Nature and its meanings. However, 'Nature' is not a clear-cut but a rather ambiguous notion, employed by both supporters and opponents of biotechnology. As has been seen, according to the optimistic view Nature is a legitimate object for investigation that will gradually reveal all of its secrets to scientific researchers. A completely different understanding of Nature draws on the distinction between the natural and the unnatural, or between the natural and the artificial. This opposition underlies many of the images described in the context of moral evaluations and criticisms of biotechnology in which biotechnology is referred to as being in contradiction to Nature. In this case, 'Nature' provides the anchor for a diaphoric understanding of biotechnology. Contrary to the common uses of metaphors, the basic concept of Nature is not equated to the new phenomenon; rather, the phenomenon is likened to the source domain's – i.e. Nature's – negation.

The dominant images of biotechnology represent attempts to construct a clear distinction between the natural and the unnatural. But this drawing of boundaries goes further, and also provides a basis for the separation of the good from the bad. In this modern world, in which people know that 'the natural' (in the sense of being 'the original') can no longer exist, identifying the 'natural' with the 'good' provides a reassuringly solid foundation for morality. The normal (i.e. that which is taken for granted) is equated with the normative. Thus, because it is usual for humans to originate by way of sexual reproduction, this is seen as the way things must be. This restores stability to the world and its 'natural kinds'.

Although people in the six countries studied do not tend to use the term 'unnatural' when talking about medical applications, they do use it when discussing reproductive technologies. This is because the latter seem to challenge a form of procreation that is taken for granted. Being the result of the procreative act, and being oneself able to reproduce, have come to be part of the definition of being human. In a similar way, humans' bodily integrity is considered to be challenged by the transplantation of organs from other species. In both respects, biotechnology is seen to be capable of subverting both gender and generation relationships and the idea of the limited lifespan. Personal identity and the sense of uniqueness are endangered, and life is demystified and replaced by 'artefacts'. As with any artefact, there is a determined outcome; there is no variation and chance ceases to be an important factor. Thus, with biotechnology, science seems to be on the verge of abolishing the last 'miracles' of life. People fear a cold, sterile future in which everything that exists does so according to preordained but worldly plans, a life that is fake without fate. To repeat, the distinction between the natural and the unnatural is made synonymous with the divide between the good and the bad.

Potentially having the power to alter basic aspects of the material and social world also implies that we will be forced into making decisions (Beck, 1986). After all, which are really more desirable, blue or green eyes, fair or dark hair? We will need to establish the criteria upon which to base our decisions. However, being able to decide also implies a responsibility for that decision, ushering in new forms of guilt. In this future scenario, being ill or having some physical imperfection would no longer be conceptualised in terms of a destiny that must be accepted, but would be seen as an error of judgement.

Further, how can we be responsible for something whose consequences we do not know? Scientists, industrialists and politicians are not considered to be capable of bearing these responsibilities. Instead, they are perceived as 'playing God', striving to create a new world out of genes, the building blocks of life. The public just wonders what sort of world this 'second Genesis'[8] will bring about.

Methodological appendix

Method

ALCESTE investigates the distribution of vocabulary in data material consisting of text (in our case, answers to an open question and interview transcriptions). The method was introduced by Max Reinert (1983) and produces a descending hierarchical classification. That is, ALCESTE separates classes of specific vocabulary that empirically co-occur without regarding the meaning of the words. In a second step these semantic word classes must be interpreted. To be able to trace the vocabulary back to its original context, ALCESTE provides a list of statements associated with each class. This helps to check if the interpretation of the single words is correct.

ALCESTE is a French program that can be used for analysing data in any language using Latin letters. To yield more precise results it is advantageous to exclude words such as articles, prepositions, pronouns and conjunctions from analyses, as well as to reduce plurals and conjugations to the word's root form. For that purpose ALCESTE needs dictionaries of these suffixes and function words. Since dictionaries are available for only a few languages, the working files of ALCESTE were modified by us to exclude a number of words (articles, prepositions and the like) for all languages required.

In order to obtain reliable results, ALCESTE computes two analyses using statements of slightly differing length. Only those statements that can be classified in a stable manner in both analyses are used further on. A stability coefficient indicates the percentage of statements that can be classified in a stable manner.

Data sets

The results presented in the chapter are based on two separate sets of textual data for six European countries. These are single and focus group interviews, on the one hand, and responses to a survey question, on the other. The two data sets were collected as part of a European Concerted Action research project, 'Biotechnology and the European Public'.

The Eurobarometer survey 46.1, which was conducted in each European Union country in 1996, was based on multi-stage random sampling methodology. It provides a representative sample of individuals aged 15 and over. In a face-to-face interview, respondents were asked the following question: 'Now I would like to ask you what comes to your mind when you think about modern biotechnology in a broad sense, that is including genetic engineering.'

Interviews with individuals were conducted in Austria and France (data from the latter country are integrated only to a small extent in this chapter); focus group interviews took place in Austria, the UK and Sweden. It was intended not to obtain a representative sample but to cover a wide range of perspectives on the topic of biotechnology. For that reason, persons associated with different socio-demographic categories were recruited. The interview questions covered the topics of medical applications, agricultural applications, genetically manipulated food, transgenic animals, reproductive genetic technology, control, risks

Table 9.2 *Main characteristics of data sets and results of analyses*

	France	UK	Sweden	Germany	Austria	Norway
Total number of responses (open question)	1,004	1,074	1,008	1,990	1,009	966
Number of analysed responses to open question (statements indicating 'I don't know' excluded)	812	973	836	1,990	814	730
Percentage of stable classified responses: open question	83	76	77	76	71	70
Number of classes	10	6	7	6	4	7
Number of interviews	20	–	–	–	18	–
Number of focus group interviews	–	5	2	–	7	–
Percentage of stable classified responses: interviews and focus groups	–	83	–	–	75	–

and consequences of biotechnological applications. Data from Austria and the UK were analysed by using ALCESTE, whereas data from Sweden were analysed according to the grounded theory approach (Glaser and Strauss, 1967).

Table 9.2 shows the number of respondents to the open question for each country, the number of responses analysed (for most countries, answers indicating 'I don't know' were excluded from the ALCESTE analysis) and the number of responses classified in a stable manner in the ALCESTE analysis. The stability of the results – ranging from 70 per cent to 83 per cent stable classified statements – can be considered as satisfactory for all countries. Table 9.2 furthermore indicates the number of interviews and focus groups conducted in each country, as well as the stability coefficient for the ALCESTE analyses of the British and Austrian interview data.

NOTES

1. Note that France was not included in the analysis of categorisations owing to the unavailability of data, but will be included in the following sections. Likewise, the Netherlands, Finland and Greece were included in the categorisation analysis, but will not be considered in the following section owing to missing content analyses.

2. One has to remember that we are dealing with different European languages here; differing vocabulary and language use can have an impact on the construction of the word classes.

3. Note that the survey took place before intensive media coverage about cloning Dolly the sheep started in 1997.

4. Throughout the text, literal quotations of responses to the open question are marked by the country code and the respondent's number. Statements from interviews and focus groups are marked accordingly. The codes are: AT = Austria, FR = France, GE = Germany, NW = Norway, SE = Sweden and UK = United Kingdom.

5. These cattle originated in central and upper Belgium and the breed was established in the early twentieth century. There is a large proportion of muscle hypertrophy in Belgian Blue, which is genetically inherited; this means that the animals are born with double thigh muscles, which give more meat and better productivity. Muscle hypertrophy also makes the internal organs of the animal smaller and caesarean delivery is frequently called for.

6. Even though the mouse bearing the human ear was not a product of xenotransplantation.

7. It is interesting to note that many of the images described in this chapter also emerge as part of the media discourse about Dolly the cloned sheep. Since the survey and interview data were collected before coverage about Dolly started, it can be supposed that media and public discourses are mutually determined. Media discourses equally reflect and influence public opinions (Wagner and Kronberger, 2001).

8. This was the title of an Austrian radio transmission about biotechnology.

REFERENCES

Bauer, M.W. and G. Gaskell (1999) 'Towards a Paradigm for Research on Social Representations', *Journal for the Theory of Social Behaviour* 29: 163–86.

Beck, U. (1986) *Risikogesellschaft. Auf dem Weg in eine andere Moderne*, Frankfurt am Main: Suhrkamp.

Geertz, C. (1973) *The Interpretation of Cultures*, New York: Basic Books.

Glaser, B.G. and A.L. Strauss (1967) *The Discovery of Grounded Theory. Strategies for Qualitative Research*, New York: Aldine.

Herzlich, C. (1973) *Health and Illness: A Social Psychological Analysis*, London: Academic Press.

Jodelet, D. (1989) 'Représentations sociales: un domaine en expansion', in D. Jodelet (ed.), *Les Représentations sociales*, Paris: Presses Universitaires de France.

Johnson, M. (1987) *The Body in the Mind. The Bodily Basis of Meaning, Imagination and Reason*, Chicago: University of Chicago Press.

Kronberger, N. and W. Wagner (2000) 'The Statistical Analysis of Text and Open-Ended Responses', in G. Gaskell and M. Bauer (eds.), *Methods for Qualitative Analysis*, London: Sage.

Lakoff, G. and M. Johnson (1980) *Metaphors We Live By*, Chicago: University of Chicago Press.

Luhmann, N. (1991) *Soziologie des Risikos*, Berlin: de Gruyter.

Moscovici, S. (1984) 'The Phenomenon of Social Representations', in R.M. Farr and S. Moscovici (eds.), *Social Representations*, Cambridge: Cambridge University Press, pp. 3–69.

Reinert, M. (1983) 'Une méthode de classification descendante hiérarchique: application à l'analyse lexicale par contexte', *Les Cahiers de l'Analyse des Données* 8(2): 187–98.

(1990) 'ALCESTE. Une méthodologie d'analyse des données textuelles et une application: Aurélia de Gérard de Nerval', *Bulletin de méthodologie sociologique* 26: 24–54.

(1998) 'Manuel du logiciel ALCESTE (Version 3.2) [computer program]', Toulouse: IMAGE (CNRS-UMR 5610).

Sloterdijk, P. (1987) *Kopernikanische Mobilmachung und ptolemäische Abrüstung*, Frankfurt am Main: Suhrkamp.

Wagner, W. (1998) 'Social Representations and Beyond: Brute Facts, Symbolic Coping and Domesticated Worlds', *Culture and Psychology* 4(3): 297–329.

Wagner, W. and N. Kronberger (2001) 'Killer Tomatoes! Collective Symbolic Coping with Biotechnology', in K. Deaux and G. Philogene (eds.), *Social Representation: Introduction and Exploration*, Oxford: Blackwell.

Wagner, W., F. Elejabarrieta and I. Lahnsteiner (1995) 'How the Sperm Dominates the Ovum – Objectification by Metaphor in the Social Representation of Conception', *European Journal of Social Psychology* 25: 671–88.

Wagner, W., G. Duveen, R. Farr, S. Jovchelovitch, F. Lorenzi-Cioldi, I. Marková and D. Rose (1999) 'Theory and Method of Social Representations', *Asian Journal of Social Psychology* 2: 95–125.

Part III

The watershed years 1996/97:
two case studies

10 Testing times – the reception of Roundup Ready soya in Europe

Jesper Lassen, Agnes Allansdottir, Miltos Liakopoulos, Arne Thing Mortensen and Anna Olofsson

In the autumn of 1996, ships carrying the annual harvest of soybeans for the European Union (EU) market were sailing from US harbours. This soya was intended partly for the European food industry, which uses soya as a raw material in the production of additives, food products and livestock feed. But, for the first time, the ships' holds contained more than traditionally bred soya. These shipments were mixtures in which up to 2 per cent comprised a genetically manipulated strain of soya. The new strain was called 'Roundup Ready'.

The soybean was the first genetically manipulated food to be marketed on a large scale on the European market. It was the culmination not only of many years of biotechnological research, but of the improving efforts of the agricultural industry as a whole, and of Monsanto (the US company responsible) in particular. The novelty of 'Roundup Ready' lay in the genetic modifications that rendered it resistant to Glyphosate, the active ingredient in 'Round-up', a herbicide produced by Monsanto and one of the most widely used agricultural chemicals in the world.

Thus, apart from being the first genetically manipulated product intended for consumption by the European market, the soybeans represented a bio-agricultural strategy with considerable economic and practical implications. This was also a technology that had already been heavily criticised for its potential environmental and social consequences (see, for example, Kloppenburg, 1988; Goldburg et al., 1990; Busch et al., 1991). As a result, the ships approaching Europe were not only transporting the first genetically altered food, they were also carrying with them the potential for serious conflicts – controversies that would not be confined to environmental, social or food-safety issues. As the controversy over 'Roundup Ready' soya escalated, critics began to argue that businesses such as Monsanto are jeopardising democratic processes. In this context, Monsanto's Roundup Ready soybean emerged as the European test-case for the food-gene technology project, and became the focus for more than a decade of public debates about the pros and cons of genetic manipulation in food and agriculture.

The introduction of a product such as Round-up Ready soya (or any other genetically manipulated food product) into the European Union was formally regulated by the Directive on Deliberate Release enforced in 1991 (EEC, 1990). This directive required manipulated organisms intended for release in the environment to acquire official approval and undergo risk assessment studies prior to import, marketing or release. Accordingly, Monsanto submitted an application to have Roundup Ready sanctioned by the UK in the winter of 1994–5. Approval from the British Advisory Committee on Novel Foods and Processes was obtained for the use of modified soya in food in February 1995 and as livestock feed in May 1995. Thereupon, in early spring 1996, an application for approval was sent to the European Commission. Subsequently the approval was considered by the Commission, submitted to hearing among the member states, and finally passed by the Commission in April 1996. Yet, despite this approval, no further progress was made with Roundup Ready soya. This was partly because the question of labelling was left open for interpretation by the national authorities, and partly because the member states were still able to demand that approval of food products containing genetically modified organisms (GMOs) could be dealt with at a national level. This situation was the result of an ongoing debate among the member states and the institutions in the EU on the so-called Novel Food Regulation – a set of rules intended to regulate novelties in the food area, including products containing or derived from GMOs.

The Commission had initiated work on novel food regulation in 1992, but resolutions were still pending in October 1996 when the soya controversy was heating up. This delay was due to disagreements among the member states in the Council, as well as among such EU institutions as the European Parliament, the Commission and the Economic and Social Committee (Behrens et al., 1997: 105ff.). Conflict centred on two fundamental issues. First, and most importantly, the question of labelling had been left open for interpretation by national authorities in the Directive on Deliberate Release. Secondly, debate focused on whether regulation should be based on an additional notification and/or approval procedure. Clearly, until the final version of the Novel Food Regulation was adopted by the Commission (on 27 January 1997), there was no clear perception of what a common policy on the marketing of genetically manipulated food products would involve (EC, 1997). Four months later, however, this new directive was put into force. Crucially, as a consequence, the door was left open for the formulation of national policies concerning the marketing of the products based on 'Roundup Ready' soya.

Seen at a societal level, it is interesting to observe how differently the soya story developed in the various EU countries during the four-month gap before the Novel Food Regulation was decided upon. At one end of the spectrum, the soya issue became a 'hot' issue in countries such as Denmark and Austria, where both public attention and political awareness were intense. At the other end of the spectrum were countries such as Greece and Italy, in which neither the public nor the political system paid the issue more than the most perfunctory attention during these months. In between these extremes were countries such as Sweden and the UK, with a moderate degree of attention directed at the soya issue.

The prospect of the introduction of modified soya struck all EU countries simultaneously, yet it elicited widely different responses in terms of both public attention and political action. This case therefore offers a unique opportunity to study the development of a controversy and political reactions to it. The two main questions this raises are, first, why did Roundup Ready spark controversy in some countries while being almost totally neglected in others; secondly, how we can explain the differences in how the political outcomes were shaped in those countries where soya *did* become an issue?

This chapter will offer some answers to these questions through an analysis of the soya issue in Denmark, Sweden, Italy and the UK. We do not claim that these four countries provide a full picture of the reception of the genetically manipulated soya in the EU. Rather, we contend that the fate of a social issue and the political reactions elicited can be understood only if we proceed from an examination of different national contexts. Explanations *must* focus upon the actors engaged in the conflict and the different material, cultural and political structures and traditions within which the potential issue evolves. These four countries have been selected in order to illustrate the variability in reception across the EU. Each country represents and embodies a different political culture; different agricultural interests in the biotechnology issue; and different public attitudes towards biotechnology in general, and food biotechnology in particular.[1]

The analytical background: the development and political impact of social issues

There is a long tradition of research into the question of how social issues develop and then disappear. It has been suggested that the evolution of social conflicts follows a systematic 'issue attention cycle' in which the social problem lasts for a relatively limited period of time, during which it passes through different stages until it eventually disappears from public view (for example, Downs, 1972). A social problem fading from view, it is claimed, will either leave a 'trace' in the shape of political action or simply melt back into oblivion. Such 'natural history' models are, however, open to criticism: first, because they lack the ability to explain the dynamics that guide the selection of potential issues and those that gain public attention and thereby become true public issues see for example, Hilgartner and Bosk, 1988); and, secondly, because their stipulation that social issues last for a relatively short period of time is arguably erroneous (Jasper, 1988).

Consequently, in attempting to explain the varying degrees of attention focused on the soya issue in different countries, this chapter will not proceed along the route preferred by 'natural historians', but will instead build on research into the development of social issues that stresses the importance of context and of actors asserting different positions (for example, Hilgartner and Bosk, 1998; Jasper, 1988). Following this line of research in explicating the development of social conflicts, the following core issues must be addressed: the discursive environment, the material setting, the political culture and the actors and their strength.

The discursive environment Here the question concerns how the new social issue relates to present fields of discourse. Those that are consistent with existing discourses are likely to benefit from this relationship and prosper at the expense of other issues. Thus, the general level and direction of the science and technology discourse will be of importance with respect to the controversy over Roundup Ready soya, as will be the more specific characteristics of the biotechnology debate. Closely related to this is the question of how the issue is defined or interpreted. Some interpretations will relate to and draw nourishment from existing discourses within the science and technology area, or even wider discourses.

The material setting This concerns the relationship between particular issues and wider economic structures. On the one hand, the general standard of living and welfare provision will be of importance in determining the amount of attention assigned to an issue, considered as a simple marginal utility relation. This view has been put forward as one explanation of different attitudes towards biotechnology in general in the EU (Nielsen, n.d.). On the other hand, the relative strengths of different national business sectors will affect the uptake of, and responses to, new social issues. In the present case, countries with powerful food and agricultural sectors are likely to produce different conditions for the emergence of the soya issue than countries with no such industry.

The political culture The development of an issue, once it has emerged as a prominent public debate, will be profoundly affected by a country's participatory tradition, as well as by its openness and, therefore, the public's access to information. Issues developing within a political culture with participatory traditions, where the views of different actors are incorporated in the policy process, are thus more likely to reflect these different views than are issues developing within countries with a more technocratic culture.

The actors and their strength Issues do not enter the public arena by themselves; they require 'social carriers' both to introduce them and to maintain their position on the public agenda. Issues will also encounter opposing actors, groups or individuals who will try to interpret them differently or even to remove them (Lukes, 1974; Bachrach and Baratz, 1962). Thus, the life of an issue depends not only on the resources (economic, knowledge, organisational, etc.) of the actors involved but also on their relations to the media and the political system. Closely related to the question of the actors is the question of 'problem definition'. Actors may use their resources to influence the way in which a certain issue is debated and thereby bring it into line with dominant discourses that either promote or reduce the longevity of the issue. Two of the opposing strategies within biotechnology have been to relate the industry to a positive image or a discourse of welfare, which includes industrial progress and prosperity, or conversely to implicate biotechnology in the discourses of social, environmental and health risks. These framing processes impact not only upon the ability of an issue to reach, and remain upon, the public agenda, but they may also influence the political reactions the issue evokes.

Of course, an analysis of the dynamics affecting the public profile achieved by the soya issue in different countries must take into account the political process targeting 'Roundup Ready' soya. On the one hand, actors may attempt to steer the political process as a means of achieving a congenial political outcome. Therefore the established political system (the government, the parliament and the bureaucracy) may be the main arena for the development of the issue, both because actors in the early stages of the conflict may want to persuade the political system to take up the issue and because, in the more mature stages, they may lobby for particular decisions to be made. On the other hand, the political system can become a dominant player in the development of the issue. This may happen either because outside pressure shifts the issue from the public into the political sphere, or because the government takes up the issue on its own initiative.

In explaining the outcome of a conflict, the first question to address, therefore, is what the political outcome was. The outcome is important because it represents the formal trace of the conflict: once the contest is over, its effects – new or altered regulation and new procedures (institutional learning) – will remain. A conflict may also leave a trace in the form of changes in everyday language, as well as by becoming enshrined in the stories or myths of the public and political spheres. In the long term, such traces will contribute to the construction of a shared frame of reference within the nation, and will thereby influence the trajectory of new controversies and political debates.

In this chapter, attention will largely be focused upon the formal traces of the 'Roundup Ready' soya issue; and particular attention will be devoted to the wording of the laws or political resolutions passed, and an investigation of whose interests these decisions advanced. This will involve, in accordance with the pluralists' tradition of political analysis, a study of the visible conflict and an identification of its winners (for example, Dahl, 1986). In addition, we will attempt to identify the processes involved in producing this actual outcome. For this reason, the analysis will also encompass an examination of the ways power was exerted in shaping the political result of the conflict. Here we make use of the concept of non-decision-making (Bachrach and Baratz, 1962, 1963), or how the interests of certain groups may be excluded by powerful actors controlling social values, political values and institutional practices in order to secure policies that minimise harm to their own interests. At the same time, we recognise that the outcome is also shaped by the more subtle and covert use of power, as manipulation or the exercise of authority, involving the deliberate shaping of the counterpart's wishes and interests (Lukes, 1974: 23ff.).

The media play a key role in the processes described by Lukes, Bachrach and Baratz, since they constitute an important stage not only in the marketing of counter-arguments but also in the attempts to control values, manipulate public opinion and demonstrate authority. Taking this a step further, we will analyse the relationships between the formulation of public policy and the activities of the media. In doing so, we accept in part the social constructionist view that a problem does not exist until it has been socially defined. This raises the central question of who determines how a problem is defined. As Rochefort and Cobb (1994) note, the definition and redefinition of problems serve as tools employed by opposing parties in order to gain advantage in the political conflict. By way of example, they claim that some actors may be inclined to define issues in narrow and technical terms since this might serve to confine the debate to just a few parties. Conversely, others may prefer a wider definition of the problem – as a matter of democracy or liberty – with the effect of involving more, potentially supportive, groups of actors.

During this process of definition, the media play a central role. This is partly because they provide important arenas for the development of the issue itself, but also because their control of these arenas enables them to influence the debate as, in effect, indirect actors. Problem definition invariably becomes a key issue for those social actors seeking to exert influence on the policy process. After all, certain definitions of the problem will significantly reduce the range of political actions considered appropriate; in this way, controlling the process of definition confers considerable influence over the political reactions to the problem. Pline (1991), for example, showed that proponents of biotechnology in the USA have successfully transformed the public image of their industry, generating a positive image that links biotechnology with economic progress. This has provided a secure platform for fighting off opponents on the political scene. Von Schomberg (1993) gives another illustration of this phenomenon in his analysis of political decision-making in relation to the release of genetically engineered organisms. He demonstrates that ecologists and biotechnologists are advancing two different definitions related to the issue of deliberate release. Ecologists maintain that 'release' is a new issue, and one that therefore demands that that new knowledge (i.e. test systems) is established before any major decisions are taken. Biotechnologists, in contrast, argue that there is nothing novel about genetically engineered organisms, that they are comparable to organisms produced by traditional methods, and that existing knowledge and methods of risk assessment should for this reason be deemed sufficient.

From the example put forward by von Schomberg, one would expect the biotechnologists to prefer issue definition to take place in a restricted technical arena. In contrast, ecologists will gain from moving the issue into the public arena via the mass media, partly because they could benefit from what Pline (1991) calls issue association (i.e. linking the issue to prominent discourses), and partly because in this way they might be able to draw other organised interests, or even public opinion, onto their side. This emphasises the key role the mass media play in issue definition as part of the democratic process, and raises the further question of the extent to which the media themselves reflect public opinion. The growing influence of the media in the political process in general highlights the importance of access to the media for those desirous of securing a particular political outcome. Moreover, as the media tend to become more concentrated, ordinary citizens are less and less likely to gain access to media expressing their views. Of course, public opinion will, to some extent, affect the political process through the activities of different non-governmental organisations (NGOs) representing the different sides of the issue. Following Schattschneider (1975), however, this raises the question of the NGOs' legitimacy. It is arguable that the media do not represent the public because they instead tend to report the attitudes of large collective actors (Schattschneider, 1975). Following this argument, any assessment of the effectiveness with which public opinions influence the political process must determine the extent to which the NGOs themselves represent the public.

This chapter does not aim to provide complete answers to all these questions. Instead, we will focus on the role of the media in this debate between early autumn and the settlement of the Novel Food Regulation. More specifically, we will show how the arguments related to the soya issue put forward in the media can be interpreted as a process of issue definition. In doing so, the media will be shown to be a basic part of the political process.

The reception of Roundup Ready soya in Denmark

Aside from occasional references to GM soya as one of the most advanced applications of biotechnology, soya was not an issue in Denmark until Monsanto delivered its application for approval to EU member states in 1995. This application prompted two questions to be asked in the parliament from the left-wing opposition Socialist People's Party (Socialistisk Folkeparti or SF) during the spring of 1995, but the story never provoked more than a few media articles. At this stage, however, the soya question became part of an attempt by SF and Enhedslisten (EL), another left-wing opposition party, to attack Danish policy regarding the Novel Food Regulation. They ensured that the government endorsed the unanimous parliamentary decision of 1994 (Folketinget, 1994) that, among other things, stated that it was Danish policy to demand, in both Denmark and the EU, the labelling of food products containing and/or produced with the aid of GMOs.

When Monsanto's application was sent for its hearing among the member states in the spring of 1996, for the first time soya became an issue – albeit a minor one – in the media as well as in the parliament. And the story was conjoined with that of Ciba Geigy's manipulated maize, for which Novel Food Regulation was also pending. Again the debate was led by EL and SF, and again the main concerns were to ensure that all soya products (as well as all other products based on GMOs) are labelled according to the parliamentary decision of 1994, and that organic products are kept free of GM organisms at all stages of production. These debates were also represented in the media during the spring and summer of 1996. Actors presented the pros and cons, and the possibility of banning genetically manipulated soya was also raised by representatives from EL and Greenpeace.

On 14 March 1996, the Danish government voted against Monsanto's EU application. Its decision was largely prompted by the absence of labelling requirements, but, in addition, the government drew attention to the fact that Danish regulations require that the Minister of Health issue an approval before such products can be released in Denmark. Monsanto applied for Roundup Ready soya to be approved by Denmark in May 1996, and permission was granted on 18 November, following the Levnedsmiddelstyrelsen's (the National Food Agency) insistence that the manipulated soybean should not be considered to be different from the traditional beans. However, the conditions of approval stated that imported beans must be accompanied by information about the content/or possible content of GM soybeans, so as to permit labelling 'in a suitable way' (Levnedsmiddelstyrelsen, 1996a).

The soya issue in Denmark: November 1996–January 1997

The Danish media closely followed the soya issue in the period analysed.[2] The media focused on basic arguments and events that promised to make good stories, and succeeded in making Roundup Ready soya one of the most prominent social issues of autumn 1996. In the first half of December, the newspaper *Politiken* alone carried seventeen longer articles, fourteen letters from readers and two leading articles dealing with Roundup Ready soya. Moreover, the issue was addressed in ten questions and one enquiry in parliament between mid-November and mid-December.

The controversy in Denmark really began, then, in mid-November, when it became known to the media that cargo ships were crossing the Atlantic carrying modified soybeans. They learned that one of them, *Hanjin Tampa*, was destined for Aarhus, Denmark. This ship was transporting 23,000 tons of soybeans to the Danish company Central Soya, a subdivision of the French company Eridania Beghin-Say, owned by the Italian Montedison. Crucially, 0.5 per cent of this cargo was derived from GM plants. It had been known since the spring of 1996 that genetically manipulated US soybeans were likely to reach European – including Danish – markets during the autumn. Yet these approaching ships were to assume immense symbolic significance. And it was this event, above all, that sparked the major debates in the media as well as in parliament. The parliamentary soybean debate ceased only after the Minister of Health issued guidelines for labelling GM soybeans on 6 December; and a parliamentary decision on GMOs and organic products was reached on

12 December. Nonetheless, the soya issue continued to appear in the media until the spring of 1997.

By mid-November, the ships were nearing European ports and GM soya was emerging as a major media and policy issue. But though GMOs had been approved in Denmark, the necessary guidelines had not been put into place. In effect, therefore, the Minister of Health, Yvonne Herløv Andersen, had been ignoring the 30 October recommendations made by her own administration (Levnedsmiddelstyrelsen, 1996b), as well as the parliamentary decisions of 1994, which had demanded the detailed labelling of GM products.

To resolve this uncertainty, the minister was called in for consultation by the Parliamentary Committee on Health on 28 November. At this consultation, she expressed her view that the labelling requirements stipulated by parliament in 1994 had been met sufficiently by the display of signs on shop shelves. She repeated this statement during a debate in parliament on 4 December (Folketinget, 1996), which resulted in both the left and the right insisting that the 1994 decision was quite unequivocal in its demand for each product to be labelled. Alongside this debate, the left-wing opposition repeatedly raised broader issues concerning the relative merits of GM organisms and organic products.

At about the same time, towards the end of November 1996, the EU parliamentary working group on the Novel Food Regulation met with the fifteen national Ministers of Health and settled upon a preliminary agreement on the labelling matter. In the process, the official Danish policy on labelling suffered a defeat. The preliminary agreement still required the approval of the EU Parliament and the Council, but demanded labelling only of products that contain live GMOs. Where appropriate, the agreement also permitted the use of labels informing consumers that a product contains no ingredients produced by gene technology. From this moment it became clear that, within a few months, Danish policy would be replaced by a rather more relaxed set of EU regulations.

Despite this development in Brussels, the Danish Social Democrats (the largest of the three parties in the minority government) proposed the adoption of a policy analogous to that agreed in the Danish Parliament of 1994. This would have made mandatory the individual labelling of all products, or ingredients thereof, containing GM organisms. This move, combined with opposition pressure on 4 December, caused the Minister of Health (from the small centrist coalition party CD) to make a U-turn when issuing labelling guidelines on 6 December. The guidelines now stated that 'Food and food ingredients produced on the basis of genetically manipulated soybeans, intended for sale to consumers, must be labelled accordingly' (Levnedsmiddelstyrelsen, 1996c). They also specified that labels must appear on the product, and further stipulated that additives, such as lecithin, are exempt from this labelling requirement. Finally, the guidelines allowed for the positive labelling of products produced without manipulated soybeans.

In formulating these guidelines, the policy-makers first produced regulations. However, subsequent controversy, and particularly the lengthy debate in parliament on 12 December, showed that there were outstanding questions demanding resolution. One question involved keeping organic products free from GM soybeans, or, for that matter, any other GMOs. Another issue concerned the likelihood of imported soya entering the food system *via* livestock feed and meat consumption: even though the greater part of soya imports were used in animal feed, the implications of this fact were not covered by the guidelines.

Meanwhile, *Hanjin Tampa* arrived in Aarhus on 11 December to be met by thirty to one hundred activists representing the newly founded organisations Økofolk Imod Gensplejsning (Ecopeople against genetic engineering) and Danmarks Aktive Forbrugere (Danish active consumers), who successfully prevented the unloading of the cargo.

The 12 December parliamentary enquiry had been scheduled by Enhedslisten on 15 November, before the soya issue had escalated. It was due to address general issues related to GMOs and to discuss the possibility of an outright ban. However, in the event, soya became one of the chief bones of contention, and the result of the enquiry was a rejection of the Enhedslisten agenda. In its place there emerged an alternative agenda, championed by the government parties and the Socialist People's Party, which called for the adoption of rules to ensure the availability of organic products uncontaminated by GM organisms at every stage of the production chain. In this way, two of the main themes of the debate were settled, while the use of GM organisms in livestock feed was unaffected and the question of a ban was rejected by most parties. The day after the enquiry there was a meeting of the parliamentary EU Committee, where all except the two socialist parties (SF and Enhedslisten) granted the government a mandate to vote for the proposed version of the Novel Food Regulation. On the morning of the same day, police moved in and broke up the blockade at Aarhus harbour, the off-loading of the cargo began and soya ceased to figure on the public agenda as a single issue.

Outside parliament, the demand for a ban was supported by a number of NGOs. After Greenpeace International had taken up biotechnology as an issue, Greenpeace Denmark became one of the most prominent NGO participants. Greenpeace's resistance is based partly on environmental considerations, and partly on health grounds. The environmental argument for securing a ban focuses on the increased use of pesticides; and their contentions were strengthened after Round-up breakdown products were found in UK, German and Dutch groundwater. On the basis of this argument, Greenpeace rejected and criticised the proposed introduction of labelling. Similar arguments were put forward by most NGOs. The only significant exception was the rather more positive National Consumers' Association, which supported the labelling solution.

The dynamics of the soya controversy in Denmark

The dramatic and newsworthy theme of a ship, laden with a highly controversial cargo, approaching Denmark was noted by all newspapers. But this never constituted the main story. At the risk of oversimplification, we argue that the main issue considered in newspaper stories was how the Danish labelling decision could possibly be upheld in the face of both direct attack from Monsanto and the introduction of EU regulations concerning novel foods. The dramatic tension in the stories was related to themes or discourses that aroused the readers' interest, such as decreasing food quality, national versus European regulation, assaults on democratic rights or the failure of politicians to solve the nation's basic problems.

There can be little doubt that the intensity of this media coverage can in part be attributed to the fact that the soya issue was framed as a question of labelling. It therefore tapped into the discourses about food quality that were so prominent during the first half of the 1990s. This food discourse arose from the prevalent tendency to perceive food quality materially as a matter of taste, safety, texture, and so on, but also as including the ethical, environmental and social aspects of the way food is produced (Lassen, 1993). In the 1990s, this was seen in the near-explosive growth in the purchase of organic products and, after the controversy over Brent Spar and the French nuclear experiments, with the emergence of a 'political consumer', prepared to express his or her politics through purchasing decisions. These historical circumstances both serve as a background for understanding why soya became a contentious issue, and also help explain why labelling per se became the dominant way in which the issue was framed both in the media and in the parliamentary debates. The ability of the consumer to identify the right commodity depends on information about the product itself. Therefore labelling is important and becomes a key issue for the political consumer.

Opposition to the unregulated introduction of GM soya also drew support from an environmentalist sub-discourse that had been prominent in Denmark for more than a decade. This sub-discourse related to concerns over the quality of groundwater. This was an especially sensitive issue in Denmark because the country is highly dependent upon this source of water, and until recently it had had a reputation for being clean and pollution free.

Behind the media stories, though not explicitly elaborated, lay a further theme: the discrepancy between the moderately positive attitudes to biotechnology held by most politicians and opinion-leading media personalities, and the generally negative attitudes of – perhaps – a majority of the population. An opinion poll initiated by *Berlingske Tidende*, one of the three leading national newspapers, confirmed the existence of this disparity: 68 per cent of respondents expressed the belief that GM food products should be prohibited by law (Gallup, 1996). This figure was not stressed by the newspapers (probably because it was considered unreliable) and was not taken up by the rest of the media, although the majority of the population unequivocally wanted a ban. None of the papers supported an outright ban, but all lodged strong support for consumers' right to choose and to be informed about possible GM ingredients in food products. This right was generally accepted and not disputed, which further explains the importance attached to the issue of labelling. In all interviews, the importation of Monsanto's soybeans was seen as a provocation and the importer was, for obvious reasons, placed in a defensive position.

The prerogative of the EU in regulating national labelling policy was not disputed. As a result, deeply rooted conflicts over the role of the EU were not explicitly aired in the media coverage. This was partly because of a widespread belief that the coming 'novel food' directive would in practice be equivalent to existing Danish labelling regulations. Alternatively, it was held that, even if the directive did not reproduce Danish regulations regarding soybeans, a strict reading of the EU decision could be imposed that would not significantly alter the Danish regulations. Where the delusional nature of both positions was remarked upon, criticism was diverted into another popular theme: the shortcomings of Danish politics.

The media debate shows two peaks in the level of coverage: between 3 and 6 December and between 12 and 14 December. On the whole, the problems related to the Danish political scene were dominant in the media coverage of the soybean case. Moreover, compared with other countries, these problems were of a special kind and serious enough to account for their prominence in the Danish media.

In November, the Minister of Health had approved Monsanto's application to begin marketing Roundup Ready soybeans as food, a decision she could hardly have avoided considering the recommendations of Danish scientific authorities. However, this meant that she was obliged to issue guidelines explaining how the earlier parliamentary decision about labelling should be implemented. Monsanto's refusal to separate genetically manipulated soybeans from traditionally bred beans made that practically impossible without her bending the formulation of the parliamentary decision. And this is what she tried first. Doing so proved to be a political miscalculation, and she was forced to replace her first guideline with a stricter version that interpreted the earlier decision in a literal fashion. This happened in the period around 4 and 6 December and produced a storm of media interest, with the event being covered by most newspapers as well as the electronic media. One week later, the arrival of the soya ship and the beginning of the parliamentary debate requested by Enhedslisten one month earlier happened to coincide; the media coverage reached its peak during 13–14 December.

What can be seen as remarkable in the representation of these two events is the difference in focus between the media and parliament. The media did not pay much attention to the fact that the parliamentary decision about organic food labelling had been almost unanimously accepted and was considered by all, even by Enhedslisten (the left-wing party that had initiated the debate), to be a demonstration of constructive political resolution. In the media, the focus was still on the labelling problem, and addressed such questions as: how can the decision be implemented? how could it be made compatible with the coming EU regulation? how might it benefit consumers? Questions such as these emerged during the following weeks as attention shifted from Denmark to Brussels, and perhaps also when consumers began wondering why they never saw any labels in the shops.

One of the remarkable results of the controversy over Roundup Ready soya in Denmark was a restructuring of organised resistance to genetic engineering. Previously, in the 1980s, organised resistance had been virtually monopolised by NOAH, the Danish Friends of the Earth. Then, as biotechnology declined in prominence after 1990/1, NOAH had also disappeared, and organised resistance with it. Yet, during the winter of 1995/6, the soya conflict dramatically altered this state of affairs. NOAH reappeared as a critic of the new biotechnologies, although not at the same level as earlier and this time sharing the spotlight with several new NGOs and with some older organisations such as Greenpeace and the Danish Society for the Protection of Nature. Despite this diversification in organised resistance, late in 1997 when eighteen different NGOs, organised by NOAH, joined together to collect signatures against genetically manipulated food it became apparent that they had common interests. In the wake of this development, for the first time in Denmark biotechnology was challenged by activists prepared to take direct action, and they sought to prevent the unloading of soybeans in Aarhus harbour. Until this episode, organised resistance had 'played by the rules' and had mainly focused upon raising public awareness.

The reception of Round-up Ready soya in the UK

After the approval of Round-up Ready soya by the US Food and Drug Administration (FDA) in September 1994, Monsanto applied to the Ministry of Agriculture, Fisheries and Food (MAFF) in the UK. This was the first in a series of applications that soon included Canada, Mexico, Argentina, Japan, the Netherlands, Denmark and eventually the European Union. The UK had already been the first country to approve another GM food, a tomato paste, which had proved to be commercially successful.

In February and May of 1995, Roundup Ready soya obtained official approval for food and feed respectively. This approval was based on recommendations from the Advisory Committee on Novel Foods and Processes, an independent scientific committee established in 1988 with the mandate to advise the government on the appropriateness of new foods for human consumption. The committee recommended that the new soya be approved without caveats. The application then moved to the European Commission for approval and further clarification on aspects such as labelling.

During this time there was no official debate in Parliament on the issue, and there were only fifteen articles in the press referring to it. However, the scene changed dramatically when it became clear that Monsanto was shipping the new strain of soya to Europe mixed in with the ordinary crop. When this news broke, various NGOs in the UK were active in trying to stop the shipment, instigating a public debate that was taken up in the media and in Parliament.

The soya issue in the UK: September 1996–February 1997

Following news of the Monsanto shipment, the debate was lively in parts of the media. During the peak period of the debate, a total of one hundred articles were written in the British press concerning the soya issue. There is an obvious difference between the coverage of the tabloid and the broadsheet daily national press. Only the broadsheet or 'quality' press – the *Daily Telegraph*, the *Financial Times*, the *Guardian*, *The Independent*, and *The Times* – tackled the soya issue, providing 81 per cent of the total press coverage. The tabloid press had a minimal input into the debate, amounting to only 5 per cent of this total; the rest of the articles (14 per cent of the total) appeared in the international press and magazines. The press covered a wide spectrum of arguments for and against the introduction of GM soya. It also represented the voices of all the main actors in the debate: industry, NGOs (Greenpeace, the Green Alliance) and the scientists.

The industry viewed Round-up Ready soya primarily as a safe, economical product and an unmistakable step towards the elimination of world hunger. The official regulatory stance over Roundup Ready soya, as stated in the reasoning points on which the FDA had approved the soya, was used to warrant the claims over its safety. The assertion that GM soya is safe was justified on similar grounds to those invoked by the FDA, with particular emphasis being placed on the argument that there is no substantive difference between GM and ordinary strains. That GM foods had acquired a negative image was acknowledged, but this was attributed to ignorance about the true status of the technology and misinformation given to the public by interest groups.

The industry clearly saw winning consumer acceptance as its objective, and it clearly acknowledged the need for an information and image campaign. Indeed, following this line of thinking, the European biotechnology industry has recently launched just such a campaign.

Moreover, it is significant that, apparently taken by surprise by the public reaction to the Roundup Ready shipments, the industry's arguments and claims changed over time. The argumentation of the early days of the debate showed considerable confidence as it evolved around the safety issue of the Roundup Ready soya, with the sole backing coming from its acceptance by regulators around the world. Yet it soon became apparent that GM soya raised doubts beyond the specific regulatory issues, and could not be separated from debates over the acceptance of biotechnology as a whole. As the focus of the dispute shifted from the specific product to the technology in general, questions of morality entered the debate. At that stage, the industry changed its main line of argument: rather than emphasising safety issues relating to Roundup Ready soya, it began to herald biotechnology as the solution to global problems.

Scientists, another main actor, approached the debate from a technical point of view. Their arguments concentrated on specific technical aspects of the Roundup Ready soya case, including the regulatory safety check procedures and past genetic research. They also questioned the naturalness of genetic engineering technology, the integrity of the regulatory procedures for the acceptance of biotechnology products, and even the ethical credentials of the scientific research itself.

Furthermore, the scientists directly challenged the official line of the regulatory authorities. In particular, on the basis of a recent discovery that foreign genes can pass into human intestinal cells, they strongly criticised claims that Roundup Ready soya is fundamentally the same as ordinary soya and therefore unlikely to produce any negative side-effects upon human consumption.

On the whole, the scientific community adopted a critical stance on biotechnology. Of course, scientists were not unanimously sceptical of

GM soya; after all, the regulators who had approved the soya also based their decision on scientific arguments. But most of the scientific arguments to reach the media, and therefore that contributed to the public debate, were strongly critical.

NGOs (environmental groups and consumer associations) took a strong and unambiguous stance in this debate. They advanced arguments at three different levels. First, at the scientific level they recapitulated the arguments put forward by sceptical scientists, especially those who called into question the safety of GM foods for human consumption. Secondly, at the level of the status of biotechnology as a whole, they portrayed the technology as unnatural and invoked technical procedures such as those involving the transferring of genes from animals to plants in attempting to evince this. Thirdly they questioned the morality of the political decision-making process, especially with respect to the labelling issue (i.e. the idea that all foods containing genetically manipulated ingredients should be clearly labelled). Labelling was at the heart of the NGOs' position during the Roundup Ready soya controversy; and this echoed the will of the majority of the public (82 per cent), who wanted genetically manipulated foodstuffs to be labelled (European Commission, 1996).

Between September 1996 and February 1997, the soya issue was debated on ten occasions by Members of Parliament in the House of Commons. The debate was mainly between members of the Opposition (at that time the Labour Party) and the Minister of Agriculture, Fisheries and Food (of the then Conservative government), whose ministry was responsible for overseeing the introduction of new foods into the food chain. In Parliament, the Opposition's case was clearly based upon the environmental risks stressed in the media debate. Accordingly, it laid particular emphasis on long-term risk assessment and the question of labelling. The Opposition was especially mistrustful of existing risk assessment schemes, and used the analogy of BSE to claim that the consumption of GM foods might have unforeseen long-term effects on health. This was reinforced by the scientists' argument that the consumption of antibiotic markers in GM foods might lead to a resistance to antibiotics. Further, the Opposition claimed that foods containing Roundup Ready soya should be labelled in view of the public's mistrust of the industry and the technology itself. Overall, the Opposition assumed an unambiguous position towards agricultural applications of biotechnology. It also displayed considerable sensitivity to lay fears, openly acknowledging the arguments advanced by organised bodies, for example religious groups, retailers and consumer organisations.

Largely in response to the Opposition's attacks, the government itself also focused its attention on the issues of risk assessment and labelling.[3] The Conservative government had set up the existing mechanisms of risk assessment, and unsurprisingly, therefore, expressed considerable confidence in their efficacy. The government dealt with the issues of labelling and GM contamination with apparent resignation. Although expressing sympathy for the critics of existing labelling policy, the government claimed that it was unable to take any action because of European and international regulations. But it is also clear that the government was unwilling to inconvenience industry by introducing constraints on its activities. Aside from expressing the government's support for industry, those government spokesmen involved in the debate adopted the biotechnology industry's preferred argument: that by increasing crop yields GM foods would help alleviate world hunger.

The dynamics of the soya controversy in the UK

In conclusion, we see that the media and parliamentary debates concerning the introduction of Roundup Ready soya in the UK followed parallel courses. As the debate developed, it was framed in two main ways: GM soya as a question of food quality and as a matter of progress and economics.

The soya issue was framed as a matter of food quality primarily by the parliamentary Opposition, and, in the media debate, by the NGOs. This framing, one that stressed safety issues and labelling, was strongly influenced by the prevalent UK discourse concerning BSE. The way in which the British government had handled the crisis over infected British beef led to a deep public mistrust of the government's competence to form reliable judgements on questions of food safety. Further, this dissatisfaction only accentuated existing discourses on food safety provoked by such issues as salmonella-infected eggs. Thus, the BSE crisis provided a point of reference against which every new issue of food quality was judged and developed, including GM soya. The question of BSE and food quality was consistently promoted as a matter of science and technology, and the discussion did not touch on ethical concerns. For this reason, the controversies surrounding GM soya were exclusively concerned with future health risks that might emerge.

Moreover, the Opposition repeatedly accused the government of acting in an irresponsible and underhand manner during the BSE crisis, on the grounds that it had ignored vital scientific reservations and withheld such information from the public. The GM soya debate was set against this background, with the parliamentary Opposition insinuating that the public no longer trusted either the government or industry.

The overall result of this was to relocate the debate's chief point of reference from the sphere of expert opinion to that of public choice. The very considerable support shown by the government for the labelling of *all* products that contain (or might contain) manipulated foodstuffs indicates a will both to regain public trust and provide a means of public regulation in the GM foods issue.

The second way in which the debate was framed, in this case mainly by the industry and government, was related to technological progress and the economic competitiveness biotechnology might bring to the UK. Here an important point of departure was the fact that the UK has the second-largest biotechnological industry in the world, with 250 active companies in 1998. In this context, industry and government saw a serious danger in that any decision that threatened the production and/or marketing of manipulated soya would threaten the viability of this commercial sector.

Having this material setting in mind, it is not surprising that the government consistently stressed the positive aspects of this technology. In a similar fashion, the industry itself focused on the alleged ability of GM crops to produce more and cheaper food. Framing manipulated soya as an economic issue limits the range of alternative arguments and solutions, because it makes it clear that any decision will have consequences for the economic competitiveness of the country – and, taking it to its logical conclusion, also for the welfare state.

In addition to these two framing devices, the issue of GM soya was less frequently framed and debated in terms of the relationship between national and European government and of notions of trust in public authorities. In the former case, the inability of the government to act on the labelling issue was attributed to the stifling effect of European legislation. Allusions to the limits placed upon national independence by the UK's membership of the European Union may be identified, but this never developed into a full-blown debate either in Parliament or in the media. The issue of how much trust people should have in public authorities, especially after the BSE crisis, was also brought into the debate. Both the media and parliamentary debates contained references to the public mistrust of official decision-making procedures. Again, however, these never escalated into major issues.

The reception of Roundup Ready soya in Sweden

In Sweden, GM soya first emerged as a public issue at the beginning of 1995, when Monsanto submitted its application for approval to MAFF in the UK. The Swedish Board of Agriculture handled the application during the hearing phase, and concluded that the import of GM soya should not be restrained by Swedish legislation. In the spring of 1996, the Board handled the application for the import of GM soya prior to the formal voting procedure of the EU member states. Once again the Board of Agriculture assumed a positive stance. The Ministry of Agriculture, which was responsible for dealing with GM plants, did, however, ensure that the matter was discussed by other relevant ministries. The Ministry of Agriculture suggested that the proposal by the Swedish Board of Agriculture be followed, but the Ministry of the Environment registered its opposition on the grounds of contrary public opinion and the potential for negative environmental consequences. Conversely, keen to avoid jeopardising trade relations with the USA, the Ministry for Foreign Affairs endorsed the Board's proposal. Yet, as a result of this round of consultations, three days later the proposed approval was changed into a rejection. The formal outcome of this was that Sweden voted against the EU proposal to approve GM soya, even though Sweden had not objected to the initial application.

This rejection was based mainly on political arguments, and was not clearly supported by scientific evidence of risks, as required in the EC Directive on Deliberate Release (EEC, 1990). An important consequence of this first stage of the issue was that the government decided to change the Swedish statute that dealt with applications for approving the release of GMOs. From then on, the government, rather than the competent authorities, was to express its views as early as the 'objection phase', and not simply at the formal voting stage. Later on, when the EU had approved the import of genetically manipulated soya, Sweden accepted this and made no further moves to prevent its import.

Roundup Ready soya shocked the Swedish into thinking more seriously about GMOs. So far, the debate over gene technology had focused mainly on issues such as medical applications, formal regulation and ethical issues, while GM food had been only a minor subject. From the beginning of 1996, however, the debate escalated and came to focus largely on GM soya. The Centre Party (former Agriculture Party) was clearly hostile to the planned import of GM soya into the EU. At this point, the Swedish food distributors and retailers entered the fray and they too expressed themselves in opposition to the import of GM food. Furthermore, on 23 October 1996, fifteen NGO spokespersons issued an appeal in the large-circulation national quality newspaper *Dagens Nyheter* (Ström et al., 1996), urging the food industry and distributors not to buy GM soya from the USA.

The soya issue in Sweden: October 1996–January 1997

After the appeal issued by the NGOs, the issue received greater attention in the media as well as in parliament. Between October and the end of 1996, about 300 articles addressing the question of GM food were published in all kinds of newspapers and magazines. Most of these, however, dealt with the EU's decision, a couple of days before Christmas, to approve the Ciba-Geigy GM maize. In parliament, there were a number of questions on this subject and MPs' bills were prepared, mainly by the Green Party, the Centre Party, the Christian Democrats and the Left Party. All of these expressed a distinctly negative attitude towards GM food, and GM soya in particular. Indeed, a majority of MPs argued for labelling regulations.

As a result of the interventions by the Centre Party and the Green Party, a short debate was held in parliament on 29 November 1996. Opponents of GM soya argued that Sweden should prevent its import by invoking article 16 in the EC Directive on Deliberate Release (EEC, 1990), on the grounds that the potential consequences for the environment and possible health risks were unknown. As an illustration of the potential hazards of GM food, opponents offered the example of the introduction of the Nile perch in Lake Victoria. The very same example had been used, a few days earlier, by a representative from Greenpeace in *Göteborgs-Posten*, one of the largest-circulation quality newspapers (Berg, 1996). This camp also claimed that, if the use of GM soya were permitted, then other products – such as GM oilseed rape and GM sugar beet – that are not grown in Sweden should also be permitted. Furthermore, it was argued, if it is impossible to prevent the import of GM soya, a proper labelling policy had to be implemented, giving consumers the option of choosing whether or not to buy GM food. Parliamentary opponents insisted that they identified with, and were representing, the consumer's position. This is quite clear from the statements of critics such as Gudrun Lindvall (MP): 'We consumers are not moderately restrictive in this case; we don't want this kind of crap' (Lindvall, 1996).

The Agriculture Minister, Annika Åhnberg, explained the official Swedish position on genetic engineering in general, and GM soya in particular (Åhnberg, 1996). She did not oppose the EU decision, and claimed that article 16 could not be used because the potential risks of releasing GM soya were insufficiently serious and because growing soya in Sweden is impossible. Her view, consistent with her previous statements, was that Sweden should be restrictive towards genetic engineering without rejecting the possible benefits it might bring. Indeed, she hoped that Swedish industry would successfully adopt this new technology. On the whole, Annika Åhnberg gave the impression of being well informed and quite positive about the future uses of genetic engineering.

An analysis of the media debate in Sweden reveals that thirty-nine articles addressed this issue.[4] Thus, GM soya and the US shipments to Europe constituted a media 'event', but not a very significant one. It is noteworthy that the majority of the articles appeared towards the end of December, and were triggered by the more intense debate over the EU's approval of Ciba-Geigy's GM maize. Looking at the actors and how their positions were represented in the media, a pattern similar to that of the preceding stage is discernible. The retailers promised to boycott food containing GM ingredients and NGOs such as Greenpeace and the LRF (the National Organisation for Farmers) expressed criticism. However, the LRF and the retailers were mostly preoccupied with consumer reactions, whereas Greenpeace was more concerned with environmental and health risks. Conversely, the politicians were fairly positive and saw an opportunity for enhancing food quality and improving Sweden's future economic performance. Actors such as scientists and industry hardly participated in the media debate at all.

Taking a closer look at the arguments put forward by the different actors, it appears that the representatives of NGOs sought to prohibit the selling of GM food on environmental and health grounds. For instance, they stressed the potential for threats to biodiversity and the unpredictable long-term consequences of an increased use of pesticides. Furthermore, GM soya was not perceived as something from which consumers could benefit. On the contrary, it was seen in a negative light because of the risk of increased allergenic reactions and the spread of antibiotic resistance.

Critics also demanded labelling as a means of protecting consumer interests. Likewise, the retailers argued that GM soya should be labelled, though they did so on the presumption that consumers do not wish to buy GM products. The retailers also perceived the need for information campaigns to give people a better foundation upon which to make well-reasoned decisions.

Journalists also appeared as actors. Most of them articulated sceptical and negative attitudes, insisting that GM soya should not be imported into Sweden because of risks to health such as antibiotic resistance and allergenic reactions. Journalists also reiterated the claim that more information needed to be supplied to the public, on the assumption that the populace was too ignorant to make its own proper judgements.

Representatives of the government (most often Annika Åhnberg) saw genetic engineering, in general, as a progressive field encompassing many opportunities in terms of both economics and increased food quality. The risks were viewed as exaggerated and the opponents of GM as reactionary. Moreover, the government stressed the fact that a common agreement in the EU on labelling should be given the highest priority since the issue was not confined to Sweden.

Many of the published articles comprised small news items or short articles from news agencies that did not tend to focus on the different actors. It is interesting to note that, in this kind of article, genetic engineering is by definition considered to be something negative. This hostile climate is exemplified by an account in *SydSvenska Dagbladet* of the NGOs' blockade in Aarhus, Denmark, which does not give the reader an explanation of why the demonstration occurred. The author baldly states, 'The ship contains 23,000 tons of soya beans, of which some are genetically manipulated' (Nilsson, 1996), and feels able to assume that the reader will understand the justification for the demonstration.

There was almost no significant reaction after the adoption of the Novel Food Regulation by the Commission on 27 January 1997. On 7 April 1997, the food industry and the retailers voluntarily agreed upon labelling certain products that contained GM soya. At this point, the debate faded to some extent. However, a new issue had now been placed on the public agenda, and from then on the public was aware of what had been dubbed 'gene food'.

The dynamics of the soya controversy in Sweden

Overall, the amount of attention paid in Sweden to the issue of GM soya has been rather modest, especially at the political level. This relatively low level of debate can partly be explained by the fact that none of the ships sailing from the USA were bound for Sweden. In addition, for simple biological reasons, soya cannot be grown in Sweden.

When the issue was discussed by actors other than the government, debate was dominated by Greenpeace, the Green Party and the Centre Party. As demonstrated above, environmentalist and consumer groups did show a fair amount of interest in the issue. Their comments were solicited during the early phase of the debate, but their overall impact on the political process seems to have been marginal. The actors who opposed GM soya did so for different reasons: retailers, politicians and NGOs (such as LRF) were sensitive to the views of consumers and concluded that the public does not want GM food; journalists and NGOs such as consumer organisations and Greenpeace, on the other hand, perceived the potential for environmental and health hazards and pointed to the fact that consumers have nothing to gain from the import of GM soya.

The dominant framing of the issue in parliament and in the media seems to have been very similar, with GM soya being identified in both arenas as a possible threat to the environment and, to some extent, to public health. As in Denmark, this way of framing the issue, although dominant, did not result in a rejection of GM soya in the end, when Sweden accepted the EU decision. This is probably because these arguments were always countered by a more positive discourse in which GM soya was framed as a source of prosperity. At the political level, this is reflected in the fact that restrictive policies were preferred to the introduction of prohibitive measures. Thus, existing legislation was retained and the EU's decisions were accepted. At a more practical level, the labelling of products containing GM ingredients was debated, and then implemented, by the food industry and retailers in late spring 1997.

The framing of the issue as an environmental question must be seen against the background of the pronounced environmental discourse in Sweden. As in most other northern European countries, the environmental question has ranked high on both the public and the political agendas. In debates concerning food products, this discourse has been expressed through a general conviction that, in general, Swedish products and farmers are environmental friendly. In relation to the more general debate over genetic engineering, the environmental discourse fostered the view that genetic engineering implies (unacceptable) interference with nature.

As indicated earlier, the positive, mainly economic, framing of the soya issue cannot simply be explained in terms of the interests of the Swedish food industry. The domestic food industry is of minor national importance and mainly supplies the home market. A possible explanation for the promotion of the positive framing of GM soya could, on the one hand, simply be a general optimism about technology among Swedish politicians. But the explanation could, on the other hand, also be a fear about the indirect consequences the soya issue might have for uses of biotechnology within, for example, the pharmaceutical sector. Were the application of GMOs within the food sector questioned, this could have the potential of reopening more general debates over the industrial use of biotechnology. This might threaten the pharmaceutical industry and other biotechnological industries. In addition, the importance of the USA as a market for Swedish industrial products seems to have been a consideration. Sweden is highly dependent on exports from its large and successful industrial sector and was therefore reluctant to harm relations with the USA.

Another element, which adds to our understanding of the relative success of the critical framing of the soya issue, is of a cultural nature. Economic prospects and profits are not always seen as something positive, especially not if those benefiting are (foreign) multinational companies. In the media debate, the view that only multinational industry – and not consumers – would benefit from the introduction of manipulated soya predominated.

The reception of Roundup Ready soya in Italy

GM food has traditionally been an issue of only minor importance in Italy and, symptomatic of this, the Monsanto Round-up Ready soya did not generate public concern in Italy prior to the arrival of the first cargoes containing the product.

The soya issue in Italy: October 1996–January 1997

The soya story was relatively slow to take off in Italy. In late October, representatives of NGOs (Greenpeace Italia and Lega Ambiente) raised questions with the Interministerial Commission for Biotechnology (CIB) about the actual levels of residual pesticides in the Monsanto soya, and in November Greenpeace called upon the Minister of Health to block the use of GM soya and take the matter up at the European level. This appeal did not lead to any direct policy responses. In mid-November, the media coverage of naked women protesting against Monsanto soya at the FAO summit in Rome caught the public attention for a fleeting moment. At this point, the issue of GM soybeans was clearly not a question for public debate in Italy. But, by the second half of November it had become practically impossible to separate the issue of GM soya from that of the imminent importation of the Ciba-Geigy maize.

The Novel Food Regulation elicited a very critical reaction among the media, which criticised the EU for giving in to pressure from the biotechnology industry at the expense of the well-being of the public. The new regulation was seen as a compromise that could jeopardise public safety. In the time-frame under study here, there is little evidence of any kind of policy activity regarding the soya case per se. This perhaps reflects a tendency to delegate responsibility in such matters to European rather than national authorities. In fact, there were no discussions in parliament over the issue until March 1997 when news of Dolly, the cloned sheep, rekindled the whole debate about biotechnology in Italy.

Within our time-frame, the Italian media coverage of the arrival of the Monsanto soybean hardly constitutes a debate. The intensity of the media's coverage of the issue was weak, and would be more aptly described as news reporting in response to events that for the most part happened far away. Articles that dealt exclusively with the soya issue did not amount to more than a handful in the whole of the national press. However, from the second half of November, particularly after the decision in December to allow the importation of GM maize, the soya story became fused with the maize issue. The approval of GM maize was covered by all the leading national papers. The real peak in media coverage of the soybean, and other GM foodstuffs, was in the following spring after Dolly the cloned sheep lowered the threshold of media attention to all aspects of biotechnology. With the exception of the financial paper, the tone of the press coverage started out sceptical and ended up being generally negative.

The main actors with a voice in the debate or in media coverage were NGOs (Greenpeace and Lega Ambiente), the Italian biotechnology industry and a number of leading scientists in the field. Some of the main actors – or rather standard commentators – involved in the debate at the global level were notably absent in Italy, as were governmental actors. The three main actors involved advanced differing arguments and had a varying influence upon the debate.

Overall, the arguments advanced by the NGOs were clear and consistent throughout the three months under study. Their general argument was concerned with the long-term and indeterminate risks to human health, and they questioned the ability of government institutions to ensure the safety of GM products. In this context they referred repeatedly to the well-known BSE crisis. This argument is closely related to the scientists' concerns regarding the impossibility of guaranteeing a product to be risk free. The political, or perhaps moral, dimension to this argument was linked to appeals for an adequate labelling of GM products that would allow the public, as customers, a choice. The environmental movement also put forward a more specific technical argument that found its way into the media. The actual level of residual pesticide in the soybeans, they claimed, was too high and imports should therefore be banned.

The initial arguments put forward by industry mostly concerned the safety of the product. The industry claimed that GM food is safe on the grounds that it is not substantially different from ordinary strains. In this context, the earlier FDA approval was widely cited. Industry also advanced a more global argument in favour of the application of biotechnology, insisting that GM crops represent the possibility of a new green revolution that promises to benefit the environment and reduce world hunger.

The dynamics of the soya controversy in Italy

The Roundup Ready soybean on its own did not instigate a public debate in Italy. Media attention to this issue was relatively weak in intensity, while the tone of coverage was clearly negative. Furthermore, no policy initiatives emerged that were directly related to the soya case or the issue of labelling GM products.

The issue of GM plants and their use in foodstuffs was hard to reconcile with the way in which the biotechnology debate was conventionally framed in Italy – the benefits of medical progress versus a number of ethical concerns. Initially, the question of GM soya was put on the agenda by environmental groups as a technical regulatory issue focusing on the unacceptably high levels of pesticide in the soya, and not as an issue within the wider biotechnology debate. In 1996, advanced biotechnology was not a major issue in the Italian public arena and the intensity of media coverage was even lower than in preceding years. In the Italian context, environmental impacts and risk assessment issues traditionally rank second to medical and ethical questions. This is reflected in the fact that the issues that tend to trigger public debate over biotechnology in Italy concern not plants but the delicate connections between biotechnology and reproductive technologies, or any intervention in human life.

As time went on, the soya issue became increasingly linked to other discourses. By mid-November, attitudes towards GM soya had become inseparable from the sceptical reactions to the modified maize. At this point, the soya and maize discourses became part of an EU discourse. This introduced an anomalous theme into the Italian political handling of biotechnology. Following the EU decision to allow the import of GM maize, the issue focused on the claim that the EU had yielded to pressure from the industry, placing the interests of North American or multinational companies above the well-being of its citizens.

A discourse of diminishing trust in scientific institutions adequately to assess risks also appeared. Frequently linked to the BSE crises, this argument was discussed only with reference to other countries, but it stood as a lesson on the dangers of the authorities not responding adequately where public health was potentially at risk.

The issue of labelling was discussed only after this. In Italy, food safety was not a sufficiently prevalent issue in 1996 for it to be strongly implicated in debates concerning GM foods. Hence, the parliament did not discuss this question until the Novel Food Regulation was in place. Whereas the soya issue failed to provoke a public debate, let alone a political reaction on its own, the issue was catapulted back onto the political agenda in the wake of the controversy over Dolly, the cloned sheep.

The principal framing of biotechnology was initially that of economic progress and development. The application of biotechnology to agriculture was seen as a positive step forward, and any opportunity to further the reportedly weak Italian biotechnology industry was welcomed by the authorities. This framing was particularly evident in the financial press. In the media coverage as a whole, in contrast, the issue was most commonly framed in terms of safety and risk, and GM food became increasingly associated with the theme of trespassing on the boundaries of nature with unknown consequences. The third main framing strategy was that of regulation and public accountability. This became dominant after the decision to allow the import of the Ciba-Geigy maize in December 1996. This discourse stressed the allegation that the EU had betrayed its citizens by yielding to pressures from industry and multinationals. This story took a further turn when the Novel Food Regulation was introduced; this was depicted as an inadequate response and a further betrayal of the public.

A knowledge of the political culture in Italy enhances our understanding of the fate of the soya issue. When soya was on its way to Europe, Italian political life was preoccupied with economic and fiscal reform. Issues such as food safety, consumer rights and biotechnology were not of prime concern in the midst of an already hectic political agenda. Questions of this kind, defined as technical in nature, are typically handled behind closed doors by appointed governmental experts. Moreover, the issue was seen as an EU responsibility, on which little national policy activity was called for. In all matters concerning the regulation of modern biotechnology, except those considered to be of an ethical nature, Italy has conformed to European-level decisions.

The main actors in the soya debate in Italy were industry, scientists and environmental NGOs (mainly Lega Ambiente and Greenpeace Italia). Relations between the environmental NGOs and their European sister organisations had some impact on the debate in Italy, but international NGOs more directly influenced the Italian debate, Greenpeace International being the first to stimulate concern and then to organise opposition.

In contrast, the industry emphasised economic aspects and the potential benefits to be derived from the use of biotechnology in modern agriculture and food production. Scientists, mostly governmental experts, also played an important role in the debate. Scientific, or rather technical, arguments were used by all actors but, considering the unknown risks involved with GMOs, the scientists remained somewhat sceptical of the use of the new biotechnologies.

All in all, the case of Roundup Ready soya in Italy turned into a story about the general acceptability of GM foods, and in particular the competence with which the issue had been handled at both a European and a national level. The most interesting feature of the Italian reaction to the Monsanto soybean is how slowly and late it came. The debate really started after the event, indicating a low prior awareness of these issues among actors. Indeed, in the policy arena, most of the action took place at a very late stage.

Conclusion

Impacts on the discursive environment

Roundup Ready soya served to bring biotechnology back onto the public agenda, with the possible exception of Italy. Yet it played different roles in the disparate discursive environments of the four countries studied.

Whereas its effect in Denmark and the UK was largely to reopen dormant debates on biotechnology and cause qualitative developments within these discussions, the impact in Sweden and Italy was more profound, with the controversy introducing new discourses. In Sweden, a new and critical discourse concerning GM foods emerged from both the soya and the maize issues, where the dominant biotechnology-related discourses had previously been medical, ethical and regulatory. In Italy, the pattern is somewhat different, because the soya issue itself was of only minor consequence. However, combined with the maize issue, it has had a more or less similar effect to that in Sweden. In addition to this, there are subtle traces of a change in the Italian EU discourse, from a rather positive attitude towards a more critical position triggered by the GM maize and soya controversies. Thus, Italy differs from the three other countries, where EU scepticism was an existing and viable discourse that the soya issue could benefit from being associated with.

USA–Europe relations

The USA's reaction to the EU's critical reception of soya largely comprised allegations that the EU was using the soya issue as a pretext for setting up protective walls around itself with the aim of prohibiting US imports. The USA's interpretation of the soya issue is therefore consistent with Pline's finding that biotechnology was defined as an economic issue (1991). However, the analysis of these four countries indicates that this was not the case. There is no evidence that the GM soya issue was framed in terms of US imports threatening the European food and agricultural sectors. Such a framing would have been part of an economic argument for protectionism, which does not seem to have any pertinence here. Only in the UK was an economic frame dominant. But even in this case, as in the USA, economic arguments stressed the positive economic rewards biotechnology might bring and emphasised the need for a low level of restrictions.

Furthermore, for there to have been a protectionist discourse in the EU, European countries would have to have been producers of soya, or of crops with similar functional characteristics. There is, however, no European soya production to protect, nor is there a European surplus of plant proteins from other sources that could be used in feed production and thus be endangered by US exports.

It should be noted that critiques of the way Monsanto was handling the soya issue were developed in all four countries. But this was not so much a question of EU/US relations as a critique of the way in which multinational companies attempt to dictate politics and bypass fundamental consumer rights, such as the freedom of choice. This was, in other words, a moral statement and not a rejection of Monsanto's right to sell products on the European market per se.

The Eurobarometer survey in 1996 (see chapter 7 in this volume; Durant et al., 1998) indicates a widespread negative attitude towards GM food in the four countries studied here. Around 60 per cent of the respondents in Denmark and Sweden disagreed or tended to disagree with the statement that the genetic manipulation of crops should be encouraged; the percentage in Italy and the UK was around 45 per cent. Unfortunately, Denmark is the only country in which a survey was carried out during the controversy. However, the survey's results are consistent with those of the Eurobarometer survey, showing that as many as 68 per cent of respondents wanted a ban on GM foods. It is reasonable to expect that the same picture would have been found in the other three countries. The political consequence of accepting the public's reluctance to support or encourage biotechnology – in this case, soya – would be to impose a ban. Yet none of the countries has taken this step. This raises the issue of the extent to which political decision-makers should follow public opinion. Ultimately this is a question of populism. Closely related to this, it raises a more fundamental question of democratic influence on decision-making.

Political decisions that do not reflect the desires of large sections, or even the majority, of a population should not necessarily be interpreted as an expression of an undemocratic political process. So long as the different arguments present within the population are given an opportunity to influence the decision-making process, democracy is not subverted. The problem in the soya case, however, is that the critical argument and its ultimate consequence – a ban – are very weakly represented in the political process. Denmark provides perhaps the strongest evidence of this, since the survey showed that around two-thirds of the population wanted a ban. And, despite its persistence, the only political party prepared to promote this position (Enhedslisten) never really succeeded in getting the issue of a ban onto the political agenda. The result is that the question of 'if soya' has never been allowed to enter the process of decision-making; instead, the question has become one of 'how soya'. This provides a concrete example of the observation, made in chapter 2 of this book, that the political process has been directed at the goal of 'making biotechnology happen' and not seriously at questioning its application.

The same pattern is observable in Sweden and the UK, although the Swedish parliament saw rather stronger representations of the critical attitude. Conversely, in Italy, GM soya as a single issue never managed to provoke a political process in parliament.

A logical consequence of framing the GM soya issue as a question of 'how' is that labelling becomes a key issue in the debate. The problem is thereby transferred to the individual sphere and the decisions consumers make in the supermarket. Individuals can choose not to buy products containing Roundup Ready soya if they are labelled properly. From this perspective, the industry is likely to have gained most from the struggle over how the GM soya issue was framed. Crucially, it avoided what was, for it, the worst-case scenario: the discussion of a total ban of food biotechnology.

At present, consumers are left with the role of individuals in the marketplace. As a group they are heterogeneous and weak, and will remain so until consumer organisations succeed in taking up the issue and encouraging consumers to act collectively in the form of a boycott; or until retailers decide to obey the voice of consumers and remove GM products from their shelves, as major European supermarkets tried to do during the spring of 1999.

If no collective action is taken by organisations or retailers representing consumers, Roundup Ready soya will continue to be produced. To put it crudely, in such circumstances, individual consumers will be able to gain no more from proper labelling than a clear conscience and a certainty that their health will not be adversely affected by eating manipulated soya.

NOTES

1. For a detailed account of the different approaches towards the new biotechnologies in Denmark, see Jelsøe et al. (1998); for Sweden, see Fjælstad et al. (1998); for The United Kingdom, see Bauer et al. (1998); and, for Italy, see Allansdottir et al. (1998).
2. The analysis covered the period between 1 November 1996 and 31 January 1997. Most national newspapers in Denmark were included (*Politiken*, *Berlingske Tidende*, *Aktuelt*, *Information*, *Kristeligt Dagblad*, *Børsen*, *Ekstrabladet* and *Weekendavisen*). The total number of articles addressing soya in this period was 208, including 10 editorials, 138 longer articles, 25 short notices and 25 letters to the editor.
3. The political landscape in the UK changed dramatically after the elections of May 1997. The Opposition (Labour Party) won with a large majority of parliamentary seats. As a government, Labour has shown a willingness to stand by the attitude towards GM soya it adopted when in opposition, and it formed a Select Committee on Agriculture to examine the GMO issue. The committee recognised previous mistakes concerning food-quality issues and suggested more openness to the public and the introduction of consumer-friendly legislation. Indicative of the change in government policy is the wording of the committee's report: 'It is our view that consumers have the right to know if foods contain genetically-modified organisms, or if there is a possibility that they may contain them, and we fully support the Government's labelling policy for such

foods. GMOs also have potential environmental consequences, but these matters lie outside the scope of the Report. We strongly support Dr Cunningham's [Minister for Agriculture, Fisheries and Food] call for continued vigilance both on the food safety and environmental consequences of GMOs' (22/04/98).
4. The analysis covered the period between 21 October and 31 December 1996. All newspapers in Sweden with a circulation greater than 90,000 per day were included.

REFERENCES

Åhnberg, A. (1996) 'Svar på interpellation 1996/97:72 "om genmanipulerad soya", 29 November 1996'. Rixlex, http://rixlex.riksdagen.se.

Allansdottir, A., F. Pammolli and S. Bagnara (1998) 'Italy', in J. Durant, M.W. Bauer and G. Gaskell (eds.), *Biotechnology in the Public Sphere. A European Sourcebook*, London: Science Museum, pp. 89–102.

Bachrach, P. and M.S. Baratz (1962) 'Two Faces of Power', *American Political Science Review* 56.

—— (1963) 'Decisions and Nondecisions: an Analytical Framework', *American Political Science Review* 57.

Bauer, M.W., J. Durant, G. Gaskell, M. Liakopoulos and E. Bridgman (1998) 'United Kingdom', in J. Durant, M.W. Bauer and G. Gaskell (eds.), *Biotechnology in the Public Sphere. A European Sourcebook*, London: Science Museum, pp. 162–76.

Behrens, M., S. Meyer-Stumborg and G. Simonis (1997) *Gen food. Einführung und Verbreitung, Konflikte und Gestaltungsmöglichkeiten*, Berlin: Edition Sigma, Rainer Bohn Verlag.

Berg, A. (1996) 'Den nya maten. Gentrixandet – katastrof eller framtidshopp?' *Göteborgs-Posten*, 18 November.

Busch, L., W.B. Lacy, J. Burkhardt and L.R. Lacy (1991), *Plants Power and Profit. Social, Economic, and Ethical Consequences of New Biotechnology*, Oxford: Basil Blackwell.

Dahl, R. (1986) 'Power as Control of Behaviour', in S. Lukes (ed.), *Power*, Oxford: Basil Blackwell, pp. 37–58.

Downs, A. (1972) 'Up and Down with Ecology: the Issue Attention Circle', *The Public Interest* (New York), No. 28: 38–50.

Durant, J., M.W. Bauer and G. Gaskell (eds.) (1998) *Biotechnology in the Public Sphere. A European Sourcebook*, London: Science Museum.

EC (1997) Regulation of the European Parliament and of the Council of 27 January 1997 Concerning Novel Foods and Food Ingredients, 97/258/EC, *Official Journal of the European Communities* L43, 14.2, p. 1.

EEC (1990) Council Directive 90/220/EEC of 23 April 1990 on the Deliberate Release into the Environment of Genetically Modified Organisms, *Official Journal of the European Communities* L117, 8 May, p. 15.

European Commission (1996) 'Proposal for a Parliament and Council Regulation on Novel Foods and Novel Foods Ingredients', *Bulletin* EU 11-1996, point 1.3.30.

Fjæstad, B., S. Olsson, A. Olofsson and M.-L. von Bergmann-Winberg (1998) 'Sweden', in J. Durant, M.W. Bauer and G. Gaskell (eds.), *Biotechnology in the Public Sphere. A European Sourcebook*, London: Science Museum, pp. 130–43.

Folketinget (1994) 'Forespørgselsdebat nr. F13: Forespørgselsdebat til Sundhedsministeren', *Folketingets Tidende*, p. 1440.

(1996) Spørgsmål nr. 707, 4 December 1996, *Folketingets Tidende*, pp. 2061 ff.

Gallup (1996) 'Gallup Instituttet for Berlingske Tidende: Opinionsundersøgelse. Forbrugernes holdninger til gensplejsede fødevarer', Projekt 9363.

Goldburg, R., J. Rissler, H. Shand and C. Hassebrook (1990) 'Biotechnology's Bitter Harvest. Herbicide-Tolerant Crops and the Threat to Sustainable Agriculture', a report of the Biotechnology Working Group, New York, March.

Hilgartner, S. and C.L. Bosk (1998) 'The Rise and Fall of Social Problems: a Public Arenas Model', *American Journal of Sociology* 94(1): 53–78.

Jasper, J.M. (1988) 'The Political Life Cycle of Technological Controversies', *Social Forces* 67(2): 357–77.

Jelsøe, E., J. Lassen, A.T. Mortensen, H. Frederiksen and A.W. Kamara (1998) 'Denmark', in J. Durant, M.W. Bauer and G. Gaskell (eds.), *Biotechnology in the Public Sphere. A European Sourcebook*, London: Science Museum, 9 pp. 29–42.

Kloppenberg, J.R. (1988) *First the Seed. The Political Economy of Plant Biotechnology*, Cambridge: Cambridge University Press.

Lassen, J. (1993) 'Food Quality and the Consumers', MAPP Working Paper No. 8, Aarhus School of Business.

Levnedsmiddelstyrelsen (1996a) 'Letter of approval from National Food Agency to Monsanto', J.no. 571.1065-0001, 18 November.

(1996b) 'Notat til Folketingets Sundhedsudvalg om godkendelsen i henhold til §11 i lov om miljø og genteknologi af genetisk modificeret soja til anvendelse i levnedsmidler', J.no. 571.1065-0001, 30 October.

(1996c) 'Vejledning om mærkning af levnedsmidler og levnedsmiddelingredienser fermstillet på grundlag af gensplejsede sojabønner', Sundhedsministeriet.

Lindvall, G. (1996) 'Interpellation 1996/97:72 om genmanipulerad soya (29 November 1996)', Rixlex, http://rixlex.riksdagen.se.

Lukes, S. (1974) *Power: a Radical View*, London: Macmillan.

Nielsen, T.H. (n.d.) 'Bioteknologi i "Anderledeslandet". Politik, etik og opinion i norsk lovgivning om bioteknologi'.

Nilsson, G. (1996) 'Sojademonstranter kördes bort av polisen', *SydSvenska Dagbladet*, 20 December.

Pline, L.C. (1991) 'Popularizing Biotechnology: the Influence of Issue Definition', *Science Technology and Human Values* 16(4): 474–90.

Rochefort, D.A. and R.W. Cobb (1994) 'Problem Definition: an Emerging Perspective', in D.A. Rochefort and R.W. Cobb (eds.), *The Politics of Problem Definition – Shaping the Political Agenda*, Kansas City: University Press of Kansas, pp. 1–31.

Schattschneider, E.E. (1975) *The Semisovereign People. A Realist's View of Democracy in America*, Hinsdale, IL: Dryden Press; first published 1960.

Schomberg, R. von (1993) 'Political Decision-Making in Science and Technology. A Controversy about the Release of Genetically Engineered Organisms', *Technology in Society* 15: 371–81.

Ström, T., B. Jonsson, G. Axell, B. Thunberg, L. Andreasson, B. Bergnér, M. Persson, C. Elmstedt, U. Paulsson, C. Axelsson, E. Lundmark, K. Nilsson, N. Carlshamre, L. Nolte and M. Ekman (1996) 'Vi kräver stopp för genförändrad mat. Nu kommer de första lasterna med genmanipulerade sojabönor från USA – Bertil Jonsson och 14 andra organisationsledare kräver förbud', *Dagens Nyheter*, 23 October, p. A2.

11   Brave new sheep – the clone named Dolly

*Edna Einsiedel, Agnes Allansdottir, Nick Allum,
Martin W. Bauer, Anne Berthomier, Aigli Chatjouli,
Suzanne de Cheveigné, Robin Downey, Jan
M. Gutteling, Matthias Kohring, Martina Leonarz,
Federica Manzoli, Anna Olofsson, Andrzej Przestalski,
Timo Rusanen, Franz Seifert, Angeliki Stathopoulou
and Wolfgang Wagner*

Her name is Dolly. She is a Finn Dorset sheep that happens to be the most famous farm animal in the world. It was in late February 1997 that the world was stunned by the announcement of her birth, and she was proclaimed to be the first animal cloned from an adult cell. Other animals had been cloned before, but the creation of a sheep from a single cell from a six-year-old ewe electrified both the scientific community and the general public.

The piece in the journal *Nature* was given the rather prosaic title, 'Viable offspring derived from fetal and adult mammalian cells', with not a mention of the word *clone* or *cloning* (Wilmut et al., 1997). The implications did not escape the science journalism elite, however, and their coverage raised the alarm for the rest of the world's press corps to follow.

We consider the Dolly story to be the first real global[1] and simultaneous news story on biotechnology.[2] In this chapter, we will compare the varied ways in which this story unfolded across eleven European countries and Canada. We focus on the elaboration of the global trigger, recognising that there will be variations in its assimilation and accommodation to local contexts. Although a unique story on its own, our comparative analysis of the media coverage of this event is also intended to further our more general examination of the continuing public elaborations of biotechnology. The question we pose is, on the face of it, a simple one: how was the story of Dolly, the cloned sheep, elaborated in media texts across these twelve countries?

We use the term 'story' intentionally, recognising that journalistic accounts are indeed narrative constructions, or storytelling (Darnton, 1975; McCombs et al., 1991). The story of Dolly is no different from other media stories in the sense that it is subject to similar, basic journalistic routines and news values, reflecting the operating logic of the media.

313

Nonetheless, within the panoply of journalistic frames, there develops a story-line that may be both unique and timeless. The cloning of Dolly, we intend to show, has its own distinctive narrative elements but, at the same time, elements of the story appeared in a wide range of countries, apparently irrespective of differing national contexts. How this story unfolded in different national contexts may reveal the existence of a common narrative around the complex of biotechnology.

Our examination of the early elaborations surrounding this event attempts to reconstruct and analyse the narrative thread that evolved, as well as its accompanying themes, images and metaphors. All of these elements are tied within a storytelling frame or the dominant story-line used to organise a story (Gitlin, 1980). These elements, in turn, may be utilised as a way of further understanding the dynamics of social representation. We suggest that these representations may provide a window into our continuing reflections on technology in the context of modernity.

Why Dolly?

Indeed, this question must be raised and addressed. Dolly is doubly significant in our large-scale examination of biotechnology. First, the event represents one of the true breakthroughs in the biotechnological path.[3] As a scientific feat that captured international attention, her birth, public announcement and reception represent a milestone that deserves attention in our attempt to portray the life of this technology. Secondly, and more importantly, Dolly as a social phenomenon deserves further examination. As we intend to show, it is a phenomenon around which larger questions of the place of technology in society coalesce.

Dolly's science

Dolly was the first mammal to be cloned from adult tissue, even though there had been many previous attempts at cloning. In the past, successful attempts had usually involved a donor cell taken directly from an embryo in a process involving 'nuclear transfer'. This donor cell, containing all of its DNA, is then fused with an egg cell from which the DNA has been removed. Once fused, the developing embryo is implanted into a surrogate mother.

Dolly was the first instance of a mammal resulting from a stem cell (in this case, from the mammary tissue of a six-year-old ewe), and it took the Roslin Institute team 277 attempts to clone a sheep successfully. Their technique involved the careful coordination of the states of the donor cell and recipient egg. In the early phase of cell division, the Roslin team artificially induced a state of quiescence by starving the cell of its nutrients. This kept the donor cell in tune with the egg cell. When the two cells fused, development then proceeded normally.

The scientific significance of Dolly was three-fold. First, she brought closer the distinct possibility of cloning humans. Second, the technology responsible for her creation made the alteration of the genetic make-up of animals much simpler. It was in this context that the Roslin team was doing its work, as they attempted to clone sheep for the production of pharmaceuticals. Third, she expanded scientists' understanding of the role of DNA in the development of animals to adulthood. In this connection, the implications for understanding processes associated with ageing were evident.

Dolly in a social context

In many ways, Dolly's story may also be a tale of technology in modernity. The American cultural historian, Marshal Berman, has described being 'modern' as finding ourselves in an environment that promises us adventure, power, joy, growth and transformation of ourselves and the world, 'and, at the same time, that threatens to destroy everything we have and everything we know' (Berman, 1982: 15). Modernity, he maintains, means constant change that is both promising and threatening, tempting yet terrifying, exhilarating as well as exhausting. Living in this world of high modernity is in essence 'riding a juggernaut' (Giddens, 1991: 28).

One of the hallmarks of this world, particularly in late modernity, is the double-edged nature of science and technology, offering beneficent promises for humankind at the same time as they create new definitions of risk and danger at every turn (Giddens, 1994). In this 'risk society' (Beck, 1992), what we previously saw as 'natural' we now subject to control, making us, in turn, worry about what we are doing to nature rather than what nature could do to us (Beck, 1998; Giddens, 1998). In this society, in which risk is an everyday concern and where the flaws of technology have been experienced, a disenchantment with experts ensues and, with it, an erosion of the authority of science.

In this context, we intend to show that the social conversations and questions that might surround a technological event such as Dolly may also represent the way we continually negotiate the place of technology in our world.

The context of social representations

We consider events such as Dolly as phenomena of social representations (see, for example, Moscovici, 1984; Farr, 1987; Wagner, 1996; Bauer and Gaskell, 1999), a theoretical framework that may be helpful to structure our analysis. 'A social representation is the collective elaboration of a social object by the community' (Moscovici, 1984: 25). Such an elaboration becomes a social reality for the community. That is, one cannot distinguish between subject and object; an object exists only in the context of how a community constructs this object through talk and action (Moscovici, 1976: 251). The media offer one forum in which such community conversations take place. These conversations, in turn, may focus on stories of the moment that capture community attention. In this sense, we view media texts (used broadly to include not just words but also images) as an outcome of social processes. That is, they are social in their production, their circulation and their consumption (Wagner et al., 1999).

Two processes are important to the elaboration of social representations. The first is a process of *anchoring* – making the unfamiliar familiar by means of classifying or naming so that new ideas are put into a familiar context. This is not an arbitrary process, nor is it simply an attempt to pigeon-hole for the sake of order or clarity. Anchoring, while facilitating interpretation by classifying and naming, can also communicate a social attitude (Moscovici, 1984: 35).

The second process is called *objectification* – making the abstract explicit. Objectifications unfold in explicit arguments in public or, more implicitly, in pictorial or other materials, conferring clarity on an abstract notion. The process is an attempt to capture the essence of a phenomenon by making it comprehensible, at the same time as it plays on social schemas already in place (Moscovici, 1976; Wagner et al., 1999).

In the context of the present study, we compared newspaper coverage of Dolly over a period of eleven days across twelve countries. In this instance, we were observing the phenomenon of representation in the making in media texts. We considered two modes of representation: the linguistic and the pictorial modes. We further compared emotional associations and metaphors with respect to the anchoring processes and looked at pictures of Dolly and their contexts to elucidate some of the objectification at work in the public discourse of cloning.

We examined newspaper stories from the following countries: Austria, Canada, Finland, France, Germany, Greece, Italy, the Netherlands, Poland, Sweden, Switzerland, and the UK (although some of the material is incomplete for the Netherlands and Poland), covering an eleven-day time-frame from 23 February to 5 March 1997. The sample of newspapers and a more detailed description of the method are described in the methodological appendix to the chapter. A qualitative description of the day-by-day coverage was provided by each country team to allow the charting of patterns in the story's evolution. Themes and metaphors were also identified and described. A quantitative analysis of a set of variables was employed to further describe thematic structures, actors, consequences in terms of risks and benefits, and general portrayals of science. Finally, visual images were analysed separately.

Anchoring: from the esoteric to the common

The basic feature of anchoring is naming the unfamiliar thing or event. By giving it a name, we link the 'thing' to a multitude of references, out of which it takes a shape that is familiar and understandable, identified as enemy or friend, or with ambivalence until further clarification.

The translation of both the scientific technique and the journal's language began almost immediately. The first step in this translation was to name or label the event by means of its most obvious implication, with a term that was already familiar. We had already noted that 'cloning' was not part of the vocabulary of the research paper but, keen to highlight its most important article, *Nature* itself was quick to use as its cover theme 'A flock of clones'. The press release from the Roslin Institute and the newspaper headlines trumpeted cloning from the very beginning, with the term 'clone' or 'cloning' nailing down the event in a network of references that unfolded in the next ten days. Cloning is a word loaded with connotations reaching far into the world of fiction. It is a word that evokes negative expectations, if not fear and loathing.

'Dolly' introduced a second name into the event. The product of the scientific process, an animal, was christened after an American pop singer and movie star, Dolly Parton, perhaps in a moment of whimsy among the Roslin team.[4] At once, the sheep that emerged from a cold laboratory process was humanised.

Once these key anchoring markers were in place – the sheep and the cloning that brought her about – the process of making their meanings concrete was woven into the tapestry of the narrative that subsequently unfolded. We will first trace this narrative development, then identify the attributes of the representational field that emerged.

The evolution of the story

Evolution is used metaphorically. It points to the fact that the social representation of the cloning event takes on a different shape with time and as contexts differ.

The public story of Dolly began with the submission to the journal *Nature* by Ian Wilmut and his colleagues of the account of their experiment. It was written in typical dry scientific prose but, even without referring to cloning, there was no mistaking the import of their account. Scheduled for publication on 27 February, the journal imposed its usual embargo on an important story.[5] Given the fierce competition among news outlets, it was not surprising that the embargo was broken. Two

Italian newspapers and a British newspaper, the *Observer*, relying on its own sources, published news of Wilmut's feat.[6] The *Observer* proclaimed the feat thus: 'Scientists clone adult sheep: triumph for UK raises alarm over human use.' Perhaps missing the import of the story on their hands, the Italian newspaper, *L'Unita*, buried the story in its inside pages and factually declared, 'A cloned lamb is born' (*L'Unita*, 22 February 1997).

Following on the heels of the *Observer* were the world's press. Here, we provide a synopsis of the story as it evolved in written text and visual images over the eleven-day period in our study newspapers. Preliminary analysis of the pictures indicated a remarkable homogeneity across the countries in this study, linked to the practices of the distribution of images through international press agencies. The commentary on the pictorial material is based on an initial determination of the homogeneity of pictures across the twelve countries. A more detailed analysis of the complete set of images from Canada, Greece, Germany and Italy, with references to images from other countries, was then made.

Day 1 (22/23 February)

The Italian newspapers broke the embargo on Saturday, 22 February. *L'Unita* provided a calm and composed descriptive account of the cloned lamb. *Il Giornale* echoed the announcement: 'The first cloned sheep is born.' Given the tone and lack of play, it is likely that the full significance of the story was missed by these Italian newspapers (see Wilkie and Graham, 1998). The UK's *Observer* account appeared on Sunday, the 23rd, and was instantly picked up by the international wire services. Robin McKie, the science editor, wrote a 635-word piece, opening his report with remarks on the triumph of UK science, a breakthrough that could lead to further advances in work on ageing, genetics and medicine. But, with a clear sense of the cultural imaginary behind cloning, he quickly provided references that countered this scientific optimism: Huxley's *Brave New World* and *Boys from Brazil*, a horror movie about producing clones of Adolf Hitler that had been adapted from a book of the same title (Levin, 1977). These references clearly framed the dark side of the technique's possibilities for humans. The main body of the article then elaborated on what had happened at the Roslin Institute, going in some detail into the processes involved and mentioning the more salutary implications: the possibility of new drugs and new insights into the ageing process. The article closed with the image of flocks of medicine-producing sheep and the observation that the vision of creating armies of dictators would attract most attention because the technique, in principle, was suitable for cloning humans.

Day 2 (24 February)

This was really the first day of coverage for the rest of the newspapers in most of Europe and North America. Whereas most newspapers trumpeted the announcement as a major breakthrough, some remained restrained in their coverage, focusing on the announcement of the event. Some countries seemed to miss or downplay the significance of the event. Finland simply reported the announcement, while Sweden buried the story in the inside pages. Perhaps because the achievement was considered a national triumph, and because it had been following the studies more closely, the UK *Guardian* downplayed the fears and highlighted its positive aspects: 'Scientists scorn sci-fi fears over sheep clone: a British breakthrough which has brought new hope to the study of presently incurable genetic diseases.'

Other countries, however, immediately coupled the announcement of the breakthrough with alternating amazement and trepidation. In Canada, Dolly's announcement was heralded as 'a dazzling technological leap and a conundrum for ethicists'. Another called it 'a genetic marvel spawning an ethical nightmare'. The spectre of human cloning was raised right from the start in all the countries under study, as the headlines suggested: 'Will cloned humans follow?' (*Kurier*, Austria); 'Research breakthrough involving female sheep may mean that humans can be duplicated as well' (*Globe and Mail*, Canada); 'Cloning, yes, but not on humans!' (Italy, *Il Giornale*). Coincidentally, in Greek, the word 'shocking' is almost indistinguishable from the word 'cloning' ('klonismos' and 'klwnismos', respectively), and *Apogeumatini* made much of this juxtaposition.

The metaphors that accompanied this announcement and naming process were pronouncements of doom. Although the magnitude of the achievement was recognised, this scientific breakthrough was also equated with the development of the nuclear bomb and continued to draw on the imagery of Huxley's *Brave New World* and the film *The Boys from Brazil*.

Day 3 (25 February)

Without exception, the focus on human cloning in newspapers became even more pronounced. The Finnish as well as the British, Greek, Swedish and Swiss press developed this theme even further: 'In this year, someone will try to clone a human' (Finland, *Ilta Sanomat*); 'First a sheep, then a human?' asked the Swedish paper *Aftonbladet*. Suggestions for who might be cloned varied: in the UK, it could be 'the rich'; in Greece and Switzerland, it would be 'some women'.

On this day, a crescendo of ethical concerns was heard. Many papers interviewed ethicists and philosophers for their views. *Figaro* quoted a psychiatrist and ethologist, Boris Cyrulnik, who said the ethical ramparts man had constructed for moral guidance were collapsing. A Viennese gene expert and philosopher, Johannes Huber, observed in *Kurier*: 'It does not help if we create a new Einstein and remain Stone Age people in ethical terms.'

After a day to catch their collective breath, media storytellers further drew from their bag of metaphors, and this time those related to 'copies' and 'duplication' were widely used. Related to this, another feature emerged, one that seemingly contradicted the moral outrage expressed over the prospect of cloning humans: puns, cartoons and other witticisms began to appear. Cartoons of cloned politicians and sports and entertainment figures were featured. Was this a way of relieving the anxieties of the moment, like gallows humour or jokes around a surgical operating table?

The articles in the early days were accompanied by photographs of Dolly, two of which were widely printed. The first was a picture of Dolly looking straight into the camera (see plate 11.1). This photo became the standard image in the press coverage. The photograph was remarkable for its neutrality, simply picturing a sheep standing on barnyard hay, its body filling the frame with nothing else to catch the eye. A cloned sheep is still

Plate 11.1 Dolly, the cloned sheep, facing the world's media cameras

much like a normal sheep, the photo suggested, but the visual shot also attested to the reality of a clone as an embodiment or the objectification of the abstract idea of cloning.

The other common image in the first days of newspaper coverage was that of Dr Ian Wilmut, standing in a pen with non-distinct sheep in the foreground and talking to Ron James, the managing director of PPL Therapeutics on his right. The connotations primarily concern the business of sophisticated animal husbandry and breeding. All the conventional references to science were notably absent from this picture: both men were without white coats; the scene was set outside, without a glimpse of anything related to the laboratory; and they appeared to be talking business. We might say that the picture anchors the cloner in the world of business and breeding, but not in the world of science.

Day 4 (26 February)

The then US President Bill Clinton's first appearance on the scene was made with his announcement that he had ordered his National Bioethics Advisory Commission to report to him on the issue of cloning. British physicist, and Nobel peace prize winner, Joseph Rotblat was also highlighted when he made a call for an international ethics committee to review the cloning issue. Added impetus to these expressions of concern about the possibilities of human cloning came from a variety of reputable actors at the national and international level. Domestic journalists were finding their own local sources to interview (generally, scientists and government officials) but, at the same time, prominent sources' views began circulating beyond their domestic markets. Officials in the German Bishops' Conference condemned cloning and their views were carried in German and other countries' papers (for example, Austria and Switzerland). Clinton's action was widely reported by all papers.

The science of cloning received further attention. This happened in Canada (*Globe and Mail*), France (*Le Monde*, *Libération*), Greece and Sweden (*Aftonbladet*), with the steps involved in the cloning process being diagrammed or outlined. Positive aspects of cloning in agriculture, particularly with respect to animal breeding, were discussed in Canadian, French, Greek and Swedish papers.

At this stage, close-ups, or the equivalent of facial portraits, of the cloned sheep were widely published. These images are a part of the functional process of identifying protagonists in newspaper stories. The effect is not only to humanise or personalise the cloned sheep but also a way to attribute status to a protagonist or a celebrity. The existence of the sheep had already been visually testified to and it was taken for granted that the reader would recognise the sheep; but at this point Dolly, the object, took

Plate 11.2 Professor Wilmut contemplating a test tube

on new meanings through associations. The media were visibly reflecting upon the irony of attributing celebrity status to a sheep and there were various images where Dolly is being photographed with a microphone held up to her, as though she were being interviewed. The world's media were paying homage to a sheep, another reflection upon the irony of modernity.

At the same moment in the story, cartoons and photomontages began to proliferate. Most of the cartoons contained representations of politicians and cultural celebrities, but also common were armies of fictional clones, or a multiplicity of identical politicians or actors from the world of sport or fashion. We can postulate that one of the primary functions of cartoons is to use irony to reflect on, if not subvert, reality. The same holds for the explicit manipulation of images in photomontages. Both kinds of visual storytelling fall within the realm of narrating possible worlds.

Another important development in the visual storytelling is the picture of Wilmut in profile, holding a test tube at eye level (plate 11.2.). Wilmut, the cloner, has now moved out of the pen and into the laboratory. Although clearly a photograph, it also produced a somewhat nightmarish effect. There were two important signs in this image: the profile of Wilmut and his hand holding the test tube. This image was the first to assert strong associations between the cloner and science, and the test tube as a sign can be read as a metonym, as a part that stands for the whole, evoking associations with life sciences. In particular, the suggestion of 'test-tube babies' was implicit. The emphasis given to his hand

and to his eyes gazing into the liquid in the test tube further strengthened associations with the manipulation of life.

Days 5 and 6 (27–28 February)

Demands for ethical guidelines and laws against human cloning were now being articulated very clearly. Legal regulations should be introduced at an international level, suggested a number of newspapers from Austria, particularly after doubts were expressed regarding the integrity of scientists. Despite Wilmut's denouncement of human cloning, he was portrayed by some as self-interested and motivated by greed. 'After all, this story is not just about scientific knowledge, it is about cash as well' (*Die Presse*). Swiss papers pointed out that human cloning is forbidden at the national level, while female Swedish parliamentarians called for an ethical debate on cloning and embryo experiments. The Vatican also weighed in, calling for immediate laws prohibiting human cloning, an event reported in France, Switzerland, Austria and Italy.

On days 5 and 6, the dominant metaphorical images revolved around humans having crossed boundaries and science spinning out of control. Metaphors that suggested humans were 'playing God' or fooling around with nature evoked fears of humankind overstepping its proper limits. The metaphor of Frankenstein was especially prevalent, and portrayed scientists and the scientific enterprise as having gone wild.

Days 7 and 8 (1 and 2 March)

Over the weekend, a number of articles appeared reflecting on Dolly and the possibilities of human cloning. 'Are there some doors that science should leave closed?' asked the *Calgary Herald* (Canada). In Germany, weekend reflections discussed research limits, regulatory considerations and economic advantages (*FAZ*). The alternative intellectual paper *Taz* devoted two articles to calls for further debate and discussion. Attention was drawn to the media circus around Dolly and the potential implications for human independence and self-determination, as well as potential redefinitions of 'mother', 'family' or 'parenthood'. Greek papers referred to domestic cloning research already under way.

The theme of regulation was also evident in the French coverage, especially after President Jacques Chirac was cited as seeking advice from the country's Ethics Committee, and the UK's 'Nine Wise persons' committee was spotlighted. Greek coverage continued to reflect shock over the cloning feat and speculation about what this might mean for the future.

At this juncture, the announcement was made by the UK Ministry of Agriculture, Fisheries and Food that funding for the Roslin cloning project was to be cut, a reaction to the continuing furore caused by the event and widespread calls for further controls. This was reported in Canadian, Greek, French, Swedish, Swiss and British newspapers.

Perhaps because the weekend provides more time for reflection and extended debate, the metaphors at this time covered the entire gamut of emotions. A few alluded to science marching inexorably onwards and the difficulty of containing the search for knowledge. However, the predominant imagery again involved copies, monsters and magic, the crossing of boundaries into the realms of the unnatural and the Divine, and science unbridled.

The pictorial story became more complex over this weekend, partly reflecting the conventions of the weekend editions of newspapers. The main new development was the appearance of images of science, such as pictures of scientists in their white coats going about their chores in laboratories, or pictures of laboratory equipment. The effect of such images was to move the locus of action back into the world of science. At the same time, most of the press published more or less elaborate diagrams that explicated the technique of cloning. The imagery at the weekend moved the cloning of Dolly out of the pen, away from the political context and celebrity attention, and into the heart of science.

Days 9 and 10 (3 and 4 March)

'Monkey embryos cloned.' The news was carried by virtually every newspaper, and everyone pointed to this step as being inescapably closer to human cloning. From a scientific point of view, the cloned monkeys were in no way pioneering, since the method of embryonic cell cloning had already been applied many times in the past. Even human cloning was attempted back in 1993 with human embryonic cells.[7] Though not all texts discussed the difference between the monkey cloning approach via embryonic cells and Dolly's cloning from an adult somatic cell, virtually all emphasised the diminishing gap between cloning in 'lower' forms and cloning humans. 'The example of Oregon and the manipulation made on primates show that there is basically no biological barrier to undertaking cloning with humans' (Switzerland, *Tages-Anzeiger*). 'Clone-sheep, clone-monkeys, soon, perhaps, clone-humans' (Austria, *Neue Kronen Zeitung*). 'This time, it's about primates; man's turn seems close' (France, *Libération*). 'It is not possible to come closer to humans than this' (Sweden, *Aftonbladet*).

In addition, four countries – Austria, Finland, Greece and Switzerland – printed extracts from an interview Wilmut gave to the German magazine *Der Spiegel*. Wilmut had received numerous calls, primarily from women who wanted to be cloned, and he warned: 'The fear of misuse is justified.

With our technique, you can produce genetic copies of humans. Only clear laws can prevent misuse' (Switzerland, *Le Matin*). On day 10, the same theme continued to dominate coverage: 'Closer to the mass production of humans after the cloning of the sheep and the monkey in the USA', warned Greece's *Eleutherotypia*.

Pope John Paul II expressed concerns regarding 'dangerous experiments' and criticised 'the merchants of life'. Italian and Austrian papers and Canada's French-language newspaper reported this event, as did the German and British press.

Photographs of the Oregon monkeys were printed in most countries. Again, the primary function of the image was to attest to the reality or veracity of the clones by showing the embodiment of an earlier idea, but the most significant aspect of this picture was the difference from the standard Dolly image. The monkeys are huddled together in an otherwise empty corner and the image bears strong emotive connotations through the apparently frightened look on their faces, whereas the standard Dolly image depicts a neutral-looking sheep. The embodiment of the cloning technique was taking on new meanings and associations along the way. The message appeared to be that cloning was really getting too close for comfort, but, at the same time, images were also being used that seemed to suggest that cloning had already moved some way towards acceptability. An interesting indication of this emerging process of naturalisation was the images of natural identical twins that some newspapers employed as illustration, associating clones with what nature was already doing without any manipulation.

Day 11 (5 March)

The fear of misuse continued to be articulated in the newspapers. Would scientists stick to laws? Pros and cons were highlighted and Wilmut's statements about scientists needing boundaries and laws were a focus of all papers except Finland's press. Doubts and cynicism about scientists' integrity were expressed: 'What's technically possible will sooner or later be done' (Switzerland, *Appenzeller-Zeitung*); 'Someday, somewhere, some madman will do cloning experiments with humans, that's for sure' (Austria, *Kurier*).

President Clinton announced the imposition of a ban on federal funds being used for human cloning experiments and called for a voluntary moratorium on human cloning research, a development reported in all of the countries studied.

An analysis of the issues most emphasised in the various countries showed similarities and distinctions (see table 11.1).[8] Human cloning

Table 11.1 *Dolly day by day, 24 February – 5 March 1997*

Issues	Day 2	Day 3	Day 4	Day 5	Day 6	Weekend: Days 7/8	Day 9	Day 10	Day 11
Peak day: production of articles		Can	CH I DE	F	Fin	Gr S UK			A
Human cloning	A Can CH DE F Fin Gr S UK	A Can CH DE F Fin Gr I S UK	A Can CH DE F Fin Gr I S UK	A Can CH DE F Fin Gr I S UK	A Can CH DE F Fin Gr I S UK	A Can CH DE F Fin Gr I S UK	A Can CH DE Fin Gr I S UK	A Can CH DE F Fin Gr I S UK	A Can CH DE Gr I S
Human fallibility and fear of misuse	Can DE S UK	Can CH F Fin Gr I UK	A CH F Fin Gr I UK	A Can CH DE F S UK	A Can CH Fin Gr I UK	A Can CH F Fin I S UK	A Can CH DE Fin Gr I UK	A F Fin Gr I	A Can CH DE Gr I
End of sex	Gr	Can F S	Can	Can F Fin	Can F Gr	Can	DE Gr		Gr S
Scientists have no morals	UK	A I UK	A Can DE I UK	A I	A Can DE I UK	A Can I	F DE I S UK	A Can I S UK	Can CH I S
Dolly started ethical debate		F	Can DE F I	I	CH I	CH I	Gr	Can F	CH
Scientists are ethically responsible	Can CH UK	CH DE F Gr S	Can CH DE F Gr	F UK	CH I UK	I	CH DE F Fin Gr I	A DE	CH Gr
Dolly a media event	CH	Can	I	A Can CH DE F	CH I UK	Can Gr		A Can I	A Can Gr
International actors		Can DE	A Can CH DE F Fin Gr I UK	A CH F I S UK	A CH DE F Fin I S UK	Can DE Gr UK	Can DE F I S UK	A CH F Gr	A Can CH F Fin I
Religious sources	UK		A DE I	A CH F	A DE F I UK	Gr I S	Can DE F I	A DE I UK	A I
The monkey story						Can	A Can CH F Fin Gr I S UK	A CH F Gr I UK	A CH Fin
Call for international laws	Fin	A Can UK	A CH DE F Fin Gr I UK	A CH F I S	A CH F I UK	Gr I UK	F I S UK	A CH F S	A CH Fin I

References to national themes, laws	Can	A Can DE F	Can DE F CH I	Can DE F S	Can CH DE Gr I S UK	A Can F I	Can CH Gr I S	CH DE F I	A CH DE F I UK	A Can CH DE Gr I
Roslin funding cut		Can F UK	Can CH DE F UK	F Gr S UK	A F	Can	Fin S UK I UK	Gr CH UK	F UK	CH
Dolly: business aspects	A Can CH F S UK	Can CH DE F UK	A Can CH DE F Fin Gr I UK	Can F I S UK	A Can DE F I UK	A Can DE S UK	Can Gr I UK	Can F I	F Gr I UK	A DE Gr I
Agricultural and medical benefits	A Can CH DE F Gr I S UK	A Can CH DE F Fin Gr I UK	A Can CH DE F Fin Gr I UK	Can F I S UK	A Can DE F I UK	Can CH Fin I S UK	Can Gr I S	CH DE I UK	A F Gr I S UK	A CH Gr I
How does cloning work?	A Can CH DE F Gr UK	A Can DE F Gr S UK	CH DE F Fin I	CH F Fin I S	A CH DE Fin Gr I UK	A F I	Can Gr I S	A Can DE F I UK	F I UK	A Fin I
Scientific limitations	Can F	F S	A Can CH DE F	Can F Fin	Can CH F I	Can F I	Can S UK	A Can DE F I UK	Can F	A CH I
Do not reduce human to DNA	CH	CH S	CH	Can F Gr	Can DE I UK	Can CH DE	Can Gr S	I	Can I	A DE Gr I
Scientific meaning of Dolly: totipotent cells		F	CH F	F	CH DE	A CH	I	F	F	A
History of cloning	Can CH Gr UK	A F S	Can DE F	F	A CH DE Gr	Can I	Can Gr S UK	F UK	F Gr I	A CH Gr
Cloning has nothing to do with biotech		A CH I	A CH I	CH	CH	A			CH	A

Note: A = Austria; Can = Canada; CH = Switzerland; DE = Germany; F = France; Fin = Finland; Gr = Greece; I = Italy; S = Sweden; UK = United Kingdom.

was mentioned in every country in the course of our sample period. Although the usefulness of the cloning technique was described (usually in areas of medicine and animal breeding), this was often accompanied by concerns about potential misuse. These worries were underlined by arguments concerning human fallibility (references to the dark side of human nature or the inability of resisting the temptation to clone). The trustworthiness of scientists (or lack of it) was also tied to this theme.

Most countries discussed the science behind Dolly, although it was less prominent in Finland and Greece. Also generally covered were the scientific limitations around cloning, including the inefficiency of the technique. At the same time, Dolly's immediate positive impact in economic terms was duly noted by most countries.

An examination of the story's evolution suggested the emergence of a distinct story-line: although triggered by the cloning of a sheep, the story was essentially that of human cloning and the horrifying possibilities it presented. In parallel to these early discussions of how the scientific 'genie in the bottle' could be contained, more positive interpretations emerged (for instance, that the bottle contains a beneficial magic potion). Yet, at this stage, these voices were overwhelmed by the voices of those who forecast danger. Although the story of Dolly obviously did not end here, the outlines of the tale are sufficiently discernible. In the next section, we will elaborate on the representations of this story that have emerged.

Framing the narratives

The meanings of Dolly were fixed by major social actors, including the media, in argumentations and iconography. Themes are major threads in the arguments concerning the cloning event, while metaphors are tools that help to amplify these themes.[9] These elements, in turn, tend to be organised within a 'frame'. The concept of 'frame' refers to the patterns of interpretation, presentation, selection and emphasis (or exclusion) employed in organising stories, which are all aspects of developing a general story-line (Gitlin, 1980; Gamson and Modigliani, 1989). 'To frame is to select some aspects of a perceived reality and make them more salient in a communication text, in such a way as to promote a particular problem definition, causal interpretation, moral evaluation, and/or treatment recommendation' (Entman, 1993: 52). These frames arise from media practices, as well as from the claims-making activities of other social actors. These framing practices, in turn, are part of the processes of social representation, illustrating our earlier point about the social nature of news production.

Actors

Who are the relevant social actors who help in framing a technology's emergence in the public sphere and who provide suggested interpretations of its meanings and implications? Our comparative quantitative analysis of sources showed that the scientific voice was dominant in all countries except Canada and Germany, where actors from industry were most frequently cited. Politicians were second in prominence in Switzerland, Germany, Greece, Sweden and the UK, and industry came close behind.

While newspapers highlighted local actors (scientists, ethicists, industry representatives), prominent international actors circulated their messages, which, in turn, were amplified as a result of recirculation through international news agency channels. These actors included the US President, the Pope and well-known Nobel prize-winning scientists, all proclaiming the dire consequences of this event and calling for a moratorium and/or additional controls.

The media arena as a forum for these claims-making activities can, in turn, be viewed as the site of a competitive struggle within and between communities of interest. Scientists, for example, do not necessarily belong to a unified community. The competing claims-making activities even among scientists is suggestive of the process of 'boundary work' that characterises controversial areas of science (Gieryn, 1995). Boundary work refers to claims-making activities occurring around the institution of science – who its practitioners are, what are considered its appropriate methods, its accepted stocks of knowledge, its recognised values. The objective is to construct a social boundary to distinguish science from non-science, or to make claims about acceptable versus unacceptable science, as part of a legitimation process (Gieryn, 1983: 782). These activities occur as different social actors contend for, challenge or validate the cognitive authority of science, including the power, credibility, prestige, other material resources or public acceptance that accompany this authority.

If science is viewed as a cultural map, with each new scientific discovery resulting in new demarcations and reconfigurations, the process of drawing boundaries to define or explain each new 'discovery' constitutes a cultural and political exercise. The view that human cloning would be abhorrent to the general public was marked by one set of claims – made, for example, by Ian Wilmut and his colleagues – which separated the work on Dolly (defined as acceptable) from human cloning (defined as unacceptable and abhorrent). On the other hand, claims by other scientists, from industry or other institutions, that the potential benefits of this technology are too invaluable to humanity to justify constraints, expressed the conviction that science and progress should not – and could not – be contained.

Calls for the regulation of cloning work illustrated further attempts to draw these boundaries. Reminders that certain countries already had rules in place to govern or restrict human cloning were part of this effort. The drawing of moral boundaries similarly evoked attempts to delimit how far science might go or how far it was seen to have gone.

Frames

Two general frames pervade the Dolly narrative: a frame of doom and a frame of progress, with the former initially more dominant and pervasive than the latter.[10] The frame of doom is characterised by thematic concerns that centre around threats to identity, the dangers of crossing boundaries (specifically the step into domains hitherto identified with 'nature' and 'God') and runaway science. The last of these themes refers to the gap between science and the rest of society, and the lack of social control over the exercise of scientific power.

The frame of progress, on the other hand, revolves primarily around specific utilitarian arguments. However, an identity theme also runs through this discourse but with a different take – rather than loss, it speaks of the retention of unique identities. A third strand of this frame is that science is a predictable enterprise comprised of incremental steps toward a laudable goal. Underlying this is the suggestion that it is folly to try to contain scientific activity; science, it is claimed, has its own momentum and its own set of controls. Yet these frames are not always mutually exclusive, and at times they overlap.

The analysis of themes and metaphors helps to elaborate further the way the narrative was framed. Metaphors function to anchor unfamiliar notions, events or things, to include them in ordinary categories and place them in a familiar context. They transfer what we know about some area of life onto the unknown one (Lakoff and Johnson, 1980).

Threats to identity Identity is the very essence of self. We equate identity with uniqueness, with a singular personality, with the core of our humanity. The word 'clone' is the antithesis of identity: it evokes the sub-human, the zombie-like state of the replicant. These ideas were captured in the chorus of dismay that greeted the announcement of Dolly. 'Each life is unique, born of a miracle that reaches beyond laboratory science' (Clinton, quoted in a number of newspapers). Austria's *Kurier* declared: 'Life is arbitrarily copied and the respect for individuality and dignity of man is lost'; France's *La Croix*: 'Cloning goes against the certainty that every human is unique, issuing from the singular meeting of one man and one woman, each with their own particular history'; Canada's *Toronto Star*. 'There should only be one of any of us.'

Identity is also conferred by parentage; and in Dolly's case the notion of lineage was certainly questioned. 'This is Dolly the clone, daughter of none', declared the *Toronto Star* (Canada). Austria's *Die Presse* suggested: 'Dolly's father is no ram but the scientist Ian Wilmut.'

Metaphorical extensions of this theme on identity could be found in allusions to *mechanical reproduction* of two types: references to the photo-copying process and references to the assembly-line. For example, 'Life by production line' (UK); 'Cloning, the factory of life', 'Photocopied mammals' (Italy); 'Carbon copies' (Canada); 'Mass human production is not far' (Greece); 'The singular individual is put into question by ge-netic copy machines' (Austria); 'Mass-produced nature', 'The molecular copying machine rotates to spew out duplicates like the devil' (Germany); and 'We turn animals into factories' (Switzerland).

Crossing boundaries If God is omniscient and represents perfec-tion, being human means accepting imperfection. It also means occu-pying a particular place in the perceived 'natural order' of things. This, in turn, presumes the presence and acceptance of boundaries. Dolly's cloning and intimations of human cloning blurred these boundaries. Thus, playing God or meddling with creation and nature were common expressions of discomfort: Canada's *Calgary Sun*, 'Morally, [cloning] ruf-fles all sorts of religious feathers. . . . Will doing such an end run around God's divine order result in a final and ignominious end to the game?'; Austria's *Salzburger Nachrichten*, 'This technique is an unauthorized intrusion into creation.' 'An absolute transgression', cried France's *Le Figaro*. The German newspapers suggested that dreams of immor-tality and resurrection could now be realised.

Other concerns were expressed in terms of the disappearance of normative sexual reproduction. 'Sperm is passe; so are men. Only eggs are needed. It's as if the birds and the bees have suddenly been rendered irrelevant' (Canada, *Calgary Herald*)

'Cloning bares a spectacular and symbolic aspect: the disappearance of sexual reproduction' (France, *Le Figaro*)

'Has the time come to move from Homo sapiens to Homo xerox? The only sure thing is that we are mathematically driven to the abolishment of sex as the only means of reproduction' (Greece, *Eleutherotypia)*

'A nation of amazons gets closer!' (Germany, *Taz*)

References to God and religion constituted another metaphorical cat-egory reflecting dismay at humans overstepping the proper boundaries of intervention and existence:

'Compared with man, God is just a beginner' (Italy)
'This is meddling with creation. God's sacrosanct make-work project' (Canada)
'Virgin birth' (UK)
'God created man and man created the clone' (Austria)
'Science keeps on taking a bite from the forbidden fruit in paradise' (Italy)

References to lambs, or little lambs, in the English papers were clear allusions to William Blake's 'little lamb, who made thee?' Equally likely, playing on the religious symbol of innocence, a quality now perhaps lost, was emblematic. In Germany, *Der Spiegel* was blunt in this regard when it declared that 'the lamb, epitome of devout nativity, has become the symbol of the lost innocence of science'.

Runaway science A large part of modernity involves living with the benefits – and risks – of science and technology. The authority of science, however, is most often questioned when we are first made aware of some startling or profound discovery, or an event that generates challenges and calls for control. The commentary from various newspapers serves to illustrate this.

'Gene research out of control: cloned sheep and then?' (Austria, *Die Presse*)
'Man is being left behind by the technology he created' (Greece, *Apogeumatini*)
'Scientists and the general public have gone in very different directions. Morality is always one step behind technology' (Canada, *Globe and Mail*)
'Bestialisation of science' (Germany, *FAZ*)
'We are treating nature as a continuous guinea pig without appropriate thought or reflection'(Germany, *Taz*)

Eleutherotypia denounced scientists as 'crazy people now able to make their dreams come true!'
The metaphors depicting this theme evoked images of the scientist or science as *monstrous*, with Frankenstein the dominant image:

'Projects out of Dr Frankenstein's horror cabinet' (Austria)
'Frankenstein's monster' (UK)
'The ghost of Frankenstein' (Finland)
'The preliminary stage of Frankensteinian labour creatures' (Germany)
'Human Frankensteins' (Greece)
'We do not know if we wish to play God or Dr Frankenstein' (Canada)

The press also had recourse to several other popular cultural resources, most notably *The Boys from Brazil*, a best-selling novel-turned-film about the Nazi doctor Joseph Mengele and his attempts to raise clones of Hitler, and Huxley's *Brave New World*. More current references to *Jurassic Park* were also made.

In addition, military metaphors were commonly invoked to depict the unbridled power of science. The development of the nuclear bomb was frequently used as an analogy, with the same allusions to the double-edged sword of ultimate power and destruction:

'Dolly, the winning stage of gene technology . . . with the drastic effect of the atom bomb' (Germany)
'Like having an atom bomb in the house' (Italy)
'The bio-engineering equivalent of the first nuclear weapon' (Canada)

Other military metaphors were also prominent, especially in the context of replicating armies of robot-like foot soldiers.

'Whole armada of identical humans could be produced' (Austria)
'Slave armies' (UK)
'Phalanxes of identical Hitlers' (Canada)
'Army of clones' (Finland)
'Legions of cloned Rambos and dictators' (Switzerland)

The world of myth and magic provided a further wellspring of imagery with which to suggest that science was out of control:

'The human clone stands on the horizon like Chronos who eats his own children' (Austria)
'Genetic voodoo'; 'the genie is irretrievably out of the bottle' (Canada)
'Sorcerers' apprentices of science' (France)

The progress frame It was noteworthy that, in virtually all the coverage in our sample, the progress frame, although present, was secondary to the frame of impending doom. However, this frame was still much in evidence in the shape of four different themes: the utilitarian argument; the defence of identity; the limitations on science and scientists; and the more general argument of the inexorable process of discovery in science.

Claims for the utility of cloning focused on agricultural and health benefits, themselves deriving from the production of pharmaceuticals (which was the Roslin team's initial concern); a better understanding of cell development (with consequences for understanding the ageing process); and the development of more efficient animal husbandry practices. All countries in the sample, at one point or another, mentioned these potential advantages.

The frame of progress was also promoted by way of attempts to diminish the concerns related to cloning. This was particularly evident for the theme of personal identity. In response to the chorus of voices that alleged cloning to be antithetical to the individual's sense of identity, a few sources made the counter-argument that environment plays an equally important role in human development; they evinced their claim by observing that identical twins are still different in many ways.

Arguments about the limits already imposed upon science and scientists were also heard. In eight of the ten countries, these varied from pointing to controls already in place to mentioning ethical norms to which science routinely adheres. The unlikelihood of human cloning was also discussed by reference to the inefficiency of the cloning process ('it took 277 attempts').

Finally, the progress frame was also discernible in the occasional – but noteworthy – attempts to place Dolly in the context of previous scientific advances. For example, German, British, Canadian, French and Swiss papers provided brief historical forays into earlier cloning attempts, with the milestone of Dolly portrayed as simply another step in the inevitable progress of science.

At the level of metaphors, this theme was sometimes couched in terms of the military ('a battle won') or sport ('race'). The metaphors of the 'wheel of progress' or of discovery ('opening doors') were employed and the term 'leap' was also often used.

On the whole, however, progress was characterised in direct terms rather than with the aid of metaphorical imagery. The references to the production of more health care products, the greater potential for studying and understanding genetic diseases, and improvements to animal husbandry were straightforward rather than allusive. This is possibly a sign that technological advancement, equated with progress, is taken to be the norm and no longer needs 'translation' and familiarisation.

Consequences Discussions of the potential consequences of cloning included several basic themes. In order to assess their relative importance we content analysed the different risks and benefits emphasised. From this evaluation, it emerged that health benefits predominated, followed by research gains. Economic benefits were touted in only three of the ten countries analysed (table 11.2). The perception of risk, on the other hand, is more homogeneous, with moral risks underlined in every single country. This ubiquity supports the frame of doom within which the story of Dolly is usually couched.

Portrayals of science The preceding discussion of frames was complemented by an additional quantitative analysis of the perceptions of science contained within the press coverage of the Dolly event. This involved quantifying, by means of semantic differential scales, the overall image of science projected in news stories.[11] As table 11.3 notes, science as an

Table 11.2 *Dolly: type of benefit and risk evaluations*

| | Benefits | | | Risks | | | | |
Country	Economic	Health	Research	Health	Inequality	Moral	Other	Environment
UK		58	20			50		16
Germany		35	48			43	28	
Netherlands		34	44			87		
Austria	50	20	20			95		
Sweden		33	20			87		13
Greece		37	40		13	53		
Finland		27	33	18	18	37		
Switzerland	28	46				94		
Poland		62	21			82		
Canada	37	49				71		

Note: Only scores larger than 10 per cent are listed. Data for France and Italy are not available.

Table 11.3 *Dolly: the image of science*

| | Image of science | | | | | |
Country	Creative vs. uncreative	Successful vs. unsuccessful	Moral vs. immoral	Conscious vs. unconscious of responsibility	Constructive vs. destructive	Public welfare vs. egoistic
UK	6.2	6.6	**4.7**	**5.2**	**5.6**	**4.3**
Germany	7.3	8.1	**4.0**	**4.0**	**5.2**	**4.1**
Austria	6.0	6.7	**3.9**	**4.2**	**5.7**	**4.7**
Sweden	5.6	6.1	**4.2**	**5.1**	**5.4**	**5.1**
Greece	8.0	7.1	**4.5**	**5.1**	5.8	6.0
Finland	7.5	7.0	**4.8**	**5.0**	**4.9**	**5.0**
Switzerland	**5.1**	6.4	**5.5**	5.8	**5.0**	**5.5**
Poland	7.3	7.3	**5.3**	**5.4**	**5.5**	**5.5**
Canada	7.5	7.3	**4.4**	**5.4**	**4.8**	**4.9**

Note: Scores represent rating of image on a semantic differential scale of 1–10.

enterprise was rated by different country teams according to criteria relating to scientific and social orientation. Aspects of the former included notions of creativity and success, while the latter incorporated morality, responsibility, consideration of public welfare, and constructiveness. The picture of science that emerged was of a highly creative and successful enterprise, but at the same time a socially irresponsible and unresponsive institution (table 11.3).[12] This finding, which was confirmed by a factor analysis of the entire sample of stories,[13] showed that, though science was framed in different ways, in this context negative imagery was clearly dominant.

The use of humour We have identified the use of humour as another element in the evolution of the story on Dolly. It was notable that no country was immune to the need to joke. Cartoons and photomontages werc frcqucnt vehicles for humour (see plate 11.3), but wordplay was another humour tool. Humour was a common approach to anchoring the Dolly event for readers, and we discuss it separately here if only to point out another unique attribute of the Dolly story.

Textual humour was often conveyed through the use of puns or other forms of wordplay. In Canada's *Globe and Mail*, these puns were illustrative: 'Are "ewe" ready?' or 'Send in the clones' (a play on the popular song 'Send in the clowns'). Cartoon subjects focused on the theme of copying by duplicating politicians or sports or entertainment figures. The Swedish papers suggested, for instance, that Ian Wilmut might make a good house doctor for the Swedish hockey team by replicating the best players.

What was it about this cloning story that so easily – and so frequently – lent itself to joking? Was it because the dark possibilities of human cloning seem so distant that one can have easy recourse to reassuring humour? Or was it a reflection of unease or nervous fascination at what humankind had wrought? Was it the protective cover that joking so often affords, providing a release from the anxieties and paranoias evoked by the dangers cloning might generate, which place in jeopardy our most cherished notions – of who we are, the sacredness and uniqueness of life, the importance of sex? Going to an extreme, were we feeling perhaps 'that this joking in the face of a new possibility for mass-produced life is in fact joking in the face of death, death of the spirit and death of the (male) body' (Miller, 1998: 80)?

Reflecting on the same theme of joking in the context of cloning, Miller (1998: 81) maintains that 'the intense presence of the urge to joke is as sure an indication as there is that we are approaching the dangerous, the sacred, and the magical. Pious and grave talk about human dignity is so often untrustworthy..., so unfelt, so by rote, so safe and predictable that some feel it necessary to retreat to the joke to pay serious homage.'

Different motivations may be at play at different times, for different papers or countries, but it is not inconceivable that the varied dimensions of cloning and its implications could evoke a range of humorous reactions – from the laughter of conceit to that of resistance, from the laughter of amusement to laughter in the face of doom.

Plate 11.3 Cover page of *Der Spiegel*, March 1997
Note: The captions read 'Science marching towards the cloned human being' and 'The Fall of Man'.

Iconography

All contemporary news media rely heavily upon visual communication through pictorial representations. These images, along with headlines, are important seduction devices to capture readers' attention and to guide their reading of the text. They are also important tools in fixing meaning and anchoring elements in more global contexts. Finally, images capture, crystallise and render visible the abstract or ambiguous. Out of all potential images of cloning and clones, the standard picture of Dolly was really the only constant in the story told through iconography. This image can be understood as an instance of objectification by which the abstract, ambiguous or threatening idea of cloning was embodied in the image of the sheep. It was present throughout the period, at times subjected to explicit manipulation in photomontages (such as on the cover of *The Economist*) but always recognisable as embodying the core of the issue. The sheep came to embody the image of all clones. By its omnipresence in the process of storytelling, this image absorbed other meanings and associations made in the course of narration, and it maintained and carried with it traces of the political and ethical debate. It can be argued that the image of Dolly became a symbol for cloning and, moreover, a motivated symbol for biotechnology by encapsulating ambivalent public attitudes. The image of an innocent-looking sheep captured the fleeting moment of moral consensus when the world seemed to agree on the boundaries of acceptable interference in life. Its frequent appearance in subsequent stories was a reminder of that fleeting moment. The image of Dolly retained the connotations of an innocent creature and of the necessity for, but at the same time the impossibility of, drawing stable boundaries between manipulation and unacceptable interference in the nature of life (see Kolata, 1998).

From a historical point of view, images invoked in relation to science and scientists displayed considerable constancy. Turney (1998), for example, traces the evolution of biologists' image in the media and shows that shifts in the way they are portrayed correspond to changes in biologists' practices. As they moved from description and classification to experimentation, from a focus on the organism to the more abstract realm of molecular biology, the image evolved from biologists' looking through microscopes to posing with molecular models or a double helix. Turney maintains that life has become less and less recognisable and, over time, has more frequently been reduced to abstractions (1998: 43). In the Dolly story, Wilmut with the test tube may be viewed in this latter context.

Local differences in elaborations

Although we have focused on many commonalities across the twelve countries, there are, of course, notable distinctions as well. Not surprisingly, national pride is clearly evident in British coverage, with frequent emphasis on Dolly being a *British* scientific accomplishment. The usefulness of the cloning procedure was also frequently highlighted in the UK, as were the economic benefits.

Finland was noteworthy for its almost unquestioning representation of the event. Scientists escaped the censure evident in other countries' coverage and calls for control or regulation were muted. This non-critical representation corresponds with more positive Finnish attitudes toward biotechnology in general (see chapter 7 in this volume). At the other extreme, Italy and Greece focused enormous attention on its negative aspects and connotations of aberrancy.

In the case of Germany, the shadow of not-too-distant history may account for the absence of allusions to Hitler (as a symbol of the potential for 'copying' evil) that were so prominent in all the other countries. *Der Spiegel*'s photomontage displayed in plate 11.3 is the exception.

A strong emphasis on regulatory control was conspicuous in the Swiss coverage. This was both a reaffirmation of the ban on cloning already in place but also a reflection of the debates surrounding the impending national referendum on genetic engineering research.

In Canada, Italy and Sweden, some of the rhetoric on cloning was framed in the context of recent domestic debates about reproductive technologies and human embryo experiments. In the case of Italy, the announcement of the birth of a healthy cloned sheep was made in a very particular discursive environment. In the weeks and months before, a heated political debate on the status of the embryo and embryo research had been waged, involving criticisms of the absence of existing legislation on reproductive technologies. It did, however, take a few days for the two stories to fuse. In Canada, a Royal Commission's three-year-long examination of reproductive technologies had already proposed, among other things, a ban on human cloning. In the Canadian press this received renewed attention.

Conclusions

In tracing the evolution of this narrative, we have shown the chief story-line emerging as that of human cloning. Dolly the sheep became the motivated symbol for humans as clones. In this story, the initial consensus was one of moral outrage and condemnation.

Why did Dolly hit the headlines worldwide? It is not the scientific event as such that explains the story's popularity, but the fact that the Dolly issue had profound cultural resonances. That it occurred *across* national imaginations reveals a social construction process of a large-scale mosaic.

One can point to journalistic routines that are evoked in any major media story, and Dolly was no exception. Journalists drew on similar metaphorical images; they highlighted the same key events and actors; they tried to 'balance' their accounts of those who denounced the breakthrough and those who exalted it, though the former were clearly in the ascendant. In terms of media practices, it is interesting that the account of Dolly was treated as a unique event, an enormous surprise, a 'technological leap', despite the fact that the idea behind her creation had been around for more than half a century. This is perhaps because journalism is encoded in short-term memory. Various cloning efforts over several decades, from frogs to mice to cattle, have been detailed in historical accounts of modern biology and its portrayals in the popular press (see, for example, Kolata, 1998; Turney, 1998). With few exceptions, however, this historical context went unutilised.

In viewing the Dolly story as an episode of social representation, the activities of anchoring and objectifying emphasised the media's role in the processes of diffusion and the propagation of representations. The acts of highlighting the scientific advance and classifying it in terms of its human cloning implications helped to anchor the issue, while the uses of imagery, the patterns of thematic emphasis and argumentation, and the elucidation of the scientific processes involved assisted in making concrete or objectifying the event. The questions Moscovici (1984: 23) posed earlier remain highly relevant: how do we explain these patterns and why do we create such representations? In this instance of cross-national comparison, why has there been such a convergence of narratives and representations?

Social representations are said to reveal themselves with greater clarity in instances of crisis or during an upheaval of thought (Moscovici, 1984). Dolly as a technological event provides an important blip on the evolutionary landscape of biotechnology that has instigated – even necessitated – an important restructuring of mental maps. Where the notion of cloning ourselves had been strictly confined to the realm of fiction, Dolly transposed it into the real world.

If the purpose of representations is to make the unfamiliar familiar, what the story of Dolly has done is to acquaint us with a new dimension of our technological prowess. One function of this is to advance a preferred vision of some ideal world, what Moscovici has called 'the hypothesis of desirability' (1984: 23). One version of this desirable world – the world of scientific progress, of conquered ills – is often packaged in 'the rhetoric of hope' (Mulkay, 1993). It is a tale that has been told countless times before and employs devices common to audiences across cultures.

At the same time, this vision has to compete with a version that unequivocally rejects it. This rival frame is replete with intimations of doom and employs an uncompromising 'rhetoric of fear' (Mulkay, 1993). In this version, the evocations of monsters, myth and magic and of 'mad science' have been found across various retellings and across different technologies (Toumey, 1992; Turney, 1998). The use of this 'rhetoric of fear' is functional when science and technology can be represented as violating basic cultural categories and moral codes (Mulkay, 1993: 724). Mulkay argues that the voices of fear represent 'a culturally subordinate discourse of science' (1993: 736). Images of monsters (commonly Frankenstein), of mad scientists and the need for controls, of the distasteful aspects of industrial assembly-line and cookie-cutter uniformity, and of the violation of identity and humanity are all highly serviceable weapons in the struggle to define a technology and its proper limits.

This struggle to define the more enduring representation of a technological advance is at the heart of the story of Dolly. It is a story that illustrates a fleeting moment of moral consensus on this particular scientific endeavour. At the same time, it richly encapsulates the continuing ambivalence we have about the project of modernity and its notion of 'progress'. It encompasses our railings against the increasing chasm between a runaway science and the rest of society, and it involves the projection of our worst fears about the unbridled power of scientists. It incorporates our ambivalence about science and embodies our deepest fears about threats to our identity. Through the allusions to the crossing of boundaries between the 'natural' and the 'artificial', it has brought out in sharp relief our deepening anxieties about the usurpation of divine power. In short, this may be the narrative of our ongoing, reflexive project of ourselves in modernity.

Postscript

About a year after the Dolly story broke, she was reported as giving birth (a natural pregnancy) to a healthy lamb, named Bonnie. This event declared the cloned sheep to be no different from a 'normal' sheep. Not long after, scientists at the University of Hawaii replicated the Roslin effort, this time on mice. The mice were successfully cloned by transferring donor nuclei (not whole cells) into eggs. With this, *Time* magazine declared: 'Dolly, you're history.'

Methodological appendix

Sample

Austria   Neue Kronen Zeitung (mass readership, opinion leader); *Die Presse* (small readership, quality, conservative Catholic); *Kurier* (mass readership); *Salzburger Nachrichten* (high quality, local readership).

Canada Globe and Mail (quality, conservative, national newspaper); *Calgary Herald, Toronto Star, Calgary Sun* (all local city English newspapers); *Le Devoir* (quality French paper).

Finland Helsingin Sanomat (independent, centrist); *Savon Sanomat* (independent, centrist); *Demari* (left wing, represents Social Democratic Party); *Suomenmaa* (centrist, Centre Party); *Kotimaa* (religious, Lutheran); *Ilta Sanomat* (independent afternoon paper); *Kauppalehti* (business paper).

France Le Monde (centre, liberal); *Libération* (centre-left); *Le Quotidien du Médicin* (specialist press); *Ouest France* (regional); *France Soir* (popular); *Le Figaro* (conservative); *La Croix* (Catholic).

Germany BILD (national daily, 5 million circulation); *Frankfurter Allgemeine Zeitung* (*FAZ*, largest national daily, right); *Tageszeitung* (national daily, left wing); *Leipziger Volkszeitung* (regional east German daily paper); *Der Spiegel* (national weekly news magazine); *Taz* (alternative, intellectual).

Greece Ethnos (socialist); *Apogeumatini* (popular, right wing); *Ta Nea* (independent, socialist); *Eleutherotypia* (opinion leader); *Kathimerini* (quality Sunday paper).

Italy Corriere della Sera (opinion leader, centrist, largest-circulation daily); *La Repubblica* (alternative opinion leader, centre-left); *L'Unita* (centre-left, formerly Communist Party daily); *Il Giornale* (right wing); *La Nazione* (semi-popular, regional, right wing).

The Netherlands Volkskrant (quality, left of centre); *NRC/Handelblad* (quality, right of centre); *Trouw* (quality, left of centre); *Algemeen Dagblad* (popular, right of centre); *Telegraaf* (popular, right of centre).

Poland Gazeta Wyborcza (quality daily, liberal); *Rzeczpospolita* (quality daily, centre); *Zycie* (quality daily, right); *Trybuna Demari* (left, represents Social Democratic Party); *Slowo* (Catholic daily); *Przeglad Tygodniowy* (weekly, centre); *Zycie Warszawy* (Warsaw daily, national circulation, centre to apolitical); *Gazeta Poznanska* (local Poznan daily, centre to apolitical); *Glos Wielkopolski* (local Poznan district daily, apolitical); *Expres Poznanski* (local Poznan daily, apolitical).

Sweden Aftonbladet (tabloid).

Switzerland *Blick* (German popular daily); *Tages-Anzeiger* (German, opinion leader); *Neue Zurcher Zeitung* (elite, international reputation, German); *Basler Zeitung* (German daily, regional); *Le Matin* (French daily); *Le Nouveau Quotidien* (French daily, urban readership); *Journal de Genève* (French daily, quality); *Appenzeller Zeitung* (German daily, rural).

The UK *Guardian* (left, quality), *Independent* (centrist, quality), *The Times* (conservative, quality).

Procedure

A census of all stories about Dolly in the selected newspapers was made by the country teams. The selection of newspapers for analysis was left to each country team, with an eye to having a diverse set for analysis.

Both qualitative and quantitative analyses were carried out. The qualitative analysis was conducted through close reading of the textual material by each country team. A day-by-day synopsis was then prepared in English for each paper. The second coding stage involved developing a list of themes from the text. From this data set, a summary matrix was created showing which themes were covered by which countries, providing a picture of similarities and differences across countries.

A third stage involved listing all the metaphors employed in each story and categorising this data set. The fourth stage incorporated analysis of the images (photographs, diagrams, cartoons) on the subject.

Quantitative analysis was also carried out. A coding sheet was drawn up with a variable set that included actors, risks and benefits, and an overall image of science as determined from the tone of the article. This data set was analysed with SPSS.

NOTES

1. To put the term 'global' in context, there is considerable evidence of coverage of Dolly in many parts of the world. Our study, however, focuses on a small sample of Western industrial countries.
2. It should be noted that the term 'biotechnology' is not exclusive to the process of recombinant DNA, more commonly known as genetic engineering. Its broader definition includes processes such as protoplast fusion and cell and tissue culture. It is in this respect that the notion of cloning can be considered to be a biotechnological activity. Nor is the term 'cloning' entirely precise, covering a number of different scientific processes. For example, it has been used to refer to molecular cloning (the duplication in a host bacterium of DNA strings containing genes); cellular cloning (where copies of cells are made indefinitely to create 'cell lines'); embryo twinning (where a sexually developed entity, an embryo, is split into identical halves); and nuclear somatic transfer, the process by which Dolly was created (see Pence, 1998).

3. We make this judgement not just in terms of the media attention accorded to the event, but also on the basis of the judgement from within the scientific community itself. The journal *Science* deemed the cloning story one of the major 'breakthroughs' at the end of 1997 (*Science*, 19 December 1997). It described a breakthrough as 'a rare discovery that profoundly changes the practice or interpretation of science or its implications for society' (p. 2029).

4. Dolly, the sheep, was created from a mammary cell. This udder lineage inspired a moment of wry frivolity, with the well-endowed Dolly Parton providing the naming inspiration, according to Wilmut (Kolata, 1998: 3).

5. An 'embargo' (literally, an official suspension of an activity) is a press relations tool to hold a story announcement or keep its contents from public dissemination until a specified date. Important journals such as *Science* or *Nature* will send out press releases on an important upcoming article and will sometimes put an embargo on the story until the day of publication, ensuring maximum publicity for the journal.

6. The UK *Observer*'s science reporter was able to break the embargo because he used a different source – an upcoming TV documentary on the Roslin research (see Wilkie and Graham, 1998; also Kolata, 1998). As for the Italian newspapers, the national news agency, Anza, may have provided the initial story (see Wilkie and Graham, 1998), but it is not clear where the agency's information originated. In any case, the sensationalised coverage in Italy did not begin until Monday, the 24th, when everyone else raised the human cloning theme in alarm.

7. In 1993, American scientist Jerry Hall was reported to have cloned seventeen human embryos into forty-eight. This was an attempt to increase the embryo supply in fertility clinics. This was not a nuclear somatic transfer procedure, however, but a case of embryo twinning (see Pence, 1998).

8. We recognise that there are obvious differences between newspapers within any country (for example, the differences between quality and popular newspapers). However, our analysis glossed over these differences, emphasising instead between-country distinctions or similarities. More in-depth analyses of media coverage of this event have been carried out in Italy, Canada and France (see Berthomier, 1999; Di Palma, 1998; Downey, 1999; Manzoli, 1998; and Rizzo, 1998).

9. We rely on the work of Lakoff and Johnson (1980), who maintain that our conceptual system is largely metaphorical, that is, our ways of thinking and experiencing and doing are structured by metaphor. When metaphors are used as linguistic expressions, they succeed largely because these are usually present in the recipients' conceptual systems.

10. In identifying the general frames of doom and progress that have surrounded representations of the Dolly story, our findings correspond with similar framing approaches that emerged around the Human Genome Project. According to Durant and colleagues (see Durant et al., 1996), a discourse of progress or hope and a discourse of concern were embedded in the British media coverage, but were similarly echoed in focus group discussions with the general public.

11. Semantic differential scales are rating scales using polar adjectives to tap evaluative dimensions of an object, particularly quality and intensity dimensions

(Osgoode et al., 1971). These measures are used to analyse 'meaning' by tapping representational processes in language. In this instance, 'science' as an enterprise might be rated according to how it is portrayed in a news story as 'socially responsible' at one end of the scale and 'socially irresponsible' at the other end.

12. This observation is restricted to science mentioned in respect to Dolly and covers only the first ten days of the newspaper coverage.

13. Two components explain about 60 per cent of the variation: 'ingenious science' and 'morality of science' (these terms are derived from the factors with the highest loading).

REFERENCES

Bauer, M.W. and G. Gaskell (1999) 'Towards a Paradigm for Research on Social Representations', *Journal for the Theory of Social Behaviour* 29: 163–86.

Beck, U. (1992) *Risk Society: towards a New Modernity*, London: Sage.

(1998) 'Politics of Risk Society', in J. Franklin (ed.), *The Politics of Risk Society*, Cambridge: Polity Press, pp. 9–22.

Berman, M. (1982) *All That Is Solid Melts into Air: the Experience of Modernity*, New York: Simon & Schuster.

Berthomier, A. (1999) 'Discours médiatiques sur la biotechnologie en France, 1973–1996', unpublished PhD thesis, Ecole Normale Supérieure de Fontenay/Saint-Cloud, France.

Darnton, R. (1975) 'Writing News and Telling Stories', *Daedalus* 104: 175–94.

Di Palma, V. (1998) 'Tra Sacro e Profano: biotecnologie e clonazione nell stampa cattolica q in quella laica. Tesi di Laurea in Scienze della Communicazione', University of Siena.

Downey, R. (1999) 'The Social Construction of Technology in Late Modernity: a Case Study of Cloning in the Canadian Media', unpublished MA thesis, University of Calgary.

Durant, J., A. Hansen and M. Bauer (1996) 'Public Understanding of the New Genetics', in T. Marteau and M. Richards (eds.), *The Troubled Helix: Social and Psychological Implications of the New Human Genetics*, Cambridge: Cambridge University Press.

Entman, R. (1993) 'Framing: towards Clarification of a Fractured Paradigm', *Journal of Communication* 43 (4): 51–8.

Farr, R. (1987) 'Social Representations – a French Tradition of Research', *Journal for the Theory of Social Behaviour* 17: 343–70.

Gamson, W. and A. Modigliani (1989) 'Media Discourse and Public Opinion on Nuclear Power: a Constructionist Approach', *American Journal of Sociology* 95: 1–7.

Giddens, A. (1991) *Modernity and Self-Identity: Self and Society in the Late Modern Age*, Stanford, CA: Stanford University Press.

(1994) 'Living in a Post-Traditional Society', in U. Beck, A. Giddens and S. Lash (eds.), *Reflexive Modernization: Politics, Tradition and Aesthetics in the Modern Social Order*, Stanford, CA: Stanford University Press.

(1998) 'Risk Society: the Context of British Politics', in I.J. Franklin (ed.), *The Politics of Risk Society*, Cambridge: Polity Press, pp. 23–34.

Gieryn, T. (1983) 'Boundary work and the demarcation of science from non-science: strains and interests in professional ideologies of scientists', *American Sociological Review* 48: 781–95.

(1995) 'Boundaries of Science', in S. Jasanoff, G. Markle, J. Petersen and T. Pinch (eds.), *Handbook of Science and Technology Studies*, Thousand Oaks, CA: Sage Publications, pp. 393–443.

Gitlin, T. (1980) *The Whole World Is Watching: Mass Media in the Making and Unmaking of the New Left*, Berkeley: University of California Press.

Kolata, G. (1998) *Clone: the Road to Dolly and the Path Ahead*, New York: William Morrow.

Lakoff, G. and M. Johnson (1980) *Metaphors We Live by*, Chicago: University of Chicago Press.

Levin, I. (1977) *The Boys from Brazil*, New York: Random House.

McCombs, M., E. Einsiedel and D. Weaver (1991) *Public Opinion: Issues and the News*, New Jersey: Lawrence Erlbaum.

Manzoli, F. (1998) *Divulgazione o finzione? La clonazione rappresentata sui quotidiana. Tesi di Laurea in Scienze della Comunicazione*, Siena: University of Siena.

Miller, I.W. (1998) 'Sheep, Joking and the Uncanny', in M. Nussbaum and C. Sunstein (eds.), *Clone and Clones: Facts and Fantasies about Human Cloning*, New York: W.W. Norton, pp. 78–87.

Moscovici, S. (1976) *La Psychanalyse – son image et son public*, 2nd edn, Paris: PUF.

(1984) 'The Phenomenon of Social Representations', in R. Farr and S. Moscovici (eds.), *Social Representations*, Cambridge: Cambridge University Press, pp. 3–70.

Mulkay, M. (1993) 'Rhetorics of Hope and Fear in the Great Embryo Debate', *Social Studies of Science* 23: 721–42.

Osgoode, C., G. Suci and P. Tannenbaum (1971) *The Measurement of Meaning*, Urbana, II: University of Urbana Press.

Pence, G. (1998) *Who's Afraid of Human Cloning?* Lanham, MD: Rowman & Littlefield.

Rizzo, B. (1998) 'La construzione sociale della notizia scientifica: Il caso di Dolly, pecora clonata. Tesi di Laurea in Scienze della Comunicazione', University of Siena.

Toumey, C.P. (1992) 'The Moral Character of Mad Scientists: a Cultural Critique of Science', *Science, Technology and Human Values* 17: 411–37.

Turney, J. (1998) *Frankenstein's Footsteps: Science, Genetics, and Popular Culture*, New Haven, CT: Yale University Press.

Wagner, W. (1996) 'Queries on Social Representations and Construction', *Journal for the Theory of Social Behaviour* 26: 95–120.

Wagner, W., G. Duveen, R. Farr, S. Jovchelovitch, F. Lorenzi-Cioldi, I. Marková and D. Rose (1999) 'Theory and Method of Social Representations', *Asian Journal of Social Psychology* 2: 95–125.

Wilkie, T. and E. Graham (1998) 'Power without Responsibility: Media Portrayals of Dolly and Science', *Cambridge Quarterly of Healthcare Ethics* 7 (2):
150–9.
Wilmut, I., A.E. Schneike, J. McWhir, A.J. Kind and K. Campbell (1997) 'Viable
Offspring Derived from Fetal and Adult Mammalian Cells', *Nature* 385:
810–13.

Part IV

The transatlantic puzzle

12 Worlds apart? Public opinion in Europe and the USA

George Gaskell, Paul Thompson and Nick Allum

Introduction

Through the 1990s research and applications of modern biotechnology were making impressive strides in the United States. Viewed from the perspective of the biotechnology lobby in Europe, the United States was enjoying an enviable and apparently effortless assimilation of the fruits of the technology of the twenty-first century. By the middle of the decade, the idea of 'Life Sciences Conglomerate' integrating agri-chemicals and genetically modified (GM) seeds, foods and pharmaceuticals became a reality. Monsanto emerged as the Microsoft of gene technology and appeared to be on the threshold of global domination. New GM strains of soya, maize and cotton, along with a number of other GM crops, were planted across millions of hectares of the United States and were taking larger and larger market shares. After the failure of the 'Flav'r Sav'r Tomato[®]' (1995–6) tomato, it seemed as if the potential of modern biotechnology to revolutionise food production was to be realised.

On the other side of the Atlantic, to the chagrin of the biotechnology lobby, Europe was in turmoil as biotechnology became increasingly controversial and political. In 1997 the cloning of Dolly the sheep (see Einsiedel et al., chapter 11 in this volume) became an international, and largely polemical, news event. In response to public hostility over the importing of GM crops, Monsanto launched a European public relations campaign in 1998. The company's own research indicated that the campaign failed to persuade the sceptical public.

Following widespread controversy, six European member states, in contravention of EU regulations, introduced a moratorium on the commercial planting of GM crops. Prompted by commercial rather than scientific considerations, supermarkets in a number of countries announced boycotts of GM foods. In 1999, prompted by the Pusztai research,[1] which claimed to demonstrate that rats fed on a diet of GM foods suffered ill-health, the media turned against modern biotechnology.

As a result a fault-line opened between Europe and the United States in a World Trade Organisation (WTO) dispute, bringing threats of a trade war. The US companies and the Department of Agriculture asserted that GM products were safe, judged on the criterion of scientific risk assessment. In Europe, confronted by a sceptical public operating along the lines of an intuitive version of the precautionary principle, many national governments dithered.

The United States, however, could not insulate itself from the impact of the European revolt against GM crops and foods. Confronted by a consumer opposition in Europe that spread to other parts of the world, US farmers worried about markets for their produce and the shareholders of the life sciences companies took flight. As stock market valuations fell, the integrated life science concept was in tatters and there followed the break-up of agri-chemicals and GM seeds from the still successful pharmaceutical divisions.

With the benefit of hindsight, a shift in the European public's response to biotechnology can be identified. Before the 'watershed years' of 1996 and 1997 it was an esoteric issue, the concern of specialists and of relatively little interest to the majority of the public. From 1997, biotechnology entered a second 'political' phase of development. As products came into the market, the traditional criteria for technological assessment of risk and safety were challenged on both scientific and ethical grounds. Bioethics, environmental impacts and public participation had featured in the European debate since 1975, but now these issues achieved greater prominence.

In this chapter we contrast public perceptions of biotechnology in the United States and Europe and investigate whether the controversy of the late 1990s might have been anticipated. Can we find at least partial accounts for the dramatic differences between the reception of biotechnology in Europe and the United States in the mid-1990s and for the subsequent change in fortunes of the US life science companies at the close of the decade? Were there harbingers in the pre-political phase of the problems to come?

Our analysis starts with a comparison of public perceptions of five applications of modern biotechnology in Europe and the United States and then explores the roots of these differences in the context of media coverage, scientific knowledge and trust in the regulatory processes.

Public perceptions of biotechnology in Europe and the United States

Our primary data source is the 1996 Eurobarometer survey (46.1) on biotechnology (Durant, Bauer and Gaskell, 1998). Many of the questions

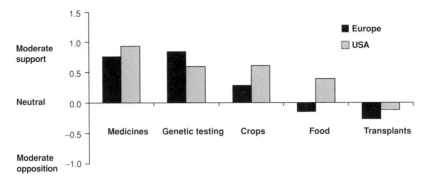

Figure 12.1 Mean support for five applications of biotechnology: Europe and USA

Note: Support is measured on a scale from −2 to +2. The USA and Europe differ significantly for each application (f-values from one-way ANOVAs for each application were all significant with $p < .05$).

from the Eurobarometer survey were also used in a US survey in late 1997.[2] These surveys provide a historical snapshot of public perceptions in 1996–7. Of course, with the rapid advance of food biotechnologies and other developments in the life sciences we would not expect to find the same opinions and attitudes today. But the use of similar questions in the surveys makes it possible to look at comparative structural differences in the pattern of public perceptions that may hold clues to understanding the contrast between the USA and Europe in more recent years.

Respondents were asked whether they thought each of five biotechnologies – genetic testing, GM medicines, GM crops, GM food and GM animals for use in human transplantation ('xenotransplantation') – was useful, risky, morally acceptable and to be encouraged. Figure 12.1 shows the mean levels of support (encouragement) on a scale from +2 to −2 for all the applications.

People in Europe and the USA showed varied levels of support across the different applications. GM medicines and genetic testing received the highest levels of support, GM crops and GM foods received intermediate levels of support, and xenotransplantation received least support. And there was not always strong support for biotechnology in the USA: for example, the average US respondent was opposed to xenotransplantation. Furthermore, they were not always more supportive than Europeans; for example, Europeans were more supportive of genetic testing. However, people in the USA were significantly more supportive of GM crops and GM foods than were people in Europe.

When the surveys were conducted, biotechnology was a relatively unfamiliar topic. On the questions about the five applications, 19 per cent

Table 12.1 *Three common logics in relation to attitudes to five applications of biotechnology*

Logic	Attitude			
	Useful	Risky	Morally acceptable	Encouraged
1. Support	YES	NO	YES	YES
2. Risk-tolerant	YES	YES	YES	YES
3. Opposition	NO	YES	NO	NO

of people in the USA and 27 per cent of Europeans did not give a complete set of responses. With this level of unfamiliarity we can assume that some people responded to the questions with poorly informed and unintegrated 'non-attitudes' (see Midden et al. in chapter 7 of this volume). Such responses would be likely to be volatile if, for example, the issue became more controversial. In the absence of a filter question allowing us to exclude people with 'no opinion', the following analysis uses only those who gave a full set of responses, on the assumption that they were more likely to have better formed opinions. Judgements of use, risk, moral acceptability and encouragement were each collapsed into a dichotomy (useful/not useful, etc.) in order to model patterns of response (henceforth 'logics') over the four dimensions of attitude. This produces sixteen possible combinatorial 'logics', but empirically only three were widely used (see table 12.1). Logics 1 and 2 are similar in being supportive, but they display different perceptions of risk. For the 'supporter', risk is not an issue. The 'risk-tolerant supporter' sees, but then discounts, the risk. Opponents take a position exactly opposite to that of supporters.

Table 12.2 shows the distribution of these three prevalent logics for each application. For GM medicines and genetic testing, supporters constituted the single largest category. Levels of risk-tolerant support were also relatively high, and levels of opposition were relatively low. Higher opposition to genetic testing in the USA ($p < .05$) than in Europe may indicate a sensitivity about genetic privacy in the context of work, credit or insurance. In contrast, for xenotransplantation, supporters and risk-tolerant supporters totalled only 36 per cent in Europe and 42 per cent in the USA, with about 33 per cent in opposition.

Turning to GM crops and GM foods, we see a considerable contrast between Europe and the USA. Both GM crops and GM foods were better supported in the USA than in Europe (for both contrasts, $p < .05$). The contrast is greatest in the case of GM foods, to which 30 per cent of Europeans with a logic were opposed, compared with only 13 per cent

Table 12.2 *Distribution of logics for five applications of biotechnology: Europe and USA*

| Application | Europe | | USA | | |
	% of respondents with complete set of responses[a]	% of total sample[b]	% of respondents with complete set of responses[c]	% of total sample[d]	T-value[e]
Medicines					
Support	41	30	54	44	4.76
Risk-tolerant	37	27	29	23	1.52
Opposition	8	6	5	4	
Genetic testing					
Support	50	37	51	41	−6.08
Risk-tolerant	33	24	21	17	−9.38
Opposition	7	5	14	11	
Crops					
Support	35	26	51	41	8.17
Risk-tolerant	26	19	22	18	3.07
Opposition	18	13	10	8	
Food					
Support	22	16	37	30	11.89
Risk-tolerant	21	15	24	19	8.13
Opposition	30	22	13	11	
Xenotransplants					
Support	16	12	23	19	2.86
Risk-tolerant	20	15	19	15	−1.47
Opposition	33	24	35	28	

Notes: Loglinear modelling on each application, with 'opposition' as the reference category, shows that the probability of being a 'supporter' or 'risk-tolerant supporter' differs significantly ($p < .05$) for the USA and Europe, with the exception of 'xenotransplants' and 'medicines' where there is no significant difference in the probability of risk-tolerant support.
[a] $N = 12,178$.
[b] $N = 16,500$.
[c] $N = 863$.
[d] $N = 1,067$.
[e] T-values of >1.96 indicate significance at $< .05$.

of Americans. By the same token, whereas over 60 per cent of people in the USA were supporters or risk-tolerant supporters of GM foods, the comparable figure for Europe is just over 40 per cent.

That, as early as 1996, such a large percentage of Europeans were concerned about the usefulness, risks and moral acceptability of GM foods is of particular note. What is clear is that there was a groundswell of people opposed to GM foods, for whom assurances about the absence of scientific risks would have been unlikely to alleviate their concerns. For them the tortuous discussions and long delay in introducing a labelling scheme for GM foods may well have created further anxieties. People may have thought 'If *they* say it is safe, why should *they* hesitate to label it? *They* must be hiding something' – *they* being a mixture of the industry and anonymous regulators.

A fourth possible logic – 'moral opposition' (in terms of table 12.1, answers = yes, no, no, no) – counts for no more than 3 per cent on any applications. That so few people adopted the converse of the logic of risk-tolerant support implies that respondents with concerns about gene technology tended to think principally in terms of moral acceptability rather than risk – a significant difference from the way in which experts normally judge the acceptability of new technologies. But we must not assume that the opposition was purely of a moral kind without considerations of benefits and costs. In the public mind the representation of potential danger is a complex of perceived benefits and risks. The first generation of GM foods offered producer rather than consumer benefits, yet any health risks were to be borne by the consumer. In these circumstances it seems not unreasonable to take the position, 'Why should I take any risk with this GM food, if there is no apparent benefit for me?'

In summary, with 30 per cent of Europeans opposed to GM foods there is clear evidence that the foundations of the consumer revolt were laid before 1997. This raises the question about the origins of the contrasting views of GM foods in the United States and Europe. In some respects the perplexing issue is why Europe was relatively more negative – perplexing because, when the surveys were conducted, there was relatively little penetration of GM food products in Europe (save for the introduction of GM tomato puree in Britain in 1996, voluntarily labelled as genetically engineered and greeted with little interest).

It seems as if negative opinions were based on individual cognitions and preferences rather than being the product of debates in the public domain. In this sense the opposition may have been implicit, articulated only in the context of a survey question, rather than reflecting the explicit contents of prior conversations and discussion.

Exploring the bases of the transatlantic differences

We consider three factors that may help to account for the differences in public attitudes between Europe and America.

Media coverage

First, consider the influence of the press. One popular view suggests that the content (either positive or negative) of press coverage shapes public perceptions in the corresponding direction. Another hypothesis suggests that in technological controversies it is the sheer quantity of press coverage that is decisive: the greater the coverage, the more negative the public perceptions (Leahy and Mazur, 1980).

In order to compare US and European press coverage, we analysed a longitudinal sample of articles drawn from elite national newspapers in twelve European countries and the USA (see Gutteling et al., chapter 3 in this volume). We assume not that these newspapers are widely read, but rather that they inform politicians and other journalists and, over time, reflect the tone of the national debate. In order to compare Europe and the USA (and because there is no trans-European press), figure 12.2 shows the average of the twelve European national newspapers compared with the *Washington Post*. Since we are exploring post hoc explanations, strict comparability of measures is not essential.

Between 1984 and 1991 there is a broadly similar trajectory in Europe and the USA. Thereafter, however, the European trajectory rises more

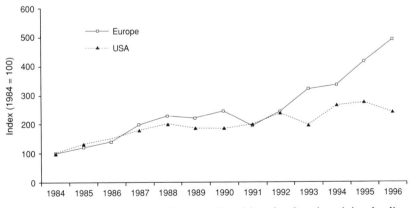

Figure 12.2 Number of articles about biotechnology in opinion-leading press, 1984–1996 (index: 1984 = 100)
Note: $N = 12$ newspapers for Europe; $N = 1$ newspaper for USA.

steeply than that in the USA. The comparison is consistent with the hypothesis concerning the importance of the quantity of media coverage. The relatively greater increase in coverage in Europe goes together with greater public concern.

Based on coding categories designed to facilitate systematic comparison of media coverage (Durant et al., 1998), table 12.3 shows the content of coverage in Europe and the USA. From 1984 to 1990, there are relatively few differences between the European and the US press. 'Progress' and 'economic prospect' are the dominant frames in both cases, and the important themes are 'health', 'basic research' and 'economics'. From 1991 to 1996, differences between Europe and the USA are evident. The *Washington Post* moves from 'progress' to 'economic prospect', while in Europe 'progress' remains dominant. The emerging frames are 'public accountability' and 'nature/nurture' in the USA, but 'ethics' in Europe. In the USA, we see fewer 'benefit' stories and more 'risk and benefit' stories. There is no evidence of increasing 'risk' stories in Europe.

These results do not confirm the view that public perceptions reflect the content of press coverage. On the contrary, whereas the trend in European press coverage was more positive than that in the USA, by 1996 public opinion in Europe was more negative. Instead, our evidence supports the hypothesis that increased amounts of press coverage of technological controversies are associated with negative public perceptions. What may have happened is that, on reaching a critical threshold, the press coverage acted as a catalyst, transforming private concerns into more visible public attitudes.

Scientific knowledge

Another factor is the role of knowledge in public perceptions. A common belief is that scientific literacy generates support for science and technology. Two types of knowledge of biology and genetics were tested by nine items. Six true/false items tested general knowledge:

1. More than half of human genes are identical to those of chimpanzees.
2. The cloning of living things produces exactly identical offspring.
3. Yeast for brewing beer consists of living organisms.
4. It is possible to find out in the first few months of pregnancy whether a child will have Down's syndrome.
5. Viruses can be contaminated by bacteria.
6. It is possible to transfer animal genes into plants.

Three true/false items tested images of food biotechnology:

7. Ordinary tomatoes do not contain genes while genetically modified tomatoes do.

Table 12.3 Content of press coverage in the USA and Europe (%)

	Frames[a]	USA	Europe	Themes[b]	USA	Europe	Risk/benefit	USA	Europe
1984–1990[c]	Progress	50	49	Health	29	24	Benefit	39	43
	Economic prospect	17	18	Basic research	14	12	Risk and benefit	34	30
	Nature/nurture	10	1	Economics	13	10	Risk	6	12
	Ethical	8	12	Regulation	11	9	Neither	21	15
	Public accountability	8	13	Safety and risk	11	7			
1991–1996[d]	Progress	30	50	Health	37	30	Benefit	27	38
	Economic prospect	29	15	Economics	12	9	Risk and benefit	44	24
	Nature/nurture	15	4	Regulations	12	10	Risk	9	10
	Public accountability	15	10	Safety and risk	11	7	Neither	20	30
	Ethical	6	16	Basic research	8	10			

Notes: The figures show the average of twelve European national newspapers compared with the *Washington Post*.
a Frames are the perspectives in which biotechnology is discussed.
b Themes are specific topics within the area of biotechnology.
c Europe: N = 1,769; USA: N = 117.
e Europe: N = 2,861; USA: N = 89.

8. By eating a genetically modified fruit a person's genes could become modified.

9. Genetically modified animals are always bigger than ordinary ones.

For the textbook items, an incorrect answer is presumed to reflect a lack of scientific knowledge. For the image items, an incorrect answer is presumed to reflect both a lack of scientific knowledge and the willingness to entertain an image of threatening possibilities of food adulteration, infection and monstrosities.

The textbook and image items formed two scales. For the textbook items, the scale records the number of correct responses (0–6). For the image items, the scale records the number of threatening images (0–3).

On textbook knowledge the mean score for Europe is 2.9, which is not significantly different from the US score of 3.0.

Thus, textbook knowledge does not explain the more positive attitudes of people in the USA. By contrast, the mean score for threatening images of food biotechnology in the USA is 0.24, significantly lower than the European mean score of 0.88 (T = −36.24, $p < .0005$). The lowest score for threatening images in any European country is more than twice as great as the US score. If more Europeans think that GM foods are the only foods containing genes, that eating GM foods may result in genetic infection and that GM animals are always bigger, it is hardly surprising that they approach modern food biotechnology with greater suspicion.

There are several possible explanations of the greater prevalence of menacing images of agricultural and food biotechnology in Europe. First, food has strong cultural connotations; it is a part of national identity. It is possible that European culinary traditions are more resistant to technological change. Secondly, the recent series of food safety scares in Europe, most notably BSE, may have sensitised large sections of the European public to the potential dangers inherent in industrial farming practices.

Beyond these agriculture and food-related issues, it is possible that other factors shape contrasting perceptions of biotechnology in Europe and North America. The new genetics touches upon deep-rooted beliefs about the boundaries between the natural and the unnatural and about the differences between 'good genetics' and 'bad genetics'. In this context, GM crops and GM foods may be hard to classify. Are these 'natural' or 'unnatural', 'good genetics' (like medicines) or 'bad genetics' (like eugenics)? As a result of different histories and different assumptions about the boundaries of the natural, it may be that Europeans are less inclined than Americans to embrace food biotechnology.

Trust in regulatory systems

The third putative explanation for the differing levels of support for biotechnology in the United States and in Europe concerns public trust in the processes of regulation. In an increasingly complex world, trust functions as a substitute for knowledge (Luhmann, 1979). Essentially people need to act on the assumption that systems of regulation and control supporting everyday life will not fail. With trust in the regulatory process, people can behave, more or less, as if the future were certain, and with confidence that unforeseen problems will be sorted out. Under conditions of distrust, however, the future is uncertain, perceived risks and resulting anxieties may be accentuated, and there may be doubts about the ability of the authorities to counter emergent problems. For this reason, the extent to which the public has trust and confidence in the regulatory processes may be a further factor contributing to public opinion on the new developments in biotechnology.

To put the issue of trust into context it is necessary to understand the rather different histories of biotechnology regulation in Europe and the United States. The European history is outlined in detail by Torgersen et al. in chapter 2 of this volume. In Europe there has been a protracted public debate and great difficulty achieving a viable multi-level consensus. Biotechnology has been treated as a novel process requiring novel regulatory provisions, leading to a complex series of national and European initiatives embracing a wide range of both known and unknown risks, including risks to the environment. The development of regulatory arrangements has been further complicated by the competing agendas of different directorates of the European Commission (Cantley, 1995). Although it is fair to conclude that biotechnology was not as controversial in the United States as in Europe during the 1990s, for most of the period since 1975, policy activity in the USA has been lively and hotly contested.

Excursion: US biotechnology policy, 1975–2000

Since the history of US policy-making on biotechnology is rather different from that of Europe, this section provides a brief overview. Following a summary of the general trends and events of the period, three subsections trace policy decisions undertaken by the US government.[3]

Few Americans outside the leading graduate research programmes in biology would have known much about recombinant DNA (rDNA) in 1975. Although Watson and Crick's discovery of DNA was, by this time, part of every well-educated American's knowledge set, the technological possibilities and the attendant risks and benefits were wholly unknown. This changed with the Asilomar Conference, called to debate whether transfer of genes from one organism to another – something that was already known in 1975 as genetic engineering – was an inherently dangerous activity. The conference was widely covered by US news media and, within a few years, popular books on genetic engineering began to appear.

One of these was *Who Should Play God?* by Howard and Rifkin, published in 1977. The book did not create a sensation. It was one of many that attempted to raise awareness about ethical questions concerning the prospects for human eugenics in the coming world of biotechnology. It is primarily noteworthy as the first entry for Rifkin, who went on to publish a bestseller entitled *Entropy* in 1980. *Entropy* was a highly accessible introduction to a broad range of environmental issues, and it was published with a postscript by Georgescu-Roegen, the distinguished theorist who is given credit for the first effective integration of economics and ecology. *Entropy* gave Rifkin prestige and an audience, and it laid the basis for the formation of an environmentally oriented non-governmental organisation (NGO) entitled the Foundation on Economic Trends (FET).

It was from this platform that Rifkin launched a series of attacks on biotechnology, beginning with *Algeny* in 1983. Though not as widely read as *Entropy*, the new book characterised genetic engineering as the natural extension of a reductionist philosophical programme launched by Bacon and Descartes some three hundred years earlier. The book linked environmentalists' interest in holistic thinking with postmodern critics of science, and opened the door for an interest in matters of the spirit. The link to spirituality was made more explicitly in Rifkin's 1985 book, *Declaration of a Heretic*, in which the Darwinian elements of ecology were replaced with a Biblical account of creation and a broad denunciation of modern biology. The specific content of these three books may be less important than the fact that they allowed Rifkin to assemble a loose coalition of environmentalists, postmodern intellectuals and fundamentalist Christians – strange bedfellows indeed – in a focused attack on the nascent products of biotechnology.

Rifkin's public activism during this period consisted in college speaking engagements, popular magazine and radio interviews, and coalition-building with his constituency groups. During these activities, the theme was frequently to highlight the ethical concerns associated with human eugenics that he had first raised in 1977, and to suggest that American science was on a slippery slope that would lead inevitably to this result. Rifkin told his audiences that they would be tempted by extremely attractive possibilities – drugs and therapies that would provide hope for people afflicted with horrible disease. Nevertheless, these temptations would tend toward ecological collapse and corruption of our respect for the integrity of life.

Behind the scenes, the Foundation on Economic Trends was launching a series of activities that would bedevil the scientists and companies that were hoping to develop new products of biotechnology. A coalition of religious leaders remarkable for its breadth was induced to sign a statement protesting against animal patenting. Meanwhile, FET filed an application for a patent on human beings (with the full expectation that the patent would be denied), partly for publicity, partly to make a point and partly to establish some legal precedents in patent law. The most successful action was a lawsuit (discussed below) that delayed agricultural experiments on ice-nucleating (so-called 'ice minus') bacteria for several years, and tested the US government's entire regulatory apparatus for agricultural biotechnology.

Rifkin found himself with a number of allies when it came to the environmental and agricultural implications of biotechnology. In 1985, Doyle published *Altered Harvest,* a book that alerted the attentive educated public to the importance of genetic diversity in maintaining an ecologically robust agriculture. In 1986, noted veterinarian and animal activist Fox published *Agricide.* In 1987, the Biotechnology Working Group, a consortium of environmental and agricultural activists, published 'Biotechnology's Bitter Harvest' (Goldburg et al., 1990). In 1993, Vandana Shiva, the well-known advocate for women of the developing world, published *Monocultures of the Mind.* Although none of these publications captured the attention of more than a fraction of Americans, all of them were sharply critical of agricultural biotechnology. Cumulatively, they created a climate of suspicion about the likely social and ecological impact of biotechnology in agriculture.

Thus, although concerns about human eugenics were never too far distant, the decade of the 1980s was a time when a number of Americans were feeling sceptical about biotechnology. It was, however, also a time when there were no products of agricultural biotechnology on the market. As such, it was difficult for this sceptical minority to raise much concern among ordinary Americans. In the meantime, a number of organisations undertook efforts to deflect the brunt of the criticism being levelled by sceptics. The first large-scale effort was conducted by the Keystone Foundation, a US NGO oriented toward finding non-violent consensual solutions to contentious issues. Keystone funded and conducted a series of workshops around the USA between 1995 and 1998 involving leadership and representatives from environmental organisations, as well as scientists and representatives from the US biotechnology industry. These workshops allowed issues and concerns to be aired, and became the forum in which many environmental leaders learned about biotechnology.

Following the completion of the Keystone project, a consortium of US and Canadian universities and not-for-profit research centres formed the

National Agricultural Biotechnology Council (NABC), which held its first meeting in Ames, Iowa, in 1989. This meeting also brought critics together with university and industry scientists, as well as government regulators. NABC has held similar meetings every year since. Ralph Nader, US Green Party presidential candidate and frequent critic of biotechnology, was the keynote speaker in 2000. NABC meetings have been structured with ample small-group discussion time, and a process that generates a report on areas of agreement, areas where research or education efforts are needed, and some recommendations, which may be directed to US government officials or to NABC members themselves. In addition to holding annual meetings, NABC publishes a newsletter and an annual report and holds a congressional briefing on consensus recommendations each year in Washington D.C.

The outcome of the Keystone and NABC projects, beyond a parade of books critical of biotechnology, has been that US policy on agricultural biotechnology has been formed amidst a rich and often contentious atmosphere of debate. Other activities, such as series of workshops on teaching ethical issues in biotechnology conducted by Iowa State University, augmented the Keystone and NABC efforts. Issues in human genetics and medicine have been in the background of those debates, but there have arguably been fewer policy decisions in the human/biomedical area in any case. By the mid-1990s, US government agencies had made formal policy decisions for regulating biotechnology, which are described below. Although many of the individuals who were most active in criticising biotechnology between 1985 and 1995 were deeply dissatisfied with the direction of US policy, the decade of workshops and hearings had apparently exhausted the energy and interest of the broader community of politically active US citizens.

Undoubtedly, many environmental leaders who participated in Keystone, NABC and other activities decided that, whatever environmental issues might be associated with biotechnology, other problems were more pressing and more worthy of their limited time and effort. Similarly, few NGOs representing consumer interests placed biotechnology on their agenda for public action. By 1995, when products began to appear in large numbers, Rifkin's odd coalition of environmentalists, advocates of social justice, animal activists, postmodern intellectuals and fundamentalists had apparently had enough of working together. When the US Food and Drug Administration (FDA) announced its intent to implement a policy that would discourage labelling of genetically engineered foods in 1996, few of the groups even bothered to comment. Rifkin himself had gone on to write books about the beef industry and the end of work. All apparently had other fish to fry.

By 2000, this situation had changed in a manner that surprised many scientists, government regulators and the biotechnology industry itself. Rifkin was back with a new book, *The Biotech Century*, in 1998. In 1997 and 1998, a US Department of Agriculture proposal to allow genetically engineered foods to be labelled as 'organic' was soundly criticised – the mirror image of the FDA's experience in 1996. In 1999, controversy in Europe began to be reported in US newspapers, and this spawned stories critical of biotechnology in the widely read Sunday supplements of the *Washington Post* and the *New York Times*. Protestors at the 1999 WTO meetings in Seattle got biotechnology on the nightly news, and leading news outlets gave sensational coverage in 2000 to a report that a leading brand of taco shells contained samples of genetically engineered maize that were not approved for human diets.

Although US policy on biotechnology seemed stable and complete in 1995, it is not at all clear that it will remain so. No key policy changes have been made during the period of controversy, but the apparatus for policy change is beginning to move. Hearings are being held and politicians are expressing outrage. However, it is important to have a general appreciation of the US policy process in order to understand the significance of these events. Some elements of current US policy on biotechnology were developed through normal procedures – the assimilation of the new technology into existing policies – and this is likely to continue in the future. Other elements were developed as a response to controversy of all sorts – the processes of accommodation to challenges – and the revived controversy of the waning years of the 1990s may lead to a new round of policy change as a form of political reaction.

Here, 'normal procedures' include activities of review and regulation that occur as a matter of course under US law. Following Kuhn's distinction between normal and revolutionary science, we can classify policy developments that occurred in response to the novelty of the new biotechnology as 'revolutionary'. However, normal procedures are not wholly distinct from revolutionary ones. Indeed, the most contentious issue of the late 1980s and early 1990s concerned whether or not existing regulatory procedures were adequate for biotechnology. During that period, regulatory agencies adapted policies and procedures that had not been previously applied to research and products of gene transfer. Whether this adaptation was 'normal' or 'revolutionary' is largely a matter of interpretation. The value of the normal/revolutionary dichotomy consists primarily in providing a succinct way to characterise two strands of policy development in the USA, and in suggesting an analysis of how and why policies continue to be contested. The following sections build on this framework, and discuss the development of both biomedical and agricultural policy.

Normal regulatory policy in the United States

On the normal side, the United States has fairly well-established regulatory procedures for research and development of activities having the potential for impact on human, animal or environmental health. Pure research activities are self-regulated by both non-profit research organisations and commercial firms, though these self-regulatory activities are themselves overseen by the primary research funding agencies of the US government. Drugs and food technologies are regulated by the US Food and Drug Administration, while chemicals or organisms with a potential for environmental impact are regulated by divisions of the Environmental Protection Agency (EPA) and by the Animal and Plant Health Inspection Service (APHIS). These federal agencies operate under a suite of laws with varying degrees of specificity regarding both the methods used to ascertain risks and the criteria for determining acceptability of risk.

In some cases, such as drugs or pesticides, firms must register a product with the regulating agency; in other cases, such as the development of new foods or food additives, submission of a product for regulatory review is optional. Thus, drugs developed using recombinant techniques must receive FDA approval before being marketed, whereas foods, such as the Flav'r Sav'r Tomato®, are reviewed only on request. In either case, however, the review process requires that organisations requesting approval must generate data demonstrating the safety and efficacy of the product. These data are reviewed by FDA officials, who frequently raise additional questions and request additional studies. Drugs must offer the chance for substantial benefit in exchange for risks in order to be approved, and must be marketed in a manner that informs end users (prescribing physicians, pharmacists and their patients) about the risk of side-effects. Generally speaking, foods must meet a *de minimus* standard of risk, though risk standards for foods have been evolving in the past twenty-five years.

Policies for product approval apply to any developer of a new technology, whether that be an individual, a for-profit firm or a non-profit research organisation (including universities or the US government itself). Traditionally, public programmes such as the US Department of Agriculture and US agricultural universities have obtained approvals for new crop varieties and have made these products available to the public at a nominal cost. However, it would be very unusual for a publicly funded organisation to undertake the extensive testing needed to demonstrate safety for drugs, pesticides or novel foods. It is only because most new crop varieties have been exempt from regulatory review on food safety grounds that public research institutions have been able to develop and register new seeds.

Whenever US agencies take action to approve a product or to alter their procedures for product approval, notice of the action is published in the US Federal Register, and parties that are potentially affected by the action are given an opportunity to comment before the action becomes final. Agencies are required to respond to public comments, also in the Federal Register, providing a public record of the basis on which decisions are made. Clearly, the US public is oblivious to most of what is published in the Federal Register. However, non-governmental organisations representing special interests do monitor the Federal Register for actions that affect their members. These organisations often express dissatisfaction with the US government's responsiveness to their concerns, and this has particularly been true for decisions regarding products of rDNA technology in the food and agricultural arena. Nevertheless, the normal US procedure for setting policy with respect to science and technology is both transparent (one may discern the basis on which decisions are made) and responsive to the interests of affected parties, when compared with that of many other industrialised countries. US agencies have made normal regulatory decisions on hundreds of drug and agricultural products since 1980.

It is important to recognise that Americans also have recourse to the courts to pursue issues of interest to them. Two venues merit special note. First, it is possible to bring a civil action against a regulatory agency in an attempt to overturn or reverse an agency's action on a particular issue (though the US government limits the opportunities to bring such action). Agencies permit and even encourage such actions, especially in cases where agency administrators believe that there may be flaws in existing legislation or procedures for conducting regulatory review. Secondly, Americans have extraordinarily broad access to legal venues for product liability actions, and settlements can be quite large. As such, firms operating in the United States have significant incentive to limit their exposure to product liability lawsuits. This leads them to seek regulatory approval whenever there could be any question about the safety of a product and to be assiduous in submitting the data needed to demonstrate the safety of a product.

Revolutionary biomedical policy and rDNA in the USA: 1975–2000

In normal circumstances, the US press would not take an interest in an issue unless agency decision-making (including its checks and balances) had clearly failed. As noted, many products (especially pharmaceutical and diagnostic devices) have moved through the normal process without exciting comment during the past twenty-five years. Thus, the fact that an issue receives coverage in national media outlets is evidence that something out of the ordinary is going on. It is fair to say that many biotechnology policy decisions had become normalised in the USA by 1997, and this may have led to a general perception that biotechnology was uncontroversial and accepted by the US public. However, biotechnology made the news regularly between 1975 and 1995, and the resurgence of reporting on biotechnology issues in 1999 and 2000 suggests that the revolutionary period of US policy may not be entirely over.

Controversy over policy on biotechnology research began with the Asilomar Conference in 1975, a meeting of scientists working at the frontiers of science of recombinant DNA. The immediate policy result of this meeting was that scientists voluntarily agreed to a moratorium on genetic transformation of organisms until procedures that would limit the risks of such research could be established. A set of guidelines for conducting such research was developed under the auspices of the US National Institutes of Health (NIH), a government agency that both conducts and funds research in biomedical science and public health. NIH also established the Recombinant DNA Advisory Committee (RAC) to review proposals for research using rDNA and to advise administrators on agency-wide policy for future research (Krimsky, 1982).

The Asilomar Conference sparked several years of speculative debate about the risks and likely applications of gene transfer. Much of this debate centred on the question of whether such research is inherently dangerous or immoral (Goodfield, 1977). The city of Cambridge, Massachusetts, home of Harvard University and the Massachusetts Institute of Technology, attempted to ban rDNA research within its boundaries, initiating what some have called the first serious attempt to regulate science in the United States (Krimsky, 1982). The Cambridge controversy also led to the 1983 creation of the Committee for Responsible Genetics, the first US NGO focusing on biotechnology policy. However, by the early 1980s most US biotechnology research was required to meet only the least onerous of NIH safety guidelines. By the end of the 1980s, review for the safety of rDNA had been thoroughly normalised, being conducted primarily by local institutional review boards at the organisations (universities, government laboratories and private firms) employing principal investigators.

As rDNA research led to the development of valuable drugs and diagnostic procedures, media coverage of the science shifted to the financial pages and toward stories that portrayed the science in non-controversial terms. Through the 1980s the NIH RAC debated the scientific and ethical issues associated with new protocols for gene therapy (Anderson, 1987). However, these debates neither engaged a large segment of the American public nor created an environment in which biotechnology policy evolved in response to controversy or public pressure. It would be more accurate to say that, by 1990, the NIH process for soliciting advice through the RAC and for making judgements about the risks and ethics of genetic research had evolved into an entirely normal process.

One key (and hence *somewhat* revolutionary) development in the normalisation of policy at NIH was the creation of the Ethical, Legal and Social Issues (ELSI) program of the Human Genome Initiative in 1992. NIH initiated this program to support both social science and philosophical research on issues in human genetics that would become problematic or controversial as more information about the location and function of genes became available. The original proposal was to dedicate 5 per cent of human genome funding to ELSI projects, though the figure dropped to considerably less than half that amount. Even at this level, however, the ELSI program alone exceeded the total amount of federal grant money that had been available for research on ethical topics before its inception. Over its history, ELSI has funded dozens of research projects and conferences on topics such as genetic discrimination, genetic screening, privacy, implications for insurability and access to health care, and the genetic basis of behavioural traits. The result is a fairly large scholarly literature on these topics and frequent coverage of these issues in the US press. A compendium of this scholarly work, *The Encyclopedia of Legal and Ethical Issues in Biotechnology*, was published in the summer of 2000 (Murray and Mehlmann, 2000).

A final spurt of revolutionary policy-making for biomedical biotechnology was sparked by the announcement that a sheep had been cloned at the Roslin Institute in February 1997. Within days of the announcement, President Clinton commissioned the National Bioethics Advisory Commission (NBAC) to study the ethical issues raised by the possibility of human cloning and to issue a report within a ninety-day time horizon. The NBAC is not a standing committee of the US Executive Branch, but was in fact an ad hoc committee of experts in reproductive medicine, embryology and medical bioethics that had been convened to review the federal government's policy on research using human stem cells. However, the NBAC launched into an intensive study on likely uses of human cloning and its risks and on ethical issues raised both by secular bioethicists and by representatives of the principal religious denominations. In addition to hearings conducted by the NBAC, the US House and Senate both conducted hearings on cloning, in which they heard from many of the same people who had prepared testimony for NBAC.

Some of the religiously oriented respondents expressed strong opposition to cloning based on their views on human embryos – views that had already surfaced in the debate over stem cell research. However, the NBAC ultimately decided not to revisit these issues, and the result was a consensus on a somewhat narrow set of issues concerning the safety of human cloning research. Specifically, the NBAC noted that many questions remain about the safety and efficacy of the procedure that produced Dolly, and recommended a moratorium on any attempt to use adult cell nuclear transfer for the purpose of producing a human child (National Bioethics Advisory Commission, 1997). As of 2000, the US Congress had not acted on the NBAC recommendations.

There is no US law that prohibits or regulates mammalian cloning of any kind. The NIH would rule on any proposal to use federal funding in such a research project, hence there is a 'normal' mechanism for considering the ethical implications of human cloning when public funds are involved. It is clear that NIH would apply many of the same considerations recommended in the NBAC report. However, NIH policies do not apply to research conducted solely with public funds, hence it is possible that human cloning could be undertaken in the USA without violating any laws.

Revolutionary agricultural biotechnology policy: 1975–2000

Immediately after the initial controversies associated with Asilomar, the next round of public controversy over biotechnology was associated with agriculture. A University of Wisconsin researcher submitted a proposal to conduct experiments using bacteria that had been transformed in a manner that could affect the freezing point for crops such as strawberries or potatoes. In 1994, Lindow, the scientist who proposed this research on so-called 'ice-minus' bacteria, duly sent his research protocol to the NIH RAC, which approved the experiment. However, as noted already, activist Rifkin and the Foundation on Economic Trends successfully sued NIH and the University of Wisconsin to block the experiment, arguing that the NIH – a biomedical research agency – lacked the scientific expertise to assess environmental risks. Rifkin's lawsuit sparked public protests and destruction of field plots where ice-minus research was being conducted. The lawsuit and public controversy eventually led to the formation of the Agricultural Biotechnology Research Advisory Committee (ABRAC) in 1986 within the US Department of Agriculture (USDA). ABRAC was intended to perform an advisory function similar to that of the NIH RAC, but focusing on applications of food and agricultural biotechnology (Thompson, 1995).

As controversy over ice-minus was subsiding, controversy was beginning over the next product of agricultural biotechnology, recombinant bovine somatotropin (rBST), which can be used to stimulate milk production in dairy cows. As an animal drug, rBST clearly came under the approval process for veterinary drugs administered by the FDA. However, a 1995 paper on the socio-economic impact of rBST had predicted that it would exacerbate economic trends that were affecting both the farm size and geographic location of dairy production in the United States (Kalter, 1985). The FDA took an unusually long time to conduct its review of rBST. The agency needed to be satisfied that the drug was efficacious in stimulating milk production, that it posed no risk to human health and that it posed acceptable risks to animal health.

Only the first of these decisions was straightforward. With respect to human health, the FDA ultimately developed one of the first versions of the policy that has come to be known as 'substantial equivalence'. Based on a comparison of the structure of the rBST protein with that of natural BST, and on the fact that the proteins are indistinguishable in any of the studies that FDA scientists were able to devise, the agency decided that evidence for the safety of natural BST (present in all milk) could be extended to rBST as well (Juskevich and Guyer, 1990). That this should be regarded as a revolutionary rather than a normal policy decision is supported by the fact that FDA officials went well beyond standard agency practices to publish the basis for this evaluation (Krimsky and Wrubel, 1996: 173). Debate over the animal health implications of rBST became protracted. High-yielding dairy cows are at greater risk of mastitis, whether or not they have been treated with rBST, but rBST can clearly increase milk yields, moving animals into this higher-risk group. In the end, the FDA decided that, since careful herd management could control the risk of mastitis, there was not sufficient reason to withhold approval of rBST, and rBST was approved for use in US dairy herds in November 1993.

However, the US Congress intervened in the process, declaring a ninety-day moratorium on the use of rBST and requiring the White House to complete a study on the impact of rBST. Although rBST did come on the market after the ninety days had expired, the fact that Congress took this extraordinary measure testifies to the level of public outcry and controversy that surrounded rBST during the long, nine-year history of FDA review. Although debate over rBST may never have risen to the level of public awareness associated with so-called 'Frankenfoods' in the UK during the late 1990s, rBST was extremely controversial in the USA for well over a decade. Debate over rBST split the US farm community, with some farm NGOs coming out strongly against, and others equally strongly in favour. Animal protection NGOs joined with farm groups – an unprecedented and unrepeated event in US NGO politics – to oppose 'bovine growth hormone' (or BGH), the term that opponents of rBST preferred to use (see Krimsky and Wrubel, 1996). The debate surfaced in political cartoons and in the comedy routines of American television performers. Retailers such as Ben and Jerry's ice cream continue to print anti-BGH statements on their product labels as of this writing.

The US debate over rBST has both formal and informal policy implications. The formal implications are largely confined to the key years of 1993 to 1996, while the informal implications cover a fifteen-year period (continuing to this day) in which US NGOs, the biotechnology industry and scientific organisations such as universities, disciplinary groups and government agencies jockeyed for control. In terms of formal policy decisions taken by the US government, the significance is three-fold. First, the fact that the FDA decision was ultimately allowed to stand indicates a decision in favour of 'normalising' regulatory decision-making for biotechnology by the Clinton administration. The ABRAC was closed in 1996 and biotechnology policy was given over wholly to the normal processes of review under the auspices of the FDA, the EPA and APHIS. Secondly, the White House explicitly rejected the legitimacy of regulating a biotechnology on the basis of its socio-economic impacts. None of the agencies listed above is allowed to consider socio-economic impact when making a regulatory decision. Thirdly, the FDA and other US government agencies adopted policies that placed heavy legal and economic burdens of proof on anyone who attempted to impose labelling requirements on food products in which modern biotechnologies had been used (Thompson, 1995).

Public trust in the regulatory processes

With this background we can now explore trust in the regulatory systems in Europe and the United States as evidenced in the surveys conducted in 1996 and 1997.

European respondents were asked to select, from a list of national and international institutions, the one best placed to regulate biotechnology. The results show most confidence in international organisations such as the United Nations or the World Health Organisation (34.5 per cent), followed by scientific committees (21.6 per cent) and national public bodies (12 per cent). Secondly, Europeans were asked, 'Which of the following sources of information do you have most confidence in to tell you the truth about genetically modified crops grown in fields?' Here the vote of confidence went to environmental, consumer and farming organisations (23 per cent, 16 per cent and 16 per cent, respectively), whereas national public bodies (4 per cent) and industry (1 per cent) commanded little support. These results appear to confirm the trend, observed by others (Samuelson, 1995), of an increasing lack of confidence

Table 12.4 *Trust in the US Department of Agriculture and the federal Food and Drugs Administration (%)*

	Lot of trust	Some trust	No trust	Don't know
FDA	22.9	61.1	15.4	0.6
USDA	21.8	68.3	9.3	0.6

Note: $N = 1,067$.

in national political institutions. They also suggest that biotechnology is seen by many Europeans as having transnational consequences that national bodies are powerless to influence.

US respondents were asked, 'If the US Department of Agriculture and the Food and Drugs Administration [separate questions] made a public statement about the safety of biotechnology, would you have a lot, some or no trust in the statement about biotechnology?' The USDA carried the support of 90 per cent of respondents, the FDA 84 per cent (see table 12.4).

These findings clearly indicate that trust in the regulatory system is considerably greater in the USA than in Europe. However, the American public's confidence in its regulatory organisations should probably be interpreted to mean that people are satisfied with the performance of this system *in toto*. The normal process for policy formation in the United States involves much more than a simple process of scientific judgement. The fact that organisations such as the American Diabetes Association or the Union of Concerned Scientists can and do monitor the Federal Register should be viewed as part of the total system. In addition, Americans' access to the courts reinforces the accountability of agency performance, as well as providing economic incentives for industry compliance. Hence the US system of checks and balances in the regulatory process inspires confidence that applications of biotechnology will be appropriately monitored. It is perhaps Europe's failure to establish a similarly coherent and stable system that has created the conditions of distrust.

Conclusion

In conclusion, no single explanation accounts for the greater resistance to food biotechnology in Europe. Media intensity, knowledge of biotechnology and trust in the regulatory process are implicated and interrelated. Different histories of media coverage and regulation go together with different patterns of public perceptions; and these in turn reflect deeper cultural sensitivities, not only towards food and novel food technologies but also towards agriculture and the environment. It is also apparent that the European public's concerns about GM foods were brewing some years before the consumer revolt of 1999. What is perhaps of note is that many of the European concerns about agricultural biotechnologies were also expressed, sometimes much earlier, in the United States. But differences between the European and American political systems led to different outcomes. The two-party system in the USA filtered out the sceptical voices once the regulatory regime had been established. By contrast, the unsettled pluralism of multi-party state systems and the evolving integration of Europe offered more effective channels to those challenging the scientific and industrial actor network.

The events of the watershed years of 1996 and 1997 point to an uncertain and contentious future for biotechnology on both sides of the Atlantic. For Europe, Torgersen et al. (chapter 2 in this volume) see the prospects of more conflicts arising from national diversity and European integration. In the USA, an increasing level of controversy over biotechnology may be a harbinger of difficulties to come. Events such as the USDA's difficulties with organic labelling (1999), a rise in critical media reportage spilling over from Europe, the protests at the WTO meetings in Seattle (1999), expressions of concern by politicians and the calling of public hearings suggest that Europe and the USA may not be worlds apart. How European and American policy and regulation respond to the likely challenges will be watched with interest around the globe.

NOTES

1. See Pusztai's website: http://www.freenet pages.co.uk/hp/A.Pusztai.
2. The US survey of public perceptions of biotechnology was conducted under NSF award 9732170, Principal Investigator: Jon D. Miller.
3. Note that the review omits one extremely important policy domain, that of intellectual property. There is little doubt that the US Supreme Court decision in *Diamond vs. Chakrabarty* in 1980 and the US Animal Patent Act of 1986 have had a crucial influence over the direction of US biotechnology research and development. Furthermore, US religious groups have taken a particular interest in these issues (Lesser, 1989).

REFERENCES

Anderson, W. French (1987) 'Human Gene Therapy: Scientific and Ethical Issues', in R. Chadwick (ed.), *Ethics, Reproduction and Genetic Control*, London and New York: Routledge, pp. 147–63.

Cantley, M. (1995) 'The Regulation of Modern Biotechnology: a Historical and European Perspective', in D. Brauer (ed.), *Biotechnology*, vol. 12 (New York: VCH), pp. 505–81.

Doyle, Jack (1985) *Altered Harvest: Agriculture, Genetics, and the Fate of the World's Food Supply*, New York: Viking Press.

Durant, J., M.W. Bauer and G. Gaskell (eds.) (1998) *Biotechnology in the Public Sphere: a European Sourcebook*, London: Science Museum Publications.

Fox, Michael W. (1986) *Agricide: the Hidden Crisis That Affects Us All*, New York: Shocken Books.

Goldburg, R., J. Rissler, H. Shand and C. Hassebrook (1990) 'Biotechnology's Bitter Harvest. Herbicide-Tolerant Crops and the Threat to Sustainable Agriculture', a report of the Biotechnology Working Group, New York: Environmental Defense, March, at http://www.environmentaldefense.org/pubs/Reports/biobitter.html, accessed November 2001.

Goodfield, June (1977) *Playing God: Genetic Engineering and the Manipulation of Life*, New York: Random House.

Howard, Ted and Jeremy Rifkin (1977) *Who Should Play God? The Artificial Creation of Life and What It Means for the Future of the Human Race*, New York: Dell Books.

Juskevich, J.C. and C.G. Guyer (1990) 'Bovine Growth Hormone: Human Food Safety Evaluation', *Science* 264: 875–84.

Kalter, Robert (1985) 'The New Biotech Agriculture: Unforeseen Economic Consequences', *Issues in Science and Technology* 13: 125–33.

Krimsky, Sheldon (1982) *Genetic Alchemy: the Social History of the Recombinant DNA Controversy*, Cambridge, MA: MIT Press.

Krimsky, Sheldon and Roger Wrubel (1996) *Agricultural Biotechnology and the Environment: Science, Policy and Social Issues*. Urbana, IL: University of Illinois Press.

Leahy, P. and A. Mazur (1980) 'The Rise and Fall of Public Opposition in Specific Social Movements', *Social Studies of Science* 10: 259–84.

Lesser, William H. (1989) *Animal Patents: the Legal, Economic and Social Issues*. New York: Stockton Press.

Luhmann, N. (1979) *Trust and Power*, Chichester: Wiley.

Murray, T.M. and M.J. Mehlmann (eds.) (2000) *Encyclopedia of Legal and Ethical Issues in Biotechnology*, New York: Wiley.

National Bioethics Advisory Commission (1997) *Cloning Human Beings*, 2 vols. Rockville, MD: NBAC.

Rifkin, Jeremy (1985) *Declaration of a Heretic*, Boston: Routledge & Kegan Paul.
(1998) *The Biotech Century: Harnessing the Gene and Remaking the World*, New York: Jeremy P. Tarcher/Putnam.

Rifkin, Jeremy, with Ted Howard (1980) *Entropy: a New World View*, New York: Viking Press.

Rifkin, Jeremy, with Nicanor Perlas (1983) *Algeny: a New Word; a New World*, New York: Viking Press.

Samuelson, R. (1995) *The Good Life and Its Discontents: the American Dream in the Age of Entitlement*, New York: Times Books.

Shiva, Vandana (1993) *Monocultures of the Mind: Perspectives on Biodiversity and Biotechnology*, London: Zed Books.

Thompson, Paul B. (1995) *The Spirit of the Soil: Agriculture and Environmental Ethics*, London and New York: Routledge.

Part V

Towards a social theory of new technology

13 The biotechnology movement

Martin W. Bauer and George Gaskell

Preliminary considerations

In this chapter we outline a framework for a social scientific analysis of the changing relations between science, technology and the public. Our concern is with modern biotechnology, which emerged in the early 1970s with the development of recombinant DNA (rDNA) techniques. We regard biotechnology as the third strategic technology of the period since the Second World War, following nuclear power and information technology. The framework outlined in this chapter reflects the research presented in this book on biotechnology in the European public spheres. Because it is not yet possible to talk of an integrated European public sphere, we pragmatically assume a multitude of public spheres, more or less related to the nation-states.

Our framework is built on various concepts: the biotechnology movement, the scientific-industrial complex, primary and secondary objectification, activity and the symbolic environment, actor networks, challenge and response, arenas of the public sphere, social milieus, public conversations and perceptions, media coverage and the regulatory process, and assimilation and accommodation. This terminology is not an end in itself, but an attempt to add clarity to the development of new technologies under the conditions of late modernity.

In the first section we refine a model formulated in 1994 in the so-called 'Hydra paper' (Bauer et al., 1994). This served as the rationale for our research activities with our colleagues in the multinational project, the outcomes of which are presented in this volume. The model proposes a triangular model of the public sphere of biotechnology comprising public perceptions, media coverage and regulation. Secondly, we integrate insights from science and technology studies. The analysis of the social shaping of technology gives us a better understanding of the 'system' that, because of our foregrounding of the public sphere, remains in the background of our research activities. However, we find in the research on the social shaping of technology only limited ideas and empirical research regarding the operations and functions of the public sphere. In short, in this chapter we develop a language for grasping the relationship between a technology trajectory and the public sphere within the context of post-industrial societies. We aim to build a framework theory that is capable of integrating middle-range hypotheses on public perceptions, media coverage and the regulatory process and the technological trajectory.

We start with a distinction between 'biotechnology' as a development of the routine activities of production and consumption – the primary objectification – and 'biotechnology' as it is represented or commonly understood in various societal domains and milieus – the secondary objectification. Primary objectification involves putting ideas and visions into products and services, while secondary objectification involves the appropriation of these products and services into everyday life in the form of ideas, symbols and behaviour. It is inappropriate to conceive of objectification as a simple causal process from primary to secondary forms. In our framework, primary and secondary objectification constitute mutual challenges, setting mutual opportunities and constraints. Representations and actions are co-determined.

The producers, observers and consumers of biotechnology, the 'insiders' of the movement, represent the technology in a variety of different ways. Others, for a variety of reasons, are either unaware of it, or ignore it or are indifferent to it; they are the 'outsiders'. But, because biotechnology is a strategic technology and, as such, increasingly challenges different aspects of everyday life, the scope for remaining an outsider is diminishing. The move from outsider to insider is accompanied by the adoption of a position towards the technology: support or rejection, or possibly a conditional approach of weighing up the options for the development of the technology – the 'third way'. We consider outsiders and insiders less as individuals and more as classes of individuals either as actors, as in primary objectification, or as arenas and milieus, as in secondary objectification, for whom the new technology takes on a particular symbolism.

Three ideas are central to the research presented in this volume. First, the development and exploitation of genetic engineering techniques is the focus for a growing 'biotechnology movement' at the core of which is a scientific-industrial complex. This is dominated by a tendency that inherits some of the significance of the older military-industrial complex, 'big science' (Price, 1965) or the techno-structure (Galbraith, 1968). However, in the case of biotechnology, the controls are not in the hands of a secretive state operating with a Cold War mentality, but are dispersed across international business and small to medium-sized companies with R&D functions operating and competing in the global market.

In its development of products and services, the scientific-industrial complex of biotechnology creates a variety of challenges to society that lead to responses. The idea of 'challenge and response' is attributable to the historian Arnold Toynbee (1978: 298), who employed these terms to understand the clashes between and the growth of civilisations, or, in other words, the encounter between large-scale projects. A challenge arises when a change to the status quo affects or threatens to affect the project or rationale of a social group. The response to a challenge is unpredictable. If the challenge is too small, it may go unnoticed; or it may be too great, leading to a crisis and deleterious effects. A response to a middle-range challenge will be creative, and exert a momentum for future development. The challenges of biotechnology may be seen by a particular interest group as a threat, leading to responses that in turn constitute counter-challenges towards the scientific-industrial complex. On the other hand, a challenge may be seen as an opportunity, leading to enhanced pro-tendencies. In our framework, the biotechnology movement comprises actors from not only the scientific-industrial complex but also the groups for whom the challenge of biotechnology leads to a response, whether of a 'pro' or 'anti' nature.

Beyond this biotechnology movement is a heterogeneous public, the 'outsiders' to whom the various 'insiders' in the biotechnology movement appeal for support. Over time this heterogeneous public is increasingly drawn into the movement, not least because the technology expands into more and more areas of life. People may be asked to adopt new products and services as promising investments or consumer goods, or they may be confronted by claims that the technology is a threat to the environment and human dignity. In other words, the 'outsiders' among the public are 'invited' to join the bandwagon, and have to decide whether to join the movement and on which side of the fence to sit.

Secondly, we assume that national public spheres are involved in constituting the counter-challenge of biotechnology by taking a 'position' vis-à-vis biotechnology. Without an understanding of the esoteric scientific principles, the public must rely on other forms of knowledge. This is the process of representation, whereby the primary activities of biotechnology are re-presented in terms of common sense, taking the symbolic form of ideas, stories of key events, historical analogies, metaphors and iconic images. We expect that these secondary objectifications of biotechnology are an eclectic mixture of modern and pre-modern mythological elements (Blumenberg, 1990).

In our research we distinguish three key arenas of the public sphere: public perceptions and everyday conversations, mass media coverage and regulatory processes. In the public sphere, we conceive of public opinion on biotechnology as comprising perceptions and conversations and also mass media coverage. We distinguish these two elements of public opinion from the regulatory processes. An objective of our research is to monitor and to disentangle the complex relations between the elements of public opinion and between these elements and regulation. There are, of course, other arenas of the public sphere that represent biotechnology in their own terms: the capital markets and stock exchanges, for example. But with our focus on the public sphere we are mainly concerned with the symbolic capital, and as such temporarily ignore the areas of business and commerce (e.g. Ernst and Young, 2001).

The representations of biotechnology in the public sphere are not epiphenomenal, but constitute the symbolic environment in which biotechnology operates (Boulding, 1956). This environment forms the 'developmental space', sometimes enabling and at other times constraining the course of biotechnology. To illustrate the point, during the 1980s many people working in the scientific-industrial complex thought that the regulatory process was the primary hurdle or constraint to their activities, only to discover in the 1990s that a second and problematic hurdle of public opinion could no longer be ignored and had to be taken into account.

Thirdly, as biotechnology, the primary objectification, develops and changes in form over time, and in so doing creates new challenges, so there is a parallel and dynamic process attached to the secondary objectifications. In other words, there is an evolution of symbolic representations in the three arenas. This may be related to a variety of contingencies: the changing cycles of public attention; mutual influences among the arenas, e.g. convergence, disjunction and 'causal linkages' at different phases of development; or the emergence of new frames of discourse, such as the emergence of bio-ethics as an issue in the 1990s.

The final issue is in the nature of a question concerning the relations between science, technology and the public sphere. What are the forms of and conditions for collective learning, in the biotechnology movement and in the public sphere? The various actors in the biotechnology movement, whether the pro or anti tendencies, try to influence and react to one another's positions and to what is happening outside the movement in the arenas of the public sphere. Having identified the 'second hurdle', the scientific-industrial complex must decide whether, and by what means, it should assimilate and/or accommodate its strategic project to other actors in the biotechnology movement and to the public sphere. In moving from ignoring to representing biotechnology, the public sphere assimilates and/or accommodates biotechnology in a particular way, thus conditioning the likely development and reception of future technologies.

Technology: politics by other means

Technology, defined as a complex of know-how, people, artifacts and services, is a relatively neglected area within the social sciences. This is rather strange given that technology is the prime example of a socially constructed reality, understood in a literal sense. Any technological innovation is a novel piece of reality. It has been constructed out of the ideas of designers and engineers, and carries with it both intended and unintended consequences for society. In this constructivist perspective a technology must be analysed like social norms and institutions. Any technically constructed device, simple or complex, embodies a 'behavioural script' that tells people what to do and what not to do (Joerges, 1988). This social fact produces a likely pattern of behaviour that is represented by things or services that become affordances for actions (Gibson, 1977; Norman, 1988). People are socialised into using things and services through the acquisition of tacit knowledge or skills or 'know-how' and through the acquisition of explicit symbolic knowledge or 'know-that'.

However, socialisation is never perfect. People may not use certain things, or they may use them in ways that were not intended or even imagined. In a Durkheimian sense, social norms and technology are functionally equivalent 'social facts' constraining behaviour into recurrent patterns (Linde, 1982), making some actions more likely while discouraging others. Whereas norms are enforced by social sanctions, the use of things is channelled by affordances, the action adequate for the thing or the opportunity costs from not using the thing. For example, the rise of the personal computer has increased keyboard writing and has probably decreased the intensity and quality of handwriting. Thus technology structures behaviour, both individually and collectively, bringing both opportunities and threats to human pursuits. This is the basic ambivalence of new technology, and it is why it constitutes a permanent challenge to social life. This is not to imply that technology is independent from social norms. Technology creates the context for reasserting social norms or the creation of new norms. Furthermore, existing social norms influence the process of technology through both informal processes (voice and exit decisions) and formal processes (regulation).

A number of models have been proposed to explain the trajectory of technology.

Technology as rational design sees the process as the outcome of rational choices between alternative designs on the criterion of maximising desired outcomes. The decision-taker may also consider the unintended consequences of his or her earlier decisions, which may carry implications for decisions to ameliorate such problems. This elite model privileges a professional class of experts, who by dint of their higher form of 'rationality' can make better decisions between design options. This elite theory considers interference either from capitalistic owners or from public politics as a limit to efficiency and effectiveness. This is variously described as the technocratic or engineering model (Winner, 1977: 135ff.).

By contrast, *technology as politics* sees the trajectory of a technology as a Machiavellian process of strategic survival based upon the managerial logic of local optimisation of returns (Simon, 1969). The main task for the promoter of a technology is to design things flexibly in order to sign up more and more people and to maintain their support; in other words, to incorporate more and more people. This is achieved by transforming the original designs and by making appropriate concessions to attract resistant users. Thus 'user-oriented' design (Norman, 1998) sets out to make friends, but at the same time operates to keep enemies or sceptics at bay. This may be the work of an actor network that conspiratorially expands its realm into a monopoly of power, or of a multitude of actor networks muddling through in an opportunistic fashion, making the best out of opportunities as they arise. This perspective considers technology as the continuation of politics by other means (Latour, 1988).

In recent years social scientific studies of technology have undergone a paradigm shift. The old paradigm focused on assessing the technology 'after the fact' by monitoring intended and unintended consequences (see Merton, 1936). The technology is a fait accompli; the research questions concern its impacts after the event. Feedback, at most, leads to fine-tuning of the packaging of the product or the service, or the design of the context as considered in diffusion and acceptance research (Rogers, 1996). The new paradigm, by contrast, investigates the processes contributing to the development of the content of a particular technology as it moves in a sequence of stabilised forms (Hughes, 1989; Mokyr, 1992; Pinch, 1996). In the new paradigm the monitoring of impacts is moved to ever-earlier stages of the design process. Resistant users are canvassed at the conceptual or at the proto-typing stage, thus lifting the 'afforded user' from the designer's imagination into empirical reality. This approach considers the shaping of a technology as a continuous feedback loop from relevant but resistant groups onto technical designs, and raises the questions of technological options and alternative trajectories. The historical, sociological or psychological reconstruction seeks to explain why certain trajectories prevailed while others failed, and why sub-optimal solutions succeed in a world of rational aspirations. Why, for example, do all English keyboards follow a QWERTY layout? Or why do bicycles have pneumatic tyres? (See Pinch and Bijker, 1987). Technology is a political learning process, in which actors aim at securing support either by pretend play or by accommodating resistance into the redesign of their project (Bauer, 1995).

In these perspectives, the concern is with a technology and its relevant actors. As yet, however, they provide a limited conceptualisation of the public sphere as a constitutive context of a technology movement.

Specific changes in the context: post-industrial society

The biotechnology movement exists, like previous technologies, in a historical environment that enables and constrains its future. We conceive of this state of affairs as a system–environment relationship (Luhmann, 1995). The colonising system is the biotechnology movement; the environment comprises relatively autonomous spheres of activities such as the economy, the sciences, the law, the polity, religion, education and the arts, each one operating according to a particular logic (Luhmann, 1995). Each sphere of activity represents biotechnology in its particular way. For example, in the sciences, genetic engineering is the latest frontier of progress for the twenty-first century. In law, it is an issue of risk regulation or of intellectual property rights; in business, a high-risk stock, a market opportunity or a means to rationalise production. In the polity it appears as concerned public opinion; in the religious domain as a potential threat to the moral fibre of society and to the sanctity of God's creation; and in the arts world it may be a new medium of expression.

Luhmann presents a long-term view of secular changes in societies away from hierarchical structures to functional differentiation. Science, which during the course of modernisation displaced religion from the apex of the hierarchy, is no longer taken for granted as the dominant form of knowledge about the world. Today it is one among others. Nested within this perspective of social change, a number of sociological imaginations postulate more recent transitions of societal structures that bear on the development of technology in contemporary times. These characterise axial changes in the public sphere within which the actors of the biotechnology movement must operate.

Bell (1973) and Stehr (1994), for example, allude to the increasing importance of knowledge as a productive sector of the economy. And with this comes the privatisation of knowledge as its production moves from public universities and state research institutions to the private sector of corporate R&D (Cohen and Noll, 1994). This development carries consequences for the operations of the public sphere – for example, in public access to knowledge and the availability of independent expertise.

Inglehart (1990) identifies a secular shift in value orientations, particularly among the affluent young generation in advanced countries, away from a commitment to materialistic progress towards a post-materialist focus on a more humane society. This value shift cultivates concerns for sustainable development. For Beck (1992), the axial change is the uncontrollable and globalised risks attributable to human activities; for Giddens (1991), it is the uncertainties of late capitalism without traditions. The success of past technologies undermines the very foundations on which they are based. The unintended consequences of modern technologies go beyond the controls of the traditional institutions, becoming increasingly global and unmanageable. Whereas, in the past, technology was a means to control local dangers and to extend human capacities, technology increasingly has perverse effects with global impacts (Tenner, 1996). Witnessing these trends, modernity turns reflexive, which undermines its foundational belief in 'progress'.

Durant et al. (2000) speculate on the emergence of different cultures of public understanding of science. As scientific knowledge increases among the public with progress in education and economic development, so is there a specialisation even in popular scientific knowledge, a decline in interest in science and technology, and an increasing ambivalence towards science and technology. The potential negative consequences of scientific and technological developments loom larger, giving rise to support for the 'precautionary principle'.

These trends – the private knowledge sector, value shifts, global risk and diversity of popular scientific cultures – bring a change in the dominant representation of technology. Technology is no longer presumed to be equated to *progress* (Touraine, 1995), and is instead part of the problem that it purports to solve.

A consequence of these nested axial changes is that the relationship between specific technology movements and their representations in societal arenas is also changing. From a system theoretical point of view, we expect a plurality of representations of biotechnology in the public sphere arising from autonomous domains of society other than science. From forms of sociological imagination we expect increased challenges to the autonomy of the scientific-industrial complex and calls for national and international regulation (the primary hurdle), and increasing legitimation of resistance by interested groups and the wider public (the second hurdle). These counter-challenges are likely to increase the significance of public representations of biotechnology to the actors in the biotechnology movement. The overall outcome will be the expansion and professionalising of proactive and reactive public relations activities of all actors of the biotechnology movement as they attempt to secure the support of various social milieus for their particular vision of biotechnology – whether utopian, dystopian or some 'third way'.

The activities of the biotechnology movement: the primary objectification

A technology at any moment in history is a complex network of actors, materials and images that objectify ideas for a particular future of society. The anthropological significance of technology is the extension of control over the environment and the shaping of natural processes into some human design (Ropohl, 1979). Every technology represents a particular vision of society through the past and future activities that are embodied in devices and services. It is hardly surprising that there is a genre of literature that explicates these imaginations in the 'atomic state', the 'space society', the 'information society' or the 'bio- or genetic-society'. Ideas are institutionalised into 'object scripts' for dealing with things or services, which people either adopt spontaneously as affordances, or into which people are socialised symbolically as aspirations. Some people will resist this process because their ideas are at variance with the affordances or the aspirations of the new technology. As resistance to a technological project arises, so is the question raised: what does resistance contribute to the process of development? The detached analysis has to see beyond the polemic that diagnoses resistance as technophobia or Luddism, and ignores the deficiencies in design and implementation (Bauer, 1995).

Like its predecessors nuclear power (Weart, 1988; Radkau, 1995) and computerisation (Kling and Iacona, 1988), the biotechnology movement is not free of controversy. All sides in the controversy attempt to cultivate representations in the public sphere in order to rally support. Like any technological movement there is a growing network of interested and concerned insiders. It excludes those who are indifferent to the technology, either because they have been kept ignorant or by their own deliberate choice, but these outsiders are potential members of the movement and constitute the pool of mobilisation.

In the context of the biotechnology movement, three distinctions regarding groups of actors may be made. First, there are those concerned with the creation and exploitation of the new knowledge: the scientific-industrial complex (figure 13.1). This includes scientists, engineers, bankers, managers, marketing executives, stockholders, communicators, patent lawyers and many other actors. All are more or less unified by an economic interest in biotechnology, albeit in competition over market segments for particular products and services.

Secondly, there are those who see opportunities for the project of their group, without being directly involved in the production of biotechnology: for example, patient groups who have high expectations that genetic engineering will eliminate diseases or alleviate the suffering of their members.

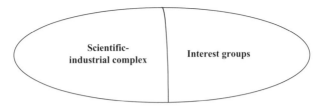

Figure 13.1 The biotechnology movement

Other examples are those who would benefit from genetic testing and a 'genetic information society', such as the police and forensic services in catching and convicting criminals, and insurance companies in their risk assessment strategies.

Thirdly, there are those who voice reservations and cultivate concerns about these new developments. For example, the environmental, consumer and other specialist interest groups mobilise religious believers, entrepreneurs, retailers, scientists, lawyers, protesters, activists, patients, consumers and others to express concerns about genetic drift in the environment, food adulteration, civil liberties, animal rights and threats to privacy or human dignity.

The scientific-industrial complex is at the centre of the biotechnology movement. By 'scientific-industrial' we refer to the ever-closer integration of knowledge production and corporate exploitation of molecular genetics since the 1980s. This is evidenced by various international trends (Haber, 1996; Ernst and Young, 2001): the number of small companies emerging since the mid-1980s with an overwhelmingly R&D orientation; the number of joint ventures between university departments and private companies; and the trends in international mergers of pharmaceutical, agri-chemical and agri-seed companies around a vision of the 'life sciences'. By 'complex' we stress its principal heterarchical nature: there is no necessary national or international unified line of command. However, at times the conspiracy theorist may be close to the truth in assuming an integrated global strategy. The complex comprises ongoing and multifaceted negotiations among various actors over the control of genetic processes in the pursuit of economic imperatives.

The biotechnology movement may be schematised as a field of forces in continuous shifts in the equilibration of demands and concessions, resistance and accommodation (Lewin, 1952; Latour, 1988). 'Progress' is depicted as the meandering movement along a non-linear curve that links alterations to the technology to formerly resistant participants. Each point on this trajectory represents new actors that signed on to a particular development of the original design. Additional users are 'bought' by changes to the original design. The strategy is called 'user-oriented design' (Norman, 1998). For example, biotechnology is developing an increasing range of products and services. In the course of this development certain products are discontinued, such as the 'terminator gene' for agricultural seeds, while others are altered, such as the removal of antibiotic marker gene.[1] The appropriate mix of flexibility and rigidity for the technology is an uncertainty of the process. If a design is inflexible it will fail to sign up new participants, but if it is too fluid few will be in a position to recognise its identity as a technology and be able to integrate it into daily routines. The objectification of an idea 'closes' a technology into a temporary match between materials and actors. Closure represents an equilibrium of the forces within the social-technical system at any moment in time. Progress and development arise through a series of 'closures'. In responding to perturbations, the design is unfrozen, changed and then refrozen.

The process can be considered a Machiavellian one: actors are working both within current regulations and public opinion, and on the limits in order to obtain competitive advantages on multiple frontiers. 'Working on limits' may involve moves to change existing regulations and public opinions, to prevent regulation in line with or contrary to public opinion, or to set selected limits where there are currently none to allay public concerns.

Challenge and counter-challenge

Where a development in biotechnology affords an alteration to the modus operandi of a particular group in society, it is seen as a challenge (figure 13.2). A challenge may be positive in the sense of an emerging new opportunity to be exploited, or negative in the sense that it threatens the status quo or the world-view of the group. Such challenges lead to collective action either to embrace the opportunity or to resist the perceived threat. These responses, which, as they unfold, constitute counter-challenges, are in turn 'internally represented' or 'incorporated' within the biotechnology movement through the activities of interest groups.

All these groups of actors are competing by forming coalitions with the objective of imposing their aspirations for the future of biotechnology

Figure 13.2 Challenge and response within the movement

onto the biotechnology movement. In order to achieve this, actors will attempt to mobilise and incorporate the wider world. In the context of our research, the wider world is the public sphere of regulation, media coverage and public perceptions.

Of the many actors engaged in the biotechnology movement, some have been specifically founded for this purpose. For example, there is the proliferation of small biotechnology companies, often joint ventures with university departments; there are the groups that coordinate and support such efforts ('Bingos', or business NGOs); and there are interest groups that scrutinise the developments in and around biotechnology, such as the group that coordinated the opposition against gene technology in the two Swiss referenda of 1992 and 1998 ('Pingos', or public NGOs).

Beyond these interest-specific groupings are existing actors/groups that were formed for other purposes but discover that biotechnology offers new opportunities. Thus they become issue entrepreneurs using biotechnology as a way of sustaining and building their general project. For example, environmental organisations, consumer groups and human rights groups adopt the emerging biotechnology as an issue within their remit.

The public arenas and social milieus that make up the public sphere of biotechnology represent the new developments in a variety of different ways, that is, as different challenges. It is crucial to recognise that these representations are not epiphenomenal and of merely passing interest. They constitute the symbolic environment of biotechnology, and symbolism is the core of any culture; and culture both constrains and enables activities.

Traditionally, social milieus have channelled their visions for society through party political activities. Traditional segmentation separates the population into distinct milieus of *Weltanschauung* with particular orientations towards the past, present and future of society and with a project for society as a whole. These structures have been the basis for party political affiliation such as conservatism, socialism or liberalism. Nowadays, social milieus constitute the new reserve army for pressure groups emerging within the biotechnology movement. It is unclear when issues raised by specialist actors will become aligned to the party political cleavages. Post-industrial society is in transition, with a mix of hierarchical milieus and modern functional differentiation. Modern segmentation increasingly reflects the spheres of productive activity such as business, law, education, religion, politics or the arts. Although interest groups have emerged from the functional differentiation of society, they appeal to traditional milieus for the mobilisation of support (figure 13.3).

Analysing the structure and function of these varied symbolic activities is the analytic challenge. In our analysis, we focus on the environment

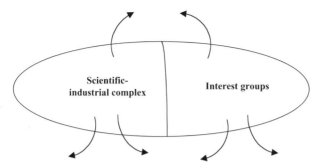

Figure 13.3 The technology movement reaching out to the public

of biotechnology defined by the public spheres of various European countries.

The symbolic environment of the biotechnology movement: the secondary objectification

The birth of modern technology is accompanied by wild imaginings and as yet unwarranted claims as to its significance. The symbolic elaboration of a new technology into utopian or dystopian *Leitbilder*, what we call secondary objectification, is a crucial element of its reality and its future potential (Dierkes, Hoffmann and Marz, 1996). Such imaginings are an integral part of the biotechnology movement. Our research focuses on secondary objectifications of biotechnology as found in public arenas and in social milieus, and the effects that these may exert on the trajectory of biotechnology.

The actors of the biotechnology movement are not only talking among themselves; they are also mobilising sections of the public in support of their views. In this way, a fait accompli, such as the exporting to Europe of genetically modified soya beans by the US company Monsanto in the autumn of 1996 (see Lassen et al., chapter 10 in this volume), offered much scope for reactive mobilisation in society.

Activists of the biotechnology movement are more or less professionally involved in their activities, whereas the public's concerns are much broader. The public at any moment in time needs to be convinced of the relevance of the issue. Advocates and opponents have one thing in common: they push the significance of biotechnology as an issue in the public domain. In consequence they combine forces to draw others into the biotechnology movement.

This leads to tangible representations that can be researched in the three arenas of the public sphere. Representations are likely to be structured as frames that organise the issues into imagery and a range of positions vis-à-vis the technology (Gamson, 1988). A frame delineates what to agree or to disagree about: biotechnology may be a matter of morals, risk, democracy, utility, globalisation, public accountability, etc. Particular sponsors promote particular frames; they match the routines of media production, and resonate more or less with public perceptions and with the regulatory process. The contest is two-fold: the choice of which frame and the position within the chosen frame.

These secondary objectifications familiarise the novelty of the fait accompli in public and co-determine the future of the 'fact'. They anchor the new technology in resonating images and objectify the uncertainties in arguments about the technology. They anchor the challenge of novelty by offering historical analogies and familiar images of the future. In the public sphere, representations are cultivated in and carried by particular social milieus.

Representation involves different modes and mediums. Modes refer to habits, cognition, informal conversations or formal communication such as in mass media or public speaking. Mediums refer to language, images, sounds or movements that embody and familiarise the new technology and establish a relation for or against it (Bauer and Gaskell, 1999).

The public sphere of modern societies is structured and dynamic. On the one hand it comprises audiences, actors and historically established forums and arenas, the typical occasions for public exchanges (Neidhardt, 1993), while on the other hand its quality of functioning may improve or decline (Habermas, 1989). We focus on three arenas of the public sphere, thus considering different modes and mediums of representation: biotechnology represented in the formalised mass media coverage, in everyday conversations and perceptions, and in regulatory routines and regimes (figure 13.4).

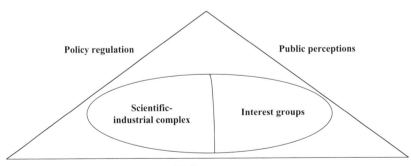

Figure 13.4 The three arenas of the public sphere

Within the public sphere these arenas serve different functions: media coverage and public perception constitute the public opinion of biotechnology, while the regulatory process constitutes the governance of biotechnology. From the point of view of the biotechnology movement, these are the primary and the secondary hurdles, and cannot be ignored.

Advocates and opponents of biotechnology attempt to make inroads in the three arenas in order to make their representation of biotechnology become the 'real future' of routine activity. The outcome for biotechnology is uncertain, as was the case with previous technologies, and is likely to be different from that which is imagined by any of the actors.

We presume that the representations arising from the challenge posed by biotechnology are beyond the control of the scientific-industrial complex or other activist groupings. This is because representations link spheres of action and as such are located in between biotechnology and other societal domains. The law, education, the media, politics, the political public, religion and the arts, although subject to the influences of the scientific-industrial complex, enjoy relative autonomy in their symbolic activities.

The triangular model of the three arenas is a heuristic device that achieves two objectives simultaneously. First, it captures the complexity of the public sphere without privileging, in the sense of uni-directional causal relations, any of the three arenas. Secondly, it allows us to simplify the complexity within the context of research. Each side of the triangle may be taken separately as a perspective on the public sphere. Thus we can ask how public perceptions relate to the policy process and to media coverage; or how media coverage relates to policy process and public perceptions; or, finally, how the policy process is related to public perceptions and to media coverage. Each of these perspectives on the public sphere calls attention to a variety of middle-range models from the social sciences, the formation of public opinion, theories of media production, and models of policy-making. The 'ideal framework' keeps the debate on a certain level of complexity and invites a synthesis of perspectives. However, at present, this may be beyond the scope of current social science theory. The triangular model also works as a shield against simplistic reduction of the 'public sphere' to surveys on public perceptions, or analysis of media coverage, or the analysis of documents on policy-making. Each arena must be understood in the context of the other two and in the context of the activities of the biotechnology movement.

Table 13.1 provides a characterisation of the modus operandi, achievements and functions of the three arenas of the public sphere. The mass media work to a short production cycle. One or two weeks is a long time

Table 13.1 *The operating principles of the three arenas of the public sphere*

	Public arenas		
	Public conversations and perception	Mass media coverage	Regulation
Modus operandi	Medium cycles Some consistency Memory	'Media logic' Short cycles News values Imperative of novelty Little consistency Hardly any memory Framing	Long cycles Bias against novelty Consistency Long memory
Achievements	Opinion Attitude Stereotype Schemata Awareness Skills	Dissemination Propagation Propaganda Advertising Education and training Agenda-setting	Regulatory regimes

Functions	Being able to act on it
	Communicating with
	    other people
		Information
		Entertainment
		Linking domains
		    of societal action
			Allocation of
			    responsibility
			Assimilation and
			    accommodating
			    public opinion
			Enabling technology

in the media world. News values and routines guide the selection of science and technology stories from a large pool of possible stories (Hansen, 1994). Consistency in reportage over time is not a key feature, which indicates a limited memory span; novelty is imperative. The main functions of the media are information, entertainment and the linking of concerns across different societal domains, for example business with politics or science with business (Kohring, 1998).

From the point of view of our research, the mass media act as a 'window' on to biotechnology for the majority of the public. We do not assume a direct influence on public perceptions of media coverage. This relationship is one of mutual constraint. Mass media coverage is a source as well as a constraint for everyday conversation and the regulatory process. Although opinions of the audience or regulations set limits to the coverage in the mass media, we may hypothesise about possible linkages between media styles and public perceptions. For example, diffusion of information forms opinions; propagation, by virtue of its qualified and value-based judgement, fosters attitudes; and propaganda, characterising a black-and-white world in terms of friends and enemies, is likely to cultivate stereotypes (Moscovici, 1976). While advertising seduces awareness and stimulates desires, training by observation establishes new skills and routines where educational functions are retained.

Media coverage for or against biotechnology does not generally map directly onto public perceptions or into policy-making. Most of the time we expect the mass media to disseminate information and/or to entertain ('info-tainment'). Beyond this, we may expect the media to synchronise attention to biotechnology across spheres of life in terms of agenda-setting (McCoombs and Shaw, 1972) and cultivation (Morgan and Shanahan, 1996). Figure 13.5 shows the increasing coverage of biotechnology across Europe aggregated from elite newspapers in twelve countries between 1973 and 1996, and this can be regarded as an index of reduced public indifference and, under certain conditions, of public controversy.

Public conversations and perception may focus on an issue such as biotechnology. But, to the dismay of many in the biotechnology movement, the public at large does not share their level of concern with the issue. The issues of biotechnology are often non-salient, because other concerns are more pressing. Conversations and perceptual topics have a certain degree of consistency and work with some memory of what went on before, and concerns come and go in a medium-term cycle. Awareness, skills, values and stereotypes tend to be rather stable and consistent, whereas opinions and attitudes may change more easily. Attitudes bring general values to bear on concrete issues, and provide categories and evaluative judgements regarding biotechnology. We expect awareness, opinions, attitudes, stereotypes or skills to enable people to communicate and, to different degrees, to influence everyday actions.

The effect of public perception on policy-making and media coverage is uncertain. Although some individuals may have direct access to the regulatory process, most opinions count only in mass. Aggregate opinion may legitimate political action; it may support or resist biotechnology, or remain ambivalent.

Public perception is achieving greater significance for technological trajectories. This is because it is the key legitimisation for both pro- and anti-activism within the biotechnology movement. Without claims of public support, interest groups would wield little influence. However, the relative autonomy and unpredictability of public perceptions is a source of great concern among activists. Hence, the attempt to take control of public opinion. Think of the effort and subsequent failure

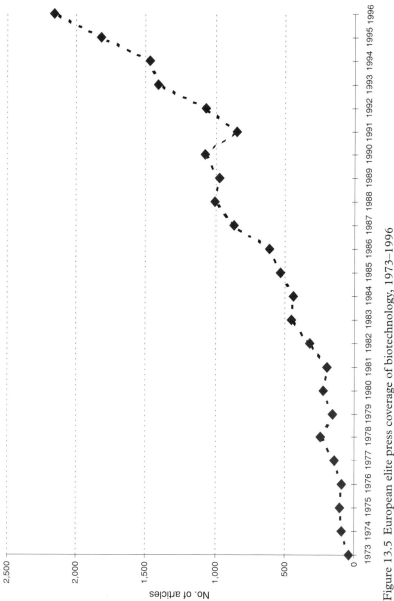

Figure 13.5 European elite press coverage of biotechnology, 1973–1996
Note: N = 12 newspapers.

of the 1998 'charm' campaign following the introduction of Round-up Ready soya into the European market (see Lassen et al., chapter 10 in this volume).

The regulation of technology in policy- and law-making is slow, with a long memory of past activities and a strong concern for consistency between levels and across domains of regulation. There is an inherent bias against novel regulation; the working assumption is that existing procedures cover the new things. The new is assimilated into existing regimes and frameworks. There is an inherent resistance to accommodation, in the sense of changing the regulatory processes or regimes in response to challenges from various actors (Hood and Rothstein, 1998). Within this broad operating logic, different regulatory regimes may be identifiable. For example, Hood et al. (1999) identify three ideal-types of policy regimes for risk management. 'Responsive government' reflects public and media opinion, 'client politics' reflects vested organised interests, and 'minimum feasible response' is oriented towards the correction of failures in the market or tort-law processes, in effect avoidance of blame. A regime is a temporarily stable modus operandi. As the biotechnology movement expands, the policy regime may move or switch back and forth between these types of risk management.

Regulation is the primary constraint from the point of view of the biotechnology movement. The representation of biotechnology in the regulatory process reflects the degree of autonomy or incorporation of the policy process by the biotechnology movement, on the one hand, and by public opinion, in terms of media coverage or public perceptions, on the other hand. The main functions of the regulatory process are to enable a new technology to serve economic imperatives, to avoid blame for future failures and to alleviate concerns expressed in the media and in public perceptions. Each of the above regimes seems to highlight one of these functions of the regulatory process.

Our final theoretical concern is how to account for the multiple representations of biotechnology that appear in different societal arenas in many societies. For this we need to go beyond the idea of social groups as the locus of activism to the concept of the social milieu. Social milieus are vector combinations in the triangular framework in which a resonance links media coverage, public perceptions, regulatory regimes and particular actors. The concept of social milieus acknowledges the heterogeneity of the three key arenas, and links divisions of the three arenas to particular representations. A milieu links particular public perceptions and media resources, and projects these into the policy arena. In so doing, a milieu supports a particular regulatory regime advocated by an actor in the biotechnology movement. In their resonance between positions

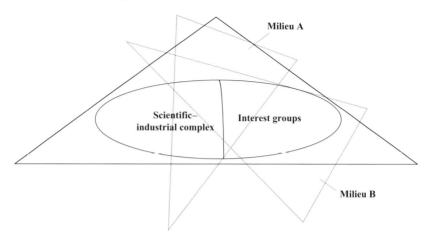

Figure 13.6 The movement mobilising milieus in the three arenas

in the three key arenas, social milieus can be said to carry particular representations of biotechnology in the public sphere (figure 13.6).

Researching the public arenas

With our description of the biotechnology movement, we expect to find multiple representations of biotechnology in the public sphere – that is, a variety of representations in public perceptions, media coverage and regulation. From the point of view of the biotechnology movement, milieus constitute established segments of the public, which are often associated with particular media outlets. In the jargon of public relations, they are target groups for whom effective means of access need to be identified.

Linkages, in terms of content, across the three arenas would be expected to be associated with a particular milieu. Let us illustrate this idea. Consider environmentalism as a relatively new but established milieu in many countries. It is characterised in its relationship with biotechnology by particular media frames and outlets, by particular attitudes and stereotypes and by support for 'responsive government' on the basis of the precautionary principle. Or consider the commercial milieu: this is characterised by other media frames and outlets (growth potential, the business press), different attitudes and stereotypes (investment opportunities) and a preference for minimal response risk regimes. These milieus legitimise actors, for example environmental or commercial, who cultivate their particular representations and seek to incorporate them into the growing biotechnology movement.

A milieu may be observed from the perspective of three viewpoints. Looking in from the regulatory arena, the milieu is the voters; from the media it is readers, listeners or viewers; and from public perceptions it is like-minded people, the community of common interests and lively conversations.

This discussion opens up the following implications and principles for research. Secondary objectifications are structures of common sense cultivated in communication and spanning the three arenas of the public sphere. With the concept of 'social representation', research focuses on the comparative analysis of milieus and their common sense rather than on the representations-for-action developed by actors in the biotechnology movement. The latter would be a different focus of research and require a different methodology (Bauer and Gaskell, 1999).

Empirically, in the research reported in this volume, informal communications are observed in everyday conversations during focus group sessions and in surveys of public perceptions. The formal communications of the mass media are assessed by the analysis of press coverage. And finally the regulatory routines are analysed by recording regulatory activities and by scrutinising official reports and documents (see Gaskell et al., 1998, for methodological details).

Representations of biotechnology vary across the three arenas in two basic dimensions: intensity and content. Each arena makes reference to biotechnology with a certain degree of intensity. The public is more or less aware of these new developments and pays attention to them; the media carry a certain number of biotechnology stories; and the regulatory sphere produces more or fewer codes of practice or statutory regulations. In all arenas, we have to assume a limited capacity of attention, where different issues compete for limited resources. The finding of simultaneous attention and activity across the three arenas may constitute an indicator of a 'crisis'. Secondly, we find different contents, that is references, frames and themes, on the topic of biotechnology. Similarities and differences of representations in time and space, across arenas and across different milieus, are empirical matters.

We attempt to relate intensity or content in one arena to intensity or content in another on the basis of middle-range hypotheses. For example, the 'intensity of coverage hypothesis' relates increasing media coverage (intensity) to the rise in negative public perceptions (particular content) of a new technology. The 'agenda-setting hypothesis' of media effects relates changes in contents of media to changes in public perception. The 'responsive government hypothesis' suggests that public concern (content of public perception and media coverage) will antecede policy-making (content).

Observing learning by assimilation or accommodation

Learning is a useful metaphor for the processes and consequences of socio-technical change. Change may be associated with progress, accumulation and improvement, or with problems, decay and collapse. Learning is based on a dual process of assimilation and accommodation. In assimilation or 'pretend play', changes are absorbed into current routines. Continuity is achieved with minimum disruption as the new is anchored in the old and familiar. By contrast, accommodation is the often painful process of readjustment, involving structural change. The new resists classification into the old categories and resists management within the old routines. When assimilation reaches its limit, the system is forced to change its categories and routines in order to accommodate the new.

This model of learning was developed to explain the cognitive development of children coming to terms with the world of things and adults (Flavell, 1963; Piaget, 1972). In the case of the biotechnology movement or the public sphere, we are working with a system and its environment without a single fixed referent; a co-evolution, with no fixed world of things with which to come to terms. Rather, things as well as people are on the move. In this process of co-evolution the participants are not equally prone to learning. All participants can choose to ignore, assimilate or accommodate the other.

On the side of the biotechnology movement, efforts to assimilate can be observed in evolving actor networks, their increased specialisation and diversification, their lobbying on regulation, public relations activities, and the use of established forms of activism. Accommodation can be observed in new forms of activism, for example the confluence of activism and public relations (such as 'no action without a TV camera or an audience of journalists'), or the setting up of NGOs on a business basis (the so-called 'bingos'); in the abandonment of product and service lines such as the 'terminator gene'; or in the alteration of product and service lines to attenuate public concerns, such as eliminating antibiotic markers.

On the side of the public process, we may observe a growing sensitivity to public opinion in the regulatory process. In some countries such as the USA, assimilation prevails in the regulatory process, as modern biotechnology is considered to be nothing new; whereas in Europe a gamut of new laws and regulation accommodate the development within the different legal frameworks (see Torgersen et al., chapter 2 in this volume). Public accommodation may also be observed in experiments with novel arenas of public debate such as 'consensus conferences' and 'deliberative democracy', or in the introduction of new regulatory regimes and frames such as 'ethics', the precautionary principle or wider notions of 'risk' such as social risk or phantom risks.

Public perceptions are informal, much more diffuse and less organised. However, in some countries we have seen a steady decline in the belief in the traditional links between science/technology and progress. With this has come greater interest in the control of science and technology, and new concerns about labelling and the right to choose, and about issues of a moral and ethical nature. There is a growth in the expectation – whose seeds were sown in the area of conflicts over civil nuclear power – of increased public participation. This public unrest has been called 'the second hurdle'. Much of the activity of actors in the biotechnology movement is directed towards managing the second hurdle, either to lower or to raise it.

The media are probably more likely to assimilate new technology into their existing modus operandi. The scientific or political journalists are covering just another technology within the established news routines of novelty and human interest. It may be that, as in the case of the environment some years ago, a particular journalistic specialisation may appear: the biotechnology correspondent. This would constitute accommodation.

Conclusion

In this chapter we have characterised a model of the reception of a new technology in the public sphere that has informed our research on biotechnology in the European context during the period 1973–96. It has served as a background from which to reflect on what has been achieved, and, as it has been extended and developed, it has pointed to possible new research questions. The model sets out a broad framework within which to understand the many interrelated processes that shape the public sphere. The public sphere at the end of the twentieth century constituted the symbolic environment for current technological developments and for those to come. This environment is not epiphenomal to an autonomous technological process, but constitutes a counter-challenge that will influence the future trajectory of biotechnology. The chapters in this book explore some, but by no means all, of the elements of this environment. Furthermore, we have restricted our gaze to the three arenas of the public sphere – public perceptions, media coverage and regulations – and have not investigated how the biotechnology movement also changes in this process. Although we are aware that both the scientific-industrial complex and other actor groups are in constant change, the dynamics of the biotechnology movement are beyond the scope of this research. In the course of this project we have been systematically exploring the 'learning conditions' up to 1996, rather than the 'learning outcomes' for the biotechnology movement. To document the outcomes we may have to wait some years to assess them with the necessary circumspection. The scientific-industrial complex may be well advised to learn from and to accommodate to public concern and calls for new forms of regulation. Otherwise the public may simply assimilate modern biotechnology to nuclear power, which once had a seemingly unlimited future but whose trajectory radically changed in the last quarter of the twentieth century.

NOTE

1 Monsanto's declaration of intent not to commercialise terminator gene technology was reported by Associated Press, 7 October 1999.

REFERENCES

Bauer, M. (1995) 'Towards a Functional Analysis of Resistance', in M. Bauer (ed.), *Resistance to New Technology – Nuclear Power, Information Technology, Biotechnology*, Cambridge: Cambridge University Press, pp. 393–418.

Bauer, M.W. and G. Gaskell (1999) 'Towards a Paradigm for Research on Social Representations', *Journal for the Theory of Social Behaviour* 29: 163–86.

Bauer, M., G. Gaskell and J. Durant (1994) 'Biotechnology and the European Public. An International Study of Policy, Media Coverage and Public Perceptions, 1980–1996', unpublished manuscript, June, Hydra, Greece.

Beck, U. (1992) *The Risk Society*, London: Sage; German original, 1985.

Bell, D. (1973) *The Coming of Post-Industrial Society*, New York: Basic Books.

Blumenberg, H. (1990) *Work on Myth*, Cambridge, MA: MIT Press.

Boulding, K. (1956) *The Image*, Michigan: University of Michigan Press.

Cohen, L.R. and R.G. Noll (1994) 'Privatising Public Research', *Scientific American* 271(3): 72–7.

Dierkes, M., U. Hoffmann and L. Marz (1996) *Visions of Technology: Social and Institutional Factors Shaping the Development of New Technologies*, Frankfurt am Main: Campus.

Durant, J., M. Bauer, G. Gaskell, C. Midden, M. Liakopoulos and L. Scholten (2000) 'Two Cultures of Public Understanding of Science and technology in Europe', in M. Dierkes and C. von Grote (eds.), *Between Understanding and Trust. The Public, Science and Technology*, Amsterdam: Harwood Academic Publishers, pp. 131–56.

Ernst & Young (2001) *Integration. 8th Annual European Life Sciences Report, 2001*, Cambridge: E&Y International.

Flavell, J.H. (1963) *The Developmental Psychology of Jean Piaget*, Toronto: van Nostrand Corporation.

Galbraith, K. J. (1968) *The New Industrial State*, New York: American Library.

Gamson, W. (1988) 'The Constructivist Approach to Mass Media and Public Opinion', [The 1987 distinguished lecture], *Symbolic Interaction* 11: 161–74.

Gaskell, G., M.W. Bauer and J. Durant (1998) 'The Representation of Biotechnology: Policy, Media, Public Perception', in J. Durant, M.W. Bauer and G. Gaskell (eds.), *Biotechnology in the Public Sphere*, London: Science Museum, pp. 3–14.

Gibson, J.J. (1977) 'The Theory of Affordances', in R.E. Shaw and J. Bransford (eds.), *Perceiving, Acting, Knowing*, Hillsdale, NJ: Erlbaum, pp. 67–82.

Giddens, A. (1991) *Modernity and Self-Identity*, Cambridge: Polity Press.

Haber, E. (1996) 'Industry and the University', *Nature Biotechnology* 14: 441–2.

Habermas, J. (1989) *The Transformation of the Public Sphere*, Cambridge: Polity Press; German original, 1961.

Hansen, A. (1994) 'Journalistic Practices and Science Reporting in the British Mass Media: Media Output and Source Input', *Public Understanding of Science* 3: 111–34.

Hood, C. and H. Rothstein (1998) 'Institutions and Risk Management: Problem Solvers or Blame-Shifters', in Opening Plenary address 'Risk Analysis: Opening the Process', Paris, 11–14 October, pp. 13–22.

Hood, C., H. Rothstein, M. Spackman, J. Rees and R. Baldwin (1999) 'Explaining Risk Regulation Regimes: Exploring the "Minimal Feasible Response" Hypothesis', *Health, Risk and Society* 1: 151–166.

Hughes, T.P. (1989) *American Genesis – a Century of Technological Enthusiasm*, New York: Penguin.

Inglehart, R. (1990) *Culture Shift in Advanced Societies*, Princeton, NJ: Princeton University Press.

Joerges, B. (1988) 'Technology in Everyday Life: Conceptual Queries', *Journal for the Theory of Social Behaviour* 18: 219–30.

Kling, R. and I. Iacono (1988) 'The Mobilisation of Support of Computerisation: the Role of the Computerisation Movement', *Social Problems* 35: 226–43.

Kohring, M. (1998) 'Der Zeitung die Gesetze der Wissenschaft vorschreiben? Wissenschaftsjournalismus und Journalismus-Wissenschaft', *Rundfunk und Fernsehen* 46(2/3): 175–92.

Latour, B. (1988) 'The Prince for Machines as well as for Machinations', in B. Elliot (ed.), *Technology and Social Process*, Edinburgh: Edinburgh University Press, pp. 21–43.

Lewin, K. (1952) *Field Theory in the Social Sciences*, London: Tavistock.

Linde, H. (1982) 'Soziale Implikationen technischer Geräte, ihrer Entstehung und Verwendung', in R. Jokisch (ed.), *Techniksoziologie*, Frankfurt: Suhrkamp, pp. 1–31.

Luhmann, N. (1995) *Social Systems*, Stanford, CA: Stanford University Press.

McCombs, M.E. and D.L. Shaw (1972) 'The Agenda-Setting Function of the Mass Media', *Public Opinion Quarterly* 36: 176–87.

Merton, R.M. (1936) 'The Unintended Consequences of Purposive Social Action', *American Sociological Review* 1: 894–904.

Mokyr, J. (1992) 'Technological Inertia in Economic History', *Journal of Economic History* 52(2): 325–38.

Morgan, M. and J. Shanahan (1996) 'Two Decades of Cultivation Research: an Appraisal and Meta-Analysis', *Communication Yearbook* 20: 1–45.

Moscovici, S. (1976) *La psychanalyse – son image et son public*, Paris: Presses Universitaires de France.

Neidhardt, F. (1993) 'The Public as Communication System', *Public Understanding of Science* 2: 339–50.

Norman, D.A. (1988) *The Psychology of Everyday Things*, New York: Basic Books. (1998) *The Invisible Computer*, Cambridge MA: MIT Press.

Piaget, J. (1972) *The Principle of Genetic Epistemology*, London: Routledge & Kegan Paul.

Pinch, T. (1996) 'The Social Construction of Technology: a Review', in R. Fox (ed.), *Technical Change. Methods and Themes in the History of Technology*, Amsterdam: Harwood Academic Publishers, pp. 17–32.

Pinch, T. and W. E. Bijker (1987) 'The Social Construction of Facts and Artifacts: or How the Sociology of Science and the Sociology of Technology Might Benefit Each Other', in W.E. Bijker, T.P. Hughes and T. Pinch (eds.), *The Social Construction of Technological Systems*, Cambridge MA: MIT Press, pp. 9–50.

Price, D.K. (1965) *The Scientific Estate*, Cambridge MA: Harvard University Press.

Radkau, J. (1995) 'Learning from Chernobyl for the Fight against Genetics?' in M. Bauer (ed.), *Resistance to New Technology – Nuclear Power, Information Technology, Biotechnology*, Cambridge: Cambridge University Press, pp. 335–55.

Rogers, E.M. (1996) *Diffusion of Innovation*, 4th edn, New York: Free Press.

Ropohl, G. (1979) *Eine Systemtheorie der Technik. Zur Grundlegung der allgemeinen Technologie*, Munich: Hanser.

Simon, H.A. (1969) *The Sciences of the Artficial*, Cambridge MA: MIT Press.

Stehr, N. (1994) *Knowledge Societies*, London: Sage.

Tenner, E. (1996) *Why Things Bite back. Predicting the Problems of Progress*, London: Fourth Estate.

Touraine, A. (1995) 'The Crisis of Progress', in M. Bauer (ed.), *Resistance to New Technology – Nuclear Power, Information Technology, Biotechnology*, Cambridge: Cambridge University Press, pp. 45–56.

Toynbee, A. (1978) *A Selection from His Work* [parts 6 and 7: Challenge and Response], Oxford: Oxford University Press [originally published between 1934 and 1954].

Weart, S.R. (1988) *Nuclear Fear. A History of Images*, Cambridge, MA: Harvard University Press.

Winner, L. (1977) *Autonomous Technology. Technology-out-of-control as a Theme in Political thought*, Cambridge, MA: MIT Press.

Index

Media Journal

Second Edition

Media Journal

Reading and Writing about Popular Culture

Joseph Harris
University of Pittsburgh

Jay Rosen
New York University

Gary Calpas
University of Pittsburgh

Allyn and Bacon
Boston • London • Toronto • Sydney • Tokyo • Singapore

Vice President, Humanities: Joseph Opiela
Editorial Assistant: Mary Yarney
Marketing Manager: Lisa Kimball
Editorial Production Service: Chestnut Hill Enterprises, Inc.
Manufacturing Buyer: Suzanne Lareau
Cover Administrator: Linda Knowles

Internet: www.abacon.com

Between the time Website information is gathered and published, some sites may have closed. Also, the transcription of URLs can result in typographical errors. The publisher would appreciate notification where these occur so that they may be corrected. Thank you.

Library of Congress Cataloging-in-Publication Data
Media journal : reading and writing about popular culture / [compiled
   by] Joseph Harris, Jay Rosen, Gary Calpas. — [2nd ed.]
      p. cm.
   Includes bibliographical references and index.
   ISBN 0-205-27483-8
   1. Mass media criticism.   2. Feature writing.   I. Harris, Joseph
(Joseph, D.)   II. Rosen, Jay   III. Calpas, Gary
P96.C76M4   1998
808'.0427—dc21   98-27417
                                                                    CIP

Printed in the United States of America
10 9 8 7 6 5 4 3 2   RRD-VA   03 02 01 00

Contents

Contents

Preface

One of the pleasures of revising a book is that it allows you to return to a text, to rethink and rework its ideas and phrasings, and, then, to claim that this is what you meant to say from the very start. And so, as we have worked with the first edition of *Media Journal* and talked with others who have used it in their classes, it has indeed become clear to us that there are at least two things we've wanted to say all along: The first—and most important—is that our focus here is on the teaching of writing. Our belief as teachers is that a writing class should take the experiences and perspectives of students seriously. Using the media as the subject for a course in writing is thus a *strategic* move that allows students to center their work on a subject they know and care about. But our goal is not for students to learn more about the media *per se,* but rather to *use* their experiences with the media in learning to write more thoughtfully, effectively, and critically.

The second thing we want to say has to do with a term we have used twice already: *experience.* For some years now a debate has raged among college writing teachers who believe that students should write from personal experience and others who insist that students need to learn how to write about books and ideas. We approach this argument from what may seem an odd angle, since, although we believe strongly that good writing stems from experience, we are most interested in students' experiences as *intellectuals.* Almost all of us are immersed every day in a world of texts—of newspapers, magazines, radio, television, movies, videos, billboards, flyers, menus, pop songs, buttons, logos, printed T-shirts and caps, memos, directions, homework assignments, web pages and e-mail postings, and even, on occasion, books—that we constantly use and interpret but are rarely asked to think critically about. It is this experience of decoding and manipulating a *textual* world that we want students to draw on and examine in their writings. The journal assignments in this book thus ask students to look closely at their everyday interactions with the media, with the ways they use popular texts in navigating their worlds and in forming a sense of who they want to be. And the assignments for writing criticism ask students to draw on their experiences with the media in responding to the readings in this book, to imagine that they too have something to say about our common culture. The sort of prose we teach is thus personal but not confessional—writing that is concerned with the texts, voices, images, and ideas of the media as they enter everyday life, with our common culture as it is lived and experienced.

In this book we ask students to do three things: (1)To keep a **media journal** in which they reflect on the uses they make of the voices and images of popular culture; (2) to **read and respond** to the work of other media critics, to test their own views and experiences against those of the writers included in these pages, and (3) to try their hands at **writing media criticism** themselves. All three kinds of work ask students to find and write about texts from the media culture around them, to think critically about what they see and hear on their television sets and radios, in magazines and newspapers, on city streets and shopping malls, at the movies, and at concerts and clubs. To put it another way, we believe that a book such as this can provide only *some* of the materials for a course on writing about popular culture, that the remaining materials must always come from the media themselves and the experiences students have with them. Our aim is not to inculcate students with a certain set of critical methods or terms or to introduce them to the academic study of popular culture, but to offer them opportunities to rethink and write about their own experiences with the media, to come to their own understandings of our common culture.

In preparing this second edition of *Media Journal*, then, we have made three major changes. First, we have updated many of the readings to better address the diverse media culture of the 1990s. Nineteen of the 39 readings in this volume are new—including pieces on the Internet, O. J. Simpson, shock journalism, the culture of celebrity, the design of everyday appliances, music and technology, talk shows, office zaniness and "casual days," pop revivals, trendspotting, reality TV, greeting cards, and more. Second, we have restructured the book itself so that it begins with all of the media journal assignments grouped together, followed by the readings listed alphabetically by author, and concluding with some longer projects for writing criticism. We hope that this new structure not only supports our interest in having students reflect critically on their own experiences with the media but also allows teachers more flexibility in designing their own courses using *Media Journal*. And, third, we have made the questions following the readings more systematic and (we hope) useful. Each reading is now followed by suggestions for *Coming to Terms* (understanding the author in your own words), *Reading as a Writer* (looking at style and strategy), and *Writing Criticism* (making an author's words and ideas part of your own thinking and writing).

Gary Calpas joins us as a coauthor of this edition, and has helped us tremendously both in expanding the range of issues and texts we've been able to consider and in forging connections between essays and assignments. We owe many other people thanks as well: An exceptional group of teachers at San Diego State University—led by Ellen Quandahl, and including Melody Kilkrease, Laura Headley, Amy Kelly, and Holly Bauer—shared their experiences working with the first edition and offered us wonderful advice for the second. At the University of Pittsburgh, Barbara Edelman and Heidi Walter also worked with the first edition and were generous with their time and help. We also benefited from the comments of several perceptive reviewers of the first edition, including: Patrice Dunmire, Kent State University; Donna Qualley, Western Washington University; Lisa A. Stolley, University of Chicago, Illinois; Susan Seyfarth, Valdosta State University. Our editor at Allyn & Bacon, Joe Opiela, showed both persistent faith in this project and excellent judgment and tact in suggesting changes to it. And we owe Carol Mysliwiec and Jeremy Jacobs special thanks for their help preparing the final version of the manuscript. Nor can we forget the many people who helped us prepare the first edition of *Media Journal*, including: David Bartholomae, Kathleen L. Bell, Rashmi Bhatnagar, John Clifford, William Costanzo, Joel Dinerstein, Bianca Falbo, Angie Farkas, Barbara Heinssen, Lee Jacobs, Joseph Janangelo, Constance Mayer, Lori Nielson, Christine Nystrom, Neil Postman, Robert L. Root, Jr., Annette Seitz, James Seitz, Geoffrey Sirc, Stephen Sutherland, Robert Thompson, Donna H. Winchell, and Eva J. Winkle.

Media Journal

Introduction

Reading Culture

"To write this stuff, you have to read like hell and watch a lot of television." So says Matt Wickline, a comedy writer on the staff of *Late Night with David Letterman*, explaining how he gets ideas for Letterman's jokes and routines. Those of us who, like Wickline, were born after World War II have probably been raised on popular culture. We're likely, that is, to have watched a lot of television, seen a lot of movies, heard a lot of pop songs, glanced at a great many ads. As college teachers and students, we are now also asked to read continuously, seriously, and widely. This combination—media culture plus a habit of reading—can be a powerful one, not only for comedy writers but for critics. And that is what we ask you to become in this book: critics of the culture that has produced David Letterman, the Beach Boys, James Dean, and microwave popcorn.

To consume the products of popular culture requires no instruction. We already know how to watch television, operate a tape player, scan the pages of a magazine, or go out to the movies. But criticism differs from consumption. A critic is concerned with appreciating, understanding, connecting, and talking back to the media. A consumer simply uses the products of the media—for enjoyment, for information, or as part of a daily routine. There would be nothing wrong with this if the media were not also trying to use us. But as audiences for the mass media we are often quite useful to others. When we're watching the evening news, or flipping through the pages of a magazine, or listening casually to the radio in the morning, we are part of the production of a valuable commodity—a mass audience that can be packaged and sold to advertisers. Ordinarily we think of packaging as something done to products. But we ourselves are packaged when, for example, *Rolling*

Stone advertises itself as a good place to reach the "young spenders," those 18-to-35-year-old college-educated men and women whose buying habits its advertisers wish to shape. To live in a culture where the mass media are heavily commercialized is to be on the receiving end of innumerable acts of packaging and persuasion—of attempts to convince us to accept this image or that notion of ourselves. And so car commercials sell us on speed, power, freedom, and mobility; movies sell us on action; sitcoms promote particular notions of family life; magazines present beauty, glamour, and fame as worlds we might inhabit. Often these propositions are harmless; sometimes they can be fun, or even inspiring. Some of them we may wish to accept as true—or at least as important to ourselves and our dreams. Others we might revise or transform for our own purposes. Yet others we reject. And still others we may protest and seek to change. But this sorting-out process begins with criticism—with an awareness of what our culture is saying when it talks to us through the mass media. This is what "reading" culture is all about.

By *reading,* then, we mean something more than simply lifting information out of books or articles. To read a text or event is to do something to it, to *make* sense out of its signals and clues. And so, for instance, a poker player reads the faces of the others around the table; a quarterback reads the defense arrayed before him; a professional guide on a rafting trip reads the river and its (possibly treacherous) currents. Reading is thus not something we do to books alone. Or, to put it another way, books and other printed surfaces are not the only texts we read. Rather, a "text" is anything that can be interpreted, that we can make meaning out of or assign value to. In this sense, all culture is a text and all culture can be read. Advertising slogans can be read, but so can the images in ads. Song titles can be read, but so can the singer's attitude. We can "read" a movie for the messages it is sending us. We can "read" a TV show for the values it promotes. Machines can be "read" for the ideas or propositions they embody. Reading, in this sense, is an intentional act—it is to have a certain purpose in mind in examining an object or experience. It is to be on the lookout for something. What sort of thing to look *for* is part of what we hope to teach you in this book.

Reading is also often an act with consequences. It is not neutral but *interested*—in the sense of being engaged, involved, committed. We read to size things up, to form our own ideas, to decide where we stand, or whom we want to stand with or against. "To read is to undo," says Mark Crispin Miller, a critic of television.[1] What he means, we think, is that all texts, and perhaps especially the texts of a commercial culture, are trying to "do" things to us or with us. This is perhaps most obvious in the case of advertising, which is trying to get us to buy something. But even texts that seem to

---

[1] *Boxed In: The Culture of TV* (Evanston, IL: Northwestern University Press, 1988), p. 19.

do nothing more than entertain can also be seen as trying to get us to buy certain images of ourselves or our world.

Consider a television game show like *Jeopardy*. We can watch it just to pass the time, or we can play along by trying to answer the questions with the contestants. We can also read *Jeopardy* as a text. That is, we can ask what sorts of claims the show might be said to make about the world and also what claims it makes on us as its viewers—what it tries to get us to accept or believe. We might notice, for instance, that *Jeopardy* celebrates a view of knowledge as the simple ability to recall random and disconnected facts—and we could then begin to ask why we find this mastery of trivia so entertaining. Or we could grant that there's always a certain joy in knowing the right answer to a question, however foolish or inane it may be, and we might then try to connect this pleasure to patterns of schooling that stress the memorization of facts or to the experience of being routinely bombarded with vast amounts of (mostly useless) information nearly every day of our lives—as we move from the morning paper to the car radio to billboards and magazines and memos and junk mail and supermarket displays to finally the TV and the news at eleven.

In thus looking at the products of popular culture as *texts*, we can start to subvert the intentions of their makers and to replace them with our own. This is not as unusual as it might seem. Products are often used in unexpected ways by the people who buy them: Recipes and formulas get changed; texts get misread on purpose. Safety pins become earrings; the knees of blue jeans are intentionally ripped; sweatshirts are worn several sizes too big. Bart Simpson shows up on T-shirts as an African American. TV ads and sitcom jingles are parodied; soaps are talked back to. The faces of celebrities become masks on Halloween; homemade dance tapes segue wildly and happily from pop to funk to rap to rock, violating all marketing categories along the way. Yet this skeptical knowledge of (and pleasure in) the voices and images of our common culture rarely finds a place in college classrooms, which means that it also rarely gets questioned or pushed or changed. Instead, our lives are too often divided in two—with friends, family, work, clothes, music, movies, TV, magazines, advertising, and almost everything else on one side, and academic study on the other.

Keeping a Media Journal

Our aim in this book is to overcome that split, to make the media culture outside the classroom one of the things that gets talked about inside it. The best way to do this is to begin with experience—that is, with ways in which the media figure in our own lives. And so we decided to organize this book not around types of criticism or forms of media but *categories of experience*—or common ways of using and interacting with the media. Throughout the book we ask you to reflect on what you do with the mass media: the shows you watch, the movies you like, the music you enjoy, the ads you see, the news you hear, the performers you admire, the habits and routines you've developed over time.

In most courses offered at a university, the challenge is to assimilate new material, to master new knowledge by becoming familiar with its uses. In this book, we ask you to engage in a kind of reverse process: to begin with the familiar territory of your own experience and to make this experience seem new or "strange." By "strange" we simply mean "in need of explanation" or "worth thinking (and writing) about." The challenge here, then, is not to master unfamiliar material but to "read" familiar experiences in a new or productive way. This means noticing things you have always done but rarely questioned—habits and preferences, likes and dislikes, important moments and recurrent patterns in your use of media.

To help with this process, we often present you with lists of questions you can ask yourself to investigate the media's place in your life. Not all of these questions will seem relevant to you; not all will produce interesting results for your own thinking and writing. That's to be expected. The point is not to answer them all, the way you would in a standardized exam, but to use the questions you find in this book to improve your own understanding of media experiences. That means selecting the questions that seem most interesting or relevant to you—the ones that best illuminate your own experience. After all, a question in a textbook is itself a "text" that can be read in a variety of ways. The best way to "read" the questions in this book is simply to seize on those that cause you to think or suggest a new way of looking at things.

As a way of noticing and reflecting on your media experiences, we recommend that you keep throughout the term a **media journal**—a notebook in which you jot down ideas, work out your responses, record your observations, and generally conduct the intellectual work required of you by the assignments in this book. (It is possible to keep this notebook in electronic form—that is, on a computer disk. But it is probably just as easy to make it a handwritten exercise. Either way, the journal should be useful to you, and organized according to your own working style.)

The journal is intended to accomplish three kinds of tasks. The first, and most structured, involves a series of **journal assignments.** These are thinking and writing tasks, presented at the beginning of each chapter. Most of them ask you to reflect on your own experience with the media. For instance, almost everybody uses the media in some way to "keep in touch" with people or events, and, conversely, just about everyone also uses the media to "escape," to be entertained or distracted. So those are two categories of experience we ask you to think about and write about in your journal. Others include feeling hyped or pressured by the media, connecting with celebrities, identifying with a certain sort of music, and watching TV. The journal assignments are built around questions we ask you to ask yourself. Keep in mind that it is not necessary to answer all the questions—just those that get you thinking.

Another task for which your journal will be useful is responding to the **Coming to Terms Questions.** By coming to terms we mean a set of questions we propose to help you in your initial encounter with an author—some suggested places to start taking the author on your "own terms." If you keep your journal in a notebook, you may want to set aside a few pages for each of your file categories. Or if you work on a computer, you may want to set up a separate document for each one. But there is no right or wrong way to keep such files; the point is to record significant aspects of your media experience for possible use throughout the term. Here are some ideas for lists you may want to create and add to as the term proceeds. These are only suggestions. The actual categories you develop will have to grow out of your own experience and working style.

- **Incidents from memory:** Important moments in your life when the media played a key role.
- **Firsts:** Moments of discovery when you first realized the power or joy of a particular medium or media experience.
- **Favorites:** Lists of performers, films, shows, songs, books, and images that have developed some sort of hold on you.
- **Strong reactions:** Particular encounters with the media that moved you deeply, or changed the way you looked at things, or caused you to feel outraged, or otherwise disturbed you in a way that may be worth examining
- **Questions to yourself:** Things you don't understand, or would like to explore about your own history with media or your current habits and preferences.
- **Comments from others:** Things friends, acquaintances, interviewees, or classmates have said that struck you as interesting or in need of exploration.
- **Routines:** Habits you may have developed in your use of the media, or recurrent patterns that show up when you examine how the media enter into your life.

In addition to your responses to the **Coming to Terms Questions,** you will probably also want to respond to the **Reading as a Writer Questions** in your media journal. These are reading tasks we ask you to undertake after you have read through the essays in this book at least once. In most cases these questions ask you to do something to or with the author's text—to take notes, make a list, or apply your experience to what is said in the piece. We suggest that you do so in your media journal. Answers to the **Reading as a Writer** sections are meant to be discussed in class. That means, of course, that you should bring your journal with you whenever your class meets. You may often find that what is said there will prompt you to add some new notes or observations to those already in your journal.

Keep your journal in whatever form suits you. Again, some people write their entries in a notebook they carry with them from place to place, while others type theirs on a personal computer. The point of journal writing is to jot down your thoughts and observations, not to write a finished or elegant essay. We think that you will soon find yourself speaking from these notes in class, that you will use them in contrasting your own experiences and ideas with those of the critics you read, and that you will also draw on them in drafting your own more formal writings.

Writing Criticism

In addition to keeping a media journal, we also ask you to read and respond to the work of other cultural critics, to match your experiences and test your ideas against theirs. What we looked for when selecting the readings for this book was, first of all, writing that we liked, that we found in some way intriguing, lucid, compelling, witty, or original. This does not mean that the writings presented here are easy. Quite the opposite. In trying to stretch the boundaries of criticism, such pieces often prove as difficult as they are interesting. The **Reading as a Writer** sections ask you not just to understand what the authors are saying but also to use what they have said to clarify your own experience with the media and popular culture. You are again asked to use the readings in the sections on **Writing Criticism** that conclude each chapter. Here, we usually ask you to do two things: (1) To draw on what you've read in interpreting some other media text; and (2) to go out and do what the critics presented here are doing, to engage in the kind of criticism on display in the essays you've read.

You will find that such work rarely ends up with answers that are clearly right or wrong. Instead, the aim of the critic is to form his or her own stance toward a subject in relation to what others have thought and said about it. The readings in this book are meant as provocations for such criticism, to get you to think through and reinterpret your experiences as part of our media culture, to become a more skeptical and attentive reader of its many voices and images, to become a critic yourself.

Criticism begins in looking at something as a text—as an object to be given meaning, interpreted, handled by your own intelligence. Often the products of media culture must somehow be detached from their usual contexts before they can be read in this way. Sometimes simply bringing them into the classroom can do the trick. The cover of the *National Enquirer* looks "natural" and familiar when we encounter it on a newstand or supermarket rack. But in a college classroom it can start to look very strange—in need of some new sort of comment and response. Or play some rock music in class, and observe how people are suddenly not quite sure what to do with their bodies. Should they move with the beat? Tap their feet or fingers? Take notes? Or look at a TV sitcom together, and note the uneasy glances that follow the chuckles and guffaws from the laughtrack. People are no longer sure whether they should be laughing along. In each of these cases, a familiar object has suddenly been made "strange." It has been transformed into a "text," and thus made available not simply for consumption but for criticism as well.

There are other ways of altering how you experience a text. One way, for instance, of defamiliarizing the imagery of TV is to turn off the sound. Without the soundtrack there to tell you how to read the images on the screen, you may find yourself looking at them in new ways. The mere act of taking notes on what you're looking at or listening to can also help, for the act of criticism starts as soon as you begin to select those details that strike you as significant or worth recording. Similarly, simply talking over your experiences with the media with friends or classmates can prove useful, since such talk can show that their responses are not at all the same as your own. And, of course, reading what another critic has had to say about a text can often help you notice new things about it when you go back to review it.

The important thing about a text is that it can be *reread*. To look again at a text with a different notion or question in mind is one of the most basic acts of criticism. So it is always an advantage if you can revisit the object of your attention—whether it is a magazine, a TV show, a movie, a place, a style of music, or a popular image. If you can, then, you should try to get a copy of the text itself, have it before you to review and cite when you are thinking over your own ideas, revising your own text. VCRs and other recording devices have now made it possible for us to "have" many texts of popular culture in a way that literary critics have always "had" in their possession the books they've considered. But even where this is not possible—as in the case of a current movie not yet available on video—you will find that taking notes can both prompt your memories of the text and allow you to rethink your first responses to it. Media culture tends to rush by us like the traffic on the street, as a continuous flow of images, sounds, propositions, famous people, and the like. You will need to find some method of arresting this flow, of stopping the action so you can apply your intelligence to it, in order to come to grips with what this culture is saying and doing. Keep this in mind when deciding what to write about. A text you can obtain and reread, whose images or phrasings you can isolate and quote, almost always makes a better subject for writing than one that's simply floating around "out there" or that exists only in your memory.

Drafting, Revising, Editing

If criticism begins in reading and rereading, noticing and note-taking, talking and arguing, it almost always ends, at least at a university, in the act of writing. As a critic, you respond to a text by creating one of your own, by writing out your "reading" of it in the form of a paper or article. There is no single correct way to do this. Some people draft their writings quickly from beginning to end, others (perhaps most) plod along in bits and pieces, starts and stops, adding a few sentences here, fiddling with a line there, reworking the first part after they've written the last, and so on. Some people pretty much know what they're going to say before they start writing, others (again, probably most) figure it out as they go along. (One of the few things that can be reliably said about the process of writing is that almost nobody begins with an outline and sticks to it.) So we won't try in this book to offer you a method or formula for writing. We argue instead that the point is not to find yourself stuck with rigid rules and inflexible plans—like always making an outline, or never using words whose spelling you're not sure of—that can get in the way of your attempts to write.

Still, we have found a few terms and ideas useful in thinking about what goes on when we sit down to write. As we see it, most of us experience three sorts of difficulties in writing.[2] First there are problems of *fluency*—of finding something to say, getting words and ideas down on paper. Then there is the question of how to rework those words and ideas in a *persuasive* way—so they make sense to you and your readers. And finally there are worries about *correctness*, about making sure that your finished text looks and sounds okay. We can think of these three problems of fluency, persuasion, and correctness as corresponding with certain rough stages of composing a text. First, a writer must *draft* what she has to say, move from the thoughts in her head to words on a page or screen. Her next task is to work with and on the rough and approximate text she has just created, to *revise* what she has written to make it more compelling and clear. Finally, she must proof-read and *edit* her text for points of style, punctuation, grammar, spelling, and the like. And, of course, in working through all three of these activities—*drafting, revising,* and *editing*—a writer can draw usefully on the *responses of readers.* The whole process can be mapped out something like this:

WHAT TO WORK AT	STAGES OF THE COMPOSITION PROCESS	
Fluency Getting words and ideas down on paper	Drafting	
Persuasion Reworking your text to make its phrasings effective and clear	Revising	Response from readers
Correctness Making sure your finished text looks and sounds okay	Editing	

---

[2]Much of what we have to say in the next few pages draws on the work of John Mayher, Nancy Lester, and Gordon Pradl in *Learning to Write/Writing to Learn* (Upper Montclair, NJ: Boynton/Cook, 1983).

Again, we stress that the activities of drafting, revising, and editing often blur into one another. We are not trying to define a lockstep formula or method for writing. We have found these terms useful, though, in approximating the things that happen both as a writer composes a single draft of an essay and as she works through several drafts of the same piece. That is, a writer may go through one round of drafting, revising, and editing in the course of an evening before a piece is due, and then several times more again in the next few weeks as she redrafts her piece in response to the comments of her readers. In either case, as soon as you look at the process this way, one of the most frequent mistakes made by beginning writers becomes clear: They worry about issues of correctness before they have even begun to figure out what they want to say. They start at the bottom rather than at the top of our map, trying to edit and correct what they write before they have even really drafted it. The result is that they often become blocked, unable to write because they fear that anything they put down on paper is likely to be marked as an error, panicked by the apparent need to figure out at once both *what* to say and *how* to get it down in the proper form. But in writing, you can work at the two separately. You can first make a rough draft of what you want to say, one that is more for you than for your readers, and then go back later to work again on its ideas and phrasings. The first step, then, in becoming more fluent as a writer is to distinguish between the acts of drafting, revising, and editing— and, in doing so, to learn how to get words on the page (or screen) without being blocked by worries about correctness. One reason we suggest that you keep a media journal is to give you practice in this sort of writing.

But if drafting involves moving from thought *to* text, ideas *to* words, revising is work *with* text—changing and reshaping and adding to what you have written. How you go about doing this is likely to vary from piece to piece. You may decide simply to abandon your first draft and start over from scratch (though, of course, it won't really be from scratch since you will have already done a good bit of work on your subject). Or you may find yourself adding some sections and cutting others, perhaps reordering paragraphs as you go, or reframing what you have to say with a new opening or closing. And sometimes you may make only small or stylistic changes to some parts of your text, even as you recast others entirely. In any case, actual revision (as opposed to editing or proofreading) always involves more than just some tinkering with words and phrases here and there. It means reworking and rethinking a text. There will be time later to edit your prose for style or grace. But it makes no sense to polish a text that needs first to be rethought.

This does not mean that the issue of correctness disappears. No one wants to feel as though he is the first person to have looked over a text that has just been handed to him. Proofreading is a courtesy to your reader (and one that is often necessary if you want to have your work read at all). You should try to make the final draft of any piece you write as clean and free of mistakes as you can. We are simply suggesting here that you think of proofreading as something you do *after* you've drafted and revised a piece, that you approach the issue of correctness as largely a problem in *reading* what you have written. Rather than trying to correct your prose as you write it, learn to set aside time to proofread your text once you have (almost) finished it, so you can then catch and correct whatever mistakes you may have made while drafting it. You might also want to ask a classmate or friend to help you proofread your work. This is not cheating. Professional writers have editors who help them in similar ways.

Entering the Conversation

But all this still leaves us with what is most interesting and difficult about writing. For you can write a text that is fluent and correct but still empty and dull—a text that is not worth reading because its author has nothing much to say. The most important problem any writer faces is that of being interesting. It can be fairly easy to learn how to catch mistakes and errors in your writing. But it can be extremely hard to find a way of talking that others think is worth listening to. How do you talk about texts and ideas (since they are the subject of our work here) in a way that does not simply repeat what others have had to say, but that adds something to the conversation?

One way of imagining what you will be doing with the materials of this book, then, is to think of yourself as entering through them into an ongoing conversation about the shape and character of our common culture—a conversation that includes both our own voices and those of the writers we have read. As the critic Kenneth Burke once put it:

> *Imagine that you enter a parlor. You come late. When you arrive, others have long preceded you, and they are engaged in a heated discussion, a discussion too heated for them to pause and tell you exactly what it is about. In fact, the discussion had already begun long before any of them got there, so that no one present is qualified to retrace for you all the steps that had gone on before. You listen for a while, until you decide that you have caught the tenor of the argument; then you put in your oar. Someone answers; you answer him; another comes to your defense; another aligns himself against you, to either the embarrassment or gratification of your opponent, depending on the quality of your ally's assistance. However, the discussion is interminable. The hour grows late, you must depart. And you do depart, with the discussion still vigorously in process.[3]*

---

[3]*The Philosophy of Literary Form: Studies in Symbolic Action* (Berkeley: University of California Press, 1973), pp. 110–11.

The name usually given to this sort of talk and writing is *criticism*—meaning not a kind of discourse that is persistently negative but simply one that takes as its subject other texts and writings. To write as a critic is to deal with and comment on the works and ideas produced by others. The trick of doing criticism well is to interpret a text in ways that will interest other readers, to find a way of talking about it that does not merely repeat what its author said, but that also allows you to make a point of your own. To be a critic is not to go around pointing out the flaws in everything you see. The best critics are deeply interested in their own enthusiasms. They want to explain why they like what they like—explain it, first of all, to themselves. At the same time, they reserve the right to object, to be disturbed, disappointed, even mad at their culture. To be a critic is to be both curious and articulate about your own responses—whether they are positive or negative, outraged, excited, or amused.

How you learn to do this is hard to say—since more often it involves acting on hunches and intuitions than it does following a list of explicit rules and strategies. And so the best advice we can give you now, at the start of a course on writing, is to enter as best you can into the "interminable discussion" called criticism and see what you can learn as part of it, rather than wasting time looking for a guru or taskmaster who can "tell you exactly what it is about" and give you fixed and unchanging rules for participating in it. The point is that you don't get better at reading and writing by memorizing rules or formulas. You have to try things out for yourself and see what works and what does not. We hope you will use this book to do just that: To write on texts and issues that matter to you; to read what others have said on those subjects; to listen to what your own readers think about what you've written; and to revise your work in response to what you have read and heard.

How This Book Is Set Up

We have tried in this book to offer a set of materials for a course in reading and writing about popular culture. But we have not tried to fix in advance what the shape of such a course might be or what directions it might take. Our aim is to raise a set of questions and issues, to provide some starting points for writing and talk, not to suggest what we think are the right ways of answering those questions. We are not advocates (at least not here) for a particular way of looking at popular culture or for a certain approach to doing cultural criticism.

The next section of this book offers a series of ideas for writing in your media journal. What distinguishes these assignments from those following the readings in this book is their direct focus on your *experiences with and uses of* the media. Our aim in composing them is to provide you with a set of ways to begin collecting, reporting, and interpreting your day-to-day encounters with the texts, voices, and images of our culture. These assignments ask you to gather observations from your own experiences and memories, to jot down key insights and information, to ask questions of yourself and others, and to reflect critically on the ways the media have entered your life. They ask you to think about your media routines, for instance, or about the ways you use the media both to connect with the outside world and to escape its demands, to identify with various forms of music or with the celebrities that populate our culture, and to read the meanings embedded in places and everyday devices. You may want to use these assignments as a way of forming your own responses to the readings in this book, or as the start of longer essays themselves.

The third and main section of this book is a collection of writings by media critics. We have followed each reading with three sets of questions: First we ask you to **Come to Terms** with the ideas of the critic you have just read, to translate his or her phrasings into language of your own, to begin to make their insights part of your own repertoire of strategies for interpreting the media. Then we ask you to **Read as a Writer,** to look at some of the distinctive moves and gestures these critics make in composing their essays, and to locate moments in their work that you may want to imitate in your own prose. And, finally, we close each reading with ideas for **Writing Criticism,** in which we try to suggest some ways you can make use of what you have read in your own work as a writer—perhaps through building on the ideas of another critic to interpret some other text or event, or by drawing on your own experiences with the media to point out some problems with or limits to her approach, or by making connections between (or defining the conflicts among) the work of several critics. In any case, the aim of these questions is not simply to see if you've understood the work of the writers in this book but to give you the chance to *do something with* what they have to say, to push their ways of reading our culture, to make their methods, ideas, and phrasings part of your own. Finally, we end *Media Journal: Reading and Writing about Popular Culture* with ideas for some longer **Projects in Criticism** which ask you to do some research of your own into the ways the texts of our culture are used and transformed by their readers, listeners, and viewers.

Media Journal
Assignments

The following series of assignments ask you to reflect on the ways the media have entered your life and on how you have made your own uses of the texts, images, words, and sounds they have offered you. They ask you, for instance, to think about how you use media to keep in touch with the world, or to escape it, to identify with various celebrities or styles, or to resist them. The style of writing these assignments call for tends to be informal. They ask you to gather observations from your own experiences and memories, to jot down key insights and information, to ask questions of yourself and others, to collect, list, describe, take notes. Use these journal assignments, then, to begin to think about your own tastes and commitments in the media, to form a position that you can speak from in responding to the views of your class-mates and the work of the other cultural critics collected in this book. The sign of good journal writing is that it is *generative*—that it leads to more writing, talk, reflection, criticism. You may find yourself turning to your journal, then, as a way of getting talk about a text or issue started in class, or for notes and ideas for longer papers and projects.

You should probably give yourself about two hours to respond to most of the journal assignments in this book. You don't have to answer every question asked in each assignment, and you don't have to write an "essay." The point of keeping a journal is to have a place to put your thoughts down, not to put them in finished form. So don't fuss over spelling, style, punctuation, and so on. If you want to switch topics, just start a new paragraph and do so. Err on the side of inclusion. That is, if you're not sure whether you should say something, or exactly how to say it, then just get it down whatever way you can for now. There will always be time later to go back to it. If you feel more comfortable working on a typewriter or personal computer, then do so, but don't bother retyping or recopying your writing. (Even if you work on a computer, you will still probably also want to keep a notebook that you can carry around with you and write in when needed.) So long as your text is legible and of some use or interest to you, it's okay.

A Day in the Life

Throughout this book we ask you to keep a media journal in which you think and write about how various forms of popular culture—music, movies, TV, books, newspapers, fashion, advertising, and the like—have entered into your life and the roles they have played in it. In asking you to keep such a journal we are, of course, assuming that all of us have been influenced by our experiences with the media, that the ways we think and act and feel and talk and dress often show traces of the movies and shows and ads that we have seen, the music we listen to, the images and personalities that we have been shown by magazines, TV, and newspapers. But we are also assuming that we all make our own individual uses of these media—that each of us remembers, interprets, and enjoys different things for different reasons. And so we will often ask you to write not only about the *effects* the media may have had on you but also about the important *uses* you have made of the images, characters, myths, ideas, and stories that the media have offered you.

Your Task

Here, as a way of getting started, we would like you to document and reflect on your media habits and routines.

Possible Approaches

From the time we get up in the morning, when many of us habitually turn on the radio, until the time we retire in the evening, when we may have the television on, we are often in contact with the media in one form or another. Some of these experiences are freely chosen—putting a tape or compact disk in the stereo, for example. Others are chosen but become more or less unconscious because they are so routine—as when we leave the TV on as we do other things about the house or room. For this assignment, we would like you to begin to reflect on your media routines and preferences—what you are in the habit of doing that involve the media in some way.

For the next few days, note down any time you catch yourself interacting with some sort of media text—when you find yourself reading a paper or magazine, or watching TV, or listening to the radio, or scanning the ads on a bus, or windowshopping for clothes or music, or the like. (Our guess is that you will be struck by how much of your daily routine involves the media in one way or the other.) Then, once you have this list, see if you can use it in jotting down some thoughts on your media habits, routines, and rituals. Here are some questions to help you in your thinking:

As a media user, what do you find yourself doing a lot of? Are you a habitual reader, TV watcher, music listener, journal writer, telephone talker? Would you go so far as to say that you're "addicted" to any particular communication experience? Which medium, if removed from your life, would leave a significant absence? Which media experiences do you find most pleasurable, or, to put it another way, the easiest to have? If you could have your way and spend your time as you please, would you spend a significant portion of it at the movies, listening to music, watching TV, reading trashy novels, or flipping through magazines? What media experiences do you find most rewarding? Which ones touch you most deeply?

When and how were your media routines established? Why have you made them your routines? If they were suddenly unavailable to you, what would be missing from your life? What do your media habits say about you? That is, what kind of person do your daily media experiences show you to be? What relationships can you spot among the various kinds of media encounters you have? Is there any common pattern? Or do different media serve different needs? What are those needs? Where do they come from? And what do you think you get from each of the routines and preferences you've established? What do they offer you? Finally, are there any media routines you've abandoned as you've grown older? Why?

Keeping in Touch

One common motive for using the media is to remain *in touch* with the world, or with a part of the world that you choose to follow. We would like you to do some thinking in your journal here about the ways you put yourself into contact with what's going on beyond your personal horizon. To get an idea of what's involved in the assignment, imagine that you had no access to the media—that is, suppose that you could see, hear, and know only those things that happened directly in front of you, only the events you personally witnessed or participated in. Most of us would quickly find this situation intolerable. We would feel cut off in some way, "out of touch." Or we might find life too boring to endure. Or imagine yourself on a hiking trip for weeks with no opportunity for contact with the rest of the world. This situation might well offer a certain pleasure, at least at first. But over time, the chances are that being out of touch—with friends, lovers, human society, world events—would begin to seem more of a deprivation or loss.

Your Task

Compose some thoughts in your journal about the ways in which you remain in touch rather than out of touch, and your uses of the media for this purpose.

Possible Approaches

(Here are some questions to ask yourself in approaching this journal assignment. There's no need to answer them all. Make use instead of the ones that help get you thinking and writing.) Keeping in touch with the world, or that part of the world you keep up with, is a motive we usually associate with the *news* media. You will want to attend, then, to the manner in which you get your news. How does news come to you? Through what sources? And what kind of news is most important to you? Some people, for example, follow the world of baseball with a faithful intensity throughout the season. Others keep up with developments in the stock market, in politics and public affairs, in Hollywood, or in their hometowns. Think about the different worlds or realms of activity you try to stay informed about. How does news of these worlds enter into your world? Of course, there are many different kinds of news, among them "news" of what's happening in your circle of family and friends. How does news of this type come to you?

Once you've established the different devices through which you stay in touch—and the different worlds you try to stay in touch with—you can begin to think more carefully about this experience. What accounts for your desire to follow events, to stay informed, to remain in touch in whatever way you do? When did this desire begin? What is satisfying about staying informed, up on things? Is being in touch a matter of knowing specific things? Or is it more like a feeling of being in contact? Describe this feeling and why it may appeal to you; or, to put it slightly differently, describe what may be disturbing in the feeling of being out of touch. How do you know when you are out of touch? Is there anything pleasureable in *that* feeling? Do you ever prefer to be out of touch? Why? Note that the whole issue of staying in touch has a moral dimension. That is, we sometimes feel that we should stay in touch with certain things, that we have a kind of obligation to do so, and that we are somehow guilty if we neglect to do so. Do you ever feel this way? Where do you think this feeling comes from? What do you do about it?

Try to connect these observations to the media, to the means of communication through which you remain in touch. What properties of the media allow them to serve this function for you? What can you learn about the media (as well as yourself) by looking at the ways they let you stay in touch?

Escaping

In the first journal assignment we asked you to consider the uses of the media to remain in touch. Another common motive for using the media is essentially the opposite of staying in touch. It is usually described by the word *escape.* It is often said, for example, that people read romance novels or watch TV sitcoms in order to "escape." Or, to use a more neutral term, we often say that we go to the movies or watch TV for purposes of "entertainment." We may even describe a media product as "mindless" entertainment.

Your Task

To think about how you use the media for purposes of escape or entertainment, and to consider what those terms mean in your own experience.

Possible Approaches

(Again, there's no need to answer all these questions. They are simply meant to help get you thinking. Respond to the ones that work for you.) You might begin by asking yourself: Which media experiences seem to provide you with an opportunity for escape? When or at what moments do you feel the need to escape? The word *escape* implies a release from a confining circumstance and a flight into a more positive or open one. Thus whenever we talk about the media and escape it is possible to ask: Escape *from* what *to* where?

If you were unable to escape through the media, where would you have to remain? And what's wrong with remaining there? If you do escape through the media in one way or another, where exactly are you after you have succeeded in doing so? What is the new territory like, and what is pleasurable, satisfying, or interesting about being there?

Why is escape desirable? Does the effort to escape typically succeed? That is, in what sense is escape really possible? What features of the media seem to make it possible? Or does the escape typically fail? And if it does fail, how does it fail, and why?

Similar questions can be posed about *entertainment.* Which media do you use for entertainment purposes? When do you typically feel a desire to be entertained, and what is the origin of this desire? How do you know when entertainment has occurred? That is, what's happening to you when you are being successfully entertained? What are the properties of different media products that allow them to serve as entertainment? And what is entertainment, anyway? If you had to define it, what sort of definition would you give? Is there an opposite to entertainment? What brings on this opposite feeling? Is it possible to overdose on entertainment? Or, to put it another way, does entertainment always work? If not, what happens when it fails to work?

Try here to talk about your own experiences with the media. That is, try to give answers to these questions that are true for you, not for the "masses" or some imagined media user. In other words, work from your *own* dealings with the media rather than from your assumptions about what *other* people might do.

Finally, consider that, like the issue of staying in touch, the question of entertainment and escape can have a moral dimension. Is escape something you feel you should (or should not) do? Where does this moral dimension originate? And when does it enter into your experience with the media? Is entertainment something we all should (or should not) seek? How and when does this question of *should* enter into your uses of the media?

Identifying with Music

Few things in our culture have more to do with our sense of who we are (or who we want to be) than the music we listen to and the clothes we wear. Of course the two often go together. Punk, disco, metal, rap, funk, country. The words conjure up not only sounds but images; they evoke not only certain kinds of music but certain kinds of people. This chapter asks you to think about the sorts of music you call your own, and the uses you make of it in forming and expressing who you are.

Your Task

To think about the ways in which you use music, and to try to define what you "gain" by listening to the music you prefer. Obviously this means analyzing your own tastes. But it also means understanding the cultural power of music—its ability to say something to a lot of people while also allowing those people to say something back through their musical preferences. You can think of this assignment, then, as a case study of how music can enter into and affect people's lives—especially your own.

Possible Approaches

(Here are a few ways of thinking about the roles music plays in your life. Use those that seem best suited to your own interests and experiences.) A good place to begin is by trying to determine not only the music you like but the music that you think of as your own. By *your* music we mean those styles, traditions, performers, or recordings that are meaningful to you, that in some way express who you are or suggest who you might be, or that move you in a special way. How did this music come to be yours? When and where did you first encounter it, and how did it find a place in your life?

How does your music fit in with the rest of your life? Does it have anything to do with the clothes you wear, or the places you go, or the people you spend time with? Think about how and where you listen to your music. On the radio? Cassette player? Stereo? MTV? At home? In the car? In the dorm? Are you usually by yourself, or with others? Do you have friends who listen to the same music and talk with you about it? Do you ever read about your music? Do you go to concerts or clubs? Do you play an instrument? Are you a member of a band?

What is it about the music you like that makes it not only rewarding or fun or pleasurable but meaningful to you as well? Is it something in the music itself? Is it something about a particular performer? Does it have to do with image and style? With clothes? videos? places? Or is it a matter of the emotions and attitudes evoked by the music? Often there is something elusive, even mysterious, about music we find compelling. Try to locate and describe this quality in the music you prefer, whatever it might be.

Once you have considered what's attractive or compelling about the music you prefer, ask yourself why you, in particular, are attracted or compelled to it. In other words, what is it about *you* that draws you to this music?

Connecting with Celebrities

This week we want you to think about fame. But rather than approach it from the perspective of the famous person, we want you to think about an ordinary person's relationship to someone famous. That ordinary person is you.

Your Task

To reflect upon your relationship with a public figure or performer, any sort of famous person, whom you "know" primarily through your experience with the media. The figure you select could be an artist, writer, actor, musician, political leader, sports star—anyone toward whom you've had a strong response, or in whom you have a continuing interest. You may consider yourself a fan of this person, or it may simply be someone who has intrigued or provoked you for one reason or another. Your response may be positive, negative, both, or neither, but it must be a reaction you find meaningful enough to explore in this assignment. In other words, choose someone who interests you.

Possible Approaches

(Here are some strategies and questions that may be useful in approaching this assignment. Focus on those that help you most with your writing.) Once you have selected your subject, you can begin to think about your reactions to him or her. What distinguishes this person from others of a similar type? What's attractive, compelling, or noteworthy about him or her? Be as precise and as detailed as you can about this: What is it about this particular person that interests or provokes you? Is it their achievements—the things they have done? Or is it more a matter of style, the manner in which they go about things? Does it have to do with what the person stands for or represents to you? Is it something physical? Or perhaps it's none of these things. What is it, then?

When you think you have a grasp of what it is about your subject that draws you in or provokes your interest, consider the question from the other side: What is it about *you* that causes you to react in the way that you do toward this person?

Next, you will want to consider the means through which you have come to "know" your subject. This means asking questions about the media. Through what medium have your encounters with your subject taken place? How has the nature of this medium shaped the nature of that encounter? (Obviously there are many famous people we know through their appearances in many media—TV, papers, tabloids, books—but it will be useful for you to ask these questions about each medium in which they appear.)

A famous person can often seem both near to and distant from us. You will want to consider both of these aspects. In what ways does this person seem familiar to you, a member of your world? In what ways is your subject a distant figure you can't possibly know, someone who dwells in "another world?" What are the attractions of the nearness we seem to feel towards to the famous? What are some of its signs? That is, how do you experience this sense of nearness or familiarity? Conversely, what are the effects of the distance that always separates us from the famous? And what are the reminders of that distance? When is that distance most obvious? Is there anything attractive or alluring about it?

Finally, try to think of your subject as someone with whom you have a relationship that might be compared, at least in some ways, to those you have with friends, classmates, coworkers, lovers, family members, and the like. What kind of relationship is this that we strike up with a famous figure? What makes it real, human, meaningful? What makes it artificial or empty?

Reading Places

As you continue working in your journal, we would like you to shift your attention outward—toward the places you live, work, and study in, as well as those you simply travel through. We want you to try to "read" some of these places as "texts" (or as collections of texts) in order to see what they tell you about how you should speak and act when you are in them. A schoolroom, for instance, is supposed to offer a different set of messages to its users than is a shopping mall, or a public square or park. And it goes without saying that not all schoolrooms or shopping malls or parks are alike. Rather most places must somehow announce their similarity to others like them (this is a place to study; this is a place to shop) at the same time they also distinguish themselves from those other places (this is an "open" classroom; this is an "expensive" boutique).

Your Task

To reflect in your journal on *places as communication environments.* You can do this by focusing on a particular place in your environment that "means" or "says" something to its users. Then try to "read" the messages of that place.

Possible Approaches

(Use these questions to help get you thinking and writing.) There are any number of places you might want to examine—ranging from the private (bedrooms, dorm rooms, car interiors) to the public (restaurants, subways, malls, stores, bus stops), from places of leisure (ballparks, dance clubs, backyards) to places of work (factories, offices, kitchens), from landmarks (museums, historical sites, municipal buildings) to the barely noticed (elevators, sidewalks, gas stations). We suggest that you pick a place to study that you know fairly well, that interests you in some way, and that you can spend some time observing.

A useful model for analyzing any form of communication is to ask: *Who* is saying *what* to *whom* through what *medium*? For what *reason* and with what *effect*? It can be interesting to ask these questions of a place, even though we don't usually include places among the media of communication. You might begin by looking around a place and asking yourself a simple question: what does this place "say" to those who enter it? What messages does it send? How would you put those messages into words? What devices are used to send those messages? What are those messages about? And who (or what) is doing the speaking? Who is listening? How do the messages sent by a place influence or determine the activity that occurs in that place?

Once you start thinking about these questions, you'll probably discover that the answers get complex. Often, there are many messages, a number of media, and a variety of speakers. Keep this complexity in mind, but try to focus on how the place *itself* sends a message to its inhabitants. For example, the books in a library say millions of things. But the library itself clearly sends messages to its users, among them "be quiet," a message that is not sent by a country-and-western bar. We want you to think not only about the kinds of messages a place can send to the people in it but also about the kinds of communication it encourages them to have among themselves. Try to describe how people interact with each other in the place you've chosen to study, and how the place itself shapes those interactions.

As you read through the notes you've taken on various places, see if you can isolate one or two aspects of communication that seem the most interesting about or distinctive of each. These are likely to be the least obvious aspects of what you've noticed, those that required the most intelligence and work for you to get at. In short, make sure you write down what you've seen that a more casual observer might have missed.

Interpreting Technologies

In this assignment your focus will be on an object—a media device or some other technology that you consider essential to the way you live or to the person you are. Technologies are valued for their practical uses—they help us get things done. Media technologies are particularly valuable because they allow us to see more than we would otherwise see, hear more than we could otherwise hear, learn more than we could otherwise learn. In short, through media we experience things we could encounter in no other way. We also get to manipulate and rearrange our own experiences. But in addition to their practical uses, technologies may cause us to see things differently. Add remote control to your television set and suddenly you may see TV differently—as something you can control continuously by "flipping" through the channels. The very meaning of "television" changes, as does your TV experience. Technologies are like that. They are not only ways of doing, but ways of seeing and defining things.

Your Task

To examine the manner in which a media device or technology shapes not only your behavior but your *outlook*. This means you must consider not only what the device does but what it means or "says" to you—the significance it has in your life.

Possible Approaches

You should begin work on this assignment by selecting for study either a media device or some other everyday technology that you find essential. By a "media device" we mean things like the telephone, the television, the radio, a VCR, a CD player, a portable cassette player, a camera, and so on. Other "everyday" technologies it would be useful to examine include the computer, the fax machine, the automobile, or the microwave oven. But you need not select one from this list. Choose a device that interests you and plays an important role in your life. (It will probably be useful for you to have this device at hand as you write.)

After you have selected a device you find essential and worth examining you can start to ask yourself what you do with it. Initial answers to this question tend to be rather obvious. What you do with a car is drive it, what you do with a microwave oven is to heat or reheat food, and so forth. But if you think a little more deeply about such devices you'll find that much of what you do with them is more complex. It is true, for example, that people drive their cars, but they may also see their cars as an extension or statement about themselves—in other words, they communicate with their cars. Owning a car can also change your relationship to your surroundings. It makes you look at the landscape in a certain way (as something to be crossed), just as owning an audiotape recorder makes you look at your record collection in a certain way (as something to be raided, rearranged, manipulated, etc.). This is the next step in your inquiry: To ask yourself how the device you've selected has changed the way you look at certain things, or altered the meaning of an experience. What is exactly that you get from your device? Is it simply something it *does for* you, or also something it *means to* you? That is, in what ways does the appeal of this device stem not only from what it does but also from what it signifies, not simply from its practical uses but also from the feelings you associate with it? Finally, what's different in your life because of the presence of this device?

Forming a Style

Clothes are a form of communication, and in this assignment we ask you to consider what is being communicated by the way you dress. This is also an opportunity to think about style—your own. Specifically, we want you to examine the way you dress and the significance of the forms of dress you prefer.

Your Task

To think about the ways in which you use clothes to form a personal style—that is, to express yourself and to communicate with others.

Possible Approaches

Obviously clothes have purposes simpler than communication. They keep us warm and cover our bodies. But we rarely choose the clothes we wear for these reasons alone. What you wear is also usually an expression of something—and to understand what this "something" is lies at the heart of this assignment. Of course, clothes don't mean what they mean simply because the person wearing them decides that they do. A quick glance around any classroom will sort people into various groups—preppies, yuppies, punks, dweebs, jocks, wanna-bes, slackers, GQers, and so on. There's no escaping such categories; they're social facts. Even people who "don't care about clothes" invariably care enough about them *not* to look like members of certain fashion groups. "Normal" and "casual" are categories like any other. They have their codes and rules, too, and they require a certain amount of work to be part of. And like any other category or fashion, these codes and rules can be played with, flaunted, resisted, revised, and redefined by individuals. In thinking about your own style, then, consider both the various fashion groups and categories that you may fall into, and also what you do to make certain looks or fashions "your own," expressions not only of group belonging but identity.

You might want to begin simply by cataloguing the clothes you have worn in the past few days. Don't worry if all you've done in those days is go to school or work. Your clothes are still likely to reflect a set of decisions that you've made, at one time or another, about how you want to present yourself to others. So what do your clothes tell other people about you? What do they purposely not tell? How do you think other people "read" your style? What fashion group or category are they most likely to assign you to? What would they miss out on or misunderstand about you in doing so? What part of your style strikes you as the most individual, the most expressive of you as a person? Which clothes do you feel the most or least comfortable wearing? If you could change something about your "look," what would it be?

You might also want to compare your own style to those of other people you know. Who else on your hall or your block or in your class dresses like you? Who doesn't? In what ways do these similarities and differences in dress reflect various ties of friendship or belonging? In what ways don't they? Are there people you *don't* want to look like, styles you avoid? Why?

Or you may want to think about how your clothes express some link to the culture at large. Have you ever come across an article of clothing in a store and thought, hey, that's me? How do you account for this match between a mass-produced product and your sense of your own individuality, your own style? Or have you ever seen anyone in a movie, or a magazine, or a video, and thought, I'd like to look like him (or her)? Or, conversely, have you ever encountered a figure in the media and thought, why are they dressed that way? What's wrong with this person? What's going on in such moments, besides a simple recognition of good or bad looks or good or bad taste? Or, to put it slightly differently, how can you tell that someone has good or bad taste, and what else does that say about them? Even people who "don't care about clothes" usually don't want to be told that they dress badly. So what's the difference between "casual" and "bad"? What's at stake in this difference?

Finally, you may want to consider this paradox: To be "in style" is by definition to be dressed like other people. Yet it is also supposedly a sign of individuality. How can this be? How have you negotiated this tension yourself, in the clothes you choose to wear? Or, to look at it from the other side, one way of being an "individual" would be to dress in ways that are completely *out of style*. Yet almost no one does this. Why not?

Seeing Gender

One way of putting the difference between *gender* and *sex* is to say that the one is learned behavior while the other is simply a given fact. That is, we are all born either male or female, but that does not seem to preclude a nearly unending series of lessons on how to be a man or to act like a woman. Many of these lessons come through the media.

Your Task

To think (in writing) about some of the ways the media are involved in shaping your sense of what it means to be a woman or a man, a girl or a boy, "feminine" or "masculine."

Possible Approaches

Start by locating a text—an ad, a magazine, a song, a book, a TV program— that claims to target you directly and specifically by sex, that is, as either a woman or a man. One approach would be to think about the *information* this text provides you with, the "content" of the "lesson" it offers you in being a man or a woman. What does the text have to say about women or men that strikes you as useful, interesting, or accurate? What seems misleading, exaggerated, false? What qualities or attributes does the text mark off or suggest as being specifically "feminine" or "masculine" ? What emotions, values, or attitudes does it associate with being male or female?

You could also think about how the text *addresses* you as part of its intended audience. How can you tell that this text is aimed specifically at your sex? What does it seem to assume about you and your concerns as a man or woman? How does it want you to respond to it; what does it want you to do or feel or believe? What *tone* would you say it takes in addressing you? Does it make promises about who you might become? Lecture you about who you are? Offer you rules or suggestions about how to do something better? Hint at secret indulgences or escape? Perhaps all of these, or something else entirely? In short, what sort of *role* does this text offer you to play as its reader or audience? What sort of role does it propose for the opposite sex? What kind of person does it ask you to imagine yourself to be? To what degree are you comfortable in doing so? In what ways are you not?

Finally, you may want to consider a more general question: in what ways have media experiences influenced your understanding of gender, your sense of what it means to be "feminine" or "masculine," a man or a woman?

Discovering Politics

In this assignment, your subject is politics, but we do not mean your political opinions or party affiliation. Rather, we want you to examine the ways in which you've experienced politics through the media—not just the news media, but all forms of communication, including reading, word of mouth, and face-to-face conversation.

Your Task

To discover how you've made sense of politics as you've experienced it through the media, and to reflect on that process in your journal.

Possible Approaches

(Here are some strategies and questions that may be useful in approaching this assignment. Not all of them will apply to you. Focus on the ones that help get you thinking.) You might begin by trying to recall your early experiences with politics and politicians. What were these experiences like? Who were the main figures in it? The memorable events? What images or phrases stick in your mind? What did you think politics was about when you were a child? What meanings did it have for you? Did it draw your interest, or did it repel you? Or did it fail to move you at all?

What was the role of the media in forming your early responses to politics? What was the role of family members and friends? What was the role of the events themselves?

How have your attitudes about politics changed as you've grown older? How do you account for these changes? Can you recall a particular moment or event that affected the way you thought about politics? What role did the media play in these changes? Or, to ask a slightly different question, how have your views and uses of the media changed as you've revised your outlook on politics?

When you read reports about politics in the paper or magazines, or hear them on the radio, or see them on TV, how do you typically react? Do you feel yourself a part of, or apart from, the world of politics? Or perhaps a better way to ask this would be: *When* have you felt yourself a part of politics, and *when* have you felt yourself apart from, or indifferent to, or untouched by politics? How did the media figure in these different experiences?

What political figures have you responded strongly to? Are there any political leaders who have inspired or meant something special to you? If so, how have your encounters with this figure been shaped by the media? How would you describe your attitude toward politics and politicians in general? Can you trace these attitudes back to particular figures or events? Again, what role do you feel the media has played in this?

Do you ever find politics absurd, ridiculous, impossible to take seriously? Do you see any ways in which the media encourages such responses? Or do you ever resent politics and its claim on your attention? Has politics ever seemed to break into your experience without permission? How have the media figured in this intrusion?

Feeling Hyped

In most of your journal assignments so far you've been asked to find meaning or pleasure—some sort of positive significance—in your encounters with
the media. But here we would like you to reflect on a negative experience.
By this we mean not simply an experience in which you felt in some way
bored or let down, but one in which you felt that your own identity or value
as a person was somehow being slighted by the media or by something in
popular culture. The slight may be subtle and hidden, or it may be open and
direct, but it should be something you have felt personally—that is, something that "hit" you and left an impression. The point here is not simply to
denounce those media experiences you would prefer not to have—the bands
you dislike, the TV shows you think are dumb, the movies that you don't care
for, and so on. Rather your aim should be to find an experience that bothered
you because it "got" to you in some negative way—not something in the media you'd rather ignore, but something you found hard to ignore. This could
be a film, something you saw on television, an ad campaign, a popular image, a slogan, a magazine cover, a popular style or fashion—any product of
the media that disturbed or continues to disturb you.

Your Task

To reflect (in writing) on a meaningful but negative encounter with the media—
one that invites your attention, but in a way that insults or reduces or simply
bothers you.

Possible Approaches

First you must locate a particular text or media experience that "gets" to you.
Why do you find it difficult to ignore or forget? Is there anything compelling,

perhaps even fascinating about it, despite your negative response? What names might you give to the feeling this experience produces in you? Does it provoke anger, shame, self-doubt, mistrust? Does it reveal a side of yourself you'd rather not see? Is it the feeling of being ridiculed, having your intelligence insulted, your values doubted, your beliefs trashed? Or perhaps it is something more subtle—the suggestion that you're good (but somehow not good enough), that you're smart (and yet ignorant in some way), that you're being flattered (but also manipulated), that you're included (and yet also left out), that your "type" is okay, but not the cool type, the best type, the most glamorous type. Or it may be none of these things. Your challenge is to describe your encounter in a way that defines and clarifies the negative feeling it left in you. What was it that disturbed you so, and what can you learn about yourself and the media from the experience?

Watching TV Watching

Almost everyone has, at one time or another, wondered about the "effects" of watching television. But those effects may differ depending on the circumstances in which television is encountered. In this assignment, we want you to examine what's happening when a group of people watches TV together. Your challenge is to watch the watchers, while at the same time participating in what they watch.

Your Task

To observe a group of people as they interact with each other and the set while watching television, and to reflect on your observations.

Possible Approaches

To do this assignment you must find a group of people (three or more) with whom you can watch television. Your family, a friend's family, a group of your own friends, fellow students in a dorm are obvious possibilities. It might also be interesting to observe a group of children watching a favorite show, or fans at a sports bar during an important game. Any setting where three or more people are watching together is acceptable.

This assignment requires you to switch your attention back and forth from the television set to the group arranged around it. Try to keep both "dramas" in mind at once—what's happening on the screen, and also what's happening *in the room* among the people attending to the screen. Take note, first, of people's behavior as they watch TV. What are they doing besides watching the screen? In what ways do they interact with one another? In what ways do they interact with the set, or with the people on TV? How are their interactions as a group influenced by what they're watching? What sort of talk do you hear as the group watches? How much of this talk relates to what's on television? What is the purpose of this talk?

Once you've observed such a group of television watchers, you might consider some questions about the possible meanings or implications of their behavior. How does watching in a group change the experience of watching television? What does a group "get" by watching together that the individuals might not get by watching alone? What happens to the relationships among people when the television set is on? How are those relationships changed by the presence of a television set in the room?

Readings

In one sense this section forms the heart of this book. The readings collected here offer you a varied and provocative set of approaches to the media, of ways of thinking about their uses and effects. We've chosen these essays because we think they're unusually smart, unpredictable, and stylish—that they're the work of writers who are both worth imitating and worth arguing with. And so the first two kinds of questions that follow each of these selections are designed to help you better understand and make use of what you have read. In **Coming to Terms,** we ask you to take the ideas and phrasings of these critics and make them your own, to begin to make their insights part of your own repertoire of strategies for interpreting the media. In **Reading as a Writer,** we then ask you to look at some of the distinctive moves and gestures these writers make in composing their essays and to locate moments in their work that you may want to imitate in your own prose. But in the third set of questions, **Writing Criticism,** we challenge you to go beyond what these critics have done, to test their views against new texts and experiences, to show the limits and gaps in their work, to push against what they have to say, to add something of your own. In that sense, these readings are not the center or focus of the work we want you to do as a writer and a critic, but only its beginnings.

We imagine that most of your responses to the questions in **Coming to Terms** and **Reading as a Writer** will take the form of **Media Journal** entries, that you will most often use them as a way of thinking through the readings and preparing yourself to discuss them in class. (Although your teacher may also at points decide to use one of these assignments as the basis for a more extended or formal essay.) The **Writing Criticism** assignments are meant, though, as prompts for more sustained, formal, and critical essays—which you may need to take through a number of drafts and which you will be expected to edit and proofread with care.

I Saw God and/or Tangerine Dream

Lester Bangs

I decided it would be a real fun idea to get fucked up on drugs and go see Tangerine Dream with Laserium. So I drank two bottles of cough syrup and subwayed up to Avery Fisher Hall for a night I'll never forget. For one thing, emerging from the subways into this slick esthete's Elysium is like crawling out of a ditch into Jackie Onassis's iris—a mind-expanding experience in itself. A woman there told me that the management had quite soured on rock clientele, and it was easy to see why: here's this cornersteel of cultural corporations, and what staggers into it but the zit-pocked lumpen of Madison Square Garden. And when worlds collide, someone has to take the slide.

What kind of person goes to a Tangerine Dream concert? Here's a group with three or maybe even four synthesizers, no vocals, no rhythm section; they sound like silt seeping on the ocean floor—and this place is sold out. Freebies are rife, yet I don't think that kid in front of me wiped out in his seat got in for nothing. So I ask some of the Tangs' fans what they find in their music, and get a lot of cosmic, Todd Rundgren mulch-mouth. I tell one guy I think they're just a bunch of shit, a poor man's Fripp and Eno, and he looks me over and says: "Well, you gotta have *imagination*..."

Everyone is stoned. Some converse re the comparative merits of various items in the Tangs' oeuvre—one guy declares the double album *Zeit* a masterpiece, another is an *Alpha Centauri* man. Three times as many males as females at least. A thirtyish guy sitting next to me in ratty beard and ratty sweater reminisces about 1968 forerunner Tonto's Expanding Head Band, and tells me about the time the Tangs played the Rheims Cathedral in France. ("6000 people cram the ancient building with a 2000 capacity," boast the program notes.) "They didn't have any bathrooms in the cathedral," he laughs, "so the kids pissed all over it. After it was over the high fathers, monsignors or whatever, said it was the devil and asked for an exorcism of the church."

DJ Alison Steele comes out, a fashion-modelish silhouette in the dimmed green light, and says that the management does not allow smoking in the theatre. As soon as she says her name, people around me scream out, "Eat shit!" and, "You're a prune!" The microphone she spoke through will stand there unused for the rest of the evening, a thin, black line cutting into the pyschemodal otherness of Laserium.

The music begins. Three technological monoliths emitting urps and hissings and pings and swooshings in the dark, little rows of lights flickering futuristically as the three men at the keyboards, who never say a word, send out sonar blips through the congealing air. Yeah, let's swim all the way out, through the Jell-O into the limestone. I close my eyes and settle back into the ooze of my seat, feeling the power of the cough syrup building inside me as the marijuana fumes sift through the cracks in the air, trying to conjure up some inner-eyelid secret movie. Oh lawd, I got the blues so bad I feel just like a cask of Amontillado. Yes, there it is, the swirls under the surface of my life are reconfiguring into: Daniel Patrick Moynihan, caricatured by Ronald Searle. He dissolves like a specter on a window shade, and is replaced by neon tubing writhing slowly into lines and forms until I think it is going to spell out a word, but no, it doesn't quite make it. Goddamn it, I guess I'll have to try harder. On the other hand, maybe no news is good news.

I open my eyes again. Now the Laserium, which I had forgotten all about in my druggy meanderings, has begun to arise from the deep and do its schtick on the screen above the synthesizers. First, a bunch of varicolored clots slowly sludging around each other; they could be anything from badly seeded clouds to cotton-candy cobwebs to decomposing bodies. Then two pristine laser circles appear afront the muck, one red and one blue, expanding and contracting and puckering at each other. They get larger and larger until they are gyrating and rubberbanding all over the place with a curiously restful freneticism. The synthesizers whisper to them as they bounce. The music goes on for a long time, seems to ebb rather than end.

Intermission. Many audience members seem uncertain whether it actually *is* intermission or if they should just pick up their stethoscopes and walk.

Back for more of the same, but more aggressive this time, if that's a way to describe quicksand. The Laserium begins to flash more violently, exploding in dots and points and lines that needle your retinae as the synthesizers suck you off and down and the towering mirrors at the sides of the stage turn slowly, reflecting beams of white light that are palpably irritating but by and gone and by again in a flash. I close my eyes to check into home control, to see if any little twisted-wax visions might be coagulating. Nothing. Blank gray. I open them and offer myself up totally to the Laserium. Flash, flash, flash—the intensity grows until I am totally flattened; I feel like an eight-track cartridge that has just been jammed home. After that, I become slightly bored and restless, although the other bodies around me are rapt. I have seen God, and the advantage of having seen God is that you can always look away. God don't care.

So, finally, picking up my coat and lugging my clanking cough-syrup bottles, I push my way through the slack and sprawling bodies—out, out, out into the aisle. As I am walking up it, I am struck by an odd figure doddering ahead of me, doubled over under raggedy cloth and drained hair. I don't trust my Dextromethorphaned eyes, so I move closer until I can see her, unmistakably, almost crawling out the door. . .*a shopping-bag lady!*

What's *she* doing at a Tangerine Dream concert? Did someone at CBS give her a ticket, or did she find one cast off by a jaded rock critic in some 14th Street garbage can? Never mind—there will be a place for her in the wiring of this brave, new world. I myself had earlier considered giving one of my extra tickets to a wino so he could get a little sleep in a comfortable chair. Look, there's got to be some place to send these whipped dogs so we don't have to look at them, and where better than Avery Fisher Hall? Let them paw through the refuse of a better world, listening to the bleeps and blips and hisses and amusing their faded eyes with the test patterns and static that our great communications combines have no better use for anyway. Just before I left, I turned around for one last taste of the Tangs and Laserium, and by gum, I had my first real hallucination since drinking the Romilar that afternoon: I saw a whole audience of shopping-bag ladies.

COMING TO TERMS

In this short piece, Lester Bangs offers us a report on a concert by a relatively obscure synthesizer band called Tangerine Dream. In addition to the music, which included no vocals, the concert featured a laser light show (courtesy of the "Laserium"). One of the reasons why we chose Bangs' piece is that it's a piece of writing that, we think, compels readers to examine and rethink their reading practices and possibly their expectations of what students may be asked to read in a college composition or writing course. We are interested in the ways in which this text might be said to *frustrate* your previous training as an "academic" reader or to call upon you to read a text in alternative ways. From past educational experiences, readers are often trained to analyze stories or narratives for *plot* or *universal themes,* for instance, or to talk about the value of a piece of writing to the degree a reader might *identify with* certain characters or the author. These are traditional and helpful ways to talk about reading, although to apply them successfully to "I Saw God and/or Tangerine

Dream," we think, might be problematic, to say the least. First, in your journal, retell what happens to Bangs as his report proceeds. How would you break down this story into significant events? Does this text have a beginning, middle and an end? Or how is this piece of writing organized? What is the theme or moral of Bangs' report? What is the plot, the conflict in the story? You may find, in attempting to answer these questions, that this approach to Bangs is not particularly helpful. So, what's the point of Bangs' report, then? Does this text remind you of anything else you have previously read?

READING AS A WRITER

Bangs attempts to describe an inner experience—what was going on in his mind and what he felt like during the concert. The experience Bangs wants to describe is a private one—it happened to him and to no one else—even though Bangs is part of an audience at a concert and is surrounded by people. As he escaped into his mind—with the help of the music, the lights, and the cough syrup he says he consumed as a drug—Bangs drifted further and further from the conscious world, including the world of language. In writing about the concert, he thus faced a difficult task: to describe in language an experience intended, in part, as an escape from language, to bring to the conscious world of the printed page a "trip" into the unconscious realm of the mind, where, as the title states, he "saw God and/or Tangerine Dream."

Go back through the text with a pen and mark those places where you think Bangs is attempting to describe in words an experience that is supposed to be "beyond" words. Which passages, in your view, succeed at describing in language something beyond language? Why do they succeed? Initially you might focus your re-reading of those places in Bangs' text where he tries to render his impressions of the sights, sounds, and smells he is experiencing. As you think about these places in the text, what must Bangs do with or to his own language to make it convey the feel of these sensory experiences? Next, you might look at how dialogue or speech is represented in this text. Who speaks? When people do speak, why might it be significant that Bangs does not respond by talking? How does he respond? Or what does Bangs choose to hear as far as others speaking? In writing about this inner experience, Bangs must refer to the outer world—what was happening around him in the concert hall. How successful has Bangs been in being able to use what was happening around him to suggest something of the quality of his inner or private perceptions?

WRITING CRITICISM

To write or talk about the media is very often to confront the gap between *experience* and *analysis,* between what something *feels like* and what it *means.* It is easy enough, for instance, to quote and write about the lyrics of a song—but it is not always the words being sung or their meanings that most draw us to a piece of music. It is instead something about the voice, the beat, the way the music makes us want to move, to sing along, the feelings it gives rise to. But what is that "something"? How do you describe it or evoke it in writing? Similarly, one way of writing about a movie is to tell what it was "about," to recount its plot, to talk about interesting characters and the theme, or to quote memorable lines from its script. But that often tells us little about what it felt like to actually watch it—about the wash of images, gestures, sounds, cuts, colors, movements, and glances—the very sensation of being in a darkened and silent room, aware that people are in the seats around you, yet somehow alone in an inner world. All that makes up so much of the experience of the film. But, again, how do you go about putting into words an experience that may have little or nothing to do with language? This is the problem Lester Bangs confronts in his piece, "I Saw God and/or Tangerine Dream."

We'd like you to take that as the problem you address in this writing. In most of the other assignments in this book, we ask you to try to gain a critical distance from and perspective on your experiences with the media. But it is hard to criticize or come to terms with your response to a text if you haven't articulated what it is yet, if you can't say what sort of impact the text had on you. So here we want you to try to render or evoke an experience you've had with the media in as full and intense form as you can, to recreate for your reader what it felt like to listen to *that* song, or see *that* performance or event, or to become caught up in *that* movie or book or video. Or you might want to write about a more ambiguous experience, to describe what it's like to have mixed or conflicting feelings about a text—as is the case, for instance, when you feel at once drawn to what an advertisement offers and skeptical of its promises, or when you feel both aroused and repelled (or embarrassed) by the depictions of sex or violence or love in a movie or book. In either case, the challenge and test of your writing will be how strong a sense you are able to give not only of the text you are dealing with—of its meaning and structure—but of your experience of it, of what it felt like to read or watch or listen to it.

Imaginary Social Relationships

John Caughey

On 14 June 1949, there was a shooting in a Chicago hotel. Eddie Waitkus, first baseman for the Philadelphia Phillies, was staying in the downtown hotel with his team for a series with the Chicago Cubs. Late in the evening Waitkus received a call in his room from Ruth Steinhagen, an eighteen-year-old girl who had just checked into the same hotel. She had never spoken to Waitkus, never met him, never communicated with him in any way. But she had to see him, she said, about "something important." When Waitkus knocked on her door, Ruth told him to come in. When he entered, there she was—with a .22 rifle.

"For two years you have been bothering me," she said. "And now you are going to die." And then she shot him in the stomach.

Waitkus was rushed to the hospital. He survived and lived to play professional baseball again. But who was Ruth Steinhagen and what was her motive? At first the press was mystified. Then it emerged that although she had never met Waitkus, she had a strong emotional attachment to him. She was "one of his greatest fans." She had been "in love" with him for two years. In a psychiatric report to felony court, the events leading up to the shooting were reconstructed as follows.

Steinhagen had first noticed Eddie Waitkus in April 1947, two years before the shooting, when she and a girl friend had attended a baseball game at Wrigley Field. At the time Waitkus was playing for the Chicago Cubs and Steinhagen became very interested in him. Her interest developed into a strong emotional attachment or "crush," a highly intense, totally one-sided fantasy romance. Steinhagen frequently went to baseball games in order to see Waitkus play and, with other girls, sometimes waited afterward to watch for him when the players left the stadium. However, she hid herself and never actually approached him to talk or get his autograph. She collected all kinds of media information about him, including press clippings and photographs, and at night was accustomed to spreading these out in a kind of shrine. She slept with one of his pictures under her pillow. She became preoccupied with 36, his uniform number, and because of his background developed a strong interest in things associated with Lithuania and Boston (she wanted baked beans "constantly"). This artificial romance also affected many of her actual relationships. She talked about "Eddie" sentimentally and incessantly with her family and girl friends.

She also engaged in a variety of fantasy relations with him. She talked out loud to pictures of him, dreamed about him, held imaginary conversations with him, and had daydreams about meeting, dating, and marrying him. When he was traded to the Phillies she cried for a day and a night, saying she could not live if he went away.

The relationship went on for many months, but as Steinhagen put it, "After a year went by and I was still crazy about him I decided to do something about it." She decided to resolve the situation by murdering Waitkus: "I knew I would never get to know him in a normal way, so I kept thinking I will never get him and if I can't have him nobody else can. And then I decided I would kill him." Despite shooting Waitkus, she did not stand trial for attempted murder. Instead she was ruled insane—"schizophrenic"—and committed to a state mental hospital.

This shooting was not unique. The years since have seen dozens of attacks on celebrities, ranging from the assassinations of the Kennedys, through the beating of rock star Frank Zappa, to the shooting of Larry Flynt, the publisher of *Hustler*. There have also been thousands of death threats against sports figures, politicians, popular musicians, and other public figures. The problem has become a pattern in American society. Sometimes a desire for personal fame seems to have motivated the assailants, and in other cases political considerations were involved. But in many cases there is also evidence that a fantasy relationship influenced the attackers. This is apparent in two recent shootings.

John Lennon was killed 8 December 1980 by a man he did not know, Mark David Chapman. The only previous meeting between the two had occurred earlier the same day on the sidewalk outside Lennon's apartment. It was a typical fan-celebrity encounter. Chapman stepped from a group of fans and asked Lennon to autograph a copy of his new album, *Double Fantasy*. Lennon complied. "John Lennon signed my album," [Chapman] exulted to Goresh (another fan)...after the Lennons had left. "Nobody in Hawaii is going to believe me." Six hours later, as Lennon returned from a recording session, Chapman shot him dead on the same sidewalk.

Lennon knew nothing about Chapman. They had never communicated, never talked, never met. But to Chapman, Lennon was very familiar. Evidence gathered from newspaper interviews with former friends and acquaintances clearly shows that Chapman had a strong relationship with Lennon—he had been a major figure in Chapman's life for fifteen years. His fantasy relationship has some parallels to Steinhagen's involvement with Waitkus. Here, however, the basis of the imaginary relationship was not romantic love, but admiration and hero worship. It began when Chapman was a teenager collecting Beatles albums and pictures. Lennon became his favorite. As a friend reports, "He tacked John Lennon's pictures on his wall and played the 'White Album' over and over."

Later, the relationship developed into patterns of emulation and identification. Chapman developed a mystical fascination with Lennon's lyrics. At least once he claimed to know Lennon. Several times he signed his name as "John Lennon." He also imitated Lennon. He wore Lennon-style wire-rimmed glasses and took up the guitar. He "yearned to be a rock star like Lennon but lacked the talent." According to a close friend, "I have a feeling that in his mind, he wanted to become Lennon." Others who knew him cited additional evidence: they noted "how he used to play Beatles songs constantly on his guitar, how he taped the name 'John Lennon' over his own on the ID badge he wore as a maintenance man at a Honolulu condominium, how he emulated Lennon by marrying a Japanese woman several years his senior."

Citing the interpretations of psychiatrists and psychologists, the media presented the details of Chapman's relationship with Lennon as symptoms of mental abnormality. Chapman's concern for Lennon was characterized as a "lethal delusion" and an "obsessive identification." Chapman was said to have obscured his own identity by making Lennon his alter ego.

This identification with Lennon was not only seen as evidence of mental pathology; it was also presented as the motive behind the shooting. Because of the identification, several psychological experts suggested that the murder was really a substitute for suicide, a "suicide turned backward." Given Chapman's frustrated musical career, resentment was also said to have been a motive. "Lennon with his expensive apartments and happy marriage represented a symbol of stability and success that the killer himself wanted."

On Monday, 30 March 1981, there was another shooting, this time in Washington, D.C. In an attempted assassination, President Ronald Reagan was shot in the chest and Press Secretary James Brady, a Washington policeman, and a Secret Service agent were gunned down with him. The gunman was described as the "son of a wealthy Colorado oil executive" and a "drifter." At first no motive was apparent. By Wednesday, however, the press reported that the assailant, John Hinckley, had a strange connection to eighteen-year-old movie starlet Jodie Foster. "He did it for her," one source reported. "She's the key."

While Hinckley never met Foster, he developed an interest in her through her movies and other appearances in the media. According to some sources, Hinckley saw *Taxi Driver*—a movie in which Foster plays a teenage prostitute who is saved by a gun-wielding anti-hero—at least fifteen times. He is also known to have assembled media information about her various movie roles and to have carried several pictures of Foster in his wallet.

Six months before the shooting, in September 1980, Hinckley traveled to Connecticut and spent several days in New Haven, where Foster had just enrolled as a freshman at Yale. He hung around her dormitory with other fans and, like many others, sent her love letters. He evidently hoped to meet Foster, but she threw his letters in the garbage. Later Hinckley wrote to Paul Schrader, scriptwriter on *Taxi Driver*, and asked him to arrange an introduction. Again he was unsuccessful. In early March, Hinckley returned to New Haven. During his stay he told a bartender and two Yale students that Jodie Foster was his girl friend. Again Hinckley hung around Foster's dorm. This time he left several love notes, but again she did not respond and again he was unable to meet her. After several days in New Haven, Hinckley returned to Colorado. There, apparently, he developed his plan to assassinate Reagan.

Before the shooting, Hinckley's life had been marginal but unremarkable. He was an ordinary high school student and a college dropout. For a time he saw a psychiatrist in Colorado. He became an unemployed wanderer, alienated from his family, disconnected from any close relationships, and given to traveling about the country and living in cheap hotels.

On the basis of such sketchy and limited information, the media developed a psychological portrait of Hinckley. Through their own descriptions and the quotes of laymen, the press immediately characterized him as mentally ill. He has a "mental condition." He was a "desperate man," a "kook," a "screwball," and a "lone nut." Many experts quickly agreed. Psychiatrically he was speculatively diagnosed as a "pathological case" suffering from "sexual inadequacy," "self-loathing," "isolation," "schizophrenia," and "sexual and psychological confusion." His imaginary relationship with Foster was seen as prime evidence of his mental abnormality. His fantasy relationship was a "fanatic relationship" which went far beyond the normal fan's aesthetic appreciation of a star. It was an "obsession," a "monumental obsession," an "erotomanic obsession." Later, of course, this interpretation was officially sanctioned. At the conclusion of his trial, Hinckley was found "not guilty by reason of insanity."

In fact, the precise nature of Hinckley's mental condition remains controversial and unclear. What is evident is that he had a strong fantasy relationship with Foster, and that this was an important motive in the shooting. Shortly before he shot Reagan, Hinckley wrote a "last" letter to Foster. He made it clear that his attempt to "get Reagan" was designed to impress Foster and to persuade her to change her mind about him. The letter ended as follows: "Jody, I'm asking you to please look into your heart and at least give me the chance with this historical deed to gain your respect and love. I love you forever. [signed] John Hinckley."

All three of these shootings developed out of an intense emotional connection between the assailant and someone he or she did not actually know. This connection was not an actual social relationship—there had been no real interaction—but rather something that parallels a social relationship, a connection which we will here call an *imaginary* or *fantasy relationship*. In each of the cases, the common interpretation is that the assailant was mentally disturbed. Primary evidence for this interpretation is the description of the imaginary relationship which had preceded the shooting. On one level there is justification for this interpretation. On another it is erroneous.

Ironically, these brutal shootings are seen as "pathological" largely because they do not seem to have been motivated by "normal" criminal thinking. If Waitkus had been shot by a jealous girl friend, if Lennon had been killed in a robbery attempt, or if Reagan had been shot by political rivals, the shootings would have been no less tragic, but they would not have been characterized as "abnormal" or "crazy." In the absence of such "normal" motives, the shootings seem senseless and bizarre.

And yet there is a pattern here—one that is not confined to these three cases. As we shall see, imaginary relationships of the kind described above are characteristic of many of those persons who are diagnosed as mentally disturbed in American society. In this sense these fantasy relationships are not merely individual symptoms but part of a widespread psychiatric complex, or *cultural syndrome*, characteristic of many American psychotics.

But there is another, more significant way in which the characterization of these three cases as "abnormal" is problematic. A primary source of evidence for the diagnosis of "pathology" is the imaginary relationships which these three individuals had before they decided to shoot their victims. Would these fantasy relationships have been viewed as pathological if they had *not* led to a shooting? The conventional answer, implicit in the reports quoted above, would be yes. Each of these relationships is described in the language of pathology—as "obsessive," "abnormal," "delusionary," "fanatical," and "schizophrenic." In fact, such an interpretation is untenable. It is probable that these three individuals were in some sense psychiatrically ill before they decided to shoot their victims—but this cannot be determined solely by reference to the prior imaginary relationship.

Some evidence of this can be found within the commentary accompanying the above reports. Hinckley was but one of many fans outside Jodie Foster's Yale dormitory. His love letters were merely a fraction of the many she received each week from unknown admirers. Chapman was but one of the many fans outside John Lennon's apartment, and those he talked with sensed nothing amiss. Much of the background on Chapman is supplied by his long-time friend Gary Limuti; not only did Limuti share Chapman's teenage admiration for Lennon and the Beatles, but he himself became a Beatles-style rock musician. While Ruth Steinhagen's crush on Waitkus was irritating and disturbing to some members of her real social world, especially her father, it was considered entirely acceptable by others, including her sister and girl friend. Not only did they know all about it—and see nothing amiss—but they were involved in similar celebrity love relationships themselves.

This is not to claim that Ruth Steinhagen, Mark Chapman, and John Hinckley were "sane," but only to raise the question whether or not, in this society, fantasy relationships such as these can properly be characterized as abnormal. As we shall see, an examination of the structure and dynamics of such imaginary relationships shows that they are an important, powerful, and pervasive aspect of contemporary American life. Indeed, intense imaginary relationships through fantasy, media, dreams, and the stream of consciousness are characteristic of contemporary American society.

• • •

Interviewed about their interest in a given figure, many fans attempt to explain their attachment by specifying an actual social relationship whose emotional quality is similar to that which they feel to the star—one celebrity will be seen as a "father figure," another as a "sister type" or a "friend." This leads us in the right direction. The basis of most fan relationships is not an esthetic appreciation but a social relationship. Fans have attachments to unmet media figures that are analogous to and in many ways directly parallel to actual social relationships with real "fathers," "sisters," "friends," and "lovers." I will consider two other major varieties below, but I will begin with the most dramatic example, "artificial romance." Here the fan casts the media figure in the role of imaginary lover.

Ruth Steinhagen had a romantic attachment to baseball player Eddie Waitkus. She was not engaged in an "esthetic appreciation" of Waitkus; she was in love with him. The psychiatric report argues that her relationship was abnormal from the start, but this is not the case. All the patterns of behavior that Steinhagen engaged in prior to the time she decided to murder her imaginary lover are standard among those many—probably millions—of normal Americans who have love relationships with unmet media figures. The data for the analysis that follows come partly from published sources, but most of the detailed information comes from my own research with seventy-two Americans who are or have been engaged in such artificial love relationships. Fifty-one of these relationships involved females (ranging in age from eight to sixty-two), all of whom had intense romantic attachments to male celebrities they had not actually met. The objects of their affections included Paul Mc-

Cartney, John Lennon, John Travolta, Donny Osmond, Elton John, Roger Daltrey, Gary Cooper, Frank Sinatra, O. J. Simpson, Bill Bradley, Marlon Brando, Rudolf Nureyev, Richard Chamberlain, Dick Cavett, James Dean, Charlton Heston, Rhett Butler, Bobby Sherman, Jonathan Hart, Clint Eastwood, Robby Benson, Richard Dreyfuss, Cat Stevens, Burt Reynolds, and Robert Redford.

Like Steinhagen, most of my informants explicitly described their relationships in romantic terms. They were "infatuated with," "fixated on," "obsessed with," "crazy about," or (most commonly) "in love with" the favored media figure. Erotic attraction is a basic part of the appeal. "At this time I did not have a boyfriend, so Elton John filled that role. . . . I was in love with him. Even though he was far from what people would view as sexually appealing, I was highly attracted to him." For one seventeen-year-old, it was the eyes. "I fell in love with him [John Travolta] when he appeared in *The Boy in the Plastic Bubble*. I liked his beautiful blue eyes. . . . Although I have somewhat outgrown my teenage fetish, I can't help but think of him from time to time. When I see those blue eyes, I still melt." Another informant, age twenty-one, is even more explicit.

> *It almost takes my breath away to look at him or even think about him, and seeing him magnified on the movie screen makes him look even more like a Greek God. . . . He's elegantly dark, with short, thickly waving hair pushed hurriedly away from his face. . . . His nose is precisely formed, his lips thin but soft and sensitive. Coarse, dark, curling hair covers his chest, arms and legs, and this masculine look is heightened by the contrasting softness of his back and his fingers.*

The intensity of these love relationships is often as strong as that of a "real" love affair. As an Elvis Presley admirer put it, "No one will ever understand how I feel. I love him! He has given me happiness and excitement in my life that will never die."

But aside from these intense feelings, what do such relationships involve? First, they involve regular, intense, romantically structured media consumption. "I saw the film [*Saturday Night Fever*] not only once, twice, but six times." In viewing a dramatic production, the individual sometimes projects herself into the female lead and plays out romantic interactions with her media lover. "I put myself in the wife's role; my personality took over her part." In other cases, the individual personally responds in the role of romantic partner.

> *I lust for him in a daydreamy way when I watch him in the show.*

> *He sang a song he wrote with such feeling and conviction that I was enchanted. The fact that he wrote it indicated to me that those were powerful emotions he personally felt. I wanted to climb right through the screen.*

From the moment he first walked on stage, I was completely enchanted. His movements and speech were spellbinding, and I found that I couldn't take my eyes off him.

One twenty-year-old with a crush on Roger Daltrey of The Who describes her first concert as follows:

As I sat in my seat waiting for the concert to begin, my heart was pounding with excitement. Then the lights went out and The Who appeared on stage. From that point on, I was totally absorbed in the concert. I didn't speak to my friend, who was sitting next to me, nor did I move my eyes away from the stage. I felt as though The Who were playing their songs just for me. I was surrounded by their music. When I left the concert, I felt exhilarated, as though I had actually met The Who in person.

This reaction is described again and again. The individual feels that the TV singer is singing directly to her. Mass media productions—including those in magazines and posters—are taken as *personal* communications.

I would go grocery shopping with my mother just so I could leaf through the various teen magazines that contained heartthrob stories about Donny, who at twelve years old was made out to be a loving, sensitive and caring young man. I, too, covered my walls with posters of Donny, which were inscribed with personal statements such as, "To you, from Donny" or "I love you, Donny." I secretly thought that the messages on his posters and on his records were directed towards me, and often I would become openly emotional over them. For instance, in one of his records he cried out, "Help me, help me please," and I played it repeatedly while crying for poor Donny.

Such relationships are not confined to media consumption. Typically they develop into elaborate patterned forms of behavior that constitute symbolic substitutes for actual interaction. Several parallel the behavior of separated real lovers. We have noted that Ruth Steinhagen collected and treasured all sorts of mementoes and media information about Eddie Waitkus, including photographs and press clippings, and that at night she spread these out in a kind of shrine. Such collecting behavior is standard.

At this time, I decided that I was going to live and die for my man Elton John. I started collecting every album he made and memorized every word of every song. Whenever I found an article in a newspaper or magazine, I cut it out and placed it in a scrapbook. Of course, this was the scrapbook that I would present to him once he met me and fell in love with me. My world became Elton John. I had four huge posters of him in my bedroom.

Steinhagen was preoccupied with Eddie's uniform number and, because of his background, she developed a special interest in Lithuania and Boston. My informants report the same kind of interest in seemingly insignificant details. A Beatles fan put it this way:

> *I relished every little detail I could find out about Paul. If I read that he ate scrambled eggs and bacon for breakfast, I developed a sudden love for scrambled eggs—one food I had always hated. Details of his life became extremely important to me. I wanted to know when he woke up, when he went to bed, what color socks he wore, and if he liked french fried or mashed potatoes.*

More important aspects of the beloved's life assume critical significance. The figure's career successes and setbacks are taken seriously and emotionally, as are personal events such as a sickness in the celebrity's family. Of particular concern are events that affect the individual's role orientation to the beloved. Several informants were distraught by media reports that Elton John was bisexual.

> *Since I was emotionally involved with Elton John, his announcement of his bisexuality was devastating to me. I no longer could view him as a suitable boy friend, lover, or marriage partner.*

> *While reading the newspaper I came upon an article stating that he was gay. No way, I thought. Not my Elton. He just hasn't met me yet. If he knew me, he would become "normal." But then the articles and stories started pouring in from every source about him. Could it be true? The love of my life, gay? I took down all but one poster of him in my room. . . . He had let me down. . . . How could he do this to me!*

A similar, if slightly less distressing, problem occurs when the media figure marries someone else. "I still remember the hurt and loss I felt in my senior year of high school when I was told by a friend that Donny Osmond was going to get married. . . . No, he's a Mormon. He'll *never* get divorced. It really dashed my hopes."

People are not only interested in the beloved's life events; they typically take up parallel activities. A 1974 letter to "Dear Abby" offers an illustrative, if extreme, example.

> *Dear Abby: How can I meet Prince Charles? I have always admired him, and it has been my dream to meet him one day, but I'm not having any luck. I've written him several letters, and each time his secretary has answered saying: "The Prince of Wales regrets that he is unable to meet you."*

> *I am a normal, intelligent 20-year-old college girl. I'm told I am pretty and have a pleasant personality. I've read everything I could find about the royal family in general and Prince Charles in particular. I'll bet I know more about the royal family than most people living in England and the rest of the United Kingdom.*
>
> *I hope you won't think I'm crazy, but I have been taking horseback riding lessons, and I plan to take flying lessons when I can afford it because I know those are Prince Charles' favorite sports. Also, if we ever meet I will have something to talk to him about.*
>
> *Abby, you're supposed to have all the answers. Can you help my dream come true?*

Such activity is often keyed to hopes of an actual meeting. Indirect communications are also often attempted. Many celebrities receive thousands of fan letters from their adoring lovers, but these letter writers probably constitute only a fraction of the star's actual romantic following. While several of my informants thought about writing to their imaginary lovers ("I wanted to write to him to ask him to wait for me until I got older"), few actually did so.

However, *pseudo communications* were standard. Ruth Steinhagen talked out loud to pictures of Eddie Waitkus. Such behavior is typical.

> *This poster [of Bobby Sherman] soon reigned alone on my closet door. I played the album constantly, singing along while staring into Bobby's eyes. Sometimes I would talk to the poster as if it was Bobby Sherman alive in my room.*

> *He [Donny Osmond] had this song about pulling a string and kissing you. Every time I'd hear this song, I'd pull a purple thread and kiss his poster.*

Such external symbolic interactions are paralleled by something even more significant. Steinhagen had fantasies about meeting, dating, and marrying Eddie Waitkus. Far from being extraordinary, such fantasy interactions are *characteristic* of this kind of social relationship.

Certain fantasies occur again and again. One involves "the meeting," an imagined social situation in which the beloved first interacts with his fan.

> *I began to fantasize about meeting Roger Daltrey. I'd imagine myself sitting in a bar or at a pool and seeing Roger Daltrey walk over and sit next to me. We would start small talking and I would casually mention that he was one of the members of The Who. He'd smile shyly and say, "Yes."*

Of course, the celebrity does not just notice the fan. He is smitten. One girl imagined how her beloved would spot her in a restaurant and, struck by love at first sight, would "send me a single rose with a note."

Another young woman, a law student, regularly conjures up a series of alternative meetings with O. J. Simpson.

> *Prince Charming crashes into the back of my brand new silver sports Mercedes with his sleek, red Jag. He races to the front of my car with every intent to curse me out, but once he notices the twinkle in my eyes, he asks me out for dinner instead.*

> *This time O. J. Simpson is still active in the NFL. As he runs a fly pattern and makes the game-winning touchdown, he crashes into me and my cameras in the end zone. As he helps me up, he notices the twinkle in my eyes and I notice his twinkle. We fall in love and live happily ever after.*

> *We sometimes meet because he needs help. On these occasions, I am the best lawyer in town. . . . a willing soul ready to do battle.*

A fourth scenario involves an even more dramatic rescue. Ironically, this fantasy was related to me less than two weeks before John Lennon's death:

> *As a bystander in a crowd listening to O. J. Simpson give his farewell football speech, I notice a suspicious-looking man. He is about to shoot Simpson. Just in the nick of time I leap between Simpson and the bullet, saving his life but critically wounding my own. As I fall to the ground the six-foot-two-inch former running back catches me in his arms and gently places his coat under my head as he rests me on the ground. As a tear runs down his lean, smooth face, he says nothing and kisses me just as I close my eyes. Now I turn into the suffering heroine. Others realize the bravery and courage it took for me to risk my life. People from all over the world visit me in the hospital, and of course, immediately after a full recovery, O. J. Simpson and I get married and live happily ever after.*

While many fantasies focus on the meeting and courtship aspects of the relationship, others picture the marriage that follows.

> *At that time I imagined I could communicate with Paul. I had long conversations with him. I imagined meeting him and having him fall in love with me. Of course, he begged me to marry him and we lived happily ever after. . . . He had to go away for concerts but he was always true-blue and loyal to me, of course. . . . He always came home from trips and told me how much he missed me.*

Sometimes people seek to translate such fantasies into real meetings. Several of my informants not only hoped but expected to meet their lovers. Occasionally people actually succeed. "Groupies" represent a culturally recognized category of successful celebrity seekers who have managed to turn a fantasy attraction into a "real" relationship. Several of my informants made token efforts to meet their lovers by attending a "personal appearance." One of my Elton John fans gives a characteristic example.

> *Somehow time flew, and the night of the concert was here. I looked beautiful as I left the house, smiling from ear to ear, knowing that somehow this would be the night. Maybe he would see me in the audience and fall magically in love and invite me on stage to do a song with him. Or maybe he would spot me and send one of his guards for me. It didn't matter how, it just had to happen!*

Relatively few people seriously attempt a meeting. Tacitly, at least, they often seem to realize not only that an actual meeting is virtually impossible, but that the fantasy is better than an actual encounter.

> *Although I dream of Prince Charming falling in love with me, I know in my heart that he never will, that I will never meet him and much less hold him. All this doesn't matter. What does matter is the creations of my imagination.*

Most, while recognizing the unlikelihood of an actual meeting, retain a ray of hope. Recalling her adolescent crush, one young woman said, "I'd still like to meet him; I'd be perfect for him."

Romantic fan relationships are common among adolescent girls. An informant who grew up with the Beatles noted that in her day it was considered odd *not* to have a crush on one of the Beatles. Tacit cultural rules make it more acceptable for young females overtly to express such relationships than for people of other age or sex categories to do so. Nevertheless, such relationships are not confined to adolescent girls. A letter to a TV "answer man" makes this point forcefully: "I'm an adult so don't treat this question lightly, please. I am absolutely in love with Richard Thomas.... I don't mean a crush; I mean love. I think about him every waking moment of every day and I must write and tell him. Please, his private address so I can tell him how serious I am." This letter is not unique. Many of my informants were well out of their adolescence, and one of my John Travolta fans was in her fifties. The following enjoyed by aging celebrities such as Frank Sinatra and Liberace demonstrates that such fantasy love relationships can extend far beyond adolescence. An elderly suburban matron with grown children still has a "special thing" for Frank Sinatra. When depressed, she pours herself a drink and listens to his romantic songs. Her attraction began some forty years ago. Like many of her age mates, she has had, in effect, a lifelong affair with Sinatra.

Men also have such relationships. John Hinckley is not an adolescent girl, nor was his relationship to Foster unusual. Although they tend to be more secretive about them, few of my male informants completely denied ever having had such relationships. Some of them described such attractions in detail. Among the female stars they have "loved" are Bonnie Raitt, Jane Fonda, Donna Summer, Princess Caroline of Monaco, Brooke Shields, Diane Keaton, Ann-Margret, Cheryl Tiegs, Joan Baez, and Yvette Mimieux. One man reported a series of such relationships extending through much of his adolescence. As this example suggests, most of the patterns described for female love relationships are paralleled among males.

> *I would read all I could about the actresses I had crushes on in* TV Guide, Life, Look *and whenever I visited my grandmother, in* TV Screen *and* Movie Mirror. *Yet all these crushes pale in comparison to my love for Yvette Mimieux. I had a crush on YM for a long time—at least four or five years. It all started with an episode of* Dr. Kildare *called "Tyger, Tyger," in which YM played a devil-may-care Southern California surfer girl. From that time on, I watched every show that YM was in that I possibly could. I would imagine meeting her in one way or another. I imagined her coming to a basketball game in which I played really well, or would imagine heroically saving her life, sometimes sacrificing my own in the process—but never before she had a chance to kiss me and thank me.*

Again, such relationships regularly extend beyond adolescence. Another Yvette Mimieux fan was only "turned off" in his mid-thirties, when he was disappointed by her "cruelty" in the movie *Three in the Attic*.

Homosexuals, too, engage in artificial love relationships, such as the TV newswriter who had a long-term crush on a black ballad singer. He found the singer "physically attractive and appealing," followed his career with interest, and kept several scrapbooks on him. He bought all the singer's records and collected information from friends who knew him personally. He was "overjoyed" to learn that the singer was gay, and then began "to fantasize in earnest."

Ethnographic investigation shows that elaborate love relationships with unmet media figures are not characteristic just of American schizophrenics. On the contrary, such relationships represent a significant, and pervasive, culture pattern in modern American society. But *why* are Americans given to these relationships?

Both the forms and the contents of the contemporary American media are conducive to the development of such relationships. Especially through the vehicle of the electronic media, the individual is regularly transported into the midst of dramatic social situations involving intimate face-to-face contact with the most glamorous people of his time. The seeming reality of this experience naturally engenders emotional reactions—especially since these figures are deliberately and manipulatively presented in the roles of sexual objects and lovers. Given their intimate, seductive appearance, it would be peculiar if the audience did *not* respond in kind.

But in addition, these relationships often fill gaps in the individual's actual social world. As Elihu Katz and Paul Lazarsfeld observe, escapist media often serves as a direct substitute for socializing activity. If the social situation is dissatisfying, an individual may compensate with artificial companions. When a person accustomed to company at dinner must dine alone, he or she typically substitutes artificial beings by way of a book, newspaper, or TV program. This suggests that an individual would be most likely to engage in a media lover relationship when he or she is without a real or satisfying actual lover. While sometimes valid, such an interpretation does not fully suffice. It does not explain the suburban grandmother who had a lifelong "affair" with Frank Sinatra despite forty years of marriage. It does not explain why, in adolescence and in later years, artificial love relationships often persist after an actual lover is found. Take one of the fans introduced earlier. This young woman, like several other informants, explicitly offered the "substitute lover" interpretation herself:

> *Donny Osmond served to replace a missing element in my life. At this time I had little relationships with other boys. My sister and many of my friends did have these relationships. Donny and I had such a relationship, and I had yearned for one.... My feeling eventually began to fade as I did begin to have crushes on real boys in sixth and seventh grade.*

However, as she herself remarked, the relationship with Osmond "faded" but did not end. It persisted through her senior year in high school and even now, at age twenty-two, she still buys Donny Osmond magazines and retains her fan club card. While she is now engaged in serious relationships with real lovers, she still has various artificial love relationships. "Now I have the same kind of relationship with Buck Williams [basketball star]. He's sooooo great, wow!"

In its simplistic form, the "substitute lover" theory assumes that the imaginary lover will disappear when the real lover appears. Tacitly it assumes that real love relationships are better than imaginary love relationships. The first assumption is often untrue, and the second may also be unwarranted. In some ways fantasy relations are often *better* than real love relationships. Media figures are more attractive than ordinary mortals, and they are carefully packaged—through makeup, costuming, camera angles, and film editing—to appear even better. Even when portrayed in a tired or disheveled condition, they seem cute, humorous, and "sexy." The media figure's prowess is typically almost supernatural. The hero is so strong and brave that the villains are always overcome ("His shoulders and arms are massive—secure comforters that are capable of doing away with any problem..."). The media figure's personality is also carefully sanitized, glamorized, and perfected. The continually brave, kind, interested, patient, and passionately devoted lovers of many media worlds are not to be found in reality. Celebrities are also more "successful" than ordinary people. This connects to a powerful American value. Extreme material wealth taps yet another basic American interest. Finally, since the figure is worshipped by millions of others, the star has a legitimacy and appeal lacking in a person who is not "somebody."

But it is fantasy that helps to make these relations superior to actual social interactions. To love a glamorous rock star through the media is to make a kind of intimate contact with a powerfully appealing figure—but you remain, in this dimension, only one of millions of other fans. However, through fantasy, the rock star moves from the public realm and picks you out of the crowd to be his special friend, lover, and wife. Landing such a widely adored figure confirms your self worth; it makes *you* somebody. It is analogous to landing the most popular boy or girl in high school, only better. One Donny Osmond fan put it this way:

> *I fantasized and dreamed about him constantly, thinking that if Donny ever bumped into me somewhere or met me, he would immediately fall in love with me and whisk me away. I wasn't really jealous of the other girls who liked him because I knew that if he met me he would forget about them.*

One of her rivals had similar plans.

> *I felt we were meant for each other, and if only we could meet, all my dreams would come true. Since Donny was a famous superstar, this gave me a feeling of superstar worth also.*

Sometimes this gratifying elevation of the self is based solely on becoming the beloved figure's chosen lover or spouse. In other cases the individual's fantasy includes stardom for himself or herself as well. Like several other informants, an Elton John fan pictured herself literally sharing the stage with her idol:

> *I saw Elton as the necessary contact I needed to break into the music business. I would frequently fantasize that Elton would hear me play the piano, fall in love with me, then take me on tour with him as an opening act. These fantasies were always very elaborate and intense and provided me with a way to fulfill all my dreams and desires.*

Through fantasy, a media love relationship is exquisitely tuned not to the needs of the celebrity, but to the needs of the self. Imaginary lovers unfailingly do what you want with grace, enthusiasm, and total admiration. The whole course of the relationship is under your control. The relationship runs—and reruns—its perfectly gratifying course from dramatic and glamorous first meetings, through courtship and consummation, to happily-ever-after marital bliss. And through it all a fantasy lover smiles fondly, never complaining, never burping, never getting a headache, never wrecking the car or making you do the dishes. You owe her no obligations. He is there when you want him and gone when you do not. Real love relationships include all sorts of unfortunate realities: fantasy love relationships do not. It is not surprising that Americans sometimes prefer fantasy lovers to ordinary mortals.

A second media figure relationship, which also directly parallels an actual social relationship, is based on antagonism. Most informants can readily list media figures they despise. Controversial public figures like Howard Cosell, Jane Fonda, William Buckley, George Steinbrenner, Richard Nixon and villainous fictional characters like Alan Spaulding ("Guiding Light"), or JR ("Dallas") are commonly mentioned examples. Considering that the individual has never met these people, and that many of them are fictional beings, the level of hostility is often astonishing. Sometimes these negative feelings lead the individual to elaborate an artificial social relationship that is the inverse of the stereotypic fan relationship. Here the basis of the relationship is not esthetic appreciation, admiration, or love, but hatred, anger, and disgust.

As with love relationships, media consumption is often intense. Several informants have developed antagonisms to local talk show hosts, and sometimes they watch such shows for the pleasure of hating the celebrity. Soap opera fans are often as interested in characters they dislike as those they admire. "He is one of the most self-centered, arrogant, selfish, uncaring persons I have ever come across," said one informant. "His main purpose is to have power at the cost of others. . . . I watch him with feelings of hatred. . . . I am overjoyed to see other characters beat him at his own game." One set of three young men used to come together out of their mutual dislike of a TV evangelist. Their mutual relationships, both with each other and with this media figure, suggested the opposite of a fan club. They used to watch this evangelist's television show regularly because they thought he was an amusing "farce." When he appeared on the screen they would laugh and ridicule him.

Such relationships often have an important fantasy dimension. Serious participants conjure up fantasy meetings with their enemies and carry out imaginary arguments with them. Sometimes these pseudo interactions become violent. An otherwise peaceful informant described fantasies about torturing and killing "evil" politicians. These pseudo interactions, both in media consumption and fantasy, allow the expression of a hatred that is more extreme and presumably more satisfying than that which can safely be expressed in real social relations. The media figure's humanity can be denied in a way that a real person's physical presence makes difficult. One runs no risks of legal punishment or retaliation since the hated figure cannot fight back.

These relationships sometimes goad people to attempts at actual consummation. It may seem amusing that an actor who plays a villainous soap opera husband is regularly stopped on the street and berated for treating his "wife" so badly. Unfortunately, serious attacks sometimes result from such imaginary relationships.

A far more common and significant group of relationships are those in which the media figure becomes the object of intense admiration. Such relationships approximate the general stereotypic conceptualization of the fan, but much more is involved than esthetic appreciation. Characteristically, the admired figure comes to represent some combination of idol, hero, alter ego, mentor, and role model.

The media figures around whom informants have built intense admiration relationships are surprisingly diverse. Consider the following examples: John Wayne, Judy Garland, Loren Eiseley, Steve Carlton, Betty Ford, Neil Young, Ralph Ellison, Charles Manson, Frank Zappa, Jane Fonda, Arnold Palmer, Hawkeye Pierce, Jack Kerouac, Barbra Streisand, Woody Allen, Anne Frank, Bruce Springsteen, James Dean, Olivia Newton-John, James Bond, Diana Rigg, Tony Baretta, Isak Dinesen, Clint Eastwood, and Mary Tyler Moore. People express strong emotional orientations to such figures, speaking not only of "admiration" and "sympathy," but also of "worship" and (platonic) "love." Again people frequently characterize the attraction by comparing it to a real social relationship. They speak of their hero as a "friend," "older sister," "father figure," "guide" or "mentor." As with the love relationships, the general source of the appeal is clear. Media figures are better than ordinary people. They have godlike qualities that are impossible for mortals to sustain. Furthermore, the emotional attachment is not complicated by the ambivalence that characterizes actual relationships; admiration is unchecked by the recognition of faults and limitations.

One man pointed out that the "father figure" he admired—a fictional John Wayne-type TV cowboy—outshone his real father in every respect. His father has several admirable qualities and he "loves him very much." But as a child he "needed someone to identify with," and his father did not measure up. A young woman described the perfection of her "TV mother," Mary Tyler Moore, and showed how Moore shifted from parent figure to role model.

> *It all began when I used to watch "The Dick Van Dyke Show" about eleven years ago, so I was ten years old at that time. The first thing I can remember from that time is how much I admired Mary; after all, she was slender, feminine, funny, talented, intelligent, cute, attractive, a kind mother, a loving wife, a caring friend, a good cook, a clean housekeeper, and more. . . . Mary became the person that I wanted my mother to be. . . . I would watch my mother cooking, for example, and I would imagine that my mother was just like Mary.*
>
> *Things really took a turn when I got into junior high school about eight years ago. "The Mary Tyler Moore Show" had been on TV for about three*

years by that time, but I didn't pay much attention until I realized that I wanted to learn how to become a woman for myself. . . . and I had the perfect person to model myself after: Mary. On her show she was a career woman and still as perfect as before; she dressed well, she was slender, she knew how to cook, she was independent, she had a nice car, she had a beautiful apartment, she never seemed lonely, she had a good job, she was intelligent, she had friends, etc. There was still nothing wrong with her.

From out of thousands of glamorous alternatives, why does the fan seize on one particular figure rather than another? The appeal is often complex, but the admired figure is typically felt to have qualities that the person senses in himself but desires to develop further. The admired figure represents an ideal self-image. Of course, it is sometimes difficult to establish whether the similarity existed prior to the "meeting" or whether it developed after the media relationship flowered. Sometimes, as in the following account, both factors are involved. As a high school student at an overseas international school, this informant did not have access to the most current American television fare.

I became familiar with the character Tony Baretta of the TV police series when newly arrived friends from the United States started to call me "Baretta." I learned that I looked a good deal like him—short, stocky, and black hair. The fact that I was from New York and had a tough guy reputation also helped.

Eventually the series came to Belgium and I saw who this character was. I was not displeased at our apparent similarity. He portrayed a hard but sensitive character, a heavily muscled tough guy who at the same time could be counted on to help people through thick and thin. I also enjoyed his speech, attitudes towards smoking, and life in general.

Once I started looking at the series, my behavior, largely because of peer expectation, started to resemble that of Baretta's more closely. . . . The next day after the series had been on friends and fellow students would say things like, "Man, I caught you on TV last night." Soon most people called me "Baretta" or "Tony" and I was sort of expected to play this character. . . . Often I would dress like him. Tight, dark blue T-shirts were a favorite during the warmer months. Naturally I was soon anxiously awaiting the weekly airing of the series and soon had assimilated many of his mannerisms. Although I never sounded like him, I do say many of his phrases: "You can take that to the bank," "Don't roll the dice if you can't pay the price," etc. . . . The TV character probably helped cement my character into the mold that is now me.

As this case also suggests, the media figure's appeal may be linked to the individual's actual social relationships. Here others made and supported the initial identification ("You are Tony Baretta"). One young woman's fifteen-year fixation on a soap opera character began with her social situation as a first grader. Arriving home from school she wanted to be with her mother, so she would sit down with her and watch "Guiding Light." A young golfer's attraction to Arnold Palmer developed in part because Palmer came from a neighboring town and because his father, also an avid golfer, deeply admired Palmer. Many people have been turned on to a given novelist or musician by a friend who admired the figure and gave the person his first book or record as a present. In some cases the initial interest may be insincere. One young man, charmed by a female Bob Dylan fan, feigned appreciation of this musician in order to ingratiate himself with the woman. In the process he gradually developed a genuine interest in Dylan that continues up to the present—long after the woman has gone.

Once the initial identification has been made, patterned forms of behavior typically develop. As with artificial love relationships, the individual typically collects totemlike jewelry, T-shirts, locks of hair, photographs, posters, first editions, records, tapes, newsclippings, and concert programs. Media consumption is also likely to be intense. The individual reads the author's work repeatedly ("I read the novel five times"), or travels long distances to attend a concert. During media consumption, personal involvement tends to be sympathetic and emotional. "From the minute she stepped on stage I was in a trance. I was mesmerized by everything she did and I could actually see myself doing the same singing.... It actually gave me the chills." As this passage implies, identification is usually significant. Sometimes it is partial ("When I watch the show I think to myself how much I would like to be like her"); sometimes it is complete ("When I see her I don't see a TV character, I see myself").

Admiration relationships with media figures also have an important fantasy dimension. As with love relations, fantasies link the individual *socially* with the admired figure. Three types of fantasies are especially common. In the first the individual meets the idol, in the second the person becomes someone like the idol, and in the third he or she becomes the idol.

Fantasy encounters are not mere meetings. One does not just shake the celebrity's hand and move on. A close and intimate social relationship is established. Several informants played out elaborate scenarios with writers whose work they admired. One man imagined journeying to the reclusive writer's home, becoming fast friends, and going out on drinking bouts full of brotherly adventure. A young woman liked to imagine herself visiting Loren Eiseley. "I visualized the two of us drinking coffee while sitting at a comfortable old kitchen table in his house. We would talk about philosophy, time, exploring damp caves, and reactions to nights in the country."

A standard variant of this fantasy involves establishing a professional role relationship with the admired figure.

> *Steve Carlton is a six-foot five-inch 220-lb. pitcher for the Philadelphia Phillies. Since I myself am a frustrated athlete, I admire Carlton. He was given a*

great body and he has made the best possible use of it. Carlton possesses all the physical attributes I wish I had.... Carlton is also an expert in martial arts and is intellectual, having studied psychology and Eastern philosophy. I admire the man because he is the type to accomplish tasks and achieve goals.

This fan has several fantasies about establishing a social relationship with his idol.

One has me in the big leagues pitching for the Phils and Carlton becomes my friend, takes me under his wing, and I become his protégé. I have also imagined that one day I will become Carlton's manager and that we have a great relationship. He admires me for my managing and coaching ability.

Another variant involves becoming, imaginatively, a close relative of the admired figure. One middle-aged woman, now herself a professional writer, described her attachment to Isak Dinesen. As a child she read and reread Dinesen and sought out all the biographical information she could find. She not only spent much time "reliving" scenes from the writer's adventurous life—"art school in Paris, her marriage to a Swedish count, the beginning of her African adventure..."—but she also constructed fantasies about growing up in Africa as Isak Dinesen's daughter. The dynamics of such fantasy interactions involve several different components but, again, a crucial dimension is the elevation of the self. Acceptance by the admired figure is the vehicle for indirect self-acceptance. (I admire them—I am like them—they accept me—I am a good person.)

In the second type of fantasy, the individual becomes someone like the admired figure and lives out experiences similar to those of the idol. An admirer of Jack Kerouac frequently imagined himself hitchhiking across the United States and Europe. Often a curious new self emerges, an imaginative combination of the fan and the admired figure. The following description comes from an admirer of *Gone with the Wind*.

In my fantasies I see myself dressed in one of those beautiful hoop skirt dresses complete with parasol, hat, and fan—making me your typical Southern belle. I would speak in a sweet Southern accent. I would belong to a very wealthy Southern family with a large plantation stretching out for miles. My family would entertain frequently and I would grow up becoming very well raised in the social graces. As a woman, I would be very at ease with others, hospitable, and charming in my relationships with men. In my fantasies I look like myself, and I still have some of my peculiar habits and personality traits, but I do take on most of Scarlett's charming attributes—her vivaciousness and way of dealing with people, the way she carries herself, etc.

One step further, in the third type of fantasy, the fan abandons his or her self and *becomes* the media figure. This is a common desire ("I would change places with him in a moment," "If I could be anyone else it would be her"). In fantasy the desire is realized. A young man studying to be a doctor reported a special relationship with James Bond, agent 007. In reality he sees himself as very different from Bond. The young man is "afraid of decisive action," Bond is cool and calculating; he "thrives on security," Bond gambles; he "feels compelled to ask for advice and permission," Bond is independent; he is "sexually naive," Bond is a Don Juan. However, through his artificial relationship a transformation of self occurs. In his media consumption and fantasy he sloughs off his own "inadequate" personality and turns into James Bond. Here, in fantasy, the individual's consciousness is "possessed" by the media self; it colors perception, patterns decision making, and structures social behavior.

Such an influence is not confined to fantasy. In the actual world as well, the admired figure sometimes guides actual behavior. First, the fan may adopt the idol's appearance. One teenage fan went to the barbershop with a picture of his hero and asked the barber to "cut my hair just like Fabian's." Writer Caryl Rivers describes a series of imitative identifications: "I had a Marilyn Monroe outfit—off-the-shoulder peasant blouse, black tight skirt and hoop earrings.... When Grace Kelly came along, I swept my hair back and wore long white gloves to dances and practiced looking glacial. I got my hair cut in an Italian boy style like Audrey Hepburn. For a while I was fixated on lavender because I read that Kim Novak wore only lavender." She goes on to describe how facial expressions were also borrowed and incorporated.

> I wanted the kind that I saw in movie magazines or [on] billboards, the kind that featured THE LOOK. THE LOOK was standard for movie star pictures, although the art reached its zenith with Marilyn Monroe. Head raised above a daring décolletage, eyes vacant, lips moist and parted slightly—that was generally agreed to be looking sexy. I practiced THE LOOK sometimes, making sure the bathroom door was locked before I did.

Such relatively superficial influences are often part of a deeper identification in which the media figure's values and plans are incorporated into the fan's social behavior. This is mentally accomplished in ways that directly parallel fantasy interactions. Operating in his or her own identify as someone who wants to act like the ideal figure, the individual may employ the media figure as a mentor or guide. Probably this is often unconscious. But sometimes the individual deliberately turns to the guide for help.

> Especially when I am upset, I think about her [Anne Frank's] outlook. I look to her as a reference. Then I try to act like she did.

Woody Allen plays the role of my pseudo conscience. He discriminates what is silly and irrelevant from what is worthwhile.... In awkward situations, I tend to rely on what I've seen him do or say.

A Mary Tyler Moore admirer shows how such a process can move beyond advice to deeper forms of self-transformation.

I used her personality in my... anticipations about myself, especially when I had a problem. I would find myself thinking "What would Mary do?"... I would imagine myself in the situation that I wanted to be in... and also that I was exactly like Mary, that I had her sense of humor, her easygoing manner, etc.... This was a daily occurrence for me until about five years ago.... It was a way to solve my problems or to guide my behavior according to a model of perfection; a natural result was that I was pleased about how I conducted myself, especially with other people. I ended up regretting my behavior a lot less when I behaved as I thought Mary would behave. As a teenager I was very concerned about meeting role expectations, but with Mary inside my mind, it was much easier to meet those expectations.

As in certain fantasies, the fan here *becomes* the idol. Occasionally the fan consciously seeks to induce this self-transformation. "Sometimes I find myself saying, 'You are Carlton. Believe you are and so will everyone else.'" Here the fan overtly acts according to the hero's values, goals, and plans.

Because of such identification, media figures exert pervasive effects on many different areas of actual social life. Sports provide a good example. Many—probably most—American athletes are affected by media exemplars. Even the most successful professional athletes regularly report such role modeling. O. J. Simpson, for example, deliberately patterned himself after Jim Brown. This kind of imitation is so common that sports interviewers regularly ask their subjects about this aspect of their sports development. Many of my own informants reported on this process. The young golfer whose hero is Arnold Palmer reads all the Arnold Palmer how-to-do-it books, studies his games on television, practices his style of shots, and pretends to be Arnold Palmer while actually playing golf.

A successful college basketball player described her close relationship to a favorite professional player.

Many times I sit and stare at a basketball court and imagine Bill down there warming up with a series of left- and right-handed hook shots...and still there are other times when after seeing him play on TV I imagine myself repeating all his moves.... On the court I pattern many of my offensive moves from Bill Walton.

As is typical with such artificial relationships, the imitation of sports behavior is part of a more profound influence. A young tennis star shows vividly how his pattern of play is but one manifestation of a larger imitation of values and lifestyle.

> *I'm extremely aggressive on the court, I guess. I really like Jimmy Connors's game. I model myself after him. I read somewhere that he said he wants to play every point like it's match point at Wimbledon. . . . I like the individualism of tennis. It's not a team sport; it's an ego trip. You get all the glory yourself. That's what I thrive on, ego. . . . I want to become No. 1 in the world and become a millionaire. . . . I want to become like Vitas Gerulaitis, with the cars, the shopping in Paris, and the girls.*

As American idols, media figures do not influence merely one narrow aspect of life. They influence values, goals, and attitudes, and through this they exert a pervasive influence on social conduct.

Given such evidence, it may seem odd that the existence of imitative media effects is still debated. The focal area of research here has been antisocial conduct in general and violence in particular. While most researchers now conclude that violent media promotes imitative actual aggression, a few deny this. They argue that violent TV fare "evaporates" or that it has a "cathartic," pacifying effect on the audience. One problem involves the behavioristic methods by which much of the research has been carried out. Many researchers seek only to assess the immediate observable effects of media consumption, as by experiments in which test children are shown a film of violence and then observed and rated as to level of aggressive action afterward. Some such studies reveal imitative effects, others do not. However, this approach ignores a crucial intervening variable, the individual's consciousness and mind set. How a viewer reacts to a given media figure's example may critically depend on the kind of artificial-imaginary social relationship he or she has established with the figure. A closely related variable is the effect the communications have not on immediate overt behavior but on the individual's consciousness and knowledge. Just because the individual does not immediately engage in imitative violence does not mean that he or she has not been affected. When the media figure is a personal hero, the individual's tendency toward violence may have been confirmed or increased. The individual may also have learned new ways of attacking, fighting, shooting, stabbing, or torturing other people.

People vividly remember the aggressive techniques of their heroes and they see an astonishing number of examples of aggression. One estimate suggests that between the ages of five and fourteen the average American child has witnessed the violent destruction of 13,000 human beings on television alone. Furthermore, Americans often rehearse what they have witnessed in the media. American children regularly imitate the violent antics of their media heroes in play, sometimes immediately after media consumption, sometimes days, weeks, or even months after the original exposure. Playing *Star Wars*, playing "Guns," playing "Cowboys"—these media-derived games are among their most common pastimes. Children and adults also "practice" violent routines by adopting the hero's persona and playing out scenes of violence in their fantasies. The violence of American inner experiences can often be directly linked to particular media productions. Media violence does not evaporate. It affects attitudes, enters the stock of knowledge, and is acted upon in fantasy. It is thus readily available—in practiced form—for use in actual conduct as well. One American criminal who has spent fifteen years in jail spoke as follows:

> *TV has taught me how to steal cars, how to break into establishments, how to go about robbing people, even how to roll a drunk. Once, after watching a* Hawaii Five-O, *I robbed a gas station. The show showed me how to do it. Nowadays [he is serving a term for attempted rape] I watch TV in my house [cell] from 4 P.M. until midnight. I just sit back and take notes. I see 'em doing it this way or that way, you know, and I tell myself that I'll do it the same way when I get out. You could probably pick any ten guys in here and ask 'em and they'd tell you the same thing. Everybody's picking up on what's on TV.*

The following case shows how an antisocial media figure may guide an unmet fan. At the time, the informant, then thirty, was teaching at a California college. His mentor was Charles Manson.

> *At the time Manson was busted for the Tate-LaBianca murders, I was very much interested in his group and especially the power control he exerted over the women. At the time Manson was being tried in the Hall of Justice in Los Angeles, several of his followers were holding a vigil on the streets outside the courthouse. They had shaven their heads and carved swastikas on their foreheads. I had seen them on TV and wanted to talk with them. So I drove to L.A. and did just that.*
>
> *I was very curious as to how such an insignificant person could garner such devotion from the women in his clan. Through brief talks with the willing group on the sidewalk, I soon got an inkling. The girls loved Charles because he was a forceful, dominant man whom they thought was godlike and was the savior for their kind: middle-class, confused, drugged, unstable, etc. youths.... Charles used biblical quotes, song lyrics, and drugs to partially control his flock; the real element of power, according to the girls, was the hypnotic spell his eyes could achieve in any face-to-face experience.... At the time I had been teaching a course in folklore and my-*

thology and been caught up in the occult, mind games, mysticism, sexual myths, therapeutic processes, etc. Also at this time I was doing drugs heavily. . . . I was also experiencing a breakdown in my marriage. The point is that I was on the edge and very impressionistic. When I talked to the Manson girls who were convinced of the power of Charles's eyes, I acknowledged that sense of power in myself.

I began to experiment with "eye psychology" in my social relationships. I used it on colleagues, strangers, and especially on young . . . women in my classes. For the most part it worked. In my sensitive state, I saw positive control of others result through my hypnotic gaze. To wit: one young woman I concentrated on fell under my spell. She confessed her love for me and was convinced that my eyes, etc., brought her under my control. She would do anything for me. At this point I pulled a Charles Manson. I gave her drugs, told her what was good and what was bad, completely controlled her social behavior. She was obligated to me and I was in no way obligated to her. I was on a power trip; she was on a slave trip. I, the manipulator, was in control. . . . After I got off heavy drugs, and after my marriage broke up, I was guilt-ridden. In my drugged-out state I probably had had the same perverted dreams as Charles.

But it is not just relations with the antisocial types, the real or imaginary media villains, that promote antisocial conduct; it is relations with heroes as well. And not just the James Bonds and Barettas, but the all-American good guys, the Gary Coopers and the John Waynes. Their teachings on violence are clear. Violence may be regrettable, but it is often justified and required (only a coward would back down). Violence is the proper solution for difficult and threatening social situations. Vengeful violence is satisfying, and successful violence will be admired and rewarded (after the killing you get the girl). The mass media did not create these values. They have their roots deep in the American value system. But our mass media promotes these teachings in a more pervasive, glamorized, insistent, and unrealistic way than does any other social world the individual is likely to visit. Like Ruth Steinhagen, Mark Chapman, and John Hinckley, we are the children of this world.

COMING TO TERMS

Caughey writes: "Every time an American enters a movie theater, turns on a media machine, or opens a book, newspaper, or magazine, he or she slips mentally out of the real social world and enters an artificial world of vicarious social experience." What might Caughey mean by calling our media experiences "social," especially since so many of our encounters with the media take place while we are alone—reading, watching TV, listening to music, and the like? How does it change our view of these encounters to think of them as social rather than private? And what role does the media play in making them social? Respond to these questions in your journal. Also, if Caughey was present when you were reading his discussion, what questions would you ask him about "imaginary social relationships"? At what places in his analysis would you stop and ask him a question? What would these questions be? Jot them down in your journal and explain why you're asking these questions.

READING AS A WRITER

An important and difficult task a writer sometimes faces is the work of fairly representing an author's theory, ideas, analysis, point of view, etc. First, we would like you to represent Caughey's way of analyzing "imaginary social relationships." As you reread Caughey, select passages from his essay which you feel fairly represent his ideas. Write these passages in your journal. We are interested in having you show how you are reading these passages. What does the passage mean? Are there key words or phrases in the passage which deserve further comment? Why do you think a particular passage is important? Record responses to these questions for each passage you cite. This kind of intellectual work is sometimes called critical commentary or close reading, and it's a kind of work that academic audiences expect you to do. You might look at this audience expectation in the following way: quotations or paraphrases don't speak until you, as a reader, make them speak.

Another kind of academic work you need to be aware of is the application of an author's ideas. We would like you to apply Caughey's ideas to an "imaginary social relationship" which can be found in one of the selections in *Media Journal.* We think John Wideman's *Michael Jordan Leaps the Great Divide* would be a good choice for this work of application. One of the intriguing features in Wideman's piece is the way in which Wideman's complex and ambivalent relationship with Michael Jordan evolves. Wideman has a real and an interactive relationship with Jordan, but we might also say that Wideman has an "imaginary social relationship" with Jordan. In your journal, record passages from Wideman's essay where both kinds of relationships occur, real and imaginary. How might Caughey explain Wideman's relationship with Michael Jordan in the passages you've chosen? Look to your representation of Caughey's ideas: What has Caughey said that might hold some kind of relevance in explaining this relationship? Wideman admires Jordan for many reasons. Where does Wideman want to be *like* Jordan? Where does Wideman see differences between himself and Jordan? Why are these differences important to this relationship? Having done this work of representation and application, in what ways does the Wideman–Jordan relationship confirm what Caughey has to say about "imaginary social relationships"? What are the aspects in the Wideman-Jordan relationship which Caughey's ideas can't explain?

WRITING CRITICISM

See if you can extend Caughey's ideas by talking with several friends or fellow students about the sorts of "relationships" they have with the media figures they care most about. Take notes on or record these interviews on tape. Then write an essay in which you compare their comments with those of Caughey's informants. How well does his analysis of the workings of "imaginary relationships" explain the remarks of the people you talked to? You need to give these remarks attribution; in other words, who is talking? You might also ask some questions that help explore the backgrounds of the people you interview. How old is the person? Academic major? General opinions on popular culture? When you analyze what these people have to say about their relationships to celebrities, does the language they use sound like the people Caughey's quotes? If not, how do you account for the differences between them, and how might those differences affect our understanding of the "imaginary relationships" Caughey describes?

Is Anything for Real?

Travis Charbeneau

A few years ago "Saturday Night Live" did a takeoff in which Elvis Presley's gold lamé suit toured on a coat hanger "playing" to packed houses long after his death. Today this doesn't seem so far-fetched. Not after the Milli Vanilli scandal where two Grammy winners were caught not only singing along with their records in ostensibly live concerts, but having failed even to sing on the original recordings. And technological "intrusions" are hardly restricted to modern pop.

George Gershwin recorded a number of exquisite piano rolls early in his career which have been laboriously transcribed for MIDI, the musical instrument digital interface, a sort of player piano for computers. The results are extraordinary.

Music technology, like technology generally, is now advancing at such a furious pace that we are increasingly confronted with philosophical questions.

Can art be automated?

Should art be automated?

As with most technological innovation the "should" only gets asked long after the "can" has been answered in the positive. This has certainly been the case in the world of music.

Today, when you purchase an audio recording or attend a concert, a Broadway show, even an opera or ballet, do you know how much of the music is automated by sophisticated synthesis and computers? Do you care? Should you? Or is "the play the thing," whatever the machinations going on in the studio or backstage?

Travis Charbeneau. *World Monitor*, March 1991, pp. 74–78. Reprinted with permission of the author.

Even before Milli Vanilli, the pop world had been shaken by a wide variety of artists lip-synching, or moving their lips along with studio recordings of their music, during "live" shows.

Older audiences and critics seethe at the notion of forking over $50 for what amounts to a celebrity pantomime of a $10 recording. Younger audiences, used to MTV's lip-synched videos and arena-sized "concerts," where spectacle has long reigned supreme over music, seem less indignant. From the bleachers you can't tell if the singer's lips are moving at all—in synch or out.

From Lip-Synch to Robotics

The lip-synch flap is merely the most recent and visible of an old and ornery squabble over technology's role in music. In recent years, even many players of acoustic (nonelectrified) instruments have had at least part of their stage acts electronically "sequenced." That is, they use a computer driving synthesizers and digital samplers (more about them later) to add, say, the punctuating clusters of notes known as "fills" in the sound of Latin percussion instruments.

It's only a small step to having the computer supply the string section for occasional "sweetening," instead of hiring a group of (unionized) humans. "Techno-pop" brings perhaps most of the music under direct control of the computer. Finally, in the ragingly popular realm of rap music, it's almost Japanese *karaoke* time; a literal sing-along with recorded music, much of that itself *re-recorded* (digitally sampled) from old records! Lip-synching merely closes the circle.

And these examples relate just to live concerts. In the recording studio, the computer has long been king.

Is all this something to be tolerated, not just legally but aesthetically? There's a stiff whiff of fraud in a lip-synched "concert." That sequenced incredible breakthrough for artistic expression, courtesy of the vacuum tube. The technology of the electrified guitar—an acoustic guitar with a microphone on it—was accepted, just like that of the piano.

But then came the *electric* guitar. Merely sticking a microphone onto a hollow-body, acoustic guitar invites ear-piercing feedback. It's just the nature of acoustics. In 1941–42, innovator Les Paul was the first to stick a microphone on what he unapologetically called "The Log," a solid slab of wood (with curvy cutouts stuck on the sides to make it at least *look* like a guitar). The "solid body" electric guitar was born. It resisted feedback even when turned up quite loud.

Ready to Rock

Which is exactly what happened next. When rock-and-roll replaced big band music in the '50s, "the older generation" had more to object to than that overtly Africanized beat: those awful electric guitars! "Do you have to play that loud?" Well, no, as it turned out.

By the '60s, thanks to the solid-body electric Les Paul had invented for jazz, heavy rock-and-rollers found they could play even *louder*. There is a perhaps apocryphal but telling anecdote concerning amplifier maker Jim Marshall's first encounter with guitarist Eric Clapton of the rock supergroup Cream in the late '60s. Mr. Marshall had designed a new line of powerful, efficient amplifiers with Chet Atkins and other low-key musical stylists in mind. Inquiring as to what settings on the exquisitely calibrated new Marshall amp the British rock guitar hero might be finding best for any given occasion, the engineer was dismayed to hear Clapton say he just twisted all the knobs to "10" and left 'em there.

The resulting sound was *loud*. So loud it could actually produce ear-piercing (if expertly controlled) feedback even with the solid-body Les Paul guitar. But, worst of all, it was distorted! Vacuum tubes driven to the point of "clipping," or distortion, populated the nightmares of every electronics expert. But Clapton, who had started out as a traditional blues artist with more of a B. B. King, "clean" guitar sound, was playing larger and larger clubs and halls. He, too, wanted to be heard. Cream, after all, was only a trio: guitar, bass, and drums. Along with other (mostly British) guitarists of the period, he found that by turning both his Les Paul and his Marshall (soon to be "a wall of Marshalls") to "10," he not only rocked the hall but was enabled by the resulting distortion to sustain, or hold, a note—again, just like a sax or horn player—to be bent, perhaps modulated like a violin, or chorded and held as long as an organ. The combination also produced sounds never before heard on this planet. "Everything on 10" was, in fact, an entirely new timbre or instrumental tone color, to be added to the world's musical palette, just like those of the piano and "electrified" guitar. And it was due again to technological intrusion, this time perverse intrusion at that.

Echoing the feelings of the music establishment, classical guitar virtuoso Andrés Segovia was disgusted. "Sacrilege" would have conveyed insufficient scorn: To his dying day The Master would not even allow himself to be miked to a standard public address system, no matter how large the hall. It was just he, his classical guitar, and many in the audience straining to hear.

But by the late '60s new technologies were helping artists create new timbres every day. Jimi Hendrix took electric guitar to heights still universally acknowledged as unequalled.

Then, in 1969, Robert Moog marketed the first commercially viable synthesizer, and the dam truly burst. The Minimoog could crank out a new color with the mere crank of a knob.

In 1982 Yamaha introduced its famous DX-7 synthesizer, which not only could produce sounds from Mars but proved very adept at simulating the sound of the electric Hammond organ, the Fender Rhodes piano, as well as synthesized but fairly convincing varieties of brass, strings, reeds, and other acoustic instruments.

Within another couple of years, digital samplers appeared, which played actual digital recordings of these instruments that were even more than fairly convincing: drums, human chairs, even to the horror of purists, the concert grand piano. A digital sampler is like a CD player with a keyboard attached. Press a key and the sampler will play back a CD-quality recording of the real instrument playing that note.

The Blindfold Test

Today, if blindfolded, the average listener, and many a professional, cannot tell the difference between a digital grand piano and a *miked* acoustic playing in the same hall. On recordings, there is virtually no instrument—possibly excepting the acoustic guitar—that cannot be convincingly simulated, especially now that the computer has entered the loop, providing microscopic, surgically precise control of every aspect of sound creation, composition, and performance.

It's a long way from the 400-pipe organ of Winchester Monastery (980) to the computer-driven digital piano, but "parts is parts," and only the "parts" distinguish these technologies. If the bottom line is music, if indeed "the play's the thing," if the fans are happy, why all the fuss?

Even regarding lip-synch, we should recall that in the '50s Les Paul had Mary Ford sing along with herself on tape for harmony parts, surreptitiously—in concert. Audiences were delighted. To his credit, Mr. Paul teased early audiences by challenging, "I'll give ten bucks to anyone who can guess how Mary did this." He was eventually caught out by a little girl, who simply asked, "Where's the other lady?" That was good enough for Les.

Obviously, in the case of "live" concerts that aren't, some "truth in labeling" might be appropriate. But new music technologies raise far broader issues. Not least among them is the displacement of traditional musicians, striking not only the soul, but the pocketbook. Dedicated talented people who have trained all their lives to master difficult instruments increasingly find "one human and a computer" taking over the world of film, television, and advertising, dominating both rehearsal and recording, elbowing them out of the Broadway orchestra pit and even off the concert stage.

If Robots Do It, Is It Art?

This raises unprecedented cultural questions. "Robots" were never supposed to be able to make "art," remember? This was the forever safe domain of real, live humans. In addition to everything else I've described, today you can purchase algorithmic compositional software that not only orchestrates and produces sounds but literally composes tunes, reducing even the composer to the status of "editor": Keep the good stuff, delete the junk. And yet: Bach and Mozart experimented, no one knows exactly how extensively, with then-available musical algorithms and mathematical equations that generated melodies, counterpoint, chords, and retrograde and/or inverted successions of notes.

"Parts is parts?" Is it merely a matter of degree?

It is no accident that Benny Goodman chose the word "sacrilege." Where do we draw the line? Can we? *Should* we?

This is, of course, where we came in, focusing on the discomforting aspects of technological change, especially discomforting when it impacts upon something as emotional as music, which Plato called "the language of the soul."

But, for all the discomfort of adapting appropriately to the new technology, we might take comfort in the potential long-term empowering aspects of this same technology.

I played guitar for 20 years until I lost the necessary mobility of my fingers. I am now an electronic composer: a human and a computer. In the most vital respects, plus many I simply never imagined, music has been given back to me.

What of the millions who love music, who hear new music every day in their heads, but simply lack the physical skills, time, or money to realize their creations?

Today, an inexpensive computer, the right music sequencing software, and a few black boxes are midwifing the birth of a "new folk music," enabling common "folks," without access to music attorneys or big record companies, to explore their talents.

And what about children, who take to using computers as though born for it? What will succeeding generations do with the ancient instinct to sing, given *these* vocal cords?

The technology of fire was a two-edged sword. (So was the technology of the two-edged sword, now that I think about it.) The joys of the computer music workstation—and tribulations like the "is it live or is it lip-synch?" concert performance—also cut two ways.

Still, the human has to control the computer, no less than any other machine. As technology continues to advance and challenge in every department of human endeavor, awareness is the only lasting tool we can develop to maintain humanity; wisdom the only continuing craft that can save us from genius.

COMING TO TERMS

Like several other writers in *Media Journal,* Travis Charbeneau carefully weighs the changes wrought by technological progress through a personal and cultural prism of gains and losses. Maybe you have read, for example, M. Kadi's piece on the Internet ("The Internet Is Four Inches Tall") or Neil Postman's discussion of communication technologies ("The Peek-a-Boo World") or Joshua Meyrowitz's critique of television ("The 19-Inch Neighborhood"). All of these writers have entered into a similar critical conversation and tell cautionary tales of technological progress. Charbeneau enters this critical conversation from a slightly different perspective: In "Is Anything for Real?" Charbeneau wants to argue *aesthetics,* to argue about the role (and perhaps the dangers) of technology in art. As you come to terms with Charbeneau's text, we'd like you to consider and respond to several questions in your journal. First, mark those places in the text where Charbeneau chronicles the major changes in the history (according to Charbeneau) of musical technology. What are the changes that he chooses to include? What instruments does he talk about? What countries are included in this short history? Beyond the obvious, what do you see as linking these changes together? What do they, for example, tell us about what Charbeneau thinks is important in a discussion of music, of art? What issues remain a constant for Charbeneau? What issues does he return to again and again?

READING AS A WRITER

Perhaps more than most writers, Charbeneau makes an extensive use of questions in writing his article. Not only is Charbeneau's title a question—"Is Anything for Real?"—but his writing also incorporates questions posed for stylistic, rhetorical, and strategic purposes. One type of question Charbeneau uses is often referred to as a *rhetorical question,* a question offered for style or effect, but a question not intended to be answered by the author. Sometimes, Charbeneau does ask a question and then provides an answer. Other times, we see Charbeneau doing something quite striking with questions posed in his article: at times, it appears that several different "voices" are asking questions. Charbeneau is making use of devices to make his writing seem more like speech; he shifts intonation and inflection, giving "voice" to some questions. Here, Charbeneau seems to be trying to give special emphasis to his questions. (At first, you might not read and *see* the shifts among these "voice" questions, so you might want to read the essay aloud to *hear* these shifts.) We'd like you to re-read Charbeneau's text and analyze how he uses questions to set the limits of his discussion of music technologies and to further his critical argument. First, we'd like you to restate Charbeneau's argument in your own words in your journal. Then respond to the following questions: What are examples of rhetorical, "real" (the kind that the author answers), and "voice" questions? Record examples of each kind in your journal. Why do you think Charbeneau doesn't stop to answer some questions?

How does this contribute to his critical argument? In the case of "voice" questions, who might be asking this question? What kind of person or group of people might this particular "voice" represent? In what ways might this question represent you? All of us? How do these questions presume to be important in or crucial to a discussion of music technology?

WRITING CRITICISM

Charbeneau speaks here to the problem of "authenticity" in art and how to identify it in a commercial culture, in which products are not only promoted and sold all the time, but are also created exclusively for this purpose. Of course, some of the products of that culture still strike us as more "commercial" than others—in their look, feel, design, or execution (in the way that they are presented). In this assignment we would like you to explore what we mean when we label a media product "commercial" and, conversely, what we mean when we perceive something as *not* commercial.

Select two media products—one in each category—and write a piece comparing them to each other. One of these products should exhibit, in your view, the qualities that make something worthy of the label *commercial.* This product might be a commercial-sounding album, a commercially-oriented radio station, a commercial-looking film, a work of commercial fiction, a magazine that has the feel of commercialism, a commercial newscast, etc. (We don't want you to conclude that a media product exhibits this quality of commercialism simply because it is for sale.)

Likewise, a product, such as a television program, a radio program, or a magazine, shouldn't be classified as commercial just because it is interrupted by commercials or advertisements. Contrast this media product with another in the same medium that strikes you as not commercial—an album that doesn't have that commercial sound, a radio station that's non-commercial, a non-commercial work of fiction or non-fiction like an autobiography, a magazine that strikes you as non- or even *anti-commercial,* a newscast that doesn't exhibit the trails of commercial broadcasting.

The purpose of your writing should not be simply to denounce one product as "commercial" and to uphold the other, but rather to understand commercialism as a quality that a media product may or may not have. As you do the work of this comparison you might address several questions. How do you identify this quality of commercialism? What tells you it's there in a media product? What do commercial products look like, sound like, feel like? How do these products satisfy the needs of a particular audience, group of users or reading, viewing, or listening culture? How would you define this group identity? What's positive about commercialism in media? In what ways does it affect aesthetics, such as artistic "authenticity" in the media? What might be the negative effects? How would you define that quality in opposition to commercialism, the non-commercial product? How do you identify it or how does it work or manifest itself in a product? What names would you give to it? What are the advantages of a non-commercial approach to art in the media? Does such a position or designation require special responsibilities of the person or people who produce something non-commercial? What might be the disadvantages? These are some of the questions to keep in mind in making your comparison between two kinds of media product.

There is one more step you need to take for this assignment. Writers do the work of comparison/contrast for a reason to advance a critical argument, to take a position on an issue or make some kind of conclusion. Sometimes, often in high school, students are asked to write a comparison/contrast essay. We don't see this work so much as an end in itself, but rather as a kind of intellectual work, which, in turn, puts a writer in a strong position to do other kinds of work. Now that you have done this comparison of the commercial and non-commercial product, you'll need to put this work to *work*. Is this distinction between two kinds of media products a helpful distinction, a valuable way of reading media culture? What have you realized about media now that you have done this work? What are your new insights? Has this comparison been problematic? In what ways?

Ideal Homes

Rosalind Coward

There's a wide, light sitting-room with French doors opening on to a well-tended garden. The room has a pine ceiling with down-lighters; the walls are pale coloured and on them hang a few framed pictures. There are lots of big plants and two modern sofas. In the centre of the room is a glass table; in one corner, an ornate antique cabinet. This is the home of an architect, we are told, a home "noted for the particular brand of elegance in furniture and decoration to be seen throughout."[1]

The pictures are there to give us "furnishing ideas." If we can find out where to buy these furnishings (and can afford them), we too can acquire a "stylish home." "Style" is something which some people have naturally, but thanks to such photographs all of us can find out what it is and where to get it.

Magazines like *Homes and Gardens* and *Ideal Home,* which deal in such images, are specialist magazines, like those aimed at men—*Custom Car* and *Hi-Fi* for instance. Home magazines, too, carry specialist advertising aimed at an interest group, those involved in the process of home-buying, home-improvement and decoration. But the general tenor of the articles makes it clear that women are taken to be the main consumers of such images. The Do-It-Yourself section of *Homes and Gardens* was at one time a detachable section, easily removed from the main bulk of the more feminine concerns: fashion and

---

[1]This and all subsequent quotations come from one of the following magazines: *Ideal Home,* December 1977, October 1980, December 1980, and May 1982; *Homes and Gardens,* July/August 1980, November 1980, March 1980; *Options,* July 1982; February 1983; *Good Housekeeping,* June 1982; *Company,* May 1982.

beauty, food, gardening and general articles like, "Why do women always feel guilty?" And if specialist home magazines carry general articles for women, it is also the case that general women's magazines carry articles about homes and furnishings remarkably similar to those of the specialist magazines.

Here, then, is a regime of images and a particular form of writing aimed largely at women. And, as with fashion, home-writing encourages a narcissistic identification between women and their "style." The language of home-improvement in fact encourages an identification between women's bodies and their homes; houses like women are, after all, called stylish, elegant and beautiful. Sometimes the connections become more explicit, as when someone refers to "Mr. X and his decorative wife." Advert[isements] too play with these connections. One company advertised its new range of bath colours with the caption "Our recipe calls for Wild Sage"; it showed a woman applying her make-up in a beautiful bathroom. Enforcing these connections makes one thing clear; the desire for the beautiful home is assumed to be women's desire.

The images and articles relating to home-improvements don't just show any old picture of any old house. There's a preferred kind of image and a very definite style of writing. Lifestyle is a term much favoured by home-improvement writers, and the activity of reading about and looking at other people's homes is essentially offered as a chance to peep into these lifestyles. Home-literature is essentially voyeuristic, a legitimate way of peeping through keyholes in a society where the private realm is kept so separate from public life. Through these images, women can look in on other people's decor, their furnishings, and pick up clues about how other people live. This is one of the few routes of access which women have into the private lives of people other than their immediate circle.

This keyhole voyeurism has two characteristic forms, forms which in fact turn up in all sorts of magazines, not just the specialist home magazines. One common mode is sneaking glimpses of the rich and famous, those alleged "personalities" spawned by our society. In "The Los Angeles home of Veronique and Gregory Peck" we get a quick peek at their "elegant dining-room" which, "strong, cool and dramatic," reflects "something of the personalities behind it." Indeed, more than all the writings about the love-lives of these personalities, ideal-home writing seems to unearth the most intimate form of revelations. After all, if you had seen the puce-coloured bedroom of Terry Wogan, would there be anything more to know?

The other kind of voyeurism, though, is rather more common. This is the scrutiny of the homes of the unknown, the ordinary people who have done up their homes in such exemplary ways. Unknown these people may be but unsuccessful they certainly aren't. Here we have hordes of successful individuals, "a rising star amongst dress designers," or heterosexual professional couples, called Sue and Nick, or Jackie and Mike, who have worked miracles on their terraced London houses. These unknowns have actually achieved the ultimate accolade. Just by the sheer force of their "personal style," their houses have been selected to teach the rest of us how.

The discourses of home-improvement have a very precise language without a precise referent. It is the photographic images which have to flesh out the meanings for the vague language of "individuality," "style," "personality" and "flair." The articles talk of "homes full of warmth and colour," "elegant town houses," and describe how "antique furniture combines happily with modern hand-crafted pieces, reflecting catholic tastes." Home-improvements get praised when they achieve "Today's look for Yesterday's houses." Houses are "lovingly restored to elegance and comfort" by their owners and admired for being "a delightful combination of comfort and elegance, delicacy and crispness, that produces a relaxed formality." The aims are "colour and comfort," "homely and handsome" but above all, "stylish."

Style, as with the fashion magazines, is always the name of the game. Style lies like a blanket across all the descriptions of houses and their decorations; "Leonard Rossiter: we visit his stylish family town house"; "Ceilings: Round off your room in style"; "Strong on Style"; "Stylish Alternatives." We are told about a Victorian terraced house ready for demolition which "a young couple" transformed into a "tasteful and stylish home." All the loving hard work, it seems, is worth it if they achieve the ultimate accolade, "personal style." Hard graft on "daunting prospects" pays off if it adds up to a "home of individual flair and style": "It's been a lot of hard work but the final result is an original home with a very personal stamp on it. As she says, 'It's so nice to come home to.'"

Personal style—a strange paradox; individuals have it but we can all copy it. Personal style is, in reality, nothing other than the individual expression of general class taste and the particular ideals promoted by that class. No wonder the sign of an individual's uniqueness is so easily copied. The photographic images make it quite clear that there are consistent ideals even if the language remains stunningly vague. The kind of pictures taken, the kind of homes shown, offer very definite ideologies as to what domestic and private life is like. The reference to *ideal* homes is no light reference. Everything shown is at an ideal moment. The rooms shown are always curiously vacated: there's never a trace of a mess, of dishes left unwashed, of beds unmade. Miraculously, the only signs of ubiquitous children are "the children's bedroom, featuring a complex blue, black and white frieze by Osborne and Little" or "a homework area for children in a spare room, with light-stained fitted louvred cupboards."

In spite of the offer of an intimate glimpse into a private home, all traces of life in that home tend in fact to be obliterated. The owners are evicted by the photographic regime. The house is photographed as it probably never is—tidy, sparkling clean, free of persons and their ephemera. These are not *homes* but *houses*. They are the finished products, the end of the long years of planning or "loving restoration." These are the houses that exist in the imagination during the years of painting, scrubbing, and hammering. The photographic regime captures the illusory moment of an "after," that moment frequently imagined during all the hard work. The lure during such work is precisely this, a finished product. But even if such work were ever finished—and for most of us, it never is—a house that is truly lived-in could not sustain such perfection. To represent a lived-in home would destroy for ever more the illusion that a house could ever be finished and perfect. In a lived-in home, the satisfying moment of "after" never occurs; "after" is endlessly postponed until the fuzzy felt farm has been peeled off the carpet and the socks put into the dirty linen basket.

This regime of imagery represses any idea of *domestic* labour. Labour is there all right but it is the labour of decorating, designing and painting which leads to the house ending up in this perfect state. We hear about how much the wallpaper cost and how long it took to get the underlying wall into good condition. We don't hear about how long it took some woman to get the room tidy, or who washed the curtains. But then the photographs only show the ever-tidy, clean and completed home. The before and after imagery endorses the joint work of couples, the husband and wife team who plan, design and decorate the house. The suggestion is that together men and women labour on their houses. The labour is creative and the end product is an exquisite finished house-to-be-proud-of. Domestic labour, the relentless struggle against things and mess, completely disappears in these images. The hard and unrewarding ephemeral labour usually done by a woman, unpaid or badly paid, just disappears from sight. Frustration and exhaustion disappear. Instead a condition of stasis prevails, the end product of creative labour.

The only suggestion of the actual life of the inhabitants is in a vision of an empty dining-table. Inhabitants of ideal homes, it seems, do a lot of entertaining. Indeed, the empty dining-table surrounded by empty but waiting chairs is often presented as the hub of the house, an empty stage awaiting the performance. Could it be that "entertaining" is the main way in which these inhabitants display their ideal homes? And could it be mere coincidence that cooking meals is also an activity which relies heavily on women's labour?

The centrality of the dinner party has been made more explicit by one magazine, *Options*. Here a "lifestyle feature" caters for the less subtle voyeurs. In these features, we get to see the house and hear all about the home-improvements; we see the person (or couple) at work in their tastefully decorated studies; and then, *pièce de résistance*, we see pictures of their dinner parties *and* hear about their favourite menus. Such features virtually tell us the names of the guests; "says Jeremy, 'Since a lot of our friends are in the media or theatre, our house can look like a rest home for weary London celebrities.'" And the articles leave us in little doubt about how enviable such lifestyles are meant to be; "you can't spend long with the Pascalls without thinking they're the kind of sociable couple you wouldn't mind having on your Christmas card list."

Not much thought is required to realize that we are being offered the styles and lifestyles of a very precise class fraction. It's not just that Nick and Jessica, Jeremy and Ann are middle-class. They are a very particular middle-class grouping. Indeed, my suspicions aroused, I found on closer examination that virtually everyone whose home appeared was involved in some way with the media. Publishing, advertising, and television are high on the list for favoured subjects, but far and away the most likely target for an ideal-home spread were designers—graphic designers, fabric designers, architects. We see the home of Mary Fox-Linton, "at the top of the tree as a decorator," and of "Lorraine, a graphic designer of very definite tastes." Then we learn about John: "although he is a successful graphic designer running his own business, he and his wife...have made their home on a ten-acre farm high on the North Downs." Even little professional subtleties are explained to the unitiated: "If you have ever wondered what the difference was between a designer and a decorator you can see it in Fanny's home, an artist's studio in Kensington." And just as you begin to wonder why this group of people, you are told: "It is, of course, easier to be original if you have the sort of design flair that makes Patrick Frey and his firm Patifet famous for its exquisite furnishing materials."

Far be it from me to suggest that there may be an element of self-interest at work here. Perhaps it is just coincidence that these designers get such good publicity in the magazines? Perhaps information about these people just happens to find its way into the home-making magazines with only the smallest amount of help from PR and communications companies? However it gets there we can be sure of one thing. The groups presented are presumably remarkably like the people who produce the magazines. Self-referential always, sycophantic sometimes, the journalism and images promote a self-evident world where everyone knows what good taste and style are. Here of course is the solution to the homogeneity of style. No brightly coloured and large patterned wallpaper, no souvenirs, no cheap mass-produced reproductions on the wall; the privileged glimpses are all of one kind of person. In so far as working-class homes do crop up, they are the subject of ridicule, material for an easy joke: "An index of proscribed examples of bad taste will be regularly published. I promise that the following obvious candidates will have fallen to the axe, hammer or incinerator—replica Victorian telephones; onyx and gilt coffee tables...large red brandy glasses with tiny porcelain kittens clinging to the side...Cocktail cabinets...spare toilet roll covers...doorbells chiming tunes...Crazy paving." And as if to confirm my theories of a particular group defining and setting the ideals, I discover that the writer of this *Company* magazine article is none other than Jeremy Pascall, featured in the *Options* article quoted above, he of the dinner parties and the media personalities.

This self-reflexive and self-congratulatory group has the hegemony over definitions of design, taste, style and elegance. The group is not exactly the ruling class; they do not own the means of production and they are by no means the most wealthy or financially powerful people in the country. Yet they control the means of mental production; they are the journalists, designers, graphic designers, furnishers and publishers who can tell us what we should think, what we should buy and what we should like. This grouping clearly has enormous powers within society because the communications media have enormous potential to decide our beliefs and tastes. But it isn't even a matter of this group deciding our views. More nebulous and perhaps more influential, these are the people who design our homes, who show us how to decorate these homes, whose material is on sale in the shops. These are the people whose tastes dictate the very possibilities for much of our everyday lives.

And the standards set by this group are remarkably consistent. The houses are all geared toward a conventional living unit. And the decors are mere variations on a basic theme. The walls are plain; there is minimal furniture; an absence of what is seen as clutter; and light, open rooms. Indeed the ideal home is very much directed toward a visual impact, and within this visual impact, toward a display of possessions. Furniture and decorations are chosen with an eye to how they match each other. Walls are painted with an eye to how to display an original painting or a framed print. Shelves and tables are arranged to show off expensive objects to their greatest advantage. Above all, the light colours and plain walls tend to demonstrate constantly how clean these walls are.

What is dismissed as bad taste in working-class homes is merely the arrangement of the home according to different criteria. In working-class homes, the pictures and colour are often on the walls, as wallpaper, not framed as possessions. Items are often displayed not to demonstrate wealth but because they have pleasurable associations. Here are souvenirs— memories of a good holiday; snapshots—memories of family and friends; and pieces of furniture chosen, not for overall scheme, but because they were liked in someone else's home. This is a different modality of furnishing, not necessarily concerned with the overall visual effect. Items of furniture are not always chosen to match but for different reasons. Here, you often find a his and hers corner—an ancient armchair which *he* refuses to get rid of, or a chiming clock which drives him mad but which *she* won't part with.

There's almost certainly an element of Oedipal drama in the obsessive ridiculing of working-class homes which goes on among the ideal-home exponents. Usually this media group are not from middle-class backgrounds but are the first "educated" members of working-class or lower-middle-class backgrounds. Doubtless the determination to ridicule such homes arises from a determination to be different, to reject their background and all it represented. The class basis for this taste is always disguised in the writing, which insists there are such things as absolute good taste and good design. But class isn't the only thing disguised by the vagueness of labels of "style" and "elegance."

The economic investment in home-restoration is also disguised by the language of loving restoration. The idea that home-improvement is merely the expression of individuality through good taste obscures the way in which this kind of restoration is a very real economic activity. People able to buy and restore houses, or build their own houses, are acquiring valuable possessions. These are the possessions which—as this group gets older, dies and leaves the houses to their children—will be creating a new elite—those who have no rent to pay as opposed to those who spend enormous proportions of their income on housing. For the house-owner, even their current property represents the creation of profit out of housing. Doing up one house, moving on to a bigger and more valuable one, with tax incentives on mortgages—home-restoration is certainly also gaining economic advantages over those who can afford only rented accommodation.

But perhaps the most impressive concealment effected by this particular style is women's relation to domestic work. We've already seen how the photography obscures domestic labour. But it is also the case that the style itself sets a goal which is the obliteration of any trace of labour or the need for labour. The style emphasizes the *display* of the home, its visual impact, which will reveal things about the personality which owns it. Any house requires intensive domestic labour to keep it clean, but plain walls, open fires and polished floors probably require more than most. To keep it spotless would either involve endless, relentless unrewarding labour or using another woman's low-paid labour. Because the ideal is so much that of *absence* of clutter and mess, of emphasis on visual impact, any sign of mess is a sign of failure.

The work which women usually put in on the home is obscured in other ways. Unlike the working-class home which sometimes visibly manifests the difference between the male and female personalities in the house, the ideal-home taste obliterates differences between men and women. The house is expected to be a uniform style. If there are two people living together, the house has to express the joint personality of the couple. And not only does the style obliterate evidence of two personalities but the articles are positively congratulatory that traditional divisions between the sexes have now gone. "Peter and Alison Wadley," we are told, "have units made to their specification. 'In design terms,' Peter believes, 'the kitchen represents an interesting problem because its status has changed over the last few years. It's no longer purely a working room—in most houses, all the family tend to gravitate there so it has virtually become a living-room.'" Women, we are informed, are no longer consigned to the kitchen which becomes instead a living-room where everyone mucks in. Men, on the other hand, are just dying to get in there to try out their creativity. In short, the styles and articles about them are all about the abolition of conflict. The home is not a place where women are subordinated but a shared space, with domestic chores split happily between the sexes.

I don't wish to decry such a vision; it's just that I don't believe it. Nothing in fact could be more mystifying about the real relations of the home, the minute and the major ways in which women continue to take responsibility for domestic life. Our society is rigidly divided on sex lines and this extends even into the home. While women's employment prospects are limited by male prejudice and by taking primary responsibility for child care, the home cannot be this fine place for all. However uncritical women may be of bearing the responsibility for the home, it is a rare woman who has *never* experienced home as a sort of prison. Confined there through limited possibilities, and bearing the awesome responsibility for the survival of young children, or torn by commitments to work and children, the home is often a site of contradiction between the sexes, not a display cabinet. Even in the most liberated households, women are well aware of who remembers that the lavatory paper is running out and who always keeps an eye on what the children are up to.

Because the home has been made so important for women, the decoration of the home matters a lot to women, perhaps more than it does to men. In a world of limited opportunities, there can be no doubt that in the construction of the house there are creative possibilities offered in few other places. It is also crucially important to women that they feel all right about where they live. But the creative aspects in women's wish to determine their environment have been submitted to a visual ideal whose main statement is the absence of the work they do, and absence of conflict about that work.

COMING TO TERMS

"Personal style: a strange paradox: individuals have it but we can all copy it." Rosalind Coward's reading of personal style, in reference to the interior of homes, surfaces an intriguing paradox or tension that several writers in *Media Journal* have confronted. While media culture sometimes promises or advertises individuality through the products it promotes, that sense of individuality or personal style needs to be weighed against the limitations a culture might place on an individual's personal expression. For example, maybe you have read Daniel Harris' piece, "The Conformity of Office Zaniness," or elsewhere in *Media Journal,* in which he considers the complications of achieving and maintaining a sense of personal style within a culture. As you come to terms with Coward's article, we'd like you to explore her way of working through this paradox by answering a set of questions for your journal. What evidence does Coward offer for her belief that style is not only the expression of an *individual* but also of an entire *class* of people—that it is, in effect, a kind of badge we put on to show that we are members of a certain

group? According to Coward, what are some of the limitations that our culture places on the expression of individual style? And why does Coward see
this as a particularly troubling issue for *women*? See if you can answer these
questions both by marking relevant passages in the text and by recording
your responses to these questions in your journal.

READING AS A WRITER

It could be said that Coward's own strategy as a writer here is much the same
as those of the journalists in the magazines she criticizes, since, like them, she
reads the interiors of homes for clues to the character and beliefs of the people
who design and live in them. She and the journalists clearly differ on what they
think those clues mean and on what they tell us about the values embodied
by the "ideal home." What Coward must do as a writer, then, is somehow to
suggest and be convincing that she knows what the ideal home *really* means
in a way that the magazine journalists do not (or are trying to conceal). One
way that Coward enters into a slightly different critical conversation here is that
she attempts to read the interior of homes and their inhabitants and designers
in terms of *ideology*. (Although it is a term that can have multiple definitions,
ideology might be understood, in the sense of how Coward uses the term, as
a version of society and a version of reality.) This version or picture of society
defines the class and gender roles people play in the society. Go back through
Coward's article and try to locate places in her text in which she offers an *ideological* reading of the interior of homes or offers an alternative *ideological* reading of another journalist's work. You might look for certain words—*class,
working-class, domestic* or *private, regime,* or *gender*—which often key the appearance of this kind of analysis. In your journal respond to the following questions: According to Coward, what characteristics (values, beliefs, roles, etc.)
define the ideology of the "ideal home"? And then what is it that defines the
"working-class" home? What are the important differences between these two
ideologies? One might say that part of Coward's critical argument is that certain magazines (Coward doesn't, but we might include television programs as
well) hide the truth and deceive us about reality in the American home. But,
based upon her reading, is there something that Coward fails to see in her
ideological reading of the American home?

WRITING CRITICISM

Rosalind Coward compares the uncluttered visual space of the "ideal home"
with the more crowded and functional arrangement of a working-class house,
and in doing so tries to read the ideology of both places for what they mean
to those who live in them, for the values they try to make those spaces em-

body and express. We'd like you to take on a similar project here which offers a critical reading of two places or areas for which you can show, like Coward, contrasted meanings or ideologies.

One approach would be to select two places that you know well and can revisit for the purposes of this writing. These places should have similar uses but should also seem to you somehow to mean or stand for different things. You might choose, for instance, to study a corner grocery and a chain supermarket, or a city block and a suburban lane, or the campuses of different schools. Take some notes in your journal on the particular ways each place you visit tries to suggest or reflect the character, tastes, interests, or values of the people who live in or make use of it. (You might re-read the journal entries you've already made where Coward has made a similar kind of analysis to generate some specific areas of study.) You might also want to consider how much you feel "at home" there or not, and why—what signals the place sends you about the ways you do or don't belong there. Try to be as exact in your observations as you can. One of your aims in writing will be to define the "feel" of the places you've visited, and you will want to be able to suggest this through a use of telling and specific detail. Then draw on your notes to write a piece that defines the contrasting meanings of the two places you've studied.

Another approach for this project would be to study the contrasted meanings or ideologies present in a number of print advertisements in which the American living room, family room, or kitchen is represented. The ads might be for any number of products. Your primary focus here would be the way the interior of the "average" American home is represented in the ads. We'd like you to select print ads which might be said to cover several decades of domestic scenes. (You might begin with selecting ads which appear in *Media Journal* and supplement these ads with ones you find in other magazines or the ads that appear online for some magazines.) Your work here would be to map out the changes you observe in the ads (or argue for the consistency you see). As Coward has done, you'll need to critically read the domestic scenes which the ads portray. In addition to the product advertised, your analysis should also consider what values, beliefs, gender roles, etc. these domestic scenes appear to be promoting.

For either project, think of yourself as working in the spirit of Coward, and try to show in your writing either how you've made use of her analysis in forming your own, or how your approach to reading places differs from hers.

Visions of Black-White Friendship

Benjamin DeMott

At the heart of today's thinking about race lies one relatively simple idea: the race situation in America is governed by the state of personal relations between blacks and whites. Belief in the importance of personal relations reflects traits of national character such as gregariousness, openness, down-to-earthness. It also reflects American confidence that disputes can be trusted to resolve themselves if the parties consent to sit down together in the spirit of good fellowship—break bread, talk things out, learn what makes the other side tick.

But there's rather more to faith in black-white friendship than off-the-rack Rotarianism. There are convictions about the underlying sameness of black and white ways of thinking and valuing, and about the fundamental causes of racial inequity and injustice, and about the reasons why the idea of addressing race problems through political or governmental moves belongs to time past.

One leading assumption is that blacks and whites think and feel similarly because of their common humanity. (Right responsiveness to racial otherness and full access to black experience therefore require of whites only that they listen attentively to their inner voice.) Another assumption is that differences of power and status between whites and blacks flow from personal animosity between the races—from "racism" as traditionally defined. (White friendship and sympathy for blacks therefore diminishes power differentials as well as ill

The Trouble with Friendship: Why Americans Can't Think Straight about Race. New York: Atlantic Monthly Press, pp. 7–23. Copyright © 1995 by Benjamin DeMott. Used by permission of Grove/Atlantic, Inc.

feeling, helping to produce equality.) Still another assumption is that bureaucratic initiatives meant to "help" blacks merely prolong the influence of yesteryear. (The advent of good interpersonal feeling between blacks and whites, on the other hand, lessens yesteryear's dependency.)

Each of these closely related assumptions surfaces regularly in print media treatment of the friendship theme—material promoting interracial amity and weaving together concern for "the disadvantaged" and the "underclass," anecdotal evidence of the mutual affection of blacks and whites, and implicit or explicit disparagement of politics and politicians. And traces of the same assumptions appear in fraternal gestures favored by campaigning political candidates.

White candidates attend services at black churches, socialize at black colleges, play games with blacks (as when, during Campaign '92, Jerry Brown took gang leaders rafting). And candidates speak out in favor of black-white friendships, venturing that such ties could be the answer to race riots. On the second day of the Los Angeles riots, Candidate Clinton declared: "White Americans are gripped by the isolation of their own experience. Too many still simply have no friends of other races and do not know any differently."

But fantasies about black-white friendship are dramatized most compellingly for large audiences in images. Movies, TV, and ads spare us abstract generalizing about the isolation of the races. They're funny and breezy. At times, as in *Natural Born Killers,* they deliver the news of friendship and sympathy in contexts of violence and amorality. At times they deliver that news through happy faces, loving gestures, memorable one-liners. Tom Hanks as Forrest Gump loses his beloved best buddy, a black (Mykelti Williamson), in combat and thereafter devotes years to honoring a pledge made to the departed (*Forrest Gump,* 1994). A rich white lady (Jessica Tandy) turns to her poor black chauffeur (Morgan Freeman) and declares touchingly: "Hoke, you're my best friend" (*Driving Miss Daisy,* 1989). Michael Jackson pours his heart into a race-dismissing refrain: "It doan matter if you're black or white" (1991). Scene and action hammer home the message of interracial sameness; mass audiences *see* individuals of different color behaving identically, sometimes looking alike, almost invariably discovering, through one-on-one encounter, that they need or delight in or love each other.

> *Item:* The black actor Danny Glover sits on the john in *Lethal Weapon,* trousers around his ankles, unaware of a bomb ticking in the bowl; Danny's white buddy, Mel Gibson, breezily at home in Danny's house, saves his life by springing him from the throne to the tub.

> *Item:* A commercial set in a gym finds two jock-side-kicks, one black, one white—Kareem Abdul-Jabbar and Larry Bird—chummily chaffing each other. Kareem bets Larry that he can't eat just one Lay's potato chip. The little devils are too tasty. Kareem wins the bet and the old cronies exit

together, clever camera work and a makeup cap on Larry's head redoing them into interchangeable, dome-headed twins.

Item: The Tonight Show re-creates itself in the image of interracial bonhomie, replacing a white band with a black band, encouraging chat between white host and black second banana music director, and casually alluding to hitherto unpublicized black-white associations. (Jay Leno to Branford Marsalis: "I toured with Miles Davis.")

Item: Pulp Fiction (1994) draws together three sickeningly violent narratives by means of an overarching theme of sacrifice—blacks and whites risking all for each other. At the pivotal crisis, luck offers a white, mob-doomed prizefighter (Bruce Willis) a chance to escape; the black mobster (Ving Rhames) who's after him, is himself entrapped—raped and tortured, within Willis's hearing, by two white perverts. Spurning self-interest, Willis risks his life and saves the mobster. (The rescued black closes the racial gap in a phrase: "There's no more you and me.")

A key, early contribution to the mythology of black-white friendship was that of *The Cosby Show.* Without actually portraying blacks and whites interacting, and without preaching directly on the subject, this sitcom lent strong support to the view that white friendship and sympathy could create sameness, equality, and interchangeability between the races. Under the show's aegis an unwritten, unspoken, *felt* understanding came alive, buffering the force both of black bitterness and resentment toward whites and of white bitterness and resentment toward blacks. *Race problems belong to the passing moment. Race problems do not involve group interests and conflicts developed over centuries. Race problems are being smoothed into nothingness, gradually, inexorably, by goodwill, affection, points of light.*

The Cosby family's cheerful at-homeness in the lives of the comfortably placed middle class, together with the fond loyalty of their huge audience, confirmed both the healing power of fellow feeling and the nation's presumably irreversible evolution—as blacks rise from the socioeconomic bottom through the working poor to the middle class—toward color blindness. In the years before the show, black-white themes, in film as well as TV, had passed through several stages of development. One of a half-dozen milestones was the introduction, in *The Jeffersons,* of the first blacks to achieve middle-class affluence via entrepreneurship. Another milestone was the introduction, by adoption, of charming black children into white families—as in *Webster* and *Diff'rent Strokes.* (The "white foster parents," wrote Jannette Dates, "could then socialize the youngsters into the 'real' American way.")

And in the wake of the success of *The Cosby Show,* the eradication of race difference by friendship became an ever more familiar on-camera subject. Closeness between the races ceased to be a phenomenon registered indirectly, in surveys documenting the positive reaction of white audiences to the Huxtables; it moved to the center of mass entertainment. Everywhere in the visual media, black and white friendship in the here and now was seen erasing the color line. Interracial intimacy became a staple of mass entertainment story structures.*

Consider *White Men Can't Jump* (1992), a movie about a white quester—a dropout eking a living on basketball courts in Los Angeles—surviving, with black help, on ghetto turf. Working first as a solitary, the young white hustles black ballplayers on their own turf, trading insults with blacks far more powerful, physically, than himself. He chides black athletes to their faces for being showboats, concerned about looking good, not about winning. He flashes rolls of bills and is never mugged. Accompanied only by his girlfriend, he walks the most dangerous ghetto streets at night, once making his way uninvited into an apartment filled with black ballplayers. He mocks black musical performers to their faces in a park, describing the hymns they sing as "shit." That an arrogant, aggressive, white wiseass can do all this and more and emerge unscathed means partly that his behavior is protected under the laws of comedy.

But the armor that counts more, here as in numberless black-white friendship tales, is provided by the black buddy. The acquaintance of white Billy Hoyle (Woody Harrelson) and black Sidney Deane (Wesley Snipes) begins badly: each hoaxes the other. Later they communicate through taunts. Black taunts white for incapacity to appreciate the black musicians whom white claims to admire. "Sure, you can listen to Jimi [Hendrix]. Just, you'll never hear him." Black taunts white for dreaming that he can slam-dunk: "White men can't jump." Black mockingly offers technical aid to white: pumping up

---

*A monograph on these structures in the movies would reach back to breakthrough works such as *The Defiant Ones, Brian's Song, Guess Who's Coming to Dinner,* and *Hurry Sundown.* Some notion of the quantity of relevant recent "product" can be gained from a more or less random list of recent films—TV and junk action movies mingling with more pretentious work—that treat race friendship themes for part or the whole of their length: *The Shawshank Redemption, Lethal Weapon I–III, The Waterdance, Ghost, The Last Boy Scout, 48 Hrs. I–II, Rising Sun, Iron Eagle I–II, Rudy, Above the Law, Sister Act I–II, Heart of Dixie, Betrayed, The Power of One, Crossroads, White Nights, Clara's Heart, Storyville, Clean and Sober, Doc Hollywood, Cool Runnings, Places in the Heart, Grand Canyon, Trading Places, Gardens of Stone, The Saint of Fort Washington, Dutch, Fried Green Tomatoes, Q & A, Passenger 57, Skin Game, That Was Then...This Is Now, Platoon, The Last Outlaw, A Mother's Courage: The Mary Thomas Story, Off Limits, The Unforgiven, The Air Up There, Cop and a Half, Made in America, The Pelican Brief, Losing Isaiah, Corrina, Corrina, Tyson, Everybody's All American, The Little Princess, Diehard with a Vengeance, Angels in the Outfield, Samaritan: The Mitch Snyder Story, Searching for Bobby Fischer, Soapdish, Homer and Eddie, Running Scared, Little Nikita, The Stand, An Officer and a Gentleman, American Clock, Yanks, State of Emergency, Poltergeist, Dr. Detroit, P.C.U., Flight of the Intruder, Ghostbusters, Lionheart, The Blues Brothers, Prayer of the Rollerboys, The Client, The Abyss, Showdown, Robin Hood: Men in Tights, A Matter of Justice, Smoke, Under Siege II, Clueless,* and so on.

his Air Jordans for dream-flight. White jabs back hard, charging black with ex-hibitionism and sex obsession.

Yet the two make it as friends, form a team, work their scams in harmony. More than once the buddies save each other's tails, as when a black ballplayer whom they cheat turns violent, threatening to gun them down. (The two make a screaming getaway in the white quester's vintage ragtop.) And the movie's climax fulfills the equation—through sympathy to sameness and interchange-ability. During a citywide, high-stakes, two-on-two tournament, Billy, flying above the hoop like a stereotypical black player, scores the winning basket on an alley-oop from his black chum, whereupon the races fall into each other's arms in yelping, mutual, embracing joy. Cut to the finale that seals the theme of mutual need and interdependency; black Sidney agrees to find quasi-honest work for white Billy at the floor-covering "store" that he manages:

Billy (helpless)*:* I gotta get a job. Can you give me a job?

Sidney (affectionately teasing)*:* Got any references?

Billy (shy grin)*:* You.

Like many if not most mass entertainments, *White Men Can't Jump* is a vehicle of wish fulfillment. What's wished for and gained is a land where whites are unafraid of blacks, where blacks ask for and need nothing from whites (whites are the needy ones; blacks generously provide them with jobs), and where the revealed sameness of the races creates shared ecstatic highs. The precise details of the dream matter less than the force that makes it come true for both races, eliminating the constraints of objective reality and redistributing resources, status, and capabilities. That force is remote from political and economic policy and reform; it is, quite simply, personal friendship.

Another pop breeding ground of delusion is the story structure that pairs rich whites and poor blacks in friendship—as in *Regarding Henry* (1991), a Mike Nichols film about a white corporation lawyer and a black physical therapist. The two men meet following a holdup during which a gunman's stray bullet wounds the lawyer, Henry Turner (Harrison Ford), in the head, causing loss of speech, memory, and physical coordination. The therapist, Bradley (Bill Nunn), labors successfully at recovering Henry's faculties.

In outline, *Regarding Henry*—video store hit—is a tale of moral transfor-mation. Henry Turner is a corporate Scrooge who earns a fortune defending insurance companies against just suits brought by the injured and impecu-nious. Between the time of the gunshot wound and his return to his law firm, he experiences a change of heart—awakens to the meanness and corruption of his legal work and begins a movement toward personal reform. The sole influence on this transformation is Bradley, who shows the lawyer a persuasive example of selfless concern for others.

Bradley is called upon, subsequently, to give further guidance. Back in his luxo apartment and offices, Henry Turner, aware finally of the amoral selfishness of his professional life and of his behavior as husband and father, sags into depression—refuses to leave his bed. His wife, Sarah (Annette Bening), summons the black therapist, the only man Henry respects. Over beer in Henry's kitchen, Bradley tells his host of a crisis of his own—a football injury which, although it ended his athletic career, opened the prospect of the more rewarding life of service he now leads.

But does Bradley really believe, asks Henry Turner, that he's better off because of the accident? His black friend answers by citing the satisfactions of helping others, adding that, except for his football mishap, "I would never have gotten to know you."

The black man speaks as though fully convinced that his own turning point—his unwanted second choice of life—and Henry Turner's are precisely similar. Nothing in his voice hints at awareness either of the gap between riches and privation or of the ridiculousness of the pretense that race and class—differences in inherited property, competencies, beliefs, manners, advantages, burdens—don't count. Wealthy white lawyer and humble black therapist speak and behave as though both were Ivy League clubmen, equally knowledgeable about each other's routines, habits, tastes. The root of Bradley's happiness as he sings his song of praise to his white buddy is that, for Henry Turner, difference doesn't exist.

The predictable closer: a new Henry Turner launches an effort at restitution to the poor whom his chicanery has cheated; black-white friendship not only makes us one but makes us good.

When crime enters, fellow feeling should in theory exit. But visions of the force of friendship challenge this rule, too. They thrust characters and audiences into hitherto unexplored passages of self-interrogation and self-definition, obliging whites to clarify, for themselves, the distinction between humane and racist responses to troubling black behavior. And they present the process of arriving at a humane response—i.e., one that doesn't allow a criminal act to derail black-white sympathy and friendship—as an act of personal reparation.

As in John Guare's highly praised *Six Degrees of Separation* (1990; movie version 1993). This work alludes to a real-life episode involving the victimization of a Manhattan family by a young black hustler, and it develops two themes. The first is that of black hunger for white friendship. The second is that of white readiness to suffer any injury if doing so is the only means of of sympathy and maintaining a one-on-one experience of sympathy and sameness with the other race.

Like Bradley the therapist, Paul, the hustler in the story, poses no black counterview—at the level of beliefs, tastes, feelings, or aspirations—to the values of rich whites. A man without a history, he studies and apes white, upper-bourgeois manners—speech, accents, clothes—not for the purpose of better cheating his models but out of desire to make those manners his own. When the heroine he's hustling asks him, "What did you want from us?" he answers, honestly, "Everlasting friendship." He tells her husband, an art dealer, "What I should do is what you do—in art but making money out of art and meeting people and not working in an office."

He begs the couple to "let me stay with you," tells them, again truthfully, that an evening he spent with them—before his scam was discovered—was "the happiest night I ever had." The heroine sums it up: "He wanted to be us. Everything we are in the world, this paltry thing—our life—he wanted it. He stabbed himself to get in here. He envied us."

This black idolater of fashionable white "grace," "style," and "tolerance"— this youngster obsessed with being one of *them*—defines the world as the majority might wish it: a place wherein the agency of interracial friendship can create, in itself, with minimum inconvenience, conditions of relative sameness between blacks and whites. And the relationship between Paul and the heroine dramatizes the imagined eagerness of whites—once they as individuals discover the rewards of one-on-one friendship—to go to any limit to preserve and nourish the friendship.

The heroine seldom wavers in her determination to indemnify the hustler for the injuries he does her. She knows the man to be a liar, knows him to be a thief, knows he's caused an innocent man's suicide and has betrayed her husband. Beyond all this she knows that, although her attachment to him is moral-imaginative, not sexual (the hustler is a homosexual), it's endangering her marriage.

Yet, in keeping with the overarching vision (sympathy heals all), she denies him nothing—not understanding, not kindness, not the goods of her house. Word that he may have killed himself brings her to the edge of hysteria. And at the crisis of the work her forbearance is dramatized as joy. The hustler, wanted for thievery, agrees to turn himself in to the police—and at once begins interrogating his white friend and victim about what she'll do to ease his imprisonment. Will she write to him when he's in jail? Yes, she promises. Will she send him books and tapes? Yes again. Will she visit him in jail? Yes. Will she wear her best clothes on these visits? "I'll knock them dead."

As her pledges come faster and more freely, the pace of the hustler's importunacy quickens; he pushes her—his victim-friend—harder, ever harder, as though intent on driving her to the borders of her generosity, the far reaches of her sufferance. His demands carry no hint of racewide protest at general injustice, no suggestion of any motive except that of one individual black's personal longing for the unconditioned, unconstrained friendship of whites.

Will they—the white woman and her husband—help him find work when he's out? Will they let him work for them, learn from them, learn the whole trade of art dealing, "not just the grotty part"? "Top to bottom" is the answer. The scene mounts to a climax at which the white woman's exuberant realization that her generosity has no limits tears a burst of wild laughter from the center of her being:

Paul: You'll help me find a place?

Ouisa: We'll help you find a place.

Paul: I have no furniture.

Ouisa: We'll help you out.

Paul: I made a list of things I liked in the museum. Philadelphia Chippendale.

Ouisa (bursts out laughing): Believe it or not, we have two Philadelphia Chippendale chairs—

Paul: I'd rather have one nice piece than a room full of junk.

Ouisa: Quality. Always. You'll have all that. Philadelphia Chippendale.

You'll have all that: a personal covenant, a lone white woman's guarantee, to a black hustler, that she'll sacrifice endlessly to ensure his well-being. Audiences are bound into that covenant by the final blackout, identifying with the passionate sacrament of the heroine's pledges, vicariously sharing the ideal of self-abnegating beneficence to black countrymen who long to become their friends. At its climax *Six Degrees of Separation* is about the "truth" that individual whites and blacks can scarcely bear not being each other. And the play's lesson is twofold: first, when whites are drawn into friendship and sympathy, one on one, with blacks, they will go to extreme lengths to suppress vexations and the sense of injury; second, once blacks are awarded unconditional white friendship, *as individuals,* they cease to harbor any sense of vexation or injury that would need suppressing.

The message confirms, for the right-minded majority, that racism is one-dimensional—lacking, that is, in institutional, historical, or political ramifications. And the quantity of similar confirmation elsewhere in popular entertainment is, speaking matter-of-factly, immense.

Incessantly and deliberately, the world of pop is engaged in demonstrating, through images, that racism has to do with private attitudes and emotions—with personal narrowness and meanness—not with differences in rates of black and white joblessness and poverty, or in black and white income levels, or in levels of financing of predominantly black and white public schools. The images body forth an America wherein some are more prosperous than others but all—blacks as well as whites—rest firmly in the "middle income sector" (the rising black middle class encompasses all blacks), where the free exchange of kindness should be the rule.

This America is of course remote from fact. One out of every two black children lives below the poverty line (as compared with one out of seven white children). Nearly four times as many black families exist below the poverty line as white families. Over 60 percent of African Americans have incomes below $25,000. For the past thirty years black unemployment rates have averaged two to three times higher than those of whites.

But in the world of pop, racism and fraternity have to do solely with the conditions of personal feeling. Racism is unconnected with ghetto life patterns that abstractions such as income and employment numbers can't dramatize. Racism has nothing to do with the survival strategies prudently adopted by human beings without jobs or experience of jobs or hope of jobs. It has no link with the rational rejection, by as many as half the young black men in urban America, of such dominant culture values as ambition, industry, and respect for constituted authority. Pop shows its audiences that racism is *nothing but* personal hatred, and that when hatred ends, racism ends. The sweet, holiday news is that, since hatred is over, we—blacks and whites together, knit close in middleness—have already overcome.

COMING TO TERMS

What are the assumptions, which according to DeMott, define the ways in which black–white friendships are presented in advertising, on television, and in films? Find these assumptions and record them in your journal. Where does DeMott say these assumptions come from? Would you agree with De-Mott that these assumptions are commonplace? Would *you* make these assumptions? Having read what DeMott has to say about black–white friendships, as they are portrayed in popular media, do you think it's possible to represent black–white friendships in a more realistic way? DeMott doesn't do this work, but in your journal explain how you think a black–white friendship *might* be presented.

READING AS A WRITER

One way of reading "Visions of Black–White Friendship" is to see DeMott as a writer presenting an argument or position, then organizing and presenting evidence to support the claims he makes about how media culture represents black–white friendships. Where in DeMott's article does he state his position or argument? Record this passage in your journal. Next, mark those places in the text where you think DeMott is presenting evidence to support his position, to show what he has observed is true, valid, or important. We then would like you to think about the following set of questions and record a response to them in your journal: Has DeMott convinced you that what he has said is true, valid or important? Why or why not? DeMott seems to be very deliberate in presenting only those examples or illustrations of ads, films, and television programs that fully support his viewpoint or position. That is, he offers no exceptions or complications. As a writer who has made arguments before, what's your position on this approach? What might be the consequences of this approach? Would you take a different one? Why or why not?

WRITING CRITICISM

Using DeMott as a place to begin, we would like you to think about the ways in which popular culture represents friendships, antagonism or significant relationships between or among people of different races, social groupings, economic classes, ethnicities, ages, sexual orientations, or even professions. In his reading of popular culture, DeMott seems convinced that media-generated black–white friendships do not acknowledge the profound historical, economic, cultural, and social differences to be found between these two races. Popular culture, according to DeMott's way of seeing things, erases differences and covers up serious problems that need to be addressed.

We want you to select some popular culture medium and write an essay in which you discuss the friendship, adversarial, or significant relationship between two or more people who have formed a relationship of difference based on race or any of the other differences mentioned earlier. You might select a relationship from a film, television program, commercial or print advertisement, or celebrity biography or autobiography. First, you'll need to describe this relationship of difference by talking about what was said and what was done when these people interact. For printed ads, you'll have to imagine the relationship from what is presented in the ad. After you discuss and provide details of this relationship, we would like you to test DeMott's assertion that ethnic and cultural differences are erased or forgotten through the representation of difference in popular culture. Based on your reading of DeMott, and your reading of a popular culture medium, would you concur with DeMott, seeing moments of truth in what he has said or would you challenge what DeMott has to say? Like DeMott you'll need to present supporting evidence for your position.

Blue Jeans

Umberto Eco

A few weeks ago, Luca Goldoni wrote an amusing report from the Adriatic coast about the mishaps of those who wear blue jeans for reasons of fashion, and no longer know how to sit down or arrange the external reproductive apparatus. I believe the problem broached by Goldoni is rich in philosophical reflections, which I would like to pursue on my own and with the maximum seriousness, because no everyday experience is too base for the thinking man, and it is time to make philosophy proceed, not only on its own two feet, but also with its own loins.

I began wearing blue jeans in the days when very few people did, but always on vacation. I found—and still find—them very comfortable, especially when I travel, because there are no problems of creases, tearing, spots. Today they are worn also for looks, but primarily they are very utilitarian. It's only in the past few years that I've had to renounce this pleasure because I've put on weight. True, if you search thoroughly you can find an *extra large* (Macy's could fit even Oliver Hardy with blue jeans), but they are large not only around the waist, but also around the legs, and they are not a pretty sight.

Recently, cutting down on drink, I shed the number of pounds necessary for me to try again some *almost* normal jeans. I underwent the calvary described by Luca Goldoni, as the saleswoman said, "Pull it tight, it'll stretch a bit"; and I emerged, not having to suck in my belly (I refuse to accept such compromises). And so, after a long time, I was enjoying the sensation of wearing pants that, instead of clutching the waist, held the hips, because it is a characteristic of jeans to grip the lumbar-sacral region and stay up thanks not to suspension but to adherence.

After such a long time, the sensation was new. The jeans didn't pinch, but they made their presence felt. Elastic though they were, I sensed a kind of sheath around the lower half of my body. Even if I had wished, I couldn't turn or wiggle my belly *inside* my pants; if anything, I had to turn it or wiggle it *together with* my pants. Which subdivides so to speak one's body into two independent zones, one free of clothing, above the belt, and the other organically identified with the clothing, from immediately below the belt to the anklebones. I discovered that my movements, my way of walking, turning, sitting, hurrying, were *different.* Not more difficult or less difficult, but certainly different.

As a result, I lived in the knowledge that I had jeans on, whereas normally we live forgetting that we're wearing undershorts or trousers. I lived for my jeans, and as a result I assumed the exterior behavior of one who wears jeans. In any case, I assumed a *demeanor.* It's strange that the traditionally most informal and anti-etiquette garment should be the one that so strongly imposes an etiquette. As a rule I am boisterous, I sprawl in a chair, I slump wherever I please, with no claim to elegance: my blue jeans checked these actions, made me more polite and mature. I discussed it at length, especially with consultants of the opposite sex, from whom I learned what, for that matter, I had already suspected: that for women experiences of this kind are familiar because all their garments are conceived to impose a demeanor—high heels, girdles, brassieres, pantyhose, tight sweaters.

I thought then about how much, in the history of civilization, dress as armor has influenced behavior and, in consequence, exterior morality. The Victorian bourgeois was stiff and formal because of stiff collars; the nineteenth-century gentleman was constrained by his tight redingotes, boots, and top hats that didn't allow brusque movements of the head. If Vienna had been on the equator and its bourgeoisie had gone around in Bermuda shorts, would Freud have described the same neurotic symptoms, the same Oedipal triangles? And would he have described them in the same way if he, the doctor, had been a Scot, in a kilt (under which, as everyone knows, the rule is to wear nothing)?

A garment that squeezes the testicles makes a man think differently. Women during menstruation; people suffering from orchitis, victims of hemorrhoids, urethritis, prostrate and similar ailments know to what extent pressures or obstacles in the sacroiliac area influence one's mood and mental agility. But the same can be said (perhaps to a lesser degree) of the neck, the back, the head, the feet. A human race that has learned to move about in shoes has oriented its thought differently from the way it would have done if the race had gone barefoot. It is sad, especially for philosophers in the idealistic tradition, to think that the Spirit originates from these conditions; yet not only is this true, but the great thing is that Hegel knew it also, and therefore studied the cranial bumps indicated by phrenologists, and in a book actually entitled *Phenomenology of Mind*. But the problem of my jeans led me to other observations. Not only did the garment impose a demeanor on me; by focusing my attention on demeanor, it obliged me to *live toward the exterior world*. It reduced, in other words, the exercise of my interior-ness. For people in my profession it is normal to walk along with your mind on other things: the article you have to write, the lecture you must give, the relationship between the One and the Many, the Andreotti government, how to deal with the problem of the Redemption, whether there is life on Mars, the latest song of Celentano, the paradox of Epimenides. In our line this is called "the interior life." Well, with my new jeans my life was entirely exterior: I thought about the relationship between me and my pants, and the relationship between my pants and me and the society we lived in. I had achieved hetero-consciousness, that is to say, an epidermic self-awareness.

I realized then that thinkers, over the centuries, have fought to free themselves of armor. Warriors lived an exterior life, all enclosed in cuirasses and tunics; but monks had invented a habit that, while fulfilling, *on its own*, the requirements of demeanor (majestic, flowing, all of a piece, so that it fell in statuesque folds), it left the body (inside, underneath) completely free and unaware of itself. Monks were rich in interior life and very dirty, because the body, protected by a habit that, ennobling it, released it, was free to think, and to forget about itself. The idea was not only ecclesiastic; you have to think only of the beautiful mantles Erasmus wore. And when even the intellectual must dress in lay armor (wigs, waistcoats, knee breeches) we see that when he retires to think, he swaggers in rich dressing-gowns, or in Balzac's loose, *drôlatique* blouses. Thought abhors tights.

But if armor obliges its wearer to live the exterior life, then the age-old female spell is due also to the fact that society has imposed armors on women, forcing them to neglect the exercise of thought. Woman has been enslaved by fashion not only because, in obliging her to be attractive, to maintain an ethereal demeanor, to be pretty and stimulating, it made her a sex object; she has been enslaved chiefly because the clothing counseled for her forced her psychologically to live for the exterior. And this makes us realize how intellectually gifted and heroic a girl had to be before she could become, in those clothes, Madame de Sévigné, Vittoria Colonna, Madame Curie, or Rosa Luxemberg. The reflection has some value because it leads us to discover that, apparent symbol of liberation and equality with men, the blue jeans that fashion today imposes on women are a trap of Domination; for they don't free the body, but subject it to another label and imprison it in other armors that don't seem to be armors because they apparently are not "feminine."

A final reflection—in imposing an exterior demeanor, clothes are semiotic devices, machines for communicating. This was known, but there had been no attempt to illustrate the parallel with the syntactic structures of language, which, in the opinion of many people, influence our view of the world. The syntactic structures of fashions also influence our view of the world, and in a far more physical way than the *consecutio temporum* or the existence of the subjunctive. You see how many mysterious paths the dialectic between oppression and liberation must follow, and the struggle to bring light.

Even via the groin.

COMING TO TERMS

Eco distinguishes here between clothes that seem to allow or encourage "the interior life" and those that force one instead "to live towards the exterior world." In your journal explain what you think Eco means by these two classifications. Then draw a line down the middle of a page in your journal. Label one side of the line "interior" and the other "exterior." Now go back through Eco's piece in order to see what examples he gives of both kinds of dress. What patterns do you see forming on the two sides of the page? For example, what might dictate whether someone dresses for the "interior" or the "exterior" world? Or do you see any pattern which might be understood in terms of gender? How do your lists allow you to question Eco's terminology and his belief that clothes can affect how we act and even how we think?

READING AS A WRITER

Eco is writing about one of the most mundane subjects imaginable: blue jeans. Yet the same could hardly be said for his style of writing—which is dense with technical terms, learned phrases and asides, and allusions to philosophy, history, and literature. How might you account for this disparity between style and subject? Do you think Eco is simply trying to raise blue jeans to a properly serious and high-minded level of discussion? Or perhaps he is mocking those who would talk about such things seriously or academically? Or maybe he is doing both at the same time?

These are questions a reader might ask in attempting to get a hold of an author's tone or attitude toward a subject. How would you describe the tone of "Blue Jeans"? Is the tone pretentious? funny? clever? smart? Choose some passages from Eco's work and in your journal explain how you might read these passages as forming a consistent tone. Or do the passages you have selected show an inconsistency of tone? Once you have determined the tone, how do you respond to Eco's tone? In what ways does the tone make you sympathetic to Eco's argument? Or does the tone make you suspicious or even hostile? Have you read other authors in *Media Journal* who have written pieces in which there seem to be incongruities among style, tone, and subject? How would you represent their success or failure?

WRITING CRITICISM

To get at what he sees as the meaning of blue jeans, Umberto Eco considers not only what they look like to others, or signify in some abstract or semiotic sense, but also draws on the actual experience of *wearing* them. In a very real way, then, his essay is concerned with the ways the meanings of fashion are felt or experienced on the body. We'd like you to test and extend Eco's project here. First, you would need to represent Eco's way of seeing clothes as either part of an "interior" or "exterior" world. Find passages that might help you explain Eco's means of analysis here.

Next, see if you can add some examples to his from your own collected experiences with clothes or fashion. Are there other clothes, besides blue jeans, in your closets or drawers that make you somehow more aware of the "exterior world" when you put them on? When do you wear them? For what occasions? To be seen or noticed by whom? How would you explain this "exterior world" to others? Conversely, do you have clothes that let you (more or less) forget about the "external world"? How do the clothes accomplish this? When do you wear these clothes?

Next, take a look around you. Do the other students in your classes seem to dress for an "interior" or "exterior" life? What about your teachers? Parents and siblings? The people on the bus or subway? Those you meet working in offices or shops or the mall? Those you see at the movies or in libraries or museums? Do you see any patterns to who dresses in what way? Might these patterns be understood in terms of gender, age, profession, race, or social status? Having now applied Eco's way of reading clothes to yourself and others, what might you add to what Eco has to say about fashion?

Playboy *Joins the Battle of the Sexes*

Barbara Ehrenreich

*I don't want my editors marrying anyone
and getting a lot of foolish notions in their heads
about "togetherness," home, family, and all that jazz.*
—HUGH HEFNER

The first issue of *Playboy* hit the stands in December 1953. The first centerfold—the famous nude calendar shot of Marilyn Monroe—is already legendary. Less memorable, but no less prophetic of things to come, was the first feature article in the issue. It was a no-holds-barred attack on "the whole concept of alimony," and secondarily, on money-hungry women in general, entitled "Miss Gold-Digger of 1953." From the beginning, *Playboy* loved women—large-breasted, long-legged young women, anyway—and hated wives.

The "Miss Gold-Digger" article made its author a millionaire—not because Hugh Hefner paid him so much but because Hefner could not, at first, afford to pay him at all, at least not in cash. The writer, Burt Zollo (he signed the article "Bob Norman"; even Hefner didn't risk putting his own name in the first issue), had to accept stock in the new magazine in lieu of a fee. The first print run of 70,000 nearly sold out and the magazine passed the

From *"Playboy* Joins the Battle of the Sexes" by Barbara Ehrenreich, 1983, *The Hearts of Men.* © 1983 by Barbara Ehrenreich. Reprinted by permission of Doubleday, a division of Bantam Doubleday Dell Publishing Group, Inc.

one-million mark in 1956, making Hefner and his initial associates million-aires before the end of the decade.

But *Playboy* was more than a publishing phenomenon, it was like the party organ of a diffuse and swelling movement. Writer Myron Brenton called it the "Bible of the beleaguered male." *Playboy* readers taped the centerfolds up in their basements, affixed the rabbit-head insignia to the rear windows of their cars, joined *Playboy* clubs if they could afford to and, even if they lived more like Babbits than Bunnies, imagined they were "playboys" at heart. The magazine encouraged the sense of membership in a fraternity of male rebels. After its first reader survey, *Playboy* reported on the marital status of its constituency in the following words: "Approximately half of PLAYBOY'S readers (46.8%) are free men and the other half are free in spirit only."

In the ongoing battle of the sexes, the *Playboy* office in Chicago quickly became the male side's headquarters for wartime propaganda. Unlike the general-audience magazines that dominated fifties' newsstands—*Life, Time, the Saturday Evening Post, Look,* etc.—*Playboy* didn't worry about pleasing women readers. The first editorial, penned by Hefner himself, warned:

> *We want to make clear from the very start, we aren't a "family magazine." If you're somebody's sister, wife or mother-in-law and picked us up by mistake, please pass us along to the man in your life and get back to your Ladies' Home Companion.*

When a Memphis woman wrote in to the second issue protesting the "Miss Gold-Digger" article, she was quickly put in her place. The article, she wrote, was "the most biased piece of tripe I've ever read," and she went on to deliver the classic anti-male rejoinder:

> *Most men are out for just one thing. If they can't get it any other way, sometimes they consent to marry the girl. Then they think they can brush her off in a few months and move on to new pickings. They ought to pay, and pay, and pay.*

The editors' printed response was, "Ah, shaddup!"

Hefner laid out the new male strategic initiative in the first issue. Recall that in their losing battle against "female domination," men had been driven from their living rooms, dens and even their basement tool shops. Escape seemed to lie only in the great outdoors—the golf course, the fishing hole or the fantasy world of Westerns. Now Hefner announced his intention to reclaim *the indoors for men.* "Most of today's 'magazines for men' spend all their time out-of-doors—thrashing through thorny thickets or splashing about in fast flowing streams," he observed in the magazine's first editorial. "But we don't mind telling you in advance—we plan spending most of our time inside. WE like our apartment." For therein awaited a new kind of good life for men:

> *We enjoy mixing up cocktails and an* hors d'oeuvre *or two, putting a little mood music on the phonograph and inviting in a female acquaintance for a quiet discussion on Picasso, Nietzsche, jazz, sex.*

Women would be welcome after men had reconquered the indoors, but only as guests—maybe overnight guests—but not as wives.

In 1953, the notion that the good life consisted of an apartment with mood music rather than a ranch house with barbecue pit was almost subversive. Looking back, Hefner later characterized himself as a pioneer rebel against the gray miasma of conformity that gripped other men. At the time the magazine began, he wrote in 1963, Americans had become "increasingly concerned with security, the safe and the sure, the certain and the known...it was unwise to voice an unpopular opinion...for it could cost a man his job and his good name." Hefner himself was not a political dissident in any conventional sense; the major intellectual influence in his early life was the Kinsey Report, and he risked his own good name only for the right to publish bare white bosoms. What upset him was the "conformity, togetherness, anonymity and slow death" men were supposed to endure when the good life, the life which he himself came to represent, was so close at hand.

In fact, it was close at hand, and, at the macroeconomic level, nothing could have been more in conformity with the drift of American culture than to advocate a life of pleasurable consumption. The economy, as Riesman, Galbraith and their colleagues noted, had gotten over the hump of heavy capital accumulation to the happy plateau of the "consumer society." After the privations of the Depression and the war, Americans were supposed to enjoy themselves—held back from total abandon only by the need for Cold War vigilance. Motivational researcher Dr. Ernest Dichter told businessmen:

> *We are now confronted with the problem of permitting the average American to feel moral...even when he is spending, even when he is not saving, even when he is taking two vacations a year and buying a second or third car. One of the basic problems of prosperity, then, is to demonstrate that the hedonistic approach to his life is a moral, not an immoral one.*

This was the new consumer ethic, the "fun morality" described by sociologist Martha Woffenstein, and *Playboy* could not have been better designed to bring the good news to men.

If Hefner was a rebel it was only because he took the new fun morality seriously. As a guide to life, the new imperative to enjoy was in contradiction with the prescribed discipline of "conformity" and *Playboy*'s daring lay in facing the contradiction head-on. Conformity, or "maturity," as it was more affirmatively labeled by the psychologists, required unstinting effort: developmental "tasks" had to be performed, marriages had to be "worked on," individual whims had to be subordinated to the emotional and financial needs of the family. This was true for both sexes, of course. No one pretended that the adult sex roles—wife/mother and male breadwinner—were "fun." They were presented in popular culture as achievements, proofs of the informed acquiescence praised as "maturity" or, more rarely, lamented as "slow death." Women would not get public license to have fun on a mass scale for more than a decade, when Helen Gurley Brown took over *Cosmopolitan* and began promoting a tamer, feminine version of sexual and material consumerism. But *Playboy* shed the burdensome aspects of the adult male role at a time when businessmen were still refining the "fun morality" for mass consumption, and the gray flannel rebels were still fumbling for responsible alternatives like Riesman's "autonomy." Even the magazine's name defied the convention of hard-won maturity—*Playboy*.

Playboy's attack on the conventional male role did, not, however, extend to the requirement of earning a living. There were two parts to adult masculinity: One was maintaining a monogamous marriage. The other was working at a socially acceptable job; and *Playboy* had nothing against work. The early issues barely recognized the white-collar blues so fashionable in popular sociology. Instead, there were articles on accoutrements for the rising executive, suggesting that work, too, could be a site of pleasurable consumption. Writing in his "*Playboy* Philosophy" series in 1963, Hefner even credited the magazine with inspiring men to work harder than they might: "...*Playboy* exists, in part, as a motivation for men to expend greater effort in their work, develop their capabilities further and climb higher on the ladder of success." This kind of motivation, he went on, "is obviously desirable in our competitive, free enterprise system," apparently unaware that the average reader was more likely to be a white-collar "organization man" or blue-collar employee rather than a free entrepreneur like himself. Men should throw themselves into their work with "questing impatience and rebel derring-do." They should overcome their vague, ingrained populism and recognize wealth as an achievement and a means to personal pleasure. Only in one respect did Hefner's philosophy depart from the conventional, Dale Carnegie-style credos of male success: *Playboy* believed that men should make money; it did not suggest that they share it.

Playboy charged into the battle of the sexes with a dollar sign on its banner. The issue was money: Men made it; women wanted it. In *Playboy*'s favorite cartoon situation an elderly roué was being taken for a ride by a buxom bubblebrain, and the joke was on him. The message, squeezed between luscious full-color photos and punctuated with female nipples, was simple: You can buy sex on a fee-for-service basis, so don't get caught up in a long-term contract. Phil Silvers quipped in the January 1957 issue:

A tip to my fellow men who might be on the brink of disaster: when the little doll says she'll live on your income, she means it all right. But just be sure to get another one for yourself.

Burt Zollo warned in the June 1953 issue:

It is often suggested that woman is more romantic than man. If you'll excuse the ecclesiastical expression—phooey!... All woman wants is security. And she's perfectly willing to crush man's adventurous, freedom-loving spirit to get it.

To stay free, a man had to stay single.

The competition, meanwhile, was still fighting a rearguard battle for patriarchal authority within marriage. In 1956, the editorial director of *True* attributed his magazine's success to the fact that it "stimulates the masculine ego at a time when man wants to fight back against women's efforts to usurp his traditional role as head of the family." The playboy did not want his "traditional role" back; he just wanted out. Hefner's friend Burt Zollo wrote in one of the early issues:

Take a good look at the sorry, regimented husbands trudging down every woman-dominated street in this woman-dominated land. Check what they're doing when you're out on the town with a different dish every night... Don't bother asking their advice. Almost to a man, they'll tell you marriage is the greatest. Naturally. *Do you expect them to admit they made the biggest mistake of their lives?*

This was strong stuff for the mid-fifties. The suburban migration was in full swing and *Look* had just coined the new noun "togetherness" to bless the isolated, exurban family. Yet here was *Playboy* exhorting its readers to resist marriage and "enjoy the pleasures the female has to offer without becoming emotionally involved"—or, of course, financially involved. Women wrote in with the predictable attacks on immaturity: "It is...the weak-minded little idiot boys, not yet grown up, who are afraid of getting 'hooked.'" But the men loved it. One alliterative genius wrote in to thank *Playboy* for exposing those "cunning cuties" with their "suave schemes" for landing a man. And, of course, it was *Playboy,* with its images of cozy concupiscence and extra-marital consumerism, that triumphed while *True* was still "thrashing through the thorny thickets" in the great, womanless outdoors.

One of the most eloquent manifestos of the early male rebellion was a *Playboy* article entitled, "Love, Death and the Hubby Image," published in 1963. It led off with a mock want ad:

TIRED OF THE RAT RACE?
FED UP WITH JOB ROUTINE?
Well, then…how would you like to make $8,000, $20,000—as much as
$50,000 and More—*working at Home in Your Spare Time? No selling!
No commuting! No time clocks to punch!*
BE YOUR OWN BOSS!!!
*Yes, an Assured Lifetime Income can be yours now, in an easy, low-pressure,
part-time job that will permit you to spend most of each and every day as you
please!*—*relaxing, watching TV, playing cards, socializing with friends!…*

"Incredible though it may seem," the article began, "the above offer is completely legitimate. More than 40,000,000 Americans are already so employed…"
They were, of course, wives.

According to the writer, William Iversen, husbands were self-sacrificing
romantics, toiling ceaselessly to provide their families with "bread, bacon,
clothes, furniture, cars, appliances, entertainment, vacations and country-
club memberships." Nor was it enough to meet their daily needs; the heroic,
male must provide for them even after his own death by building up his sav-
ings and life insurance. "Day after day, and week after week the American
hubby is thus invited to attend his own funeral." Iversen acknowledged that
there were some mutterings of discontent from the distaff side, but he saw
no chance of a feminist revival: The role of the housewife "has become much
too cushy to be abandoned, even in the teeth of the most crushing boredom."
Men, however, had had it with the breadwinner role, and the final paragraph was a stirring incitement to revolt:

> *The last straw has already been served, and a mere tendency to hemophilia
> cannot be counted upon to ensure that men will continue to bleed for the plight
> of the American woman. Neither double eyelashes nor the blindness of night
> or day can obscure the glaring fact that American marriage can no longer be
> accepted as an estate in which the sexes shall live half-slave and half-free.*

Playboy had much more to offer the "enslaved" sex than rhetoric: It also
proposed an alternative way of life that became ever more concrete and vivid
as the years went on. At first there were only the Playmates in the centerfold
to suggest what awaited the liberated male, but a wealth of other consumer
items soon followed. Throughout the late fifties, the magazine fattened on advertisements for imported liquor, stereo sets, men's colognes, luxury cars and
fine clothes. Manufacturers were beginning to address themselves to the adult
male as a consumer in his own right, and they were able to do so, in part, because magazines like *Playboy* (a category which came to include imitators like
Penthouse, Gent and *Chic*) allowed them to effectively "target" the potential syb-
arites among the great mass of men. New products for men, like toiletries and
sports clothes, appeared in the fifties, and familiar products, like liquor, were

presented in *Playboy* as accessories to private male pleasures. The new male-centered ensemble of commodities presented in *Playboy* meant that a man could display his status or simply flaunt his earnings without possessing either a house or a wife—and this was, in its own small way, a revolutionary possibility.

Domesticated men had their own commodity ensemble, centered on home appliances and hobby hardware, and for a long time there had seemed to be no alternative. A man expressed his status through the size of his car, the location of his house, and the social and sartorial graces of his wife. The wife and home might be a financial drag on a man, but it was the paraphernalia of family life that established his position in the occupational hierarchy. *Playboy*'s visionary contribution—visionary because it would still be years before a significant mass of men availed themselves of it—was to give the means of status to the single man: not the power lawn mower, but the hi-fi set in mahogany console; not the sedate, four-door Buick, but the racy little Triumph; not the well-groomed wife, but the classy companion who could be rented (for the price of drinks and dinner) one night at a time.

So through its articles, its graphics and its advertisements *Playboy* presented, by the beginning of the sixties, something approaching a coherent program for the male rebellion: a critique of marriage, a strategy for liberation (reclaiming the indoors as a realm for masculine pleasure) and a utopian vision (defined by its unique commodity ensemble). It may not have been a revolutionary program, but it was most certainly a disruptive one. If even a fraction of *Playboy* readers had acted on it in the late fifties, the "breakdown of the family" would have occurred a full fifteen years before it was eventually announced. Hundreds of thousands of women would have been left without breadwinners or stranded in court fighting for alimony settlements. Yet, for all its potential disruptiveness, *Playboy* was immune to the standard charges leveled against male deviants. You couldn't call it anti-capitalist or un-American, because it was all about making money and spending it. Hefner even told his readers in 1963 that the *Playboy* spirit of acquisitiveness could help "put the United States back in the position of unquestioned world leadership." You *could* call it "immature," but it already called itself that, because maturity was about mortgages and life insurance and *Playboy* was about fun. Finally, it was impervious to the ultimate sanction against male rebellion—the charge of homosexuality. The playboy didn't avoid marriage because he was a little bit "queer," but, on the contrary, because he was so ebulliently, even compulsively heterosexual.

Later in the sixties critics would come up with what seemed to be the ultimately sophisticated charge against *Playboy*: It wasn't really "sexy." There was nothing erotic, *Time* wrote, about the pink-cheeked young Playmates whose every pore and perspiration drop had been air-brushed out of existence. Hefner was "puritanical" after all, and the whole thing was no more mischievous than "a Midwestern Methodist's vision of sin." But the critics misunderstood *Playboy*'s historical role. *Playboy* was not the voice of the sexual revolution, which began, at least overtly, in the sixties, but of the male rebellion, which had begun in the fifties. The real message was not eroticism, but escape—literal escape, from the bondage of breadwinning. For that, the breasts and bottoms were necessary not just to sell the magazine, but to protect it. When, in the first issue, Hefner talked about staying in his apartment, listening to music and discussing Picasso, there was the Marilyn Monroe centerfold to let you know there was nothing queer about these urbane and indoor pleasures. And when the articles railed against the responsibilities of marriage, there were the nude torsos to reassure you that the alternative was still within the bounds of heterosexuality. Sex—or Hefner's Pepsi-clean version of it—was there to legitimize what was truly subversive about *Playboy.* In every issue, every month, there was a Playmate to prove that a playboy didn't have to be a husband to be a man.

COMING TO TERMS

"The real message," Barbara Ehrenreich writes, "was not eroticism but escape." What might she mean by this comment? What does she see the readers of *Playboy* as trying to escape? Why? What role do the images of naked women play in this escape? What other parts of the magazine become important in thinking of it as "escapist" rather than simply "erotic"? As you come to terms with these questions, we're interested in your take on the word "escapist" or "escapism." Often these words have negative connotations, seen as representing just daydreams, not worthy of intellectual consideration. What is your sense of how Ehrenreich is using the word "escape"? (Have you read other authors in *Media Journal,* who, like Ehrenreich, see escape through media culture as worthy of serious discussion and cultural explanations?) Go back through her essay with these questions in mind and jot down some notes that respond to these questions in your journal. Then, after you work out *what you* think Ehrenreich is trying to say, we'd like you to figure out why she might want to make this sort of critical argument. Is she simply trying to defend *Playboy* against those who would criticize or dismiss it as pornography? Or does she have some other kind of criticism to make of the magazine? If so, what is it?

READING AS A WRITER

There's an old joke about men who claim to read *Playboy* "*for* the articles," but that is, in fact, exactly what Ehrenreich does here, looking closely at what Hugh Heffner and his editorial staff actually had to say about what they were trying to do in their magazine. But what's also important here is that Ehrenreich is also concerned *when* Heffner and his staff were speaking in *Playboy,* in other words, with its historical and cultural context. Ehrenreich is doing a historical and cultural reading of the writers, editors, advertisers and readers of *Playboy.* We often think of the materials of history, its archives and sources, in terms of important national documents, the letters and papers of famous people, and the collections of dates and facts. In her text Ehrenreich challenges us to re-think the materials of history, to include the texts of media or popular culture. Keep this kind of work in mind as you re-read her text. Go back through Ehrenreich's essay and pay close attention to the passages she quotes from *Playboy* and how she comments on them. We might call this response to quotations *critical commentary,* and it's a kind of important academic reading or interpretation which might be heard to say: quotations don't speak until a reader makes them "speak." How does Ehrenreich make these quotations from *Playboy* speak? What do these passages tell her that simply looking at the pictures would not? How does she then connect them to other parts of the magazine, to its ads, and graphics? In what ways does Ehrenreich's reading of these passages reveal aspects of American culture and history of the 1950s? Significant historical events? Cultural values? Political beliefs or movements? Roles that women and men might be said to play? Respond to these questions in your journal. How compelling (or not) do you find the textual evidence for Ehrenreich's counter-reading of *Playboy* to be?

WRITING CRITICISM

In her piece on *Playboy,* Ehrenreich writes as a woman about a magazine aimed at men. Part of what's interesting in her piece involves this clash of gender perspectives: How does a woman view a media or popular culture product that is aimed at men, a product that includes portrayals of women produced for the pleasure of men? In this assignment we would like you to take on a project similar to Ehrenreich's. We want you to look at the other gender as shown through the media. If you are a man, select a media project aimed at or typically enjoyed by women. If you are a woman, select a media product aimed at or typically enjoyed by men. Then write a piece on the view of men and/or the view of women suggested by this product.

An effective way of finding material for this assignment is to visit a magazine stand. There you will find a larger number of women's magazines, and a smaller but still significant number of magazines aimed at men. Simply choose one that appeals to the opposite gender and that you find interesting, odd, outrageous, fun, or somehow in need of a critical reading. (Or you might make some appropriate preliminary choices here and see if the magazines are available in your school or a public library. You might also check the magazine to see if it is available, or a version of it, on the Internet. This website information often appears as part of the "letters to the editor" section of a publication.) But while magazines are the most likely material for this essay, there are other media forms you may want to consider. For example, there are categories of popular fiction (romances) aimed at women and other categories aimed at men (spy novels). Non-fiction books, like self-improvement or self-help books or even autobiographies, can be said to appeal to a specific gendered audience. You might use your experience with the Internet here as well. Websites, chatrooms, and home pages might also be read as being designed for a specifically male or female audience. Network television programs might be aimed at either a male or female audience. The same holds true for cable television: while channels like *Lifetime* or *Romance Classics* might be targeted at women, channels like *ESPN* might generate a predominantly male audience. Likewise, programs like QVC or The *Shopping Channel* might be selling a line of products for either men or women. You can choose any media form you want, but whatever you select you must be able to observe it first hand. Ideally you should have a copy of your text available to re-examine or re-read—which is another reason we suggest the use of a popular magazine.

Once you have found your material, there are a number of different approaches you might take. One is to analyze the appeal of the product for the gender it appeals to. For example, as a man, what seems to you to be the appeal of a women's fashion magazine like *Vogue?* Or, as a woman, how do you explain the American male's love of televised pro football? Another kind of approach would seek to describe the view of men offered to men by a man's media product, or the view of women offered to women by a woman's product. How are men portrayed, what is assumed about them, how is masculinity defined in a men's magazine like *GQ*? How are women discussed, what is assumed about men, and how are they addressed on a show like Oprah Winfrey's? A third approach would attend to the portrayal of women in a product aimed at men, or the portrayal of men in a product aimed at women. How are women pictured, who are they assumed to be, for example, in Playboy? What role are men given, what is assumed about them, in a typical romance novel?

These three approaches are not mutually exclusive. You may well want to combine them to create a piece that says everything you have to say about gender. (For example, you may want to analyze how the relations between genders are represented in a magazine like *Cosmopolitan* or *Esquire*.) In general, your task is to employ the advantage of the cross-gender situation that this assignment deliberately creates. That means your essay should profit from the fact that you're examining something ordinarily experienced by the opposite gender. Any method you choose to convert this fact into a series of interesting observations will work for the purposes of this assignment. As always, you should write about what provokes you in the material you are examining. In this particular essay, though, it is important to remain aware that you are writing as a member of your gender—as a woman or a man—in a particular culture, in a particular time. Just exactly what *that* might mean is part of what we'd like you to think about in this assignment.

Wrestling with Myself
George Felton

It's Saturday morning, 11 A.M., right after the cartoons: Time for "The NWA Main Event." As I watch the ringside announcer set up today's card, a wrestler—huge, topless and sweating, wearing leather chaps and a cowboy hat, carrying a lariat with a cowbell on it—bursts into frame, grabs the announcer by his lapels, and, chunks of tobacco spraying out of his mouth, begins to emote: "Well lookee here, this is just what eats in my craw.... I don't care if you're the president or the chief of police, it don't matter. I'm gonna do what I wanna do," and what he mostly wants to do is wrassle somebody good for once—enough nobodies in the ring, enough wimps running the schedule. As quickly as he spills into the camera, he veers out, having delivered exactly the 20 second sound bite required. Our announcer blithely sends us to commercial, and another Saturday's wrestling hour has begun. I feel better already.

I soon find out this cowboy's name is Stan Hanson, he's from Border, Texas, and lately he's been getting disqualified in all his matches for trying to kill his opponents and then "hogtying" them with his lariat. We get to watch a recent match in which he kicks some poor guy's stomach furiously with his pointed-toe cowboy boots and drop-slams his elbow into his neck and, after getting him down, hits him over the head with the cowbell, and first whips, then strangles him with his lariat. It's great stuff, with the bell ringing madly and the referee waving his arms, but Stan's already yanked the guy outside the ring onto the apron and he's still on top, trying to kill him.

Why do I love this? Why am I crazy about Stan Hanson, who's old and fat and a man the announcer warns us "ought to be in a straitjacket and chains"? Because he personifies the great redemption of pro wrestling, the way it delivers me from civilization and its discontents. Not only is Stan Hanson mad as hell and not taking it anymore, but he's doing it all for me—getting himself disqualified so that I won't run the risk myself, but inviting me to grab one end of the rope and pull. He is my own id—the hairy beast itself—given a Texas identity and a push from behind, propelled out there into the "squared circle" where I can get a good look at it: sweat-soaked, mean, kicking at the slats, looking for an exposed neck. My heart leaps up, my cup runneth over.

Obviously I can't tell my friends about too much of this. If I even mention pro wrestling, they just stare at me and change the subject. They think I'm kidding. I am not supposed to like pro wrestling—its demographics are too downscale, its Dumb Show too transparent. They complain that it's fake and it's silly, which to me are two of its great charms. If it were real, like boxing, it'd be too painful to watch, too sad. I like knowing it's choreographed: The staged mayhem lets me know someone has studied me and will toss out just the meat the dark, reptilian centers of my brain require to stay fed and stay put. Sadomasochism? Homoeroticism? I am treated to the spectacle of Ric "The Nature Boy" Flair, astride the corner ropes and his opponent. His fist may be in the air, triumphant, but his groin is in the other guy's face, and he keeps it there. For once the ringside announcers are speechless as we all stare, transfixed, at the clearest of symbolic postures. Consciously I am squirming, but my reptilian center feels the sun on its back.

Racism? Ethnocentrism? Am I unsettled about Japanese hegemony? No problem. There is, in the World Wrestling Federation, a tag-team of scowling, unnervingly business-oriented Japanese toughs—the Orient Express, managed by Mr. Fuji—who invite me to hate them, and of course I do. Their failure is my success, and I don't even have to leave the living room. Two oversized, red-trunked Boris types used to parade around the ring under a red flag and insist, to our booing, on singing the Russian national anthem before wrestling. Since the Cold War has become passé, I notice that an upcoming match pits the Russians *against each other,* and that, as my newspaper tells me, is not passé. I hear groans of delight from below, as this reprise of Cain and Abel croons its libidinal tune.

I mean where else can I take my id out for a walk, how else to let it smell the sweaty air, root its nose through the wet leaves? Cartoons? No amount of Wile E. Coyote spring-loaded bounces, no pancakings of Roger Rabbit, none of the whimsical annihilations of Cartoonville can approximate the satisfactions of a real boot in a real belly, a man's head twisted up in the ropes, the merry surfeit of flying drop kicks, suplexes, sleeper holds and heart punches, all landed somewhere near real bodies. Pro sports? I get more, not less, neurotic rooting for my teams—my neck muscles ache, my stomach burns with coffee, after enduring a four-hour Cleveland Browns playoff loss on TV. The Indians? Don't even get me started. The violence of movies like *RoboCop 2* and *Total Recall*? Needlessly complicated by story line.

No, give it to me straight. Wrestling may be a hybrid genre—the epic poem meets Marvel Comics via the soap opera—but its themes, with their medieval tone, could hardly be simpler: warrior kings doing battle after battle to see who is worthy, women pushed almost to the very edges of the landscape, Beowulf's heroic ideal expressed in the language of an after-school brawl: "I wanna do what I wanna do. You gonna try to stop me?"

I also appreciate the pop-culture novelty of pro wrestling, its endearing way of creating, a little smudged and thick-fingered, but with a great earnest smile ("Here, look at this!") new betes noires for our consumption. One of the newest is something called Big Van Vader, a guy in a total upper torso headgear that looks like Star Wars Meets a Mayan Temple. He carries a stake topped with a skull and can shoot steam out of ventricles on his shoulders, but it looks like all he can do to keep from toppling over. He's horrifying and silly all at once, an atavistic nightdream wearing a "Kick Me" sign.

Such low-rent Show Biz, this admixture of the asylum and the circus, is central to wrestling's double-tracked pleasure. Its emotional reductio ad absurdum taps my anger like a release valve, but its silliness allows me to feel superior to it as I watch. I can be dumb and intelligent, angry and amused, on all fours yet ironically detached, all at the same moment. It's a very satisfying mix, especially since life between Saturdays is such an exercise in self-control, modesty and late 20th-century Angst. To my students I am the helpful Mr. Felton. To my chairman I'm the responsible Mr. Felton. To virtually everybody and everything else I'm the confused, conflicted Mr. F. My violence amounts to giving people the finger, usually in traffic. When I swear I mutter. To insults I quickly add the disclaimer, "just kidding," a move I learned from watching David Letterman temper his nastiness. I never yell at people, threaten them, twist my heel into their ears, batter their heads into ring posts, or catch them flush with folding chairs. I don't wear robes and crowns and have bosomy women carry them around for me, either. In short, I never reduce my life to the satisfying oversimplification I think it deserves.

I'm a wimp. Just the sort of guy Cactus Jack or old Stan himself would love to sink his elbows into, a sentiment with which I couldn't agree more. And that brings us to the deepest appeal of pro wrestling: It invites me to imagine the annihilation of my own civilized self. When Ric Flair jabs his finger into the camera and menaces his next opponent with, "I guarantee you one thing—Junkyard Dog or no Junkyard Dog, you're going to the hospital," when another of the Four Horsemen growls, "I'm gonna take you apart on national television," the real thrill is that they're coming for me. And when Stan offers me one end of the rope, we both know just whose neck we're pulling on. Ah, redemption.

COMING TO TERMS

In "Wrestling with Myself" George Felton writes about the satisfaction he receives from watching professional wrestling on TV. His essay is about pleasure and his need to explain and defend this sense of pleasure to others in viewing the wrestling spectacle. "Why do I love this?" Felton asks himself. Go back through his text and make two lists in your journal of all the answers he gives to this question. One list would comprise personal answers or explanations; the other might be titled cultural answers, answers that somehow connect Felton's love of wrestling to his place and position in society, to his reading of certain cultural roles, rules, and constraints he lives with everyday. Of these various answers, which make the most sense in explaining the appeal of wrestling to someone like Felton? And which illuminate the ritual of wrestling for you, allow you to see something about it you might otherwise have missed?

READING AS A WRITER

Successful writers have a good sense of the audience for whom they are writing. One way to make sense of George Felton's relationship with his subject, televised professional wrestling, and to an imagined group of readers (his audience) is to look at the ways in which Felton uses *allusion* in his writing. Allusions are direct or implied references to other texts, printed or otherwise, historical events, and well-known persons within a culture. Why do you think that Felton brings together allusions from what might be called "media culture" (Wile E. Coyote) and from "high culture" (Sigmund Freud's term id)? Go back through Felton's writing, and in your journal jot down and explain as many allusions as you can from both categories. (A good college dictionary will help you with many of the allusions.) What do you think the use of these allusions has to say about Felton's relationship with his subject and audience? How would you describe this relationship? How might your work of understanding allusion help you to rethink the title, "Wrestling with Myself"? That is, what might be some of the issues Felton is "wrestling" with?

WRITING CRITICISM

Most of Felton's piece is an attempt to explain his love of wrestling, a media experience which is usually looked down upon. In the course of doing so, he must say a good deal about wrestling and what it means to him. He must also say something about himself, about the societal and cultural codes, rules and expectations under which he lives, works, and interacts with others, and why he as an individual and as a member of a particular culture and society finds meaning in the ritual of wrestling. Finally, he must defend wrestling against the views of those who don't see in it what he sees. In this assignment, we'd like you to do something like what Felton does when he asks, "Why do I love this?" That is, take some media experience that you enjoy that you know is considered somehow unacceptable, unimportant, or incomprehensible by many others. It might be something that is commonly looked down upon, that is deemed trashy or juvenile or dumb or even dangerous, something that people like yourself are not supposed to enjoy. It might be something that you feel guilty about doing (but do anyway), or something that you take seriously while others do not. Or it could be something that is meaningful for you but obscure for others—that is simply hard for them to understand. You can write about almost any sort of media experience that you want—a TV show you watch, a magazine you read, a performer or celebrity you admire, a band you like, a kind of music you prefer, a certain type of movie you know you'll love, an Internet chat room, a form of fiction you avidly consume as science fiction or romance novels or a media ritual you've developed. These are only suggestions, and they do not cover all the possibilities.

In trying to account for the pleasure or meaning you find in this media experience, you will want to ask yourself three basic kinds of questions. First, what is it about this *experience* that you find so attractive? Here the challenge is to be as specific and detailed as you can in describing the attraction. Second, what is it about you that causes you to be so attracted? Here, the challenge is to be revealed in an interesting way—to talk about yourself in a way that helps the rest of us to see why the experience you describe is compelling, worthwhile or just plain fun. And third, how would you explain your attraction in terms of a larger cultural background? Here, the challenge asks you to understand your attraction to a media experience because your place, position, and identity are somehow defined and regulated by certain rules, beliefs, or expectations. Implied in Felton's question, "Why do I love this?" is another, related question, "Why don't others love what I love?" You may want to address this directly. That is, you may want to say something about what it is that others can't see, fail to appreciate, or don't understand about the experience you enjoy.

Confessions of a TV Talk Show Shrink

Steven Fischoff

To audience applause, Geraldo Rivera runs down the center aisle of the studio. Three women are seated next to me on stage, ready to talk, their chests nervously heaving, faces frozen in polite smiles. They listen as Geraldo details their shocking biographies to the audience. My pulse accelerates as Geraldo moves beside me, his hype escalating: "Dr. Stuart Fischoff, a clinical psychologist from Los Angeles, is here to help answer the $64,000 question: Why would a woman want to marry her rapist?"

Throughout the show, Geraldo tosses questions at me. Whys, whys, and more whys. But I don't know these women. I am armed with only bits and pieces of their self-justifying explanations. People are authorities on women, on marriage, on rape. But no one is an authority on why women marry their rapists.

I offer only general comments, about low self-esteem (it's always about low self-esteem, isn't it?) and about the illusory bond between rape and the romantic myth of being taken because one is so needed, so desirable. "It's not about love and desire," I say. "It's about anger and dominance." My words go past the three women like whistles down a wind. They don't hear me. They can't hear me. They are not there to be helped. They have come to be validated.

The focus shifts. The studio mikes are activated. The tension rises. The women in the audience are openly furious. Validation is not on their agenda. Their questions and accusations speak of betrayal, of pandering to the odious

Steven Fischoff, "Confessions of a TV Talk Show Shrink," *Psychology Today*, Sept/Oct 1995, pp. 38–45. Reprinted with permission from *Psychology Today* Magazine. Copyright © 1995 (Sussex Publ., Inc.).

stereotype that women secretly want to be raped. The women on stage try to defend the indefensible. It is a battle they cannot win. It is another rape. Only this time it's by a gang of women who are angry and want to dominate.

In the show's last segment, Geraldo stands on stage, points to me, and says, "In thirty seconds or less, Dr. Fischoff, give us your impression of these women."

My mind gulps, "Thirty seconds? Is he kidding?"

No, he's not. My thoughts run wild. "What the hell am I doing here? Open your mouth, say something, something smart, clever, incisive. Don't embarrass yourself."

My mouth starts before I know what I'm going to say. Out spill the words; glib, facile words. I do my job. I perform.

"What the hell was I doing on *Geraldo*?" I asked myself a few weeks later, after watching a videotape of the show. As a university professor and a clinical psychologist in private practice for over 25 years, appearing on television to discuss the human condition was something I was comfortable with. More accurately, I loved it. Teaching and TV commentary are both performance arts.

Public speaking ranks first among phobias; not every academic nor every psychologist relishes the opportunity to speak to millions of people via the electronic pulpit. But if you like offering insightful *bon mots* on camera and do it well, you get ample opportunity. The mass media have taken psychologists and their ilk to their collective bosom; we turn up everywhere, on television, in magazines, in newspapers, opining away. And if you live in Los Angeles, one of the top news and media markets given its identification with Hollywood, your media exposure opportunities are increased to the tenth power.

I began in the 1970s, discussing assertiveness training. Over time, I evolved into what is now officially known as a media psychologist. There is even a division of the American Psychological Association devoted to media psychology. Media psychologists research and write about the impact of the media on society. Those who can talk easily and succinctly and explain complex social events and psychological issues in nonjargonized language also appear on or in the media themselves.

My love of media exposure notwithstanding, I had adamantly refused to do interviews with tabloid publications like *The National Enquirer* or such magazines as *Hustler*. Whatever the merit of a particular article involved, the context of the publication as a whole was off-putting. So when a producer from *Geraldo* called (she got my name from the public affairs department of the APA), I approached with caution on the invitation to appear on the show.

I had done some local talk shows and my experiences on them had been good. But Geraldo was another matter entirely. He epitomized the tabloid format of contemporary talk shows. First there was *Donahue*. Then there was *Oprah*. Then there were *Geraldo* and *Sally*, and the platform of talk show taste standards collapsed completely.

Despite my misgivings, I went on the show. It was network television. It would be a learning experience.

It was. But so is falling off a cliff. Looking at the *Geraldo* tape, I was embarrassed. Forget what I said about these rapist-marrying women and the neurotic lock in which they were embraced. The words were fine. Doing the show—*that* was the embarrassment! I was caught up in the talk show juggernaut that later I would see trap other "experts." Worse, Geraldo's unexpected request to describe these women in 30 seconds put me in the role of shoot-from-the-hip psychologist. I didn't like it. I liked even less my attempt to deliver the goods.

It will get better, I reassured myself. And, in fact, there were times over the course of the numerous daily talk shows when the show's tempo or the host's integrity offered me opportunities. The *Montel Williams Show* was a prime example. Early in his career, when the show originated in Los Angeles, Montel dealt with such issues as interracial dating, the impact of gangster rap lyrics on society, and parent–child conflicts. Generally these shows were done with a purpose—to educate, not simply titillate. Things got rowdy at times, but Montel never seemed to lose his vision of accomplishing something worthwhile, especially in the arena of race relations.

Unlike Geraldo, Sally, or Oprah, Montel Williams let an expert talk in more than sound bites so that delicate and complicated issues could be explored without sacrificing light for heat. Indeed, after appearing on a show in which I voiced my support of interracial dating as a way of breaking down racial barriers, people came up to me in stores and restaurants to thank me for making their lives a little easier. Such experiences reinforced my belief that reaching millions of people via the forum may indeed be worth some of the silliness an "expert" occasionally encounters.

But things change. The *Montel Williams Show* became corrupted by its own modicum of success and its quest for ratings. The day I got a call from one of the show's producer-bookers to mediate battles between couples in a boxing ring while wearing a striped referee costume, I knew the wind had shifted. Montel had joined the talk show circus in earnest.

Before Montel's fall from grace, did I really believe talk shows could be reliable forums for imparting psychological wisdom? For a while I did, so I kept doing them. I was like a pigeon in a Skinner box, pecking away on a schedule of partial reinforcement, convinced that the next peck, the next show, would reward me with the experience of having done something worthwhile.

Then finally, after appearances on *Oprah* and *Sally,* the simple, stupid truth of it all sunk in: Talk shows exist to entertain and exploit the exhibitionism of the walking wounded. If you want to explore your problem, you go to counseling. If you want to exhibit your life, attack and humiliate your spouse, or exact revenge for some misdeed, you go on a talk show.

Proof came to me in a phone call late one night. A woman's voice asked if I was the doctor she had seen earlier that day on *Oprah*. I was.

"Would you help me get on *Oprah?*" she asked. "I need to talk about how I was sexually abused as a girl by my brother. I told my parents. They never would listen. Now I'm in Seattle, living with a woman. My life is screwed up and I need to get better."

"What about therapy?" I suggested.

"No," she replied, "I must go on *Oprah*."

"Why *Oprah*, why a talk show?" I asked. "Surely therapy would be a better place to work out your anger and resentment."

"No" she screamed. "My parents have to pay for what they did, for not believing me, for ruining my life. If I went into therapy, only the therapist would know. If I go on *Oprah*, millions will. My parents couldn't ignore me then. And their lives would be ruined like mine was."

After that call, I began to turn down show invitations. Most, but not all. Their narcissistic allure still whispered in my ear, albeit more faintly.

Help, I've Tuned In and I Can't Tune Out

Like cancer cells, talk shows multiply. In the 70s there were three. Now there are 20 and counting. They have surpassed soap operas as the number one draw of daytime TV. Their appeal is obvious. These shows are a source of electronic gossip, of safe scandal. They provide endless opportunities to compare one's own life with those on the screen and breathe a superior sigh of relief. If you feel like one of life's ciphers, how uplifting to see people make fools of themselves on talk shows.

Talk shows also offer vicarious revenge. If you seethe inside because you have been betrayed or deeply disappointed in relationships, how pleasured you are by watching infidels try to justify their actions and get clobbered in a pincer movement by guests, hosts, and audiences.

Like the soaps, shopping networks, and endless women-in-jeopardy movies of the week, talk shows owe their popularity primarily to women. They constitute over 70 percent of the viewing audiences.

Talk shows are relationship shows. But because they're steeped in gender stereotypes, they polarize relationships between the sexes. Women come on talk shows mostly to discuss betrayals and victimizations. Women in the audience either attack or embrace. They attack when the women on stage live up to the weaknesses to which they are heir—needing a man to feel legitimate and validated, and sacrificing their dignity on the altar of that need. Women in the audience embrace when the enemy, the male, is on stage betraying the female guests in the same way women in the audience feel they have been betrayed.

If, as Bette Davis once said, old age is no place for sissies, then the talk show is no place for men. You wonder why men who won't commit, who sleep with their girlfriend's best friend, or who abuse their wives come on such shows to get predictably garroted. From what I've observed, most do it to pacify their partner's desire to simply get on one of the shows.

These men will perform on cue, say dumb things, reveal dumb attitudes about fidelity or inattention to their partner's needs. They get pummeled— but they leave the stage unscathed. For them it's a joke. "I hope you get off my back now," I heard one young man say to his girlfriend as they left the stage at the end of the show. She had just ripped him apart on stage but now was all warm and cuddly...for the moment.

There are men in the audience, but most are recruited by magazine ads and audience brokers. They sometimes ask questions and make pronouncements, but their hearts aren't in it. That's obvious when you sit on stage and watch them squirm as Oprah or Sally walks by: "Not me, please don't ask me to speak." Women, on the other hand, line up, stand up, raise their hands, shout, "Me, let me speak!" This is their chance to take out a few of the Neanderthal men or viperous women who inhabit every woman's nightmare and probably most women's autobiography of woe.

Circus Gratuitus

Unless they are from Mars or Venus and have never seen these talk shows, never seen the guests behave like *Gong Show* alumni or the experts pander away their professional prestige to the theatrical demands of the burlesque format, why do people continue to go on them? Why do they risk damaging personal life or professional limb? Because guests and experts alike whisper to themselves, "I can do better." But they don't. Because they can't. The iron-maiden format of the talk show ensures that.

Talk shows occupy two realities. There's the reality of witnessing a talk show on television in the familiar, benign environment of your home—the "passive reality." Then there's the "active reality" of actually being a guest or expert confronted by the kaleidoscope of glaring studio lights, perambulating cameras, charismatic hosts, stares of the studio audiences, and the mesmerizing fact of being on television. Moreover, the studio is far smaller and more intimate in actuality than it appears on television. The sheer psychologically coercive power of the host and of the people in the audience and on stage, in such close proximity, is invisible to the home audience. And its effects cannot be anticipated, only experienced.

Many times before a show, guests have confidently told me they are sure the show will be a positive experience, even when they have painful topics to discuss. But it is precisely the poor control that guests have over the proceedings (or themselves) that makes the unsophisticated, "Look Ma, I'm on TV" guests such attractive prey for the talk show. Once on stage, a guest's self-restraint evaporates in the hot glare of lights.

Media psychologist colleagues have shared with me similar post-show shellshock. They rarely have the chance to say what they thought they would when they agreed to do the show. Being chastened by the host when one's explanations are too prolonged is jolting. The experts find themselves with two choices: be glib or be ignored.

The Acrobatics

Calm intellectual discourse is unwelcome to most talk show viewers—they want action! Emotions and conflict are the two critical ingredients of the talk show recipe. They give it the tang that is so viewer-addictive. Conflict is king! And producers, hosts, and studio audiences use the guests to sow the seeds.

Certainly conflict is the essence of good storytelling. But the conflicts provoked on stage by hosts and studio audiences are the not scripted fictions of a made-for-TV movie. The tumultuous dramas talk show guests enact are *their* lives, *their* wounds, the crimes of *their* hearts and *their* loins. Unlike actors, talk show guests must ultimately answer for their on-camera confessions when they return to their everyday lives.

It's one thing to recount your troubles and misdeeds to a stranger in a bar. It's quite another to do it in front of 20 million people. But it's the predictable unpredictability of the dirty-linen flaunting that viewers find so irresistible.

The Dancing Bears

My talk show experiences make one thing painfully clear: Most guest are drawn from America's abundant population of *have nots*. They *have not* high intelligence. They *have not* high income. They *have not* many opportunities. They *have not* any way to snatch the brief celebrity that television confers except to exhibit themselves. They sell their misery the way hookers sell their bodies.

Talk show tradition is like a limbo bar. The lower it goes, the lower people who follow must go to play the game successfully. Guests will say the most intimate things precisely because they have watched others do it before them. If guests discuss their sex lives, other guests will do the same. If guests attack their spouses, other guests will do the same. And if guests admit to incest, incredibly, other guests will go on to do the same. Like some revivalist tent show, once guests have fallen to the ground, touched by the spirit, speaking tongues, others will follow, tongues wagging, shame and privacy shunted aside.

Some guests, of course, are more canny, intent on exploiting the exploiters. One woman called after seeing me on a talk show. She told me she had
the disease *du jour,* multiple personality disorder. She wanted to go on *The
Home Show* or *Geraldo* (she had seen me on both) so she could tell her story
and maybe get a producer to option her life for a television movie. She had
heard I was also a screenwriter. If I would help her get on, she said, she
would let me write the screenplay.

Other guests have humbler exploitative goals—merely to get on a show.
For one *Oprah* show the bookers found four couples who alleged they were
struggling with various strains of sexual jealousy. A man and woman in their
middle 20s got Oprah's attention. He was an accountant, she a homemaker.
She felt jealous all right, but not the kind of jealousy they had promised the
booker—"bikini-clad women at work." The wife was jealous of his work, period! Bikini-clad women had nothing to do with it.

When this tidbit of truth finally tumbled out of the wife's mouth, Oprah
rendered a sublime imitation of Mount Vesuvius. "This is why you came on
the show, to discuss your jealousy over his working long hours to get
ahead?" she sputtered with anger—then turned to me. "What do you have
to say about that, Dr. Fischoff?"

"About what in particular?" I asked, searching for something to focus on.

"About anything. Just talk," she hissed. So I spoke while Oprah brooded
and paced. The housewife was shocked and hurt. "Why was Oprah so upset?"
she asked me later. "After all, I just wanted to get on *Oprah.* Doesn't everyone?"

Others who grope for air time are not quite as naive. There are a handful
of guests, often transvestites or people with multiple personality disorder,
who somehow hop from show to show, flaunting their oddities.

The Jeering Crowd

Spectators they are not. If you take the studio audience out of the picture,
you take away the talk show spectacle as we know it. The audience provides
tribal impact, people provoking people to say and do things they would
never say or do unless they were drunk or assured anonymity.

The audience is laced with sharpshooters and soapboxers who all too
often use a guest to draw themselves into the limelight, to engage not in dialogue but in inquisition. The more their questions make the guest squirm
and lose control, the more powerful the audience members feel. That makes
a guest something of a defendant. The audience members are judge and jury.
And they will try to take your head off, for they sing the executioner's song.

Audiences may not walk in the studio with their fangs bared, but they
are soon salivating vampirically. Before the show goes on they are exhorted
by a warm-up staff to "Say what's on your mind," "Don't hold back," "You

don't like what they're saying, tell them!" When the guests come on stage and climb out on a ledge, the audience is already whipped up—transformed into a crowd of Manhattan pedestrians, looking skyward, urging the pathetic wacko on the fifteenth-floor window ledge to jump.

As an expert, you sit on stage and feel the negative charges of electricity hurtling out of the audience, enveloping the guests. But your expert status is by no means a safe-house. The audience's opinions are deliberately placed on even footing with those of the expert.

The expert may have the training, the clinical experience, or other *bona fides* to offer an educated opinion about a topic under discussion. But the audience members come armed with their personal experiences, the equalizing power of the studio mike, and the encouragement of the host to "let it rip." A meeting of the minds it isn't. If you go against the audience's strident opinions, they go against you.

The show was *Sally Jessy Raphael*. The subject: "Men Who Won't Commit." The audience kept spewing questions at the hapless males (are there any other kind on most talk shows?) on the stage. "Immature, immature," yelled women (and some men) in the audience, pointing fingers like Winchester rifles. I jumped in and noted that sometimes women confuse a man's general fear of commitment with an unwillingness to commit to a specific woman. And rather than maturity, it is often social conditioning and other less romantic agendas that compel many women to push for a marriage commitment.

Big mistake! The audience fell on me like children smashing a piñata. I might just as well have accused Mother Teresa of being a transsexual. Sally loved it, of course. What she didn't love was when one of the "uncommitted" proposed marriage on stage and placed an engagement ring on his girlfriend's finger. The exploiters got exploited.

The Ringmaster

You have seen him or her on television a hundred times nurturing those with righteous, sympathetic causes or ripping into the misfits, opportunists, and freaks. Phil? You think you know him. Sally? You think you know her. But you don't. Most hosts' sympathies and concerns are all too often mere contrivances to seduce the guest into self-exposure and beguile the television audience. Sympathy bonds freeze during commercial breaks and reanimate when the taping resumes. Caring? There are rare circumstances of on-camera concerns and off-camera follow-ups. But these tender moments are often later publicly exploited by the shows's spin doctors.

And rarely (Oprah is the occasional exception) is the expert treated any differently. For most talk show hosts, when the show ends, so does the existence of experts. I've been on tabloid (*Geraldo, Oprah, Sally*) and non-tabloid

(*Sonya Friedman, Larry King*) talk shows dozens of times. Except for Montel, a host has never chatted with me before or after a show. Many colleagues relate similar tales of off-camera invisibility.

Call in the Clowns

Who are the "experts" who appear on these shows? Usually psychologists or, more often, non-Ph.D. psychotherapists who do the talk show circuit regularly because they have a book to sell or a private practice to nurture and can speak in sound bites. Many veterans offer workshops to other psychologists who want to get on television. They even get together to trade war stories and get feedback on a recent "performance."

In theory, the experts could offer sound advice on a show's general topic. They could even offer useful information to the guests. But this is not what generally happens. The "formula" gets in the way. Even experts with the best intentions get caught in the talkshow undertow, the hurried rush to judgment, and do misguided on-air counseling or mediation between warring guests. But that's like singing to a deaf man. Guests are not there to humble themselves and gain therapeutic insight. Not in front of 10 million people. They are there for validation or, like the expert, to get some TV exposure.

So what function do psychological experts serve in actuality? In part, they give the talk show a frisson of legitimacy. But in the main, experts are the laugh track to help audiences identify whom to blame, whom to side with, and who "just doesn't get it." That may not be why the experts think they're there. But that's why they are there.

Some experts will say publicly that, "on balance," talk shows are worthwhile, that they help the viewing public, if not the guests. But for most experts I know, the only "balance" they are thinking about is their own, in the form of self-promotion. The dirty little secret most media psychologists know is that, with rare exceptions, if a psychologist truly wants to educate the public, the last place to do it is on a contemporary tabloid talk show.

The Barker

With very few exceptions, those who book the guests must be con artists and ambulance chasers. They get the names and phone numbers of prospective guests from a variety of sources: viewers who call in response to an announced theme ("If you have had a boyfriend betray you with your best friend, call us"); those who read an ad for prospective guests in the classified section of local newspapers; those who list themselves in publications devoted to specific oddities of human behavior. Or the bookers call psychotherapists or other personal-service specialists and ask them to bring their patients on as guests for particular theme shows. For ethical reasons respectable therapists must refuse such requests; the requests keep coming nonetheless.

The bookers need social misfits to feed the beast. Guests are given no warning that the electrified climate of the set will loosen their tongues and obliterate their self-protective sensibilities. That would spoil the fun.

Prospective guests are offered only a forum for personal advocacy ("obese women deserving love," "transvestites deserving women who understand them") and encouragement to tell all. The promise of an opportunity to meet Phil or Sally, of planes and first-class hotels, and of a night on the town can be tempting to people who usually have little access to such luxuries.

There are no warnings about surprise guests, as a grandmother discovered on one *Montel Williams* show on which I was the expert. Thinking she was on stage only to discuss her misgivings about interracial marriages, this grandmother was totally blindsided when the mixed-race grandchild she had refused to see or even acknowledge was brought on stage and placed in her arms. She had no choice but to submit or else be seen as the racist Ice Queen of the century. The gimmick played well to the audience. But the on-stage reunion had little to do with grandmother's off-stage fury about being set up. The staff of the show was subjected to her tirade.

A psychologist or psychotherapist is there to feed the beast as well. When a booker calls, he or she needs to determine whether you can talk without resorting to psychobabble and whether your point of view is compatible with the show's topical focus. A booker once called me to appear on a show to discuss the pain of recovery from repressed memories of abuse. I told her I was unconvinced of the legitimacy of many so-called recovered memories. She thought for a minute, said that wasn't the point of view needed for this particular show, but they were planning another show on "the false memory syndrome," and she would keep me in mind. The sands of principle shift easily.

The less experienced you are as an expert, the easier it is to be misled by bookers. If you tell them that you don't want to be part of a circus, that you need to be able to seriously explore the topic, they tell you they agree and that their show is *different*. But if you watch an episode of the show before deciding, most of the time you realize you have been lied to.

It doesn't take a brain surgeon to figure out why. Their show is a circus because circuses get ratings. In March, the *Jenny Jones Show* garnered some unplanned publicity when, a few days after their appearance on the show, one guest shot another. The alleged murderer had been asked to appear on a show about secret admirers. Unaware that the show's real theme was about men who have secret crushes on men, he obliged. To his shock, a neighbor appeared with his heart on his sleeve. Three days later the admiring neighbor was found dead and the homophobic guest surrendered to police. The next week, the show's ratings jumped 16 percent.

Sweeping Up

Do guests go on, do their strip show, and walk off into the sunlight? Some apparently do. According to published research, few will admit to being devastated by an appearance. Unless it becomes the first round in a lawsuit. Montel Williams encountered that surprise after he ambushed (yet) another reluctant guest with the revelation that her boyfriend had been sleeping with her sister. She subsequently sued the show for pain and suffering.

I was contacted by her attorney to review a tape of that show. I was asked to judge whether the woman did indeed seem caught off guard and traumatized by her sister's betrayal. I was also asked to prepare for possible expert testimony regarding such matters as springing cruel surprises on guests. The woman later collected an undisclosed sum in an out-of-court settlement.

There are less extreme instances, of course. But even then, most guests will simply say it was an experience, it was somewhat disappointing, or that they had to do it to help correct a terrible social prejudice—even if it was a wretched forum in which to do it. Such benign public admissions may be misleading.

After one *Sally* focusing on "women with bad reputations," I uncomfortably shared an airport limo with one of the guests. She had just been shredded by the audience and the host. Nor did my on-air comments validate her self-abusive lifestyle. She insisted the experience was wonderful. "I was on *Sally*," she gushed. Denial insulates one from regret and humiliation.

Denial insulates, that is, unless your parents saw you on *Geraldo* and heard—for the first time—that your husband is the man who raped you on your second date. Then things don't go so well. Or unless your son saw you on *Sally* and learned that you habitually go to neighboring towns on the weekends, end up in bars doing stripteases, and go home with strangers. Guests quickly learn that they paid a big price for their brief stab at celebrity.

I have unsettling concerns about the people who volunteer as guests on talk shows. You might argue that if guests want to go on, let them. If they get burned, that's the price their choice exacted. No one held a gun to their heads.

True. But it is equally true that people don't always have the sophistication to make the right choices or grasp the consequences of their decisions. I would argue that until people fully understand the risks of parading their life-flaws for a few moments of cheap celebrity, until they understand that a talk show exerts an intoxicating pull on self-divulgence they can't grasp sitting at home wishing for a chance to get on *Geraldo,* then they are far less responsible for the degrading spectacle than are the savvy producers of the talk shows.

Circus freak shows are no longer the cruel attraction they once were. As a society we have become more sensitive to the feelings of "freaks" and to the dehumanizing stigma the term implies. We have also come to roundly condemn the promoters who exploited freaks for profit. Perhaps guests on these televised freak shows should be accorded the same compassion and the shows's producers the same condemnation. If I had understood this earlier in my talk show career, I would have dropped it sooner.

Will television talk shows run their course? Will producers be brought to their senses and pull back from the class warfare of the *haves* exploiting the *have not* exhibitionists for the amusement of voyeuristic audiences? Not until shame and privacy reassert themselves in the pantheon of social values. The "I am victim, hear me whine" chorus has a seemingly inexhaustible supply of chanters to parade across television screens and inspire fascination in viewers.

In time, I learned that I could exercise more control over what I said instead of obligingly feeding the beast. But controlling my presentation was just not enough. My presence still legitimized the circus. In the end, there was no way I could further delude myself into believing I was serving any informative function for audiences or guests. The forces at work are simply too powerful. The talk show circus will no doubt continue, but with one less clown.

COMING TO TERMS

As an expert guest appearing on the television talk show circuit, Stuart Fischoff's relationship with the talk show has become complex, undergoing several changes. His initial hopes of using this medium as a way to inform and educate studio and national audiences about serious personal and social problems seem to have been replaced by a cynical attitude: the talk show is a kind of insatiable, hungry "beast" which devours guests (and experts like Fischoff). What has happened to change Fischoff's opinion of the potential and worth of the talk show? In your journal collect some of the significant events Fischoff experienced as a media psychologist which caused him to change his mind and retell some of these key events as a story. List specific events as part of this history you are retelling.

READING AS A WRITER

In one way, stepping back and writing about his experiences with television talk shows allows Fischoff the time and distance for careful analysis: he can now speak and write from a position of expertise, carefully choosing his words, using an appropriate language, doing a certain kind of analysis and asking certain kinds of questions. We might call this way of speaking and talking a *discourse.* Often academic and professional ways of speaking and talking are called discourses. But we might also say that students use a specialized discourse in talking, for example, about the experiences of being a student or that the ways late-night talk show hosts talk and ask questions of guests comprise another kind of discourse.

Return to specific passages in "Confessions" in which you see Fischoff speaking the discourse of an expert, a clinical psychologist. What are the important words and phrases he uses? What does he choose to analyze? There seems to be another discourse at play in Fischoff's article. Are there occasions where Fischoff uses language which surprises you or denies your expectations of how a psychologist might speak? You might look at the title and subtitles of Fischoff's article. How would you describe this discourse? What might be the effects of using more than one discourse in a piece of writing?

WRITING CRITICISM

When Fischoff observes that "teaching and TV commentary are both performance arts," he is drawing our attention to the fact that both activities occur before an audience and are part of a social dynamic. While a performance does affect an audience, we might also say that the audience helps to define and limit the performance. As part of his analysis of talk show culture, Fischoff invests considerable time in studying the role audience plays. In fact, he identifies two audiences: the studio audience and the national "at-home" audience.

We'd like you to do a similar kind of work and write an essay in which you describe and analyze the experience of being the member of an audience at some kind of media event. Choose any media event in which the audience plays a significant and dynamic part in the experiencing of an event. For example, you might write about the audience experience of seeing a play or a film or attending a musical event. You could also write about the "audience" of an Internet chat room which you visit on a regular basis. You could, in fact, be experiencing a sense of audience where you and a group of people are watching a television program or listening to a radio call-in program.

We would like you to describe this audience in terms of gender, race, age, and mannerisms. Did people enter the audience as a group or a couple? Alone? Why might this be significant? But as Fischoff did, we also want you to analyze the audience at a media event on several levels. Does the audience as a whole suggest or reveal a personality of its own? How would you describe and explain this personality? From its appearance, conversations, questions, comments, what observations can you make about the audience's political beliefs? Values? Likes and dislikes? How might you describe and explain the "psychology" of this audience?

Movies and History

Eric Foner and John Sayles

John Sayles: Why do people make movies about history? That's a question with a lot of different answers, and the answer depends on who the muscle behind the movie is. Sometimes historical movies get made because they're just good stories, and it's easier to begin with a story that already exists than to pick one up out of your head. Somebody's already done the living and the plot. Very often, if the story is fairly old history, if it's more than fifty years old, there's been some shaping over the years—the story has become legend—so a lot of the details that aren't necessarily dramatic have fallen away. A lot of the work has already been done for you.

Very often, though, historical films get made because the muscle is a star actor who wants to play a particular character. "I always wanted to be Davy Crockett," "I always wanted to be Wyatt Earp," whatever. There's something about the character, historically or mythically, that attracts the actor, and the line between the two of them can get pretty dim. If a very popular actor says he wants to play a historical part, it's not very hard to get the movie made.

Eric Foner: But your movies aren't star vehicles in the way that many historical films are. One of the strengths of your films, from a historian's point of view, is that you focus on ensemble situations, whether it's the team in *Eight Men Out* or an entire community in *Matewan*. Your films aren't like *Hoffa*, which concentrates on one guy, with a star playing him, and his name is the name of the picture. Presumably that's a conscious choice on your part.

Sayles: I've often had the experience of seeing a historical movie and then reading some history—and thinking that the history is a better story, a more

Foner, Eric and Sayles, John, "History according to the Movies: A Conversation between Eric Foner and John Sayles," from *Past Imperfect: History according to the Movies* (2nd ed.), Mark Carnas, Ed., 1995, pp. 11–28. New York: Henry Holt and Co. 1995. Reprinted with permission of the publisher.

interesting story, and certainly a more complex story. I feel that history, especially the stories we like to believe or know about ourselves, is part of the ammunition we take with us into the everyday battle of how we define ourselves and how we act toward other people. History is something that is useful to us and that people feel we need. I'm always interested in getting something new into the conversation that may have been overlooked, that may have been forgotten, and so sometimes I may err in not including something obvious that everybody knows about and instead going toward the stuff that people have forgotten or that was tilted in a totally different way when it was first put on the screen.

Foner: Now maybe I'm wrong about this, but my impression is that until rather recently most films about American history have been basically celebratory, whether they're *Gone with the Wind* or westerns of the traditional kind. I think you wrote in your book about *Matewan* [*Thinking in Pictures*] about the gap between American myth and American reality. I know you're not a Hollywood filmmaker, but do you think people in Hollywood are driven to make films that cheer audiences and make them feel good just because they've got to make back their fifty-million- or hundred-million-dollar investments?

Sayles: I would say so. Certainly there was no idea at the beginning of the movie industry that it was anything other than an industry. The stories that were filmed were supposed to be popular; they were the equivalent of stage shows, which were celebratory as well. Popular history was just more grist for the mill. But you have to remember that things tend to show up in movies about third: First, historians start working on something and take a look at the record. Their work usually stimulates novelists, and the novelists often stimulate movie people. Finally, things end up on television. Now we're getting Ted Turner making all these revisionist Indian stories. Sometimes the white people are a little one-dimensional, but the Indians at least get to be three-dimensional characters.

Historians started doing that during the 1940s and 1950s. Then Thomas Berger wrote *Little Big Man* in 1964, and when the countercultural thing hit the movie screen—when they stopped making *The Sound of Music* and started making *Easy Rider*—that revisionism crept into historical movies, too, and you got Dustin Hoffman as Little Big Man. Now it's reached television. What was done in *Little Big Man* may have been very broad and one-dimensional, but it was one of the first times those ideas were presented to a film audience.

I think there's actually a kind of time curve after a major war during which the enemy becomes more and more human. It was a good fifteen years after World War II before you started seeing Japanese in films who were not just spectacled, buck-toothed, evil-empire kind of guys. The same thing has happened with Vietnam: Oliver Stone's *Heaven and Earth* came along fifteen or twenty years after the fact, and people may not have been quite ready for that film because it had some problems, but they *were* at least ready to entertain the possibility that a Vietcong could be a three-dimensional person.

Foner: On the subject of things going into novels first, it's probably worth mentioning that you are a writer and were a writer before you became a film-maker, which I assume is unusual among filmmakers. Although my own connection with the film industry is much less intense, of course, I have been an adviser for a number of historical docudramas, documentary projects, etc., and I always tell the filmmakers I work with that a word is worth a thousand pictures. What I mean is that you can put so much more information into a book than into a film. That doesn't mean film can't teach in a very effective way, but I'm just wondering whether as a writer you feel constrained when you make a film. Film presents things in dramatic, visual, and direct ways, but doesn't the medium also force omissions and oversimplification because you can't put as much into a film as you can into a five-hundred-page book?

Sayles: That's why I choose different formats for different projects. When I wrote *Los Gusanos,* I was dealing with 120 years of Cuban and U.S. history through this one family. Because it was a novel, it could be complex. Each chapter in the book was told from a different character's point of view. In a two-hour movie, though, it's very, very difficult for an audience to accept or follow more than three points of view. You have the omniscient point of view; then there's the protagonist's point of view—he's the one in the closet waiting for the chain saw to rip through the door; and finally there's the antagonist's point of view—he's on the other side of the chain saw. With those three viewpoints, you can do a lot, but for an audience that's used to making an emotional connection with a film, it's very, very alienating to have too many points of view—and that fact mitigates against complexity.

One of the things that you have to do is say, "Okay, am I going to recreate this entire historical world, or am I going to take one episode that stands for it?" In making *Matewan,* I chose to focus on the Matewan Massacre because it seemed to me that this episode epitomized a fifteen-year period of American labor history. To make it even more representative, I incorporated things that weren't literally true of the Matewan Massacre—such as the percentage of miners who were black—but *were* true of that general fifteen-year period. I wanted to be true to the larger picture, so I crammed a certain amount of related but not strictly factual stuff into that particular story. That was a simplification.

Foner: Okay, it was a simplification, but *Matewan* has more than three points of view, and that's one of its strengths.

Sayles: It's a more complex story than you usually get in a strike movie, but I'm always aware that I'm simplifying certain things.

Foner: In *Matewan* and your other films, if I'm not mistaken, you abjure not only Hollywood money and Hollywood control but also Hollywood techniques of, you might say, melodrama—I mean, the usual ways in which films hook people. There's no sex in *Matewan,* no affair between Joe Kenehan and some woman in the town—

Sayles: Nothing tacked on.

Foner: You know that if *Matewan* had been made in Hollywood, there would have been. It's also a fairly slow moving film, and it's not entertaining in the Hollywood sense. Why is that? How could you afford to make those decisions?

Sayles: The decision to make a film and the ability to make it put you on very thin ice, which is why filmmakers have more freedom when they work with low budgets. You're not carrying the responsibility of having to make money for a large corporation and the three people in that corporation who just gave you the license to spend fifty million dollars of their money, not to mention another fifteen million dollars they're going to use to advertise the film. If you make a movie in a country as big as this one, with as many people in it who watch movies as this one has, and you're spending only three and a half million dollars, which is what *Matewan* cost, then you feel the people who invested in your film have a decent chance to make their money back—and that's all you really can tell anybody who's investing in a film. On the other hand, if you have to make back fifty million dollars, you know so many people have to see the film that you have to reach the mainstream of people who go to the movies to look for who's the good guy and who's the bad guy and to see the good guy win. If your movie isn't that kind of story, those people are going to be disappointed, the word of mouth is going to be bad, and you're going to fall short of the audience you need to pay back your budget

Foner: I imagine you've had opportunities to make studio films, but you don't. Can you imagine any circumstances under which you might make a big-budget Hollywood film?

Sayles: I've actually made a couple of movies for studios. I made *Baby, It's You* for Paramount which was a bad situation at the end of the day. I finally got the cut that I wanted, but there was a lot of me getting kicked out of the editing room, put back in, and all that kind of thing. And *Eight Men Out* about the 1919 Black Sox scandal was made for Orion. The interesting thing about that one is that I got to make it eleven years after I wrote the first draft of the script, which was based on Eliot Asinof's book.

One of the reasons it took eleven years—not just because I was a new film-maker and it was a period movie that would cost a certain amount of money to make—was that I wanted to tell that story, *Eight Men Out*. Not *One Man Out* or *Three Men Out and a Baby*—different versions, as the studio would say. The history said these eight guys threw the World Series, and I wasn't going to have a scene at the end where I froze the frame on somebody going, "We're number one!" These guys didn't win—unlike *The Natural*, which wasn't history but was a book. In the book, you know, Bernard Malamud has the guy strike out at the end. But in the movie, not only does he hit a home run, he hits lights that probably didn't even exist in that era, and there's a shower of sparks. I think the filmmaker said he wanted to be more in tune with the Eighties. Or he may have meant, "I want to make more money, and that's what we're doing."

I don't think Bernard Malamud was too thrilled with the new ending. You know, there's a story that Elmore Leonard called him up and said, "So how did you feel after *The Natural* came out?" And Malamud said, "I stayed in my apartment for a week." And Elmore Leonard said, "Well, I just saw the movie they made of my last book, and I'm leaving the country for a year." So you don't have to be a historian to feel the record has been distorted.

But *Eight Men Out* was interesting in that all I had to pay—if you think you have to pay something for the price of admission, to get the budget for the film—was some casting. The studio that finally made it had turned the script down three times before, and they still weren't that crazy about the story. But at that particular time, there were a lot of young actors around. So how do you get more than two or three young actors into the same movie? It's got to be either a war picture or a sports movie. And the studio said, "Well, who do you want to cast in this?" And we made up parallel lists, and I said, "Here are all the people I would like to have in this movie. Write down your list of the people you would like to have." It turned out that there were a number of people on both lists, and we actually got three or four of them to do the film. So I was happy and they were happy, and I didn't have to sacrifice the story in order to—I didn't have to burn the town in order to save it.

Usually what happens when somebody falls in love with a historical story of some sort is that a struggle develops, and you start giving away pieces of it. Your take on it may not be accurate to begin with, but you feel as though it's your story, and you resent having to make compromises to make it more dramatic and get it financed. The lie of the movie may be only the casting. Everything else might be totally historically accurate, but the person you are forced to cast just doesn't have it in him to do historical justice to the role. I read William Manchester's very psychological review of *Young Winston,* and he comes to that conclusion. The facts may have been accurate, but they missed the point of the man.

Foner: To get more into the mechanics of historical filmmaking, what sort of things did you do to re-create the period of *Eight Men Out*?

Sayles: You do the things that you think are going to strike people. For instance, I got hold of the rules of baseball from that era and the records of that World Series. We knew what happened on every play of every game because they kept those kind of line scores then. If somebody hit a ground ball to the shortstop in the third inning of the fourth game, that was the way it was shown on the screen. We had a lot of arguments, though, about the number of games. For three years beginning in 1919, major-league baseball went to a nine-game World Series, instead of a seven-game Series, to make extra money. Because the movie was running long, the studio said, "Couldn't we just have a seven-game World Series?" But I said no. I told them it was an interesting thing and that I'd make it work for the story.

Foner: Is the portrayal of White Sox owner Charles Comiskey in *Eight Men Out* accurate, or was it just Comiskey as the players saw him, if you know what I mean? I'm prepared to believe that Comiskey was a totally selfish man—

Sayles: Actually, what I tend to do is make these guys a little bit more appealing than they really were so that people will believe them. The stuff that's in the movie, he pulled all that and a lot worse. After *Matewan,* for instance, I was criticized a little for the two guys who were Baldwin-Felts agents—

Foner: I thought you made them much more appealing than they probably deserved to be.

Sayles: I made that attempt. I mean, these were guys who used to stop Red Cross milk wagons, put kerosene in the milk, and send them off so the kids would drink the kerosene. It was hard not to dehumanize these guys.

Foner: This might seem like a leading question, but do you and other film-makers care what historians think of your films or is it of no particular concern one way or another?

Sayles: I think it's generally of very little concern. I like to think of it more as— Let me compare it to what happened in the 1960s with the New Journalism. Tom Wolfe was one of the first guys to do it at *Esquire;* Gay Talese was another. The idea they had was that if you were true to the spirit of the story, you didn't have to get all the facts exactly right. Their work led to a lot of what I consider creative writing. When it's bad, of course, you start reading it and say, "Wait a minute, this isn't about this guy, this horseshit is just here so the writer can do some moves." Take Albert Goldman's books: All of a sudden when you're reading them you realize, "Wait a minute, he wasn't there. How does he know this conversation happened?" He's making it up, but he feels he's being true to the spirit of it. I think most filmmakers feel the same way about history.

Foner: It's a resource.

Sayles: It's a story bin to be plundered, and depending on who you are and what your agenda is, it's either useful or not. You may read six books about the story you're filming. Maybe you find some of what you read useful, and you get rid of the rest: characters, ideas, countries—

Foner: I'm sure that's absolutely right, but why not just present a historical fiction? Why did Spike Lee make a film called *Malcolm X?* Why didn't he create a fictional black leader and call him Joe Something Else? A film like *She Wore a Yellow Ribbon* doesn't claim to be an accurate presentation of a particular event in history—it's a generalized picture of what happened in the West— but to name a film *Malcolm X* after a real person is a claim of some kind of truth. So is calling a movie *Matewan,* because there really was a Matewan Massacre. Why not just create a fictional town and entirely fictional characters?

Sayles: There's a certain power that comes from history. I mean, I've heard producers say many, many times that the only way a movie is going to work is if the ad says "Based on a true story." Audiences appreciate the fact that something really happened. Whether it did or didn't, they're thinking that it did or knowing that it did. That gives the story a certain legitimacy in the audience's mind and sometimes in the filmmaker's mind, whereas if you make something up out of whole cloth, it's not the same. William Goldman, the screenwriter, once said that it's not important what is true, it's important what the audience will accept as true. Barbara Kopple, who made *Harlan County U.S.A.* and *American Dream*, also made a wonderful documentary about Mike Tyson for network television. If you took that documentary and fictionalized everything word for word, nobody would believe it. Nobody would believe Mike Tyson; nobody would believe Don King; nobody would believe Tyson's mentor who's the gay handball champion of the world and also the greatest curator of fight films who's ever lived. Even *I* don't buy it; it doesn't make any sense. But if you call it a documentary, people will believe it and say it's amazing, amazing stuff.

Foner: When the film *Glory* came out, the producers called and asked me to come to a screening, which I did, and then they asked me to write a statement about the film. I wrote one saying that the film revealed a little-known feature of the American past and blah-blah-blah.... They called me back and said, "Well, this statement is of no use to us." And I said, "Well, what do you want?" They said, "We want a statement that says the film is accurate from a historian's point of view." And I said I couldn't do that because what *I* mean by accurate is not exactly the same thing as what *they* mean by accurate. I thought the film was accurate in a general way, but there were many historical inaccuracies in it. I didn't necessarily want to criticize all of them, but I wasn't going to give the film my Good Housekeeping Seal of Approval.

Sayles: So they found someone else.

Foner: So they found someone else. I guess "This is true" has a really big appeal. Have you ever been attracted to making a documentary to get around that question? Or maybe you don't want to get around it.

Sayles: The minute someone sits down in an editing room, even for a documentary, choices are made. Some footage is kept; other footage is not. I've seen some extremely biased documentaries—some well made, some not—so making a film a documentary doesn't really solve the problem, and documentaries are a lot more work. Barbara Kopple spent seven years on *American Dream*. I think Fred Wiseman comes as close to the ideal of accuracy as you can—he just turns on his camera and sees what people will do—but even he uses only some of his footage.

There's also that principle of physics—the Heisenberg Uncertainty Principle—which says that just by your presence you change the reality. Some filmmakers love that, and they actually start talking to the people: "Say, why don't you do this, and why don't you do that? Will you come and repeat that thing you did yesterday or you said yesterday?" And others really hate it. I think Wiseman hangs around for a while without pulling the trigger on the camera so people can get used to him. He figures, "They're not good enough actors, and I'm going to be here forever. They're going to get tired of me, and they're going to start acting like themselves sooner or later." That's pretty much what happens in most of his films.

Foner: Have you ever used historical consultants on your films? I mean, when historians go to see a film, we always complain, "Oh, damn, if that guy had just spoken to me or had shown me the script, I could have pointed out seventy-five errors." I know that when Richard Attenborough did the film *Gandhi,* he sent the script to Ainslie Embree, who used to teach the history of India at Columbia. It was full of little historical errors that were of no importance to the drama—you know, the years Gandhi was actually in South Africa, or did he actually meet this guy, little things. Embree wrote them all down and sent them to Attenborough—and, of course, not a single one of them got changed. Not even the date. Now correcting some of them might have altered the film, but fixing others wouldn't have changed the film at all. Attenborough didn't care what the historical—

Sayles: Or it got in the way of his storytelling.

Foner: Or it got in the way of his storytelling. You know, you can read books about history and distill their essence, but that still doesn't make you a professional historian. Do you ever feel the desire or the need to talk to a historian, or do historians just get in the way?

Sayles: I would say that I probably use historians the way most directors use them: I tend to use people who are well versed in historical details, very specifically in the details, but not in the big picture.

Foner: People who know what kind of uniform a particular guy wore in the Civil War?

Sayles: That's right. You call up people who know about trains and say, "Do you have any pictures of what a sleeper car looked like in 1920?" Or you call the telephone company and say, "Were there phone booths in 1920? Did people dial first?" I wasn't around then, but amazingly there are people who know that stuff.

Foner: But they're not so easy to find.

Sayles: What you find is that people who work in your art department have sources, because they've worked on historical films before and these problems have come up. There are also some researchers who work on films just as historical fact checkers.

Foner: I know. They're always calling me and asking me for all this information. If it's one or two questions, I say sure, that's fine. After half an hour, though—

Sayles: They think you're the public library.

Foner: And then they say they don't have any budget for this.

Sayles: We had a tent guy for *Matewan.* He told us that the miners would still be using white tents because the tents would probably be war surplus from the Spanish-American War. Back then, troops didn't have to worry about balloons or airplane attacks, so their tents could be white. The tent guy wore the right clothes and put the pegs in and everything. We also had coal miners working with us who had mined with the kinds of equipment they used in 1920.

Foner: But you didn't consult with a labor historian or someone who might have written about the history of the mine workers' unions?

Sayles: I would have read their books but not necessarily brought them on for that particular story. I think part of the fear and resistance to hiring historians is the same thing I encounter when I rewrite other people's movies. I can always sniff out a movie that's the writer's own story. You know, very often it's about his divorce. It might even be a little more interesting than that but there are things about the story that don't work dramatically, and the author is sitting on them because that's what happened to him. It doesn't make a good story, so you have to say to the guy, "Unless this is a documentary about you, and nobody knows who you are, we have to change the story."

I was just working on this film about Apollo 13, and the producers were being very, very technically accurate. I got to talk to astronauts, and my job was to translate what they were saying into language that nonscientists could understand. I had to find ways to get in the technical information without simplifying it to the point where it wasn't true anymore. But I had to make it simple enough so that nontechnicians could understand. The problem is, you can't have the head of NASA saying, "Oh, by the way, we don't point the rocket at the moon. We point it where the moon is going to be." So you find ways to do that: You create characters who didn't exist but who are commentators. You combine some characters. With the Apollo 13 script, there was quite a bit of fudging around to make it a better story, which the astronauts resisted at first. Then they started calling up with ideas. What was interesting about their ideas was that sometimes they were very melodramatic but they usually did help you solve the story point without being untrue to the spirit of the people, because our astronauts knew those people.

On other pictures, I tend to work a lot with the Chicago Historical Society. One thing about Chicago is that sociology was kind of born there, so there are records in Chicago about poor people that are older than records from almost any other part of the country. I've written two or three gangster movies, and they've been very helpful with the details. But that's just the letter of the law. Preserving the spirit of what happened is really up to the filmmaker's abilities and intentions, though his work can be changed enormously by the studio and later by the test audience. If the test audience doesn't like the way the Civil War came out, maybe the studio will release another version for Alabama.

Foner: Do you think we as historians ought to be concerned when we see the "truth" altered somewhat for dramatic purposes? Should we be concerned about that as historians? I mean, after all, many more people learn their history from watching the film *Malcolm X* than from reading some academic tome about *Malcolm X*. And *JFK*—God knows how many people now think Jim Garrison had the assassination all figured out. Or *Glory*, which is a very good film in many ways but still makes some very striking historical errors—like the idea that this unit was largely made up of former slaves. In fact, they were all free people from the North. It may be good dramatically to have a guy take off his shirt and show his lashes, but it's not who those people really were.

Sayles: There were other units that had former slaves, but that one—

Foner: That was a specific unit with a specific commander, Robert Gould Shaw, and it was made up of free Negroes from the North.

Sayles: Which is an interesting choice, I think, in that it reminds me of those guys who went to the Spanish Civil War to fight in the International Brigades. They didn't have to; they were already free. But they put their asses on the line for somebody else—which is almost more of a risk.

Foner: Absolutely. It's a dramatic story, but maybe the filmmaker didn't think it was dramatic enough.

Sayles: Once again, he probably felt that to bring the bigger story into his specific story, he needed to create things.

Foner: So should we protest? Should we send letters to the producers saying, "Look here, you are distorting history by showing this—"

Sayles: I think sending a letter to the producer is the wrong thing to do. The letter should go to *The New York Times*. The letter should go to—I just saw a nice documentary about the civil rights movement, *Freedom on My Mind*, and part of the reason it was made was because of *Mississippi Burning*. The people who made the documentary wanted to set the record straight. If people know more about a story, then it's their responsibility to keep putting what they know into the conversation.

Foner: Does the movie industry as a whole have any responsibility to get the story right?

Sayles: I think using *responsibility* in the same sentence as *the movie industry*—it just doesn't fit. It's not high on their list of things to think about. You know, one of the things that's been happening in the industry lately is that the way politicians have been marketing themselves has affected the way movies are made, not just the way they're sold. There have always been famous cases of endings being changed because test audiences didn't like them, but now, more and more often, people are saying, "Why should we wait until the movie is made? Why don't we do the test marketing when we're writing the script, or when we're planning on the movie, just like they do with political campaigns?"

Foner: Let's just take a poll, see what moviegoers want to see, and then make it for them.

Sayles: Exactly. You remember what one of Reagan's press people used to say? "You don't tell us how to stage the news, and we won't tell you how to report it." Those people always talked about "scenarios." You know, "This is the scenario." And one of Bill Clinton's problems is that he doesn't stay with his scenarios.

Foner: Or he doesn't have the right scenario.

Sayles: And he starts one, then backs off it instead of saying, "Well, I'm going to take a stand here and play it out." Franklin Roosevelt had a lot of scenarios, and he thought this one would fly and that one wouldn't. He didn't get us involved in the Spanish Civil War because he didn't think it would play out. He thought he would lose the Catholics and that would be worse. But he had the right scenario for getting us into World War II, and it included some pretty high-handed stuff. But once he had made the decision, he stayed with it. Sometimes he lost and sometimes he won, but he won more often than he lost.

Foner: So what you're saying is, the way the movie industry is going now we're not likely to hear a lot of concern about historical accuracy.

Sayles: No, because coming up with the scenario that sells is the same thing as staying in office. You know, it's the difference between a leader and a politician.

Foner: And the stakes are so much higher now. You stand to lose a hundred million bucks instead of five million.

Sayles: I think the stakes were always the same, personally, for those guys: "Don't make a movie like this. It doesn't make money." If historical accuracy were the thing people went to the movies for, historians would be the vice presidents of studios. Every studio would have two or three historians.

Foner: Fortunately, or unfortunately, that doesn't seem likely to happen.

Sayles: You were asking before about the difference between fiction and movies. The one thing I feel that I can do in movies that I can't do in fiction is—I can do anything in a movie, and it doesn't have to go through your head first. When you're in a movie theater, you experience the film viscerally. It goes straight to your gut. If it's a true story—if it says, "This is a true story"—that adds a lot of weight, too.

Foner: But in your films you keep that much more under control than most directors. You don't go for the cheap shot. You don't go for the cheap emotional high.

Sayles: I tend to pull you in and push you away, pull you in and push you away. In *Matewan*, there's that really awful scene in which you see the kid's throat being cut. Now if that movie were just about pulling you in, the answer—what everybody would want—would be the *Walking Tall* answer: We're going to kill all those guys, and we know the audience will feel good when we do. The Clint Eastwood film *Unforgiven*—it's just one example—is predicated on the audience knowing that the lead character may be called something else but he's really Clint Eastwood. It's funny when he falls off a horse because it's Clint Eastwood falling off a horse. The foreboding of the movie, what everybody knows, is that at the end Clint Eastwood is going to kick ass, which is a fairly simplistic revenge sort of thing. I think a lot of movies like that added up to why people were ready for the Persian Gulf War—that and a certain amount of Arab-bashing, especially since the Arab-American community isn't large enough to put up much of a fuss.

Foner: They did complain about *Aladdin* a little bit, but—

Sayles: That was a mild, mild version. I'm talking about characters like those on "Mission: Impossible"—the slimy, rich Arab guys, the bad guys. But getting back to that scene in *Matewan* of the kid's throat being cut, the point was to push you back a little. And that's one of the things that the Joe Kenehan character did when he spoke out against more violence. He said, "Well, number one, just tactically it's a bad idea, because we're going to get creamed—maybe not tomorrow, but eventually we're going to get creamed. Also, morally, this strike isn't about killing people, it's about getting what we want. It's about not living in a world where you have to have an eye for an eye." He's pushing away the Hatfields and the McCoys, whereas many movies use a feud as their emotional release, and that's often what the filmmaker is going for: getting people so worked up that when they leave the theater, they want to sign up. Bertolt Brecht even talked about that. He said that he wanted people to leave the theater and join the revolution. As alienating as he made most things, that was one thing he wanted people to get viscerally involved in.

Look at any World War II film made between 1940 and 1948 or 1949. The movie industry worked very much hand and glove with the government to the point that, when John Huston made a documentary showing very, very accurately what fighting was like, he couldn't screen it because people felt it was a little too raw. That's a consequence of using film as an element to keep people mentally in a war.

Every filmmaker, like every historian, has an agenda. The difference is that historians read one another, and, because of the academic world in which they live, there's a little bit more checking up on the facts. There's a little bit more of, "Okay, this is your agenda, but where's your documentation? Where did you get this? What have you ignored? What have you overemphasized? What have you underemphasized?"

Foner: Well, we're trained to do that and so are most readers of history. But most moviegoers don't go to a film thinking how—They think *JFK* is true; they think *Matewan* is true. They basically think whatever they see is true.

Sayles: While the movie is happening. And I think that's an important point about movies: They exist during those two hours. If you make them, you hope that they have some echo, but the only thing you really have to do for the audience to buy in is to be true to the world you create for those two hours. If it's a world in which people have superhuman powers and can jump higher and run faster—if you set that up early and stay true to it—then people will buy it. But if you create a world in which there are only six shots in a six-shooter and somebody fires seven shots without having to reload, then people will get upset. In *JFK*, for example, one of the things Oliver Stone did was hit you with so many images so quickly—some of them familiar, some of them new information—in such a barrage of documentary styles that he was able to pass off stuff that was fairly speculative. Sometimes even he must have thought, "Well, I don't have any way to back this up, but it makes a great story, and I think it might be true."

Foner: I thought that film was a brilliant example of manipulation of the highest order.

Sayles: Sure.

Foner: And by the way—You're right, he created new footage that looked exactly like the original documentary footage, and you couldn't really figure out where the "real" documentary footage left off and the "new" documentary footage began.

Sayles: Unless you came to that film with a very strong sense of history, having done a lot of research and knowing the Kennedy years very well. Otherwise, you get swept up, and at least for the world of that film, for those three hours, you buy it. You buy that Kevin Costner is Jim Garrison, even though it's bad casting. I don't know whether you've ever seen Jim Garrison, but he looked around six-foot seven.

Foner: Is there any film you'd like to mention that you think was particularly successful in its treatment of history? And I guess we should narrow our definition of *history*. Most movies are set in history in one form or another—except *Star Trek* or some other futuristic film—but some are more insistent than others in claiming to be history.

Sayles: I think the more successful historical films tend to be smaller stories because of that complexity thing we talked about earlier. When you tell a wider story, it becomes either a lot of production value without much to hold you emotionally in the story or so simplified that it doesn't seem real. Sometimes people have taken a historical story about the Rosenbergs or whatever—you know, a very, very, very small piece of history—but time after time I see them crap out. What was the one about—*Guilty by Suspicion*—the blacklisting film. It seemed as though they had a good idea there, but finally they weren't comfortable enough with the period to really deal with the story, so it became just another Hollywood movie.

Foner: What about *The Front*?

Sayles: *The Front* was an interesting movie in that it fictionalized a real case, so it was able to play things both ways. It kind of captured one little aspect of the blacklisting, which is that some guys who were somewhat sympathetic became fronts and managed to keep their hands clean enough not to get blacklisted themselves. There's an interesting moral ambiguity there—of not really fighting against the blacklist; in fact, profiting from it, which was what an awful lot of the friends of blacklisted people did.

Now I happen to see the world in a fairly complex way—I see complexity everywhere—but that's not the direction that movies were going during the 1980s. Think of the big stars of the Eighties, like Arnold Schwarzenegger and Sylvester Stallone. They played Ur characters—you know, mythic characters. It was this simplistic and even heightened good-guys-and-bad-guys thing— *The Terminator*—as opposed to the moral ambiguity of the 1960s and 1970s. I was definitely swimming against the current during the Eighties, and it was a conscious choice. Now, getting back to that idea of filmmaking as a conversation: One of the things that I feel is useful about making a small movie is that film can provide a voice for people who are not being heard and not being seen on the big screen. Let's try to get them up there, even if it's in a stunted non-Hollywood movie. Don't historians feel that way, too? Aren't you trying to broaden the debate?

Foner: Most of us want to be relevant. We want to feel that the history we're writing has some impact on the way people think about America. I write a lot about black history, about the Civil War, Reconstruction, the history of American radicalism. Why do I choose those subjects instead of other ones? It's because I think they have relevance and they relate to the way I grew up. But there are a lot of pressures now from many directions: There are people who feel they've been left out of history, and they want a new version that may not be any more accurate. There are even some people who think that history's gone too far in the other direction; they think it's too critical now, and they're demanding a more celebratory history. Lynn Cheney, the former head of the National Endowment for the Humanities, was complaining recently that historians are too negative, that we're too depressing. Why don't we write about the greatness of America and not dwell upon all these—

Sayles: Which was the 1950s version that I got in—

Foner: A lot of people want to go back to that. As you well know, the public presentation of history is a tremendous battleground right now, whether it's the Smithsonian exhibition on the A-bomb, which you may have read about, or—

Sayles: Columbus.

Foner: Columbus. How are we going to present Columbus? I'd say there's more public pressure on historians today than there probably ever has been. There's more scrutiny on what historians are doing from political, social, and other angles than ever before. So filmmakers are not alone in having people watching very carefully what they're doing. It's a new thing for many of us and I welcome it. It would be worse if everyone thought we should just be ignored.

Sayles: Put it in a book and put it away.

Foner: You know, the hardest thing for people who don't think much about history to realize is that there may be more than one accurate version of history. In other words, there may not be just one "correct" view of something, with all other views incorrect. There are often many legitimate interpretations of the same historical event or the same historical process, so none of us can claim that we are writing history as objective fact, in the only way that it can be legitimately presented. I hope we can all accept what you said earlier, that we all have agendas, we all have points of view. My history is a point of view. On the other hand, there are limits. If my point of view was completely divorced from the evidence, other historians would know that my views were implausible, and they would point that out because the evidence is there and there are standards.

Sayles: Like "The Holocaust didn't happen."

Foner: Exactly. There are standards of historical proof and presentation—there are outer limits—but within those outer limits are many possible ways of looking at things, all of which may be plausible.

Sayles: Among the biggest difficulties in making a historical film is presenting just that: the idea that there may be more than one reasonable version of events. It's one of the reasons that my book *Los Gusanos* is told from about twenty-six different points of view. When I was down in Miami kicking it off, I told the Cuban community, "You are going to find your best friend in one chapter and your worst enemy in the next, but they may both be right." If you take what happened historically on a personal level, it's like a pool break: One guy went into the pocket and that was the worst thing that ever happened in his life, while for another guy it was the best thing that ever happened to him. And they're both right from their own points of view. If either of them were to write a history of the original event, it would be difficult to recognize the shot that started the whole thing—the Cuban Revolution. Understanding that is a very, very difficult thing for most Americans, who have been trained by popular entertainment to want an answer in half an hour, an hour, or two hours, depending on whether it's television or the movies. And they want an unambiguous answer.

Another thing that is very, very difficult to do in a movie—and only a little more possible to do in a book—is to remember that people's thought processes were different at different times. For example, being a socialist in America in 1920 was a very different thing than being one in 1970 or 1990. Being religious in Rome in 1722 was different than being religious today in very secular New York City. When you see a historical movie and it doesn't quite jibe, it's usually because the mindset is wrong. Maybe you go with it if it's *Butch Cassidy and the Sundance Kid.* The sensibility of that film was very 1970s, and it was meant to be. The filmmakers got away with it because the movie was an entertainment and you kind of bought these guys as—well, they're not much different than the guys in *Easy Rider.* They're a couple of revolutionary devil-may-care guys who happen to be outlaws. But there really were historical figures named Butch Cassidy and the Sundance Kid, and real things happened to them.

When I'm making a historical film, I often spend maybe the first twenty minutes trying as subtly as possible to get the audience into the heads of the people living at that time. Maybe there wasn't sexual freedom. Or if there was, maybe it was limited to a small group. There were certain givens as far as what people believed. When James Earl Jones came to be in *Matewan,* the first thing we talked about, and he brought it up, was that we didn't want his character to be a revisionist black man. "This is 1920 and he's a black man from Alabama behind enemy lines. If he talks back, he knows that the next thing to happen may be a noose around his neck." So big powerful James Earl Jones really put a tether on his voice and on his body language, because we were trying to get people into the mindset of where these black guys were coming from. You know, there was no "up against the wall" back then. It was "yes, sir" and "no, sir."

Foner: That was another problem with *Glory.* In *Glory,* the soldiers talked as though they were on the streets of Harlem in 1990, with all this slang.

Sayles: When they were around white people.

Foner: And even by themselves. You know they weren't saying, "Hey, brother, give me five" during the Civil War. *Glory* had no sense, as you say, of language, of thought processes, of the way people related to one another. Life was very different back then.

Sayles: And who even knows exactly what the slang was? Most black vernacular from that time was written by white guys who were lampooning it. You know, one of the things that I often do—for instance, when I wrote *Eight Men Out*—is read a lot of the period writers, especially guys who were known to have had an ear for dialogue—Ring Lardner, James T. Farrell, Nelson Algren. I also showed the actors Jimmy Cagney films because there was a speed of delivery affected by urban guys then that Jimmy Cagney used.

Actors today, since they've been through the "method revolution," take these incredibly long pauses. I showed them a movie called *City for Conquest* in which a hundred different things happen in an eighty-minute movie because everybody talks really fast I told the actors—except the one who played Joe Jackson, because he was supposed to be from Georgia—"This is your rhythm. Spit it out" That made a statement back then, spitting it out.

It's interesting now what you see in sports. There's a certain modesty that has come in, especially with the white players—I think it comes from John Wooden, who coached all those championship basketball teams at UCLA—whereas the black players are still on that "I'm the greatest," Muhammad Ali kind of thing. Back in 1919, though, the white athletes were poor kids, and they were in your face. "We're going to kill those guys tomorrow. They don't have a fuckin' chance." They didn't say "fuckin'," but there was the same kind of cockiness that you see among black players today. You've got to be aware of that stuff when you're making a historical movie. There wasn't any of that "Well, geez, I'm just going to hope the Lord Jesus Christ helps me out tomorrow when I do my job." There's been a lot of that in the Eighties and Nineties with white athletes, but it didn't exist in the Teens and Twenties.

Foner: On another point, one of the major trends in the writing of history nowadays—and it may take years for this to get into film—is the tremendous interest in women's history. With a few exceptions, not many historical films are really that interested in women. Most of these films are male worlds. *Matewan* is basically a male world. In fact, *Eight Men Out* is a completely male world. Why can't filmmakers make films about women's history? Is it the same reason people are always complaining that there aren't enough good roles for serious actresses?

Sayles: You know, if you look at the five hundred films that were released this year, you're going to find that about ninety-five percent of them were directed by men. The decision-makers who finance the big films are also ninety to ninety-five percent men. So part of it is just a lack of interest in women's stories. Another part is that audiences go to movies with certain expectations that have been raised by all the other films they've ever seen. Even something like *Reds*, which tried to work a woman's story into a larger historical picture, became in the second half just a romance, just a woman trudging across the tundra looking for her guy, instead of really saying, "Okay, let's drop John Reed and really see what Louise Bryant is doing."

There is this feeling in the industry that if people see a woman on the screen, she has to be involved in a romance—that to get women into the audience, there has to be some kind of romantic payoff for them. That can change, though, the same way that the treatment of blacks and Indians has changed. Before the 1950s, there were certain things you could do that were accepted because a large number of the moviegoing public didn't feel these people were totally human. You could demonize these people, stereotype them, and present them in one-dimensional ways. But then the consciousness changed—partly because of movies, partly because of life, partly because of history and the efforts of the people who were in the civil rights movement—and once that changed, all of a sudden you couldn't present people in the same bad way anymore. I'm sure there are things that we accept in movies right now that, twenty years from now, people will say, "I can't believe they made a movie like that."

Foner: I suppose you can never quite predict how an audience will react to a film. I was once at a conference on all this and I heard a man speak who had made a documentary about Caryl Chessman, a guy whose execution in 1960 aroused a lot of protest against capital punishment. Anyway, the man who had made the documentary said that he had made the film as a powerful statement against capital punishment. But a lot of people saw the film and said, "Great, they fried the guy at the end—and, boy, he really deserved it." They thought the film had a happy ending because the guy was executed. So there's a limit to the degree to which you can control how an audience will react to your work.

Sayles: You load the dice, more or less, but one thing I definitely do in my movies is allow people to draw their own conclusions, which means that some of them are going to draw conclusions I don't necessarily want them to draw—or come down on a side I think is the wrong one. It's better in the long run, though, to have people question things rather than just react without any thought, without any kind of analysis. The struggle that you often see in the making of a historical film is the struggle between how much of a viscerally page-turning, emotionally stirring story you want and how much you want people to think about what's going on.

Foner: What about changing the facts to make a point? Especially when the audience isn't aware of the changes, which is usually the case?

Sayles: You know, I ask a lot of the people in the audience. I try not to condescend to them, and implicit in that is a presumption that they will take some responsibility not to believe everything they see and also to see more than one thing. If you're not going to condescend to people, you do have to ask that. Because if you are going to condescend and spoon-feed them everything and simplify everything, then you're saying the people aren't capable of complexity, they're not capable of reading two versions and making up their own minds about which one to believe. That can be a very dangerous point of view.

COMING TO TERMS

The exchanges between Eric Foner and John Sayles make up a recorded and then transcribed conversation, what we might call a dialogue. Unlike a published story or article where writers have an opportunity to revise, reorganize, or amplify their thoughts, the dialogue exhibits most of the characteristics of conversation. Ideas and viewpoints are presented and developed, but within the give-and-take of a conversation in which people speak their minds. They ask questions, challenge what's been said, get back on track, and return to earlier points. And sometimes, in the course of a conversation, as it is with Foner and Sayle's dialogue, we speak for or represent people who aren't present.

As you read the conversation between Foner and Sales, we want you to imagine a third chair at the conversation table, but one that is empty. Foner is an academic historian and John Sayles is an independent film maker. One of the missing people in this conversation is the big budget, Hollywood studio film maker. In your journal we want you to create three lists or columns: independent film maker, academic historian, and Hollywood film maker. Then under each heading select passages from *A Conversation* which you think represent each person's attitude toward historical films. For example, how does each present an attitude towards historical accuracy in films? How would you characterize the major agreements or conflicts among these three ways of talking about history and films?

READING AS A WRITER

Maybe you've had the experience of writing an essay and your teacher writes next to one of your sentences that you've made an *assumption,* that *you* took something for granted, not showing or explaining why something is true, valid or important. Such statements are called assumptions; they state things that are taken for granted, unchallenged, not proven. Although all writers make assumptions, experienced and successful writers are often able to judge what assumptions an audience might accept or question.

In their conversation about history and film, Foner and Sayles make a number of assumptions. At one point early in the dialogue, Sayles says, "History is something that is useful to us and people feel we need." Having made this point, Sayles doesn't offer further explanation, provide supporting examples or argue for the truth of this statement: he probably assumes (and he's right) that Foner will accept that history is important and useful to people. You, on the other hand, might want to challenge the truth in this assumption.

As you re-read their conversation we want you to pay careful attention to what Foner and Sayles have to say and how they say it. As you re-read, what places in the text do they carefully explain, offer examples or reasons for a claim? And where do Foner and Sayles make assumptions? In your journal record three examples of statements that are explained or proven and three examples of statements, from your perspective, that represent writers making assumptions. One place you might begin is to extend the metaphor of an empty chair at the conversation table. Since Foner and Sayles make comments about people who go to films, how they read these films, and how they are affected by films, Foner and Sayles are talking about you.

WRITING CRITICISM

We would like you to think about the ways in which history is presented and revised through the media of popular culture. The way in which history is written, taught, and revised remains a highly controversial issue. We want you to select a historical event, something of national importance and significance, for which you can see and produce two different versions of that event: a text book or an established version of history and then a popular culture version of the same event. The popular culture version might be a film, television program, newspaper or magazine story (these would have been published at the time of the event), or a historical novel. The established version of history might come from library research, finding a book or an article in a professional history journal or accessing a university web-site on the Internet. You will need to take careful notes on each version of history, recording dates, places, names, and what has occurred. As you do this comparative work, we want you to notice similarities, differences or discrepancies between the established and the popular culture version.

As you begin to draft this essay, we first want you to present your own opinion or perspective on the value of reading, writing about or studying history. What is your stake in the question of how history is written or revised? Next you would compare these two versions of history, considering how one version differs from another. We would then like you to explain what these differences are and try to account for the differences. You could make an argument that one version is more historically accurate than the other. But it strikes us that in order to make that determination you would have to consult additional versions of this historical event.

Instead, we'd like you to see one version of history as a revision of another. How has history been revised in the popular culture version? What might each version of history say about the time period in which it was written? What we mean here is that history certainly tells us something about the past, but it also reveals certain information about the time period in which it was written. A history of the American Civil War written in the 1940s tells us about the Civil War but also something about the 1940s. This approach to history is called a "presentist" viewpoint. It maintains that history in some way is always concerned with the present, no matter what past it represents.

Grieving for the Camera

Neal Gabler

By now, there is a poignant familiarity to it. First comes the tragedy: the Oklahoma City bombing, the horrifying abduction and murder of 12-year-old Polly Klaas, the crash of TWA Flight 800. Then come the grieving relatives: parents, grandparents, siblings, sometimes cousins and in-laws. Some, understandably, looking as blank as automatons, others weeping copiously, they sit before the TV cameras and exhibit their pain while we uneasily look on—voyeurs in suffering. Recently it was Marc Klaas, apparently soothing his grief and anger by shuttling from one camera to another in the wake of the sentencing of his daughter's murderer. Next week, it will be some other devastated soul.

To many, the ghoulish exploitation of people at a time of obvious vulnerability, when all they want to do is talk about their trauma, is yet another example of the media running amok. Journalists need an angle to humanize the abstraction of tragedy. Teary relatives provide great visuals. The tragic event itself is over—reduced to strewn wreckage or an empty crime scene. But a camera fixed on the cracking mask of the grieving is recording high drama.

In light of this, it seems pointless to ask why the media shuffle talk with benumbed mourners. They are simply doing what they've always done, all the way back to the tabloid Sob Sisters, who reported heartbreaking tales of woe in the 1920s. The real question is: Why do people who have suffered terrible loss now routinely bare their souls for television?

The answer may say less about the media than about our own changing definition of privacy.

It wasn't so long ago that stoicism in the face of adversity was regarded as a virtue. We would speak approvingly of how people bore their grief, how

Reprinted by permission of Neal Gabler. Copyright 1996 by Neal Gabler. Originally published in the *Los Angeles Times,* 10/6/96, as printed in the Pittsburgh Post-Gazette, 10/13/96.

153

they held up. One of the most powerful allures of Jacqueline Kennedy was this very quality: the quiet dignity she displayed during and after the national mourning for President Kennedy. We admired her strength and her grace, and one imagines her stature would have shrunk considerably if she had instead granted an interview to Barbara Walters and, sniffling and red-eyed, publicly aired her distress.

That's because we all accepted the decision between public and private conduct. We assumed that, at some point, Mrs. Kennedy had broken down, torn her hair and railed at fate. She would have been inhuman if she hadn't. But we also knew that her breakdown was behind closed doors because this sort of personal exposure wasn't for public consumption. In public, she understood it was more appropriate to conceal her emotions than to expose them.

Her public decorum was precisely the kind of behavior that cultural historian Richard Sennett applauded in his landmark study "The Fall of Public Man." Once upon a time, Sennett argued, people had an admirable sense of reserve and propriety. They obeyed a set of public conventions—a code of behavior—and the conventions governed social relations. In Sennett's view, this role-playing was neither dishonest nor hypocritical. It was necessary to maintain order and community. There were certain things best kept private. A sensible world recognized that.

What Sennett lamented was the steady erosion of these public conventions under the onslaught of the cult of the self—what we call today "releasing the inner child" or "awakening the giant." Convinced by the new psychobabble that the old social conventions inhibited one's self-expression and denied one's true feelings, people began chucking those conventions. Where a person once guarded his privacy, now he seemed only too eager to reveal his inner self—his fears, needs, hopes, even his dysfunctions.

Sennett thought this narcissistic and destructive of the social order, but it was, nonetheless, irreversible. Before we were just cast members in a large social drama. The cult of the self made us all stars.

What Sennett didn't foresee is that the self would become a role, too. Sooner or later, all our forms of self-expression, all our problems, emotions and claims of individuality, would also become conventions—just as routinized as the old social conventions he had extolled. In fact, what we have seen over the last 20 years is not only the transformation of once private behavior into public conduct, but the transformation of this "publicly private" conduct into something predictable.

No doubt many of the victims' relatives who go before the cameras have been borne along by the wave of the self and believe that showing their grief is both legitimate and, one hopes, therapeutic. Riveted by their tragedy, they need to talk as a way of anesthetizing their hurt, and if the audience happens to be millions, rather than one, so much the better.

Perhaps they also believe that, in this age where nothing is private, public stoicism would seem a form of not caring for those they had loved. Sadly, they may think that only television can validate their loss and give it the requisite amplitude.

But however beneficial these televised appearances might be for the grief-stricken, one also suspects that, without even realizing it, the participants are playing out a new role of mourner in much the way Jacqueline Kennedy played out the old role. We now *expect* these appearances. We know the routine of anguish and, subconsciously or not, the bereaved have learned how to conform.

That is why Susan Smith could murder her children and then appeal to an alleged kidnapper to return them. Smith had watched enough TV news to be able to enact the whole gamut of sorrow: the tears, the screams, the clutched hands, the desperate entreaties. And she was savvy enough to know that the media and the public would attend.

Public grieving has become so formulaic that mourners go through the motions even when the media don't demand them. Many thought it peculiar that relatives of the victims on Flight 800 called their own news conferences to discuss their feelings. One didn't doubt their sincerity or agony, but it seemed odd to have them asking to face the cameras to show just how agonized they were—as if this confessional was now a necessary stage in the mourning process.

That may be the awful danger lurking within these heart-rending appearances—as well as the hidden lure. Once one is facing the cameras, one may no longer be just a mourner. One may become a celebrity, unwittingly exploiting one's own tragedy as the media exploit it—especially since celebrity promises to fill the void the tragedy has created.

Edye Smith, the young woman who lost two sons in the Oklahoma City bombing, has become just such a celebrity. She is visited by hordes of media who see her as the symbol of the tragedy, and she proudly keeps track of every journalist who comes. The National Enquirer even paid for the wedding when she decided to remarry her dead sons' father.

•

It is a macabre trade-off. They give us their pain, and the media give them short-lived celebrity. Obviously, it is not a trade-off anyone would willingly make. But in a society starved for new drama, and to people too disoriented to see clearly, it seems to work.

Again and again, they open their private wounds, hoping to do their loved ones honor by proclaiming their public grief—hoping to find peace in the media ether.

COMING TO TERMS

We've all seen it on TV (or maybe in person): Something terrible happens—a horrible accident, a brutal, disturbing crime, a plant or company closes, a child is missing—and a reporter and camera crew are there to capture the grief with camera and microphone. At least initially Neal Gabler seems to be expressing a common complaint directed toward media in these headline situations when he says, "To many, the ghoulish exploitation of people at a time of obvious vulnerability, when all they want to do is talk about their trauma, is yet another example of the media running amok." Later in "Grieving for the Camera," however, Gabler moves beyond the usual theme of media exploitation and argues that the relationship between the media and the grieving victim, family member, or friend is a complex reciprocal relationship (the media get something, the grief-stricken get something) and is created by changes in our cultural attitudes towards privacy. As a way of coming to terms with Gabler's reworking of some well-established media criticism, we'd like you to take stock of your attitude towards the news media as it represents the personal tragedy of others. Watch (and take notes) on a few episodes of the local or national televised news in which reporters interview victims, relatives, or even witnesses involved in a tragic event. Then respond to the following questions in your journal. How would you characterize the work of the news media in these situations? What does the camera see? And what does the microphone hear? The news footage and sound segments have been edited to emphasize what about the story? They tell what kind of story? How would you characterize the questions asked? Reporters sometimes claim they must pose the difficult questions they do because of professional responsibility and because the public has *the right to know.* Do you see your rights illustrated in these situations: do you have the right to know? Are there situations in which it's hard to claim a right to know? What would you say should be professional journalistic behavior in those situations? Finally, ask a few people (not members of your class, however) their take on the news media's coverage of personal tragedy. Do they see it as intrusive or insensitive? As justified, necessary? Do they see themselves as, in the words of Neal Gabler, "voyeurs in suffering"?

READING AS A WRITER

Neal Gabler asks, "Why do people who have suffered terrible loss now routinely bare their souls for television?" Gabler's use of two words in this question, *now* and *routinely,* suggests a writer who will be investigating historical and cultural changes in "Grieving for the Camera." We'd like you to re-read Gabler's article to examine a critical argument built upon a writer's perception and representation of historical and cultural change. In addition, Gabler's work gives us an opportunity to explore the difficulties that emerge in this kind of critical writing and the degree of uncertainty that can also emerge in serious writing which raises questions it cannot easily answer. We'd especially like you to be aware of the generalizations in Gabler's reading of U.S. culture and history. Respond to the following questions in your journal. How does Gabler describe the dominant cultural attitude towards expressing grief or defining privacy before the change in attitude occurred? What has changed? What kind of evidence does Gabler offer to support his claim? How does Gabler illustrate this change through his use of *historical markers,* people, or events that an audience would recognize? What does Gabler say are the origins of the change in our attitude towards privacy? How convincing do you think Gabler is here? Gabler's answer to the question he posed earlier in his article is as follows: "The answer may say less about the media than about our own changing definition of privacy." Here Gabler is carefully clarifying his answer: this change is a deep and significant one, not simply generated by the opportunities the television camera affords. How representative of a larger population is what is shown on television? Or to ask this question another way: Is what we see through the television camera an accurate barometer of how people really behave or what's important to them? Why or why not? Can you think of other popular or media culture texts which support or challenge Gabler's claims? Finally, at one point, Gabler says that "'publicly private'" behavior has become "something predictable." What are the consequences when a writer sets his sights on the "predictable"? What might Gabler be missing?

WRITING

We'd like you to write an essay in which you extend, amplify, or challenge Neal Gabler's critical argument concerning the changing cultural attitudes towards privacy—the blurring of clear distinctions between the private and the public self. You would also consider the consequences of those changes in U.S. culture and the appeal of a media or popular culture which exposes or brings to the surface emotions, attitudes or subjects once considered inappropriate for public display or discussion. We'd like you to focus your investigation on a specific media or popular culture text in which you see the collision of what's now public that was once only private, what Gabler paradoxically calls "'publicly private' conduct."

For this project you might choose an episode (or two) of a television talk show which addresses a controversial issue or a radio call-in program that offers personal advice or psychological counseling. A television news journal, like *60 Minutes* or *20/20,* often incorporates an in-depth interview that you might use. Or you might select a no-holds-barred print or Internet interview with a celebrity or personality currently in the news. Another text to consider would be an autobiography or biography—authorized or unauthorized.

We'd then like you to focus on and explain the appeal of—our continued fascination with—this kind of media or popular culture text which might, for example, transform "mourner to celebrity" or private tragedy into public discourse. As part of your project, your work might investigate answers to some of the following questions. What do the participants in such a discourse gain (or lose) from publicly broadcasting their suffering, pain, personal or family tragedies, addictions, narratives of abuse, etc? What do the viewers, readers or listeners gain or lose? Does the discourse strike you as being therapeutic in some way? Do people who "grieve for the camera" receive in exchange a kind of "media ether"? Do they celebrate or embrace or resist the classification of being victimized somehow? And why might this choice be significant? Is there a confessional quality to the discourse? And why might this be significant?

Is the appeal of the discourse one of prurient interests? Exploring a cultural taboo? Or, in the language of pop psychology (what Gabler might call "psychobabble"), does the discourse validate a certain problem or experience? Does the discourse serve an educational purpose? For example, does the discourse provide valuable information, generate new understanding, or create sympathy? If a studio audience is present, how does the audience contribute to the discourse, how does it further your understanding of the public and private self? Is someone or something being exploited in the discourse? As you think about these questions in terms of your text, is there something which draws us to the spectacle of people in pain, in trouble, in crisis? Are we, as Gabler, observes "voyeurs in suffering" who inhabit "a society starved for new drama"?

Finally, you may want to use your work with this text to explore a somewhat larger cultural context than Gabler provides us in "Grieving for the Camera." Gabler seems convinced that we "grieve for the camera" not so much because television creates an opportunity to do so, but because America has undergone a fundamental cultural change in our definition of privacy and the importance that we now give to the "cult of self." We wonder, however, if other cultural or social forces might be at play here. There might be multiple attitudes (even contradictory ones) directed toward a cultural issue, especially if the issue is a significant one like privacy. For example, in what ways might advancements in telecommunications technology affect how we view privacy? How might technology actually heighten our needs for privacy, keeping the private and public self separate? How does this technology protect the private self from being thrust into public discourse? Or how might changes in fundamental attitudes towards uncomfortable social issues, like alcoholism or abusive relationships, actually make possible their appearance in public discourse? Is it possible that groups within a culture—groups identified by race, gender, economic class, etc.—might exhibit competing attitudes towards issues of privacy?

Thirteen Ways of Looking at a Black Man

Henry Louis Gates, Jr.

"Every day, in every way, we are getting meta and meta," the philosopher John Wisdom used to say, venturing a cultural counterpart to Émile Coué's famous mantra of self-improvement. So it makes sense that in the aftermath of the Simpson trial the focus of attention has been swiftly displaced from the verdict to the reaction to the verdict, and then to the reaction to the reaction to the verdict, and, finally, to the reaction to the reaction to the reaction to the verdict—which is to say, black indignation at white anger at black jubilation at Simpson's acquittal. It's a spiral made possible by the relay circuit of race. Only in America.

An American historian I know registers a widespread sense of bathos when he says, "Who would have imagined that the Simpson trial would be like the Kennedy assassination—that you'd remember where you were when the verdict was announced?" But everyone does, of course. The eminent sociologist William Julius Wilson was in the red-carpet lounge of a United Airlines terminal, the only black in a crowd of white travellers, and found himself as stunned and disturbed as they were. Wynton Marsalis, on tour with his band in California, recalls that "everybody was acting like they were above watching it, but then when it got to be ten o'clock—zoom, we said, 'Put the verdict on!'" Spike Lee was with Jackie Robinson's widow, Rachel, rummaging through a trunk filled with her husband's belongings, in preparation for a biopic he's making on the athlete. Jamaica Kincaid was sitting in her car in the parking lot of her local grocery store in Vermont, listening to the proceedings on National Public Radio, and she didn't pull out until after they were over. I was teaching a literature seminar at Harvard from twelve to two, and watched the verdict with the class on a television set in the seminar room. That's where I first saw the sort of racialized response that itself would fill television screens for the next few days: the white students looked aghast, and the black students cheered. "Maybe you should remind the students that this is a case about two people who were brutally slain, and not an occasion to celebrate," my teaching assistant, a white woman, whispered to me.

The two weeks spanning the O. J. Simpson verdict and Louis Farrakhan's Million Man March on Washington were a good time for connoisseurs of racial paranoia. As blacks exulted at Simpson's acquittal, horrified whites had a fleeting sense that this race thing was knottier than they'd ever supposed—that, when all the pieties were cleared away, blacks really *were* strangers in their midst. (The unspoken sentiment: *And I thought I knew these people.*) There was the faintest tincture of the Southern slaveowner's disquiet in the aftermath of the bloody slave revolt led by Nat Turner—when the gentleman farmer was left to wonder which of his smiling, servile retainers would have slit *his* throat if the rebellion had spread as was intended, like fire on parched thatch. In the day or so following the verdict, young urban professionals took note of a slight *froideur* between themselves and their nannies and babysitters—the awkwardness of an unbroached subject. Rita Dove, who recently completed a term as the United States Poet Laureate, and who believes that Simpson was guilty, found it "appalling that white people were so outraged—more appalling than the decision as to whether he was guilty or not." Of course, it's possible to overstate the tensions. Marsalis invokes the example of team sports, saying, "You want your side to win, whatever the side is going to be. And the thing is, we're still at a point in our national history where we look at each other as sides."

The matter of side-taking cuts deep. An old cartoon depicts a woman who has taken her errant daughter to see a child psychiatrist. "And when we were watching 'The Wizard of Oz,'" the distraught mother is explaining, "she was rooting for the wicked witch!" What many whites experienced was the bewildering sense that an entire population had been rooting for the wrong side. "This case is a classic example of what I call interstitial spaces," says Judge A. Leon Higginbotham, who recently retired from the federal Court of Appeals, and who last month received the Presidential Medal of Freedom. "The jury system is predicated on the idea that different people can view the same evidence and reach diametrically opposed conclusions." But the observation brings little solace. If we disagree about something so basic, how can we find agreement about far thornier matters? For white observers, what's even scarier than the idea that black Americans were plumping for the villain, which is a misprision of value, is the idea that black Americans didn't recognize him *as* the villain, which is a misprision of fact. How can conversation begin when we disagree about reality? To put it at its harshest, for many whites a sincere belief in Simpson's innocence looks less like the culture of protest than like the culture of psychosis.

Perhaps you didn't know that Liz Claiborne appeared on "Oprah" not long ago and said that she didn't design her clothes for black women—that their hips were too wide. Perhaps you didn't know that the soft drink Tropical Fantasy is manufactured by the Ku Klux Klan and contains a special ingredient designed to sterilize black men. (A warning flyer distributed in Harlem a few years ago claimed that these findings were vouchsafed on the television program "20/20.") Perhaps you didn't know that the Ku Klux Klan has a similar arrangement with Church's Fried Chicken—or is it Popeye's?

Perhaps you didn't know these things, but a good many black Americans think they do, and will discuss them with the same intentness they bring to speculations about the "shadowy figure" in a Brentwood driveway. Never mind that Liz Claiborne has never appeared on "Oprah," that the beleaguered Brooklyn company that makes Tropical Fantasy has gone as far as to make available an F.D.A. assay of its ingredients, and that those fried-chicken franchises pose a threat mainly to black folks' arteries. The folklorist Patricia A. Turner, who has collected dozens of such tales in an invaluable 1993 study of rumor in African-American culture, "I Heard It Through the Grapevine," points out the patterns to be found here: that these stories encode regnant anxieties, that they may take root under particular conditions and play particular social roles, that the currency of rumor flourishes where "official" news has proved untrustworthy.

Certainly the Fuhrman tapes might have been scripted to confirm the old saw that paranoids, too, have enemies. If you wonder why blacks seem particularly susceptible to rumors and conspiracy theories, you might look at a history in which the official story was a poor guide to anything that mattered much, and in which rumor sometimes verged on the truth. Heard the one about the L.A. cop who hated interracial couples, fantasized about making a bonfire of black bodies, and boasted of planting evidence? How about the one about the federal government's forty-year study of how untreated syphilis affects black men? For that matter, have you ever read through some of the F.B.I.'s COINTELPRO files? ("There is but one way out for you," an F.B.I. scribe wrote to Martin Luther King, Jr., in 1964, thoughtfully urging on him the advantages of suicide. "You better take it before your filthy, abnormal, fraudulent self is bared to the nation.")

People arrive at an understanding of themselves and the world through narratives—narratives purveyed by schoolteachers, newscasters, "authorities," and all the other authors of our common sense. Counternarratives are, in turn, the means by which groups contest that dominant reality and the fretwork of assumptions that supports it. Sometimes delusion lies that way; sometimes not. There's a sense in which much of black history is simply counternarrative that has been documented and legitimatized, by slow, hardwon scholarship. The "shadowy figures" of American history have long been our own ancestors, both free and enslaved. In any case, fealty to counternarratives is an index to alienation, not to skin color: witness Representative Helen Chenoweth, of Idaho, and her devoted constituents. With all the appositeness of allegory, the copies of "The Protocols of the Elders of Zion" sold by black venders in New York—who are supplied with them by Lushena Books, a black-nationalist book wholesaler—were published by the white supremacist Angriff Press, in Hollywood. Paranoia knows no color or coast.

Finally, though, it's misleading to view counternarrative as another pathology of disenfranchisement. If the M.I.A. myth, say, is rooted among a largely working-class constituency, there are many myths—one of them known as Reaganism—that hold considerable appeal among the privileged classes. "So many white brothers and sisters are living in a state of denial in terms of how deep white supremacy is seated in their culture and society," the scholar and social critic Cornel West says. "Now we recognize that in a fundamental sense we really do live in different worlds." In that respect, the reaction to the Simpson verdict has been something of an education. The novelist Ishmael Reed talks of "wealthy white male commentators who live in a world where the police don't lie, don't plant evidence—and drug dealers give you unlimited credit." He adds, "Nicole, you know, also dated Mafia hit men."

"I think he's innocent, I really do," West says. "I do think it was linked to some drug subculture of violence. It looks as if both O. J. and Nicole had some connection to drug activity. And the killings themselves were classic examples of that drug culture of violence. It could have to do with money owed—it could have to do with a number of things. And I think that O. J. was quite aware of and fearful of this." On this theory, Simpson may have appeared at the crime scene as a witness. "I think that he had a sense that it was coming down, both on him and on her, and Brother Ron Goldman just happened to be there," West conjectures. "But there's a possibility also that O. J. could have been there, gone over and tried to see what was going on, saw that he couldn't help, split, and just ran away. He might have said, 'I can't stop this thing, and they are coming at me to do the same thing.' He may have actually run for his life."

To believe that Simpson is innocent is to believe that a terrible injustice has been averted, and this is precisely what many black Americans, including many prominent ones, do believe. Thus the soprano Jessye Norman is angry over what she sees as the decision of the media to prejudge Simpson rather than "educate the public as to how we could possibly look at things a bit differently." She says she wishes that the real culprit "would stand up and say, 'I did this and I am sorry I caused so much trouble.'" And while she is sensitive to the issue of spousal abuse, she is skeptical about the way it was enlisted by the prosecution: "You have to stop getting into how they were at home, because there are not a lot of relationships that could be put on television that we would think, O.K, that's a good one. I mean, just stop pretending that this is the case." Then, too, she asks, "Isn't it interesting to you that this Faye Resnick person was staying with Nicole Brown Simpson and that she happened to have left on the eighth of June? Does that tell you that maybe there's some awful coincidence here?" The widespread theory about murderous drug dealers Norman finds "perfectly plausible, knowing what drugs do," and she adds, "People are punished for being bad."

There's a sense in which all such accounts can be considered counternarratives, or fragments of them—subaltern knowledge, if you like. They dispute the tenets of official culture; they do not receive the imprimatur of editorialists or of network broadcasters; they are not seriously entertained on "MacNeil/Lehrer." And when they do surface they are given consideration primarily for their ethnographic value. An official culture treats their claims as it does those of millenarian cultists in Texas, or Marxist deconstructionists in the academy: as things to be diagnosed, deciphered, given meaning—that is, *another* meaning. Black folk say they believe Simpson is innocent, and then the white gatekeepers of a media culture cajolingly explain what black folk really mean when they say it, offering the explanation from the highest of motives: because the alternative is a population that, by their lights, is not merely counternormative but crazy. Black folk may mean anything at all; just not what they say they mean.

Yet you need nothing so grand as an epistemic rupture to explain why different people weigh the evidence of authority differently. In the words of the cunning Republican campaign slogan, "Who do you trust?" It's a commonplace that white folks trust the police and black folks don't. Whites recognize this in the abstract, but they're continually surprised at the *depth* of black wariness. They shouldn't be. Norman Podhoretz's soul-searching 1963 essay, "My Negro Problem, and Ours"—one of the frankest accounts we have of liberalism and race resentment—tells of a Brooklyn boyhood spent under the shadow of carefree, cruel Negro assailants, and of the author's residual unease when he passes groups of blacks in his Upper West Side neighborhood. And yet, he notes in a crucial passage, "I know now, as I did not know when I was a child, that power is on my side, that the police are working for me and not for them." That ordinary, unremarkable comfort—the feeling that "the police are working for me"—continues to elude blacks, even many successful blacks. Thelma Golden, the curator of the Whitney's "Black Male" show, points out that on the very day the verdict was announced a black man in Harlem was killed by the police under disputed circumstances. As older blacks like to repeat, "When white folks say 'justice,' they mean 'just us.'"

Blacks—in particular, black men—swap their experiences of police encounters like war stories, and there are few who don't have more than one story to tell. "These stories have a ring of cliché about them," Erroll McDonald, Pantheon's executive editor and one of the few prominent blacks in publishing, says, "but, as we all know about clichés, they're almost always true." McDonald tells of renting a Jaguar in New Orleans and being stopped by the police—simply "to show cause why I shouldn't be deemed a problematic

Negro in a possibly stolen car." Wynton Marsalis says, "Shit, the police slapped me upside the head when I was in high school. I wasn't Wynton Marsalis then. I was just another nigger standing out somewhere on the street whose head could be slapped and did get slapped." The crime novelist Walter Mosley recalls, "When I was a kid in Los Angeles, they used to stop me all the time, beat on me, follow me around, tell me that I was stealing things." Nor does William Julius Wilson—who has a son-in-law on the Chicago police force ("You couldn't find a nicer, more dedicated guy")—wonder why he was stopped near a small New England town by a policeman who wanted to know what he was doing in those parts. There's a moving violation that many African-Americans know as D.W.B.: Driving While Black.

So we all have our stories. In 1968, when I was eighteen, a man who knew me was elected mayor of my West Virginia county, in an upset victory. A few weeks into his term, he passed on something he thought I should know: the county police had made a list of people to be arrested in the event of a serious civil disturbance, and my name was on it. Years of conditioning will tell. Wynton Marsalis says, "My worst fear is to have to go before the criminal-justice system." Absurdly enough, it's mine, too.

Another barrier to interracial comprehension is talk of the "race card"—a phrase that itself infuriates many blacks. Judge Higginbotham, who pronounces himself "not uncomfortable at all" with the verdict, is uncomfortable indeed with charges that Johnnie Cochran played the race card. "This whole point is one hundred per cent inaccurate," Higginbotham says. "If you knew that the most important witness had a history of racism and hostility against black people, that should have been a relevant factor of inquiry even if the jury had been all white. If the defendant had been Jewish and the police officer had a long history of expressed anti-Semitism and having planted evidence against innocent persons who were Jewish, I can't believe that anyone would have been saying that defense counsel was playing the anti-Semitism card." Angela Davis finds the very metaphor to be a problem. "Race is not a card," she says firmly. "The whole case was pervaded with issues of race."

Those who share her view were especially outraged at Robert Shapiro's famous post-trial rebuke to Cochran—for not only playing the race card but dealing it "from the bottom of the deck." Ishmael Reed, who is writing a book about the case, regards Shapiro's remarks as sheer opportunism: "He wants to keep his Beverly Hills clients—a perfectly commercial reason." In Judge Higginbotham's view, "Johnnie Cochran established that he was as effective as any lawyer in America, and though whites can tolerate black excellence in singing, dancing, and dunking, there's always been a certain level of discomfort among many whites when you have a one-on-one challenge in terms of intellectual competition. If Edward Bennett Williams, who was one of the most able lawyers in the country, had raised the same issues, half of the complaints would not exist."

By the same token, the display of black prowess in the courtroom was heartening for many black viewers. Cornel West says, "I think part of the problem is that Shapiro—and this is true of certain white brothers—has a profound fear of black-male charisma. And this is true not only in the law but across the professional world. You see, you have so many talented white brothers who deserve to be in the limelight. But one of the reasons they are not in the limelight is that they are not charismatic. And here comes a black person who's highly talented but also charismatic and therefore able to command center stage. So you get a very real visceral kind of jealousy that has to do with sexual competition as well as professional competition."

Erroll McDonald touches upon another aspect of sexual tension when he says, "The so-called race card has always been the joker. And the joker is the history of sexual racial politics in this country. People forget the singularity of this issue—people forget that less than a century ago black men were routinely lynched for merely glancing at white women or for having been *thought* to have glanced at a white woman." He adds, with mordant irony, "Now we've come to a point in our history where a black man could, potentially, have murdered a white woman and thrown in a white man to boot—and got off. So the country has become far more complex in its discussion of race." This is, as he appreciates, a less than perfectly consoling thought.

"But he's coming for me," a woman muses in Toni Morrison's 1994 novel, *Jazz*, shortly before she is murdered by a jealous ex-lover. "Maybe tomorrow he'll find me. Maybe tonight." Morrison, it happens, is less interested in the grand passions of love and requital than she is in the curious texture of communal amnesty. In the event, the woman's death goes unavenged; the man who killed her is forgiven even by her friends and relatives. Neighbors feel that the man fell victim to her wiles, that he didn't understand "how she liked to push people, men." Or, as one of them says of her, "live the life; pay the price." Even the woman—who refuses to name the culprit as she bleeds to death—seems to accede to the view that she brought it on herself.

It's an odd and disturbing theme, and one with something of a history in black popular culture. An R. & B. hit from 1960, "There's Something on Your Mind," relates the anguish of a man who is driven to kill by his lover's infidelity. The chorus alternates with spoken narrative, which informs us that his first victim is the friend with whom she was unfaithful. But then:

> Just as you make it up in your mind to forgive her, here come another one of your best friends through the door. This really makes you blow your top, and you go right ahead and shoot her. And realizing what you've done, you say: "Baby, please, speak to me. Forgive me. I'm sorry."

"We are a *forgiving* people," Anita Hill tells me, and she laughs, a little uneasily. We're talking about the support for O. J. Simpson in the black community; at least, I think we are.

A black woman told the *Times* last week, "He has been punished enough." But forgiveness is not all. There is also an element in this of outlaw culture: the tendency—which unites our lumpenproles with our post-modern ironists—to celebrate transgression for its own sake. Spike Lee, who was surprised but "wasn't happy" at the verdict ("I would have bet money that he was going to the slammer"), reached a similar conclusion: "A lot of black folks said, 'Man, O. J. is *bad*, you know. This is the first brother in the history of the world who got away with the murder of white folks, and a blond, blue-eyed woman at that.'"

But then there is the folk wisdom on the question of why Nicole Brown Simpson had to die——the theodicy of the streets. For nothing could be further from the outlaw ethic than the simple and widely shared certainty that, as Jessye Norman says, people are punished for doing wrong. And compounding the sentiment is Morrison's subject—the culturally vexed status of the so-called crime of passion, or what some took to be one, anyway. You play, you pay: it's an attitude that exists on the streets, but not only on the streets, and one that somehow attaches to Nicole, rather than to her ex-husband. Many counternarratives revolve around her putative misbehavior. The black feminist Bell Hooks notes with dismay that what many people took to be a "narrative of a crime of passion" had as its victim "a woman that many people, white and black, felt was like a whore. Precisely by being a sexually promiscuous woman, by being a woman who used drugs, by being a white woman with a black man, she had already fallen from grace in many people's eyes—there was no way to redeem her." Ishmael Reed, for one, has no interest in redeeming her. "To paint O. J. Simpson as a beast, they had to depict her as a saint," he complains. "Apparently, she had a violent temper. She slapped her Jamaican maid. I'm wondering, the feminists who are giving Simpson such a hard time—do they approve of white women slapping maids?"

Of course, the popular trial of Nicole Brown Simpson—one conducted off camera, in whispers—has further occluded anything recognizable as sexual politics. When Anita Hill heard that O. J. Simpson was going to be part of the Million Man March on Washington, she felt it was entirely in keeping with the occasion: a trial in which she believed that matters of gender had been "bracketed" was going to be succeeded by a march from which women were excluded. And, while Minister Louis Farrakhan had told black men that October 16th was to serve as a "day of atonement" for their sins, the murder of Nicole Brown Simpson and Ronald Goldman was obviously not among the sins he had in mind. Bell Hooks argues, "Both O. J.'s case and the Million Man March confirm that, while white men are trying to be sensitive and pretending they're the new man, black men are saying that patriarchy must be upheld at all costs, even if women must die." She sees the march as a congenial arena for Simpson in symbolic terms: "I think he'd like to strut his stuff, as the patriarch. He is the dick that stayed hard longer." ("The surprising thing is that you won't see Clarence Thomas going on that march," Anita Hill remarks of another icon of patriarchy.) Farrakhan himself prefers metaphors of military mobilization, but the exclusionary politics of the event has clearly distracted from its ostensible message of solidarity. "First of all, I wouldn't go to no war and leave half the army home," says Amiri Baraka, the radical poet and playwright who achieved international renown in the sixties as the leading spokesman for the Black Arts movement. "Logistically, that doesn't make sense." He notes that Martin Luther King's 1963 March on Washington was "much more inclusive," and sees Far-rakhan's regression as "an absolute duplication of what's happening in the country," from Robert Bly on: the sacralization of masculinity.

Something like that dynamic is what many white feminists saw on display in the Simpson verdict; but it's among women that the racial divide is especially salient. The black legal scholar and activist Patricia Williams says she was "stunned by the intensely personal resentment of some of my white women friends in particular." Stunned but, on reflection, not mystified. "This is Greek drama," she declares. "Two of the most hotly contended aspects of our lives are violence among human beings who happen to be police officers and violence among human beings who happen to be husbands, spouses, lovers." Mean-while, our attention has been fixated on the rhetorical violence between human beings who happen to disagree about the outcome of the O. J. Simpson trial.

It's a cliché to speak of the Simpson trial as a soap opera—as entertain-ment, as theatre—but it's also true, and in ways that are worth exploring fur-ther. For one thing, the trial provides a fitting rejoinder to those who claim that we live in an utterly fragmented culture, bereft of the common narratives that bind a people together. True, Parson Weems has given way to Dan Rather, but public narrative persists. Nor has it escaped notice that the biggest televised legal contests of the last half decade have involved race matters: Anita Hill and Rodney King. So there you have it: the Simpson trial—black entertainment television at its finest. Ralph Ellison's hopeful insistence on the Negro's cen-trality to American culture finds, at last, a certain tawdry confirmation.

"The media generated in people a feeling of being spectators at a show," the novelist John Edgar Wideman says. "And at the end of a show you ap-plaud. You are happy for the good guy. There is that sense of primal identi-fication and closure." Yet it's a fallacy of "cultural literacy" to equate shared narratives with shared meanings. The fact that American TV shows are re-broadcast across the globe causes many people to wring their hands over the menace of cultural imperialism; seldom do they bother to inquire about the meanings that different people bring to and draw from these shows. When they do make inquiries, the results are often surprising. One researcher talked to Israeli Arabs who had just watched an episode of "Dallas"—an ep-isode in which Sue Ellen takes her baby, leaves her husband, J.R., and moves in with her ex-lover and his father. The Arab viewers placed their own con-struction on the episode: they were all convinced that Sue Ellen had moved in with her *own* father—something that by their mores at least made sense.

A similar thing happened in America this year: the communal experience afforded by a public narrative (and what narrative more public?) was splintered by the politics of interpretation. As far as the writer Maya Angelou is concerned, the Simpson trial was an exercise in minstrelsy. "Minstrel shows caricatured every aspect of the black man's life, beginning with his sexuality," she says. "They portrayed the black man as devoid of all sensibilities and sensitivities. They minimized and diminished the possibility of familial love. And that is what the trial is about. Not just the prosecution but everybody seemed to want to show him as other than a normal human being. Nobody let us just see a man." But there is, of course, little consensus about what genre would best accommodate the material. Walter Mosley says, "The story plays to large themes, so I'm sure somebody will write about it. But I don't think it's a mystery, I think it's much more like a novel by Zola." What a writer might make of the material is one thing; what the audience has made of it is another.

"Simpson is a B-movie star and people were watching this like a B movie," Patricia Williams says. "And this is *not* the American B-movie ending." Or was it? "From my perspective as an attorney, this trial was much more like a movie than a trial," Kathleen Cleaver, who was once the Black Panthers' Minister for Communication and is now a professor of law at Emory, says, "It had the budget of a movie, it had the casting of a movie, it had the tension of a movie, and the happy ending of a movie." Spike Lee, speaking professionally, is dubious about the trial's cinematic possibilities: "I don't care who makes this movie, it is never going to equal what people have seen in their living rooms and houses for eight or nine months." Or is it grand opera? Jessye Norman considers: "Well, it certainly has all the ingredients. I mean, somebody meets somebody and somebody gets angry with somebody and somebody dies." She laughs. "It sounds like the 'Ring' cycle of Wagner—it really does."

"This story has been told any number of times," Angelou says. "The first thing I thought about was Eugene O'Neill's 'All God's Chillun.'" Then she considers how the event might be retrieved by an African-American literary tradition. "I think a great writer would have to approach it," she tells me pensively. "James Baldwin could have done it. And Toni Morrison could do it."

"Maya Angelou could do it," I say.

"I don't like that kind of stuff," she replies.

There are some for whom the question of adaptation is not entirely abstract. The performance artist and playwright Anna Deavere Smith has already worked on the 911 tape and F. Lee Bailey's cross-examination of Mark Fuhrman in the drama classes she teaches at Stanford. Now, with a dramaturge's eye, she identifies what she takes to be the climactic moment: "Just after the verdict was read I will always remember two sounds and one image. I heard Johnnie Cochran go 'Ugh,' and then I heard the weeping of Kim Goldman. And then I saw the image of O. J.'s son, with one hand going upward on one eye and one hand pointed down, shaking and sobbing. I couldn't do the words right now; if I could find a collaborator, I would do something else. I feel that a choreographer ought to do that thing. Part of the tragedy was the fact of that 'Ugh' and that crying. Because that 'Ugh' wasn't even a full sound of victory, really." In "Thirteen Ways of Looking at a Blackbird" Wallace Stevens famously said he didn't know whether he preferred "The beauty of inflections / Or the beauty of innuendoes, / The blackbird whistling / Or just after." American culture has spoken as with one voice: we like it just after.

Just after is when our choices and allegiances are made starkly apparent. Just after is when interpretation can be detached from the thing interpreted. Anita Hill, who saw her own presence at the Clarence Thomas hearings endlessly analyzed and allegorized, finds plenty of significance in the trial's reception, but says the trial itself had none. Naturally, the notion that the trial was sui generis is alien to most commentators. Yet it did not arrive in the world already costumed as a racial drama; it had to be racialized. And those critic—angry whites, indignant blacks—who like to couple this verdict with the Rodney King verdict should consider an elementary circumstance: Rodney King was an unknown and undistinguished black man who was brutalized by the police; the only thing exceptional about that episode was the presence of a video camera. But, as Bell Hooks asks, "in what other case have we ever had a wealthy black man being tried for murder?" Rodney King was a black man to his captors before he was anything else; O. J. Simpson was, first and foremost, O. J. Simpson. Kathleen Cleaver observes, "A black superhero millionaire is not someone for whom mistreatment is an issue." And Spike Lee acknowledges that the police "don't really bother black people once they are a personality." On this point, I'm reminded of something that Roland Gift, the lead singer of the pop group Fine Young Cannibals, once told a reporter: "I'm not black, I'm famous."

Simpson, too, was famous rather than black; that is, until the African-American community took its lead from the cover of *Time* and, well, blackened him. Some intellectuals are reluctant to go along with the conceit. Angela Davis, whose early-seventies career as a fugitive and a political prisoner provides one model of how to be famous *and* black, speaks of the need to question the way "O. J. Simpson serves as the generic black man," given that "he did not identify himself as black before then." More bluntly, Baraka says, "To see him get all of this God-damned support from people he has historically and steadfastly eschewed just pissed me off. He eschewed black people all his life and then, like Clarence Thomas, the minute he gets jammed up he comes talking about 'Hey, I'm black.'" And the matter of spousal abuse should remind us of another role-reversal entailed by Simpson's iconic status in a culture of celebrity: Nicole Brown Simpson would have known that her famous-not-black husband commanded a certain deference from the L.A.P.D. which she, who was white but not yet famous, did not.

"It's just amazing that we in the black community have bought into it," Anita Hill says, with some asperity, and she sees the manufacture of black-male heroes as part of the syndrome. "We continue to create a superclass of individuals who are above the rules." It bewilders her that Simpson "was being honored as someone who was being persecuted for his politics, when he had none," she says. "Not only do we forget about the abuse of his wife but we also forget about the abuse of the community, his walking away from the community." And so Simpson's connection to a smitten black America can be construed as yet another romance, another troubled relationship, another case study in mutual exploitation.

Yet to accept the racial reduction ("WHITES V. BLACKS," as last week's *Newsweek* headline had it) is to miss the fact that the black community itself is riven, and in ways invisible to most whites. I myself was convinced of Simpson's guilt, so convinced that in the middle of the night before the verdict was to be announced I found myself worrying about his prospective sojourn in prison: would he be brutalized, raped, assaulted? Yes, on sober reflection, such worries over a man's condign punishment seemed senseless, a study in misplaced compassion; but there it was. When the verdict was announced, I was stunned— but, then again, wasn't my own outrage mingled with an unaccountable sense of relief? Anna Deavere Smith says, "I am seeing more than that white people are pissed off and black people are ecstatic. I am seeing the difficulty of that; I am seeing people having difficulty talking about it." And many are weary of what Ishmael Reed calls "zebra journalism, where everything is seen in black-and-white." Davis says, "I have the feeling that the media are in part responsible for the creation of this so-called racial divide—putting all the white people on one side and all the black people on the other side."

Many blacks as well as whites saw the trial's outcome as a grim enactment of Richard Pryor's comic rejoinder "Who are you going to believe—me, or your lying eyes?" "I think if he were innocent he wouldn't have behaved that way," Jamaica Kincaid says of Simpson, taking note of his refusal to testify on his own behalf. "If you are innocent," she believes, "you might want to admit you have done every possible thing in the world—had sex with ten donkeys, twenty mules—but did not do this particular thing." William Julius Wilson says mournfully, "There's something wrong with a system where it's better to be guilty and rich and have good lawyers than to be innocent and poor and have bad ones."

The Simpson verdict was "the ultimate in affirmative action," Amiri Baraka says. "I *know* the son of a bitch did it." For his part, Baraka essentially agrees with Shapiro's rebuke of Cochran: "Cochran is belittling folks. What he's saying is 'Well, the niggers can't understand the question of perjury in the first place. The only thing they can understand is, 'He called you a nigger.'" He alludes to *Ebony*'s fixation on "black firsts"—the magazine's spotlight coverage of the first black to do this or that—and fantasizes the appropriate *Ebony* accolade. "They can feature him on the cover as 'The first Negro to kill a white woman and get away with it,'" he offers acidly. Then he imagines Farrakhan introducing him with just that tribute at the Million Man March. Baraka has been writing a play called "Othello, Jr.," so such themes have been on his mind. The play is still in progress, but he *has* just finished a short poem:

Free Mumia!
O. J. did it
And you know it.

"Trials don't establish absolute truth; that's a theological enterprise," Patricia Williams says. So perhaps it is appropriate that a religious leader, Louis Farrakhan, convened a day of atonement; indeed, some worry that it is all too appropriate, coming at a time when the resurgent right has offered us a long list of sins for which black men must atone. But the crisis of race in America is real enough. And with respect to that crisis a mass mobilization is surely a better fit than a criminal trial. These days, the assignment of blame for black woes increasingly looks like an exercise in scholasticism; and calls for interracial union increasingly look like an exercise in inanity. ("Sorry for the Middle Passage, old chap. I don't know *what* we were thinking." "Hey, man, forget it— and here's your wallet back. No, really, I want you to have it.") The black economist Glenn Loury says, "If I could get a million black men together, I wouldn't march them to Washington, I'd march them into the ghettos."

But because the meanings of the march are so ambiguous, it has become itself a racial Rorshach—a vast ambulatory allegory waiting to happen. The actor and director Sidney Poitier says, "If we go on such a march to say to ourselves and to the rest of America that we want to be counted among America's people, we would like our family structure to be nurtured and strengthened by ourselves and by the society, that's a good point to make." He sees the march as an occasion for the community to say, "Look, we are adrift. Not only is the nation adrift on the question of race—we, too, are adrift. We need to have a sense of purpose and a sense of direction." Maya Angelou, who agreed to address the assembled men, views the event not as a display of male self-affirmation but as a ceremony of penitence: "It's a chance for African-American males to say to African-American females, 'I'm sorry. I am sorry for what I did, and I am sorry for what happened to both of us.'" But different observers will have different interpretations. Mass mobilizations launch a thousand narratives—especially among subscribers to what might be called the "great event" school of history. And yet Farrakhan's recurrent calls for individual accountability consort oddly with the absolution, both juridical and populist, accorded O. J. Simpson. Simpson has been seen as a symbol for many things, but he is not yet a symbol for taking responsibility for one's actions.

All the same, the task for black America is not to get its symbols in shape: symbolism is one of the few commodities we have in abundance. Meanwhile, Du Bois's century-old question "How does it feel to be a problem?" grows in trenchancy with every new bulletin about crime and poverty. And the Simpson trial spurs us to question everything except the way that the discourse of crime and punishment has enveloped, and suffocated, the analysis of race and poverty in this country. For the debate over the rights and wrongs of the

Simpson verdict has meshed all too well with the manner in which we have long talked about race and social justice. The defendant may be free, but we remain captive to a binary discourse of accusation and counter-accusation, of grievance and counter-grievance, of victims and victimizers. It is a discourse in which O. J. Simpson is a suitable remedy for Rodney King, and reductions in Medicaid are entertained as a suitable remedy for O. J. Simpson: a discourse in which everyone speaks of payback and nobody is paid. The result is that race politics becomes a court of the imagination wherein blacks seek to punish whites for their misdeeds and whites seek to punish blacks for theirs, and an infinite regress of score-settling ensues—yet another way in which we are daily becoming meta and meta. And so an empty vessel like O. J. Simpson becomes filled with meaning, and more meaning—more meaning than any of us can bear. No doubt it is a far easier thing to assign blame than to render justice. But if the imagery of the court continues to confine the conversation about race, it really will be a crime.

COMING TO TERMS

While Gates' essay centers on the famous O.J. Simpson trial, it is not exclusively about the trial itself, and indeed, there are several sections of his piece that hardly refer to the trial at all. In fact, this alternative way of *reporting* on this trial was one of the reasons we chose Gates' piece for *Media Journal*. Gates "reads" the O.J. Simpson trial as a *cultural text,* as a kind of index or barometer, as a place where he attempts to *come to terms* with what he and others see as the defining social issues in contemporary American life. Because Gates' work is complex and rather inclusive, we'd like you, in your journal, to make a kind of "map" of Gates' essay by first writing a brief summary of each of its eleven sections. Then use these summaries as the basis for a paragraph or two in which you restate, in your own words, what Gates has to say not about the Simpson trial itself but about what the coverage of and reactions to the trial tell us about ourselves, about our convictions and contradictions in beliefs, about our tensions and concerns as a society.

READING AS A WRITER

In the course of his essay Gates speaks with a remarkable number of people (and quotes from the writings or televised appearances of still more) about their reactions to the O.J. Simpson trial. But we think Gates' project here is more than to compile an extensive collection of quotations to give his article authority or the semblance of having his finger on the pulse of public opinion. We'd like you to examine Gates' use of quotation in light of the rhetorical purposes those quotations might reveal. In other words, how do Gates' choices here promote or limit a particular kind of critical argument or conversation? And secondly, how does Gates define his own *perspective* in a piece of writing which invites so many others to speak? We'd like you to respond to several questions in your journal. Re-read Gates' article and mark those places in the text where others are being quoted. Can you see a pattern or some logic that might explain why Gates has chosen these particular personalities to speak in his article? Do these personalities fit certain categories of expertise? Historians? Political commentators, etc? Does Gates omit someone that you might have expected to hear from in an article of this kind? How does Gates' choice of personalities reveal his sense of the important issues surfaced in the O.J. trial? Which quotations does Gates comment on? Does this commentary help you to get a sense of Gates' position? Or what passages in the text would you point to as giving a sense of Gates' own critical position among the many competing voices in this article?

WRITING CRITICISM

"Who would have imagined that the Simpson trial would be like the Kennedy assassination—that you'd remember where you were when the verdict was announced?" Gates' subject here is the impact of two events—the O.J. Simpson trial and the Million Man March—that together helped to shape public and media discourses of the mid-1990s. We'd like you to write an essay in the spirit of Gates, one that also studies the cultural meanings of a particular event as they are presented and filtered by the media. The event you select to write about does not need to have had a political importance (though it might have), but it should have had some sort of cultural impact. It should be the kind of event that was broadly commented on in the media, that one might perhaps imagine relative strangers talking about, that in some way touched on popular concerns about sex, race, fame, politics, work, family, money. . . . Your task will be to define what those concerns were and how this event you are writing about helped to crystallize and define them.

We'd like you to draw upon two moves Gates makes as a writer in your project. The first is to research the ways in which your event has been represented in the media, to catalogue and interpret the images, characters, and phrases that it made part of our cultural vocabulary. (As, for instance, "the race card," the bloody glove, the white Bronco, "the rush to judgment," side bar, Mark Fuhrman, and DNA testing all became common knowledge and contested symbols in the coverage of the Simpson trial.) You'll need to draw upon some print media texts as a source here and not rely upon your own memory of the event. You might access some online sites for magazines or newspapers or you might use *The Readers' Guide to Periodical Literature* in your campus library. In any case, you'll want to get the details of this event, examine what images, phrases or personalities emerged from this kind of historical record.

The second move here is to speak with other people about how the event entered into their lives and the meanings they drew from it. Ask several people—maybe a cross-section of people, not just your friends—what details they remember about the event and why the event holds some kind of significance for them.

As you think about putting this project together you need to consider how both the media record of the event and your interviews with others contribute to your understanding of the larger cultural impact of this event. Based upon a comparison of the media representation of your event and what others remember about the event you may find discrepancies between what was reported and what now is remembered. What is the cultural significance of what was remembered and what was forgotten here? What has passed into or become part of our cultural memory of this event?

Pop Goes the Culture

William Grimes

Those who forget the past are condemned to repeat it. So are those who remember it all too well.

Take a quick look at the cultural landscape, preferably in your rear-view mirror. It's the '60s. "Tommy" is on Broadway. "Hair" is scheduled for a revival in London. The Velvet Underground is getting together again. Jimi Hendrix and the Doors rule. So do bell-bottoms and "Right on!" politics.

It's the '70s, too. The grand amusement park of American popular culture proudly announces a new attraction, 1974 Land, a bright and shiny place where crocheted vests, platform shoes, shag haircuts and "The Brady Bunch" live on forever.

If you like, it's the '30s, '40s and '50s too.

Nearly everything that human reason says could never, ever—nohow, no way—stage a comeback has come back. An entire universe has been retrofitted and given a new lease on life.

In just a few years the '80s will return, artfully repackaged, retooled, deconstructed, gleaming with a semiotic wax job and somehow more real than they were 10 years ago. The second time around, you get a chance to think about things, to ponder some basic questions. Like: What is going on here?

•

Two things are going on. The first is easy to understand. Time passes and a new generation discovers new and exciting things in the old stuff. Frantic picking and gleaning goes on, and creative editing. This is one way tradition moves forward, by moving an old idea into a new context and discovering that it can do the work of a new idea.

Old styles can supply vital nutrients. The '60s are back in part because Americans in their late teens and early '20s like the idea of a dominant youth culture. The decade looked pretty silly from the vantage point of 1974, but now some of the good stuff is coming back.

But what about all the bad stuff? That's the second, and much more significant part of the equation. The culture is delivering unsolicited warmed-over pizzas to America's doorstep. Fleetwood Mac. "Bewitched." LSD. "Ironside." Polyester. Who ordered this?

What exactly are the rules here? And if Birkenstocks can run around the track one more time, why not Earth shoes? If bell-bottoms, why not leisure suits? If "Laugh In," why not "Sing Along With Mitch?" If Lulu, why not Tony Orlando?

"With this question," said Robert J. Thompson, an associate professor of television at Syracuse University, "you're really beginning to flirt with the Rosetta Stone for understanding American pop culture in the waning days of the 20th century."

As a case study, consider the phenomenal second life of "The Brady Bunch," a show that went off the air in 1974 after five seasons and then lay dormant, a ticking cultural time bomb. In the meantime, the original audience for the show grew up, went to college, married and absorbed the reflexive sense of irony that is widespread in American culture today.

In 1991, "The Brady Bunch" exploded. A theater company in Chicago began performing episodes of the program, following the original scripts, to standing-room-only audiences. The show, "The Real Live Brady Bunch," is now running in four cities. There are three "Brady Bunch" books. A film version is in the works.

"We used to try to identify ourselves by trying to display what we like," Thompson said. "Now there's more of a tendency to display what we make fun of. That's a major cultural shift."

Thompson proposed four rules governing a successful revival:

- The product must be must be low- to middlebrow, and must have been consumed the first time around with no sense of irony, often by people of a lower economic class than the people consuming it the second time around. (This is why "Masterpiece Theater" does not generate many revivals.)
- The product must carry as many markers as possible that link it to a particular period. ("The Brady Bunch" was a nonstop festival of early '70s fashion, speech mannerisms and hair styles.)
- The revival must be spontaneous. (When Sherwood Schwartz, who created "The Brady Bunch," produced a television revival of the show, it flopped.)
- There must be a minimum level of technical competence.

The Thompson taxonomy is, of course, debatable. (Point 4, Thompson admits, may be a little shaky). But it covers a lot of ground. And there's no reason why extra points cannot be tacked on in the search for a comprehensive theory to explain why some things happen twice and some don't.

The power of mere exposure needs to be considered, for example. In the same way that a baby duck can form a child-mother bond with a vacuum cleaner during the critical imprinting stage, children will develop an emotional attachment to anything that flickers on the tube often enough.

This applies even to fare like "Gilligan's Island," which survived a total of three seasons (1964–67) against competition like "Lawrence Welk," "Donna Reed" and "Kentucky Jones" but went on to a thriving afterlife in syndication and now threatens to become an off-Broadway musical.

But mere exposure is not enough to explain why "Gilligan" thrives and "My Favorite Martian" doesn't. The reason is that "Gilligan" although little more than a situation and a handful of little-known actors with a laugh track attached, exhibits a complex texture. "My Favorite Martian" is unidimensional, the television version of a pet rock. You see it; you get it; you forget it. It's gone. There's no point in kicking it around.

But "Dragnet" survives because it was taken seriously throughout its lifetime on television. Only later did audiences learn to hoot at Jack Webb's wooden acting style, the corny voice-overs, the portentous theme music, the awesome sincerity of the whole project.

Too much popularity, in fact, can permanently taint any second-level pop-culture phenomenon. Peter Max was so overexposed during the '60s that his revivability potential is well below zero. Trudgers who manage, by persistence and guile, to string out a career long beyond any reasonable expectation do not make good revival candidates, either. Hence, no afterlife for Tony Orlando, and it's a pretty sure bet that Kenny Rogers will not be packaged in an irony-laden five-CD set by Rhino Records.

Earnest mid-level competence is nonrevivable. A musician like Peter Frampton offers no ironic openings. You have to dive a little lower in the talent pool.

The same goes for Top 40 hits. "What makes them the most dated makes them the most durable," said Gary Stewart, a vice president of Rhino, which is now in its 22nd volume of repackaged '70s hits, "Have a Nice Day." He carefully distinguished between efforts like David Soul's "Don't Give Up On Us," or "Love Grows Where My Rosemary Goes" and the more revival-worthy "The Night Chicago Died."

Fashion is a little more complicated. It demands more commitment than listening to a three-minute song or sitting through a failed sitcom. It costs money and it actually touches your body. No one gets ironic except with accessories. A betting man might jump for a leisure suit right about now, but who wants to be first? Hair is even more difficult. Make a mistake and you're stuck with it for a long time.

One-joke fads don't travel well through time. Not even John Waters is going to bother with mood rings, pet rocks, Hula Hoops or Rubik's Cubes. The public's love affair with these choice items was a one-night stand, a mad night of passion that can never be repeated, even in an ironic mode.

The same dynamic insures that no best seller can ever return to popularity. "Peyton Place," which was to books what "Star Wars" was to films, no longer exists as a cultural artifact. Even deader are books with philosophical overtones, the ones that define, in a particularly embarrassing way, an entire generation of college freshman. Books like "Love Story" or "Jonathan Livingston Seagull" or "Siddhartha." Do not wait for a second coming of "Your Erroneous Zones."

●

The end of an age may be approaching, however. The pleasures of irony are not inexhaustible. As a strategy it has proven effective in allowing the educated to enjoy mass culture while sending out signals that they know better—and *are* better—than those who live in tract houses and see nothing funny about lawn sculpture or avocado-colored appliances.

But the strategy, which is an elaborate defense mechanism and more than a little cowardly, works only so long as the people in the tract houses don't get the joke. And that is changing. They can watch Letterman, too.

"The last decade has been the golden age of the revival of kitsch," Thompson said. "But once the regular people begin to get wind of this, and start making fun of Big Boys, then the whole purpose of revivalism is defeated. The game is up."

COMING TO TERMS

At one point in "Pop Goes the Culture," William Grimes observes, "The second time around, you get a chance to think about things, to ponder some basic questions like: What is going on here?" Grimes' comment and question are referring to the often perplexing (and, at times, apparently inexplicable) revived interest in fashion trends, television series, toys and novelties, musical groups, and other cultural texts from past decades. As you come to terms with Grimes' article, we'd first like you to restate in your journal Grimes' argument or ways of explaining why some cultural texts do indeed have a "second time around" (while others do not). Next, see if you can select one or two cultural texts from a past decade which have undergone a revival in the 1990s and which confirm Grimes' explanation for a revival. Explain your choices in your journal. Then choose one or two texts which have been revived but appear to run contrary to Grimes' explanation. (In this latter category, we think that the revival of yo-yos, popular in the 1970s, and the revived interest in the *Star Wars* trilogy would seem not to fit Grimes' rationale for a revival.) Finally, select several cultural texts—fashion trends, style of music, toys, television series, etc.—which you can argue are representative of the 1990s. Using Grimes' rationale for predicting the revival of a cultural text, explain why some of these texts might likely have a "second time around" and why others probably will not.

READING AS A WRITER

Grimes documents a "reflective sense of irony" that he claims characterizes much of current popular culture. There are many kinds of *irony* that writers use and readers need to be aware of. On one level, *irony* occurs when an author uses language to mean something other than the literal, expected meaning of that language. For example, we might comment on the "beautiful weather today" when, in fact, the weather is quite depressing. To appreciate an author's use of irony, it's important to understand the context or meaning an author might be trying to achieve. Grimes is discussing irony on a different level here. A place to begin is to think about why a cultural text might logically undergo a revival, but then consider why Grimes is arguing that the current revival of certain cultural texts from past decades is ironic.

In your journal try to define what Grimes means by a "reflexive sense of irony" and then respond to the following questions. What does he suggest are the allures or satisfactions of irony? Who does irony seem to most appeal to? What are its limits or drawbacks? What other terms, tastes, and concepts does Grimes associate irony with? Are all revivals of old texts or old ideas ironic? What is Grimes' own stance or attitude towards pop irony? How does he make this stance clear?

WRITING CRITICISM

One way of extending the project of another writer involves taking certain ideas, phrases, or examples and *elaborating* on them, thinking through their details and possibilities. In "Pop Goes the Culture," William Grimes tries to describe a general trend in 90's pop culture, and in doing so offers several quick examples of ironic revivals of songs, fashions, and TV programs—without really pausing to analyze any of these examples in much detail or complexity. That's what we'd like you to do here. Begin by locating a popular text from an earlier decade that has recently enjoyed some sort of renewed life or attention. (This may involve an actual remaking of the original text—as is the case with "covers" of old songs, or when TV shows are sometimes made into feature films—but it may also simply be a revival or recycling of an older text.) Your task then is to write an essay in which you account for the *current* meaning and appeal of that text. (In doing so, you may want to make use of the four "rules"—the Thompson taxonomy—for a successful revival that Grimes refers to, though you shouldn't hesitate to add to or argue against these rules as needed.)

The challenge will be to define the *specific* appeal of the text you are looking at. To say simply that the 70s are back in style (if they happen to be, when you are reading this) is probably not to say very much, for instance, but to be able to suggest how and why a *particular* version of the 70s—that embodied, say, by the Partridge Family rather than by The Cowsills or by Donnie and Marie—

might be to say a good deal. You may want to ask yourself: If this text "revives" something from an earlier decade, then what exactly is that something? And why is that something needed, pleasurable, and interesting now?

The Conformity
of Office Zaniness

Daniel Harris

In the modern office, the desk functions as a transitional space between the public and private lives of corporate employees, who use it to assert their personalities and to counteract the effects of institutional decor designed precisely to eliminate individual differences. In a world whose appearance is dictated by the needs and imperatives of business transactions, the desk is a small, tidy museum of selfhood, the symbolic locus of a contained insurrection of the specific over the general. Our efforts to infuse our "work stations," as they are now called, with individuality can be as radical and unsparingly expensive as those of executives who strip off the firm's uniform carpeting and inlay the floor with marble or parquet, or as nominal and unobtrusive as the display of a polished rock, a pink ribbon, a row of plastic ants meandering single file over a keyboard, or a blown-up fortune from a fortune cookie.

The office desk also reveals workers' disheartening attitudes toward labor and its relation to what we are taught to believe is its opposite: leisure, privacy, and family life. Virtually every personal item we exhibit functions to transport us—mentally or emotionally, if not physically—out of this staid professional environment. In its place, a pastoral realm is invoked: games, sports, sensuality, sex, humor, exoticism, and even extremes of meteorology (snapshots of snowbound cars, deserts, bayous, mud flats, which serve as an antidote to the office's artificial mindlessness and its deadening lack of variation in light, humidity, and temperature). Everything in, on, and around the desk works to deny the reality of the business world: the indispensable containers of hand lotion, the Curél and Revlon Aquamarine, which women furtively massage into their hands to suggest sensual acuity in an environment in which the senses are kept in a constant state of deprivation; the souvenirs of vacation spots, which encourage fantasies about places other than the one in which the desk is situated; and even the invariable cluster of family photos, that secular reliquary which, in the context of these generic and collective spaces, seems to call out plaintively to another world.

Even workers who leave little or nothing on their desks (a chocolate-chip cookie on a napkin; an empty Planter's Peanuts jar) express their individuality in their freedom *not* to decorate—a quiet act of civil disobedience that almost sullenly attests to the occupants' contempt for this world, as if they were only bivouacking here and had no intention of remaining longer than necessary.

In the absence of real freedom of behavior, we use our desks to create an illusion of liberty, even license, of a seductive but essentially false form of unruliness and autonomy, as if we were contemptuous of (rather than servile to) the annihilating strictures of professionalism imposed on us. The result is the innocuous institutionalization in conveniently inanimate office paraphernalia of something best described as controlled nonconformance. Zaniness, loopiness, weirdness, wackiness (rather than real eccentricity, disturbingly unsocialized differences of behavior) are displayed with a sense of pride, of unrestrained self-congratulation in honor of the symbolic victory the office worker often flatters himself that he has achieved over the sobriety and lifelessness of his colleagues. Salacious calendars are one of the most widely used weapons available in the worker's arsenal in this ongoing campaign to express originality and freedom. (On one, the Lone Ranger, now retired, sits in a rocking chair where he "makes an unpleasant discovery" as he thumbs through a dictionary: "Oh, here it is... 'Kemosabe: Apache expression for a horse's rear end.'") Cartoons also constitute one of the major forms of office humor, popular because they function as a convenient generalizer, a way of indicating a universal grievance common to all offices (as if the person posting the cartoon had intended to say, "Look, this problem occurs everywhere, to such an extent that there is even a cartoon about it"). While the display of the cartoon is meant to reassure the office comedian that he or she is a madcap nonconformist, its generic "zaniness" often acts to shore up conformity, neutralize dissidence, and smooth over conflict by suggesting that overwork, frustration, and unhappiness are endemic to this world, and that therefore it is pointless to resist them.

Self-infantilizing is one of the most important ways workers convince themselves that they are idiosyncratic interlopers in the business world rather than faceless bureaucrats who blend effortlessly into their drab surroundings. Both secretaries and executives take particular delight in rebelling against professionalism by projecting an image of themselves as childlike, immature, and innocent—the ultimate expression of controlled nonconformance, of their abortive efforts to suggest that they are the proverbial square peg in the round hole, the kooky, pleasantly maladjusted oddball. Indeed, taken as a whole, the office desk is a celebration of infantilism—teddy bears, stuffed animals, coloring books, and pictures of kittens for the secretaries; model cars, windup toys, gross plastic jokes, and Groucho Marx noses for the executives. Humor in the office— at least humor in the inoffensive form of lunacy and childishness—becomes a vehicle of self-assertion against an environment that discourages individuality, a way of suggesting that one hasn't sold out entirely, that one maintains a subtly antagonistic relationship with the business world, an enmity sublimated, devitalized, and ultimately stifled in calculated displays of the goofy and the puerile. Incarnated so passionately in the personal artifacts of office decor, illusions of unsuitability, of being mismatched with one's profession, dominate the inner lives of the workers of corporate America.

COMING TO TERMS

In "The Conformity of Office Zaniness," Daniel Harris writes about the ways in which workers try to safeguard their sense of individuality and freedom in spite of the "reality of the business world." The clash between the individual and the corporate world (or any other large institution or system) has been a pervasive theme in U.S. culture for many decades. In fact, the struggle for a perceived personal freedom and individuality moves far beyond being simply a popular theme in the American literary canon. It is a deeply ingrained cultural value or ideal and is celebrated as such in, among other things, film, television, advertising, and in the clothes we choose to wear. As a way of coming to terms with Harris' article, we'd like you to think about the ways in which business, industry or corporate America is represented in popular or media culture. We'd like you to choose two texts that depict the business or corporate world. These texts might include films, a television series, a print advertisement, a television or radio commercial, etc. Or you may have recently read a biography or an autobiography of a well-known business personality. Or you might even search for corporate sites on the Internet. In any case, one of these texts needs to be an example of a company, corporation, or business representing or advertising itself (you might call this one the *company* perspective). The other text you choose needs to show an *outsider's* or even a *consumer's* perspective on the company or corporation as seen through the eyes of characters in a film or television program. Then in your journal respond to the following questions. How is corporate America represented from the company's viewpoint? And how is it represented from the outsider's point of view? What kinds of images are emphasized in both of these representations? How are they different, similar? How would you characterize the relationship between worker and management in each? Do you recognize stereotypes in either one of these representations? Do you see large companies or corporations as trying to put on a human face? How is this accomplished? What does the outside representation choose to criticize about corporate America? How is the individual worker or consumer represented?

READING AS A WRITER

As a response to the "reality of the business world," the desk, according to Harris, becomes "a small, tidy museum of selfhood." The desk is seen as an expressive medium, which can be shaped and transformed to show a worker's individuality and personal sense of identity. This need to personalize office space is, for Harris, a thoroughly human response to an anonymous corporate culture that divests workers of any sense of individuality. Maybe it is that Harris assumes his readers have experienced or are well aware of cultural representations of the business world, for Harris does not invest much time in specifically detailing this world. It is also possible that Harris is drawing upon another rhetorical technique to further his critical argument, namely *defining* by *negation*. Throughout his article, Harris characterizes the business or corporate world by showing us its opposite qualities, what it is not, how it is different from "real" living. From this stance or position of negation, a reader might then *infer* (draw some conclusions not explicitly stated by an author) how Harris is reading corporate culture. We'd like you to reread "The Conformity of Office Zaniness" and find places in the article where Harris is "defining" the business world through negation. (Harris does occasionally use the textual markers of "business" and "corporate," so you might also look for them.) Then answer the following questions in your journal. What can you infer from these passages as to the ways Harris imagines the daily "reality" of business or corporate cultures? What are the specific qualities of these cultures, in a sense, conspicuous by their absence? What do you make of Harris' possible assumption that we might already know (and apparently agree with) his particular reading of the business world? How might this assumption affect your reading of his critical argument?

WRITING CRITICISM

Whatever we choose to call it—personal freedom, autonomy, individuality, personal style, self-identity, originality—Daniel Harris argues that this defining quality of our lives is jeopardized or even sacrificed when we encounter the constraints of the business or corporate world. And although workers might transform their desks into "a small, tidy museum of self-hood," Harris sees these efforts as rather marginal: "In the absence of real freedom of behavior, we use our desks to create an illusion of liberty, even license, of a seductive but essentially false form of unruliness and autonomy. . . ." (In fact, why do you think Harris uses the word museum? What does a museum suggest to you?) Even those workers who "express their individuality in their freedom not to decorate," seem to exhibit a predictable, controlled nonconformity. Harris focuses his critical attention on corporate America; but we think other important institutions in U.S. society, such as government, higher education, health care, to name a few others that might threaten individuality or self-identity, might also undergo a similar kind of scrutiny.

We'd like you to write an essay in which you examine, through a media or popular culture text, the cultural paradox of trying to maintain individuality or personal freedom within a culture that also places many kinds of restrictions on expressions of freedom and that also values conformity as well as nonconformity. We have called this dynamic a cultural paradox because it's a good example of the contradictory messages or signals a culture might send: be an individual, but join a group.

For this writing assignment you'll need to select media or popular culture texts which allow you to see this paradox at play. In other words, how is individuality being celebrated or promoted in this cultural text? And how might this individuality be circumscribed or controlled by other cultural forces or institutions, such as being accepted by a particular group as a way of achieving success? (You might call upon other writers in *Media Journal* as intellectual allies for this project. Jay Weiser, Umberto Eco, and Rosalind Coward, for instance, strike us as also confronting versions of this cultural paradox.)

Regardless of the media or popular culture text you choose, you will need to find a text in which this cultural paradox is at play. You might, for example, select a television series or program in which a character(s) seem to be forging or protecting a sense of individuality or personal freedom or in which the story involves some kind of conflict with an institution or resists some kind of dominant cultural belief. Or you might select a popular song whose lyrics make similar statements about individuality or freedom. You might also show how a current fashion is seen as an act of resistance. Or finally you might select several print ads whose products advertise or promote individuality or nonconformity as well as a particular product. We'd like you to make, as part of your project here, some observations on how the ideal of freedom or individuality might be presented differently (there may be similarities as well) from one decade to another. The products here need not be the same; instead what you're looking for are multiple ads which promise individuality by purchasing a product. *Remember:* for any of these choices, you're attempting to show a cultural paradox at play: a call for individuality and how that individuality might be controlled or modified by larger cultural values and beliefs.

The Importance of Being Oprah

Barbara Grizzuti Harrison

Looking at pictures of herself as a young girl—"A *nappy*-haired little colored girl"—Oprah Winfrey sees herself on a porch swing, "scared to death of my grandfather. I feared him. Always a dark presence. I remember him always throwing things at me or trying to shoo me away with his cane. I lived in absolute terror.

"I slept with my grandmother, and my job was to empty the slop jar every morning. And one night my grandfather came into our room, and he was looming over the bed and my grandmother was saying to him, 'You got to get back into bed now, come on, get back in the bed.' I thought maybe he was going to kill both of us. I was 4. Scared. And she couldn't get him to get back in his room. And there was an old blind man who lived down the road, and I remember my grandmother going out on the porch screaming, 'Henry! Henry!' And when I saw his light going on I knew that we were going to be saved.

"But for years I had nightmares that he would come in the dark and strangle me."

The girl who emptied out the slop jar is the woman who now wears Valentino, Ungaro, Krizia clothes; powerful, glamorous, rich. She doesn't know what got her from there to here. How does one make a self? Why? Oprah Winfrey's rushed headlong to get the answers, sometimes in advance of framing the questions.

Her audiences are co-creators of the self and the persona she crafts. Her studio is a laboratory. She says hosting a talk show is as easy as breathing.

Here she is, an icon, speaking:

"I just do what I do—it's amazing.... But so does Madonna.... Everybody's greatness is relative to what the Universe put them here to do. I always knew that I was born for greatness....

"If it's not possible for everybody to be the best that they can be, then it has to mean that I'm special, and if I'm special then it means the Universe just goes and picks people, which you know it doesn't do.... I've been blessed—but I create the blessings.... Most people don't seek discernment; it doesn't matter to them what the Universe intended for them to do. I hear the voice, I get the feeling. If someone without discernment thinks she hears a voice and winds up being a hooker on Hollywood and Vine, it is meaningful for the person doing it, right *now*. She is where the Universe wants her to be....

"According to the laws of the Universe, I am not likely to get mugged, because I am helping people be all that they can be. I am all that I can be.... I am not God—I hope I don't give that impression—I'm not God. I keep telling Shirley MacLaine, 'You can't go around telling people you are God.' It's a very difficult concept to accept."

Her amber eyes fill with tears. We are talking about the fact of human misery, and about her phenomenal success; and she has adopted a metaphysical theory that encompasses both—several metaphysical theories, actually, partaking of Eastern religion and Western religion and of what is called New Age. She says she has achieved peace and the serenity of total understanding. (She is 35.) Knotty contradictions in the fabric of her belief have not, up to now, impeded the progress of what she calls a "triumphal" life. If her comfortable truths do not entirely cohere—if on occasion they collide—they are, nevertheless, perfect for the age of the soundbite. They make up in pith for what they lack in profundity.

She is as likely to rest her beliefs on Ayn Rand as on Baba Ram Dass. She brings the baggage of her contradictions to her television talk show, which she calls her "ministry." An avowed feminist, she does not challenge or contradict when a guest psychologist repeatedly attributes lack of self-esteem (a favorite daytime talk-show subject) to a negative "tape"..."the internalized mother" (talk shows are full of language like this). Her contradictions work for her: they act to establish kinship with an avid audience, whose perplexities they reflect; they insure that she will be regarded as spontaneous, undogmatic.

Her great gift for making herself likable is married to a message smooth as silk: *Nothing is random.* Whether Oprah Winfrey is in her fatalistic mode ("if you were abused as a child, you will abuse someone else as an adult") or espousing free will ("I was a welfare daughter, just like you...how did you let yourselves become welfare mothers? Why did you choose this? I didn't"), she chases away the fear that things may sometimes happen by malicious accident, or by the evil forces of others. In Winfrey's scheme of things, the mugger and the mugged were fated to meet—*and* they chose this fate...the starving man from India chose the path that led to his death. The fact of human suffering she manages to erase by divesting it of its apparent aimlessness. This is what commercially successful television programs do, no matter what the format: in an hour or less, they *resolve.*

"I want it to be for a reason, Oprah...." a recovering alcoholic (with four alcoholic children) says, deploring the anarchy of fate. By the time the hour is over, the audience, if not the woman, will be convinced both that it had to be and that it *has* been for a reason. Oprah Winfrey—sassy, sisterly, confiding—has said so.

She brings her audiences into her life. No one who watches Oprah Winfrey—and an estimated 16 million people do—does not know about her weight loss—by her reckoning, 67 pounds on a 400-calories-a-day liquid diet (Optifast), not her meanest achievement—or about her hairdresser, her seamstress, her history of childhood abuse, her golden retriever, her boyfriend, Stedman Graham. There appears to be no membrane between the private person and the public persona.

The woman herself embodies a message, and a sanguine one. It says: You can be born poor and black and female and make it to the top.

In a racist society, the majority needs, and seeks, from time to time, proof that they are loved by the minority whom they have so long been accustomed to oppress, to fear exaggeratedly, or to treat with real or assumed disdain. They need that love, and they need to love in return, in order to believe that they are good. Oprah Winfrey—a one-person demilitarized zone—has served that purpose.

Last year, Oprah Winfrey made $25 million. Her production company, Harpo Productions Inc., gained ownership and control of her top-rated television talk show; she secured the unprecedented guarantee that ABC would carry "The Oprah Winfrey Show" on its owned and operated stations for five additional years. She also bought, reputedly for $10 million (including partial renovation costs), an 88,000-square-foot studio in Chicago, which, when renovated, will provide facilities for producing motion pictures and television movies, as well as her talk show. It will, says Harpo's chief operating officer, Jeffrey Jacobs, be "*the* studio between the coasts, the final piece in the puzzle" that will enable Winfrey to do "whatever it is she wants to do, economically, and under her own control." It is also in the path of Chicago real-estate development.

Winfrey, whose television production of Gloria Naylor's "The Women of Brewster Place" won its time-period ratings on two successive nights, owns the screen rights to Toni Morrison's "Beloved" and to "Kaffir Boy," the autobiography of the South African writer Mark Mathabane. She has bought a 162-acre farm in Indiana. She is part-owner of three network-affiliated stations. She has an interest in The Eccentric, a Chicago restaurant which she often works three nights a week, going from table to table, shaking every hand. She says she entered into partnership because she "wanted a place to dance."

...When she walks from room to room in her large apartment, which receives cool light from Lake Michigan, she talks to herself, little purring, self-approving speeches—as if to reproach the silence, as if silence were a form of inactivity, a vacuum that called out to be filled....

She brings a practiced wit, an evangelist's anecdotal flair and a revivalist fervor to the dozens of speeches she gives every year. When she spoke to 6,000 women who attended an AWED (American Woman's Economic Development Corporation) conference in New York in February, the audience—nearly half of whom were black, atypical for an AWED gathering—loved her from the moment she bounded onto the stage in form-hugging black and peach. Her speech didn't bear close scrutiny, as it advanced at least two opposing ideas, which may, in fact, be part of her appeal—her audiences are seldom called upon to follow a difficult path either of reasoning or of action; they seem to be able to draw from her words an affirmation of that which they already believe to be true.

"There is," she said, "a false notion that you can do and be anything you want to be...a very false notion we are fed in this country...there's a condition that comes with being and doing all you can; you first have to know who you are before you can do that.... The life I lead is good, it is good...people ask me what temperature I would like to have my tea.... What the Universe is trying to get you to do is...to look inside and see what you feel...all things are possible."

They loved her. Perhaps they saw that in spite of perceived anomalies—(once) fat in a world of thin, female in an industry dominated by males, black in a racist society—she is, her flamboyance notwithstanding, deeply conventional in her thinking: a born-again capitalist, she believes that goodness is always rewarded, and that the reward takes the form of money, "if you expect it to take the form of money."

When Winfrey found demands on her time and on her new money overwhelming, the author Maya Angelou told her: "Baby, all you *have* to do is stay black and die.... The work is the thing, and what matters at the end of the day is, were you sweet, were you kind, did you get the work done."

Winfrey calls upon Angelou, Bill Cosby, Quincy Jones and Sidney Poitier for counsel; they are available to discuss anything from investments to the criticism she has received from time to time from the black community. "You are their hope," Poitier says.

She calls Quincy Jones "the first person I have unconditionally loved in my whole life. He walks in the light. He is the light. If something were to happen to Quincy Jones, I would weep for the rest of my life."

Celebrities need one another—they ratify one another's myths. They are one another's truest fans.

Having refused to hire an agent, a manager and an attorney—"I don't get why anyone would want to pay 40 percent of their earnings in commissions or retainers"—she lavishly praises Harpo's Jeffrey Jacobs, who fills all those roles: "If something were to happen to him, I don't know what I would do, I don't know."

She has "weeded out" the people who "get on my nerves and just take up my time."

...Her living room is white on white—white pickled oak walls and floors and white rugs with silver threads and fat white sofas and skinny white chairs; crystal obelisks and vases, brushed chrome, white marble and white onyx and vaguely deco wall sconces, steel and Lalique; a heroic Elizabeth Catlett sculpture (a gift from Cosby) and two architectural plants, unwatered, droopy. The light that enters the room is silvery...

Winfrey embodies the entrepreneurial spirit; she is an Horatio Alger for our times. Salve to whites' burdened consciences? She shoves the idea aside. Role model for black women? For all women, she says, her aim being to "empower women."

An active fund-raiser, not necessarily for glamorous causes, she is philanthropic and fabulously generous.

She set up a "Little Sisters" program in Chicago's Cabrini-Green housing projects, to which she devotes time. She annually endows 10 scholarships to Tennessee State University, her alma mater. The story that she made her longtime and best friend Gayle King, a newscaster in Hartford, a millionaire with a Christmas check of $1,250,000 is part of the Oprah legend.

Two Christmases ago, she took three producers and her publicist to New York. At the hotel, she sent them a note saying: "Bergdorf's or Bloomingdale's—you have 10 minutes to decide."

Reliving the enacted fantasy with a breathy mix of chumminess and adulation that is characteristic of the staff of Harpo Inc. and peculiar to devotees of a guru, Christine Tardio, then one of her producers, says: "We go to Bergdorf's and she hands us each an envelope and in the envelope is a slip of paper and it says you have one hour to spend X amount of money. So we are frantically running around shopping and then she walks around and pays for everything.

"The next morning a little piece of paper is slipped under the door and it says, you have five minutes to decide where you want to buy boots and shoes—Walter Steiger or Maud Frizon. And then while she is paying for the boots at Frizon, she hands us another slip of paper and it says you have like two minutes to decide, leather or lingerie. So we all pick leather; so again we're like throwing clothes around the store, and half an hour later we get into the limo and she blindfolds us with terry-cloth sweat bands.

"We're drinking champagne, driving around New York. We get to our destination and we're helped out of the car, through a revolving door, into an elevator and through another door and she says, 'O.K., on the count of three take off your blindfolds,' and we're in the middle of a furrier's. And she hands us another note that says, 'You can have anything in the store you want, except sable.' So three of us got minks and one of us got fox. And she was a little kid through the whole thing. Oh! She's wonderful!"

The producer Mary Kay Clinton says: "I would take a bullet for her."

The woman who elicits this rapturous loyalty was born on a farm in Kosciusko, Miss. She lived there with her grandmother until she was 6; then she moved to Milwaukee to be with her mother, Vernita Lee. When she was 9, and for years thereafter, she was sexually abused by a teenage cousin, and then by other male relatives and friends.

She has talked about this on her show: "It was not a horrible thing in my life. There was a lesson in it. It teaches you not to let people abuse you." She told *People* magazine in 1987 that she found the attention pleasurable.

"When that article came out the response I got was, how dare you say that you liked it? Which wasn't what I was saying at all. If someone is stroking your little breasts, you get a sexy, physical feeling. It can be a good feeling, and so it's confusing, because you then blame yourself for feeling good, not knowing that you had nothing to do with that kind of arousal. A child is never to blame."

How did she ever learn to trust a man?

"Really," she says, "the question is, how did I ever learn to trust anybody?"

A rebellious teen-ager—she was sent to a juvenile detention home when she was 13 and turned away only because there were no available beds—she went to live with her father, Vernon Winfrey, a barber and city councilman in Nashville, and a strict disciplinarian who strongly encouraged her to read a book each week and to write a book report and to learn a word a day.

This is the story as she has crafted it and as she tells it. Her half-sister and her half-brother have no place in this story. Her candor is more apparent than real.

Miss Black Tennessee of 1971, she was hired by WTVF-TV, the CBS affiliate in Nashville, in 1973 when she was 19, as a reporter-anchor. In 1976, she joined the ABC affiliate WJZ-TV in Baltimore as co-anchor; in 1978, she became co-host of the talk show "People Are Talking." The ratings soared. In 1984, she became host of the floundering show "AM Chicago" on WLS-TV.

Quincy Jones, in Chicago on business in 1985, flipped the television dial and saw Winfrey. He arranged an audition for the role of Sofia in the screen adaptation of Alice Walker's "The Color Purple," a role Winfrey says she coveted from the moment she read the book. Her acting in the film earned her an Academy Award nomination. "Luck," she says, "is a matter of preparation. I am highly attuned to my divine self."

In September 1986, "The Oprah Winfrey Show" went national. The show has received three Daytime Emmy Awards; its star—and its real if not its touted subject—has received one.

Members of her staff—who speak in the corporate, adoring "we"—see these things as a matter of "divine convergence."

Winfrey's interest in herself is vivid. She is inclined to believe—her friend, Maya Angelou, has recently suggested this theory—that what got her from there to here is "obedience." As a child, she was obedient "to escape a whipping." Now she is "obedient to my calling. One of my favorite Bible verses is: 'I press toward the mark...of the high calling of God....'" (Philippians 3:14).

When she was a child, "church was my life. Baptist Training Union. Every black child in the world who grew up in the church knows about B.T.U. You did Sunday School, you did the morning service, which started at 11 and didn't end until 2:30, you had dinner on the ground in front of the church, and then you'd go back in for the 4 o'clock service.

"It was forever, oh, it was forever. It was how you spent your life.

"At a Sunday-school performance—all weekend—they'd say, 'And little Mistress Winfrey is here to do the recitation.'... I'd have these little, little patent leather shoes. Oh, *very* proper. 'Little Mistress.' And you're in a church with a fan going around the ceiling and there are three wasps' nests under the fan. And they called it 'being on program.'" Being on program.

"It was a very narrow view of God. Well, each person gets God at whatever level they're able to accept. That's why there are all these people—Holy Rollers, Episcopalians, Baptists—who can only accept God as a man with a long white beard and a black book checking off the things you can do. And that's really O.K., that's O.K. if that's as big as God is for you and that keeps you under control. Not everyone can be what we want them to be. You still benefit, but your benefit will be as limited as your vision is."

"I can afford anything."

She is convinced that, had she not attended to the will of the Universe, she'd still be making "a nice little six-figure income."

"Jesus says He knows how much we can bear, and He won't put any more on us than we're able to bear. The slaves used to sing it. And before I had a studio—when I was just talent—I had a slave mentality. He won't *give* you more than you can handle."

The house in which her boyfriend Stedman's aunt lived burned down. Oprah wrote her a check. Oprah chose not to build her a new house.

"The thing is, she'd spent the last 20 years taking care of her mother. And her mother died. And she refused to let go. She sits in the house all day long, holding on, walking through room after room crying, crying at dinner time 'cause there's nobody to feed, nobody to look after. And so what happens? I was going to send her on a cruise, get her out of the house. But she'd have none of that, no, no, no. So what happens? Lightning strikes the house and burns it down.

"So I wrote her a check."

In this story Oprah Winfrey is indistinguishable from the Universe.

"Would *you* like this money?" she asks. "A lot of people wish they had my money. It's jealousy and self-hatred.... You aren't saying I shouldn't have it, are you?"

Criticism leveled at Oprah is often as petty as why she sometimes chose to wear green contact lenses. She has also received substantive criticism: the N.A.A.C.P. expressed concern about Winfrey's production of "The Women of Brewster Place." Winfrey's publicist, Christine Tardio, says Winfrey had already decided to soften the portraits of black men. But the intense femaleness of the four-hour miniseries—not only a hymn, gritty and sentimental, to black women, but an implied indictment of black men—remained undisturbed.

To all professional criticism of that show, Winfrey has one answer: the ratings were good.

If money has, as she says, a "deep spiritual" meaning, perhaps ratings have spiritual import, too; that would explain why Winfrey is so remarkably casual about criticism she does not construe as personal.

She accepts Stedman's observation that black men—"I don't see them"—respond to her "very well." She is hurt by what she calls "negativity from black women"..."talks and rumors," stemming in large part from "the whole overweight thing. I was overweight and Stedman is gorgeous. It would have been easier on me if Stedman wasn't gorgeous."

He is gorgeous, as both the tabloids—which predict the demise of the relationship weekly—and Winfrey's audiences know. She talks about him a lot. She tells them—they beam—that she prayed and prayed for a good, hard-working man, "and Lord, let him be tall!" He is.

A former model, Stedman Graham is president of a North Carolina-based public relations firm called the Graham Williams Group; pressed, he says his corporation "helps people to become all that they can be." His firm "maximizes resources and helps small firms become large corporations and large corporations become multinational, multimillion-dollar corporations." Whatever this, in real terms, means, it is not surprising that he is a Republican.

Producer Mary Kay Clinton was at first wary of Stedman: "She was overweight and he was so attractive, and we wanted to know: What's he after? What's he coming here for? Where did he come from?"

Oprah frequently feels obliged to disabuse audiences of the notion that Stedman is interested in her for her money.

Winfrey's show is sometimes indistinguishable from soap opera—as when Stedman appeared to talk about the perils and pleasures of being partnered with a celebrity. The audience chanted: "When are you going to get married?"

Stedman said: "Oprah and I love each other very much. She's my woman and I'm her man.... We don't want to succumb to the pressures of the public by getting married, so we will announce happily that we are getting married when we are getting married, and not before." And a lot of people thought they saw a lump of disappointment rise in Oprah's throat.

...Oprah flings open a closet door and is for a short time lost in pleasure (perfume rises from furs), and then she tosses her head and says: "I gotta move. There isn't room in this closet for two adults. And we're getting married next year. *Next year.*" Stedman says: "We talk about it. We don't talk about it. We'll do it when we're ready."...

Oprah Winfrey's weight concerns viewers as much as her nuptials. She is now a size 8, down from a high of over 200 pounds—"but I was always shapely," she says with forgivable vanity. When she amiably carted 67 pounds of animal fat on her show in a little red kiddie wagon to illustrate her weight loss, Stedman called in to say, to millions of viewers, that her not being fat made it easier for them to walk into a crowded room. That was a bad moment for Oprah. She says now that he was misunderstood. He now says, "The only problem I had with her weight was that she had problems with it."

When Winfrey was fat, she hugged and touched her studio guests a lot. She practically cuddled. Now that she is slim and awfully glamorous, she maintains a far greater distance. The touch of a woman with perceived sexual allure is scarier, more charged, dangerous. Paradoxically, her body seemed more loose, her movements more flowing, when she was fat. When she is with Stedman, her body regains its comfortable eloquence. She vamps.

...In her hallway, past the aubergine bedroom suite with its enormous television screen, past the tiled kitchen, free of all domestic clutter and appurtenances, there is a framed letter: "Oprah, You must keep alive!" it says, "Your mission is sacramental! A nation loves you!" It is signed "Winnie Mandela."...

Stedman works with Mandela in a way he will not precisely define. He says suffering "has nothing to do with being black or white. It's not a group thing. Each individual is responsible for his own life. All lessons are personal."

When Stedman speaks, it might be Oprah speaking. Then again, it might not.

"There are more white people than black people on welfare, you know," she tells a member of her studio audience, a small flame of anger igniting.... "Why do women continue to have babies they can't afford to take care of?" she demands rhetorically of welfare mothers, her anger taking her in a different direction..."It sticks in my craw, too"—earning applause from white members of her audience.

"A small but vocal group of black people fear me," Winfrey says. "Slavery taught us to hate ourselves. I mean, Jane Pauley doesn't have to deal with this. It all comes out of self-hatred. A black person has to ask herself, 'If Oprah Winfrey can make it, what does it say about me?' They no longer have any excuse. If you don't believe that I should make it, you absolutely don't believe you should, either."

There are set pieces she offers to interviewers. One knows when one is present at the birth of a new set piece—such as The Story of the Keys:

"I lost my keys to the farm. There's a map to my house in circulation, so when I go down the road jogging, one person calls another and the next thing you know, they're out there leaning over the fences—'Hi Oprah!' So I said to Stedman, 'If the keys are found everybody will know they're my keys. I've got to find them.' He said, 'The question you ought to be asking yourself is why did you lose them.' Other people would just get aggravated, but I saw that I wasn't being aware enough. And that the keys are a way of making me see it. I think the Universe is saying something to me. And so the minute I said that—this kind of thing happens to me all the time—I found the keys. I found the lesson and I found the keys."

This story, one feels, will accrete.

She wakes up at 5 or 5:30. She runs six miles along Lake Michigan or works out in a gym daily. She gets to the studio about 8. As she is prepared for the camera by her hairdresser and makeup man, she has a "talk session"—which she terms "redundant"—with her producer. By 9, she is in front of the cameras.

She will have prepared for her show the night before for less than an hour. "She wings it," says Debra DiMaio, her executive producer. "She gets on camera and asks the questions ordinary people would ask."

The ideas for Winfrey's shows have their origins in the producer's weekly "brainstorming" sessions. (The executive producer and three of the show's female producers are white; two are black. A white male producer has recently been hired. A minority training program is in place.) Often the ideas derive from slender premises (the producers insist on the quotidian, the nonrational and the emotive; they eschew that which would require severe logic or analysis; Winfrey rarely invites politicians on because their "feelings" aren't available to her). A producer's husband was "told by a nun when he was a little boy that he was never going to be a singer." From this emerged a show called "People Who Were Humiliated in Childhood."

That many of the shows are as trivial as reruns of "Gilligan's Island" does not seem to matter to the ratings; what matters is Oprah's energetic, autobiographical presence.

On a recent show, Winfrey talked to the mothers of mass- and serial-killers.

"So…this is your son we're speaking of, who boiled people, who tortured them mercilessly…and then dismembered them…. Do you survive by going into some form of denial, even though they found 40 pounds of bones…?"

She pushes her questions through her guests' tears: "Would you want to see the tapes" of people tortured and killed? She says her shows serve a "deep, divine purpose."

Coming on the show, Winfrey has said, is the first step in turning guests' lives around. Her concern is for the hour. The guests' concern is for their whole lives. The audience's concern is to have an experience. People like to have experiences; it saves them from having to think.

Many guests are drawn from a pool of people who respond to promos broadcast locally after the show. Who, one might reasonably ask, would respond to such invitations but exhibitionists? Winfrey's answer is that televisions provides a "catharsis. People will say to 20 million people what they won't say in their dining rooms." (They are more civil in their dining rooms.)

Sometimes one feels as if one were watching theater of the absurd, or "Saturday Night Live"—"My Friend Is a Bigot," for example, during which a white woman was asked whether she would go to dinner at the home of her "best friend" (who was black). "I don't know where she lives," she answered. Thereafter a panel member asked a prune-faced white woman, identified in graphics with the words "ADMITTED BIGOT," how she'd feel if a black person were to visit and drink from one of her cups. God created her an individual with "a built-in system," she replied, and "socialwise...I'm not going to have someone in my home and make me uncomfortable."

To which Oprah later responded: "I could care.... I could bring my own cup, you know what I mean?" Winfrey—who allows herself to refer to her wealth in this manner often—also said: "You are representing a world of people who absolutely feel this, but they don't say it." Given that we know and are grieved by that, it is difficult to know what is gained, how we are enhanced, by an exchange like this. It makes one despair to think that the people who watch "The Oprah Winfrey Show" regard this kind of bizarre talk, which scarcely resembles human conversation, as a lesson in race relations.

Sometimes the consequences of haphazard preparation are embarrassing but innocuous—as when a *savant* was asked by Winfrey to play the piano and then could simply not be made to stop.

But the ante has been upped. Since the proliferation of trash-television programs, there has been an undertone of hysteria, an edge of danger, to daytime talk-TV; the potential for mischief is realized.

Last month, on a show billed as "Mexican Satanic Cult Murders," Winfrey introduced a woman pseudonymously called "Rachel," who claimed to have a multiple-personality disorder (multiple personalities being a talk-show staple). "Rachel," Winfrey said, "participated in human sacrifice rituals and cannibalism" as a child.

Rachel: "My family has an extensive family tree, and they keep track of who's been involved...and it's gone back to like 1700."

Winfrey: ..."Does everyone else think it's a nice Jewish family? From the outside you appear to be a nice Jewish girl.... And you all are worshiping the Devil inside the home?"

Rachel: "...There's other Jewish families across the country" ritualistically abusing and sacrificing children. "It's not just my own family."

Winfrey's responses to Rachel were feeble: "...This is the first time I heard of any Jewish people sacrificing babies, but anyway—so you witnessed the sacrifice?"

Winfrey several times identified Rachel as a Jew, a washing of hands that served only to distance her from the woman's mad claims, not in any real way to challenge them.

Chris Tardio attributed Oprah's flounderings to "innocence."

A spokesman for the Anti-Defamation League of B'nai B'rith called the program "potentially devastating." The director of the Religious Action Center of Reform Judaism in Washington expressed "grave concern about the lack of judgment and the insensitive manipulation" of Rachel. The president of People for the American Way said: "I think what happened here demonstrates how...freedom has to be married to responsibility."

Oprah Winfrey, for whom it is an article of faith, frequently reiterated, that "with freedom comes responsibility," was, when all hell broke loose—hundreds of angry calls came into the station—uncharacteristically silent.

Winfrey met with members of the Jewish community on May 9. Three days later, a joint statement was issued, in which Winfrey said: "We recognize that 'The Oprah Winfrey Show' on May 1 could have contributed to the perpetuation of historical misconceptions and canards about Jews, and we regret that any harm may have been done...." On behalf of the community leaders, Barry Morrison of the Anti-Defamation League said: "We are satisfied that Oprah Winfrey and her staff did not intend to offend anyone and that Oprah was genuinely sorry for any offense or misunderstanding...."

But damage had been done.

"It's not our sensitivities she ought to be concerned about," said Philip Baum of the American Jewish Congress; "it's a question of the integrity of her show."

When Winfrey was hired for "AM Chicago," the station manager was, according to her executive producer, Debra DiMaio, delighted that he had managed to find someone who wasn't an "Angela Davis type who'd picket the station with a gun in her hair." But the real danger lay not in political ideology, but in the superficial quality of Winfrey's curiosity. We are as much beguiled by easy questions as we are by easy answers.

Recently I watched again an early 1987 show of Winfrey's that many consider her finest: "On Remote in Forsyth County, Georgia." Winfrey took her crew to this county of 42,000 people after it was the scene of protests and counterprotests, not all of them peaceful. No black person had lived in Forsyth County since 1912, when a white teen-ager was allegedly raped by three black men, who were subsequently hanged.

She came, Winfrey said, not to challenge or to confront, but "to explore people's feelings...to ask why." (Outside, the Rev. Hosea Williams, a black activist, and a small group of protestors picketed the show for having an all-white Forsyth audience, and Winfrey for having "turned all white.")

What television audiences saw and heard was racial hatred, anguish, confusion and ignorance, and one very brave black woman. Her composure was magnificent—even when a hate-mongerer asked his fellows: "How many of you who welcome blacks or niggers would want your son or daughter to marry one?" Her dignity was unassailable.

But the elation (and the very high ratings) obscured the fact that "exploring people's feelings"—which Oprah Winfrey does best—is not equivalent to changing them. There are no guaranteed and automatic epiphanies.

One doesn't wish to believe, but one is forced to consider, the fact that the final achievement of a show like that in Forsyth County is to allow audiences to feel superior to blatant, uneducated racists, while cherishing their own insidious subtle racism. Simply by watching shows like this we are encouraged to believe we are doing something, embroiled in something; whereas in fact we are coddled in our passivity. This is a problem intrinsic to the medium; that an engaging Oprah Winfrey is part of a process that engenders passivity is our problem, not hers.

Oprah Winfrey has always promised her guests more than she could reasonably deliver—the opportunity for them to "release the pain" so as to feel no guilt. And she promises her audiences the opportunity to "see themselves in others" so as to sidestep guilt, and, as a result of an hour's broadcast, to "turn around." A quick fix.

She has always made things very simple: "What we are trying to tackle in this one hour is what I think is the root of all the problems in the world—lack of self-esteem is what causes war because people who really love themselves don't go out and try to fight other people.... It's the root of all the problems." Everything circles back to the prison of self, that self which Americans so relentlessly and time-consumingly seek to better and to batter into (self) submission. And it *would* be simple—if the world did not act upon us as much as we act upon it.

Winfrey endears herself to her viewers by acting as if she were just like them: "Believe me, I've been down on my knees.... If there's one thread running through this show, it's the message that you are not alone."

She is like us; and she is not. It is within her power to perform acts of genuine philanthrophy that alleviate genuine human misery. Her audiences do not have her power to act. They may, and probably do, nevertheless, model their responses on hers—and what they see is a detached, uninvolved, carefree kind of caring: "You're reeking—I love you," she tells an alcoholic in her studio audience.

How edifying is this kind of caring, the bandying about of the word love? How much, finally, do we learn from the unhappiness of others? This is a question it takes a lifetime, not a television season, to answer.

"What I'm trying to make God know," Winfrey tells me, "is that I got it already. I don't need to learn anything new today.

"I don't foresee unhappiness coming to me. The chances are, if unhappiness befalls me, it's going to be great, severe unhappiness, 'cause the test would be a very strong one, very severe. It would be God saying: "'Let's see how strong you really are.'"

On May 18, "Entertainment Tonight" broadcast a rumor that had been making the rounds for days in Chicago (and that finally surfaced in a blind item in the *Chicago Sun-Times* and in the form of insinuations by disk-jockeys): Oprah Winfrey had found Stedman Graham in bed with her hairdresser, and had shot him.

No hospital records were found to substantiate the rumor; no one came forth to confirm it in any of its particulars. On May 23, the *Chicago Sun-Times* issued a statement that they were "discontinuing the 'On the Town' column written by freelance gossip columnist Ann Gerber"—in which the blind item had appeared—citing basic differences..."regarding the journalistic standards used in producing her column." No mention of Winfrey was made; and Vera Johnson, assistant to vice president and executive editor Kenneth D. Towers, denied that any "one incident prompted the action."

Winfrey came before her studio and television audiences on May 19 to denounce and decry the "vicious and malicious lie"—the substance and nature of which, however, she left unspecified (as a result of which a large part of her audience was thoroughly mystified). "I hear a lot of rumors about myself," she said, "it comes with the territory." But she'd thought this one—"widespread, vulgar...evil..."—was "so ridiculous nobody could possibly believe it...no part of it, absolutely no part of it, is true." And, stylish trouper that she is, she performed her show with dignity and aplomb. As of this writing, Stedman Graham has been silent on the subject of this rumor.

Sidney Poitier once told Oprah Winfrey that she was "carrying the people's dreams"—and that, emulated and idolized, she would also be resented and hated.

"Violated" by the rumor, she tells me she's "stronger and better now; Stedman and I are closer than we have ever been." She is eager to distance herself from the 27-year-old Oprah Winfrey who, depressed and obsessing over a man who battered her emotionally, "sat on the floor crying—a Saturday night, 8:30—but not really serious about dying." She wrote a "suicide note" to her friend Gayle King (she asked her to water the plants). In that note, and in entries in her journal, she wrote, "I'm so depressed, I want to die." She didn't, she insists—and the difference is crucial to her—actually consider ways and means to accomplish her death: "I didn't even have the courage to end the relationship." When she read that old journal recently, she says, "I wept for the woman I used to be.... I will never give my power to another person again."

She has survived a nightmare childhood, things it bruises the mind to think of. She hasn't accomplished her own creation yet. One wishes Oprah Winfrey well at the task—for her sake, and for the sake of the millions of viewers who, lonely and uninstructed, draw sustenance from her, from the flickering presence in their living rooms they call a friend.

COMING TO TERMS

What view of Oprah Winfrey emerges from Barbara Harrison's profile of her? That is, what kind of person does Harrison seem to want us to view Winfrey as? Does Harrison change her mind or remain ambivalent or remain consistent in her representation? What might be the key moment or anecdote where, in your view, Harrison's portrait of Winfrey comes together or takes shape? Respond to these questions in your journal. And then see if you can offer a brief character sketch of Winfrey *according to Harrison* in your journal. In extending what Harrison has to say about this televised talk show, how might Harrison introduce Oprah Winfrey if she were a guest on an imaginary talk show hosted by Harrison? What might this one episode be entitled? What would be the likely topic(s) of this episode?

READING AS A WRITER

We'd like you to explore Harrison's attitude or tone towards the people who might be said to make up a typical studio or at-home audience of *The Oprah Winfrey Show* and how this representation of audience might be seen as an extension of Harrison's critical argument. Respond to the following questions in your journal. Where does Harrison place herself in relation to Winfrey's audience? Would you say that she is a Winfrey "fan"? That is, does she write from the perspective of someone who is part of Winfrey's audience, who watches the show and participates in its rituals? Or does she comment on that audience from the "outside"? How can you tell? You might ask yourself: What kinds of knowledge would a "fan" have? And what additional or different kinds of information would an "outsider" have? What clues does Harrison offer in her text about her own feelings towards Winfrey and her audience? What does she assume about us as readers of her article? Are we addressed as Winfrey watchers? Or does she speak to us as outsiders looking in on Winfrey and her fans? Finally, how does Harrison's representation of Winfrey's audience and her understanding of us as readers of "The Importance of Being Oprah" contribute to Harrison's critical argument, her way of critically reading this cultural text?

WRITING CRITICISM

Televised talk shows are almost always identified by the name of their hosts, and are as such closely identified with their personalities. We'd like you to write an essay in which you define the attitudes and values that a particular talk show host represents and enacts for his or her viewers. To do so, watch and tape an edition of *Oprah* or *Rikki* or *Jenny* or *Montel* or *Geraldo,* or any other TV talk show with a similar format—that is, in which the studio audience participates in questioning the guests and commenting on the topic at hand. In such a format, then, there are two kinds of speakers—the "guests" and the audience—which means that the role played by its host is more complex than usual.

You might begin to do so by making a list of all the different tasks the host must perform. But you will also have to do a good deal more. For the host not only keeps the show going but also sends out certain messages, exhibits certain attitudes, takes positions, voices ideas, and speaks directly to us, the viewers at home, as well as to the guests and the studio audience. So you need to describe not only what the host *does* but also what he or she *communicates.* We want you, in other words, to consider the representative function of the host—to define the attitudes or ideas or moral standards that he or she might be said to stand for or to embody. (You might draw here upon Steven Fischoff's "Confessions of a TV Talk Show Shrink," which appears elsewhere in *Media Journal.*) See if you can make explicit what the host "says" to us (and for us) in this broad sense, and relate this to what he or she does during a typical show. In what ways does the host represent the concerns and values of his or her viewers? In what ways is his or her role instead to "educate" those viewers, to offer them a new or improved set of values? How would you characterize the image of selfhood that the host projects? How does the format of the show assist the host in constructing an image?

Happy [   ] Day to You

Gerri Hirshey

Springtime in the heartland, and the funky fertilizer truck hurtles down a narrow country lane. A spaniel yaps in its pungent wake. Hawks glide over this 254-acre farm a half-hour northeast of Kansas City, Mo. And beside a small pond, two poets write of love.

Jim Howard is rangy, intense, tussling with the west wind for the pages of his yellow pad. Across the water, his lover is restless, scribbling fitfully. She is a small, curly-haired brunette with a disposition as bright as her name: Penny Krugman. This past February, the newly engaged couple appeared on Tom Snyder's "Late Late Show" under this darn cute premise: valentine writers in love. By way of introduction, Snyder honked: "Wonder who writes those *sappy* valentines? Hawwwwwwwww."

When they read aloud the heartfelt valentines they had written to one and other, the Big Fella was unmoved: *"That's the most ridiculous thing I've ever heard!"*

No respect. Howard and Krugman know it's their cross to bear, working, as they do, for Hallmark Cards, a company that has been called "the General Motors of emotion." Never mind that we trust Hallmark with our passions, grand and small, at the rate of nearly eight million cards a day; that Americans will buy nearly seven and a half billion cards of all brands this year; that we send more prepackaged sentiment per person than any society on earth save the decorous Brits. In ads, on talk shows, greeting-card makers have suffered the gleeful derision accorded, say, insurance salesmen.

Gerri Hirshey is a contributing writer to *The New York Times Magazine. The New York Times Magazine,* 2 July 1995: 20. Copyright © 1995 by Gerri Hirshey. Reprinted with permission of Sterling Lord Literistic, Inc.

Despite the snarky put-downs and the fever for electronic communiqués, we're buying more prefab passions than ever. Marianne McDermott, executive vice president of the Greeting Card Association, a national trade organization, believes her industry's robust health is largely due to the TV generation's discomfort with the written word. "Baby boomers don't know how to write letters," she says. "That's why you saw this explosion 10 to 15 years ago of the non-occasion card. New cards for every conceivable reason."

The investment is minimal—a stamp, a signature, maybe a note at the bottom. One bright sign in this verbal Sahara: blank cards are making great strides for more personal communications. "A blank piece of stationary is too intimidating to most people nowadays," McDermott says. "But a four-by-six card seems, well, doable."

McDermott, a confessed "heavy user"—"my cats send cards to the cats next door"—thinks her industry is one of the best indicators of social change. "Cards reflect things faster," she says. They're cheap, high-turnover snapshots of The Way We Are.

Writers and artists ponder our emotional requisites here in Kearney, Mo., at the rural retreat that Hallmark bought to cosset and inspire its creative staff of 700. The farm is 23 miles from the company's mammoth headquarters, a two-million-square-foot concrete mother ship of sentiment that takes up a fair corner of south central Kansas City. This sunny afternoon, a company van ferried six writers to this lovely, greeting-card perfect farmhouse surrounded by a prairie grass preserve that once trembled beneath buffalo hooves.

They are limbering up for Valentine's Day '96 with some writing workshops, led by Howard. He details the first exercise in something he calls anti-logic: absurdist twist on the love poem. Everyone disperses outdoors and, after an hour, the writers share their afternoon's gleanings.

Penny: Love is a fig Newton
　　　Love is the entire blinding sun
　　　I'd like to buy a vowel.

Jim: A bullfrog leaps inside my chest
　　　I'm smitten!
　　　My pocket's lion is no kitten.

Just kidding. Just some poetic knuckle-cracking, loosening up those iambs and thou arts. You'll never find anything so risqué, so Dada-esque, in a Hallmark card. But they like to keep 'em frisky out on the farm. Chairman Donald J. Hall Sr., son of the company's founder, Joyce C. Hall, told me that one of his first duties when he joined the family firm after wartime service was to visit the burgeoning California hive of his father's dear friend Walt Disney. At the magic kingdom, and at newspapers and ad agencies, Hall sought instruction in the care and feeding of the Hallmark's greatest asset, its creative staff.

"I learned that there was no formula," he says. "But the common thread was that you cannot leave an artist alone for too long. They need a good deal of attention and nurturing and renewal."

A few Hallmark prescriptives: grant in-house sabbaticals in state-of-the-art workshops where staff members in need of refreshment can dabble in ceramics, silk screening, sculpture or button making. Buy blocks of movie tickets and hand them out. Send them on trips to Mexico, Spain, Florence! For your edgier product lines, let them watch Letterman and "Seinfeld" tapes and tune in the O. J. trial on company time. Cultivate a flock of pop-culture vultures with the keenest zeitsight, folks who can pounce on a humor trend and drop it into the retail racks before it stiffens: Top 10 lists, court TV jokes, bungee jumping....

No one here need be convinced of the efficacy of his or her labors. Of the three large companies that control 85 percent of the American greeting-card market, Hallmark can claim about 42 percent. (American Greetings controls 35 percent, Gibson Greetings, 8 percent.) Smaller companies—about 1,500—share the crumbs. Though as a family-owned business, Hallmark releases virtually no figures, 1994 revenues are estimated at $3.8 billion. Besides cards, the company produces countless ornaments, mugs, T-shirts, gift wraps, party supplies and assorted knickknacks that are sold in nearly 40,000 retail outlets.

Diversification has been limited and cautious. In 1984, Hallmark bought the venerable Binney & Smith, makers of Crayola crayons, Magic Markers and Silly Putty. They recently divested a cable interest, deciding, Donald Hall says, that the vagaries of changing government regulations make for too much of a risk. Instead, the company spent $365 million last year for its own production company for TV and theatrical releases. Along with a few of the long-running (44 years) "Hall of Fame" dramas, the company's RHI Entertainment has produced mini-series like "Lonesome Dove" and the $40 million "Scarlett," both for CBS. Wholesome family fare will be the standard for some 200 hours of product projected for 1996.

Conservative good taste was a near-religious ethic for J. C. Hall, a man whose gut instincts could rival the gigabyte sophistications of his company's current marketing data system. One of Hall's favorite issues, a four-by-five-inch friendship card featuring a cartful of purple pansies, remains the company's most stalwart evergreen, with almost 28 million copies sold. It still moves at the rate of 780,000 a year. Original publication date: 1939.

In the farmhouse, writers, photographers and graphic artists have been known to read fairy tales aloud, to drape one another in gossamer fabric. Self-expression is relaxed, not unlike the brushed-denim comforts of Saturdays, Hallmark's new, casual-and-cozy line. The TV come-on: *"Kinda like blue jeans for your emotions."*

Here, in this living room cluttered with duck decoys and gardening books, the six people and their writing styles couldn't be more diverse. Ken Kinnamon admits he is not having an inspiring afternoon and declares his piece "too dumb and too dirty to read." (He does and it is.) Carolyn Hoppe reads a rather lush bit of verse that elicits high praise from a fellow workshopper.

"Wow!" Barbara Loots enthuses. "Image *rich!*"

Loots looks the part of career poetess: a short cap of coppery hair, dangly earrings, flowing garments in bright orange and rust. After 28 years at the company, she is one of Hallmark's favored switch-hitters, able to pull to right or left—from Christian contemporary to African-American to generic terminal illness. This week she's been knocking out the kind of traditional iambic pentameter verse she "can't seem to turn off," in the form of inspirational wedding prayers. ("*Thank you, Lord, for giving us the gift of love to share/For bringing us together in the shelter of your care....*")

When Big Moments strike us dumb, it's a comfort to secure a professional mouthpiece for just $2.25. Mourners quote from Hallmarks at gravesites. An emergency-room nurse wrote to say that she uses a few lines of Hallmark palliative when helping doctors make next-of-kin notifications. Many a trembling bride has held a Loots card in her hand at the altar and whispered a life's commitment in her softly sculptured verse. Loots gets their letters, knows her work is packed into cedar chests with crumbling dried gardenias. Before Loots wrote cards, she spent a decade on children's books, pop-up novelties and Hallmark gift books. And recently examples from that oeuvre have shown up at flea markets. "I have become a collectible!" she chirps.

This afternoon, as usual, Loots has filled the most pages; she reads her work aloud with an English teacher's poise. There's high tech love: "*I'll never reprogram, my default is you.*" And romantic whimsy: "*You make my tootsies tickle the stars!*"

Throughout this, and a second exercise, Jim Howard is gentle and encouraging to all. "Thank you," he tells them, "for taking a risk."

Majoring in Relationships at Thoughtfulness University

Spend a week at Hallmark corporate headquarters and it does not seem the sort of place where anything is left to chance. Its main building, an Inverted Pyramid, is built, bunkerlike, into a hill at the heart of an 85-acre business and residential complex called Crown Center, after the company's regal logo. The complex takes in high-rise Hyatt and Westin hotels, along with shops, restaurants, a skating rink, parks and an outdoor concert facility.

Inside the Hallmark compound, there are miles of dun-colored cubicles in a honeycomb so vast, areas are number- and letter-coded like floors in a parking garage. Conference-room walls are fitted with card racks and swinging, hinged "wings," where more than 50,000 greeting cards are displayed and meticulously coded as to customer satisfaction and sales performance.

Here, rookie retailers learn to move product at what insiders call "Thoughtfulness University." Hallmark racks are stocked with an engineer's precision: high-price point cards at eye level, lower end above and below. Retailers constantly feed sales data to Kansas City, where thousands of computerized adjustments send or redistribute product according to buying patterns. An anniversary card that might be flatlining in Dubuque but selling in Baltimore can be quickly and more heavily stocked in its hot spot.

One of Hallmark's busiest stores is right here in the main building, where workers can satisfy their own card habits at a discount. The store is just across the corridor from the Crown Room, a Brobdingnagian cafeteria that features huge crown-shaped chandeliers, crown-stamped flatware and paper napkins in perky seasonal prints. Employees can also shop at a Hallmark-run department store (Halls), pick up dry cleaning, theater tickets, customized birthday cakes and dinners-to-go at on-site facilities.

No hostile takeovers, no union cabals lurk in this manicured family preserve that is one-third employee-owned. Like its founding father, the company is pleasant, conservative and intensely private. Numbers are closely guarded, visitors strictly escorted, media interviews monitored. Pick up a card with its market ratings and performance encoded on the back—indecipherable to an outsider—and it is quickly retrieved with a gentle, "Oh, you don't need to see *that*."

Even Hallmark's darkest public days were smoothly handled. In 1981, 114 people were killed when two aerial walkways in the Crown Center's Hyatt Regency Hotel collapsed; more than 200 were hurt. But Hallmark lawyers, aided by a rift between squabbling plaintiffs, were able to negotiate settlements totaling $140 million—all covered by insurance. In 1988 Hallmark settled a copyright infringement suit that had gone to the 10th Circuit Court of Appeals, admitting no wrongdoing but quietly deep-sixing thousands of cards.

In the media and in the marketplace, the Hallmark ethos ("When You Care Enough to Send the Very Best") is to be protected at all costs. No Christmas turkeys limp off to discount outlet stores. "Unsalable products with the Hallmark name on it *disappears*," says Don Fletcher, vice president for product development. "It's recycled. Or destroyed."

More than 21,000 full-time and 15,000 part-time workers worldwide toil in what Hallmark brass refer to as the "relationship business." They work hard to make your mom happier, your love life hotter, your filial guilt smaller. Though the industry has long endured accusations of "creating" card occasions, it is we, the card "users," with our messy, breathless, complicated lives, who drive this mighty engine.

Perhaps only social-service agencies amass more data on the dicey tectonics of our emotional terrain than Hallmark. Through punctilious market research, high-volume consumer mail and a finely tuned customer data base, they know our hearts as well as any talk-radio shrink or Harvard sociologist. Long before the Dr. Ruths, Montels, and Frasiers, Hallmark was taking note of our emotional wish list and writing iambic prescriptions for what they term "situational" needs.

Hallmarkers are experts, for example, on the delicate protocols of the new "blended" families, on just how to address your father's new wife who's half your age with three toddlers from two other daddies. They have found that we are still not comfortable with the word "death" in sympathy cards, but we are modern enough to send divorce announcements. We now acknowledge imperfect childhoods with Hallmark's selection of cooler, less idealistic Mother's Day cards. We are so close to our "animal companions"—Hallmark research indicates that most Americans consider pets to be "family members"—that pet cards (sympathy, birthday, friendship, pet-to-owner, pet-to-pet) just fly off the racks. We are so stressed and guilty as to buy stick-on love notes to greet our latchkey kids, and concerned enough to lean on "To Kids With Love," a series for 7- to 11-year-olds from parents that takes on the traumas of moving schools, discipline problems, divorce and separation.

Nineties life has yielded a host of scary new Hallmark occasions: bankrupt pension plans, child custody loss, AIDS, gambling problems, chemotherapy and forced early retirement, all efficiently acknowledged by what the company calls "universal specific" cards: *"We don't always understand the things life brings our way/You're being thought of with so much love during this difficult time."*

In situations where it's tough to clean and jerk that phone receiver, when face-to-face confrontation seems untenable, a white flag is available for the price of a Big Mac. Choose from general apology cards, hey-hang-in-there cards and tough-love missives that plead, gently and generically: *"Please get the help you need."*

No Hallmark sentiment flies untested toward our hearts. A case in point: Hallmark's successful "recovery" line.

It began with another emotional tale. In 1989, Susan Giffen, a senior editor at the company, was working on "Just How I Feel," a line of "anyday" non-occasion cards—at that point the most thoroughly researched line in Hallmark History. Responding to a perceived need for greater intimacy and self-expression, Giffen's staff was creating and testing cards with "heavy I-content," very personal, me-to-you messages. And during that time, Giffen had an upsetting lunch with a friend.

"She told me that her husband was an alcoholic," Giffen recalls, "that he was trying to deal with it and that this was an awful time for them. I went back to the office thinking, Gee, I wish there was an appropriate card, just something supportive."

On Giffen's hunch, the mighty engine began to move forward. She held a brainstorming meeting with writers—a common Hallmark exercise wherein an editor or manager trolls for ideas with a familiar introduction: *I see this need.*

"Needs lists" are the marching orders for Hallmark's creative staff, handed to divisions of artists and writers with directives as specific as "need new jokes about hospitals—not nurses, gowns, food or expenses..." or exactly how many bunnies of which color should be rolling an Easter egg. Giffen's needs list dealt with demons: alcoholism, drug abuse, marital and mental cruelties, diet dilemmas.

A batch of test cards were produced, from the general "glad you're turning your life around" to the more abuse-specific ("I'm worried about your drinking"). They were sent to consumer panels handpicked by the research division in two cities. Panelists examined 50 cards and rated them on a scale of 1 to 5, then chose several favorites. Enough respondents—more than 30 percent—said they had needs met by such messages, and would buy the cards. Thus, of the 520 "anydays" produced for the line, 10 were "recovery-specific" cards, sent out as trial balloons in the turbulent skies of what Giffen calls this nation's "recovery movement."

They soared far beyond expectations for what was considered a "niche" need. Before Giffen pressed formally for a "larger offering" of recovery cards, she armed herself with statistics from the research department. The population that Hallmark now refers to as the "recovery community" extends to nearly half the American population. Fifteen million attend weekly support groups for addictions and abuses of various sorts; 100 million family members and friends are engaged in trying to help them cope. That could mean a lot of "Congratulations on the anniversary of your recovery" cards.

Giffen and her staff pored over recovery magazines and attended conventions of Adult Children of Alcoholics (A.C.A.); they consulted with professionals on 12-step program etiquette and the touchy codependencies of recovery relationships. Looking at the numbers and Giffen's keenly targeted marketing plan, the line's manager told her, "Try it."

Giffen then invited interested writers to work on the project. Since the cards were to be "message driven," the company's best lettering artists were summoned. Color directives: cheery and upbeat. In came the "specialty" team—product designers who can goose a line's revenues by applying card slogans like "100 percent alcohol-free" to mugs, key chains and T-shirts.

Rolling out this line called for alternatives to traditional advertising. Hallmark sent releases to newspapers and recovery publications and invested in a booth at an A.C.A. convention. ("If we met several hundred people there," Giffen reasoned, "They'd tell their recovery group, which would be maybe 10 or 20 more people, and so on.") In the stores, the cards would debut in 1992 under the heading: Just for Today: Cards of Support and Encouragement.

Unlike most Hallmark lines, the recovery cards bypassed the customary one-to three-month testing in 16 selected stores nationwide. But the final selection did cover a range of needs: serious apologies from the former abuser ("*I regret the pain I've caused you....*") and pastel pep talks ("*Sometimes it may seem that therapy is slow going....*"). Humor worked as well. A caped "Recovery Man" has proven a very popular 90's superhero, able to leap 12 steps "in a single bound."

After three years, the cards are holding their turf in the racks, but the specialty items were not strong enough to keep their retail space, and have been discontinued. Diet encouragement cards have remained best sellers. Given the national obsession with fat-free futures, says Giffen, "Diet cards may well be the recovery evergreen."

Is There Room for Sentiment at the Superstore?

The greeting-card business would not satisfy the lusts of an aggressive growth investor. Gray Glass, a financial consultant for Wheat First Butcher Singer in Richmond, is the nation's top expert on the greeting-card industry, producing market reports, visiting companies and lecturing at trade meetings. He watches over what he describes as "historically modest growth, 2 percent to 4 percent a year." He also says that except for a short time from 1985–7, when tough retailers forced a destructive price war between the big three companies, the greeting-card industry has never seen hard times.

Glass believes that Hallmark's biggest skirmishes will be fought not with A.T.&T. or the Internet, but in the Wal-Marts and giant drugstore chains. "The card-shop retail sector has gone from 40 percent of all card sales to less than 30 percent and is still declining," he says. Card shops, like most specialty stores, are losing ground to the big, one-stop multimarts. Harried superstore managers with thousands of products to shepherd have no room for losers.

And Hallmark has no leeway for what Don Fletcher calls operational "redundancies." Yes, he confirms, Hallmark is undergoing a reorganization to correct these, combining the expertise in their top Hallmark lines with that of their mass-market Ambassador lines. Valentine specialists, for example, will now pitch their woo for both lines at once. "Probably some jobs will be phased out," Fletcher says. "But we're not like some public company that announces we're going to cut 500 heads."

Nonetheless, Fletcher's acknowledgement of "serious reorganization" over the next 18 months may explain the Hallmark Moment I witnessed in a sprawling lounge area: a youngish male managerial type was thanking a female co-worker for a hang-in-there card (a frazzled cat blown sideways) he'd found on his desk.

"Transferring is better than you-know-what," she offered.

"Downsizing stinks," he said. "No matter what you call it." He tapped the card and smiled. "Anyhow, hey, thanks for the thought."

The Sweet Butter-Cream Center of America

"Lureen? Say, Lureen, it's like they say in those TV ads—how the heck *do* they come up with all these cards?"

In Hallmark's busy Visitors Center, a group of seniors from Ottumwa, Iowa, is marveling at the display that shows how all those perfect giftwrap rosettes are made. Push a button, watch the ribbon leap through the automated works and zap—a tiny purple bow falls smack in your palm.

"Sakes alive!"

The Iowans tell me that Hallmark was a featured stop on their trip to Branson, the sentimental Ozark theme park of country music and Tony Orlando's Yellow Ribbon Music Theater that now draws 5.8 million tourists a year to southern Missouri. Hallmark is also right on the way to the Precious Moments Chapel about two hours south in Carthage, Mo. The chapel's 50 murals are the work of a Missouri-born Michelangelo named Samuel J. Butcher—once an illustrator of greeting cards.

Butcher specializes in pastel waifs with signature teardrop-shaped eyes. They are to the 90's what those saucer-eyed Keane paintings were to the 60's. Precious Moments porcelain figurines have become the nation's No. 1 collectible, with sales of $500 million. In peak season, as many as 4,000 people a day converge on this tiny town in search of the inspirational and mantel appropriate. Hallmark, Branson, Precious Moments. Perhaps we've stumbled smack into the sweet butter-cream center of American sentiment here. The Whimsy Triangle. And maybe it's no accident that all big three card companies began as family-run concerns rooted in the Midwest—in touch with small-town feelings, free from jaded coastal cynicisms. American Greetings started in a wagon. And Hallmark Cards was born in a chilly 12-by-12 room at the Kansas City Y.M.C.A.

It was here, to this bustling, anything-is-possible cow town, that 18-year-old Joyce C. Hall ventured from his home in Norfolk, Neb., checking into Room 413 of the Y on a January morning in 1910. He had invested in two boxes of penny postcards—Christmas and Birthday greetings, silly slogans, hearts and flowers—and began wholesaling them from his room.

Americans had been mad for picture postcards since shortly after the turn of the century, sending them and collecting them in fat albums. By the time Hall hit Kansas City, the craze was at its peak, but he was doubtful of its future. He reasoned that in an increasingly mobile society, greeting cards could become a welcome middle ground between the frivolous hello from Luna Park and the time-consuming letter. In 1914, Hall Brothers (Joyce C. and his brother Rollie) opened a card shop in the Corn Belt Bank Building. World War I effectively stopped the import of European Postcards that had previously dominated the market. And Hall's greeting-card line expanded: friendship cards, get wells, Halloween. . . .

Hall grew to be a very rich, deeply conservative man who often ate in the employee cafeteria and drove the same 1963 Buick for a dozen years because he felt its transmission was nonpariel. As head of the company's Okay Committee, he approved every card until late in his 70's; in his 80's, he still piloted the Buick into work two or three days a week. (He died in 1982, at the age of 91.) During the nearly 70 years of his stewardship, Hall held a sensitive finger to the nation's emotional pulse—through the Depression

("Just a Christmas wish to your house—from the gang here at the poor house!"), through war times ("Let's kick Mussi off the map!") and into the nuclear age ("Now that we've harnessed the atom, maybe we can do something about birthdays!").

The greatest changes—the social transformations that removed pots and pans from Mother's Day cards, that took away Dad's pipe and slippers and strapped him with a full Snugli—have unfolded during the watch of Donald Hall, another pleasant, reserved man who can count George and Barbara Bush among his greatest friends. He says he doesn't pretend to understand all the changes that have moved his product. But he knows that change can be very, very good for his bottom line. After the war, Hall says: "There was a great displacement of home. Everybody had moved, which was a gift for our business. People were no longer in the town they'd grown up in. Then as time went on, the family changed. The recognition of different parts of our culture became more pronounced."

The Multi-Culti Moment: Can We Say 'Dis'?

"Uh, cornbead and collard greens. Are these things often eaten together?"

Supervising the white writers who sometimes work on Hallmark's African-American Mahogany line, Myron Spencer Davis has had some amusing moments.

"Yeeees," he'll explain patiently. "Cornbread and collards generally go together. They're relevant to our culture. We eat them on Sunday...."

Davis laughs, bouncing dozens of chin-length braids. He is a young, cheerful, former high-school teacher whose mandate is to give Mahogany a freer, more ethnic voice. The line, though small by Hallmark standards (about 800 cards), is very successful. It also has close competition from American Greetings' Baobab Tree line and Gibson's Family Collection line and scores of small enthnocentric companies. In addition to Mahogany, Hallmark markets its Primor line, for Latino cultures, and a Tree of Life line of Jewish cards.

Visiting Davis's conference room with its walls full of vibrant, bold, Afrocentric colors, I am reminded of a conversation I once had with Spike Lee's grandmother Zimmie Shelton. She told me about the years of sitting at her kitchen table applying thin watercolor washes of brown to the only greeting cards available for her children and grandchildren. Until very recently, the card racks were white only.

"We do the same seasons and everyday cards," Davis says. "But in a way that is culturally relevant." Staff members pore over Essence and Ebony for hair-style info and relationship articles. They watch "Yo! MTV Raps." Ask Davis about the parameters of Hallmark funk and he laughs. "We ask about the borders every day. Can we *do* that? Can we say 'dis'? I mean as a *verb*."

So You Think Anyone Can? Just Try It.

To a friend, I quote the little phrase from Hallmark's Shoebox Greetings humor brand that launched more than a quarter million Father's Day T-shirts: *I fought the lawn and the lawn won.*

"Eh," she says. "I could have written *that.*"

This is a common conceit about greeting-card writing, and one I shall take pains to dispel. Imagine the flop sweat, please, when the visiting Media Person is invited to participate in the afternoon brainstorming session held in Hallmark's Idea Exchange. This group of a dozen or so artists and writers, serving on a rotating basis, functions as a free-floating think tank, available to all divisions.

Theirs is not a cubicle consortium. With their slapdish configuration of antique roll-top desks and salvage yard memorabilia, the Exchange is what Hawkeye's tent was to the rest of the "M*A*S*H*" unit—a seedbed of cheery sedition. Jim Howard, who seems to be in permanent rotation here, is about to conduct one of the group's twice-weekly forced writing exercises. "We get a lot of cards accepted out of these writing sessions," he says. "We use writing sessions," he says. "We use it mainly for alternative humor lines and it works, because it bypasses the formulaic. You just go for it."

With everyone seated around a conference table, Howard holds up a stock photo or reads out an opening line from some department's current wish list. Assignment: the funny tag or caption. For about three minutes, writers scribble entries on index cards and toss them into a large baseball mitt. Someone yells *"Ding!"* Then the unsigned entries are read aloud.

First up, an opening line, "If I was French, I'd have a completely *different* way of wishing you a happy birthday...."

Tick tick. I'm getting nothing, but all around the table, cards sail into the mitt, one every 20 seconds or so. I finally toss one in at the ding, then flush miserably as Howard reads it a few minutes later: "I'd hire Jerry Lewis to slap you on both cheeks."

Happily, there are a few worse than mine, but most are better, if predictable. Brie jokes, Sadistic waiter jokes, a few more lame Jerry Lewis gags. And a possible keeper: "I'd dig a chunnel through your cake and then act resentful."

Next, Howard holds up a black-and-white photograph of a 50's housewife breaking an egg into a mixing bowl and looking somewhat disconcerted. After the ding, Howard reads this bit of inside copy: "On her 12th birthday, Martha Stewart makes an angel food cake and lays out her fiendish plan for world domination."

Another suggests a caption bubble: "Darn, was that my last egg?" Inside: "Happy 40th birthday!"

This one gets big laughs and some female moans. It's declared funny, but borderline. Depends on who's sending it. "Men are always fair game," one female writer explains. "But only women can make fun of women. House rule, you know." She laughs. "Hey, it works for me."

It's a Female Thing: Thelma and Louise and Maxine

Sisterhood is powerful at the card rack. More than 85 percent of greeting cards are bought by women. And woman-to-woman "friendship" cards have gone through the roof. "The absolute biggest thing was the women's movement," says the Card Association's McDermott. "You started to see woman-to-woman cards telling it like it is. That never existed before."

Hallmark has found that women like to send each other gentle life's-a-bitch-but-we're-in-it-together cards. They can get a tad randy: *Whenever I get depressed I just hop on my Harley and take a little spin.* (Open to a photo of a half dressed hunk.) *This is Harley.* They like to get even. A poutingly posed Adonis is stamped with a simulated F.D.A. contents label: *MAN. Ingredients: vanity, self-centeredness, arrogance, insensitivity, thoughtlessness, insincerity. Plus may contain one or more of the following: communication skills of a chimp, obsessive love for his mother, and/or an ego the size of a landfill.*

Presaged by the cranky nickel-a-session counselings of the "Peanuts" strip's Lucy, Hallmark has featured its own feisty feminists: the single, working and sardonic Cathy (from the comic strip by Cathy Guisewite) and the popular, if prickly, Maxine—a 50-plus attitude problem in saggy stockings. Maxine's delivery is Don Rickles-with-hot-flashes. Her wisdom is dubious, but it sells: *If life gives you lemons…tuck 'em in your bra. Could help. Couldn't hurt.*

The woman-to-woman card phenomenon has even prompted a handful of strident academic studies. The latest, published in the Journal of Popular Culture, takes a dim view of how "everyday female existence is being penetrated by the patriarchal corporation," i.e. Hallmark. In "Icons of Femininity in Studio Cards: Women, Communication and Identity," Melissa Schrift condemns all that "self-disparaging humor" about body parts sliding south. Yet grudgingly, Schrift does allow this: "Research on the importance of greeting cards in the aging process suggests that antagonistic birthday cards provide an acceptable outlet to communicate fears about aging."

As Maxine might say: Aw, put a sock in it. Despite the dangers to their psyches, nearly five million American women are happy to buy their sentiments with the help of a Hallmark Gold Crown card. And they know Big Daddy is watching when they present the card to the cashier to register purchases and rack up incentive points. The plastic card is similar to those used by large casino companies to track and reward heavy gaming patterns, and to stay in touch with valued consumers by mail. In Hallmark's program, many cardholders get a newsletter, The Very Best, plus free samples and discounts.

"It's called relationship marketing," explains Brad Moore, a Hallmark advertising vice president. "These are people you want to talk to directly.

These are *very* heavy users." They send an average of 120 individual cards per year, though some self-confessed addicts buy 10 to 20 cards a week. Jean Farmer, a homemaker from Garland, Tex., maxes out at 2,500 warm thoughts to friends, their pets, to hairdressers and yes—Hallmark card shop workers. Total outlay, with postage, $15,000 per annum.

The L-Word

No large corporation devotes more time and expense to parsing American expressions of love. Three years ago, Hallmark got serious enough to set up an in-house consortium on passion. They called it Val Lab. Its mandate: rolf the traditional valentine and come up with a line better tuned to the vicissitudes of 90's relationships.

Anything went, including Jim Howard's axle-grease 'n brewskie "Macho Sonnet." Printed on a rather traditional red filigreed card, it reads like one of those Iron-John-at-the-backyard-fence monologues from "Home Improvement":

"I love this woman. That's the way it is. . . . If I could show what happens in my brain when our two bodies meet, I'd count the five kerjillion ways her touch drives me insane. I'd sing like Satchmo. Hell, I'd even dance. Well, maybe not. . . ." Others were a tad less effusive: *"Be my valentine. . . . I've already told people we're sleeping together."*

Oh, the things they cranked out. "We took a big risk," says Connie Strand, Hallmark's seasonal-card programming manager, "and put a lot of it in the line." The first product line was called "Passionate Promises" and debuted in 1994. The cards sold very well, so the company has been refining the Val Lab brainstorming in the seasons that followed. I've been told that next year's theme is top secret. But after much nipping at Strand's ankles and more than a few pointed questions on the deplorable volatilities of dating these days, I get her to cough up the password for Valentine '96: *Caution.* Fewer forevers, easy on the "one and onlys" and very, very judicious use of what Hallmarkers call "the L-word." It seems that for a large segment of valentine shoppers, love futures are a risky commodity. And market research supports the retreat.

"We have found that there are people in committed relationships free to say a lot," Strand says. "But there's also a real need for a more guarded message."

This goes across the board, according to Myron Spencer Davis, who oversees African-American love missives at Mahogany. "I think singles need more help than ever," Davis says. "Men don't know how to act, women don't, everybody's liberated. Should I approach you, how much can I say?" Thus in Davis's line, love often takes a backseat to heavy like. "We might have a sentiment that's very close. But it doesn't use the L-word. I like you, I respect you for who you are *and that's as far as I'm willing to go!"*

Sex is something else, and Hallmark understands. "Our relationship might be purely sexual," Davis theorizes. "So we might have a nice little card that talks about waking up beside you, or your touch. Something like that."

No one at Hallmark is very comfortable using the G-word when it comes to love. Smaller gay greeting card companies serve that market well, I'm assured. Hallmark produced one "AIDS awareness" card that did well. And the party line is this: Hallmark is so deft in universal specifics that the most esoteric sets of lovers, from interracial lesbians to preoperative transvestites, should be able to find something "appropriate" amid the Very Best.

The Queen of Steam

"If I'm in a real hostile mood, love is the *last* thing I want to write about today," Linda Elrod says. "*I know our love will last forever—YEAH, SURE!*" She laughs. "That's the day I'll write Mother's Day or something."

This is from the valentine virtuoso known throughout the building as Hallmark's Queen of Steam. One wall of Elrod's dimly lighted cubicle is a prime-time hunkathon. Cutouts of George Clooney from "E.R.," of David Caruso, of generic rippling pecs and abs, all serve as macho muses for Elrod's palpitating oeuvre. Also inspirational: a bottle of potent after-shave in her desk drawer. Sometimes, she says, it's like acting: "Setting the stage, getting in the mood—whatever it takes."

No one does it like Elrod, who is as blessedly forthright as she is prolific. This Kansas City native began as a stringer for a commodities wire service. Her editor was exasperated by the fancy adjectives she kept tucking in among the pork bellies. Friends kept urging her to apply to Hallmark, and she says it was a perfect match. For the most part, they just let her go:

When you put your arms around me
And pull me so close
That I can feel your heart beating,
A passion rises within me
Like I've never known before.

Writing to men—wife to husband, etc.—Elrod says she gives them what they want to hear. "Men need reassurance that they still have it," she maintains. But her ego massages are never X- or even R-rated. "I never go tacky or sleazy," she says. "You can do a lot more with just implying." (Example: a Christmas card that says, "When you unwrap your presents, start with me.") Elrod admits that she was a bit disappointed when editorial nixed a swell idea she had for last Valentine's Day, a coupon book for a week's worth of hotcha fantasies. "I did them like romance novels," she recalls. "A pirate, a cowboy. . . . Well, their minds were just *too* active."

Elrod was the first Hallmark writer to receive a name credit on a card, back in 1984. The recognition thrilled her, she says, until the night she got a phone call from an agitated woman: "You wrote my husband this love letter. *Stay away from him!*" The warnings got louder and more hysterical. The woman had found the flowery card with all this lovey stuff, and it was *signed!* She knew he'd been fooling around; now she had a name to hang her suspicions on.

"NO! Turn it over!" Elrod spluttered. "Doesn't it say Hallmark?"

"No, it's *handwriting!*"

(Big points here to the Hallmark calligraphers.)

Since that distressing interlude, Elrod has enjoyed an unlisted number and a steady stream of gentler consumer response. Because of the bylines, she gets plenty of mail addressed to her personally. "This gal was telling me that she and her husband were having a rocky time and they'd just run out of words," Elrod relates. "And she found this card I wrote that said just what she wanted to say. She gave it to him and it was the first time she ever saw a glint of a tear in his eye."

The woman's sign-off nearly had Elrod reaching for the hankies: "If you never write another card, at least I got the one that saved my relationship."

Elrod is divorced and describes the current dating scene as "grisly." Like millions of American singles, like those urban careerists in hit sitcoms ("Friends," "Ellen" and "Seinfeld"), she has created her own post-nuclear family. "There are a lot of middle-age single people incorporating their friends into their family," she says. "I have a handful of really good friends with whom I can discuss *anything.*" Some of her popular coping cards come from those gabfests. Elrod looks around her cubicle at the cute forest creatures, the mugs bearing gentle sayings, the lush floral valentines that go for $5 a shot.

"It takes a lot of discipline to do this job," she says. "Everybody has days when they're miserable, they wake up hating everything. And to have to come in and get in the Hallmark mood.... Oh, *my.*"

'Like Dylan Goes Electric'

Should inspiration momentarily desert the staff at Shoebox Greetings, Hallmark's nine-year-old alternative humor line, its editorial manager, Steve Finken, is likely to hand out a clutch of movie passes and say, "Here, kids, go see 'Pulp Fiction.'" His writers churn out a total of 150 ideas a day to meet their quota of 75 to 80 new cards and slogans a week. Hallmark is banking on Shoebox to power it toward a second century with an increasingly important market segment.

It was baby boomers and their post-"Peanuts" skepticisms who fueled the bullish market called "alternative cards," the quirky, offbeat, generally humorous stuff that was once just the province of the smaller card companies that could afford to take a risk. When they caught on in the mid-70's, says Gray Glass, the industry analyst, everybody had to jump in. "You either had to get the product out or give up that huge new share of the market, the postwar boomers who had grown into the card-buying stages of their lives." Not unlike the big American car companies, who awoke from a complacent drowse to find a nation mad for Hondas, the Big Three card companies dragged their feet for almost a decade.

By the mid-80's, all three of the main players had scrambled to roll out alternative lines. Gibson created Neat Stuff. American Greetings came up with Just My Style. Hallmark added Shoebox, and for its Ambassador division, My Thoughts Exactly. They all could be sitcom titles for the audience they're beamed at: younger, college-educated, middle- to upper-income, white-collar jobs. This was a generation reared to expect *specialness*, one that browses for cards the way it shops for Saabs and coffees. As the retail bible Drug Store News explained, "Alternative card growth is fueled by the baby boomers' desire for self-expression and differentiation."

And they're willing to pay for it. Alternative cards are averaging a 25 to 30 percent growth per year, accounting for more than a quarter of the retail market. And nearly half of alternative card sales are impulse buys. Be funny or be gone.

It was this imperative that caused Steve Finken and 39 other Hallmark "radicals" to be cut from the cubicled herd to start Shoebox in 1986. It was risky, it was expensive and it worked very well. "For Hallmark it was a big blip on the screen," says Finken. "Like Dylan goes electric."

Finken, a wry 46-year-old cross between Mr. Peabody and Lou Grant, oversees the hiring, care and output of nine talented-if-twisted Shoebox writers. I find him presiding over the department's 4 P.M. meeting. Every day at this time, his writers turn in a stack of folded index cards with jokes written to an assigned theme or "needs list." They are read aloud and judged summarily with Finken's imperative shorthand: Can't do. Have. Maybe. Yeah! And "too Ted Nugent."

This week the task is Mother's Day '96. Finken is reading from the index cards: *Thanks for not naming me Amber.* It gets big laughs and a "can't do." (The writers chorus: "Too many Ambers out there.") Finken drones on: *I can't remember what it was I wanted to say to you, Mom—MINDLESS ZOMBIE GOTTA GO—since I joined the cult—200 ROBES TO LAUNDER BY SUNDAY.*

Someone hums the old "Star Trek" theme. "To boldly venture where...ooh, sorry."

Soon, Shoebox will be working on a line for the subteen set to try to instill the "personal communications" habit at an evermore tender age. Reaching the upcoming X-ers and beyond has Finken joking about his longevity at the chase. By way of example, he shows me examples from the "new stupidity" line. "It's just silliness," he says. "Silly puns, absurdist humor." One card features a tiny cartoon horse on the cover and this inside: *A pony for your thoughts.*

Such groaners are a departure from the reality-based observational humor that's been so successful for Shoebox. But it may be the wave of an even less verbal future. "It's an emerging younger market," Finken says, a trifle sadly. "They're being raised on 'Wayne's World,' 'Dumb and Dumber,' 'Beavis and Butthead.'"

Humor can change with remote-control speed. And the keenest edge can get fatally dulled in an Okay Committee. "Doing this at Hallmark is like being Mick Jagger playing with Lawrence Welk's band," he says, grinning. "There are certain creative restrictions." Kinda like a leisure suit for your blues.

Cybersentiments

As the company positions itself to care for boomers in their golden decades, to service X-ers coming into those "card buying years," no one at Hallmark is trembling about the encroachments of E-mail, the Web and the Internet. Hallmark is about to market personal-computer software for making personalized greetings, but the company is cutting back on computer kiosks that let customers personalize cards in retail stores. A partnership with America Online will sell Hallmark sentiment, but the products—albeit individualized—will still be addressed, stamped and mailed.

"I think I was in high school when I first heard that our industry was done for," says Donald Hall. "But I think there will always be a need for some hard copy deliverance of sentiment."

Real handwriting is a rarity in today's mailbox, and most often it's retrieved with pleasure and relief. It's not totting up a balance overdue. It never addresses "Occupant."

"A card is a quiet refuge amid the noise and speed of all the rapid-fire communication in everyday life," says Kristin Riott, creative director at Hallmark. "There's an intimacy about its being a folded piece of paper you have to handle."

Riott grumps, quite rightly, that no one edits E-mail; bombarded with gigabytes of blathering electrons, we may just be on the cusp of communications overload. Bestseller lists are larded with books teaching one how to talk to the opposite sex, the boss, the teen-ager. Airwaves and fiber-optics cables thrum with talk radio, call-in shows, Psychic Friends and Internet heavies all urging us to 976-CHAT. But amid all this noise, how are we at talking to those who matter most? You can now download porn, but can you find a Window for sentiment?

Doubtless Hallmarkers have been poring over the newspaper trend pieces that document on-line addictions, the compulsive excesses of computer wonks whose real social lives have shrunk every time they've upped their RAM capacity. Support groups are forming. As we natter on toward the millennium, the good folks at Hallmark will be monitoring our thoughts, dreaming, perhaps, of a new line of recovery cards, mugs and T-shirts for the electronic shut-in: Log off and *live!*

COMING TO TERMS

Hirshey offers us a "behind the scenes" look at the Hallmark Cards company, the people who work there, and the sophisticated corporate culture responsible for many of the greeting cards we have seen throughout our lives, and she also suggests some reasons why greeting cards have become such a common part of our everyday lives and culture. One reason why we chose to include Hirshey in *Media Journal* is that she takes a common text from our culture—greeting cards—and then, as if writing a history, Hirshey considers a number of cultural and sociological changes in the United States by connecting these changes to the ever-changing canon of greeting cards and their messages. In your journal, then, record some passages from Hirshey's essay that show how greeting cards have kept pace with (or resisted) major changes in U.S. culture and society. Next to each passage explain how this card or that card represents some kind of change in society. What specific change is being addressed and how does the card reflect that change? Hirshey also turns her attention to contemporary times. What does Hirshey tell us about the appeal and function of cards today? What do they offer their users that other forms of communication don't? (Why send a card, for instance, rather than make a telephone call, or post an e-mail message, or write a note?) Do we use greeting cards today in ways we have not used them in the past? How do the messages on greeting cards (and their impressive variety) today reflect our culture? Finally, as someone who probably has had considerable experience with cards (both sending and receiving), how would you question or add to what Hirshey has to say?

READING AS A WRITER

Throughout her essay Hirshey quotes a number of Hallmark cards, card ideas, and lines of development for specific sets of cards. She also quotes a number of the behind-the-scenes people involved in this industry, showing that some cards are the imaginative work of a single artist while others spring into production through compromise and committee decision. We would like you to pay special attention to the ways Hirshey makes use of quotation in her article. Sometimes, possibly from either writing a research paper in high school or doing some other kind of library research project, writers think that quotations are used only for example or to give "expert" authority to a piece of writing. These are certainly reasons to use quotation. We think, however, that Hirshey provides us with an important opportunity to explore several different ways to understand the uses of quotation as a particular kind of rhetorical strategy, a way to advance the critical argument in a piece of writing. In your journal, we would first like you to put into your own words what you think the critical argument of Hirshey's discussion is. In other words, based on your reading of the piece, what do you think is the point of this article? What is it trying to convince you of? In what ways, for example, might Hirshey be trying to show us how to "read" greeting cards in a new way? We think that you may see Hirshey's piece as attempting to argue or do several things, which is fine, as long as you say something more carefully thought about than, "She talks about greeting cards." In other words, Hirshey is using greeting cards as a popular culture text to say or to argue something about something else.

Next, we'd like you to go back to Hirshey's text and to mark those places in the article where she is using quotation. Then we would like you to re-read these quotations against a set of questions and record your responses in your journal. When does Hirshey allow people from Hallmark to speak for themselves? When does she paraphrase or summarize? Do you see some kind of pattern here? It's also important to see how Hirshey introduces a quotation or provides some kind of critical commentary after a quotation, in an attempt to show how she might be reading the quotation. In which quotations does Hirshey do this kind of work? When does she let a quotation *speak for itself*? Why might Hirshey choose not to provide a gloss on a quotation? What might this quotation or that quotation tell her readers that Hirshey cannot say in her own words? In both cases, how would you explain that these quotations are *adding to* or *complicating* Hirshey's argument?

WRITING CRITICISM

Greeting cards are "cheap, high-turnover snapshots of The Way We Are," argues Gerri Hirshey. We like Hirshey's metaphor here, but we'd like to extend it to include The Way We Were...The Way We Hope To Be...(or even) The Way We Hope To Avoid Being Seen. Write an essay in which you test out the claims of our extension of Hirshey's metaphor. Try to define the "we" that a particular line of cards tries to address and speak for. Begin by locating a set of cards that interests you and that seems to be targeted at a specific, identifiable audience. Your choice might be a "line" developed by a particular company—like one of the Hallmark series that Hirshey discusses. Or you may want to look at cards by various companies that seem to target the same audience or occasion. Or you may want to look at a set of cards that have been sent to a single person—what do they tell you about her or him? (As you can see, you may end up doing your research for this writing in a card store or supermarket, or by pulling a shoebox of cards out of a closet or from under a bed.) In any case, your task will be to define the kind of person suggested and implied by the cards you are looking at, the kind of person they seem made "for."

Ask yourself: What do these cards suggest about the age and interests of the person or people they are addressed to? About their occupation or profession? Their interests, worries, sense of humor, tastes in music, or TV? About shared experiences between sender and receiver? These are, of course, very much the sort of questions that run through our minds—though perhaps in a somewhat different or less conscious form—whenever we try to choose a card for someone we know, whenever we find ourselves saying something like: "Oh, this would be perfect for X." Here, we might say that we're using *personal knowledge* in our selection, in choosing just the right card. But we'd also like to see you extend Hirshey's project and attempt to explain your choice of a card or set of cards in *cultural terms*. For the next question to ask is: What do these cards *not* tell about their users, or even hide or distort about them? Greeting cards do say things; but, at the same time, they are careful not to say certain things. How might your choice in cards reflect certain cultural values, stereotypes, prohibitions, or taboos? In addition to the *personal* message in a card, how might this sentiment be read as a *cultural* message? Your job here is to make that familiar set of intuitions and hunches more explicit—and thus available for critics.

The Internet Is Four Inches Tall

M. Kadi

"Computer networking offers the soundest basis for world peace that has yet been presented. Peace must be created on the bulwark of understanding. International computer networks will knit together the peoples of the world in bonds of mutual respect; its possibilities are vast, indeed."
—Scientific American

"Cyberspace is a new medium. Every night on Prodigy, CompuServe, GEnie, and thousands of smaller computer bulletin boards, people by the hundreds of thousands are logging on to a great computer mediated gabfest, an interactive debate that allows them to leap over barriers of time, place, sex, and social status."
—Time *Magazine*

"The Internet is really about the rise of not merely a new technology, but a new culture—a global culture where time, space, borders, and even personal identity are radically redefined."
—Online Access *Magazine*

"When you begin to explore the online world, you'll find a wealth of publicly available resources and diverse communities."
—A 'Zine I Actually Like—so I'm not going to tell you where I got this quote

From *h2so4* 3 (November 1994), 26–30. Reprinted with permission of the publisher.

"The first time you realize the super toy you wanted is really only four inches tall you learn a hard lesson." Q: How big is the Internet? A: Four inches tall. Blah. Blah. Blah. Everyone is equal on the Net. Race, gender, sexual orientation are invisible and, being invisible, foster communication. Barriers are broken down. The global community exists coming together. (Right here, right now.)

Diversity. Community. Global Culture. Information. Knowledge. Communication. Doesn't this set your nostalgia alarm off? Doesn't it sound like all that sixties love-in, utopian, narcissistic trash that we've had to listen to all our lives? Does this sound familiar to anyone but me?

Computer bulletin board services offer up the glories of e-mail, the thought provocation of newsgroups, the sharing of ideas implicit in public posting, and the interaction of real-time chats. The fabulous, wonderful, limitless world of communication is just waiting for you to log on. Sure. Yeah. Right.

I confess, I am a dedicated cyberjunkie. It's fun. It's interesting. It takes me places where I've never been before. I sign on once a day, twice a day, three times a day, more and more; I read, I post, I live. Writing an article on the ever-expanding, ever-entertaining, ever-present world of online existence would have been easy for me. But it would have been familiar, perhaps dull; and it might have been a lie. The world does not need another article on the miracle of online reality; what we need, what I need, what this whole delirious, interconnected, global community of a world needs is a little reality check.

To some extent the following scenario will be misleading. There *are* flat rate online services (Netcom for one) which offer significant connectivity for a measly seventeen dollars a month. But I'm interested in the activities and behavior of the private service users who will soon comprise a vast majority of online citizens. Furthermore, let's face facts, the U.S. government by and large foots the bill for the Internet, through maintaining the structural (hardware) backbone, including, among other things, funding to major universities. As surely as the Department of Defense started this whole thing, AT&T or Ted Turner is going to end up running it so I don't think it's too unrealistic to take a look at the Net as it exists in its commercial form[1] in order to expose some of the realities lurking behind the regurgitated media rhetoric and the religious fanaticism of net junkies.

Time and Money

The average person, the normal human, J. Individual, has an income. Big or small, how much of J. Individual's income is going to be spent on Computer Connectivity? Does 120 dollars a month sound reasonable? Well, you may

---

[1]Techno concession: I know that the big three commercial services are not considered part of the Internet proper, but they (Prodigy, CompuServe and AOL) are rapidly adding real Net access and considering AOL just bought out Netcom...well, just read the article.

find that number a bit too steep for your pocket book, but the brutal fact is that 120 dollars is a "reasonable" amount to spend on monthly connectivity. The major online services have a monthly service charge of approximately $15. Fifteen dollars to join the global community, communicate with a diverse group of people, and access the world's largest repository of knowledge since the Alexandrian Library does not seem unreasonable, does it? But don't overlook the average per-hour connectivity rate of an additional $3 (which can skyrocket upwards of $10, depending on your modem speed and service). You might think that you are a crack whiz with your communications software—that you are rigorous and stringent and never, ever respond to e-mail or a forum while online, that you always use your capture functions and create macros, but let me tell you that no one, and I repeat, no one, is capable of logging on this efficiently every time. Thirty hours per month is a realistic estimate for online time spent by a single user engaging in activities beyond primitive e-mail. Now consider that the average, one-step-above-complete neophyte user has at least two distinct BBS accounts, and do the math: Total Monthly Cost: $120. Most likely, that's already more than the combined cost of your utility bills. How many people are prepared to double their monthly bills for the sole purpose of connectivity?

In case you think thirty hours a month is an outrageous estimate, think of it in terms of television. (OK, so you don't own a television, well, goody-for-you—imagine that you do!) Thirty hours, is, quite obviously, one hour a day. That's not so much. Thirty hours a month in front of a television is simply the evening news plus a weekly *Seinfeld/Frasier* hour. Thirty hours a month is less time than the average car-phone owner spends on the phone while commuting. Even a conscientious geek, logging on for e-mail and the up-to-the-minute news that only the net services can provide is probably going to spend thirty hours a month online. And, let's be truthful here, thirty hours a month ignores shareware downloads, computer illiteracy, real-time chatting, interactive game playing and any serious forum following, which by nature entail a significant amount of scrolling and/or downloading time.

If you are really and truly going to use the net services to connect with the global community, the hourly charges are going to add up pretty quickly. Take out a piece of paper, pretend you're writing a check, and print out "One hundred and twenty dollars—" and tell me again, how diverse is the online community?

That scenario aside, let's pretend that you're single, that you don't have children, that you rarely leave the house, that you don't have a TV and that money is not an issue. Meaning, pretend for a moment that you have as much time and as much money to spend online as you damn-well want. What do you actually do online?

Well, you download some cool shareware, you post technical questions in the computer user group forums, you check your stocks, you read the news and maybe some reviews—Hey, you've already passed that thirty hour limit! But, of course, since "computer networks make it easy to reach out and touch strangers who share a particular obsession or concern," you are participating in the online forums, discussion groups, and conferences.

Let's review the structure of forums. For the purposes of this essay, we will examine the smallest of the major user-friendly commercial services—America Online (AOL). There is no precise statistic available (at least none that the company will reveal—you have to do the research by Hand!!!) on exactly how many subject-specific discussion areas (folders) exist on AOL. Any online service is going to have zillions of posts pertaining to computer usage (e.g., the computer games area of AOL breaks into five hundred separate topics with over 100,000 individual posts), so let's look at a less popular area: the "Lifestyles and Interests" department.

For starters, there are fifty-seven initial categories within the Lifestyle and Interests area. One of these categories is Ham Radio. Ham Radio? How can there possibly be 5,909[2] separate, individual posts about Ham Radio? [There are] 5,865 postings in the Biking (and that's just bicycles, not motorcycles) category. Genealogy—22,525 posts. The Gay and Lesbian category is slightly more substantial—36,333 posts. There are five separate categories for political and issue discussion. The big catch-all topic area, The Exchange, has over 100,000 posts. Basically, service wide (on the smallest service, remember) there are over a million posts.

So, you want to communicate with other people, join the online revolution, but obviously you can't wade through everything that's being discussed—you need to decide which topics interest you, which folders to browse. Within The Exchange alone (one of fifty-seven subdivisions within one of another fifty higher divisions) there are 1,492 separate topic-specific folders—each containing a rough average of fifty posts, but with many containing close to four hundred. (Note: AOL automatically empties folders when their post totals reach four hundred, so total post numbers do not reflect the overall historical totals for a given topic. Sometimes the posting is so frequent that the "shelf life" of a given post is no more than four weeks.)

So, there you are, J. Individual, ready to start interacting with folks, sharing stories and communicating. You have narrowed yourself into a single folder, three tiers down in the AOL hierarchy, and now you must choose between nearly fifteen hundred folders. Of course, once you choose a few of these folders, you will then have to read all the posts in order to catch up, be current, and not merely repeat a previous post.

A polite post is no more than two paragraphs long (a screenful of text, which obviously has a number of intellectually negative implications). Let's say you choose ten folders (out of fifteen hundred). Each folder contains an

---

[2]Statistics obtained in June 1994. Most of these numbers have increased by at least 20 percent since that time, owing to all the Internet hoopla in the media, the consumer desire to be "wired" as painlessly as possible, and AOL's guerrilla marketing tactics.

average of fifty posts. Five hundred posts, at, say, one paragraph each, and you're now looking at the equivalent of a two hundred page book.

Enough with the stats. Let me back up a minute and present you with some very disturbing, but rational, assumptions. J. Individual wants to join the online revolution, to connect and communicate. But, J. Individual is not going to read all one million posts on AOL. (After all, J. Individual has a second online service.) Exercising choice is J. Individual's god-given right as an American, and, by gosh, J. Individual is going to make some decisions. So, J. Individual is going to ignore all the support groups—after all, J. is a normal, well-adjusted person, and all of J.'s friends are normal, well-adjusted individuals. What does J. need to know about alcoholism or incest victims? J. Individual is white. So, J. Individual is going to ignore all the multicultural folders. J. couldn't give a hoot about gender issues; does not want to discuss religion or philosophy. Ultimately, J. Individual does not engage in topics which do not interest J. Individual. So, who is J. meeting? Why, people who are *just like* J.

J. Individual has now joined the electronic community. Surfed the Net. Found some friends. *Tuned in, turned on, and geeked out.* Traveled the Information Highway and, just off to the left of that great Infobahn, J. Individual has settled into an electronic suburb.

Are any of us so very different from J. Individual? It's my time and my money and I am not going to waste any of it reading posts by disgruntled Robert-Bly drum-beating men's-movement boys who think that they should have some say over whether or not I choose to carry a child to term simply because a condom broke. I know where I stand. I'm an adult. I know what's up and I am not going to waste my money arguing with a bunch of Neanderthals.

Oh yeah; I am so connected, so enlightened, so open to the opposing viewpoint. I'm out there, meeting all kinds of people from different economic backgrounds (who have $120 a month to burn), from all religions (yeah, right, like anyone actually discusses religion anymore from a user-standpoint), from all kinds of different ethnic backgrounds and with all kinds of sexual orientations (as if any of this ever comes up outside of the appropriate topic folder).

People are drawn to topics and folders that interest them and therefore people will only meet people who are likewise interested in the same topics in the same folders. Rarely does anyone venture into a random folder just to see what others (The Others?) are talking about. This magazine being what it is, I can assume that the average reader will most likely not be as narrow-minded as the average white collar worker out in the burbs—but still, I think you and I are participating in the wide, wide world of online existence only insofar as our already existing interests and prejudices dictate.

Basically, between the monetary constraints and the sheer number of topics and individual posts, the great Information Highway is not a place where you will enter an "amazing web of new people, places, and ideas." One does not encounter people from "all walks of life" because there are too many people and too many folders. Diversity might be out there (and personally I don't think it is), but the simple fact is that the average person will not encounter it because with one brain, one job, one partner, one family, and one life, no one has the time!

Just in case these arguments based on time and money aren't completely convincing, let me bring up a historical reference. Please take another look at the opening quote of this essay from *Scientific American*. Featured in their "50 Years Ago Today" column, where you read "computer networks," the original quote contained the word *television*. Amusing, isn't it?

Moving beyond the practical obstacles mentioned above, let's assume that the Internet is the functional, incredible information tool that everyone says it is. Are we really prepared to use it?

Who, What, Where, When, and Why?

School trained us to produce answers. It didn't matter if your answer was right or wrong, the fact is that you did the answering while the teacher was the one asking the questions, writing down the equations, handing out the topics. You probably think that you came up with your own questions in college. But did you? Every class had its theme, its reading list, its issues; you chose topics for papers and projects keeping within the context set by your professors and the academic environment. Again, you were given questions, perhaps more thinly disguised than the questions posed to you in fourth grade, but questions nevertheless. And you answered them. Even people focusing on independent studies and those pursuing higher degrees, still do very little asking, simply because the more you study, the more questions there seem to be, patiently waiting for you to discover and answer them.

These questions exist because any contextual reality poses questions. The context in which you exist defines the question, as much as it defines the answer. School is a limited context. Even life is a limited context. Well, life was a limited context until this Information Highway thing happened to us. Maybe you think that this Infobahn is fabulous; fabulous because all that information is out there waiting to be restructured by you into those answers. School will be easier. Life will be easier. A simple tap-tap-tap on the ol' keyboard brings those answers out of the woodwork and off the Net into the privacy of your own home.

But this Information Highway is a two-way street and as it brings the world into your home it brings you out into the world. In a world filled with a billion answers just waiting to be questioned, expect that you are rapidly losing a grip on your familiar context. This loss of context makes the task of formulating a coherent question next to impossible.

The questions aren't out there and they never will be. You must make them.

Pure information has no meaning. I would venture to assert that a pure fact has no meaning; no meaning, that is, without the context which every question implies. In less than fifteen minutes I could find out how much rain fell last year in Uzbekistan, but that fact, that answer, has no meaning for me because I don't have or imagine or know the context in which the question is meaningful.

No one ever taught me how to ask a question. I answered other people's questions, received a diploma, and now I have an education. I can tell you what I learned, and what I know. I can quantify and qualify the trivia which comprises my knowledge. But I can't do that with my ignorance. Ignorance, being traditionally "bad," is just lumped together and I have little or no skills for sorting through the vast territory of what I don't know. I have an awareness of it—but only in the sense that I am aware of what I don't know about the topics which I already know something about in the first place.

What I mean to say is, I don't know what I don't know about, say, miners in China because I don't know anything about China, or what kinds of minerals they have, or where the minerals are, or the nature of mining as a whole. Worse yet, I don't have a very clear sense of whether or not these would be beneficial, useful, enlightening things for me to know. I have little sense of what questions are important enough for me to ask, so I don't know what answers, what information to seek out on the Internet.

In this light, it would seem that a massive amount of self-awareness is a prerequisite for using the Internet as an information source—and very few people are remotely prepared for this task. I believe that most people would simply panic in the face of their own ignorance and entrench themselves even more firmly into the black holes of their existing beliefs and prejudices. The information is certainly out there, but whether or not any of us can actually learn anything from it remains to be seen.

Fly, Words, and Be Free

The issues pertaining to time, money, and the fundamental usefulness of pure information are fairly straightforward when contrasted with the issues raised by e-mail. E-mail is the first hook and the last defense for the Internet and computer-mediated communication. I would like to reiterate that I am by no means a Luddite when it comes to computer technology; in fact, because of this I may be unqualified to discuss, or even grasp, the dark side of electronic communication in the form of e-mail. The general quality of e-mail sent by one's three-dimensional[3] friends and family is short, usually funny, and almost completely devoid of thoughtful communication. I do not know if this is a result of the fact that your 3-D friends already "know" you, and therefore brief quips are somehow more revealing (as they reflect an immediate mental/emotional state) than long, factual exposés, or if this brevity is a result of the medium itself. I do not know if I, personally, will ever be able to sort this out, owing to the nature of my friends, the majority of whom are, when all is said and done (unlike myself, I might add), writers.

Writers have a reverence for pen and paper which does not carry over well into ASCII. There is no glorified history for ASCII exchange and perhaps because of this fact my friends do not treat the medium as they would a handwritten letter. Ultimately, there is very little to romanticize about e-mail. There is decidedly a lack of sensuality, and perhaps some lack of realism. There is an undeniable connection between writers and their written (literally) words. This connection is transferred via a paper letter in a way that can never be transferable electronically. A handwritten letter is physically touched by both the sender and the receiver. When I receive an electronic missive, I receive only an impression of the mind, but when I receive a handwritten letter, I receive a piece, a moment, of another's physical (real?) existence; I possess, I own, that letter, those words, that moment.

Certainly there is a near-mystical utopianism to the lack of ownership of electronic words. There is probably even an evolution of untold consequences. Personally, I do not think, as so many do, that a great democracy of thought is upon us—but there is a change, as ownership slips away. While this is somewhat exciting, or at least intriguing, insofar as public communication goes, it is sad for private correspondence. To abuse a well-known philosopher, there is a leveling taking place: The Internet and the computer medium render a public posting on the nature of footwear and a private letter on the nature of one's life in the same format, and to some extent this places both on the same level. I cannot help but think that there is something negative in this.

Accessibility is another major issue in the e-mail/handwritten letter debate. Text sent via the Net is instantly accessible, but it is accessible only in a temporal sense. E-text is inaccessible in its lack of presence, in its lack of objective physicality. Even beyond this, the speed and omnipresence of the connection can blind one to the fact that the author/writer/friend is not physically accessible. We might think we are all connected, like an AT&T commercial, but on what level are we connecting? A handwritten letter reminds you of the

---

[3]So sue me for being a nerd. Personally, I find referring to friends one has made outside of the cyber world as one's "real" friends, or one's "objective" friends, to be insulting and inaccurate. Certainly one's cyber friends are three-dimensional in a final sense, but the "3-D" adjective is about the only term I have come up with which doesn't carry the negative judgmental weight of other terms so over-used in European philosophy. Feel free to write me if you've got a better suggestion. Better yet, write me in the appropriate context: flox@netcom.com.

writer's physical existence, and therefore reminds you of their physical absence; it reminds you that there is a critical, crucial component of their very nature which is not accessible to you; e-mail makes us forget the importance of physicality and plays into our modern belief in the importance of time.

Finally, for me, there is a subtle and terrible irony lurking within the Net: The Net, despite its speed, its exchange, ultimately reeks of stasis. In negating physical distance, the immediacy of electronic transfers devalues movement and the journey. In one minute a thought is in my head, and the next minute it is typed out, sent, read, and in your head. The exchange may be present, but the journey is imperceptible. The Infobahn hype would have us believe that this phenomenon is a fastpaced dynamic exchange, but the feeling, when you've been at it long enough, is that this exchange of ideas lacks movement. Lacking movement and the journey, to me it loses all value.

Maybe this is prejudice. Words are not wine, they do not necessarily require age to improve them. Furthermore, I have always hated the concept that Art comes only out of struggle and suffering. So, to say that e-mail words are weaker somehow because of the nature, or lack, of their journey, is to romanticize the struggle. I suppose I am anthropomorphizing text too much—but I somehow sense that one works harder to endow one's handwritten words with a certain strength, a certain soul, simply because those things are necessary in order to survive a journey. The ease of the e-mail journey means that your words don't need to be as well prepared, or as well equipped.

Electronic missives lack time, space, embodiment, and history (in the sense of a collection of experiences). Lacking all these things, an electronic missive is almost in complete opposition to my existence and I can't help but wonder what, if anything, I am communicating.

COMING TO TERMS

As you read "The Internet is Four Inches Tall," we would like you to consider the reasons why Kadi might have given her discussion about the information superhighway this unusual title. In your journal write about your take on, your explanation of this title. If you could give an alternative title to Kadi's essay, what would it be? How does Kadi's title suggest what her attitude towards or position on the Internet might be? As a way of exploring meaning in this title, you might examine those places in the essay where Kadi talks about how access to and use of the Internet has or has not changed her life.

READING AS A WRITER

One of the reasons why we've asked you to read this essay is that Kadi's work raises several important issues or problems concerning what makes for effective argument in writing. In presenting her position on whether the Internet has lived up to the expectations, promises, and power to transform lives, Kadi not only presents her position but also presents conflicting positions, viewpoints, and counter-arguments as well. We think it's difficult to read Kadi's discussion as simply as a neat demonstration of the pro and con sides of an argument. Kadi complicates this traditional approach to argument by presenting multiple positions or points of view which don't quite fit a simple dichotomy of pro and con or "I'm for it" or "I'm against it." Her positions on issues and the conflicting positions she represents appear to shift or move, at times, through a process of careful qualification. No position on an issue seems to be completely pro or con. One might say that this quality is just poor writing. We read this quality in another way, however. It seems that Kadi is using writing to do more than report on things already decided; instead, the discussion takes a more tentative tone, almost like an investigation in progress.

We would like you to analyze Kadi's argument as a journal project. First, you'll need to identify and list the issues that comprise this complex question of the Internet, issues such as education, the ownership of writing, and the promotion of diversity. These are only a few of the issues Kadi grapples with. Then you'll need to show Kadi's position on these issues as well as any conflicting positions by finding, quoting, or paraphrasing appropriate places in the discussion. Finally, we'd like you to respond to a set of journal questions: How successful is Kadi in fairly presenting multiple positions on an issue? How would you weigh the success or effectiveness of an argument which considers multiple points of view?

WRITING CRITICISM

At one point in her essay, Kadi reveals that she has deceived her readers, caught us off guard by replacing some of the words in one of the headnotes to her essay. (Headnotes are relevant quotations some writers use to introduce a piece of writing.) Kadi tells us that she has replaced the word *television* with *computer networking* in the original passage or head note from *Scientific American:*

> Computer networking *offers the soundest basis for world peace that has yet been presented. Peace must be created on the bulwark of understanding. International television networks [computer networks] will knit together the peoples of the world in bonds of mutual respect; its possibilities are vast, indeed*
>
> From *50 Years ago Today*

Kadi's response to this ruse is "Amusing, isn't it." We think that Kadi, however, may have something serious if not disturbing in mind here, especially in light of the other head notes that introduce her discussion of the Internet. What appears to link these head notes together is how each speaks about the promise of technology, the changes or transformations that will result from the Internet. In this context, Kadi's substitution of computer networking for television in the *Scientific American* passage seems to register more as a cautionary tale than as some written sleight-of-hand. Technological developments are often heralded as part of a brave new world of progress and change. While Kadi chose to replace television in the passage, we think other substitutions would be interesting to try, such as the printing press, telephone, telegraph, motion picture, satellite communication, radio, nuclear energy, or the internal combustion engine. Or as quaint as it might sound today, when the public address system was introduced into U.S. public schools in the late 1920s, it was praised as a revolutionary way to foster a sense of community among school children.

We would like you to consider this question of how the promise of technology is fulfilled or broken but specifically from the vantage point of popular culture. Thinking about Kadi's essay as a place to begin, we want you to select some popular culture medium—television, radio, film, the Internet, newspaper, magazine, print, or televised advertisement—and then present your own argument on the ways in which this medium has created, is creating, or might create a sense of national or international sense of community, progress, and change. As part of this assignment, you'll need to discuss specific examples for one of these media. You will also need to identify the important issues in your argument, which might include but are not limited to revising education, promoting understanding, encouraging democratic thinking or participation, exploring diversity, or breaking down barriers to communication.

Some places to start: Music sometimes is called the international language, a medium where no interpreter is needed. How might the broadcasting or recording of rock and roll, rap, folk, or country-western music fulfill the promise of technology in this medium? In what ways might national or international chat rooms on the Internet, radio or television talk shows, or newspapers or magazines with national or international readerships work out the promise of technology? How might films transform the way one group of people see and interact with another? Finally, as you select a medium, identify issues, provide examples or illustrations of that medium, we also want you to enact a strategy of fairly representing multiple points of view on an issue in your argument. What would be the opposition to what you have to say? How would you respond to the counter-argument?

The Malling of America

An Inside Look at the Great Consumer Paradise

William Severini Kowinski

Four survivors of the apocalypse are searching by helicopter for safety and sustenance. My God! What'll we do? Where will we go? Then, just below, they see it, spread out over one hundred acres, a million square feet of sheltered food and clothing, not to mention variable-intensity massagers, quick-diet books, Stayfree Maxi-pads, rat poison, hunting rifles, and glittering panels of Pac-Man and pinball, all enclosed in a single climate-controlled fortress complete with trees, fountains, and neon. Safe at last! Home free! The biggest, best-equipped fallout shelter imaginable, the consumer culture's Eden, the post-urban cradle, the womb, the home, the *mall.*

So begins George Romero's cult horror film *Dawn of the Dead,* the first zombie movie to be shot in a shopping mall. In the movie, the mall is a seductive place. It not only attracts passing pilgrims and lulls them into staying, even though they have to battle murderous zombie shoppers, it has acted as a magnet for the zombies themselves ("This was an important place in their lives," one of the pilgrims says). Both groups have been seduced by the mall's products and its completeness—*one-stop living,* even for the living dead.

But this completeness is also demonstrated for us when the circle is closed and the movie about the mall's power is itself *shown* in a shopping mall. For among other things, the mall is a fortress of entertainment.

Dawn of the Dead was in fact shot in the middle of the night at Monroeville Mall in western Pennsylvania, not far from Greensburg and Greengate mall. But it was only after my travels were completed that I saw Monroeville Mall as something of a model for malls across America. It isn't exactly average— it's more that Monroeville is the essential mall. It has a couple of big department stores and a reasonable selection of shops. It has all the typical mall stuff—the bathroom supply store, the fast-food dispensaries—but it also has a good French bakery and café, and a pleasant Italian restaurant.

Monroeville Mall is surrounded by a complex of other buildings, which include hotels and the Expo Mart, a convention and exhibition center. Monroeville itself is less a town now than a new downtown spread along the highway. Inside, the mall has that compact sense of being an enclosed and efficient distributor of everything. You can play the lottery there, and get quasi-religious counseling from The Talk Shop, a kiosk in the center of the mall that also functions as a living mall directory (the person there will give you directions to heaven, or to Pup-A-Go-Go). For a while there was a row of storefronts that deserved to enter into mall mythology: Funland (the game arcade) followed by the John XXIII Chapel, followed by the Luv Pub. The chapel is gone now. So is the skating rink, which was the mall's most distinguishing feature and most important community asset before shortsighted business people decided the space was too valuable. Now there's a shop that says LUV in blinking, roguish pink neon; it describes itself as being "for dating…mating and celebrating." Sure enough, inside are evening gowns, see-through blouses, and slinky, skimpy nightgowns.

If this weren't entertaining enough, on one of my visits I saw a specific link to show business on display: a traveling "Hollywood on Tour" exhibit in a shop window. Hanging behind the glass were clothes from famous movies, including a black dress worn by Marilyn Monroe in *Some Like It Hot,* the robes Greta Garbo wore in *Queen Christina,* and the rumpled khaki trench coat Humphrey Bogart wore in the Paris train station scene of *Casablanca.* "Are they the real ones?" a teenaged girl asked her mother. The real clothes worn by the real actors in the real movies, portraying fantasy characters in fantasy stories? "They could have kept them up better," her mother harrumphed. "That cape is *filthy.*"

To me the mall always felt something like a movie, a fantasy environment where it was perfectly appropriate for Bogie's coat (or Rick's) to rest overnight in a dark display case. But the ties between the movies and the malls have become more explicit as both movie studios and mall developers begin to see themselves as components of the same business, and as partners behind the stage of The Retail Drama.

It began when studios sponsored a few modest movie promotions involving poster giveaways and look-alike contests in a few mall courts. Then suddenly movie people started to see what they had here. Most of the movie theaters in America are in and around shopping centers. "That's where it really starts—where the theaters are," said Martin Levy, who handled MCA-Universal's mall promotions for *E.T.: The Extra-Terrestrial*, "—and that's a major revolution in distribution in the last ten years."

Not only that, but malls were where the teenagers of America hung out, even when they weren't headed for the movies. "The malls are where the kids are," said Bill Minot, who placed promotions for *The Return of the Jedi* in more than a hundred malls across the country. "For theatergoing demographics, twelve to twenty-four years old, that's where you've got to be. It's the modern Main Street, it's where the action is."

Not only *that*, but the mall is where a movie's licensed tie-in products are sold: the books (the novelization of *Return of the Jedi* was the top paperback best seller of 1983 and *E.T.* was second the year before; picture books and how-the-movie-was-made books sell well, depending on the movie), the records (sound tracks for *Flashdance* and *Footloose* both became major hits), the video cassettes, games, clothes, novelties, and of course the toys. This is not an inconsiderable part of the movie business, either. While the first two *Star Wars* films grossed an estimated $600 million, they generated something like $2 billion in retail merchandise sales.

It wasn't long before both studios and mall managers caught onto the synergistic possibilities, to the propensity of kids to walk out of *Return of the Jedi* with Star Wars in their eyes and to drift down the mall court and right into toy stores, bookstores, and record stores, picking up picture albums, theme music, Princess Leia lunchboxes, and Ewok dolls along the way. The movie helps to sell the merchandise; the merchandise helps to sell the movie—in effect continuously advertising it. The amount and duration of the feedback energy can even be multiplied when the movie is part of a series; the continuing characters are kept alive in the public mind by the action figures on the shelves, while the sequels keep the merchandise current and add new characters to old favorites. In fact, the trend toward sequels probably owes a good deal to the mall and its merchandise.

As movie promotions in malls proved successful, the mall industry became enthusiastic. Some developers—notably the Hahn companies in California—helped organize such promos, and both sides looked forward to the imminent day when big studios could launch national campaigns in concert with big mall developers. As the studios became more sophisticated (*Footloose*, for example, was promoted in malls together with the styles of clothes worn by characters in the movies), malls themselves added their own ideas. "You can coordinate an entire mall-wide promotion on the movie theme," explained Stacy Batrich-Smith, then marketing director for the famous Sherman Oaks Galleria. "On *The Pirate Movie* we had a 'treasure chest full of values,' and our stores had *Pirate Movie* sales, and we decorated our windows on the pirate theme. It really ties your center together. You can display all the things that are pirate-oriented, like blouson tops with wide collars and cuffs, as well as having a pirate costume contest and really involving the whole community in the promotion."

But early on, both studios and the malls decided that only certain movies could be promoted. Mall management wouldn't risk offending some of its customers by pushing the R-rated films it nevertheless shows. Studios chose films with a strong youth appeal: either films with young stars (another mall-compatible trend in movies), stars who have young followings, or films that just have *stars,* as in *Star Wars* and *Star Trek,* the spacey special-effects movies. Of course these are also the movies most likely to be marketing tie-in merchandise.

As the relationship between malls and the movies becomes better understood, the studios have powerful reasons not just to promote mall-compatible films but to make movies that will specifically sell in malls. After all, a movie can be a big-city hit, but if it doesn't sell well in malls it isn't likely to be much of a success. "There are many cities in which a first-run quality film will only open in four or five theaters, and they will all be out in [suburban] communities, and those will be in shopping centers," said marketing consultant Martin Levy. Then there is the merchandising factor. "There is no question people look at scripts for their licensing potential," Lester Borden, general manager of Columbia Pictures Merchandising, told *The Wall Street Journal.* He added that some producers commission marketing studies of merchandising potential before deciding whether or not to make the picture.

All in all, it only makes good business sense that, as an advertising and promotion executive at one studio admitted, malls have a real influence on the choice of what films the studios will back.

In any case, mall theaters are the perfect place to show movies with a suburban setting and sensibility—such as *War Games* (in which a suburban teenager would have destroyed the world with his home computer except that he had to take out the garbage), *Poltergeist* (in which two suburban fathers engage in a western shootout using television channel changers), and *E.T.* (in which suburban children do not mug a small alien but harbor him and help him home)—which link them to the mall world. Like suburbia itself, the mall is clean, new, safe, yet gently fanciful and ironic—and everyone, as Moon Zappa says, in her "Valley Girl" song, is "super-super nice."

The movies are only one of the entertainment, information, and cultural media that involve the mall in a mutually lucrative feedback loop. Malls sell merchandise that ties in with television (Smurfs, Strawberry Shortcake) and rock music (Michael Jackson dolls), as well as being prime merchandisers of books and magazines, records and tapes, video equipment and cassettes. The same synergistic effect that happens with movies also occurs with products that begin their journey to profitability at another point. An exercise enthusiast, for example, can trot through the mall picking up the *Jane Fonda Workout Book,* the Workout record, the Workout video cassette, and the Fonda line of workout clothes, without a single negative vibe from the outside world falling on a single running shoe.

The relationships can get more complicated than that. They can even get fairly bizarre. The impact of Saturday morning television has become so strong that in one case, a product it advertised (the Cabbage Patch Kid doll) was the cause of hysteria and violence in shopping centers across America. At the same time, a network children's programming producer admitted on ABC's *Nightline* that unless a new show was based on an already familiar image—more often than not, a toy or video game character, such as Pac-Man—the networks simply wouldn't put that show on the air. Meanwhile the best-selling children's books in the mall bookstores were almost exclusively spin-offs of television shows, such as *The Smurfs* and *Sesame Street.*

Television's relationship to the mall isn't restricted to children. The merchandising of *Star Trek* stuff in malls while the old TV series was in syndication led to its revival as a series of motion pictures, which were in turn promoted in malls and seen in theaters there, and eventually sold there on video cassettes. Which led to more merchandise for the mall to sell.

Daytime television in particular has a mall-shopper audience, and so personal appearances by soap opera stars are always big draws in malls. Perhaps the strangest evidence is the success of a revived *Newlywed Game*, done just as it was for years on television but no longer televised—simply taken live to shopping malls around the country as a center-court performance. According to Bob Eubanks, its host in the mall as he was on television, it became the most successful shopping-center promotion in the United States.

Even the once-rebellious image of rock music has been transformed by acts that link themselves to commercial interests, including retail companies that sell their products almost exclusively in malls. One of the first such joint ventures paired Air Supply, a soft-rock group with a clean-cut image, and Jordache jeans. As part of this relationship, Air Supply appeared at a shopping mall in Tampa, Florida, first at a clothing store—which sold 2,500 pairs of jeans that day—and then at the mall's record store, where they autographed their albums, the entire stock of which sold out.

The synergy intensified further when rock music videos became popular (and available on cassettes in the mall). Those videos also began to set fashions in clothes, which became instantaneous trends as soon as the videos were shown on such national outlets as MTV on cable, and on the broadcast networks. Such mall clothing chains as the Merry Go Round made specific efforts to sell the clothes that appeared in the videos. Movies also picked up on the music video style, which led to such films as *Beat Street,* and to Beat Street sweatshirts and tennis shoes in the mall.

Some of these relationships were organized by conglomerates with businesses in several media which could package movies, spinoff books, records, and video. When such conglomerates begin to include mall developers, then the last loops will be closed and the mall's strategy of a controlled environment can be more extensively and intensively applied.

The malling of the movies and the media may have more consequences. Many of the movies, records, and books that successfully fit into the mall world have been the work of individual vision and artistic accomplishment. But if books get published (or written), or movies, TV shows, or records get made, based principally on the results of marketing studies and how well such "products" fit into the merchandising feedback system, then the media may find that they've taken on some of the mall's characteristics as a result of their lack of spontaneity, personal statement, or moral commitment. It may be the hearts and minds of America that next become malled.

Meanwhile the mall itself is becoming a movie star. After countless cameos as the location for crime or romance, and appearances in the background of TV commercials and man-on-the-mall interviews, it played major roles in two youth-oriented movies, *Fast Times at Ridgemont High* and *Valley Girl,* both filmed at the Sherman Oaks Galleria. Romero's *Dawn of the Dead* remains the definitive mall work so far, however, and the best movie scene in a shopping center is still the car chase through a Chicago mall in *The Blues Brothers* with John Belushi and Dan Aykroyd ("It's got everything!" Belushi remarks with uncharacteristic reverence as they demolish it).

So it seemed to me that it was only a matter of time before the mall would itself become the subject of a movie aimed at the mall constituency—the citizens of everybody's hometown. In the mall's theatrical setting, scenarios of such movies often came to me. For instance, *Princess Stacey,* a TV miniseries based on a steamy novel, a *Movie of the Week* shot in a glamorous urban mall...

Stacey's plump, hot-buttered curls were spilled across the pillow. Was she a television commercial for sheets? Not today, but sometimes. Stacey—rich, a star, and beautiful—stretched her slim athletic tawny limbs into the air that thrummed with a special signature, its monogram, its label. For this was Ritz air. A Ritz Hotel bed was the host for Stacey's high-priced body, now upright and snaking across the Ritz room to the Ritz shower.

But she was not alone in the Ritz. In the bed, or in the shower. His thick chest hair, matted only moments ago with the sweat of her impassioned curls was foamy with Ritz soap. They laughed and splashed together: Stacey, her new, rich, and famous psychiatrist lover, and the Ritz. The day was just beginning. She would shave his face; he would shave her legs. Soon they would shop for diamonds.

Outside was Chicago's cold: the traffic-blackened snow, the pollution-processed slush. Also noise, crime, poverty, and desperation. But here Stacey's curls will never be moistened or dirtied or insulted. She will shop with Chicago slime outside. She will move as an angel's breath from Ritz air to the conditioned environment and comfort-controlled splendor of this titanic seven-level paradise, all agleam and a wonder, itself a kind of diamond—a diamond even bigger than the Ritz. It is called Water Tower Place.

Who would want to rescue Princess Stacey from her Tower? No one but a cad wishing to chance nasty encounters with the dragons outside Eden. "Place" had such a ritzy sound, Stacey thought, satisfied.

By afternoon Stacey was at work. She danced exactly, minimally for the cameras, for a commercial interrupted by unscripted screams. It was her brother, methodically re-forming a troop of Girl Scouts into the contours and consistency of raspberries. Sometimes medical science fails, says her psychiatrist lover evenly. We try; we succeed. But sometimes, dammit, we fail.

Then the air was pierced by an eerier scream. Stacey, her hot-buttered curls awhirl, bansheed into her brother's twisted face. "This was my story!" Stacey cried. Now you're trying to turn it into yours—Bloody Murder Nightmare at Water Tower Place!

But better yet, a movie about the more traditional suburban center, with all the *sturm und schmaltz* of a big-screen, cameo-filled epic—*MALL!*, the story of flamboyant self-made entrepreneur Byron Lord, manager of Esplanade Square Bigdale Mall, who is under pressure from the ambitious streamlined on-site development rep, a mall lifer who has never worked outside of air conditioning. Lord's affair with the chic and worldly manager of Casual Corner is getting tempestuous when...crisis! The sprinkler system is shut down following the dousing of a minor fire (and the quarrelsome county supervisor who got in the way) but another fire erupts, threatening the first floor all the way from Broadway department store past The Gap to the Piercing Pagoda. Meanwhile the parking lot is heating up as a suburban motorcycle gang, the Mall Marauders, is mixing it with a band of Hare Krishnas who've been refused permission to jingle their bells outside. Security is diverted from monitoring middle-school truants and observing car thieves and sent to these two scenes, except for young rookie Mick O'Bannion, who is delivering a baby on the terazzo tile in front of Karmelkorn. Of course all this distraction pleases glassy-eyed Dan the Dealer as he peddles his dilithium crystals near the fountain across from Tennis Lady. *But* dope-crazed Nicole is threatening to throw herself from the second level to the central-court Recreational Vehicle display because her mother (currently shopping at Big Woman) doesn't understand her, and her stoned friends are too busy riding the glass elevator to care—except Robert, who has come to the mall every day on the pretext of playing air hockey just to see Nicole. *He's in the elevator; he looks up, horrified—Nicole is perched outside the shiny railing, outlined against the bright lights of the National Record Mart....*

COMING TO TERMS

William Kowinski is writing here about shopping malls and movies simultaneously. While U.S. cinema and television have often chosen the places where people shop and congregate—from the country or general store or the five-and-ten to the big city department store and suburban mall—as a setting for comedy, melodrama and even horror, Kowinski seems to see a stronger relationship between the mall and movies than just another setting for an occasional film. We'd like you to explore this complex cultural relationship which Kowinski begins to flesh out in "The Malling of America." Maybe a good place to start would be the *representational* connection between movie and mall. What films that are set in a shopping mall does Kowinski discuss? In your journal, list the names of any film mentioned and then explain how the film uses the mall as a setting. You might include here other films which Kowinski didn't mention or have appeared after his article was written. What group or groups of people do these films seem to feature? What are the characteristics that define this group or these groups? Beyond the representational, what other kinds of connections does Kowinski make between mall and movie? Go back through his text and mark key passages where Kowinski establishes connections. In your journal, then, try to explain the connections. For example, do films and malls share similar cultural attitudes to what is important in art or entertainment? What does Kowinski see as possible commercial connections? Connections in approaches to advertising or the merchandising of products? Similar interests in history? What other connections do you see between mall and movie that Kowinski may have overlooked?

READING AS A WRITER

We see a number of places in "The Malling of America" where Kowinski seems to shift his tone or his attitude towards his subject. At times, the tone appears to be humorous, crafting plays on words or relating unusual anecdotes. Then it moves to a level of careful academic analysis and talks from a position of intellectual distance. Other times, the tone is satirical or even cautionary, warning of possible and significant social, cultural, and artistic changes. We see this *tonal ambivalence* (for lack of a better phrase) not so much as a weakness or a loss of control, but rather as a kind of testimony to Kowinski's complex personal, cultural, and intellectual relationship to movies and malls. And this relationship seems to be a continually evolving one. Take a moment to look at Kowinski's title and respond to the following questions in your journal. How do you read "inside look" in the title? Does this phrase represent a play on words? In what way? Or does this phrase remind you of a certain kind of journalistic approach to a subject or a certain kind of media investigation? How do you read the word "paradise" in Kowinski's title? If you see multiple ways of reading this title, and choose one over another, how might that interpretive choice suggest your attitude towards Kowinski's article?

To determine a writer's tone often requires careful interpretation of a text since, as readers, we don't really know an author's intent. For example, we think Kowinski uses *satire* and *parody* quite successfully in "The Malling of America." Satire and parody might be understood as an often humorous treatment of something in a culture that appears with such frequency or repetition that it is said to have become conventionalized—tough-talking hosts circulating throughout a rapt audience, unusual (to say the least) guests, experts who although qualified don't seem to get what's going on—the television talk show has become a target of parody. We laugh at a parody of the talk show because as a cultural text it's become all too familiar—we seem to know the script before everyone on the program is even introduced. With this in mind, turn to the page in "The Malling of America" where Kowinski parodies the one-word titled, big budget, star cameo, disaster film which is set in a mall: Kowinski's title: *Mall!* Hollywood's titles: *Twister, Volcano, Titanic, Airport, Hurricane, Earthquake, Ants, Piranha....* Kowinski also parodies the other conventions of this genre of film: the plot and subplots and intrigues, the stock characters, the slowly developing but inevitable disaster, etc. In your journal, we'd like you to analyze Kowinski's parody and then sketch out your own parody of a media culture production—a television talk show, a rock concert, a sitcom, televised professional wrestling match, episode of a cable shopping channel, etc.—which takes as its setting a shopping mall. You would need to analyze this medium in terms of its conventions. But like Kowinski, you will also need to write about how this media event has a connection to the mall. And finally, although your parody might be humorous in tone, in what ways might a reader interpret your parody as a kind of serious social, aesthetic or cultural criticism?

WRITING CRITICISM

Usually we write about the media by looking at the media. But in his piece on malls, William Kowinski tries a different approach. He writes about the media—in this case, movies—by looking at the shopping mall, an environment that is linked to movies (just as movies are linked to malls). In this assignment, we'd like you to do what Kowinski does: go looking for the media by visiting the mall. Take a trip to a local mall—the bigger the better. As you walk around, take special notice of the connections you spot between what's present in the mall and what's presented to you by the media. (It's a good idea to re-read Kowinski's piece before you begin this work, and as you "read" the mall, you might want to take a few minutes to sit and take some notes on your observations.) Some of these connections are direct and rather obvious. As Kowinski notes, movies are shown in the malls, and tapes and disks are sold there. You should take note of these clear connections. But other links may be more subtle and indirect, requiring you to notice small things that never stood out before. Look around you. What do you see in the mall that reminds you of what you see in media? (Reconsider the parody you have written.) How does the look or feel of the mall atmosphere compare to the look and feel of different media experiences? What media experiences most resemble the experience of visiting the mall? In what ways might the mall be considered a medium itself? Pay particular attention to the various forms of advertising, promotion, and display that you see in the mall. What do you see here that reminds you of experiences with media?

When you get home, use your notes to draft a piece on the mall and the media. Think of yourself as continuing a critical project begun by Kowinski— that is, try in your writing to see how you can add to, revise, or challenge Kowinski's way of "reading" a mall. Ask yourself: What can we learn about the mall by seeing it as a media environment? What can we learn about media by looking for the media at the mall? And finally, as someone who is now in a critical position to understand the connections between malls and the media, how might this work be of significance in explaining to others the influence of media culture on our lives?

About a Salary or Reality?—Rap's Recurrent Conflict

Alan Light

In 1990, rap dominated headlines and the pop charts as never before. Large segments of the American public were introduced to rap—or at least forced to confront its existence for the first time—through a pack of unlikely and sometimes unseemly performers. The year started with the January release of Public Enemy's single "Welcome to the Terrordome," which prompted widespread accusations (in the wake of remarks made by Professor Griff, the group's "Minister of Information," that Jews are responsible for "the majority of wickedness that goes on across the globe") that rap's most politically outspoken and widely respected group was anti-Semitic. The obscenity arrest of 2 Live Crew in June filled news, talk shows, and editorial pages for weeks. The concurrent rise of graphic, violent "Gangster Rap" from such artists as Ice Cube and the Geto Boys stoked these fires, even if their brutal streetscapes often made for complex, visceral, and challenging records.

It's easy to vilify any or all of these artists. *Newsweek* lumped several of them together and ran a cover story (19 March 1990) entitled "Rap Rage," which proved to be a savage attack on the form (including, by some curious extension, the heavy metal band Guns 'n' Roses) as "ugly macho boasting and joking about anyone who hangs out on a different block," and as having taken "sex out of teenage culture, substituting brutal fantasies of penetration and destruction." Certainly, Luther Campbell and 2 Live Crew aren't exactly the First Amendment martyrs of the ACLU's dreams; their music consists of junior high-school locker-room fantasies set to monotonous, mighty uninspired beats. The "horror rap" of the Geto Boys is deliberately shocking, and songs such as "Mind of a Lunatic," in which the narrator slashes a woman's throat and has sex with her corpse, raise issues that are a long way from the Crew's doo-doo jokes. Public Enemy, for all its musical innovation and political insight, has an uncanny knack for talking its way deeper into trouble, and leader Chuck D.'s incomprehensible waffling during the Griff incident, first dismissing Griff, then breaking up the group, then reforming and announcing a "boycott of the music industry," was maddening and painful to watch.

But there was also a different side to the rap story that was at least as prominent in 1990. M. C. Hammer's harmless dance-pop *Please Hammer Don't Hurt 'Em* became the year's best-selling album and rap's biggest hit ever. It was finally displaced from the top of the charts by white superhunk Vanilla Ice's *To the Extreme*, which sold five million copies in twelve weeks, making it the fastest-moving record in any style in five years.

Hammer has been defended by the likes of Chuck D. for being rap's first real performer, a dancer/showman/business tycoon of the first order, but his simplistic regurgitation of hooks from familiar hits quickly wears thin. Vanilla Ice not only lacks Hammer's passable delivery, he also manufactured a none-too-convincing false autobiography to validate his appropriation of black culture and subsequent unprecedented commercial success. Both are given to a self-aggrandizement so far beyond their talents that the biggest problem is simply how annoying they are.

There's nothing criminal about bad music or even simple-mindedness. And it's nothing new for the most one-dimensional, reductive purveyors of a style to be the ones who cash in commercially. But the most unfortunate result of the year of Hammer, Ice, and the Crew is that it may have determined a perception of rap for the majority of America. Any definitions of rap formed by the millions of Americans introduced to it in the last year would probably (and, so to say, reasonably) center on a simplified analysis of the genre's basest cultural and sociological components and the most uninspired uses of its musical innovation.

If, in 1990, people new to rap gave it any thought at all, they would have concentrated on the crudeness of 2 Live Crew—who may have a constitutional right to be nasty, but there is no way around the ugliness of their lyrics. Newcomers might (understandably, given much of the mainstream press coverage) have dismissed Public Enemy, self-styled "prophets of rage," as mere traffickers in hate. Whatever one thinks of the controversial sampling process, in which pieces of existing records are isolated and digitally stored and then reconstructed as a kind of montage to form a new musical track, the derivative, obvious samples of Hammer and Ice represented the triumph of the technology at its worst. And rap diehards and novices all had to contend with the cheers of "go, white boy, go" as Vanilla Ice became the biggest star yet to emerge from this black-created style.

This has made for a lot of sociocultural analysis and interpretation, which is perfectly appropriate; rap is unarguably the most culturally significant style in pop, the genre that speaks most directly to and for its audience, full of complications, contradictions, and confusion. But what gets lost in this discussion, tragically, is that rap is also the single most creative, revolutionary approach to music and to music making that this generation has constructed.

The distance between M. C. Hammer and the Geto Boys seems to be the final flowering of a contradiction built into rap from its very beginning. Though the polarity may seem inexplicable—is it progress now that we hear not just "how can both be called music?" but "how can both be called rap?"—it is actually a fairly inevitable progression that has been building for years. We can be sure that rap artists are more aware than anyone of the current condition; a press release touting the new group Downtown Science quotes rapper Bosco Money's definitive statement of purpose for 1990: "Our crusade is to fuse street credibility with a song that's accessible to the mainstream."

Rap, however, has seen problematic moments before, times when people were sure it was dead or played out or irrelevant. It has, with relatively alarming frequency for such a young art form, repeatedly found itself at seemingly impassable stylistic crossroads. Off and on for years, it has been torn between the apparently irreconcilable agendas inherent in such a radical pop creation. Throughout the decade of its recorded existence, though, rap has always emerged stronger due to its openness to musical and technological innovation and diversification.

We should not forget that 1990 was also the year that the bi-coastal post-Parliament-Funkadelic loonies in Digital Underground had a Top 10 hit with "The Humpty Dance," and Oakland's gratuitously nasty Too Short released his second gold-selling album. L. L. Cool J was booed at a Harlem rally for not being political enough, but then recorded an "old-school" style album that re-established him as a vital street force. It was a year when Public Enemy, so recently a strictly underground phenomenon, shipped over a million copies of the *Fear of a Black Planet* album; a year when the Gangster scene redefined "graphic" and Hammer and Ice redefined "crossover." It was a year when there could be no more arguing that rap sounded like any one thing, or even any two things. Despite the increasing perception among insiders that rap is a sellout or, to outsiders, that it's just a cheap, vulgar style, it is this expansion and flexibility that is the real story of rap's decade on record.

Five years before Vanilla Ice, rap had made its initial breakthrough into mainstream culture, but it looked like it had gone as far as it was going to go. Rap records had managed to cross over to the pop charts as novelty hits on occasion, beginning in 1980 with the startling Top 40 success of the Sugarhill Gang's "Rapper's Delight." By 1985, hip-hop culture, rap, graffiti, and especially breakdancing were receiving widespread popular attention; after a brief breakdancing sequence drew notice in the movie *Flashdance*, other movies like *Breakin'* and *Beat Street* quickly followed, and breakdancing became a staple in young American culture, turning up everywhere from TV commercials to bar mitzvah parties. The soundtrack to all this head spinning and 7-Up drinking was an electrosynthesized hip-hop, with lots of outer space imagery, sounds and styles borrowed from the year's other rage, video games, and featuring a burbling, steady electronic pulse (Jonzun Crew's "Pack Jam" is the most obvious example, but the style's real masterworks were Afrika Bambaataa and the Soulsonic Force's "Planet Rock" or "Looking for the Perfect Beat"). Maybe it was the rapidity of its appropriation by Hollywood and Madison Avenue, maybe it was just too obviously contrived to last, but the breakdancing sound and look quickly fell into self-parody—much of it was dated by the time it even got released.

There was, at the same time, another strain of rap that wasn't quite as widely disseminated as the break-beat sound but would prove to have a more lasting impact. Grandmaster Flash and the Furious Five's 1982 hit "The Message" offered a depiction of ghetto life so compelling and a hook so undeniable ("Don't push me/Cause I'm close to the edge") that it became a pop hit in spite of the fact that it was the hardest-hitting rap that had ever been recorded. (Harry Allen, Public Enemy's "media assassin," recently offered an alternate view when he wrote that the song was the "type of record white people would appreciate because of its picturesque rendition of the ghetto.") In its wake, "socially conscious" rap became the rage. Plenty of junk, including Flash's top rapper Melle Mel's follow-ups "New York, New York" (which was listenable if awfully familiar) and "Survival" (which was just overblown overkill), was released in attempts to cash in on this relevance chic, but the most important and influential rappers to date were also a product of "The Message." Run-D.M.C.'s first album came out in 1984.

With its stripped-to-the-bone sound, crunching beats, and such accessible but street-smart narratives as "Hard Times" and "It's Like That," in addition to more conventional rap boasting, the album *Run-D.M.C.* was the real breakthrough of the underground. One irony was that the three members of the group came from black middle-class homes in Hollis, Queens, and their "B-Boy" gangster wear was just as much a costume as the cowboy hats and space suits of Grandmaster Flash or Afrika Bambaataa. "With Run-D.M.C. and the suburban rap school," says producer and co-founder of Def Jam Records Rick Rubin, "we looked at that [ghetto] life like a cowboy movie. To us, it was like Clint Eastwood. We could talk about those things because they weren't that close to home." This "inside" look at the street, though, was enough to impress an unprecedented, widespread white audience when combined with Run and D.M.C.'s rapid-fire, dynamic delivery. To the street kids, the group's sound was loud and abrasive—and therefore exciting and identifiable. "It wasn't a hard record to sell to the kids on the street," says Def Jam's president (and Run's brother) Russell Simmons. "But it was a hard record to sell to a producer or the rest of the industry. There was no standard to compare it to."

Run-D.M.C. was the first great rap album. Even as David Toop was writing in *The Rap Attack* (now out of print, Toop's book is one of very few decent extended writings on hip-hop) that "rap is music for 12-inch singles," *Run-D.M.C.*'s release demonstrated that the form could be sustained for longer than six or seven minutes. The next year, though, as breakdance fever continued to sweep the nation, the group released *King of Rock*. It sold over a million copies, but the group sounded tired. The title song and its video were funny, crunching classics, but the in-your-face rhymes and jagged scratching style seemed stale. The tracks which experimented with a rap-reggae mix were flat and unconvincing. And though seventeen-year-old hotshot L. L. Cool J was the first and best rapper to pick up on the Run-D.M.C. style in his spectacular debut *Radio,* he was obviously second-generation, not an innovator. It was 1985, rap had been split down the middle into two camps, and it was becoming harder to care about either one.

What has sustained rap for its (ten-year? fifteen? decades-old? the argument continues...) history is its ability to rise to the challenge of its limitations. Just at the points at which the form has seemed doomed or at a dead end, something or someone has appeared to give it a new direction. If 1985 was hip-hop's most desperate hour (the rap magazine *The Source* recently referred to this era as "The Hip-Hop Drought"), in 1986 it rose triumphant from the ashes. Run-D.M.C. was at the forefront of this renaissance, but it was the group's producer Rick Rubin who inspired it. He thought that a cover, a remake of a familiar song, would be a way to make the group's new album a "more progressive" record. "I went through my record collection," Rubin recalls, "and came up with [Aerosmith's] 'Walk This Way,' which really excited me because the way the vocals worked it was already pretty much a rap song. It would be cool to have a high-profile rap group doing a traditional rock & roll song, and really not having to change that much."

Rubin brought Aerosmith's lead singer Steven Tyler and guitarist Joe Perry into the studio with Run-D.M.C. and the resultant "Walk This Way" collaboration made rap palatable to white, suburban youth across the country. It reached Number 4 on the pop singles charts and catapulted the *Raising Hell* album to multimillion selling heights. The recent massive successes of Michael Jackson and Prince had been exceptional crossover developments, opening up MTV to black artists and reintroducing these artists to pop radio. "Walk This Way" drove those opportunities home, and it did so without compromising what made rap so special, so vibrant; the beats and aggressive, declamatory vocals made it an undeniable rap record, heavy-metal guitars or no.

Rap was established as a viable pop form, at least as long as its connections to the traditional rock & roll spirit were made explicit. Nowhere was that connection more obvious than with the next major development in hip-hop. Hot on the heels of *Raising Hell* came the Beastie Boys' *Licensed to Ill,* at the time the biggest-selling debut album in history. The Beasties were three white Jewish New Yorkers who had played together in a punk-hardcore band before discovering rap and hooking up with Rick Rubin. For them, rap wasn't a way to establish racial pride or document hard times: it was a last vestige of rock & roll rebellion, a vaguely threatening, deliberately antagonistic use for their bratty, whiny voices.

Licensed to Ill sometimes rang hollow when the Beastie Boys rhymed about a criminal, decadent life beyond the fantasies of their listeners, but when it worked—as in their breakthrough hit "Fight for Your Right [to Party]"—you could hear young America laughing and screaming along. Their rhymes were simple and unsophisticated compared to those of Run or L. L. Cool J, but that made it easier for their audience of rap novices to follow. Besides, technical prowess wasn't what their yelling and carousing was about.

The strongest legacy of the album, however, was purely technical. *Licensed to Ill* introduced a newly expanded rap public to the concept of sampling. Until this time, rappers were backed up by DJs who would spin records, crosscutting between favorite beats on multiple turntables, scratching and cutting up breaks from other songs as a musical track by manually manipulating the records under the needle (the definitive scratch record was Grandmaster Flash's cataclysmic "Grandmaster Flash on the Wheels of Steel"). The advent of the digital sampler meant that a machine could isolate more precise snippets of a recording and loop or stitch them into a denser, more active backdrop. One drumbeat or a particularly funky James Brown exclamation could form the backbone of an entire track, or could just drift in for a split second and flesh out a track.

The Beasties were the perfect group to exploit this technology, because the musical accompaniment they sought was the sound they and Rubin had grown up with, heavy on Led Zeppelin riffs and TV show themes, and packed with in-jokes and fleeting goofy references. Along with a handful of dance singles, most notably "Pump Up the Volume" by a British production duo called M/A/R/R/S, the Beasties made this bottomless, careening flow of juxtaposed sound familiar to the world. After "Walk This Way" made suburban America a little more open to the idea of rap, the Beasties, with the novelty of their personalities, attitude, and sound (and, of course, their color), won huge battalions of teenagers over to the form. But just as significantly, they rewrote the rules regarding who could make this music and what, if anything, it was supposed to sound like.

Surprisingly, in the wake of the Beastie Boys, there was not a massive, immediate explosion of white rappers. In fact, within the rap community the Beasties were cause for backlash more than celebration, and the next critical development was a move toward music "blacker" than anything ever recorded. Public

Enemy emerged out of the hip-hop scene at Long Island's Adelphi University, encouraged by, once again, Rick Rubin. Their first album, 1987's *Yo, Bum Rush the Show,* didn't have the militant polemics that would characterize their later, more celebrated work, but it did have a sound and attitude as new to the listening public as the Beasties or Run or Flash had been. "When we came into the game, musicians said we're not making music, we're making noise," said producer Hank Shocklee, "so I said, 'Noise? You wanna hear noise?' I wanted to go out to be music's worst nightmare." With such open defiance, Public Enemy sculpted jarring, screeching musical tracks and wrote of the hostility they felt from and toward the white community. Public Enemy was offering an extension of rap's familiar outlaw pose, but they grounded it in the realities of contemporary urban life, with a sharp eye for detail and a brilliant sonic counterpoint that raised rap to a new level of sophistication.

If rap had crossed over, however tentatively, Public Enemy was not going to allow anyone to forget that this music came from the streets and that no rappers should feel any responsibility to take it away from there. Having found a pop audience and the kind of mass acceptance any pop form strives for from its inception, rap was now faced with the challenge of retaining its rebel stance and street-reality power. Public Enemy went on the road, opened for label-mates the Beastie Boys, and featured an onstage security force dressed in camouflage and bearing replicas of Uzis. They intimidated the young Beasties crowd and didn't really catch on with rap devotees until "Rebel without a Pause" exploded in urban black communities months later.

Rappers could now sell a million records without ever being picked up by pop radio or crossing over, and the artists were becoming aware of the kind of power that such independence meant. "Rap is black America's CNN," said Chuck D., and he would go on to claim that Public Enemy's goal was to inspire five thousand potential young leaders in the black community. This mindset ran through the work of Boogie Down Productions, who came out as hardcore gangster types (championing the honor of the Bronx in hip-hop's ongoing borough supremacy battles), but soon brought rapper KRS-One's intelligence and social concern up front in raps like "Stop the Violence." The Public Enemy mindset was there in the menacing, slow-flowing style of Eric B. and Rakim, who made the sampling of James Brown rhythm tracks and breaks the single most dominant sound in hip-hop. It was there even in the less revolutionary work of EPMD (the name stands for Erick and Parrish Making Dollars), hardly oppositional in their priorities, but uncompromisingly funky above all else.

It was an attitude that reached its fruition in 1988 when Public Enemy released *It Takes a Nation of Millions to Hold Us Back,* rap's most radical extended statement and still the finest album in the genre's history. Chuck D. castigated radio stations both black and white, fantasized about leading prison breaks, eulogized the Black Panthers, and thunderously expanded rap's range to anything and everything that crossed his mind. Shocklee and Company's tracks were merciless, annoying, and sinuous, using samples not for melodic hooks or one-dimensional rhythm tracks but as constantly shifting components in an impenetrably thick, dense collage. *It Takes a Nation* exploded onto city streets, college campuses, and booming car systems; it was simply inescapable for the entire summer. Jim Macomb wrote in the *Village Voice* that "it seemed that no one ever put in the cassette, no one ever turned it over, it was just on."

After that album, politics and outspokenness were de rigueur, simple boasting way out of style. Public Enemy's sound couldn't be copied—it still hasn't been rivaled—but that screech, once so foreign and aggressive, was soon filling dance floors in hits like Rob Base and DJ E-Z Rock's propulsive smash "It Takes Two." Public Enemy had thrown down the gauntlet, as much musical and technological as lyrical and political, and they could just wait and see when and how the hip-hop community would respond.

Public Enemy's artistic and commercial triumph crystallized two lessons that had been developing in hip-hop for some time. Number one: anything was now fair game for rap's subject matter. Number two: a rap album could sell a lot of copies without rock guitars. Rap had, after struggling with its built-in contradictions for several years, established its long-term presence as a pop style without being required to compromise anything. Terms like "underground" and "street" and "crossover" were becoming less clear as records that would never be allowed on pop radio were turning up regularly on the pop sales charts. It didn't take long for these successes sans radio and the expansion of rap's subject matter to sink in and to set in motion the forces that put rap in the unsettled, divided condition it faces today.

The most visible immediate result was geographic. Virtually all of the rappers mentioned thus far have been residents of New York City or its environs. In 1988, galvanized by Public Enemy's triumphs, the rest of the country began to compete. Most significantly, five young men from the Los Angeles district of Compton joined together and recorded their first single, "Dope Man," which may or may not have been financed by their founder Easy-E's drug sales profits. This group called themselves Niggas With Attitude (N.W.A), and demonstrated that the West Coast was in the rap game for real. Their only real precursor in Los Angeles was Ice-T, the self-proclaimed inventor of "crime rhyme" whose sharp, literate depictions of gang life were inspired by ghetto novelist Iceberg Slim. N.W.A took Ice-T's direct, balanced, and morally ambiguous war reporting to the next level with brutal, cinematically clear tales of fury and violence. The titles said it all: "Gangsta Gangsta" or even, in a gesture truly unprecedented for a record on the pop charts "[Fuck] tha Police."

Rap's single biggest pop hit in 1988 didn't come from New York or Los Angeles, however, but out of Philadelphia. DJ Jazzy Jeff and the Fresh Prince released a pleasant trifle chronicling teen suburban crises called "Parents Just Don't Understand," which hovered near the top of the pop charts all summer

(Fresh Prince went on to become the first rapper to star in a prime-time sit-com). *Yo! MTV Raps* debuted on the music-video network and not only gave hip-hop a national outlet, but quickly became the channel's highest-rated show. Miami's 2 Live Crew released *Move Somethin'*, which went platinum long before they became hip-hop's most famous free-speech crusaders, and cities such as Houston and Seattle were establishing active rap scenes.

A dichotomy was firmly in place—rappers knew that they could cross over to the pop charts with minimal effort, which made many feel an obliga-tion to be more graphic, attempt to prove their commitment to rap's street her-itage. As with the division first evident in 1985, rap had split into two camps with little common ground, and the polarization again led to a period of stag-nation. N.W.A's *Straight Outta Compton* simply defined a new subgenre—hard-core, gangster rap—so precisely that those who followed in their path sounded like warmed-over imitators. On the other side, like the breakdance electro-rap before it, poppy MTV-friendly hip-hop quickly grew stale. L.A.'s Tone-Loc, possessor of the smoothest, most laid-back drawl in rap history, re-leased "Wild Thing," the second best-selling single in history (right behind "We Are the World"), but it didn't feel like a revolution anymore.

Into the abyss in early 1989 jumped De La Soul, heralding the "DAISY Age" (stands for "da inner sound, y'all," and hippie comparisons only made them mad) and sampling Steely Dan, Johnny Cash, and French language instruction records on their debut album *3 Feet High and Rising*. After two full years of James Brown samples, with the occasional Parliament-Funkadelic break thrown in, it took De La Soul to remind everyone that sampling didn't have to mean just a simpler way to mix and cut records. It was a more radical structural development than that—sampling meant that rap could sound like anything. Michael Azerrad wrote in *Rolling Stone* that many saw the group as "the savior of a rap scene in danger of descending into self-parody." It's an unnecessarily sweeping state-ment—there were still first-rate hip-hop albums being released—but De La Soul did quickly and assuredly reenergize the form. It was the group's peculiar vision, their quirky intelligence and, just as significantly, their awareness of the true pos-sibilities of sampling technology that made possible such a leap forward.

In De La Soul's wake came a new look (colorful prints, baggy jeans, Africa medallions) and music from De La Soul's cohorts in the Native Tongues move-ment, including the Jungle Brothers and A Tribe Called Quest. The sound was varied, funny, and gentle, and the "Afrocentric" emphasis of the rhymes was warm, ecologically and socially concerned, and not too preachy. Soon there was also a whole new audience; De La Soul acquired a large collegiate follow-ing, drawn to their looser grooves, accessible hooks, and appealing goofiness. This was rap that was nonthreatening, politically correct, and didn't have the tainted feel of a crossover ploy.

In addition, De La Soul helped introduce Queen Latifah and Monie Love to the public, and thus proved instrumental in the emergence of a new generation of women rappers in 1989. Women had been part of rap from the beginning; the three women in Sequence comprised one of the first rap groups to sign a recording contract, and Salt-n-Pepa had several major hits, including "Push It," one of 1988's biggest dance hits. Latifah, however, was something new. In the face of ever-increasing graphic misogyny, she (and MC Lyte, Monie, and then in 1990 Yo-Yo and numerous others) offered a strong female alternative. For all of the attention given to 2 Live Crew or N.W.A's "A Bitch Iz a Bitch," rap was allowing a voice for women that was far more outspoken, far more progressive than anything found in other styles of pop music.

Ultimately, this new strain of headier hip-hop offered some kind of uncompromised middle ground between rap's pop and gangster sides, but it didn't slow down their ever-widening divergence. This "new school" became its own movement, while the two camps were far too entrenched to adapt the lessons of the DAISY age to their music. Los Angeles continued to dominate the East Coast in producing new sounds and styles. Young M. C., a University of Southern California economics major who had helped write Tone-Loc's hits, recorded his own raps that featured even more danceable grooves and middle-class lyric concerns, and became a star on his own. The members of N.W.A, meanwhile, took to the road with the threat of arrest hanging over their heads if they chose to perform "[Fuck] tha Police," and by the end of 1989 had received an ominous "warning letter" from the FBI. After a decade on record, the biggest challenge hip-hop faced was not survival, but avoiding overexposure and irreparable co-optation.

Hip-hop is first and foremost a pop form, seeking to make people dance and laugh and think, to make them listen and feel, and to sell records by doing so. From its early days, even before it became a recorded commodity, it was successful at these things—Russell Simmons recalls promoting rap parties with DJ Hollywood and drawing thousands of devoted New Yorkers years before rap made it to vinyl. "It's not about a salary/It's all about reality," rap N.W.A; even if they didn't claim that life "ain't nothing but bitches and money" on the same album, the fallacy would be clear. On a recent PBS rap special, San Francisco rapper Paris said that "[e]verybody gets into rap just to get the dollars or to get the fame."

At the same time, rap by definition has a political content: even when not explicitly issues-oriented, rap is about giving voice to a black community otherwise underrepresented, if not silent, in the mass media. It has always been and remains (despite the curse of pop potential) directly connected to the streets from which it came. It is still a basic assumption among the hip-hop community that rap speaks to real people in a real language about real things. As *Newsweek* and the 2 Live Crew arrest prove, rap still has the ability to provoke and infuriate. If there is an upside to this hostile response, it is that it verifies rap's credibility for the insider audience. If it's ultimately about a salary, it's still about reality as well. Asked why hip-hop continues to thrive, Run-D.M.C.'s DJ Jam

Master Jay replied "because for all those other musics you had to change or put on something to get into them. You don't have to do that for hip-hop."

At a certain level, these differences are irreconcilable. Since Run-D.M.C. and the Beastie Boys established rap's crossover potential, and Public Enemy demonstrated that pop sales didn't have to result from concessions to more conventional pop structures, the two strains have been forced to move further apart and to work, in many ways, at cross-purposes. It is a scenario familiar in the progression of rock & roll from renegade teen threat to TV commercial music. Perhaps more relevant, the situation is reminiscent of punk's inability to survive the trip from England, where it was a basic component of a radical life, to America, where it was the sound track to a fashionable life-style.

If this conflict is fundamental to all pop that is the product of youth culture, it is heightened immeasurably by rap's legitimately radical origins and intentions. But success with a wide, white audience need not be fatal to the genre. The rage directed by much of the rap community at Hammer and Vanilla Ice is ultimately unwarranted—if they make bad records, they're hardly the first, and if that's what hits, it's not going to take the more sophisticated listeners away.

Rap has thus far proven that it can retain a strong sense of where it comes from and how central those origins are to its purpose. If this has sometimes meant shock value for its own sake—which often seems the norm for recent Gangster records, such as N.W.A's *100 Miles and Runnin'*—and if that is as much a dead-end as Hammer's boring pop, the legacy of De La Soul is that there are other ways to work out rap's possibilities. Some of the best new groups, such as Main Source and Gang Starr, may have disappeared by the time this article appears, but they have been integrating melody, live instruments, and samples from less familiar sources into their tracks, learning from De La Soul and Digital Underground and company that hip-hop has other roads still left untrodden.

The paradigmatic hip-hop figure of 1990 would have to be Ice Cube, formerly N.W.A's main lyricist. He recorded *AmeriKKKa's Most Wanted* with Public Enemy's production team, the first real collaboration between the two coasts, and it ruled the streets for most of the year. The album's layered, crunching, impossibly dense sound set a new standard for rap production, the progression we've been waiting for since *It Takes a Nation of Millions to Hold Us Back.* Ice Cube's technical verbal prowess is astonishing: his razor-sharp imagery is cut up into complicated internal rhymes, then bounced over and across the beat, fluid but never predictable, like a topflight bebop soloist.

The content, though, is somewhat more troublesome. When rhyming about the harsh realities of ghetto life, Ice Cube is profane, powerful, and insightful. When writing about women—make that bitches, since he uses that a word a full sixty times on the album—things get more disturbing. He has defended "You Can't Fade Me," a first-person account of contemplating murderous revenge on a woman who falsely accuses him of fathering her expected child, by saying that he's just telling a story and illustrating that people really do think that way. It's a fair enough defense, but it's hard to believe that many listeners won't hear it as Ice Cube's own attitude. If part of rap's appeal is the "reality" of the rappers, their lack of constructed stage personae and distance from the audience, Ice Cube simply doesn't establish the constructedness suf ficiently to make the song's "objective" narrative effective.

But here's the surprise. At a press conference late in 1990, Ice Cube said that "a lot of people took mixed messages from my album, so I'm just going to have to try to make my writing clearer in the future." It's something rappers have always had a hard time doing—when a performance style is so rooted in boasting and competition, admitting that you might be wrong or even just imperfect is a risky matter. If Public Enemy had been willing to take such an attitude, of course, they could have handled the Griff affair much more gracefully. But if Ice Cube is sincere about improving his expression without compromising it (and his moving, somber antiviolence track "Dead Homiez" bodes well), he may have shown us the future. Like De La Soul's radical rewriting of sampling and hip-hop personae, gangster rap that moves beyond gore and shock is evidence that rap is not trapped in a dead-end dichotomy.

Writing about rap always has a certain dispatches-from-the-front-lines quality; sounds and styles change so fast that by the time any generalizations or predictions appear, they have often already been proven false. At any moment, a new rapper or a new attitude or a new technology may appear and the troubled times hip-hop faced in 1990 will be nothing but ancient history. This may be its first real struggle with middle age, but rap has never failed to reinvent itself whenever the need's been there.

COMING TO TERMS

"What has sustained rap...is its ability to rise to the challenge of its limitations. Just at the points at which the form has seemed doomed or at a dead end, something or someone has appeared to give it a new direction." Go back through Light's article and see if you can use these sentences as a starting point to map out the history of rap he offers. Sketch a brief outline of this history in your journal, and then, offer a response to the following questions in your journal. When and why, according to Light, did rap first head towards a "dead end"? Who or what gives a "new direction" to move in? What then proved to be the limits of this approach? How did it, in turn, give rise to a new sort of music? What does Light see as the future for rap music?

READING AS A WRITER

Once you have a sense of Light's history of rap music, we'd like you to analyze this history in some detail to see if you can identify the central themes, characters, and conflicts in the story Light tells about rap. Respond to the following questions in your journal. Who are the heroes in his narrative? What makes them heroic? What challenges do they face? What are their successes? Their failures? What battles remain to be fought? That is, as you re-read Light's article, how does Light take a jumble of facts, names, and dates, and try to make a *story* or *narrative* out of them? Is there a "moral" to the story? What does Light do to give his own writing a sense of plot and direction, to make it something more than a random string of songs and events? In what ways does Light write a narrative that you can "identify" with? Where in this story does Light stop to offer critical commentary about the story he is telling? Why does Light interrupt the story where he does to in-terpret an event or a character? How does Light find a context for this nar-rative? In other words, how is Light's narrative more than just a history of rap? What does this history tell us about American culture and society? About what we find important in life? In art?

WRITING CRITICISM

In "About a Salary or Reality," Alan Light offers a full and complex history of an entire type or genre of music: rap. Now you may not feel up to producing a history of a musical form on this sort of scale. (Certainly we don't.) What we would like you to do, then, is to define—that is, name for the first time—a form or genre or type of music that you can tell an audience something about. You can do this by noticing a number of musicians or bands who are similar enough in some way to deserve, in your view, to be grouped together; but who have not, to your knowledge, been thought of in that way. Or you can take a certain type of song that has not, to your knowledge, been dis-cussed as a type—even though it exists. This genre may be very small or idiosyncratic: groups with big hair, bands on college radio play lists, quintes-sential oldies songs, thrash-country fusion, songs that indicate that a band has "sold out," crossover singers, the ultimate dentist office songs, bands on TV shows, interracial duets, and so on.

The point is not to take one of the genres we've just suggested here, but to pick out one of your own, to identify a set of musicians (or songs) where there is something in common worth noticing and naming. This "something" should be a common factor that the rest of us might not have noticed before, and which lets us look at each of the individual members of this new group-ing in a slightly different way. In addition to giving a (fitting) name to your genre, you will have to explain, of course, why you think it deserves to be considered a genre. And, as with all of your writing on music, you will want to be able to consult recordings of the songs and artists you're considering.

Understanding Television

David Marc

In one of the few famous speeches given on the subject of television, Federal Communications Commission chairman Newton N. Minow shocked the 1961 convention of the National Association of Broadcasters by summarily categorizing the membership's handiwork as "a vast wasteland." The degree to which Minow's metaphor has been accepted is outstripped only by the appeal of television itself. Americans look askance at television, but look at it nonetheless. Owners of thousand-dollar sets think nothing of calling them "idiot boxes." The home stereo system, regardless of what plays on it, is by comparison holy. Even as millions of dollars change hands daily on the assumption that 98 percent of American homes are equipped with sets and that these sets play an average of more than six and a half hours each day, a well-pronounced distaste for TV has become a prerequisite for claims of intellectual and even of ethical legitimacy. "Value-free" social scientists, perhaps less concerned with these matters than others are, have rushed to fill the critical gap left by status-conscious literati. Denying the mysteries of teller, tale, and told, they have reduced the significance of this American storytelling medium to clinical studies of the effects of stimuli on the millions, producing volumes of data that in turn justify each season's network schedules. Jerry Mander, a disillusioned advertising executive, his fortune presumably socked away, has even written a book titled *Four Arguments for the Elimination of Television*. Hans Magnus Enzensberger anticipated such criticism as early as 1962, when he wrote,

The process is irreversible. Therefore, all criticism of the mind industry which is abolitionist in its essence is inept and beside the point, since the idea of arresting and liquidating industrialization itself (which such criticism implies) is suicidal. There is a macabre irony to any such proposal, for it is indeed no longer a technical problem for our civilization to abolish itself.

Though Mander, the abolitionist critic, dutifully listed Enzensberger's *The Consciousness Industry* in his bibliography, his zealous piety—the piety of the convert—could not be restrained. Television viewers (who else would read such a book?) scooped up copies at $6.95 each (paperback). As Enzensberger pointed out, everyone works for the consciousness of industry.

Despite the efforts of a few television historians and critics, like Erik Barnouw and Horace Newcomb, the fact is that the most effective purveyor of language, image, and narrative in American culture has failed to become a subject of lively humanistic discourse. It is laughed at, reviled, feared, and generally treated as persona non grata by university humanities departments and the "serious" journals they patronize. Whether this is the cause or merely a symptom of the precipitous decline of the influence of the humanities during recent years is difficult to say. In either case, it is unfortunate that the scholars and teachers of *The Waste Land* have found "the vast wasteland" unworthy of their attention. Edward Shils spoke for many literary critics when he chastised those who know better but who still give their attention to works of mass culture, for indulging in "a continuation of childish pleasures." Forgoing a defense of childish pleasures, I cannot imagine an attitude more destructive to the future of both humanistic inquiry and television. If the imagination is to play an epistemological role in a scientific age, it cannot be restricted to "safe" media. Shils teased pop-culture critics for trying to be "folksy"; unfortunately, it is literature that is in danger of becoming a precious antique.

As the transcontinental industrial plant built since the Civil War was furiously at work meeting the new production quotas encouraged by modern advertising techniques, President Calvin Coolidge observed that "the business of America is business." Since that time television has become the art of business. The intensive specialization of skills called for by collaborative production technologies has forced most Americans into the marketplace to consume an exceptional range of goods and services. "Do-it-yourself" is itself something to buy. Necessities and trifles blur to indistinction. Everything is for sale to everybody. As James M. Cain wrote, the "whole goddamn country lives selling hot dogs to each other." Choice, however, is greatly restricted. Mass-marketing theory has formalized taste into a multiple-choice question. Like the menu at McDonald's and the suits on the racks, the choices on the dial—and, thus far, the cable converter—are limited and guided. Yet even if the material in each TV show single-mindedly aims at increasing consumption of its sponsors' products, the medium leaves behind a body of dreams that is, to a large extent, the culture we live in. If, as Enzensberger claimed, we are stuck with television and nothing short of nuclear Armageddon will deliver us, then there is little choice but increased consciousness of how television is shaping our environment. Scripts are written. Sets, costumes, and camera angles are imagined and designed. Performances are rendered. No drama, not even melodrama, can be born of a void. Myths are recuperated, legends conjured. These acts are not yet carried out by computers, although network executives might prefer a system in which they were.

Beneath the reams of audience-research reports stockpiled during decades of agency billings is the living work of scores of TV-makers who accepted the marketable formats, found ways to satisfy both censors and the popular id, hawked the Alka-Seltzer beyond the limits of indigestion, and still managed to leave behind images that demand a place in collective memory. The life of this work in American culture is dependent upon public taste, not market research. A fantastic, wavy, glowing procession of images hovers over the American antennascape, filling the air and millions of screens and minds with endless reruns. To accept a long-term relationship with a television program is to allow a vision to enter one's life. That vision is peopled with characters who speak a familiar idiosyncratic language, dress to purpose, worship God, fall in love, show élan and naïveté, become neurotic and psychotic, revenge themselves, and take it easy. While individual episodes—their plots and climaxes—are rarely memorable (though often remembered), cosmologies cannot fail to be rich for those viewers who have shared so many hours in their construction. The salient impact of television comes not from "special events," like the coverage of the Kennedy assassinations or of men on the moon playing golf, but from day-to-day exposure. Show and viewer may share the same living room for years before developing a relationship. If a show is a hit, if the Nielsen families go for it, it is likely to become a Monday-through-Friday "strip." The weekly series in strip syndication is television's most potent oracle. Because of a sitcom's half-hour format, two or even three of its episodes may be aired in a day by local stations. Months become weeks, and years become months. Mary accelerates through hairdos and hem lengths; Phyllis and Rhoda disappear as Mary moves to her high-rise swinging-singles apartment. Mere plot suspense or identification with characters yields to the subtler nuances of cohabitation. The threshold of expectation becomes fixed, as daily viewing becomes an established procedure or ritual. The ultimate suspension of disbelief occurs when the drama—the realm of heightened artifice—becomes normal.

The aim of television is to be normal. The industry is obsessed with the problem of norms, and this manifests itself in both process and product. Whole new logics, usually accepted under the general classification "demographics," have been imagined, to create models that explain the perimeters of objectionability and attraction. A network sales executive would not dare ask hundreds of thousands of dollars for a prime-time ad on the basis of his high opinion of the show that surrounds it. The sponsor is paying for "heads" (that is, viewers). What guarantees, he demands, can be given for delivery? Personal assurances—opinions—are not enough. The network must show scientific evidence in the form of results of demographic experiments. Each pilot episode is prescreened for test audiences who then fill out multiple-choice questionnaires to describe their reactions. Data are processed by age, income, race, religion, or whatever cultural determinants the tester deems relevant. Thus the dull annual autumn dialogue of popular-television criticism:

Why the same old junk every year? ask the smug, ironic television critics after running down their witty lists of the season's "winners and losers."

We know nothing of junk, cry the "value-free" social scientists of the industry research factories. The people have voted with their number 2 pencils and black boxes. We are merely the board of elections in a modern cultural democracy.

But no one ever asked me what I thought, mumbles the viewer in a random burst from stupefaction.

Not to worry, the chart-and-graph crowd replies. We have taken a biopsy from the body politic, and as you would know if this were your job, if you've seen one cell—or 1,200—you've seen them all.

But is demography democracy?

Fortunately, TV is capable of inspiring at least as much cynicism as docility. The viewer who can transform that cynicism into critical energy can declare the war with television over and instead savor the oracular quality of the medium. As Roland Barthes, Jean-Luc Godard, and the French devotees of Jerry Lewis have realized for years, television is American Dada, Charles Dickens on LSD, the greatest parody of European culture since *The Dunciad.* Yahoos and Houyhnhnms battling it out nightly with submachine guns. Sex objects stored in a box. Art or not art? This is largely a lexiographical quibble for the culturally insecure. Interesting? Only the hopelessly genteel could find such a phantasmagoria flat. Yesterday's trashy Hollywood movies have become recognized as the unheralded work of auteurs; they are screened at the ritziest art houses for connoisseurs of *le cinéma.* Shall we need the French once again to tell us what we have?

Television Is Funny

Though network executives reserve public pride for the achievements of their news divisions and their dramatic specials, comedy has always been an essential, even the dominant, ingredient of American commercial-television programming. As Gilbert Seldes wrote, in *The Public Arts,* "Comedy is the axis on which broadcasting revolves." The little box, with its oblong screen egregiously set in a piece of overpriced wood-grained furniture or cheap industrial plastic, has provoked a share of titters in its own right from a viewing public that casually calls in the "boob tube." Television is America's jester. It has assumed the guise of an idiot while actually accruing power and authority behind the smoke screen of its self-degradation. The Fool, of course, gets a kind word from no one: "Knee-jerk liberalism," cry the offended conservatives. "Corporate mass manipulation," scream the resentful liberals. Neo-Comstockians are aghast, righteously indignant at the orgiastic decay of morality invading their split-level homes. The avant-garde strikes a pose of smug terror before the empty, sterile images. Like the abused jester in Edgar Allan Poe's "Hop-Frog," however, the moguls of Television Row make monkeys out of their tormentors. Profits are their only consolation; the show must go on.

In 1927 Philo T. Farnsworth, one of TV's many inventors, presented a dollar sign on a television screen in the first demonstration of his television system. By the late 1940s baggy-pants vaudevillians; stand-up comedians, sketch comics, and game-show hosts had all become familiar video images. No television genre has ever been without what Robert Warshow called "the euphoria [that] spreads over the culture like the broad smile of an idiot." Police shows, family dramas, adventure series, and made-for-TV movies all rely heavily on humor to mitigate their bathos. Even the news is not immune, as evidenced by the spread of "happy-talk" formats in TV journalism in recent years. While the industry experiments with new ways to package humor, television's most hilarious moments are often unintentional, or at least incidental. Reruns of ancient dramatic series display plot devices, dialogue, and camera techniques that are obviously dated. Styles materialize and vanish with astonishing speed. Series like *Dragnet, The Mod Squad,* and *Ironside* surrender their credibility as "serious" police mysteries after only a few years in syndication. They self-destruct into ridiculous stereotypes and clichés, betraying their slick production values and affording heights of comic ecstasy that dwarf their "original" intentions. This is an intense comedy of obsolescence that grows richer with each passing television season. Starsky and Hutch render Jack Webb's Sergeant Joe Friday a messianic madman. *Hill Street Blues* returns the favor to *Starsky and Hutch.* The distinction between taking television on one's own terms and taking it the way it presents itself is of critical importance. It is the difference between activity and passivity. It is what saves TV from becoming the homogenizing, monolithic, authoritarian tool that the doomsday critics claim it is. The self-proclaimed champions of "high art" who dismiss TV shows as barren imitations of the real article simply do not know how to watch. They are like freshmen thrust into survey courses and forced to read Fielding and Sterne; they lack both the background and the tough-skinned skepticism that can make the experience meaningful. In 1953 Dwight Macdonald was apparently not embarrassed to condemn all "mass culture" (including the new chief villain, TV) without offering any evidence that he had watched television. Not a single show was mentioned in his famous essay "A Theory of Mass Culture." Twenty-five years later it is possible to find English professors who will admit to watching *Masterpiece Theatre.* But American commercial shows? How could they possibly measure up to drama produced in Britain and tied in form and sensibility to the nineteenth-century novel? To this widespread English Department line there is an important reply: TV is culture. The more one watches, the more relationships develop among the shows and between the shows and the world. To rip the shows out of their context and judge them against the standards of other media and other cultural traditions is to ignore their American origins and misplace their identities.

Enter the Proscenium

The forms that came to dominate television comedy—and therefore television—were video approximations of theater: the situation comedy (representational) and the variety show (presentational). The illusion of theater is a structural feature of both. It is created primarily by the implicit attendance of an audience that laughs and applauds at appropriate moments and thus assures the viewer that the telecast is originating behind the safe boundary of the proscenium. Normal responses are thus defined. The "audience" may be actual or an electronic sound effect, but this is a small matter. The consequence is the same: the jokes are underlined.

The situation comedy has proved to be the most durable of all commercial-television genres. Other types of programming that were staples of prime-time fare at various junctures in TV history (the western, the comedy-variety show, and the big-money quiz show among them) have seen their heyday and faded. The sitcom, however, has remained a ubiquitous feature of prime-time network schedules since the premiere of *Mary Kay and Johnny,* on DuMont, in 1947. The TV sitcom obviously derives from its radio predecessor. Radio hits like *The George Burns and Gracie Allen Show* and *Amos 'n' Andy* made the transition to television overnight. Then, as now, familiarity was a prized commodity in the industry. The sitcom bears a certain resemblance to the British comedy of manners, especially on account of its parlor setting. A more direct ancestor may be the serialized family-comedy adventures that were popular in nineteenth-century American newspapers. Perhaps because of the nature of its serial continuity, the sitcom had no substantial presence in the movies. Though Andy Hardy and Ma and Pa Kettle films deal with sitcomic themes, their length, lack of audience-response tracks, and relatively panoramic settings make them very different viewing experiences. Serial narratives in the movies were usually action-oriented. The Flash Gordon movies, for example, were constructed so that each episode built to a breathless, unfulfilled climax designed to bring the patrons back to the theater for the next Saturday's resolution. The lack of life-and-death action in the sitcom makes this technique irrelevant. Urgent continuity rarely exists between episodes. Instead, climaxes occur within episodes (though these are not satisfying in any traditional sense). In the movie serial (or in a modern television soap opera) the rescue of characters from torture, death, or even seemingly hopeless anxiety is used to call attention to serialization. The sitcom differs in that its central tensions—embarrassment and guilt—are almost always alleviated before the end of an episode. Each episode may appear to resemble a short, self-contained play; its rigid confinement to an electronic approximation of a proscenium-arch theater, complete with laughter and applause, emphasizes this link. Unlike a stage play, however, a single episode of a sitcom tends to be of dubious interest; it may not even be intelligible. The attraction of an episode is the strength of its contribution to the broader cosmology of the series. The claustrophobia of the miniature proscenium, especially for an audience that has grown casual toward Cinemascope, can be relieved only by the exquisiteness of its minutiae.

Trivia is the most salient form of sitcom appreciation, perhaps the richest form of appreciation that any television series can stimulate. Though television is at the center of American culture—it is the stage upon which our national drama/history is enacted—its texts are still not available upon demand. The audience must share reminiscences to conjure up the ever-fleeting text. Giving this exchange of details the format of a game, players try not so much to stump as to overpower one another with increasingly minute, banal bits of information that bring the emotional satisfaction of experience recovered through memory. The increased availability of reruns that cable service is bringing about can only serve to deepen and broaden this form of grass-roots TV appreciation. Plot resolutions, which so often come in the form of trite, didactic "morals" in the sitcom, are not very evocative. The lessons that Lucy Ricardo learned on vanity, economy, and female propriety are forgotten by both Lucy and the viewer. A description of Lucy's living-room furniture (or her new living-room furniture) is far more interesting. The climactic ethical pronouncements of Ward Cleaver recapture and explore the essence of *Leave It to Beaver* less successfully than a well-rendered impersonation of Eddie Haskell does. From about the time a viewer reaches puberty, a sitcom's plot is painfully predictable. After the tinkering of the first season or two, few new characters, setting, or situations can be expected. Why watch, if not to visit the sitcosmos?

The sitcom dramatizes American culture: its subject is national styles, types, customs, issues, and language. Because sitcoms are and always have been under the censorship of corporate patronage, the genre has yielded a conservative body of drama that is diachronically retarded by the precautions of mass-marketing procedure. For example, *All in the Family* can appropriately be thought of as a sixties sitcom, though the show did not appear on television until 1971. CBS waited until some neat red, white, and blue ribbons could be tied around the turmoil of that extraordinary self-conscious decade before presenting it as a comedy. When the dust cleared and the radical ideas that were being proposed during that era could be represented as stylistic changes, the sixties could be absorbed into a model of acceptability, which is a basic necessity of mass-marketing procedure. During the sixties, while network news programs were offering footage of Vietnam, student riots, civil-rights demonstrations, police riots, and militant revolutionaries advocating radical changes in the American status quo, the networks were airing such sitcoms as *The Andy Griffith Show, Petticoat Junction, Here's Lucy,* and *I Dream of Jeannie.* The political issues polarizing communities and families were almost completely avoided in a genre of representational comedy that had always focused on American family and community. Hippies would occasionally appear as guest characters on sitcoms, but they were universally portrayed as harmless buffoons possessing neither worthwhile ideas nor the power to act, which might make them dangerous. After radical sentiment crested and began to recede (and especially after the first steps were taken toward the repeal of universal male conscription, in late 1969), the challenge of incorporating changes into the sitcom model was finally met. The dialogue that took place in the Bunker home had been unthinkable during the American Celebration that had lingered so long on the sitcom. But if the sitcom was to retain its credibility as a chronicler and salesman of American family life, these new styles, types, customs, manners, issues, and linguistic constructions had to be added to its mimetic agenda.

The dynamics of this challenge can be explained in marketing terms. Five age categories are generally used in demographic analysis: (1) 2–11; (2) 12–17; (3) 18–34; (4) 35–55; (5) 55+. Prime-time programmers pay little attention to groups 1 and 5; viewing is so prevalent among the very young and the old that, as the joke goes on Madison Avenue, these groups will watch the test pattern. Prime-time television programs are created primarily to assemble members of groups 3 and 4 for commercials. Although members of group 4 tend to have the most disposable income, those in group 3 spend more money. Younger adults, presumably building their households, make more purchases of expensive "hard goods" (refrigerators, microwave ovens, automobiles, and so on). The coming of age of the baby-boom generation, in the late sixties and early seventies, created a profound marketing crisis. The top-rated sitcoms of the 1969–1970 season included *Mayberry R.F.D., Family Affair, Here's Lucy,* and *The Doris Day Show.* Though all four of these programs were in Nielsen's top ten that season, their audience was concentrated outside of group 3. How could the networks deliver the new primary consumer group to the ad agencies and their clients? Norman Lear provided the networks with a new model that realistically addressed itself to this problem. In *Tube of Plenty* Erik Barnouw shows how the timidity of television narrative can be traced directly to the medium's birth during the McCarthy Era. If the sixties accomplished nothing else, it ended the McCarthy scare. The consensus imagery that had dominated the sitcom since the birth of TV simply could not deliver the new audience on the scale that the new consensus imagery that Norman Lear developed for the seventies could. Lear's break from the twenty-year-old style of the genre seemed self-consciously "hip." The age of life-style was upon us.

In the fifties and sixties, sitcoms were offering the Depression-born post-Second World War adult group a vision of peaceful, prosperous suburban life centered on the stable nuclear family. In these shows—among them, *Father Knows Best, The Adventures of Ozzie and Harriet, The Donna Reed Show, The Trouble With Father, Make Room for Daddy*—actual humor (jokes or shticks) was always subordinate to the proper solution of ethical crises. They were comedies not so much in the popular sense as in Northrop Frye's sense of the word: no one got killed, and they ended with the restoration of order and happiness. What humor there was derived largely from the "cuteness" displayed by the children in their innocent but doomed attempts to deal with problems in other than correct (adult) ways. Sometimes an extra element of humor was injected by marginal characters from outside the nuclear family. Eddie Haskell (Ken Osmond) is among the best-remembered of these domestic antiheroes. A quintessential wise guy, Eddie deviated from the straight and narrow as walked by Wally Cleaver; this was implicitly blamed on his parents. The fact that Eddie was uniformly punished by the scriptwriters made his rebellion all the more heroic.

Beneath the stylistic differences that separate a classic fifties sitcom and a Norman Lear show, the two are bound together by their unwavering commitment to didactic allegory. Lear indeed updated the conversation in the sitcom living room, but the form of his sitcoms was actually quite conservative. Like the sitcoms of the fifties, Lear's shows reinforced with Dorothy Rabinowitz has called "our most fashionable pieties." "Fashionable" is the key term. As Roger Rosenblatt has pointed out, the greatest difference between *Father Knows Best*'s Jim Anderson and Archie Bunker is that Jim, the father, is the source of all wisdom for the Anderson family, whereas Archie is more likely to be the object of lessons than the source of them.

Perhaps the reason that the sitcom has been looked down upon by critics as a hopelessly "low" or "masscult" form is that a search for what Matthew Arnold called "the best that has been known and said" is a wild-goose chase as far as the genre is concerned. R. P. Blackmur, though certainly no TV fan, commented germanely in his essay "A Burden for Critics" that the critic "will impose the excellence of something he understands upon something he does not understand. Then all the richness of actual performance is gone. It is worth taking precautions to prevent that loss, or at any rate to keep us aware of the risk." Television is not yet a library with shelves; it is a flow of dreams, many remembered, many submerged. How can we create a bibliography of dreams? Blackmur also wrote that "the critic's job is to put us into maximum relation to the burden of our momentum." Television is the engine of our cultural momentum. It has heaped thousands upon thousands of images upon the national imagination:

Jackie Gleason rearing back a fist and threatening to send Alice to the moon.

Phil Silvers's bullet-mouthed Sergeant Bilko conning his platoon out of its paychecks.

Jack Benny and Rochester guiding an IRS man across a crocodile-infested moat to the vault.

Dobie Gillis standing in front of "The Thinker" and pining for Tuesday Weld.

Carroll O'Connor giving Meathead and modern philanthropic liberalism the raspberry.

Jerry Van Dyke setting in the driver's seat of a Model T Ford for a heart-to-heart talk with *My Mother, the Car.*

In *Democratic Vistas* Walt Whitman called for a new homegrown American literary art whose subject would be "the average, the bodily, the concrete, the democratic, the popular." The sitcom is an ironic twentieth-century fulfillment of this dream. The "average" has been computed and dramatized as archetype; the consumer world has been made "concrete"; the "bodily" has been fetishized as the object of unabashed envy and voyeurism—all of this is nothing if not "popular." The procession of images that was Whitman's own art, and that he hoped would become the nation's, is lacking in the sitcom in but one respect: the technique of television art is not democratic but demographic. The producers, directors, writers, camera operators, set designers, and other artists of the medium are not, as Whitman had hoped, "breathing into it a new breath of life." Instead, for the sake of increased consumption they have contractually agreed to create the hallucinations of what Allen Ginsberg has called "the narcotic...haze of capitalism." The drug indeed is on the air and in the air. Fortunately, the integrity of the individual resides in the autonomy of the imagination, and is not curtailed by this system. The television set plays on and on in the mental hospital; the patient can sit in his chair, spaced-out and hopeless, or rise to the occasion of consciousness.

In Front of the Curtain

"The virtue of [professional] wrestling," Roland Barthes wrote in 1957, "is that it is the spectacle of excess." The sitcom, in contrast, is a spectacle of subtleties, an incremental construction of substitute universes laid upon the foundation of a linear, didactic teletheater. Even the occasional insertion of the *mirabile* or supernatural underlines the genre's broader commitment to naturalistic imitation. Presentational comedy, which shared the prime-time limelight with the sitcom during the early years of TV, vacillates between the danger of excess and the safety of consensus. The comedy-variety genre has been the great showcase for presentational teleforms: stand-up comedy, impersonation, and the blackout sketch. It is similar to wrestling, in that it too strives for the spectacle of excess. The pre-electronic ancestors of the TV comedy-variety show can be found on the vaudeville and burlesque stages: the distensions of the seltzer bottle and the banana peel; the fantastic transformations brought about by mimicry; the titillating physical, psychological, and cultural disorders that abound in frankly self-conscious art forms. But the comedy-variety show does not go to the ultimate excesses of wrestling. Like the sitcom, it is framed by the proscenium arch and accepts the badge of artifice.

While the representational genres—the sitcom and its cousins, the action/adventure series and the made-for-TV movie—have flourished to the point that they consume almost all of the "most-watched" hours, the comedy-variety show has been in steady decline since the 1950s, when it was a dominant formula for prime-time television. Since the self-imposed cancellation of *The Carol Burnett Show,* in 1978, the few presentational variety hours that have appeared in prime time have been hosted by singers (Barbara Mandrell, Marie Osmond), and comedy has been relegated to a rather pathetic supporting role. The fall of the genre took place amid increasing demand for a "product" as opposed to a "show" in the growing television industry. As the prime-time stakes rocketed upward, sponsors, agencies, and networks became less tolerant of the inevitable ups and downs of star-centered presentational comedy. The sitcom and other forms of representational drama offer relatively rigid shooting scripts that make "quality control" easier to impose. Positive demographic responses to dramatic "concepts" are dependable barometers. Performance comedy is only as good as an individual performance; the human element looms too large. Furthermore, representational drama can avoid the dreaded extremes of presentational comedy. Kinescopes of early Milton Berle shows reveal the comedian in transvestite sketches whose gratuitous lewdness rivals that of wrestling at its most intense. The passionate vulgarity of these sketches could not have been wholly predictable from their scripts. Instead, it derives directly from Berle's confrontation with the camera—his performance. Censorship of such material presents complex editing problems, which are easily forestalled in representational drama by script changes.

Berle's nova-like career is itself an illustration of the fragility of comedy-variety on television. NBC signed "Mr. Television" to a thirty-year contract in 1951; by the end of the decade he had become an embarrassment to the network and was being used as the host of a bowling show. The message was clear: Berlesque, like wrestling, had been blackballed from television's increasingly genteel prime-time circle. The late fifties, the heyday of *Playhouse 90, The U.S. Steel Hour,* and *The Armstrong Circle Theater,* was the "golden age of television drama." TV then, as now and always, was on the verge of becoming sophisticated. The NBC late-night spot passed from the wacky Jerry Lester to the neurotically urbane Jack Paar. Berle was certainly neurotic; he simply could not be urbane.

Generally speaking, the comedian has had to frame himself with the proscenium arch and don the mask of a representational character to find a place in prime-time television. The networks have thus provided themselves with a modicum of protection from the unreliability of personalities. Presentational comedy—performance art—may simply be too dangerous a gamble for the high stakes of today's market.

Interestingly, the disappearance from TV of the clown who faces the audience without a story line has occurred more or less simultaneously with rising interest in and appreciation of performance art in avant-garde circles. In "Performance as News: Notes on an Intermedia Guerrilla Art Group" Cheryl Bernstein wrote:

> *In performance art, the artist is more exposed than ever before. The literal identification of artistic risk with the act of risking one's body or one's civil rights has become familiar in the work of such artists as Chris Burden, Rudolf Schwarzkogler, Tony Schafrazi and Jean Toche.*

Burden invites an audience into a performance space where spectators sit atop wooden ladders. He then floods the room with water and drops a live electrical wire into the giant puddle. The closest thing television offers to a spectacle of this kind is Don Rickles, who evokes terror in an audience by throwing the live wire of his insult humor into the swamp of American racial and ethnic fears. Rickles, for the most part, has been prohibited from performing his intense theater of humiliation in prime time. Twice the networks have attempted to contain him in sitcom proscenia, but these frames have constricted his effect and turned his insults into dull banter. In recent years he has rarely been unleashed upon live studio audiences. The erratic quality of his occasional performances as guest host on *The Tonight Show* offers a clue to the networks' reluctance to invest heavily in the presentational comedy form.

Bernstein points to the news as the great source of modern performance art on television. She deconstructs the kidnapping of Patty Hearst by the Symbionese Liberation Army as a performance work. The SLA was a troupe formed to create a multimedia work—the kidnapping—principally for television. The mass distribution of food in poor neighborhoods in the San Francisco Bay area (in compliance with one of the SLA's demands) and the shootouts and police chases were all part of a modern theatrical art that can exist only on television. Perhaps the proliferation of the news on television can be tied to the decline of presentational comedy; the two seem to have occurred in direct proportion to each other. The schlockumentary magazine (*Real People, That's Incredible!*) is the point at which the two genres meet.

Furthermore, the bombardment of the home screen with direct presentations from every corner of the earth has created a kind of vaudeville show of history. The tensions generated by the nuclear Sword of Damocles create a more compelling package than even Ed Sullivan could have hoped to assemble. The nations of the world have become a troupe of baggy-pants clowns on TV. They are trotted out dozens of times each day in a low sketch comedy of hostility, violence, and affectation. The main show, of course, is the network evening news. Climb the World Trade Center. Fly an airplane through the Arc de Triomphe. Plant a bomb in a department store in the name of justice. Invade a preindustrial nation with tanks in the name of peace. Can Ted Mack compare with this?

The Theater Collapses

The television industry is at the heart of a vast entertainment complex that oversees the coordination of consumption and culture. "Entertainment" has been established as a buzz word for narrative and other imaginative presentations that make money. It is used as a rhetorical ploy to specialize popular arts and isolate them from the aesthetic and political scrutiny reserved for "art." An important implication of the definition of *entertain* is the intimate social relationship it implies between the entertainer and the entertained. In 1956 Gunther Anders wrote that the "television viewer, although living in an alienated world, is made to believe that he is on a footing of the greatest intimacy with everything and everybody." The technological means to produce this illusion have since been significantly enhanced. Anders described this illusion as "chumminess." Television offers itself to the viewer as a hospitable friend: Welcome to *The Wonderful World of Disney*. Good evening, folks, We'll be right back. See you next week. Y'all come back, now. As technology synthesizes more and more previously human functions, there is a proliferation of anthropomorphic metaphor: automated teller machines ask us how much money we need. Computers send us bills. Channel 7 is predicting snow. The car won't start. It is this context that television entertains. There is an odd sensation of titillation in all this service Whitman and other nineteenth-century optimists foresaw an elevation of the common man to a proud master in the techno-world. Machines would take care of life's dirty work—slavery without guilt. Television enthusiastically smiles and shuffles for the viewer's favor. Even television programs we do not like contribute to the illusion (that is, "We are not amused"). Success in tele-American life is measured largely by the quantity of machines at one's disposal. Are all the household chores mechanized? Do you have HBO? Work is minimized. Leisure is maximized. There is more time to watch television—to live like a king.

Backstage of this public drama a quite different set of relationships is at work. In a demography the marketing apparatus becomes synonymous with the state. As the quality of goods takes a back seat to the quantity of services, the most valued commodity of all—the measure of truth—becomes information on the consuming preferences of the hundreds of millions of consumer-kings. Every ticket to the movies, every book, every tube of toothpaste purchased is a vote. The shelves of the supermarkets are stocked with referenda. Watching television is an act of citizenship, participation in culture. The networks entertain the viewer; in return, the viewer entertains thousands of notions on what to buy (that is, how to live). The democrat Whitman wrote that the most important person in any society is the "average man." The demographer Nielsen cannot agree more:

> *While the average household viewed over 49 hours of television per week in the fall of 1980, certain types viewed considerably more hours. Households with 3 or more people and those with non-adults watched over 60 hours a week. Cable subscribing households viewed about 7 hours a week more than non-cable households.*

Paul Klein, the chief programmer at NBC in the late seventies, built his programming philosophy upon what he called the Least Objectionable Program (LOP) theory. This theory, expounded by Klein, plays down the importance of viewer loyalty to specific programs. It asserts that the viewer turns on the set not so much to view this or that program as to fulfill a desire to "watch television." R. D. Percy and Company, an audience-research firm, has found some evidence to support Klein's thesis. *TV Guide,* summarizing Percy's two-year experiment with 200 Seattle families, reported: "Most of us simply snap on the set rather than select a show. The first five minutes are spent *prospecting* channels, looking for gripping images." After giving in to the impulse (compulsion?) to watch TV the viewer is faced with the secondary consideration of choosing a program. In evolved cable markets this can mean confronting dozens of possibilities. The low social prestige of TV watching, even among heavy viewers, coupled with the remarkably narrow range of what is usually available, inhibits the viewer from expressing enthusiasm for any given show. The viewer or viewers (watching TV, it must not be forgotten, is one of the chief social activities of the culture) must therefore "LOP" about, looking for the least bad, least embarrassing, or otherwise least objectionable program. While I am ill prepared to speculate on the demographic truth of this picture of the "average man," two things are worth noting: anyone who watches television has surely experienced the LOP phenomenon; and NBC fell into last place in the ratings under Klein's stewardship.

I cite Klein (and Nielsen) to demonstrate the character of demographic thought, the ideology that ultimately produces most television programs and that is always employed to authorize or censor their exhibition on the distribution system. The optimistic, democratic view of man as a self-perfecting individual, limited only by superimposed circumstance, is turned on its head. Man is defined as a prisoner of limitations who takes the path of least resistance. This is an industrial nightmare, the gray dream of Fritz Lang's *Metropolis* reshot in glossy Technicolor. Workers return from their multicollared tasks drained of all taste and personality. They seek nothing more than merciful release from the day's production pressures, or as certain critics tell us, they want only to "escape."

"Escapism" is a much-used but puzzling term. Its ambiguities illustrate the overall bankruptcy of the television criticism that uses it as a flag. The television industry is only too happy to accept "escapism" as the definition of its work; the idea constitutes a carte-blanche release from responsibility for what is presented. Escapism critics seem to believe that the value of art should be measured only be rigorously naturalistic standards. Television programs are viewed as worthless or destructive because they divert consciousness from "reality" to fantasy. However, all art, even social realism, does this. Brecht was certainly mindful of this fact when he found it necessary to attach intrusive Marxist sermons to the fringes of social-realist stage plays. Is metaphor possible at all without "escapism"? Presumably, the mechanism of metaphor is to call a thing something it is not in order to demonstrate emphatically what it is. When the network voice of control says, "NBC is proud as a peacock," it is asking the recipient of this message to "escape" from all realistic data about the corporate institution NBC into a fantastic image of a bird displaying its colorful feathers in a grand and striking manner. The request is made on the assumption that the recipient will be able to sort the shared features of the two entities from the irrelevant features and "return" to a clearer picture of the corporation. Representational television programs work in much the same way. If there were no recognizable features of family life in *The Waltons*, if there were no shared features of life-style in *Three's Company*, if there were no credible features of urban paranoia in *Baretta*, then watching these shows would truly be "escapism." But if those features are there, the viewer is engaging in an act that does not differ in essence from reading a Zola novel. The antiescapist argument makes a better point about the structure of narrative in the television series. In the world of the series problems are not only solvable but usually solved. To accept this as "realistic" is indeed an escape from the planet Earth. But how many viewers accept a TV series as realistic in this sense? Interestingly, it is the soap opera—the one genre of series television that is committed to an anticlimactic, existential narrative structure—that has created the most compelling illusion of realism for the viewing audience. The survival and triumphs of an action-series hero are neither convincing nor surprising but merely a convention of the medium. Like the theater audience that attends *The Tragedy of Hamlet*, the TV audience knows what the outcome will be before the curtain goes up. The seduction is not What? but How?

Thus far, I have limited my discussion to traditional ways of looking at television. However, the television industry has made a commitment to relentless technological innovation of the medium. The cable converter has already made the traditional tuner obsolete. From a comfortable vantage point anywhere in the room the viewer can scan dozens of channels with a fingertip. From the decadent splendor of a divan the viewer is less committed to the inertia of program choice. It is possible to watch half a dozen shows more or less simultaneously, fixing on an image for the duration of its allure, dismissing it as its force disintegrates, and returning to the scan mode. Unscheduled programming emerges as the viewer assumes control of montage. It is also clear that program choice is expanding. The grass-roots public-access movement is still in its infancy, but the network *mise en scène* has been somewhat augmented by new corporate players such as the superstations and the premium services. Cheap home recording and editing equipment may turn the television receiver into a bottomless pit of "footage" for any artist who dares use it.

Michael Smith, a Chicago-born New York grantee/comedian, is among those pointing the way in this respect. Whether dancing with *Donny and Marie*, in front of a giant video screen, at the Museum of Contemporary Art, in Chicago, or performing rap songs at the Institute of Contemporary Art, in Boston, Smith offers an unabashed display of embarrassments and highlights in the day of a life with television. Mike (Smith's master persona) is the star of his own videotapes (that is, TV shows). In "It Starts at Home" Mike gets cable and learns the true meaning of public access. In "Secret Horror" reception is plagued by ghosts. The passive viewer—that well-known zombie who has been blamed for every American problem from the Vietnam War to Japanese technological hegemony—becomes a do-it-yourself artist in Smith. If, as Susan Sontag writes, "interpretation is the revenge of the intellect upon art," parody is the special revenge of the viewer upon TV.

Whatever the so-called blue-sky technologies bring, there can be no doubt that the enormous body of video text generated during the decades of the network era will make itself felt in what will follow. The shows and commercials and systems of signs and gestures that the networks have presented for the past thirty-five years constitute the television we know how to watch. There won't be a future without a past.

In *Popular Culture and High Culture* Herbert Gans took the position that all human beings have aesthetic urges and are receptive to symbolic expressions of their wishes and fears. As simple and obvious as Gans's assertion seems, it is the wild card in the otherwise stacked deck of demographic culture. Paul Buhle and Daniel Czitrom have written,

> *We believe that the population at large shares a definite history in modern popular culture and is, on some levels, increasingly aware of that history. We do not think that the masses of television viewers, radio listeners, movie-goers, and magazine readers are numbed and insensible, incapable of understanding their fate or historical condition until a group of "advanced revolutionaries" explains it to them.*

Evidence of this shared, definite history, in the form of self-referential parody, is already finding its way to the air. The television babies are beginning to make television shows. In the signature montage that introduces each week's episode of *SCTV*, there is a shot of a large apartment house with dozens of televisions flying out the windows and crashing to the ground. As the viewer learns, this does not mean the end of TV in Jerry Mander's sense but signifies the end of television as it has traditionally been experienced. *SCTV* is television beginning to begin again. The familiar theatrical notions of representation and presentation that have guided the development of programming genres are ground to fine dust in the crucible of a satire that draws its inspiration directly from the experience of watching television. *SCTV* has been the first television program absolutely to demand of its viewers a knowledge of the traditions of TV, a self-conscious awareness of cultural history. In such a context viewing at last becomes an active process. Without a well-developed knowledge of and sensitivity to the taxonomy and individual texts of the first thirty-five years of TV, *SCTV* is meaningless—and probably not even funny. In 1953 Dwight Macdonald described "Mass Culture" as "a parasitic, a cancerous growth on High Culture." By this, I take it, he meant that mass-consumed cultural items such as television programs "steal" the forms of "High Culture," reduce their complexity, and replace their content with infantile or worthless substitutes. The relationship of mass culture to high culture, Macdonald told us, "is not that of the leaf and the branch but rather that of the caterpillar and the leaf." *SCTV* bears no such relationship to any so-called high culture. It is a work that emerges out of the culture of television itself, a fully realized work in which history and art synthesize the conditions for a new consciousness of both. Other media—theater, film, radio, music—do not bend the show to televised renderings of their own forms but instead are forced to become television. The viewer is not pandered to with the apologetic overdefining of linear development that denies much of television its potential force. Presentation and representation merge into a seamless whole. The ersatz proscenium theater used by the networks to create marketing genres is smashed; the true montage beaming into the television home refuses to cover itself with superficial framing devices. The pseudo-Marxist supposition that *SCTV* is still guilty of selling the products is boring—the show is not.

SCTV is among the first tangible signs of a critical relationship to TV viewing that is more widespread than a reading of the TV critics would indicate. American TV was born a bastard art of mass-marketing theory and recognizable forms of popular culture. Thirty-five years later a generation finds in this dubious pedigree its identity and heritage. The poverty of TV drama in all traditional senses is not as important as the richness of the montage. For the TV-lifer, a rerun of *Leave It to Beaver* or *I Love Lucy* or *The Twilight Zone* offers the sensation of traveling through time in one's own life and cultural history. The recognizable, formulaic narrative releases the viewer from what becomes the superficial concerns of suspense and character development. The greater imaginative adventures of movement through time, space, and culture take precedence over the flimsy mimesis that seems to be the intention of the scripts. The whole fast-food smorgasbord of American culture is laid out for consumption. This is not merely kitsch. Clement Greenberg wrote that "the precondition for kitsch [the German word for mass culture], a condition without which kitsch would be impossible, is the availability close at hand of a fully matured cultural tradition, whose discoveries, acquisitions, and perfected self-consciousness kitsch can take advantage of for its own ends." In fact, this process is reversed in television appreciation. The referent culture has become the mass, or kitsch, culture. Instead of masscult ripping off highcult, we have art being fashioned from the junk pile. The banal hysteria of the supermarket is capable of elegant clarity in Andy Warhol's *Campbell's Soup Can.* Experience is re-formed and recontextualized, reclaimed from chaos. Television offers many opportunities of this kind.

The networks and ad agencies care little about these particulars of culture and criticism. The networks promise to deliver heads in front of sets and no more. But, as we will find in any hierarchical or "downstream" system, there is a personal stance that will at least allow the subject of institutional power to maintain personal dignity. In the television demography this stance gains its power from the act of recontextualization. If there is no exit from the demographic theater, each viewer will have to pull down the rafters from within. What will remote-control "SOUND: OFF" buttons mean to the future of marketing? What images are filling the imaginations of people as they "listen" to television on the TV bands of transistor radios, wearing headphones while walking the streets of the cities? Why are silent TV screens playing at social gatherings? When will the average "Household Using TV" (HUT) be equipped with split-screen, multichannel capability? What is interesting about a game show? The suspense as to who will win, or the spectacle of people brought frothing to the point of hysteria at the prospect of a new microwave oven? What is interesting about a cop show? The "catharsis" of witnessing the punishment of the criminal for his misdeeds, or the attitude of the cop toward evil? What is interesting about a sitcom? The funniness of the jokes, or the underlining of the jokes on the laugh track? The plausibility of the plot, or the portrayal of a particular style of living as "normal"? What is interesting about Suzanne Somers and Erik Estrada? Their acting, or their bodies? Television is made to sell products but is used for quite different purposes by lonely, alienated people, families, marijuana smokers, born-again Christians, alcoholics, Hasidic Jews, destitute people, millionaires, jocks, shut-ins, illiterates, hang-gliding enthusiasts, intellectuals, and the vast, heterogeneous procession that continues to be American culture in spite of all demographic odds. If demography is an attack on the individual, then the resilience of the human spirit must welcome the test.

"To be a voter with the rest is not so much," Whitman warned in his *Democratic Vistas* of 1871. The shopper/citizen of the demography ought to know this only too well. Whitman recognized that no political system could ever summarily grant its citizens freedom. Government is a system of power; freedom is a function of personality. "What have we here [in America]," he asked, "if not, towering above all talk and argument, the plentifully-supplied, last-needed proof of democracy, in its personalities?" Television is the Rorschach test of the American personality. I hope the social psychologists will not find our responses lacking.

COMING TO TERMS

David Marc argues that we can turn television into something much more than an intellectual wasteland if we learn to "transform cynicism into critical energy" and look at the medium in an "active" way. But how do you do this, exactly? What clues does Marc offer you about how you might actually go about "taking television on your own terms" rather than simply "taking it the way it presents itself"? Marc is setting up two significantly different ways of viewing television: one might be termed active, the other passive. To understand television, in Marc's terms, is to understand that viewers have a choice in defining their relationship with this medium. Re-read Marc's article and mark passages in the text that allow you to address the following questions in your journal. Where do you see Marc interpreting TV in different ways from the ways its producers might have intended? You need to see two versions of a television program here: what might have been intended and how Marc has chosen to see it differently, on his own terms. What does Marc seem to find most interesting about viewing television? What does he dismiss as boring or dull? What kinds of details does he notice? What sorts of questions does he ask? Take a moment: If you could comprise a "top ten" list of television programs scheduled for a station called the *Intellectual Wasteland Channel,* what programs would you list? Why did you include what you did? Select one, and ask yourself: If you looked at this program like Marc might, why would you eliminate it from your list?

READING AS A WRITER

Marc writes about a text we all know well: TV, although after reading Marc's article, we may have the sense that we don't know television as well as we thought we did, at least in the way Marc "knows" television. One challenge that Marc presents us with is that he is a heavily *allusive* writer; he makes *allusions* to many other texts—some that are on TV and some that are not, some that you may know and others that you may not. Draw up a list of allusions and references in your journal, the names and titles Marc refers to or quotes in his

article. Place a check mark by the ones you recognize, even if you're not exactly sure of who or what they are. Then look again at the ones you're less
familiar with. Who does Marc help you identify? (For example, early on he refers to Paul Klein, *the chief programmer at MBC in the late seventies*.") Who
does he seem to expect his readers to know already? At this point in your
work, think about what critical argument Marc is trying to make in his article.
In your journal, restate Marc's argument in your own words. How might Marc
be using these allusions to advance his argument? Do you see any pattern to
those figures he identifies and those he simply cites? What would happen to
Marc's piece, and his argument, if they were simply left out? What other writers
have you read in *Media Journal* who also use allusions, possibly in strategic
or rhetorical ways? In ways similar to or different from Marc?

WRITING CRITICISM

David Marc describes two different ways of watching television: one in which
the viewer passively accepts television "the way it presents itself"—that is, the
way it wants to be seen—and another, more active kind of viewing, in which
we use the medium for our own purposes, turning TV into something that is
better, more interesting, or more fun than it perhaps intends to be. Marc is talking about these two ways of attending to television when he writes: "What is
interesting about a game show? The suspense as to who will win, or the spectacle of people brought to the point of hysteria at the prospect of a new microwave oven? What is interesting about a cop show? The "catharsis" of
witnessing the punishment of the criminal for his misdeeds, or the attitude of
the cop toward evil? What is interesting about a sitcom? The funniness of the
jokes, or the underlining of the jokes on the laugh track? The plausibility of the
plot, or the portrayal of a particular style of living as "normal"?
    According to Marc, then, it is our ability to change the ways we watch TV
that saves the medium from being an intellectual wasteland. We'd like you to
do some work here in the spirit of this project. Take a television program that
asks to be seen as one sort of thing, but in your view deserves to be seen as
something else. You might select a show that presents itself as harmless fun,
mere entertainment, a blatant appeal to sexuality, an annoying collection of
laugh tracks, and demonstrate that it deserves to be seen as more complex,
or insidious, or about more serious matters than it first appears. Take something that presents itself as serious—some soap operas or talk shows might
come to mind—and show that it deserves to be seen as silly, comic, or even
pathetic. A third approach would involve describing what a program claims to
be "about" and then showing that it is actually about something else entirely.
Or you might also take a show that is supposedly aimed at one audience (children, perhaps), and argue that it is actually of interest to a different one (hip

twentysomethings, say). Cartoons, for example, can be remarkably "adult" at times. In helping you move beyond what might be intended for a program, as you take this program on your "own terms," you might also draw upon other writers in *Media Journal,* who view television in "active ways," writers like George Felton, Joshua Meyrowitz, Mark Crispin Miller, Barbara Harrison, or Alexander Nehamas, to name a few.

The subject of your writing, then, will not be so much the TV text you select as the two different readings of this text that you offer. Your focus should be on the way the program asks to be read as compared to the ways you say *it should be read.* To do this sort of double-reading well, you will need to attend carefully to the clues the program provides in support of the "official" or approved reading. First, ask yourself: What does the program do in order to suggest how it wants to be read? What are the signs that tell you the sort of reactions you're supposed to have to the program? Then consider: what makes this official or approved reading impossible, unacceptable, or unconvincing? What is going on that undermines the official reading or allows us to have a different reaction from the one intended?

Having addressed the above questions, you will be in a good position to propose a better (more interesting, novel, intelligent) reading. How do you think the text should be seen, as what kind of program? What features of the show make this alternative reading possible? What do you find preferable— more enjoyable, more defensible, more rewarding—about your reading of the text, as compared to the official or approved interpretation? Finally, what do we gain by seeing the text the way you want us to see it? What do the program's sponsors or producers gain by having us view the text in the way they want us to view it? What is at stake in the conflict between these interpretations? Why does it matter how we view the program?

A final way to describe your task, then, is to test Marc's thinking by showing what it might mean to view the same TV text in two different ways—the way it wants to be seen, and the way it *could* be seen by a viewer (such as you) who refuses to accept the terms on which television presents itself. Your writing should thus put you in a position to answer the following questions: Do you agree with Marc that watching TV in a different way saves the medium "from becoming the homogenizing, monolithic, authoritarian tool that doomsday critics claim it is"? Or are there problems with or limits to this alternative or ironic mode of viewing that he does not acknowledge?

Fortunate Son

Dave Marsh

This old town is where I learned about lovin'
This old town is where I learned to hate
This town, buddy, has done its share of shoveling
This town taught me that it's never too late
  —Michael Stanley, "My Town"

When I was a boy, my family lived on East Beverly Street in Pontiac, Michigan, in a two-bedroom house with blue-white asphalt shingles that cracked at the edges when a ball was thrown against them and left a powder like talc on fingers rubbed across their shallow grooves. East Beverly ascended a slowly rising hill. At the very top, a block and a half from our place, Pontiac Motors Assembly Line 16 sprawled for a mile or so behind a fenced-in parking lot.

Rust-red dust collected on our windowsills. It piled up no matter how often the place was dusted or cleaned. Fifteen minutes after my mother was through with a room, that dust seemed thick enough for a finger to trace pointless, ashy patterns in it.

The dust came from the foundry on the other side of the assembly line, the foundry that spat angry cinders into the sky all night long. When people talked about hell, I imagined driving past the foundry at night. From the street below, you could see the fires, red-hot flames shaping glowing metal.

Pontiac was a company town, nothing less. General Motors owned most of the land, and in one way or another held mortgages on the rest. Its holdings included not only the assembly line and the foundry but also a Fisher Body plant and on the outskirts, General Motors Truck and Coach. For a

From "Introduction" by Dave Marsh, 1985, *Fortunate Son.* Reprinted by permission of the author.

while, some pieces of Frigidaires may even have been put together in our town, but that might just be a trick of my memory, which often confuses the tentacles of institutions that monstrous.

In any case, of the hundred thousand or so who lived in Pontiac, fully half must have been employed either by GM or one of the tool-and-die shops and steel warehouses and the like that supplied it. And anybody who earned his living locally in some less directly auto-related fashion was only fooling himself if he thought of independence.

My father worked without illusions, as a railroad brakeman on freight trains that shunted boxcars through the innards of the plants, hauled grain from up north, transported the finished Pontiacs on the first leg of the route to almost anywhere Bonnevilles, Catalinas, and GTOs were sold.

Our baseball and football ground lay in the shadow of another General Motors building. That building was of uncertain purpose, at least to me. What I can recall of it now is a seemingly reckless height—five or six stories is a lot in the flatlands around the Great Lakes—and endless walls of dark greenish glass that must have run from floor to ceiling in the rooms inside. Perhaps this building was an engineering facility. We didn't know anyone who worked there, at any rate.

Like most other GM facilities, the green glass building was surrounded by a chain link fence with barbed wire. If a ball happened to land on the other side of it, this fence was insurmountable. But only very strong boys could hit a ball that high, that far, anyhow.

Or maybe it just wasn't worth climbing that particular fence. Each August, a few weeks before the new models were officially presented in the press, the finished Pontiacs were set out in the assembly-line parking lot at the top of our street. They were covered by tarpaulins to keep their design changes secret— these were the years when the appearance of American cars changed radically each year. Climbing *that* fence was a neighborhood sport because that was how you discovered what the new cars looked like, whether fins were shrinking or growing, if the new hoods were pointed or flat, how much thinner the strips of whitewall on the tires had grown. A weird game, since everyone knew people who could have told us, given us exact descriptions, having built those cars with their own hands. But climbing that fence added a hint of danger, made us feel we shared a secret, turned gossip into information.

The main drag in our part of town was Joslyn Road. It was where the stoplight and crossing guard were stationed, where the gas station with the condom machine stood alongside a short-order restaurant, drugstore, dairy store, small groceries and a bakery. A few blocks down, past the green glass building, was a low brick building set back behind a wide, lush lawn. This building, identified by a discreet roadside sign, occupied a long block or two. It was the Administration Building for all of Pontiac Motors—a building for executives, clerks, white-collar types. This building couldn't have been more than three-quarters of a mile from my house, yet even though I lived on East Beverly Street from the time I was two until I was past fourteen, I knew only one person who worked there.

In the spring of 1964, when I was fourteen and finishing eighth grade, rumors started going around at Madison Junior High. All the buildings on our side of Joslyn Road (possibly east or west of Joslyn, but I didn't know directions then—there was only "our" side and everywhere else) were about to be bought up and torn down by GM. This was worrisome, but it seemed to me that our parents would never allow that perfectly functioning neighborhood to be broken up for no good purpose.

One sunny weekday afternoon a man came to our door. He wore a coat and tie and a white shirt, which meant something serious in our part of town. My father greeted him at the door, but I don't know whether the businessman had an appointment. Dad was working the extra board in those years, which meant he was called to work erratically—four or five times a week, when business was good—each time his nameplate came to the top of the big duty-roster board down at the yard office. (My father didn't get a regular train of his own to work until 1966; he spent almost twenty years on that extra board, which meant guessing whether it was safe to answer the phone every time he actually wanted a day off—refuse a call and your name went back to the bottom of the list.)

At any rate, the stranger was shown to the couch in our front room. He perched on that old gray davenport with its wiry fabric that bristled and stung against my cheek, and spoke quite earnestly to my parents. I recall nothing of his features or of the precise words he used or even of the tone of his speech. But the dust motes that hung in the air that day are still in my memory, and I can remember his folded hands between his spread knees as he leaned forward in a gesture of complicity. He didn't seem to be selling anything; he was simply stating facts.

He told my father that Pontiac Motors was buying up all the houses in our community from Tennyson Street, across from the green glass building, to Baldwin Avenue—exactly the boundaries of what I'd have described as our neighborhood. GM's price was more than fair; it doubled what little money my father had paid in the early fifties. The number was a little over ten thousand dollars. All the other houses were going, too; some had already been sold. The entire process of tearing our neighborhood down would take about six months, once all the details were settled.

The stranger put down his coffee cup, shook hands with my parents and left. As far as I know, he never darkened our doorstep again. In the back of my mind, I can still see him through the front window cutting across the grass to go next door.

"Well, *we're* not gonna move, right, Dad?" I said. Cheeky as I was, it didn't occur to me this wasn't really a matter for adult decision-making—or rather, that the real adults, over at the Administration Building, had already made the only decision that counted. Nor did it occur to me that GM's offer might seem to my father an opportunity to sell at a nice profit, enabling us to move some place "better."

My father did not say much. No surprise. In a good mood, he was the least taciturn man alive, but on the farm where be was raised, not many words were needed to get a serious job done. What he did say that evening indicated that we might stall awhile—perhaps there would be a slightly better offer if we did. But he exhibited no doubt that we would sell. And move.

I was shocked. There was no room in my plans for this...rupture. Was the demolition of our home and neighborhood—that is, my life—truly inevitable? Was there really no way we could avert it, cancel it, *delay* it? What if we just plain *refused to sell?*

Twenty years later, my mother told me that she could still remember my face on that day. It must have reflected extraordinary distress and confusion, for my folks were patient. If anyone refused to sell, they told me, GM would simply build its parking lot—for that was what would replace my world—around him. If we didn't sell, we'd have access privileges, enough space to get into our driveway and that was it. No room to play, and no one there to play with if there had been. And if you got caught in such a situation and didn't like it, then you'd really be in a fix, for the company wouldn't keep its double-your-money offer open forever. If we held out too long, who knew if the house would be worth anything at all. (I don't imagine that my parents attempted to explain to me the political process of condemnation, but if they had, I would have been outraged, for in a way, I still am.)

My dreams always pictured us as holdouts, living in a little house surrounded by asphalt and automobiles. I always imagined nighttime with the high, white-light towers that illuminated all the other GM parking lots shining down upon our house—and the little guardhouse that the company would have to build and man next door to prevent me from escaping our lot to run playfully among the parked cars of the multitudinous employees. Anyone reading this must find it absurd, or the details heavily derivative of bad concentration-camp literature or maybe too influenced by the Berlin Wall, which had been up only a short time. But it would be a mistake to dismiss its romanticism, which was for many months more real to me than the ridiculous reality—moving to accommodate a *parking lot*—which confronted my family and all my friends' families.

If this story were set in the Bronx or in the late sixties, or if it were fiction, the next scenes would be of pickets and protests, meaningful victories and defeats. But this isn't fiction—everything set out here is as unexaggerated as I know how to make it—and the time and the place were wrong for any serious uproar. In this docile midwestern company town, where Walter Reuther's trip to Russia was as inexplicable as the parting of the Red Sea (or as forgotten as the Ark of the Covenant), the idea that a neighborhood might have rights that superseded those of General Motors' Pontiac division would have been regarded as extraordinary, bizarre and subversive. Presuming anyone had had such an idea, which they didn't—none of my friends seemed particularly disturbed about moving, it was just what they would *do*.

So we moved, and what was worse, to the suburbs. This was catastrophic to me. I loved the city, its pavement and the mobility it offered even to kids too young to drive. (Some attitude for a Motor City kid, I know.) In Pontiac, feet or a bicycle could get you anywhere. Everyone had cars, but you weren't immobilized without them, as everyone under sixteen was in the suburbs. In the suburb to which we adjourned, cars were *the* fundamental of life—many of the streets in our new subdivision (not really a neighborhood) didn't even have sidewalks.

Even though I'd never been certain of fitting in, in the city I'd felt close to figuring out how to. Not that I was that weird. But I was no jock and certainly neither suave nor graceful. Still, toward the end of eighth grade, I'd managed to talk to a few girls, no small feat. The last thing I needed was new goals to fathom, new rules to learn, new friends to make.

So that summer was spent in dread. When school opened in the autumn, I was already in a sort of cocoon, confused by the Beatles with their paltry imitations of soul music and the bizarre emotions they stirred in girls.

Meeting my classmates was easy enough, but then it always is. Making new friends was another matter. For one thing, the kids in my new locale weren't the same as the kids in my classes. I was an exceptionally good student (quite by accident—I just read a lot) and my neighbors were classic underachievers. The kids in my classes were hardly creeps, but they weren't as interesting or as accessible as the people I'd known in my old neighborhood or the ones I met at the school bus stop. So I kept to myself.

In our new house, I shared a room with my brother at first. We had bunk beds, and late that August I was lying sweatily in the upper one, listening to the radio (WPON-AM, 1460) while my mother and my aunt droned away in the kitchen.

Suddenly my attention was riveted by a record. I listened for two or three minutes more intently than I have ever listened and learned something that remains all but indescribable. It wasn't a new awareness of music. I liked rock and roll already, had since I first saw Elvis when I was six, and I'd been reasonably passionate about the Ronettes, Gary Bonds, Del Shannon, the Crystals, Jackie Wilson, Sam Cooke, the Beach Boys and those first rough but sweet notes from Motown: the Miracles, the Temptations, Eddie Holland's "Jamie." I can remember a rainy night when I tuned in a faraway station and first heard the end of the Philadelphia Warriors' game in which Wilt Chamberlain scored a hundred points and then found "Let's Twist Again" on another part of the dial. And I can remember not knowing which experience was more splendid.

But the song I heard that night wasn't a new one. "You Really Got a Hold on Me" had been a hit in 1963, and I already loved Smokey Robinson's voice, the way it twined around impossibly sugary lines and made rhymes within the rhythms of ordinary conversation, within the limits of everyday vocabulary.

But if I'd heard those tricks before, I'd never understood them. And if I'd enjoyed rock and roll music previously, certainly it had never grabbed me in quite this way: as a lifeline that suggested—no, insisted—that these singers spoke *for* me as well as to me, and that what they felt and were able to cope with, the deep sorrow, remorse, anger, lust and compassion that bubbled beneath the music, I would also be able to feel and contain. This intimate revelation was what I gleaned from those three minutes of music, and when they were finished and I climbed out of that bunk and walked out the door, the world looked different. No longer did I feel quite so powerless, and if I still felt cheated, I felt capable of getting my own back, some day, some way.

Trapped

> It seems I've been playing your game way too long
> And it seems the game I've played has made you strong
> —Jimmy Cliff, "Trapped"

That last year in Pontiac, we listened to the radio a lot. My parents always had. One of my most shattering early memories is of the radio blasting when they got up—my mother around four-thirty, my father at five. All of my life I've hated early rising, and for years I couldn't listen to country music without being reminded almost painfully of those days.

But in 1963 and 1964, we also listened to WPON in the evening for its live coverage of city council meetings. Pontiac was beginning a decade of racial crisis, of integration pressure and white resistance, the typical scenario. From what was left of our old neighborhood came the outspokenly racist militant anti–school busing movement.

The town had a hard time keeping the shabby secret of its bigotry even in 1964. Pontiac had mushroomed as a result of massive migration during and after World War II. Some of the new residents, including my father, came from nearby rural areas where blacks were all but unknown and even the local Polish Catholics were looked upon as aliens potentially subversive to the community's Methodist piety.

Many more of the new residents of Pontiac came from the South, out of the dead ends of Appalachia and the border states. As many must have been black as white, though it was hard for me to tell that as a kid. There were lines one didn't cross in Michigan, and if I was shocked, when visiting Florida, to see separate facilities labeled "White" and "Colored," as children we never paid much mind to the segregated schools, the lily-white suburbs, the way that jobs in the plants were divided up along race lines. The ignorance and superstition about blacks in my neighborhood were as desperate and crazed in their own way as the feelings in any kudzu-covered parish of Louisiana.

As blacks began to assert their rights, the animosity was not less, either. The polarization was fueled and fanned by the fact that so many displaced Southerners, all with the poor white's investment in racism, were living in our community. But it would be foolish to pretend that the situation would have been any more civilized if only the natives had been around. In fact the Southerners were often regarded with nearly as much condescension and antipathy as blacks—race may have been one of the few areas in which my parents found themselves completely in sympathy with the "hillbillies."

Racism was the great trap of such men's lives, for almost everything could be explained by it, from unemployment to the deterioration of community itself. Casting racial blame did much more than poison these people's entire concept of humanity, which would have been plenty bad enough. It immobilized the racist, preventing folks like my father from ever realizing the real forces that kept their lives tawdry and painful and forced them to fight every day to find any meaning at all in their existence. It did this to Michigan factory workers as effectively as it ever did it to dirt farmers in Dixie.

The great psychological syndrome of American males is said to be passive aggression, and racism perfectly fit this mold. To the racist, hatred of blacks gave a great feeling of power and superiority. At the same time, it allowed him the luxury of wallowing in self-pity at the great conspiracy of rich bastards and vile niggers that enforced workaday misery and let the rest of the world go to hell. In short, racism explained everything. There was no need to look any further than the cant of redneck populism, exploited as effectively in the orange clay of the Great Lakes as in the red dirt of Georgia, to find an answer to why it was always the *next* generation that was going to get up and out.

Some time around 1963, a local attorney named Milton Henry, a black man, was elected to Pontiac's city council. Henry was smart and bold—he would later become an ally of Martin Luther King, Jr., of Malcolm X, a principal in the doomed Republic of New Africa. The goals for which Henry was campaigning seem extremely tame now, until you realize the extent to which they *haven't* been realized in twenty years: desegregated schools, integrated housing, a chance at decent jobs.

Remember that Martin Luther King would not take his movement for equality into the North for nearly five more years, and that when he did, Dr. King there faced the most strident and violent opposition he'd ever met, and you will understand how inflammatory the mere presence of Milton Henry on the city council was. Those council sessions, broadcast live on WPON, invested the radio with a vibrancy and vitality that television could never have had. Those hours of imprecations, shouts and clamor are unforgettable. I can't recall specific words or phrases, though, just Henry's eloquence and the pandemonium that greeted each of his speeches.

So our whole neighborhood gathered round its radios in the evenings, family by family, as if during wartime. Which in a way I guess it was—surely that's how the situation was presented to the children, and not only in the city. My Pontiac junior high school was lightly integrated, and the kids in my new suburban town had the same reaction as my Floridian cousins: shocked that I'd "gone to school with niggers," they vowed they would die—or kill—before letting the same thing happen to them.

This cycle of hatred didn't immediately elude me. Thirteen-year-olds are built to buck the system only up to a point. So even though I didn't dislike any of the blacks I met (it could hardly be said that I was given the opportunity to *know* any), it was taken for granted that the epithets were essentially correct. After all, anyone could see the grave poverty in which most blacks existed, and the only reason ever given for it was that they liked living that way.

But listening to the radio gave free play to one's imagination. Listening to music, that most abstract of human creations, unleashed it all the more. And not in a vacuum. Semiotics, the New Criticism, and other formalist approaches have never had much appeal to me, not because I don't recognize their validity in describing certain creative structures but because they emphasize those structural questions without much consideration of content: And that simply doesn't jibe with my experience of culture, especially popular culture.

The best example is the radio of the early 1960s. As I've noted, there was no absence of rock and roll in those years betwixt the outbreaks of Presley and Beatles. Rock and roll was a constant for me, the best music around, and I had loved it ever since I first heard it, which was about as soon as I could remember hearing anything.

In part, I just loved the sound—the great mystery one could hear welling up from "Duke of Earl," "Up on the Roof," "Party Lights"; that pit of loneliness and despair that lay barely concealed beneath the superficial bright spirits of a record like Bruce Channel's "Hey Baby"; the nonspecific terror hidden away in Del Shannon's "Runaway." But if that was all there was to it, then rock and roll records would have been as much an end in themselves—that is, as much a dead end—as TV shows like *Leave It to Beaver* (also mysterious, also—thanks to Eddie Haskell—a bit terrifying).

To me, however, TV was clearly an alien device, controlled by the men with shirts and ties. Nobody on television dressed or talked as the people in my neighborhood did. In rock and roll, however, the language spoken was recognizably my own. And since one of the givens of life in the outlands was that we were barbarians, who produced no culture and basically consumed only garbage and trash, the thrill of discovering depths within rock and roll, the very part that was most often and explicitly degraded by teachers and pundits, was not only marvelously refreshing and exhilarating but also in essence liberating—once you'd made the necessary connections.

It was just at this time that pop music was being revolutionized—not by the Beatles, arriving from England, a locale of certifiable cultural superiority, but by Motown, arriving from Detroit, a place without even a hint of cultural respectability. Produced by Berry Gordy, not only a young man but a *black* man. And in that spirit of solidarity with which hometown boys (however unalike) have always identified with one another, Motown was mine in a way that no other music up to that point had been. Surely no one spoke my language as effectively as Smokey Robinson, able to string together the most humdrum phrases and effortlessly make them sing.

That's the context in which "You Really Got a Hold on Me" created my epiphany. You can look at this coldly—structurally—and see nothing more than a naked marketing mechanism, a clear-cut case of a teenager swaddled in and swindled by pop culture. Smokey Robinson wrote and sang the song as much to make a buck as to express himself; there was nothing of the purity of the mythical artist about his endeavor. In any case, the emotion he expressed was unfashionably sentimental. In releasing the record, Berry Gordy was mercenary in both instinct and motivation. The radio station certainly hoped for nothing more from playing it than that its listeners would hang in through the succeeding block of commercials. None of these people and institutions had any intention of elevating their audience, in the way that Leonard Bernstein hoped to do in his *Young People's Concerts* on television. Cultural indoctrination was far from their minds. Indeed, it's unlikely that anyone involved in the process thought much about the kids on the other end of the line except as an amorphous mass of ears and wallets. The pride Gordy and Robinson had in the quality of their work was private pleasure, not public.

Smokey Robinson was not singing of the perils of being a black man in this world (though there were other rock and soul songs that spoke in guarded metaphors about such matters). Robinson was not expressing an experience as alien to my own as a country blues singer's would have been. Instead, he was putting his finger firmly upon a crucial feeling of vulnerability and longing. It's hard to think of two emotions that a fourteen-year-old might feel more deeply (well, there's lust...), and yet in my hometown expressing them was all but absolutely forbidden to men. This doubled the shock of Smokey Robinson's voice, which for years I've thought of as falsetto, even though it really isn't exceptionally high-pitched compared to the spectacular male sopranos of rock and gospel lore.

"You Really Got a Hold on Me" is not by any means the greatest song Smokey Robinson ever wrote or sang, not even the best he had done up to that point. The singing on "Who's Loving You," the lyrics of "I'll Try Something New," the yearning of "What's So Good About Goodbye" are all at least as worthy. Nor is there anything especially newfangled about the song. Its trembling blues guitar, sturdy drum pattern, walking bass and call-and-response voice arrangement are not very different from many of the other Miracles records of that period. If there is a single instant in the record which is unforgettable by itself, it's probably the opening lines: "I don't like you/But I love you..."

The contingency and ambiguity expressed in those two lines and Robinson's singing of them was also forbidden in the neighborhood of my youth, and forbidden as part and parcel of the same philosophy that propounded racism. Merely calling the bigot's certainty into question was revolutionary—not merely rebellious. The depth of feeling in that Miracles record, which could have been purchased for 69¢ at any K-Mart, overthrew the premise of racism, which was that blacks were not as human as we, that they could not feel—much less express their feelings—as deeply as we did.

When the veil of racism was torn from my eyes, everything else that I knew or had been told was true for fourteen years was necessarily called into question. For if racism explained everything, then without racism, not a single commonplace explanation made any sense. *Nothing* else could be taken at face value. And that meant asking every question once again, including the banal and obvious ones.

For those who've never been raised under the weight of such addled philosophy, the power inherent in having the burden lifted is barely imaginable. Understanding that blacks weren't worthless meant that maybe the rest of the culture in which I was raised was also valuable. If you've never been told that you and your community are worthless—that a parking lot takes precedence over your needs—perhaps that moment of insight seems trivial or rather easily won. For anyone who was never led to expect a life any more difficult than one spent behind a typewriter, maybe the whole incident verges on being something too banal for repetition (though in that case, I'd like to know where the other expressions of this story can be read). But looking over my shoulder, seeing the consequences to my life had I not begun questioning not just racism but all of the other presumptions that ruled our lives, I know for certain how and how much I got over.

That doesn't make me better than those on the other side of the line. On the other hand, I won't trivialize the tale by insisting upon how fortunate I was. What was left for me was a raging passion to explain things in the hope that others would not be trapped and to keep the way clear so that others from the trashy outskirts of barbarous America still had a place to stand—if not in the culture at large, at least in rock and roll.

Of course it's not so difficult to dismiss this entire account. Great revelations and insights aren't supposed to emerge from listening to rock and roll records. They're meant to emerge only from encounters with art. (My encounters with Western art music were unavailing, of course, because every one of them was prefaced by a lecture on the insipid and worthless nature of the music that I preferred to hear.) Left with the fact that what happened to me did take place, and that it was something that was supposed to come only out of art, I reached the obvious conclusion. You are welcome to your own.

COMING TO TERMS

In order to describe fully the meaning that *You Really Got a Hold on Me* came to hold for him, Dave Marsh also needs somehow to evoke the impact of the song on him. But the experience of music is notoriously difficult to put into words. Go back to the moment in his essay where Marsh describes first hearing Smokey Robinson on the radio in order to see what he does to suggest what it felt like for him to listen to that particular song at that particular moment in his life. You will want to pay attention both to how Marsh talks about the music itself and to the context in which he first heard it. Ask yourself: How does Marsh try to describe the actual *sound* of the song—what moments does he cite to suggest the experience of listening to it? And what does he tell you about himself, his situation, that helps you understand why this particular song and moment stand out so strongly for him? (If you can, you may want to listen to a recording of *You Really Got a Hold on Me*. What does Marsh catch about the song? What does he miss?) Take some notes in your journal that will help you discuss what Marsh does here as a writer and (perhaps) make use of similar moves in your own work. For example, does Marsh interpret specific lyrics from the song? How does he talk about the lyrics? How does Marsh connect words in the songs to a specific event? Does he let the lyrics speak for themselves? When? Why?

READING AS A WRITER

Dave Marsh tells a story here about growing up in the "outlands" of Pontiac, Michigan, about having to move from his boyhood house to make way for a company parking lot, about the pleasures of listening to the radio and the revelations offered by Smokey Robinson's *You Really Got a Hold on Me*. We think, however, that Marsh means not only to tell us about his boyhood but also to use that story to make an argument for the value of rock and roll, to explain some of the ways rock and roll offered him and many others a needed sense of identity and purpose, "a place to stand" in our culture, and a way to participate in history.

Maybe you've heard the expression or something like it: "It's only a story." The implication here is that telling stories is something less than other kinds of discourse. But if you think about it, stories or narratives are powerful means to teach, ask questions, transfer cultural values and beliefs, or make an argument. Sacred texts in many religions impart knowledge and belief through the use of parables, apparently simple stories on the surface but stories that have yielded many volumes of explanation and interpretation over the centuries. National history is a story; yet many societies argue that this story must be told, especially to students, to perpetuate a sense of national identity. We tend to think that, more often than not, a story is probably not just a story.

We would like you to consider Dave Marsh not only as a writer who has interesting things to say about rock and roll but also as a writer of history, a historian. *Fortunate Son* might be read as a story in which a writer has very carefully located himself within several historical contexts; it is a story where multiple histories meet. The narrative strands in Marsh's story are comprised of personal, family, local and national histories, and a history of media, all in which Marsh represents himself as not only an observer but also as a participant. Marsh claims that "everything looked different" after his "intimate revelation" with the Smokey Robinson song. For you, it doesn't have to be "everything" that changed, just *something*—a valuable lesson, a surprising discovery, a shift in outlook, a new emotion, a different attitude toward yourself or the world—that you can trace back to an encounter with the media or popular culture. In thinking and writing about this experience, your goal should be to understand more clearly what it was that affected you so deeply, and what it was in you that changed as a result. You will then be able to consider how it is that the mass media or works of popular culture can have such an effect. In other words, your goal in writing should be to tell us something about the power of the media and the value of popular culture through examining a powerful personal experience and how it happened to you. To go back to Marsh once again for an example: The challenge he faced as a writer was to explain how it was that hearing *You Really Got a Hold on Me* over the radio made such an impact on him. What was it about that song? What was it about the circumstances in which he heard it? What was it about the radio as a medium that helped to produce this moment? And what was it about Marsh himself that made him open to the experience he says he had?

You will want to pose similar questions about your own experience. Perhaps you saw a film that opened you up to a new emotion or suggested a new stance toward the world. What was it about this *particular* film? What was it about the *situation* in which you saw the film? What is it about film *as a medium* that gives it the power to produce such effects? And finally, what was it about you that caused you to be open to that experience? But it doesn't have to be a film. It might be a news event that you saw unfold on television, an album that you played over and over, or a heroic performance by a performer who touched you deeply. Whatever it is, whatever it was, your goal is to become clear about exactly what happened, what made the happening possible, and what changed in you as a result.

We think that Marsh's work has much to offer us, so we'd like to offer an alternative to the assignment already explained. The previous approach has the sense of conclusion; maybe you would rather see your relationship to a popular culture medium as part of an evolving and continuing process of discovery. *Fortunate Son* can also be read as a story in which a writer makes connections between several different kinds of history: personal, family, local, national, social, economic, and a history of music, rock and roll. At his moment of "epiphany," Marsh seems to find a new way of seeing racism; it is no longer an abstruse or a distant intellectual conclusion. Rather through listening to rock and roll, and specifically in 1963 to Smokey Robinson's *You Really Got a Hold on Me,* Marsh is somehow able to see his own history as a child, student, and rock and roll fan in terms of the times and conditions of racism in the United States in the 1960s. Marsh writes, "When the veil of racism was torn from my eyes, everything else that I knew or had been told was true for four-teen years was necessarily called into question."

Marsh's moment of "epiphany" can be interpreted in many ways. One way of reading this moment is to see it as an evolving process in which Marsh has gained a critical sense of history. He is no longer disconnected from or outside of history; rather, he is now inside history, connected as a participant. You might write an essay in which you connect to or place yourself in some ongoing historical context, which has become clearer or more meaningful to you through an experience with popular culture. In what ways has listening to a certain kind of music, viewing a film, accessing a site on the Internet, following a television series allowed you, like Dave Marsh, to see yourself as *participating* in history? To be *inside* of a specific unfolding or continuation of history?

In order for you to claim a stake in some history, you'll need to name and describe what this history is. What is the subject of this history? What role are you playing in this history as an insider or participant as a result of experiencing a particular medium of popular culture? Regardless of the media event you choose, it must provide you a way of connecting personal or family history to some kind of larger historical context.

Life in the Stone Age

Louis Menand

I.

If you advised a college student today to tune in, turn on, and drop out, he would probably call campus security. Few things sound less glamorous in 1991 than "the counterculture"—a term many people are likely to associate with Charles Manson. Writing about that period now feels a little like rummaging around in history's dustbin. Just twenty years ago, though, everyone was writing about the counterculture, for everyone thought that the American middle class would never be the same.

The American middle class never is the same for very long, of course; it's much too insecure to resist a new self-conception when one is offered. But the change that the counterculture made in American life has become nearly impossible to calculate—thanks partly to the exaggerations of people who hate the '60s, and partly to the exaggerations of people who hate the people who hate the '60s. The subject could use the attention of some people who really don't care.

The difficulties begin with the word "counterculture" itself. Though it has been from the beginning the name for the particular style of sentimental radicalism that flourished briefly in the late 1960s, it's a little misleading. For during those years the counterculture *was* culture—or the prime object of the culture's attention, which in America is pretty much the same thing—and that is really the basis of its interest. It had all the attributes of a typical mass culture episode: it was a lifestyle that could be practiced on weekends; it came into fashion when the media discovered it and went out of fashion when the media lost interest; and it was, from the moment it penetrated the middle class, thoroughly commercialized. Its failure to grasp this last fact about itself is the essence of its sentimentalism.

"Life in the Stone Age," by Louis Menand, *New Republic*, 1-7, 14-91, pp. 38–44. Reprinted with permission of the publisher.

The essence of its radicalism is a little more complicated. The general idea was the rejection of the norms of adult middle-class life; but the rejection was made in a profoundly middle-class spirit. Middle-class Americans are a driven, pampered, puritanical, and self-indulgent group of people. Before the '60s these contradictions were rationalized by the principle of deferred gratification: you exercised self-discipline in order to gain entrance to a profession, you showed deference to those above you on the career ladder, and material rewards followed and could be enjoyed more or less promiscuously.

The counterculture alternative looked to many people like simple hedonism: sex, drugs, and rock 'n' roll (with instant social justice on the side). But the counterculture wasn't hedonistic; it was puritanical. It was, in fact, virtually, Hebraic: the parents were worshiping false gods, and the students who tore up (or dropped out of) the university in an apparent frenzy of self-destructiveness—for wasn't the university their gateway to the good life?—were, in effect, smashing the golden calf.

There was a fair amount of flagrant sensual gratification, all of it crucial to the pop culture appeal of the whole business; but it is a mistake to characterize the pleasure-taking as amoral. It is only "fun" to stand in the rain for three days with a hundred thousand chemically demented people, listening to interminable and inescapable loud music and wondering if you'll ever see your car again, if you also believe in some inchoate way that you are participating in the creation of the New World.

The name of the new god was authenticity, and it was unmistakably the jealous type. It demanded an existence of programmatic hostility to the ordinary modes of middle-class life, and even to the ordinary modes of consciousness—to whatever was mediated, accommodationist, materialistic, and, even trivially, false. Like most of the temporary gods of the secular society, the principle of authenticity was merely paid lip service to by most of the people who flocked to its altar, and when the '60s were over, those people went happily off to other shrines. But there were some people who took the principle to heart, who flagellated their consciences in its service, and who, even after the '60s had passed, continued to obsess about being "co-opted."

There are two places in American society where this strain of puritanism persists. One is the academy, with its fetish of the unconditioned. The other is the high end of pop music criticism—the kind of criticism that complains, for instance, about the commercialism of MTV. Since pop music is by definition commercial, it may be hard to see how pop music commercialism can ever be a problem. But for many people who take pop music seriously it is *the* problem, and its history essentially begins with *Rolling Stone*.

II.

Rolling Stone was born in the semi-idyllic, semi-hysterical atmosphere of northern California in the late '60s, an atmosphere that Robert Draper's entertaining history of the magazine does an excellent job re-creating. His book is filled with vivid sketches of many of the classic-period types who passed through *Rolling Stone*'s offices and pages during the years the magazine was published in San Francisco—from 1967 until 1977, when it was moved to New York.

The theme that Draper has selected to tie the story together, though, is the standard history-of-the-'60s theme of selling out. He chooses to illustrate it by making Jann Wenner, the magazine's founder and still its editor and publisher, both the hero and the villain of the tale—the man who seized the moment and then betrayed it. This threatens to make Wenner a little more complicated than he actually is. An opportunistic, sentimental, shrewd celebrity-hound, Wenner was the first person in journalism to see what people in the music business already knew, and what people in the advertising business would soon realize: that rock music had become a fixture of American middle-class life. It had created a market.

Wenner knew this because he was himself the prototypical fan. He was born in 1946, in the first wave of the baby boom—his father would make a fortune selling baby formula for the children to whom the son later sold magazines—and he started *Rolling Stone* (he is supposed to have said) in order to meet John Lennon. He met Lennon; and he met and made pals with many more of his generation's entertainment idols—who, once they had become friends, and with or without editorial justification, turned up regularly on the covers of his magazine. Wenner was not looking for celebrity himself; he was only, like most Americans, a shameless worshiper of the stars. "I always felt Jann had a real fan's mentality," one of his friends and associates, William Randolph Hearst III, explained. "He wanted to hang out with Mick Jagger because Mick was cool, not because he wanted to tell people that *he* was cool as a result of knowing Mick."

The person who thinks Mick is cool is the perfect person to run a magazine devoted to serious fandom. But he is an obvious liability at a magazine devoted to serious criticism. Wenner was not a devotee of the authentic, not even a hypocritical one. He was a hustler: he believed in show biz, and saw, for instance, nothing unethical about altering a review to please a record company he hoped to have as an advertiser. "We're gonna be better than *Billboard!*" is the sort of thing he would say to encourage his staff when morale was low.

Morale was not thereby improved. For the people who produced Wenner's magazine took the '60s much more seriously than Wenner did. It wasn't merely that, like many editors, Wenner demonstrated a rude indifference to the rhythms of magazine production, commissioning new covers at the last minute and that sort of thing. It was that he didn't seem to grasp the world-historical significance of the movement that his magazine was spearheading. "Here we were," Jon Carroll, a former staffer, told Draper, "believing we were involved in the greatest cultural revolution since the sack of Rome. And he was running around with starlets. We thought that Jann was the most trivial sort of fool."

Draper's view is an only slightly less inflated version of Carroll's view. "Quite correctly," he writes of the early years, "the employees of *Rolling Stone* magazine saw themselves as leaders and tastemakers—the best minds of their generation." *Rolling Stone* covered the whole of the youth culture—though it generally steered clear, at Wenner's insistence, of radical politics. ("Get back," Wenner pleaded with his editors in 1970, after the shootings at Kent State inspired them to try to "detrivialize" the magazine, "get back to where you once belonged.") But the backbone of the magazine has always been its music criticism, and its special achievement is that it provided an arena for the development of the lyrical, pedantic, and hyperbolical writing about popular music that is part of the '60s' literary legacy. *Rolling Stone* wasn't the only place where this style of criticism flourished, but it was the biggest. *Rolling Stone* institutionalized the genre.

This is what Draper responds to in the magazine, and where his sympathies as a historian lie. His principal sources are from the editorial side of the magazine, because that is his principal interest. He tells us at some length about the editorial staffs travails, but gives a perfunctory account, as though he found it too distasteful to investigate, of, for example, the business staff's "Marketing through Music" campaign—a newsletter for "Marketing, Advertising, and Music Executives," circulated in the mid-'80s, that encouraged corporate sponsorship of rock concerts and the use of rock stars and rock songs in advertising. The business deals are here, but they are generally treated from the outside, and always as inimical to the true spirit of the magazine.

From the point of view of social history, though, "Marketing through Music" is the interesting part of the story. For rock music, like every other mass-market commodity, is about making money. Everyone who writes about popular music knows that before Sam Phillips, the proprietor of Sun Records, recorded Elvis Presley in 1954, he used to go around saying, "If I could find a white boy who could sing like a nigger, I could make a million dollars." But Elvis himself is somehow imagined to have had nothing to do with this sort of gross commercial calculation, and when Albert Goldman's biography appeared in 1981 and described Presley as a musically incurious and manipulative pop star, the rock critical establishment descended on Goldman in wrath.

All rock stars want to make money, and for the same reasons everyone else in a liberal society wants to make money: more toys and more autonomy. Bill Wyman, when he went off to become The Rolling Stones' bass player, told his mother that he'd only have to wear his hair long for a few years, and he'd get a nice house and a car out of it at the end. Even The Doors, quintessential late-'60s performers who thought they were making an Important Musical Statement, began when Jim Morrison ran into Ray Manzarek, who became the group's keyboards player, and recited some poetry he'd written. "I said that's it," Manzarek later explained. "It seemed as though, if we got a group together we could make a million dollars." Ray, meet Sam.

Pop stars aren't simply selling a sound; they're selling an image, and one reason the stars of the '60s made such an effective appeal to middle-class taste is because their images went, so to speak, all the way through. Their stage personalities were understood to be continuous with their offstage personalities—an impression enhanced by the fact that, in a departure from Tim Pan Alley tradition, most '60s performers wrote their own material. But the images, too, were carefully managed.

The Beatles, for example, were the children of working-class families; they were what the average suburban teenager would consider tough characters. Their breakthrough into mainstream popular music came when their manager, Brian Epstein, transformed them into four cheeky but lovable lads, an image that delighted the middle class. The Rolling Stones, apart from Wyman, were much more middle class. Mick Jagger attended (on scholarship) the London School of Economics; his girlfriend Marianne Faithfull, herself a pop performer, was the daughter of a professor of Renaissance literature. Brian Jones's father was an aeronautical engineer, and Jones, who founded the band, had what was virtually an intellectual's interest in music. He wrote articles for *Jazz News,* for instance, something one cannot imagine a Beatle doing. But when it became The Stones' turn to enter the mainstream, the lovable image was already being used in a way that looked unbeatable. So (as Wyman quite matter-of-factly describes it in his appealing memoir) *their* manager, Andrew Oldham, cast them as rude boys, which delighted middle-class teenagers in a different and even more thrilling way.

These images enjoyed long-term success in part because they suited the performers' natural talents and temperaments. But it is pointless to think of scrutinizing them by the lights of authenticity. One reason popular culture gives pleasure is that it relieves us of this whole anxiety of trying to determine whether what we're enjoying is real or fake. Mediation is the sine qua non of the experience. Authenticity is a high culture problem.

Unless, of course, you're trying to run a cultural revolution. In which case you will need to think that there is some essential relation between the unadulterated spirit of rock 'n' roll and personal and social liberation. "The magic's in the music," The Lovin' Spoonful used to sing. "Believe in the magic, it will set you free." The Lovin' Spoonful was a self-promoting, teenybopper band if there ever was one; but those lyrics turn up frequently in Draper's book. For they (or some intellectually enriched version of them) constitute the credo of the higher rock criticism.

The central difficulty faced by the serious pop exegete is to explain how it is that a band with a manager and a promoter and sales of millions of records that plays "Satisfaction" is less calculating than a band with a manager and a promoter and sales of millions of records that plays "Itchycoo Park" (assuming, perhaps unadvisedly, that a case cannot be made for "Itchycoo Park"). Theorizing about the difference can produce nonsense of an unusual transparency.

"Rock is a mass-produced music that carries a critique of its own means of production," explained the British pop music sociologist Simon Frith in *Sound Effects* (1981); "it is a mass-consumed music that constructs its own 'authentic' audience." To which all one can say is that when you have to put the word "authentic" in quotation marks, you're in trouble.

The problem is more simply solved by reference to a pop music genealogy that was invented in the late '60s and that has been embraced by nearly everyone in the business ever since—by the musicians, by the industry, and by the press. This is the notion that genuine rock 'n' roll is the direct descendant of the blues, a music whose authenticity it would be a sacrilege to question.

The historical scheme according to which the blues begat rhythm and blues, which begat rockabilly, which begat Elvis, who (big evolutionary leap here) gave us The Beatles, was canonized by *Rolling Stone*. It is the basis for *The Rolling Stone Illustrated History of Rock & Roll* (1976), edited by Jim Miller, which is one of the best collections of classic rock criticism; and it's the basis for *Rock of Ages: The Rolling Stone History of Rock & Roll* (1986) by Ed Ward, Geoffrey Stokes, and Ken Tucker, which reads a little bit like the kind of thing you would get if you put three men in a room with some typewriters and a stack of paper and told them they couldn't come out until they had written *The Rolling Stone History of Rock & Roll*.

All genealogies are suspect, since they have an inherent bias against contingency, and genealogies to which critics and their subjects subscribe with equal enthusiasm are doubly suspect. The idea that rock 'n' roll is simply a style of popular music, and that there was popular music before rock 'n' roll (and not produced by black men) that might have some relation to, say, "Yesterday" or "Wild Horses" or "Sad-Eyed Lady of the Lowlands"—songs that do not exactly call Chuck Berry to mind, let alone Muddy Waters—is largely unknown to rock criticism.

The reason that the link between Elvis Presley and The Beatles feels so strained is because we are really talking about the difference between party music for teenagers and pop anthems for the middle-class—between music to jump up and down by and music with a bit of a brow. Even the music to jump up and down by is a long way from the blues: adolescents from Great Neck did not go into hysterics in the presence of Blind Lemon Jefferson. An entertainment phenomenon like Mick Jagger, with his mysteriously acquired cockney-boy-from-Memphis accent, surely has as much relation to a white teen-idol like the young Frank Sinatra as he does to a black bluesman like Robert Johnson. Except that Robert Johnson is the real thing. Of course some of the music of Jagger and Richards and Lennon and McCartney appropriated the sound of black rhythm and blues: that's precisely the least indigenous and least authentic thing about it.

This is not to say that rock 'n' roll (or the music of Frank Sinatra, for that matter) doesn't come from real feeling and doesn't touch real feeling. And it's not to say that there aren't legitimate distinctions to be made among degrees of sham in popular music. When one is discussing Percy Faith's 1975 disco version of "Hava Nagilah," it is appropriate to use the term "inauthentic." But the wider the appeal a popular song has, the more zealously it resists the terms of art. The most affecting song of the 1960s was (let's say) the version of "With a Little Help from My Friends" that Joe Cocker sang at Woodstock on August 17, 1969—an imitation British music-hall number performed in upstate New York by a white man from Sheffield pretending to be Ray Charles. On that day, probably nothing would have sounded more genuine.

III.

Spiro Agnew thought that the helpful friends were drugs, which is a reminder that the counterculture was indeed defining itself against something. The customary reply to a charge like Agnew's was that he was mistaking a gentle celebration of togetherness for a threat against the established order—that he was, in '60s language, being uptight. Agnew's attacks were ignorant and cynical enough; but the responses, though from people understandably a little uptight themselves, were disingenuous. Few teenagers in 1967 thought that the line "I get high with a little help from my friends" was an allusion to the exhilaration of good conversation. "I get high" is a pretty harmless drug reference. But it is a drug reference.

The classic case of this sort of thing is "Lucy in the Sky with Diamonds," also on the *Sgt. Pepper's* album. When the press got the idea that the title encrypted the initials LSD, John Lennon, who had written the song, expressed outrage. "Lucy in the Sky with Diamonds," he allowed, was the name his little boy had given a drawing he had made at school and brought home to show his father; and this bit of lore has been attached to the history of *Sgt. Pepper's* to indicate how hysterically hostile the old culture was to the new. No doubt the story about the drawing is true. On the other hand, if "Lucy in the Sky with Diamonds" is not a song about an acid trip, it is hard to know what sort of song it is.

Drugs were integral to '60s rock 'n' roll culture in three ways. The most publicized way, and the least interesting, has to do with the conspicuous consumption of drugs by rock 'n' roll performers, a subject that has been written about interminably, A. E. Hotchner's overheated book on The Rolling Stones being one recent specimen among many. Lennon eating LSD as though it were candy, Keith Richards undergoing complete blood transfusions in an effort to cure himself of heroin addiction ("How do you like my new blood?" he would ask his friends after a treatment)—these are stories of mainly tabloid interest, though they are important to rock 'n' roll mythology since addiction and early death are part of jazz and blues mythology as well.

The drug consumption was real enough (though one doesn't see it mentioned that since the body builds a resistance to hallucinogens, it is not surprising that Lennon ate acid like candy: he couldn't have been getting much of a kick from it after a while). Some people famously died of drug abuse; many others destroyed their careers and their lives. But overindulgence is a hazard of all celebrity; it's part of the modern culture of fame. That rock 'n' roll musicians overindulged with drugs is not, historically, an especially notable phenomenon.

Then there are the references to drugs in the songs themselves. Sometimes the references were fairly obscure: "Light My Fire," for instance, the title of The Doors' biggest hit, was a phrase taken from an Aldous Huxley piece in praise of mescaline. Sometimes the references were overt (Jefferson Airplane's "White Rabbit," or The Velvet Underground's "Heroin"). Most often, though, it was simply understood that the song was describing or imitating a drug experience: "Lucy in the Sky with Diamonds," "Strawberry Fields," "Mr. Tambourine Man," "A Whiter Shade of Pale."

The message (such as it was) of these songs usually involved the standard business about "consciousness expansion" already being purveyed by gurus like Allen Ginsberg and Alan Watts: once you have (with whatever assistance) stepped beyond the veil, you will prefer making love to making war, and so forth. Sometimes there was the suggestion that drugs open your eyes to the horror of things as they are—an adventure for the spiritually fortified only. ("Reality is for people who can't face drugs," as Tom Waits used to say.) The famous line in The Beatles' "A Day in the Life" was meant to catch both senses: "I'd love to turn you on." It was all facile enough; but the idea was not, simply, "Let's party."

What was most distinctive about late '60s popular music, though, was not that some of its performers used drugs, or that some of its songs were about drugs. It was that late-'60s rock was music designed for people to listen to while they were *on* drugs. The music was a prepackaged sensory stimulant. This was a new development. Jazz musicians might sometimes be junkies, but jazz was not music played for junkies. A lot of late '60s rock music, though, plainly advertised itself as a kind of complementary good for recreational drugs. This explains many things about the character of popular music in the period—particularly the unusual length of the songs. There is really only one excuse for buying a record with a twelve-minute drum solo.

How the history of popular music reflects the social history of drug preference is a research topic that calls for some fairly daunting field work. It was clear enough in the late '60s, though, that the most popular music was music that projected a druggy aura of one fairly specific kind or another. Folk rock, for example, became either seriously mellow (Donovan, or The Youngbloods) or raucous and giggly (County Joe and the Fish), sounds suggesting that marijuana might provide a useful enhancement of the listening experience. Music featuring pyrotechnical instrumentalists (Cream, or Ten Years After) had an overdriven, methedrine sort of sound. In the '70s a lot of successful popular music was designed to go well with cocaine, a taste shift many of the '60s groups couldn't adjust to quickly enough. (The Rolling Stones were an exception.)

But the featured drugs of the late '60s were the psychedelics: psilocybin, mescaline, and, especially, LSD. They, were associated with the British scene through Lennon, who even before *Sgt. Pepper's* had apparently developed a kind of religious attachment to acid. And LSD was the drug most closely identified with the San Francisco scene, especially with The Grateful Dead, a group that had been on hand in 1965 when Ken Kesey and the Merry Pranksters took their "acid test" bus trips, and whose equipment had been paid for by Timothy Leary himself. It would seem that once a person was on a hallucinogen, the particular kind of music he was listening to would be largely irrelevant; but there were bands, like The Dead, whose drug aura was identifiably psychedelic.

You didn't have to be on drugs to enjoy late-'60s rock 'n' roll, as many people have survived to attest; and this is an important fact. For from a mainstream point of view, the music's drug aura was simply one aspect of the psychedelic fashion that between 1967 and 1969 swept through popular art (black-light posters), photography (fish-eye lenses), cinema (jump cuts and light shows), clothing (tie-dye), coloring (Day-Glo), and speech, ("you turn me on").

Psychedelia expressed the counterculture sensibility in its most pop form. It said: spiritual risk-taker, uninhibited, enemy of the System. It advertised liberation and hipness in the jargon and imagery of the drug experience. And the jargon wasn't restricted to people under 30, or to dropouts. For in the late '60s the drug experience became the universal metaphor for the good life. Commercials for honey, encouraged you to "get high with honey." The Ford Motor Company invited you to test drive a Ford and "blow your mind." For people who did not use drugs, the music was a plausible imitation drug experience because every commodity in the culture was pretending to be some kind of imitation drug experience.

Psychedelia, and the sensibility that attached to it, was a media-driven phenomenon. In April 1966 *Time* ran a story on the Carnaby Street, mods and rockers, Beatles and Rolling Stones scene in London. In fact, that scene was on its last legs when the article appeared; but many Americans were induced to vacation in London, which revived the local economy, and the summer of 1966 became the summer of "Swinging London."

Swinging London was perfect mass media material: sexy, upbeat, and fantastically photogenic. So when 20,000 people staged a "Human Be-In" in Golden Gate Park in January 1967, the media were on hand. Here was a domestic version of the British phenomenon: hippies, Diggers, Hell's Angels, music, "free love," and LSD—the stuff of a hundred feature stories and photo essays. The media discovery of the hippies led to the media discovery of the Haight-Ashbury, and the summer of 1967 became the San Francisco "Summer of Love," that year's edition of Swinging London. *Sgt. Pepper's* was released in June, and the reign of psychedelia was established. The whole episode lasted a little less than three years—about the tenure of the average successful television series.

Once the media discovered it, the counterculture ceased being a youth culture and became a commercial culture for which youth was a principal market—at which point its puritanism (inhibitions are oppressive) became for many people an excuse for libertinism (inhibitions are a drag). LSD, for instance, was peddled by Leary through magazines like *Playboy,* where, in a 1966 interview, he explained that "in a carefully prepared, loving LSD session, a woman will inevitably have several hundred orgasms." This was exactly the sort of news *Playboy* existed to print, and the interviewer followed up by asking whether this meant that Leary found himself suddenly irresistible to women. Leary allowed that it did, but proved reluctant to give all the credit to a drug, merely noting that: "Any charismatic person who is conscious of his own mythic potency awakens this basic hunger in women and pays reverence to it at the level that is harmonious and appropriate at the time."

Playboy is not a magazine for dropouts; and the idea that counterculture drugs were really aphrodisiacs was an idea that appealed not to teenagers (who do not require hormonal assistance) but to middle-aged men. ("Good sex would have to be awfully good before it was better than on pot," Norman Mailer mused, presumably for the benefit of his fellow 45-year-olds, in *The Armies of the Night,* in 1968.) It was not teenagers who put Tom Wolfe's account of Kesey's LSD quackery, *The Electric Kool-Aid Acid Test* (1968), on the hardcover bestseller list. Hippies did not buy tickets to see *Hair* on Broadway, where it opened in 1968 and played over 1,700 performances, or read Charles Reich's homage to bell-bottom pants in *The New Yorker.* People living on communes did not make "Laugh-In," Hollywood's version of the swinging psychedelic style, the highest-rated show on television in the 1968–69 season. And, of course, students did not design, manufacture, distribute, and enjoy the profits from rock 'n' roll records. Those who attack the counterculture for disrupting what they take to have been the traditional American way of life ought to look to the people who exploited and disseminated it—good capitalists all—before they look to the young people who were encouraged to consume it.

IV.

After the Altamont concert disaster in December 1969, when a fan was killed a few feet from the stage where the Rolling Stones were performing, psychedelia lost its middle-class appeal. More unpleasant news followed in 1970—the Kent State and Jackson State shootings, the Manson Family trials, the deaths by overdose of famous rock stars. And even more quickly than it had sprung up, the media fascination with the counterculture evaporated.

But the counterculture, stripped of its idealism and its sexiness, lingered on. If you drove down the main street of any small city in America in the 1970s, you saw clusters of teenagers standing around, wearing long hair and bell-bottom jeans, listening to Led Zeppelin, furtively getting stoned. This was the massive middle of the baby-boom generation, the remnant of the counterculture—a remnant that was much bigger than the original, but in which the media had lost interest. These people were not activists or droupouts. They had very few public voices. One of them was Hunter Thompson's.

Thompson came to *Rolling Stone* in 1970, an important moment in the magazine's history. Wenner had fired Greil Marcus, a music critic with an American studies degree who was then his reviews editor, for running a negative review of an inferior Dylan album called *Self-Portrait* (it is one of Wenner's rules that the big stars must always be hyped); and most of the politically minded members of the staff quit after the "Get Back" episode following Kent State. There were financial problems as well. By the end of 1970, *Rolling Stone* was a quarter million dollars in debt.

Hugh Hefner, who is to testosterone what Wenner is to rock 'n' roll, offered to buy the magazine, but Wenner found other angels. Among them were record companies. Columbia Records and Elektra were delighted to advance their friends at *Rolling Stone* a year's worth of advertising; *Rolling Stone* and the record companies, after all, were in the same business.

The next problem was to sell magazines. (*Rolling Stone* relies heavily on newsstand sales, since its readers are not the sort of people who can be counted on to fill out subscription renewal forms with any degree of regularity.) Here Wenner had two strokes of good fortune. The first was a long interview he obtained with John Lennon, the first time most people had ever heard a Beatle not caring to sound lovable. It sold many magazines. The second was the arrival of Thompson.

Thompson was a well-traveled, free-spirited hack whose résumé included a stint as sports editor of *The Jersey Shore Herald,* a job as general reporter for *The Middletown Daily News,* freelance work out of Puerto Rico for a bowling magazine, a period as South American correspondent for *The National Observer* (during which he suffered some permanent hair loss from stress and drugs), an assignment covering the 1968 presidential campaign for *Pageant,* two unpublished Great American novels, a little male modeling, and a narrowly unsuccessful campaign for sheriff of Aspen, Colorado.

Thompson had actually been discovered for the alternative press by Warren Hinckle, the editor of *Ramparts,* which is when his writing acquired the label "gonzo journalism." But Thompson was interested in *Rolling Stone* because he thought it would help his nascent political career by giving him access to people who had no interest in politics (a good indication of the magazine's political reputation in 1970). A year after signing on, he produced the articles that became *Fear and Loathing in Las Vegas* (1972), a tour de force of pop faction about five days on drugs in Las Vegas. It sold many copies of *Rolling Stone,* and it gave Thompson fortune, celebrity, and a permanent running headline.

Many people who were not young read *Fear and Loathing in Las Vegas* and thought it a witty piece of writing. Wolfe included two selections from Thompson's work in his 1973 anthology *The New Journalism* (everyone else but Wolfe got only one entry); and this has given Thompson the standing of a man identified with an academically recognized Literary Movement. But Thompson is essentially a writer for teenage boys. *Fear and Loathing in Las Vegas* is *The Catcher in the Rye* on speed: the lost weekend of a disaffected loser who tells his story in a mordant style that is addictively appealing to adolescents with a deep and unspecified grudge against life.

Once you understand the target, the thematics make sense. Sexual prowess is part of the Thompson mystique, for example, but the world of his writing is almost entirely male, and sex itself is rarely more than a vague, adult horror; for sex beyond mere bravado is a subject that makes most teenage boys nervous. A vast supply of drugs of every genre and description accompany the Thompson persona and maintain him in a permanent state of dementia; but the drugs have all the verisimilitude of a 14-year-old's secret spy kit: these grown-ups don't realize that the person they are talking to is *completely out of his mind* on dangerous chemicals. The fear and loathing in Thompson's writing is simply Holden Caulfield's fear of growing up—a fear that, in Thompson's case as in Salinger's, is particularly convincing to younger readers because it so clearly runs from the books straight back to the writer himself.

After the Las Vegas book, *Rolling Stone* assigned Thompson to cover the 1972 presidential campaign. His reports were collected in (inevitably) *Fear and Loathing on the Campaign Trail* (1973). The series begins with some astute analysis of primary strategy and the like, salted with irreverent descriptions of the candidates and many personal anecdotes. Thompson's unusual relation to the facts—one piece, which caused a brief stir, reported that Edmund Muskie was addicted to an obscure African drug called Ibogaine—made him the object of some media attention of his own. But eventually the reporting breaks down, and Thompson is reduced at the end of his book to quoting at length from the dispatches of his *Rolling Stone* colleague Timothy Crouse (whose own book about the campaign, *The Boys on the Bus*, became an acclaimed exposé of political journalism).

Since 1972 Thompson has devoted his career to the maintenance of his legend, and his reporting has mostly been reporting about the Thompson style of reporting, which consists largely of unsuccessful attempts to cover his subjects, and of drug misadventures. He doesn't need to report, of course, because reporting is not what his audience cares about. They care about the escapades of their hero, which are recounted obsessively in his writing, and some of which were the basis for an unwatchable movie called *Where the Buffalo Roam*, released in 1980 and starring Bill Murray.

Thompson left *Rolling Stone* around 1975 and eventually became a columnist for the *San Francisco Examiner*. He has been repackaging his pieces in chronicle form regularly since 1979. *Songs of the Doomed* is the third collection, and most of the recent material concerns the author's arrest earlier this year on drug possession and sexual assault charges in Colorado. Having made a fortune portraying himself as a champion consumer of controlled substances,

Thompson naturally took the position that the drugs found in his house must have been left there by someone else. (The charges, unfortunately for a writer badly in need of fresh adventures, were dismissed.)

Thompson, in short, is practically the only person in America still living circa 1972. His persona enacts a counterculture sensibility with the utopianism completely leached out. There are no romantic notions about peace and love in his writing, only adolescent paranoia and violence. There is no romanticization of the street, either. Everything disappoints him—an occasionally engaging attitude that is also, of course, romanticism of the very purest sort. Thompson is the eternally bitter elegist of a moment that never really was, and that is why he is the ideal writer for a generation that has always felt that it arrived onstage about five minutes after the audience walked out.

V.

If all popular culture episodes were only commercial and manipulative, they would not matter to us. Some things are what you make them, and even the shabbiest cultures contribute to character. If you grew up in Disneyland, you would care about Mickey Mouse in spite of his artificiality, for Mickey would have been one of the presences in the world where your spirit was formed. Something like this is true for people who grew up in the '60s. For the late-'60s counterculture was not, by any means, the shabbiest episode of the postwar era, even if it now seems the most antique. It was imaginative and infectious, and it touched a nerve.

The faith in popular music, in consciousness-expansion, and in the nonconformist lifestyle that made up the countercultural ethos seems clearly misplaced today. You wonder why it didn't dawn on all those disaffected *Rolling Stone* writers and editors that Wenner was successful precisely because he wasn't the anomaly they took him for. But faith in anything can be a valuable sentiment; and what young people in the '60s thought their faith made it possible for them to do was to tell the truth. Of course, telling the truth is much harder than they thought it was; and the culture they imagined was sustaining them turned out not to be "authentically" theirs, and not really sustainable, after all. But those people had not yet become cynics.

The silliest charge brought against the '60s is the charge of moral relativism. Ordinary life must be built on the solid foundations of moral values, those who make this charge argue, and the '60s persuaded people that the foundations weren't solid, and that any morality would do that got you through the night. The accusation isn't just wrong about the '60s; it's an injustice to the dignity of ordinary life, which is an irredeemably pragmatic and open-ended affair. You couldn't make it through even the day if you held every transaction up to scrutiny by the lights of some received moral code. Radicals and youthful counterculture types in the '60s weren't moral relativists. They were moral absolutists. They scrutinized everything, and they believed that they could live by the distinctions they made.

There are always people who think this way—people who see that the world is a little fuzzy and proceed to make a religion out of clarity. In the '60s their way of thinking was briefly but memorably a part of the popular culture. Hotchner's book on The Rolling Stones is a mélange of clichés and misinformation; but it is constructed around a series of interviews with people who were around the band in the '60s, and although most of the anecdotes have the polished and improved feel of tales many times retold, a few have a kind of parabolic resonance.

One of the stories is told by a photographer named Gered Mankowitz, who accompanied The Rolling Stones on their American tours in the 1960s. It seems that there were two groupies in those days who dedicated themselves to the conquest of Mick Jagger. After several years of futile pursuit, they managed to get themselves invited to a house where The Stones were staying, and Mick was persuaded to take both of them to bed. Afterward, though, the girls were disappointed. "He was only so-so," one of them complained. "He tried to come on like Mick Jagger, but he's no Mick Jagger." The real can always be separated from the contrived: wherever that illusion persists, the spirit of the '60s still survives.

COMING TO TERMS

"Life in the Stone Age" is a complicated piece of criticism in that Menand is considering four books concerned with The *Rolling Stone* magazine, the 1960s, and the rock and roll music of that era all within the same review. Magazine editors will sometimes publish this kind of multiple book review for the complex perspective it offers on an event, a person, or a particular subject area. It's also an acknowledgment on the part of an editor that if multiple books are published within a short period of time on a subject, in this case 1960s rock and roll culture, there might be considerable audience interest in the subject.

One of the difficulties, we think, in coming to terms with this kind of criticism, however, is keeping track of who is speaking: Is it one of the authors being reviewed? Or is it Menand speaking *about* one of the authors? Making sense of a multiple book review is especially difficult if you're not familiar with the subject, event, or subject area being written about. So, in your journal, we'd like you to make four columns with an author heading for each of the books being reviewed. Then under each heading record a few passages in which Menand is quoting from each of these authors. Next beneath each list of quotations write: "Menand's response" and record how Menand is reading the quotation, responding to the author. You might find responding to the following questions helpful. Is Menand, for example, concurring with an author? Challenging what an author says about rock and roll in the 1960s? Finally, with which author (or authors) do Menand's sympathies lie?

According to Menand, what author (or authors) provides the most intelligent, accurate, or critically sound reading of rock and roll culture in the 1960s? How did you make this decision? What can you infer, from your reading of this review, is Menand's attitude towards The *Rolling Stone,* rock and roll, and the 1960s?

READING AS A WRITER

At one point in "Life in the Stone Age," Menand remarks, "All genealogies are suspect." What has captured Menand's critical attention (and what he wants to read against) "is the notion that genuine rock 'n' roll is the direct descendant of the blues, a music whose authenticity it would be a sacrilege to question." Although the term *genealogy* has rather specific meaning in current literary and cultural criticism, we think Menand's sense of this word might be understood as meaning some kind of historical analysis, an attempt to trace the "descendants" of something, for example, rock and roll, from its beginnings to the present.

Menand seems aware that, in the case of rock and roll, there might be more than one genealogy at play, competing versions of the same story which might challenge each other's sense of the past. We also think that Menand's reading of this genealogy suggests that historians or genealogists have choices to make in the ways they represent what's important from the past. And because choices are to be made, genealogies or histories work rhetorically to promote their own critical arguments, their own particular interpretation of the past.

As part of a journal assignment, we'd like you first to represent what Menand claims to be the standard genealogy of rock and roll in your journal. Then we'd like you to respond to the following questions in your journal. In what ways does Menand challenge this standard genealogy? Why might it be called "sacrilege" to question or revise this particular version of rock and roll history? And what might be lost if this genealogy was revised? What would be the consequences to the ways we see rock and roll? Finally, we'd like you to attempt to write a genealogy of your own, considering what might be the "descendants" of a contemporary media or popular culture text. We don't expect you to write the definitive genealogy of, say, the *Star Wars* films or big-budget, Hollywood disaster films or rap music or a particular fashion or a particular kind of sitcom or cop show on television. In fact, we'd like this genealogy to be exclusively yours, your entry into possible cultural mythology, your reading of culture. But you need to be able to say, "This text has a history, a genealogy." Your text needs to trace a line of textual "descendants" which came before or exist with, but are "related" to this contemporary media or popular culture text. After you have sketched out this genealogy, we'd also like you to respond to the following questions in your journal. What might be the critical argument of your genealogy? Its way of enacting a particular critical conversation about this current text and its descendants? What is your genealogy arguing that is important about this text and those places it came from?

WRITING CRITICISM

Writing to a friend in the early 19th century, Thomas Jefferson once remarked, "We may consider each generation as a distinct nation." Jefferson's metaphor of a generation as a "distinct nation" unto itself offers its own particular nuance to current discussions of multiculturalism in which generational differences are not often spoken about. And we think Jefferson's comment speaks to American politicians' endless struggle to understand (and gain the votes of) a younger generation. In the 1955 film *Rebel Without a Cause,* James Dean's character, in one of the film's climatic moments, screams out to no one in particular, "You're tearing me apart!" Dean's character comes across as a rebel without a cause, a member of a new generation (in search of something), lashing out at the perceived hypocrisy, bankrupt values, empty dreams, and false promises of his parents' generation. And in one of the 1990s film adaptations of the popular (or retro-popular) 1970s television series *The Brady Bunch,* a stranger wanders into the Brady household. After a few truly strange days—encounters with cute "bubble gum" rock and roll songs, platform shoes, love beads, leather vests, paisley and tie-dyed fabrics—this visitor from the 1990s refers to the Bradys as being "decade-impaired." If you've met the television Bradys before (or their film counterparts), you can appreciate the comment: The Bradys live in a perpetual 1970s time warp where, in the words of Greg Brady, they will forever be one "groovy, far-out, happening kind" of family.

We've pulled these very different "scenes" together for a moment as a way of suggesting the continuity of a cultural discussion whose representations in literature, film, television, music, and theater (and even politics) range from parody, to satire, to the serious, to the tragic: generation follows generation follows generation—and as each passes, in its wake it seems to leave confusion, suspicion, misunderstanding, the invention of new media forms, a sense of loss, a struggle for a new self-identity. This interplay of passing (and often colliding) generations leaves us with the sense that, in some way, one generation is always a generation of strangers to another. And this need to see, name, and describe separate generations as individual cultural moments has led to a continuing inventory of generational markers: today, we have Generation X and the Wired Generation; yesterday we had the Swing Generation, Lost Generation, Me Generation, Pepsi Generation, TV Tray Generation, Atomic Generation, Beat Generation, Woodstock Generation, Love Generation, and Latch Key Generation, to name a few.

One of the reasons why we decided to include Menand's article in *Media Journal* is that it participates in a tradition of naming or defining a generation or decade in terms of the media and popular culture texts made popular at that time. As a part of his project, Menand discusses The *Rolling Stone* magazine as a defining moment and cultural text of the 1960s, along with the rock and roll songs which are seen by many to define the 1960s counterculture. We think Menand's perspective on this subject is especially astute in the way it maintains a stance of critical distance from its subject; in other words, Menand is wary of an author's attempt to romanticize or simply celebrate the 1960s.

We'd like you to undertake a similar kind of project and write an essay in which you "define" your generation and offer your take on its own sense of self-identity or cultural independence from another generation through discussing a set of media and popular culture texts. You might choose as generational markers, for example, films, television programs, a style of music, fashions, advertisements, music videos, Internet sites, magazines, or radio programs. (If a particular text or medium is especially important in defining your generation, you might consider discussing this text by creating a genealogy of it.) You might consider other cultural texts as well in the way certain cultural artifacts are said to define, say, the 1970s—lava lamps, tie-dyed clothing, Volkswagen Beetles, etc. You might also think about how advances in telecommunications technology might be seen as a way of defining a generation. Regardless of the texts you choose, you will need to explain in some detail how you are reading these texts as a way of defining a generation, a specific cultural self-identity and show how these texts embody specific cultural values and beliefs which characterize your sense of generation.

We also have in mind an alternative approach to this writing project. Menand's discussion provides an important caveat to the kind of cultural analysis we are proposing here. To a significant degree, Menand argues that the popular images and mythology of the 1960s were created and packaged through commercial advertising, serious journalism, and academic criticism. (You might retread those sections of "Life in the Stone Age" where Menand is concerned with the genealogy of rock and roll, commercialism, and authenticity.) As an alternative approach, then, you might examine a set of popular images and generational markers already in place to describe your generation. Your project would be first to represent this conventionalized representation, say, of Generation X, as it has been presented in several magazine or on-line texts. (You'll need to cite specific passages from these texts.) What do these authors choose to speak for you in defining this generation? How do these authors characterize the cultural beliefs, values, and interests of Generation X? You might also include characters from films or television programs which you read as attempts to represent members of Generation X. Or you might even examine a set of print ads which seem to be designed for Generation X. Then the second aspect of this approach would be for you to either accept (argue for) this established version of Generation X or to challenge this representation, proposing an alternative set of media and popular culture texts and in this process explain a different set of cultural values, beliefs, or interests.

The 19-Inch Neighborhood

Joshua Meyrowitz

I live in a small New Hampshire town, but in the last few weeks I met the Lebanese leader of Amal and I was shouted at by militant Shiite hijackers. I sat beside the families of hostages as they anxiously watched their loved ones at a news conference in Beirut and as they later rejoiced when the hostages were released.

I weighed the somber questions and comments of anchorman Dan Rather as he "negotiated" with Lebanese minister Nabih Berri. I evaluated firsthand the demeanor of the news reporters, the facial expressions of the hijackers and the public comments of President Reagan and his spokesmen.

And I participated in this drama of international scope without ever leaving New Hampshire; indeed, I shared in it fully when sitting isolated in my living room in front of my television, watching the 525-line screen of flickering specks of light and color that my brain translates into pictures of people, objects and motions. The visual liveliness—like a conglomeration of thousands of flashing neon lights—and the intensity of the drama itself kept me riveted to the screen.

In contrast, the images through my window of trees, dogs and neighbors' houses are crisp and clear—tangible, real. Yet when I think of "keeping in touch" with things each day, I, along with a hundred million others, turn to the blurry television set. Recently a house in my town was destroyed by fire, and I vaguely recall reading the story in my local paper. Was anyone hurt? Is the family that lived there homeless now? Have they, too, suddenly been taken hostage by a swirl of events not of their making? I don't know. I could find out, I suppose, but I probably won't.

Reality

For I, and most of my neighbors, no longer simply live in this town; we don't live "with" each other in quite the same way our grandparents did. We, like the 98 percent of American families who own a TV, have granted it the power to redefine our place and our social reality. We pay more attention to, and talk more about, fires in California, starvation in Africa and sensational trials in Rhode Island than the troubles of nearly anyone except perhaps a handful of close family, friends and colleagues.

Our widespread adoption of television and other electronic media has subtly but significantly reshaped our world. For the first time in human civilization we no longer live in physical places. And the more we rely on our video window, the less relation there is between where we are and what we know and experience, the less there's a relationship between where we are and who we are.

Such changes affect our sense of identification with our community—and role relationships within our family. Isolated at home or school, young children were once sheltered from political debates, murder trials, famines and hostage crises. Now, via TV, they are taken across the globe before we give them permission to cross the street.

Similarly, our society was once based on the assumption that there were two worlds: the public male sphere of "rational accomplishments" and brutal competitions, and the private female sphere of child rearing, of emotion and intuition. But just as public events have become dramas played out in the privacy of our living rooms and kitchens, TV close-ups reveal the emotional side of public figures. Television has exposed women to parts of the culture that were once considered exclusively male and forced men to face the emotional dimensions and consequences of public actions.

For both better and worse, TV has smashed through the old barriers between the worlds of men and women, children and adults, people of different classes, regions, and levels of education. It has given us a broader but also a shallower sense of community. With its wide reach, it has made it difficult to isolate oneself from the informational arena it creates.

To watch TV now is to enter the new American neighborhood. The average household keeps a TV set on for 50 hours a week. One may watch popular programs not merely to see the program, but to see what others are watching. One can watch not necessarily to stare into the eyes of America, but to look over the shoulders of its citizens and see what they see.

Television has become our largest shared arena where the most important things happen. When a friend sings exquisitely, we no longer say, "You should sing in our church," but rather, "You should be on television." Our funniest friends are wished an appearance on *The Tonight Show,* not a performance at the town hall. The early presidents of this country were seen by few of the voters of their day; now it is impossible to imagine a candidate who has not visited us all, on television.

Weather

Television has replaced the local street corner and market as an important place to monitor—but, as with a marketplace, we do not always identify personally with what goes on inside. We may avidly watch what is on the news and on the entertainment and talk shows even as we exclaim, "I can't believe people watch this!" or, "What's the world coming to?"

Regardless of its specific content, then, television today has a social function similar to the local weather. No one takes responsibility for it, often it is bad, but nearly everyone pays attention to it and sees it as a basis of common experience and as a source of conversation. Indeed, television has given insularity of place a bad name; it now seems bizarre to be completely unaware or cut off. The TV set is a fixture in the recreation rooms of convents; it is even something that is sometimes watched in the formerly silent halls of Trappist monks.

Paradoxically, TV is both a hijacker and a liberator, hostage and hostage taker. It frees us from the constraints of our isolated physical locations, but flies us to a place that is no place at all. And our attention is most easily held hostage when television itself becomes a hostage of terrorists, demonstrators, politicians and other self-conscious social actors who vie for the chance to become—at least for a while—our closest video neighbor.

COMING TO TERMS

One way to come to terms with Joshua Meyrowitz's article is to become aware of two major "shifts" as he builds his critical argument: "Our widespread adoption of television and other electronic media has subtly but significantly reshaped our world. For the first time in human civilization we no longer live in physical places. "So writes Meyrowitz in "The 19-Inch Neighborhood." In any argument about a large social change, the writer must describe the situation "before" the shift in question and then go on to talk about what happens "after" the decisive shift. Meyrowitz does this several times in his piece, which is about the decisive shifts caused by television's entrance into our lives. First, re-read the passage cited above, and in your journal explain what you think Meyrowitz means by this change in our lives. Then go back through his essay and note with a pen places where he makes use of this technique of representing life "before" and "after" television's profound effects. In your journal, then, rephrase his critical argument in your own words as a series of "before" and "after" statements—as many as you can. There is another kind of "shift" we'd like you to pay attention to, which may be connected to the first. Meyrowitz begins his article by talking about himself. But, in much of his piece, he shifts from "I" to "we" and "us." We'd like you to consider this second shift by responding to several questions in your journal: What is the effect of this shift to the plural? What does it allow Meyrowitz to accomplish in his discussion of television? How might the shift in society here be somehow connected to this shift of pronouns?

READING AS A WRITER

Meyrowitz "concludes" his article in an intriguing, yet unsettling way, leaving his readers not with a neatly tied-up sense of conclusion or closure, but with a *paradox:* "Paradoxically, T.V. is both a hijacker and a liberator, hostage and hostage taker. It frees us from the constraints of our isolated physical locations, but flies us to a place that is *no place at all.*" A paradox is sometimes defined as something that on first sight is apparently contradictory, but on closer examination often reveals some kind of significant truth. Often great religious teachers or philosophers speak enduring truths through paradox. It seems as though if you're in a hurry, you miss the point; if you're patient, you get something valuable, maybe an insight. We like philosopher Kenneth Burke's paradox: "A way of seeing is a way of not seeing." But to return to the paradox that Meyrowitz leaves us with, we'd like you to consider the consequences of "concluding" a critical argument with a paradox and to respond to several questions in your journal: Is it an academic or a critical move you'd expect from an author you might study in a college writing or composition course? Why or why not? What might be the effects of leaving an audience with a paradox? Are there other possible applications of a media experience that paradoxically "frees" you from the constraints of a physical location, but transports you "to a place that is no place at all"? Through music? Film? Novels? Or the Internet? If so, how would you explain this sense of paradox and its consequences in your own media experience?

WRITING CRITICISM

One way of extending Meyrowitz's project here might be to look at some specific ways in which the "no place" of TV attempts either to (a) represent real places or (b) imagine fictional ones. In the first case, your task would be to write an essay that analyzes how a certain show works to situate itself as part of the media extension of an actual community. For instance, in contrast to the network news shows, local TV new programs usually work very hard to identify themselves with their "hometown." Your extension of Meyrowitz here would be to investigate how these local stations work to nurture this sense of "community" by creating emblems, icons, promotions, allusions, or references which help identify a certain news program as "local." How might these programs, or the various news celebrities, become an extension of the community? In what ways do the celebrities touch the lives of the viewers? In a broadcast, what local knowledge or sense of history do the newscasters employ to appear to be knowledgeable about the community, as if they too lived there? What do the conversations or remarks made among the newscasters, as a lead-in or concluding remark to a feature story, suggest that they know not only the news, but that they also know the underlying concerns, fears, hopes, or prejudices of the local community? There is one final aspect of local news coverage we'd like you to consider here. In what ways does the media experience of seeing local news represent a place or community correspond (or not) to the actual places and communities that the viewers live in, or more specifically that you live in? What are the agreements and discrepancies between these two versions of a place or a community? (Of course, the local news is only one example among many sorts of local programming; you might want to look instead at local talk shows, children's, religious, or community affairs programs—or even local radio or cable programs.)

The other approach would be to look at how TV creates a community that viewers are willing, at least for a time, to imagine themselves as being a part of. The landscape of all our memories is filled with places and cities we know through television: Mayberry, New Rochelle, Sesame Street—or the Cunninghams' or Lucy and Ricky's living room or "The Swamp" or Rosie's or—places that we've never really been to, but if were to somehow find ourselves there, we'd know exactly where we were and feel right at home. Sometimes these communities are wholly imaginary: ranging from the nameless suburbs of many sitcoms and dramas, to the various workplaces set in cities that are never quite specified, to the sinister specificity of *Twin Peaks* or the utopian Sisily of *Northern Exposure*. But often TV presents us with versions of real places: the Boston of *Cheers* or *St. Elsewhere,* the New York of *NYPD Blue* or *Law and Order* or *Seinfeld* or *Mad About You,* the Baltimore of *Homicide,* the Chicago of *Chicago Hope and ER,* the Philadelphia of *thirtysomething*. Whatever program or location that you choose, your job as a writer will be to show how TV goes about creating a sense of place, and to explain why that place is alluring enough to persuade viewers to enter it imaginatively, and what it gives them that their own towns or cities lack.

Barbara Walters's Theater of Revenge

Mark Crispin Miller

In Hollywood today, being beautiful is not enough; the stars also must exhibit "a self-mocking sense of humor." So say Hollywood's casting directors, as surveyed recently by *TV Guide*. Self-mockery, they have noticed, is "a key ingredient in star quality. 'Tom Selleck has it,' says HBO's [head of casting], 'and so do Burt Reynolds and Paul Newman.' And they *have* to have it, she says, because it allows the audience to forgive them for their looks."

That explanation points up a sobering fact about Celebrity today: The stars are obligated not just to tantalize their fans but also to appease them. The stars must seem to pay us back for the envy they've been paid to make us feel. While basking in success, they must also seem to shrug it off, or to atone for it, so as to blunt the deep resentment that is the obverse of our "love."

Such mass ambivalence is at least as old as the star system itself. In the Thirties, Nathanael West sensed the hatred simmering among the movies' most passionate fans: "The police force would have to be doubled when the stars started to arrive," he writes of the hellish "world premiere" in *The Day of the Locust*. "At the sight of their heroes and heroines, the crowd would turn demoniac. Some little gesture, either too pleasing or too offensive, would start it moving and then nothing but machine guns would stop it."

Throughout Hollywood's heyday, the press functioned to contain that great mass animus, channeling the audience's wrath according to the needs of the studio executives. Read by millions, "reporters" like Louella Parsons,

Hedda Hopper, and Sheilah Graham, among others, did not just hype the stars but also helped supervise them, by standing as the righteous tribunes of the movie-going populace. In 1950, one of Hedda Hopper's legmen gleefully called her attacks "a healthy disciplinary medium among the stars and big shots who sometimes delude themselves they're entitled to special privileges, like, for instance, adultery, taking dope, and indulging in strange sex aberrations." Through the columnists, the audience could act out its revenge against the privileged movie colony. And so the stars had to abase themselves before those hard columnists, regarding them, as Leo Rosten wrote in 1941, with "a combination of cunning, fear, hate, and propitiation."

If anything, that mass hostility seems to have increased, now that TV tantalizes everyone nonstop. Through TV, the dazzling image of Celebrity has become inescapable, a glow pervading every home and haunting every mind, with a thoroughness that was not possible when the stars were beckoning only from the screens of crowded movie theaters. The same medium that appears to bring the stars within our reach also (of course) withholds them—a kind of teasing that makes people angry. This may, in part, explain why stardom has, with the spread of TV, become more dangerous, a status that today requires not just the occasional instruments of crowd control but a staff of live-in bodyguards.

Only a few fans, of course, have vented their anger in the kind of kamikaze operation that has felled John Lennon, Ronald Reagan, Theresa Saldana, Rebecca Schaeffer. The psychotic assault on Celebrity is only the most extreme symptom of the widespread resentment, which usually finds less violent release in the same media spectacle that also rouses that resentment. While a few carry guns and keep mad diaries, millions simply feed on lurid fantasies of retribution, available at any supermarket checkout line, as well as on television.

Of all the features in today's theater of revenge, however, there is none as popular, or as subtle, as the quarterly ritual staged by Barbara Walters—an edgy televisual ceremony in which Walters acts out, and exploits, the viewers' ambivalence toward the superstars. A throwback to the likes of Hedda Hopper, Walters, on our behalf, seems to penalize the stars for their success—right after, or in the midst of, hyping it.

Thus the Walters interview usually begins with what she calls "the tour"; i.e., a bald exhibition of fantastic wealth, the camera panning reverently across stupendous lawns, vast marble foyers, bedrooms like the tombs of emperors, etc., while Walters, in voice-over, extols the property as if trying to sell it.

After this display, however, Walters must stage a punishment for the affluence she has just eyed so worshipfully. Most often, Walters will, after a little fawning, suddenly confront the subject with a very rude personal question, just as a child or imbecile would. At these moments, the interviewer actively personifies the viewing audience at its nosiest. It is, in fact, her genius to intuit the wonderings of the tactless oaf within each one of us and then to become that oaf, speaking its "mind" right in the star's home, right in the star's face.

Here, for instance, is an exchange that Walters conducted, in 1977, with Elizabeth Taylor and Senator John Warner, then newlyweds, "in the spacious kitchen of their elegant farmhouse":

B.W. [to Liz, out of nowhere]: Are you worried about putting on weight?

Liz: No. [*tense giggling*]

B.W.: Does it matter to you?

Liz: No, it doesn't. Because I'm so happy, and I enjoy eating. I like to cook—um—and I enjoy eating, and I love it—

B.W.: You wouldn't care if you got *fat?*

Liz: I *am* fat!

B.W.: I didn't want to say it.

Liz: I can hardly get into any of my clothes!

B.W.: But you don't *care?*

Liz: Not really.

B.W. [changing tactics]: Do you worry about getting older?

Liz: No, not at all. I'm forty-four years old, and I'll be forty-five next Sunday—

[*Cut to an extreme close-up, letting us enjoy the sight of Liz Taylor's laugh lines and double chin while she sits there trying to sound liberated.*]

"Why didn't you get your nose fixed?" Walters once asked Barbra Streisand. On a show featuring Bo Derek, Cheryl Ladd, Farrah Fawcett, and Bette Midler, Walters—referring to Derek's hit movie—asked Midler to rate, one to ten, her own looks. "Fifty-five!" Midler sang out genially, whereupon Walters, seeing as the oaf sees, was ostentatiously nonplussed: "Er—*really?*" she gasped, and Midler was clearly—and understandably—offended.

Obviously, Walters's purpose is not to regale the public with good conversation but to abuse the stars as that same public would, and often does, abuse them: "People can be very cruel," Barbra Streisand once told Walters. "They treat performers sometimes as if they're not alive." Tellingly, Walters used that sound bite to kick off her gala Fiftieth Special, as if to advertise the cruelty of her own approach. Her callousness shows itself not just in childish personal remarks but in even grosser moments of impertinence: "Can you have sex?" Walters asked the paraplegic singer Teddy Pendergrass.

When trying to justify such invasiveness (which she does not do often), Walters will lay the blame on us: "I *hate* to do this," she insisted to Fawn Hall (having asked her if she was dating anyone), "but *you* know people want to know." In thus voicing the abrasive curiosity of the oaf multitude, Walters recalls the Hollywood newsbags, who were also merciless inquirers. The difference is, of course, that the columnists would leave, in print, no trace of their own prying, whereas Barbara Walters asks those nervy questions right on camera—indeed, the asking often *makes* the story. Walters wants visceral responses: a gulp, a sob or two, or preferably a fit of keening. For this payoff, she will move in—often indirectly—on the most painful memories, which she will bring up with a fixed look of clammy pseudo-sympathy and in the Standard Condolatory Murmur used by people who enjoy funerals. "Betty, did you know from the first there was no hope?" Thus Walters delicately opened up the subject of Betty White's eventual widowhood.

Why do those stars go through with it? In part, of course, they want the exposure, and Walters offers a unique promotional bonanza. Moreover, it would be a sentimental error to imply that the stars are mere victims here, because Walters often stages the star's punishment with the star's own complicity—just as the Hollywood "reporters" used to do. The ambush mode is not Walters's only means of bringing down Celebrity. Just as often, she will help some star evade the mass resentment by getting her to tell us how she's Been Through Hell—a confession of past misery that permits the star to go on having all that fame and fortune ("She deserves it, with what she's been through!"), and that also reassures us oafs that we're all better off just as we are: working (if we're lucky) nine to five, waiting on line in supermarkets, watching television.

But while she generally has the women sell themselves as erstwhile victims, she tends to bring the males down in a different way. As spokesperson for the masses, Walters often moralizes, promoting the same tidy Code of Family Values that Hopper/Parsons used to push so hypocritically. Usually, however, Walters defends the code not by railing at the stars' loose behavior but by having them—the men, that is—give dewy-eyed avowals of their deep longing for domestic life. This is exactly how the stars of yore, however randy, mollified their leery public through the gossip columns. And so Sylvester Stallone told Walters what would make him happy: "I would like to be able to have a home life that is really beautiful, that is—the whole storybook, white-picket-fence bit. I really would. I would like to go for that," said the man who had married Brigitte Nielsen.

The whole pseudo-populist spectacle, all the prying and the piety, has a fundamentally elitist purpose. Although posing as a journalist, Walters is, like Hopper and the others, in the business not of public information but of popular delusion. Through her, we only seem to be allowed inside a world of peace and luxury, a world whose real inhabitants, through her, we only seem to punish for our actual exclusion. And yet there is no peace within that world, because of us. Even for the stars, in other words, the specials offer no real benefit. They sit down with Walters not only for the vast exposure but as a form of public penance, hoping thereby to pacify the eerie rage of their adoring audience. The stars are in hiding from us, and so they meet uneasily with Barbara Walters.

COMING TO TERMS

Often a critic will try to show how a text that presents itself as one thing is actually better understood as another. Explaining this complex, alternative kind of reading or interpretation is difficult to put into words. And we're not quite convinced that expressions like "reading between the lines" or "looking for hidden meaning" really hit the mark. Maybe, to put it in somewhat more algorithmic terms, a critic using this strategy might be seen as saying, in effect, something like: "If you read this text quickly, it may appear simply to mean *x,* but if instead you slow down, and look at it carefully, from the perspective I am suggesting, you will find that it means something much more interesting, something more like *y.*" Or maybe we might explain the critic's work here as making the familiar unfamiliar. Re-read what Miller has to say on "Barbara Walters' Theater of Revenge" with this scheme in mind. In your journal, we'd like you to respond to the following questions: What does Miller say that a *Barbara Walters* Special presents itself as? (What is the *x,* or what is the "familiar"?) How would you explain this reading in your own words? And then, how does Miller argue that it should actually be understood? (What is his *y,* or "unfamiliar"?) And what does Miller promise that we will "get" from reading it his way? How would you describe his way? Why does Miller think his y is more satisfying, striking, or insightful than the more obvious x? What have we gained (and lost) from viewing television in his way?

READING AS A WRITER

One aspect of Miller's writing that we think deserves close attention is the balance he achieves in his critique. While the tone in which Miller writes is sharply critical of Barbara Walters and her "theater of revenge," his essay is also about the pleasure that we, as her viewers, take in this revenge, this cruelty. In working toward this balance, Miller achieves two important effects: first, he shows himself as a writer who seeks to represent his subject as a complex one, acknowledging the popular appeal of Barbara Walters as well as critiquing her; second, Miller is imagining his readers with a kind of respect and an appreciation that they will demand a sense of intellectual fair play from his discussion. We'd like you to re-read Miller's essay to see if you can see and explain this balance in your journal. How does Miller try to reconcile these competing aims as a writer: the one being to criticize his subject, the other to understand its appeal? Find some places in the text in which you think Miller is successful at achieving a balance between *criticism* and *understanding*. What does he do to make his criticisms of Walters seem fair rather than simply ill-tempered? How does he try to show that he understands the appeal of her form of television or the programming she is set up to represent? (Some advice: You may want to observe how and when Miller uses the words "we" and "us." What is achieved by using this language? How might using these words define a writer's relationship to an audience?)

WRITING CRITICISM

Miller interprets the Barbara Walters interview as a "ritual" in which the stars are punished, in which they "pay us back for the envy they've been paid to make us feel." But this punishing of the famous is not confined to the occasional Barbara Walters special. It takes place regularly in supermarket tabloids like *The National Enquirer,* in unauthorized biographies of famous people, in magazines like *Spy, People, US,* and *Vanity Fair,* and in the gossip columns of newspapers. It also takes place on cable television, on channels like E!, and is regular fare on network television in the opening monologues of Jay Leno, David Letterman, and Conan O'Brien. The character of the acerbic Hollywood gossip columnist is often satirized or burlesqued on programs like *Saturday Night Live.* And maybe some of the most savage "punishing" of the famous occurs on the Internet, at sites like *The Drudge Report,* where libel laws remain rather fuzzy.

For this assignment we'd like you to find a popular media text in which a famous person is being "punished" or "trashed," brought down or humbled in some way, and write an essay in which you analyze the messages it sends out to (or attitudes it seems to share with) its readers about stardom and fame. Does your reading of this text support what Miller has to say about our need for "revenge" on the famous? In what ways might your reading add to or complicate his argument? A good way to begin your analysis is first to think about the usual and established image this person has or conveys to the public. Then ask: What happens to the famous person or people as a result of the treatment received in this text? Are they made to seem more human, more like us? Or are they portrayed as inhuman, cruel, even monstrous, perhaps? Is there any sense in which their problems, mistakes, or sins make them more glamorous or enviable? Or do you, at some point in the text, begin to feel sorry for them? What happened in the text to change your mind, to shift to a position of sympathy? Were people given an opportunity to defend or explain themselves? To respond to questions or charges? If your text is a printed text, how were quotations used in the text?

Once you have characterized the famous person's or people's image and then shown how this image has been recast through a popular medium, you need to address several questions in your discussion. Why does this text which "punishes" the famous appeal to those of us *ordinary* people who hear about stars, celebrities, and the famous constantly? How are we supposed to react to what it has to say? Are we supposed to be shocked, thrilled, saddened, outraged, curious, eager for more? What does your text assert about us, the audience, and our relationship to the famous? Explore Miller's assumptions about the famous person–ordinary person relationship in cultural terms, in terms of our values, beliefs, expectations?

Serious Watching

Alexander Nehamas

Traditionalist critics of recent developments in American universities are fond of comparing the current situation to the past—that is, in most cases, to their own student days. Gertrude Himmelfarb, for example, recalls a time when, in contrast to today,

> *it was considered the function of the university to encourage students to rise above the material circumstances of their lives, to liberate them intellectually and spiritually by exposing them, as the English poet Matthew Arnold put it, to "the best which has been thought and said in the world."*[1]

Nostalgia has colored not only Professor Himmelfarb's perception of the past, but also her recollection of Arnold, who actually wrote that the "business" of criticism is "to know the best that is known and thought in the world."[2] We can dismiss Himmelfarb's other inaccuracies, but we cannot overlook her replacement of Arnold's present-tense "is" by the perfect-tense "has been." For this allows her to appeal to Arnold's authority in order to insinuate, if not argue outright, that the university's concern is with the past and the present, at least in connection with the humanities, lies largely outside the scope of its function.

Such emphasis on the past does not exclude attention to contemporary works of fine art or philosophy, which can generally be shown to be part of what is often in this context called "the" tradition. But it does disenfranchise present-day cultural products that cannot be readily connected with that tradition. This is especially true of works of popular art particularly television.

From "Serious Watching" by Alexander Nehamas, 1990, *South Atlantic Quarterly, 89, no. 1* (Winter 1990). © 1990 Duke University Press, Durham, NC. Reprinted by permission of the publisher.

These are either totally overlooked, as is the case in our philosophy of art, or disparaged, not only by highbrow critics like Allan Bloom but also by the very people whose livelihood depends upon them.[3] In this way, J. R. Ewing, the hero/villain of *Dallas*, "the man everybody loves to hate," turns out to be the perfect metaphor for the medium that sustains him.

But though it is devoted to the past, this approach is historically blind. Its adherents consider Plato, Homer, and the Greek tragic poets equally parts of the tradition, but they fail to realize that Plato's uncompromising exclusion of the poets from the perfect state of the *Republic* proceeds from exactly the same motives and manifests precisely the same structure as their own rejection of contemporary popular culture. The paradigms of one age's high culture often began their life as entertainment for the masses of another.[4] Our seemingly unified tradition is significantly more complex and inconsistent than we tend to believe. And examining the popular artworks of our day is crucial for understanding the operations by means of which they are invested with value and the conditions under which they too can come to be assimilated (as some always are) into the fine arts and into high culture.

Such considerations aside, however, highbrow critics of television can also be answered in their own terms. This is the point of my essay: television rewards serious watching. Serious watching, in turn, disarms many of the criticisms commonly raised against television.

The common criticisms of television, though they are united in their disdain for medium, come from various directions and have differing points. Wayne Booth, for example, expresses a relatively traditional preference for primarily linguistic over mainly visual works:

> It is hard to see how anyone can eliminate the fundamental difference between media in which some kind of physical reality has established scene before the viewer starts to work on it, and those like radio and print that can use language only for description—language that is always no more than an invitation to thought and imagination, never a solid presentation or finished reality. . . . The video arts tell us precisely what we should see, but their resources are thin and cumbersome for stimulating our moral and philosophical range.[5]

It is worth pointing out, but necessary to leave aside, the ironic and twisted connection between this view and the famous passage of the *Phaedrus* in which books are criticized for lacking, in comparison to the spoken word, just the features for which Booth praises them:

> You know, Phaedrus, that's the strange thing about writing, which makes it truly analogous to painting. The painter's products stand before us as though they were still alive, but if you question them, they maintain a most majestic silence. It is the same with written words; they seem to talk to you

as though they were intelligent, but if you ask them anything about what they say, from a desire to be instructed they go on telling you just the same thing forever.[6]

This parallel is very important in its own right and should be studied in detail.[7] But what concerns me now is simply Booth's claim that the video arts are inherently incapable of addressing serious "moral and philosophical" issues.

A related criticism is made by John Cawelti, whose celebrated study of the arts of popular culture, particularly of formulaic literature, has led him to conclude that

formulaic works necessarily stress intense and immediate kinds of excitement and gratification as opposed to the more complete and ambiguous analyses of character and motivation that characterize mimetic literature.... Formulaic works stress action and plot.

He also considers that "a major characteristic of formulaic literature is the dominant influence of the goals of escape and entertainment."[8] The contrast here is one between the straightforward, repetitive, action-oriented, and entertaining formulaic works which by and large belong to popular culture—works which include the products of television—and the ambiguous, innovative, psychologically motivated and edifying works of high art.

Finally, Catherine Belsey, who has approached the study of literature from a Marxist point of view, following the work of Louis Althusser, draws a contrast between "classic realism, still the dominant popular mode in literature, film, and television" which is characterized by "illusionism, narrative which leads to closure, and a hierarchy of discourses which establishes the 'truth' of the story," and what she calls "the interrogative text." The interrogative text, she writes,

may well be fictional, but the narrative does not lead to that form of closure which in Classical Realism is also disclosure.... If it is illusionist it also tends to employ devices to undermine the illusion, to draw attention to its own textuality.... Above all, [it] differs from the classical realist text in the absence of a single privileged discourse which contains and places all the others.[9]

It would be easy to cite many other similar passages, but the main themes of the attack against television, to which those other passages would provide only variations, are all sounded by these three authors: (1) given its formulaic nature, television drama is simple and action-oriented; it makes few demands of its audience and offers them quick and shallow gratification; (2) given its visual, nonlinguistic character, it is unsuited for providing psychological and philosophical depth; and (3) given its realist tendencies, it fails to make its own fictional nature one of its themes. It is therefore self-effacing and constructs an artificial point of view from which all its various strands can appear to be put together and unified; it thus reinforces the idea that problems in the world can be solved as easily as they are solved in fiction and it domesticates it audience. These reasons are taken to show that television does not deserve serious critical attention—or that, if it does, it should only be criticized on ideological grounds.

And yet there are reasons to be suspicious of this view, which can all be based on a serious look, for example, at *St. Elsewhere*—a television drama that appears straightforward, action-oriented, and realistic.

My discussion of *St. Elsewhere* often concerns individual episodes or scenes. Nevertheless, my main concern will be with the series as a whole. As I have already argued elsewhere, the object of criticism in broadcast television drama is primarily the series and not its individual episodes.[10] But, of course, individual episodes are all we ever see, and it is by watching them in sufficient numbers that we become familiar with the series as a whole. A large part of the dissatisfaction with television drama, I think, is due precisely to a failure to appreciate this point. By concentrating exclusively on individual episodes, the critics of television are incapable of seeing where, as it were, the impact of the medium occurs and are therefore unable to be affected by it. Character, for example, is manifested through particular occurrences in particular episodes; but each manifestation is thin and two-dimensional, until we realize that thickness and depth are added to it if (and only if) it is seen *as* a manifestation of character which can be understood and appreciated only over time and through many such manifestations. In this respect, television is not unlike the comic strip, in regard to which Umberto Eco has written:

> *a structural fact that is of fundamental importance in the understanding of comics in general [is that] the brief daily or weekly story, the traditional strip, even if it narrates an episode that concludes in the space of four panels, will not work if considered separately; rather it acquires flavor only in the continuous and obstinate series, which unfolds, strip after strip, day after day.*[11]

There are, in fact, many similarities between cartoons and at least some television series, especially those filmed before an audience, and this is in my opinion partly responsible for the low regard in which many high-minded critics hold the latter. The stiffness of the poses held by the television actors, the necessity of their half-facing the camera, the inability of the television image to give great detail and its lack of visual texture, the narrow angles and small groups which it can only accommodate, and the staccato rhythm in which lines are often delivered with pauses for laughter or applause are all features that make of television shows cartoons that are animated in a literal sense of the term.

Whatever its connections to the cartoon, television has always been thought to be inherently realistic, not only by high-culture intellectuals but by its own creators as well. During the first years of broadcast television, documentation seemed absolutely essential for drama as well as for comedy. For example, *Medic*, the very first medical show, was, according to a recent discussion, a "highly realistic examination of surgery. The program sought to document medical case histories and used some actual hospital usage."[12] *Medic* was written by Jim Moser, a friend and ex-collaborator of Jack Webb, whose show, *Dragnet*, was also supposedly based on "actual files of the L.A.P.D."; the connection was responsible for *Medic* coming to be known as "Drugnet." The very first situation comedy, *Mary Kay and Johnny*, which opened in 1947, starred an actual married couple, whose child, born after the show was already on the air, was incorporated into the plot and thus set the pattern made famous by Lucille Ball as well as by the Nelsons in *The Ozzie and Harriet Show*. The episodes of the police show *Gangbusters* were based on "actual police and FBI files"; they concluded by airing a photograph of one of FBI's "most wanted" criminals with instructions to call the FBI or the show itself with information about them, thus anticipating the current mania for interactive programs of this sort.[13] Finally, in one of the most absurd cases of the search for verisimilitude, *Noah's Ark*, produced by Jack Webb and featuring a "messianic veterinarian," was based on "actual cases" from the files of the Southern California Veterinary Association and the American Humane Society.

Whether such a mixture of documentation and fictional narrative produces or undermines realism is a complex question. For the moment, I simply want to point out that in the 1960s and 1970s such obvious attempts to incorporate reality into television gave way to a more straightforward melodramatic mode. The mythical quality of melodrama, however, was soon infected with reality once again: a new realism from two new directions, which resulted from the intervention of two very different television authors.

Now the term "author" may well seem inappropriate here, for, among all the arts, television seems to be the most authorless. Most dramatic series are written by different people, or groups of people, each week, and it is very difficult to know precisely who is to receive the credit or the blame for a show's success or failure. In an effort to determine who is finally responsible for the character of each program, Todd Gitlin has argued that while in the film (as Bazin and others have claimed) this role, the *auteur*, belongs to the director, in television the relevant role is played by the producer, perhaps the only person who provides continuity in a show and who determines the overall look of the program.[14] This view is at least partly correct: *Hill Street Blues* (Gitlin's primary concern) and *L.A. Law* are indeed Steven Bochco's creatures, as *Miami Vice* belongs to Michael Mann. If this is a view we accept, then we can say that a crucial factor in the development of American television was the work of Norman Lear, who was responsible for the nature and success of shows like *All in the Family, Maude, The Jeffersons*, and *Mary Hartmann, Mary Hartmann* (which was actually so parodic that not even he could sell it to the networks). In his various shows, Lear introduced acute social commentary and thematized complex social and political issues through the previously innocuous format of the classic situation comedy. A particularly interesting feature of Lear's work was that it was very difficult to tell where exactly his shows' sympathies lie: "Liberals and radicals tended to interpret *All in the Family* as a left-liberal critique of bigotry and conservatism, while conservative audiences tended to identify with Archie Bunker and to see the series as a vindication of Archie's rejection of his 'meathead' son-in-law's liberalism."[15] This indeterminacy of television is to a great extent dictated precisely by the medium's immense popularity and its need to appeal to an extremely heterogeneous audience. It argues against the facile charge that television "totalizes" its narrative point of view, for, in order to be susceptible to such varying interpretations, the television "text" must be essentially incomplete and open to radical interpretation on the part of the audience (which thereby shows itself to be much more active in its reaction to the medium than our stereotypes often suggest).

Another realistic element in recent television drama is associated not with a person but with a whole production company, MTM Productions, which is behind shows like *The Mary Tyler Moore Show, Rhoda, Phyllis, The Bob Newhart Show, Lou Grant, WKRP in Cincinnati, Hill Street Blues,* and *St. Elsewhere.* MTM Productions was headed by Grant Tinker, but Tinker was not associated with the character of his many shows as directly as Lear was with his. In a serious way, credit for these programs goes to the production company rather than to any individual. And this in turn raises the interesting possibility that the television author need not only *not* be an individual but also that it need not always be an object of the same ontological order: both a concrete individual and an abstract entity—a company—can play the relevant role. Three features of MTM productions are relevant to my account.

First, these shows shifted in many cases the location of the situation comedy from the home, where the genre had been truly at home, to the workplace. But the MTM workplace—a television station, a doctor's office, a police precinct, a country inn, an inner-city hospital—always operates as the locus of an extended family within which individual characters face, defer, or resolve innumerable personal problems. The humor of these shows is less biting, less abusive, and less overtly political than the humor of Norman Lear. Part of their overall message seems to be that one's most real family consists not so much of the people with whom one lives but rather of the people with whom one works. Many characters have restrained personal lives; many are unhappy at home; and happy families, as in both of Bob Newhart's shows, are continuous with the family of the workplace.

Second, in contrast to earlier television drama and following the precedent established by *M*A*S*H,* many of these shows allow for, and depend upon, character development. Characteristically, during the opening scene of *The Mary Tyler Moore Show,* the program's title song asked of its heroine "How will you make it on your own?" In later seasons, this was changed to "You're gonna make it on your own," and eventually, in line with Mary Richards's increasing independence, any reference to this issue was dropped altogether.[16]

Finally, other MTM shows followed the lead of *Hill Street Blues*, which was in this respect influenced by the conventions of daytime soap operas and introduced multiple story lines in individual episodes. These stories would often be carried over a number of episodes but, contrary to the situation in soap operas, they would always be resolved. The possibility of containing multiple plot lines naturally depends on the existence of a relatively large cast. Accordingly, programs like *Hill Street Blues* and *St. Elsewhere* ceased to function around a single figure and developed into "ensemble shows" featuring many actors, each one of whom has relatively little time in front of the camera.[17]

With these ideas in mind, we can now turn to *St. Elsewhere*, which concerns life in a large inner-city hospital in Boston. The show seems straightforward, action-oriented, and realistic. It is full of local color. It is a serious program, addressing complex medical and moral issues (it was the first television show, for example, to present a series of episodes concerning AIDS) in the liberal manner of *M*A*S*H* and *Lou Grant*, but it is also bitingly, paradoxically funny. For example, during a title sequence a group of hospital personnel are shown hurrying a life-support machine down a corridor in what clearly seems an urgent situation. Now television doctors do occasionally fail their patients; even Dr. Welby lost a few. But, traditionally, these were always cases of nature asserting itself over technology. Here, however, the attempt fails because, in all their dispatch and intensity, the interns clumsily stumble and end up, along with their machine, sprawled across the floor. The fact that this is a scene in the title sequence, and that it is repeated week after week, fixes it in the mind of the program's audience and allows the incident to manifest a feature not only of St. Eligius but of hospitals in general, so that the scene appears realistic as well as funny: things like that do happen, more often than we like to think, in hospitals.

St. Elsewhere lacks the unrelenting technological and humanistic optimism of *Marcus Welby, M.D.* Patients die there, and often there are no lessons in their deaths. The physicians not only help but also cheat and seduce each other. None of the ultimate positions of power in the hospital is occupied by a woman. This, in fact, becomes one of the show's themes, especially when a woman—and a Vietnamese refugee at that—replaces the obnoxious, racist, sexist but technically superb chief of surgery, Mark Craig, when he smashes his hand in a fit of pique and self-doubt. Craig's sexism, which causes a breakdown in his marriage, is consistently addressed in the show, along with his wife's efforts to find a job, a life, and a voice of her own. One of the residents is a former nurse who realized that she could do a physician's job as well as or even better than many of the men who practice medicine. The program features a very successful black doctor, a *summa cum laude* graduate of Yale and chief resident at St. Eligius; a highly motivated black orderly who moves up to paramedic and then to physician's assistant before he realizes that that is as far as he can go; and a friend of his who is content to remain an orderly. For these and many other similar reasons, *St. Elsewhere,* compared to *Marcus Welby, M.D., Dr. Kildare,* or *Ben Casey,* appears to be much more accurate to life within the medical profession.

It is imperative, however, to note the terms of this comparison. The realism of *St. Elsewhere* is measured by comparing it not to life within a hospital, of which most of us know almost nothing, but to the standards and features of earlier shows on the same general subject. On the other hand, there does seem to be something inherently more realistic in a show that features a hospital not in some idealized suburban setting but in the middle of Boston. The fact that the Red Line runs right next to St. Eligius, moreover, places the hospital in the location of Massachusetts General Hospital, which adds a further touch of verisimilitude—until one realizes that far from being based upon the latter, St. Eligius is constantly being contrasted with it under its fictional name of "Boston General." St. Eligius in fact, seems modeled on Boston City Hospital. Boston is a real presence—a character—in this program: the governor of Massachusetts appears in one episode, the city's racial conflicts are often addressed, Harvard looms large. And yet this is a very peculiar Boston. For one thing, it contains a bar named "Cheers." And, on one occasion, the show's three patriarchal figures go to this bar, whose fictional existence also involves a very "real" Boston, to drink and talk things over. To complicate matters further, in one episode a resident of St. Eligius passes by the actual bar in Boston which advertises itself as the place which inspired "Cheers," and takes his little son in after asking him, in an allusion to the theme song of that show, whether he wants to eat "where everybody knows your name."

This is, then, an impossible Boston, however realistic its representation appears. Realism in a case like this is indeed measured not by proximity to reality but by distance from fiction whose conventions we have come to see as conventions. At the time, of course, the conventions of early medical shows were invisible, just as many of the conventions of *St. Elsewhere* will only become visible in the future, and the shows certainly seemed realistic. But even here the situation is complicated: the relationship between program and reality may be more ambiguous, and the television audience may be more aware of this ambiguity, than we are apt to suppose.

George Gerbner and Larry Gross, for example, report that over the first five years of Dr. Welby's television practice the show received roughly 250,000 letters, most of them requesting medical advice.[18] Their conclusion is that viewers consider television characters "as representative of the real world." But consider the fact that people still write letters to Sherlock Holmes at 221B Baker Street (in fact, the firm that occupies that address employs someone just to answer them); yet surely no one aware of Holmes believes that he is an actual person: rather it is more plausible to suspect that the people writing Holmes engage in a game that exploits Holmes's ambiguous status, his fictional genius and his "actual" address. The same idea is also suggested by the practices of television fan magazines, which explicitly mix information about the various characters of the soap operas with information about the actors who portray them in such a away that it is difficult to separate one from the other. The television audience seems to be enjoying the equivocal interpenetration of fiction and reality. As John Fiske writes in regard to the fan magazines,

> *we must be careful not to let the "cultural dope" fallacy lead us to believe that the soap fans are incapable of distinguishing between character and player.... This is an intentional illusion, a conspiracy entered into by viewer and journalist in order to increase the pleasure of the program... The reader [is encouraged into] the delusion of realism not just to increase the pleasure of that delusion, but also to increase the activeness and sense of control that go with it.*[19]

If the Boston of *St. Elsewhere* contains both the Red Line and the Red Sox on the one hand and Cheers on the other, it is and it is not a real city. And if this is so, then it is difficult to agree with Belsey's view that "illusionism" and lack of self-awareness are deeply characteristic of television drama. In fact, *St. Elsewhere* mixes fiction so thoroughly with life and is so sensitive to what is now being called "intertextuality" that only ignorance of the medium of television could ever have suggested that the program is naively and straightforwardly realistic. This becomes more obvious when we realize that only someone familiar with television—a literate viewer—can understand that Cheers is explicitly fictional and that the realistic episode involving the bar and its characters is doubly impossible.

Pierre Bourdieu, who objects to paying television "serious" attention, argues that aesthetic approaches to the medium mystify its cultural role and conceal its real importance. In all popular entertainment, as opposed to the fine arts, Bourdieu writes,

> *the desire to enter into the game, identifying with the characters' joys and sufferings, worrying about their fate, espousing their hopes and ideals, living their life, is based on a form of investment, a sort of deliberate "naievety," ingenuousness, good-natured credulity ("we're here to enjoy ourselves") which tends to accept formal experiments and specifically artistic effects only to the extent that they can be forgotten and do not get in the way of the substance of the work.*[20]

Similarly, Herbert Gans writes that members of lower "cultural taste groups" (to which the television audience by and large is supposed to belong) choose their form of entertainment "for the feelings and enjoyment it evokes and for the insight and information they can obtain; they are less concerned with how a work of art is created."[21] Such reactions to entertainment, according to Bourdieu, are

> the very opposite of the detachment of the aesthete who...introduces a distance, a gap—the measure of his distant distinction—vis-à-vis "first-degree" perception by displacing the interest from the "content"...to the "form," to the specifically artistic effects that are only appreciated relationally, through a comparison with other works which is incompatible with immersion in the singularity of the work immediately given.[22]

Based on this distinction between "investment" and "distance," between "immediacy" and "relationality," Bourdieu repudiates aesthetic interpretation and criticism because he considers them self-deceptive:

> Specifically aesthetic conflicts about the legitimate vision of the world...are political conflicts (appearing in their most euphemized form) for the power to impose the dominant definition of reality, and social reality in particular.[23]

But Bourdieu's distinction between the immediate enjoyment of popular art by the lower classes and the comparative attitude of the distant aesthete cannot be maintained. The television audience is highly literate (more literate about its medium than many high-culture audiences are about theirs) and makes essential use of its literacy in its appreciation of individual episodes or whole series. Its enjoyment, therefore, is both active and comparative. Consider the following case.

A regular secondary character on *St. Elsewhere* during the 1985 season was an amnesiac, referred to as John Doe. Having failed to regain his memory and find out who he is, Doe, who is a patient in St. Eligius's psychiatric ward, turns obsessively to television. But though his conversation is riddled with lines derived from commercials, his real interest is in the news programs: "Newscasters—*they* know who they are," he insists to his psychiatrist, Dr. Weiss. Weiss, who is concerned with Doe's state of mind, finally tells him not to watch the news any longer: "It's too depressing. I want you to watch shows that lift your soul and put a smile on your face."

Another character on this particular program is the passive-aggressive Mr. Carlin, who loves to torture Doe. Mr. Carlin, portrayed by the same actor, was a regular character in *The Bob Newhart Show,* in which Newhart, a psychologist, was treating him for his (at the time) milder disorder. But Bob apparently failed, and Mr. Carlin has been committed, finding himself in a hospital in a different show.

Doe and Mr. Carlin fight over the television set in the ward lounge, incessantly switching channels. Doe finally gets reconciled to fictional shows, and even tells Carlin that television is "filled with real people." "And they're only *this* tall," Carlin replies, placing his thumb and forefinger six inches apart. "Television is the mirror of our soul," Doe insists; "we look in and we see who we really are." And as he switches from one famous program to another, he catches for a moment the very end of *The Mary Tyler Moore Show* and the logo of MTM Productions (itself a parodic reference to MGM's famous trademark, and, in its substitution of a kitten for MGM's lion, a whole parable of the relationship between film and television). MTM Productions, of course, is responsible not only for *The Mary Tyler Moore Show* but also for *The Bob Newhart Show* as well as for *St. Elsewhere*. As soon as he sees the MTM kitten, John Doe loudly claims that he now knows who he is: he is Mary Richards, Mary Tyler Moore's character. He instantly goes into character, dons a beret like Mary's, identifies various patients and physicians with characters from the show (Mr. Carlin, for example, becomes Rhoda, Mary's friend, though he nastily refuses to play along; Dr. Weiss, naturally enough, is Mr. Grant; Dr. Auschlander, the senior figure in the hospital and quite bald, becomes Murray, and so on), and develops, like Mary, a profound devotion to his new extended family for whom, in Mary's manner, he immediately prepares a party: "Sometimes the people you work with aren't just the people you work with," he tells Dr. Weiss, echoing the main theme not only of *The Mary Tyler Moore Show* but also of *St. Elsewhere*.

Auschlander is worried about Weiss's decision to go along with Doe's fantasy, and seems slightly embarrassed at having to appear at Doe's party as Murray. He is disdainful of the television audience in general, echoing at least some of the complaints I have introduced into this discussion. "People sit in front of their televisions," he says, "believing the characters they see there actually exist, eat, breathe, sleep." But when Weiss asks him which character *he* would like to be if he had the choice (which of course he doesn't, since he already is one), he unhesitatingly replies that he would like to be Trapper John, M.D.—a doctor with a reassuring manner who invariably saves his patients, the very kind of doctor *St. Elsewhere* will not allow to exist within its own fictional space and which, from within its own fiction, is thereby asserted to be more "real" than its competitors.

As part of an independent subplot, an astronaut is being treated in St. Eligius for a case of paralysis. The astronaut, however, has also announced that on his next space mission he will walk hand in hand with God. The Navy has sent one of its medical officers to bring the astronaut to earth and to Bethesda. This Navy doctor, who has already appeared in an earlier, unrelated episode of *St. Elsewhere* and who has thus established her "independent" identity, is portrayed by Betty White. As she is on her way to visit Dr. Weiss, she runs into Doe, who immediately exclaims, "Sue Anne! The Happy Homemaker!," recognizing actress Betty White as the character she was in *The Mary Tyler Moore Show*. Betty White, naturally, responds with a blank stare and a vague "I'm afraid you have me confused with someone else."

Doe's party turns out to be a success, but this causes him to start doubting his new identity: "Mary always throws lousy parties," he confesses; "maybe I am not Mary." At that point, Mr. Carlin gets into a fight with another patient, and Doe runs to help him out, attacking this other patient, whom he identifies as "Mr. Coleman," the station manager who, when *The Mary Tyler Moore Show* was cancelled, was supposed to have fired its main characters. "I've committed a violent act," Doe says. "Mary would never do that." Carlin, moved by Doe's friendship, abandons his nastiness and decides to play along with him: "Call me Rhoda," he suggests. But Doe responds (in a way that still mixes fiction and life), "No. I am not Mary. We've just been cancelled."

The next morning, Doe goes for a walk with Dr. Auschlander, who reassures him that his many friends will help him find out who he is. At the hospital's main entrance, Doe, calm, peaceful, and happy, says "I'm gonna make it after all," and in an exact parallel to the final shot of the title sequence of *The Mary Tyler Moore Show,* tosses his beret in the air, replicating Mary in word and deed in the very process of liberating himself from her. Television and reality, fiction and life, character and person are intermixed through and through.

Only a literate and active audience could ever appreciate or even get the point of this ingenious use of intertextuality. Its point is not necessarily deep, though it does ask whom television characters are supposed to resemble, to what reality they correspond, and to whom—actor or character—one is responding in watching and enjoying a program. All these are questions important to ask and difficult to answer, and they make of this episode as "interrogative," self-conscious, and self-reflexive a work as any high-culture might possibly wish. But to be part of the high culture is not necessarily to be cultured, and to know much about literature is not necessarily to be literate.

St. Elsewhere lives by confounding fiction and life. During one of the show's last seasons, in an effort to improve ratings, its producers seemed to have decided that the old-fashioned hospital's seedy beige-and-brown background was failing to attract an audience getting used to the pastels made popular by shows like *Miami Vice.* They brought the look of their show in line with that of other high-profile programs by means of a brilliant move: they had the problem-ridden St. Eligius bought by a private hospital chain. And the first thing the chain did was to renovate the building, which provided the show with a postmodernist set and a whole new narrative dimension.

The fictional hospital company was called "Ecumena" and was immensely interested in artificial heart transplants. The Humana Corporation, on which Ecumena was obviously based, objected to the whole idea, especially because Ecumena was depicted as a cold, impersonal, profit-obsessed enterprise. They succeeded in getting a disclaimer added to the closing credits, and they finally won an injunction against the word "Ecumena." The name had to go. *St. Elsewhere* characteristically responded by mixing fact and fiction. Within the show, a nameless hospital company sued Ecumena on the grounds that their name was too close to theirs and won. The chain is renamed "Weiggert Hospitals" and as the Ecumena sign is being removed from the entrance to St. Eligius, it slips from the workers, falls to ground, splinters into countless fragments, and almost kills the hospital administrator, a devoted and often heartless employee of the chain, who looks in disgust and mutters, echoing the show's producers, "It's been that kind of day from the beginning."

In order to know where to look in order to locate the psychological power of broadcast television, we must concentrate on two features on account of which the medium has often been criticized. The first is that broadcast television works by repetition. We see the same characters in the same general circumstances though in varying specific situations week after week over a long period of time. The second is that the television camera can cover only a small visual angle and this, together with the low resolution of the television image requires a large number of close-up shots. For many, this is equivalent to saying that television is visually elementary and intellectually boring.

And yet some of the medium's greatest achievements depend on these two features. For the first allows us to become acquainted with television characters gradually and over a long period of time and the second enables us to come, in an almost physical sense, very close to them. And this closeness is not only physical. Our continuous exposure to these characters also brings us close to them in a psychological sense. Just as the characters of *St. Elsewhere* interact with one another every fictional day, so the audience comes to know them slowly, routinely, in a more or less controlled situation, not unlike the way in which they know the people they themselves work with. To a serious extent the relationship of the characters to one another replicates the relationship of many members of that audience to the people they in turn work with. We come to know these characters, and many of the characters of television drama generally, *intimately*—in both a physical and psychological sense. But to say that we come to know them intimately is not to say that we come to know them deeply. Their innermost nature, unlike the nature of the characters of novels, is not exposed; better put, television characters have no innermost nature. And yet, I want to suggest, the intimacy with which these not deep characters are revealed is one of the medium's glories.

Many of the people we know best in life often move or infuriate us by some particular gesture or action the significance of which is very difficult to communicate to others because, as we say, "it has to be seen in context." The same is often the case when we try to recount a funny, moving, or nasty moment in a television program to someone who is not familiar with it. In one episode of *St. Elsewhere*, for example, Donald Westphall, St. Eligius's chief of medicine, gets fired by the company representative. Mark Craig, who has always derided Westphall for being weak, boring, a do-gooder, and stubborn, continues to mock him for not apologizing and asking for his job back. When Westphall asks him why he cares whether he leaves or not, Craig, who is at the point of leaving the room, stops, turns, and, in a tone as supercilious as it is confessional, replies, "I'll miss you. Ridiculous as it may sound, you're the best friend I have." This is a moving and poignant moment, but it is difficult to say why, precisely because it is a *moment*, a small part of a complex relationship, and it is only within that relationship that it acquires whatever significance it has.

St. Elsewhere works through the accumulation of such moments, and allows its audience to know its characters intimately but not deeply—just as these characters themselves know one another in the extended familial space of their workplace. Such knowledge can be extremely fine-grained, but it depends essentially on long exposure. One must have learned, over time and through a large number of isolated incidents, precisely what a prude Westphall is in order to understand exactly what he does when, just before leaving the hospital and upon being told that he can have his job back provided he becomes a "team player," he turns his back and drops his pants in the face of the hospital administrator.

We might be tempted to say that even without knowledge of Westphall's character we do know what he does, though we may not know exactly what it means. But this is misleading, because it suggests that there is one level on which action is described and another on which it is interpreted, because it separates description from interpretation. It is much more nearly correct to say that we literally do not know what Westphall does on this occasion unless we see it in light of everything else he has ever done. Television, because of its serial character, highlights the essential interconnection of human actions—a psychological point—and the interdependence of their description and interpretation—a philosophical issue: Is it really true to say that it is "thin and cumbersome in stimulating our moral and philosophical range"?

The serial unfolding of character and the ability of individual characters (at least in some programs) to change and develop with time have an important consequence. They render character ambiguous. By this I do not mean that television characters are difficult to understand. Rather, the point is that television can present various aspects of its characters without offering a single, all-encompassing judgment about their ultimate nature or worth. This is another sense in which television characters have no depth.

Mark Craig, the chief of surgery at St. Eligius, for example, is a terrific surgeon and a horrible sexist. *St. Elsewhere* does not account for his sexism in any way that justifies it. Craig is also brusque, selfish, insensitive, and competitive, but also fiercely loyal to, and proud of, his residents. He is, in addition, insecure and more in need of others than he could possibly admit. He both loves and detests his son, who was addicted to drugs, married beneath him, and got killed in a car accident. We learn all this, and more besides, about him over a long period of time in a way not unrelated to that in which we come to know many of our friends and acquaintances. The net result of this gradual accumulation of detail is that there is no net result about Craig's character. It is difficult, perhaps impossible, and certainly not fruitful to say of him, or of most of the characters in this program, whether we like or dislike him. We may be devoted to him in the sense of wanting to know what he will be doing in the next episode, but approval or disapproval are not at issue. Do I *like* him? How can I, given his sexism, his crassness, his lack of sensitivity? Do I *dislike* him? No, because I do like his clipped manner, his pride in his work, his frightening straightforwardness. His character has too many sides for me to make a general evaluation of it. But where I see the absence of "totalization," others may not. And they may well like or dislike Craig, often for exactly the same reasons in each case. The television text is, in this sense, indeterminate. It allows its viewers to focus on different aspects of the characters it depicts and to see the same character in radically different ways depending on their own preferences and values. It is, as John Fiske, echoing Roland Barthes, characterizes it, a "producerly" text—subject to various operations on the part of its viewers.

Similar things can be said about the show's female characters. Nurse Papandreou, for example, can be an absolute terror, nasty and full of invective. She is also unquestionably a superb nurse. She terrifies the obnoxious Victor Ehrlich, but she also brings out the best in him (the little of it there is) when she relaxes in his company at a Greek feast, invites him up to her apartment afterward, and eventually marries him. And, once married, she shows a perceptive and mature side in her relationship to him which, precisely because it does not carry over into her other interactions, makes it impossible—for me, at least—to make a general judgment about her. Is she a good or a bad character? What about most of the people with whom we live and work on a daily basis? We live and work with one another, and toward most of them we have no single unequivocal reaction.

Classical realism, according to Catherine Belsey, always creates an overarching point of view from which all the pieces of each story can be seen to fall together. I have just argued that, at least on a psychological level, *St. Elsewhere* undermines any effort to occupy such a point of view. And just for this reason, *St. Elsewhere* reveals that its medium has the resources for presenting unusual aspects of human character—unusual enough to pass completely unnoticed if we are not willing to watch seriously.

Is *St. Elsewhere*, however, realistic in the further sense that its aim is to achieve "closure," to provide on a narrative level a final settlement of all the details of its plot and to put its viewers in the comfortable position of having finished *with* the story as well as simply having finished it? Jane Feuer has argued that *All in the Family* aimed at that goal: "The Lear family, however much they were divided along political lines, would each week be reintegrated in order that a new enigma could be introduced."[24] In fact, however, this was not quite true of *All in the Family.* Often, the show's episodes end with an extreme close-up of Archie Bunker, who has just been speechless by losing an argument. But speechlessness is not accommodation. And the look on Archie's face—part admission of defeat, part stubborn reassertion of his inner conviction that he is always right—allows viewers of different political orientations to draw their own different conclusions about the very nature of the episode they have just watched.

St. Elsewhere, however, appears to provide just the kind of closure Belsey associates with classical realism. The program was cancelled at the end of the 1987–88 season. Since this was known in advance, the show's final four episodes were devoted to constructing a complete resolution of its various subplots and to disposing properly of every single one of its regular characters. Manifesting a remarkable and unusual single-mindedness, *St. Elsewhere*, with the exception of those who died, created a future for all its characters and left absolutely no loose ends. Or so it seemed until the show's final scene.

Throughout the last episode, which is supposed to occur in the spring, the characters keep remarking that the temperature is dropping and that it is about to snow. At first these remarks are so out of place that they pass unnoticed until their cumulative weight makes them as impossible to ignore as they are to understand. Indeed, snow begins to fall. We see the hospital in the middle of a snowstorm, and the camera pans in order to take in the whole building—a shot which strongly disposes us to expect that the show has come to its end. But as the pan continues, and the building gets progressively smaller, it also inexplicably, begins to shake. And suddenly we realize that what we are seeing is not at all the "real" hospital but only a cardboard cutout enclosed in a glass paperweight and surrounded by "artificial" snowflakes.

What toy is this? It belongs to Tommy, Dr. Westphall's autistic son, who was looking at the snowfall from inside the hospital in the scene immediately preceding. Tommy, completely absorbed, is sitting on the floor shaking the paperweight. But can this be Westphall's son? He is sitting in a shabby room and not in Westphall's suburban house, and he is being watched over by Dr. Auschlander—who cannot be Dr. Auschlander, since Dr. Auschlander died of a stroke earlier on in the episode. At that point, Westphall enters: he wears a hardhat and carries a lunchbox. But, of course, this is not Westphall either: he turns out to be a construction worker and the son of the man we had known as Auschlander up to that point. Having greeted his father, "Westphall" looks at his sons and says, "I don't understand this autism thing, Pop.... He sits there all day long, in his own world, staring at that toy. What's he thinking about?" He then lifts the boy, places the toy exactly in the middle of the top of the television set, and leaves the room. The camera now closes in for the truly final shot of St. Eligius, encased—frozen—in its glass container.

Turning a story into a dream or a fantasy at the last minute is one of the most uninteresting ways of accounting for loose ends that could not be coherently pulled together. In this case, the whole show we have been watching for five years or so is made the content of the mind of an autistic boy—but only after every single one of its strands has been carefully, obsessively pulled tight. It is impossible for a story of such complexity to have been conceived by an autistic eleven-year-old. The ending is unbelievable. And it is also, since no loose ends had remained, unnecessary. Why, then, is it there?

It is there, I think, as a final reminder that the story was after all a fiction, as much a fiction as the fiction that an autistic boy could ever spin such a fiction. It is a reminder that just as the story of the boy's spinning such a fiction cannot be true, so every part of the show itself, everything that we have seen has been fiction, though it was fiction that, as we have seen, took its shape, its color, its plot, and its very end—its death—from the demands of life. What is real, this ending asks, and what is fiction?

And the toy in which the boy is absorbed, left on top of the television set, now emerges as a metaphor for television and for its viewers' relation to it. It is not very flattering, if one does watch television, to see oneself described as an autistic eleven-year-old. Yet this character, the show tells us, is the show's creator, and is acknowledged as such by the other characters' awareness of the snow for which the boy is directly responsible. Who is it, this ending asks, who, along with life, gave the show its shape, its colors, and its end? How much has the viewer contributed? Is it a good or a bad thing to watch television, and to be part, and in part a creator, of its fiction?

These are heady questions, and there are many others like them. The fact that they are raised by a program like *St. Elsewhere* shows that the literate opponents of the popular media have no monopoly on literacy, and that the very notion of literacy needs to be examined anew. "Whoever begins at this point, like my readers," Nietzsche wrote, "to reflect and pursue his train of thought will not soon come to the end of it—reason enough for me to come to an end"—and for me as well, but not before I cite one more attack on his time by an author who considered it "an age, wherein the greatest part of men seem agreed to convert reading into amusement, and to reject every thing that requires any considerable degree of attention to be comprehended."[25] Thus David Hume. But Hume's complaint is both older and more recent: it was first made by Plato and is being repeated today by countless educated people who are unaware of its provenance. It suggests that even those with the greatest knowledge of history are not necessarily the most historical of people, and that the gesture of rejecting "every thing that requires any considerable degree of attention to be comprehended" is not peculiar to "the greatest part of men"—or rather, that it is, except that the greatest part of men, and women, includes us all.

Notes

1. Gertrude Himmelfarb, "Stanford and Duke Undercut Classical Values," *New York Times,* 5 May 1988.

2. Matthew Arnold, "The Function of Criticism at the Present Time," in *Critical Theory Since Plato,* ed. Hazard Adams (San Diego, 1971). 588.

3. Allan Bloom, *The Closing of the American Mind: How Higher Education Has Failed Democracy and Impoverished the Souls of Today's Students* (New York, 1987), 58. An article in *Applause* (February 1987), the magazine of the public television station in Philadel-

phia, plugging Patrick McGoohan's series *The Prisoner,* which WHYY was about to air again, closes as follows: "Watch the show when you get the chance. It's not perfect—the fashions, for example, are horribly dated. But it's very good. And for a television show, that's saying a lot" (*"The Prisoner* Returns," 23).

4. The argument for this view, stated dogmatically here, can be found in my "Plato on Imitation and Poetry in *Republic* 10," in *Plato on Beauty, Wisdom, and the Arts.* ed. J. M. E. Moravcsik and Philip Temko (Totowa, N.J., 1982), 47–78: and, with more immediate relevance for this paper, in "Plato and the Mass Media," *The Monist* 71 (Spring 1988): 214–34.

5. Wayne Booth, "The Company We Keep: Self-Making in Imaginative Art," *Daedalus* 111 (Fall 1982): 42.

6. Plato *Phaedrus* 275D, trans. Roy Hackforth (Cambridge, 1952).

7. Nehamas, "Plato and the Mass Media," 221–22.

8. John Cawelti, *Adventure, Mystery, and Romance* (Chicago, 1976), 14, 13.

9. Catherine Belsey, *Critical Practice* (London, 1980), 68, 70, 92.

10. Nehamas, "Plato and the Mass Media," 228–30.

11. Umberto Eco, "On 'Krazy Kat' and 'Peanuts,'" *New York Review of Books,* 15 June 1985.

12. Robert S. Alley, "Media Medicine and Morality," in *Understanding Television,* ed. Richard P. Adler (New York, 1981), 231.

13. David Marc, *Demographic Vistas* (Philadelphia, 1984), 73.

14. Todd Gitlin, *Inside Prime Time* (New York, 1983), 273–324.

15. Steven Best and Douglas Kellner, "(Re)Watching Television: Notes Toward a Political Criticism," *Diacritics* 17 (Summer 1987): 104.

16. On this and other instances of character development in MTM shows, see Jane Feuer, "The MTM Style," in *Television: The Critical View,* ed. Horace Newcomb (New York, 1987), 60–62.

17. See Thomas Schatz, "*St. Elsewhere* and the Evolution of the Ensemble Series," in Newcomb, ed., *Television,* 85–100.

18. George Gerbner and Larry Gross, "The Scary World of TV's Heavy Viewer," *Psychology Today,* April 1976, 44.

19. John Fiske, *Television Culture* (London, 1987), 121, 123.

20. Pierre Bourdieu, *Distinction: A Social Critique of the Judgment of Taste* (Cambridge, Mass., 1984), 32.

21. Herbert Gans, *Popular Culture and High Culture* (New York, 1974), 79.

22. Bourdieu, *Distinction,* 34.

23. Pierre Bourdieu, "The Production of Belief," In *Media, Culture and Society: A Critical Reader,* ed. R. Collins et al. (London, 1986), 154–55.

24. Jane Feuer, "Narrative Form in American Network Television," in *High Theory/Low Culture,* ed. Colin MacCabe (Manchester, 1986), 107.

25. Friedrich Nietzsche, *On the Geneology of Morals,* trans. Walter Kaufmann (New York, 1969), 1: 17; and David Hume, *A Treatise of Human Nature,* ed. L. A. Selby-Bigge (Oxford, 1888), bk. 3. pt. 1, sec. 1, 456.

COMING TO TERMS

Nehamas says that his goal in writing is to "disarm many of the criticisms commonly raised against television" by showing how the medium can indeed reward "serious watching." But to disarm such criticisms, Nehamas first needs to say what they are. He thus writes much of his piece in a point/counterpoint form, in which he first repeats the various complaints that critics have made against television, and then tries to show why he thinks they are wrong. See if you can restate this exchange of points between the critics of TV and Nehamas in your own words. Begin by listing in your journal what Nehamas says are the three usual complaints made against television. Make sure you can state the reasons for these criticisms in your own language. Then explain why Nehamas rejects or challenges these criticisms, what mistakes he thinks the critics of TV have made. Finally, see if you can link Nehamas' reading of *St. Elsewhere* to both the argument he is trying to make for the value of "serious watching" as well as to the usual complaints made against television. In what ways does Nehamas think his reading of this show "disarms" the critics of TV? How successful has Nehamas been in this work? What might be some issues Nehamas has neglected to address?

READING AS A WRITER

Most of what Nehamas has to say is about television and the values of "serious watching." But he begins and ends his piece by talking about intellectual figures who rarely find their way into discussions about TV: Gertrude Himmelfarb, Matthew Arnold, Plato, Pierre Bourdieu, and David Hume. We see Nehamas making an important move here as a writer. He is a writer who is taking an apparently current subject (television) and then imagining this subject as part of an ongoing discussion, a "critical conversation" which is centuries old, a context which invites the participation of some widely acknowledged writers and thinkers.

We would like you to name this "critical conversation" which engages both current and historic participants. What would you call this conversation? It's not just about television; it's about something else. What is this *something else*? What is the subject of this conversation? What are the issues? For instance, how might what Plato said about writing connect up with what David Hume said about reading, or what Nehamas wants to argue about TV? For your journal, form some answers to these questions, not by necessarily doing extra research in the library (you might find many of these figures in a good college dictionary), but by noting how Nehamas quotes these people and what he says about them. (You might want to do so in the following fashion: Jot down in your journal the name of each "critic" that you come across in Nehamas' piece. Then briefly restate what Nehamas tells you that this person believed or argued. Then see if you can determine what Nehamas himself thinks about what this person had to say.) Think of your task as defining, in your own words, the intellectual *context* that Nehamas wants to situate his remarks in, the ongoing "critical conversation" in which he imagines himself an active participant.

WRITING CRITICISM

"The realism of *St. Elsewhere* is measured by comparing it not to life within a hospital, of which most of us know almost nothing, but to the standards and features of earlier shows on the same general subject." Alexander Nehamas begins here to sketch out an interesting theory in which the realism of texts on TV (and perhaps other media as well) are measured "not by proximity to reality but by a distance from fiction whose conventions we have come to see as conventions." Or, to put it another way, what we call "realism" has less to do with how much a particular show mirrors the actual world than with our sense that it is somehow *unlike* most other programs.

In writing an essay, then, we'd like you to test this theory against the evidence of a particular text. You would first want to represent Nehamas' theory in your essay, drawing upon the critical terms and phrases which you feel show the kind of intellectual analysis Nehamas is doing. Next, pick a television program that you feel strives for a kind of "realism" and that you find in some way interesting or compelling. Review it closely in order to determine, for example, which "standards and features" of earlier programming it tries to set itself against and which it still seems to subscribe to. In doing this work, you will want to look both at the content and the form of the program—that is, both at the messages it offers about work, family, gender, love, friendship, and so on, and at the ways those messages get presented, at the look and sound and pace and feel of the show. In what ways does the program seem most "conventional"? Does this approach hamper its attempts at realism? In what ways does it seem least conventional? Do these breaks with tradition always work to make it seem more "real," or do they sometimes have other (perhaps ironic or comical) effects? Are there any other strategies beside a "distance from convention" that the show uses to make itself more realistic? How would you say whether your reading of this program supports Nehamas' claim that "realism" in TV is largely an "intertextual" phenomenon, a result of comparing or setting one program against others? Or whether your reading calls for an alternative way of explaining realism? Then, what would be the substance of this alternative way of explaining realism on television?

The Home Magazine Kitchen

Donald Norman

Articles and advertisements in home magazines always show bright happy families in bright happy kitchens. They excite the imagination, but not always in the ways the magazine intended.

How come there are never any:

Dirty dishes?
Dishes in the drainer?
Drainers?
Spills?
Clutter?
Long rows of appliances?

And how come there always seem to be enough electric outlets?

Every year the busy manufacturers of the world toil and labor to produce yet another kitchen appliance. Would you believe that even now, deep in their secret caverns and laboratories, scientists are plotting to take over the kitchen with their diabolical means?

The intelligent home system is coming. Ah yes. We will have communication lines connecting all our appliances. The refrigerator will talk to the stove. The stove will talk to the washing machine. The washing machine will talk to the heating system. All the appliances will talk with one another. And the worst part of all is that they may try to talk to us.

"Hi Don. This is Fred, your friendly dishwasher. I am ready to do a load of dishes. Would you like me to wait until 3 AM, when you are asleep and when utility rates are low?"

"Don? Don? Don, I am talking to you."

"Hi, Don. This is Fred, your friendly dishwasher. I am ready to do a load of dishes. Would you like me to wait until 3 AM, when you are asleep and when utility rates are low?"

"Don? I know you are there, Don. I can sense your infrared emissions."

"Hi, Don. This is Fred, your friendly dishwasher. I am ready..."

Hour after hour after hour. And then Susan, my friendly refrigerator, will ask if she can defrost herself at 4 AM And let's not forget Tom, my friendly oven. Oops, Tom is also my friendly videocassette recorder. Uh-uh, those manufacturers didn't synchronize their naming practices. What happens if I ask Tom to turn on from 2 AM to 4 AM for channel 10?

"This is Tom, your friendly oven. What temperature should I use?"

Or what if Susan asks if she should defrost the chicken, but before I can say "No," Tom says, "Yes, I am ready to record." Susan hears the "yes," so she defrosts the chicken. And maybe Fred starts the dishes.

Ah yes, the high-technology house. The same folks who brought you the digital watches and VCRs that you can't program, who brought you washing machines and dryers with unintelligible dials, and refrigerators whose temperature controls you can't set will now bring you the high-tech house. Welcome to the world of the future.

The Four-Questions Test

Whenever I see photographs of kitchens in those home magazines, especially ones that look appealing, I peer carefully for any sign that the kitchens are actually used. I look for answers to the questions I have about my own kitchen. I never find them.

Each appliance accomplishes some wonderful task that you never realized you needed but that now you will not be able to live without. How can you ever resist? Well, my family has figured out how. We invented a test—the four-questions test—that each prospective new purchase must pass. We don't care how valuable, useful, efficient, or essential the new appliance is; it has to pass this test or we will not buy it.

Question one, the most basic of all, defeats most new appliances right from the start: *"Where would we store it?"* Since we can't even figure out where to store the appliances we already own, what would we do with new ones? Put them on the floor? In fact, in my family this issue alone has become so serious that it has almost put an end to buying. I hope.

Now for question two: *"Where would we use it?"*—even if we could find a place to store it. Wherever we try, there is something else already there. Look again at those magical kitchens in the glossy magazines. Where are the mixers, toasters, blenders, grinders, coffee makers, knife sharpeners, pencil sharpeners, juice makers, . . . , what have-you?

Question three follows logically: *"Where would we plug it in?"* The real trick is to find an available outlet that has some relationship to the place where we would like to use the appliance. These new kitchen gadgets always require plugging in. I think that if we bought another breadboard, it would require plugging in. Either that or it would use batteries—and, of course, it would require some assembly.

Question four brings up a critical issue: *"How much work would it be to clean?"* We have managed to escape several appliance fads simply by thinking about the cleanup problem. Sure, a kitchen atomic vaporizer will save us thirty seconds every time we need to do some atomizing, but at a cost of five minutes of disassembling, washing, drying, and reassembling. If we could figure out where to put it, and use it, and plug it in.

Designing Kitchens for Real Life

I suspect that kitchens are designed to look good, not to be used. If kitchen designers had their way, they wouldn't let ordinary people into kitchens, only photographers. The problem with kitchens is a general problem with design. How many people really worry about the *usability* of their kitchen—or their home or office or computer or the ever more fancy gadgets in their automobile?

The proper way to design something is to take into account what people really do and then construct things appropriately. Sound logical and sensible? Yup, but it is amazing how infrequently such common sense is followed.

Industrial Psychology and the Kitchen

Kitchens were actually one of the very first workplaces to receive careful, scientific study. In the early 1900s Lillian Gilbreth, one of the world's first industrial psychologists, studied work habits in home and industry. Among her many activities, she taught industrial psychology as a professor at Purdue University and did scientific studies of work patterns in kitchens. Gilbreth may be familiar to some of you as the mother of the family in the books and films *Cheaper by the Dozen* and *Belles on their Toes*. She and her husband, Frank Gilbreth, did indeed have a dozen children, and they practiced and polished their studies of industrial efficiency on their own family. It was two of the children who later wrote about their life.

Lillian Gilbreth studied the work patterns in the kitchen. She and her associates watched and measured, timing operations with a stopwatch—the Gilbreths were one of the pioneers in the development of "time-and-motion" studies, especially through the use of motion pictures of the activities. From this work, Lillian Gilbreth developed the concept of the "work triangle"—the proper arrangement of sink, refrigerator, and stove to make the normal activities of usage more efficient and less tiring. Gilbreth made sure there was a place for every object and every activity, a place that was carefully selected to minimize effort and increase efficiency and ease of use. To my knowledge, no equivalent studies have been conducted in recent times.

It is about time we updated Gilbreth's principles. Work patterns have changed dramatically in the kitchen since those times. Just as today it is rare to have a family of fourteen, other changes have come about in the home. For one, the kitchen used to be populated mainly by the cook, whether a member of the family or hired. In Gilbreth's own family the cook would not let other family members into the kitchen. Today almost everyone invades the kitchen, and kitchens need to be designed to accommodate them. Sometimes this population pressure is deliberate, as when two or more people collaborate in preparing a meal. Sometimes it is accidental, as when family members enter the kitchen for this or that while the main food preparer is at work. Are kitchens designed for these practices? Not very often. Moreover, we have many new appliances and completely different kinds of foods today: supermarket packaged foods, foods meant for the microwave oven, and foods that are frozen, canned, dried, and otherwise specially prepared. There are new kinds of cooking devices, new kinds of blenders, grinders, and mechanized appliances, and new cooking practices.

Even with all these changes, Gilbreth's fundamental observations and principles still apply, and still are woefully ignored. Where do we find room for all our gadgets? Where is the proper workspace to support our activities? What design principles have been applied to the work patterns within the kitchen? These are the aspects of kitchen design Gilbreth emphasized—the aspects that are lacking today.

A good kitchen makes it possible for others to enter to get a snack, a cup of coffee, or a glass of water or milk without interfering with the cooking process. Or for two people to work at once without continually bumping into each other. Or for someone to answer the phone, or just to hang around to talk.

What about cleanup, perhaps the most neglected part of the kitchen: Are appliances designed so as to be easy to clean? Are countertops easy to clean, with enough room beneath them so that you can scoop crumbs into your hand? Is there sufficient room near the sinks and dishwasher to place dirty items in preparation for washing them? Can you open the dishwasher to fill it or empty it without blocking other activities in the kitchen? And what about newly washed items? Where do they go while drying?

Once more I look through kitchen design books in vain for some signs of a rack or holder for drying dishes. Nope. Actually, I have found some clever ideas from Scandinavia. That figures: Scandinavian designers have always led the way in combining functionality with aesthetics. Too bad others have not followed. Unfortunately, those Scandinavian kitchens seem better designed for small families and small apartments. They wouldn't work in my household.

Design Neglect

The kitchen is really the high-technology center of the home. It is easy to make fun of the kitchen, but it stands as a microcosm of a lack of consideration for others in the entire design community. Benign neglect was once a popular political concept: Deliberately leave something alone because it will probably be better off without any attention, without the type of aid that politicians typically provide.

Well, perhaps what we have here in the kitchen is Design Neglect. Notice that benign neglect failed as a political strategy and so too does design neglect fail as a design strategy. The kitchen—or any other workplace for that matter—will not be an efficient, convenient place to work unless someone spends time, effort, and thought at the task. Worse, if effort is put only into aesthetics or cost, the result is almost guaranteed to hinder usability.

Beautiful kitchen cabinets are nice to look at, but many do some impractical things—like hide the handles for the sake of beauty, which means that you can't figure out how to open them. Appliances have moved toward computer controls and touch-sensitive keyboards, so that now it is impossible to figure out how to heat a cup of coffee or make a piece of toast. Can you use your microwave oven in all its glory? Do you know how to program for defrost followed by a wait cycle followed by a cooking cycle? Can you even set the temperature of your refrigerator?

There are enough problems in the kitchen to fill a book. That's fine for me. I write books. But what about the user of the kitchen? In my earlier book *The Design of Everyday Things*, I talked about how difficult it was to adjust the temperature of my home refrigerator. The labels on the temperature controls and their actual functions were completely unrelated. I was not alone: Many people wrote to me saying that they had the same refrigerator and they too could not adjust the temperature. One person was even an engineer who worked for the same company that made the refrigerator.

And what about the need for electric outlets? You would think that any modern kitchen designer would understand the need for sufficient outlets, or for room to store all those appliances, a place to put them during use. No,

I think that designers try to make everything look good. And outlets are simply not thought of as a part of the kitchen. Outlets are for the electricians to worry about. But electricians do what they are told to do, and if nobody tells them how many outlets to put in, they will follow the electrical code, which probably specifies two outlets per wall, regardless of whether it is a kitchen or a bedroom. Design neglect.

The Proper Way to Design

The proper way to design anything is to start off understanding the tasks that are to be done and the needs of the users. In a kitchen, don't start with the appliances and counters, start with the people and their needs. This is how all things should be designed, not just kitchens.

What do we do in the kitchen? It isn't hard to discover: just observe some families. Patterns probably differ depending upon the kind of family, but I suspect that there are not that many different kinds of usage patterns—a dozen perhaps? That wouldn't be too hard to study and catalog.

Here is what is required for my family, and maybe for many users of a kitchen. First of all, we use our kitchen for multiple purposes, sometimes several at the same time. One important activity is that of loading up the kitchen, bringing in newly purchased groceries. For this activity, we need places to unload the supplies and places for the storing. Cooking is the most obvious function of the kitchen, and this too has many different phases and activities, often spread out over considerable time. For these, we need places and equipment for food preparation and the actual cooking. We also need places for casual food preparation—for snacks—and places for food in the process of cooking, perhaps simmering, perhaps defrosting, or perhaps prepared ahead of time and now awaiting the final steps. An important but often neglected activity is cleaning up and taking out the garbage. Can we transport packages easily through the necessary doors? Finally, our kitchen has become the modern "communication and control" center of the home, the place where everyone congregates for information, talk, and activities.

All these different activities have to be accounted for, especially when performed by several people simultaneously. One of the goals of a successful design should be to make all the activities go smoothly, so that each of us is properly accommodated, so that we do not get into one another's way

The Kitchen as Communication and Control Center

Where would the modern family be without the refrigerator? Its glistening sides provide an ideal location for the graffiti of living. What a serendipitous design accident. Refrigerator doors are made of steel, and steel is magnetic.

Refrigerator doors are large, flat, unadorned. Those surfaces of white-painted steel cry out for a use, and a use has been found.

Sometimes I wonder which is the more important function of the refrigerator: storing the food on the inside or keeping the messages on the outside. The family refrigerator has become the major center of the home, in part through the magic of magnets. In fact, in the United States, the making of "refrigerator magnets" has become a minor industry. And where would we be without the most impressive invention of the era: Post-it Notes?

It all had to be accidental: I can't imagine anyone in the refrigerator business or the kitchen design business being clever enough to design the kitchen message board on purpose. Even in countries where the refrigerator door is not used for messages and announcements, families typically do have a need for some central message board. Everyone must know of this need for centers. Everyone must know except the refrigerator manufacturers and kitchen designers.

Designer kitchens, of course, do not allow for such an activity. Look again at those home magazines: Never would any appliance be sullied by a magnet or taped note. Moreover, if you follow the designers, you will pay twice or even three times as much money and buy a designer refrigerator with wood panels; you will pay twice as much in order to have a door that has no use at all, except, sigh, as a door. Pretty, elegant, but so dull.

The refrigerator door, quite by accident, turns out to be a nice, functional place for messages. In part this is because in many homes its location, not so accidentally, turns out to be perfect. Home design has evolved over the years. Today, especially in the United States, the kitchen is often the center of family activity and the refrigerator the center of the kitchen: well lit, central to the other activities.

I can imagine a typical kitchen inhabitant during a between-meals forage: glass of milk in one hand, cookies in the other, telephone between shoulder and chin, checking for scheduled events and messages on the refrigerator. Do designers have such a picture in mind when they design? This is real kitchen usage but it is never planned, never thought about: It just happens. But because it is so essential to kitchen life, it ought to be a major part of kitchen design. It should be thought about, incorporated into the kitchen plans. I fear that such practical uses will be destroyed by the prettiness of the home magazine kitchen. Aesthetics before function. Ah yes. The story of modern design.

COMING TO TERMS

In "The Home Magazine Kitchen," Donald Norman writes about how the design of the contemporary American kitchen and its assorted appliances has departed from common sense and has lost touch with the principle of functionality or practicality. One of the problems Norman returns to on several occasions is that the early designs of the American kitchen (still popular today) appear to dominate the work of domestic and appliance designers— that even though the American kitchen has become over the decades more than just a place to prepare and store food, the design of the kitchen has not kept pace with changes in American society. Re-read Norman's article and underline those passages in which you see him documenting sociological or technological changes that have not been reflected in the practical design of the American kitchen. Then in your journal respond to the following questions: What are these changes which Norman observes? Can you think of other changes that Norman may have overlooked? What are other functions of the American kitchen now beyond food preparation? Think of the activities in which you have been involved in your family kitchen. What can you add to a revised list of the functions of the American kitchen?

READING AS A WRITER

Norman makes several important moves as a writer, transforming what might appear on the surface to be a rather common subject—the design of the American kitchen—into a subject of considerable depth and multiple contexts. Like other writers in *Media Journal,* Norman sees a familiar and maybe mundane cultural text in unfamiliar ways: the American kitchen becomes a prism through which to examine American culture and society. One way in which Norman extends the context of his subject is to draw upon subject areas not usually associated with a discussion of the kitchen, such as demographics, psychology, and history. We'd like you to re-read Norman's article and locate places where you think he is taking his readers to an *unexpected* area of discussion. And then answer the following question in your journal. How does Norman use demographics, psychology, and history to complicate his study and our reading of the American kitchen?

WRITING CRITICISM

"The proper way to design anything is to start off understanding the tasks that are to be done and the needs of the users.... This is how all things should be designed, not just kitchens." So writes Donald Norman in "The Home Magazine Kitchen." He then spends much of his time in showing the ways in which modern designers have failed to put this principle to reasonable use. In fact, towards the end of his article, Norman seems to be concluding that "the story of modern design" might be summed up in a kind of motto: "Aesthetics before function."

We'd like you to write an essay in which you test out Norman's conclusions about the current state of design, that designers don't really consider or "start with the people and their needs." Although Norman's primary focus here is the poor design of the American kitchen, he seems quite comfortable in seeing this illustration as compelling evidence for a rather large generalization about "the story of modern design."

Continuing with Norman's metaphor of a "story" here, we'd like you to add a few additional chapters to this story of modern design. For this project, you will need to select at least two cultural texts which have functions, a sense of aesthetic appeal, and have been professionally designed. These texts might include but are not limited to cars, food preparation machines, communication or entertainment technologies, fast food franchises, dormitory rooms, bathrooms, waiting rooms in professional offices, work stations, etc. You'll need to select cultural texts against which you can fairly test Norman's ideas. You might, for example, choose one text which supports and another which challenges Norman's generalization about design: "aesthetics before function." You might also choose a text which through your reading finds a workable balance between aesthetics and function. As you analyze these texts, give other criteria besides the two Norman identifies that might have influenced the design of these texts.

The Peek-a-Boo World

Neil Postman

Toward the middle years of the nineteenth century, two ideas came together whose convergence provided twentieth-century America with a new metaphor of public discourse. Their partnership overwhelmed the Age of Exposition, and laid the foundation for the Age of Show Business. One of the ideas was quite new, the other as old as the cave paintings of Altamira. We shall come to the old idea presently. The new idea was that transportation and communication could be disengaged from each other, that space was not an inevitable constraint on the movement of information.

Americans of the 1800's were very much concerned with the problem of "conquering" space. By the mid-nineteenth century, the frontier extended to the Pacific Ocean, and a rudimentary railroad system, begun in the 1830's, had started to move people and merchandise across the continent. But until the 1840's, information could move only as fast as a human being could carry it; to be precise, only as fast as a train could travel, which, to be even more precise, meant about thirty-five miles per hour. In the face of such a limitation, the development of America as a national community was retarded. In the 1840's, America was still a composite of regions, each conversing in its own ways, addressing its own interests. A continentwide conversation was not yet possible.

The solution to these problems, as every school child used to know, was electricity. To no one's surprise, it was an American who found a practical way to put electricity in the service of communication and, in doing so, eliminated the problem of space once and for all. I refer, of course, to Samuel Finley Breese Morse, America's first true "spaceman." His telegraph erased

From "The Peek-a-Boo World," pp. 64–80, by Neil Postman, 1985, *Amusing Ourselves to Death.* © 1985 by Neil Postman. Reprinted by permission of Viking Penguin, a division of Penguin Books USA Inc.

state lines, collapsed regions, and, by wrapping the continent in an informa-tion grid, created the possibility of a unified American discourse.

But at a considerable cost. For telegraphy did something that Morse did not foresee when he prophesied that telegraphy would make "one neighbor-hood of the whole country." It destroyed the prevailing definition of infor-mation, and in doing so gave a new meaning to public discourse. Among the few who understood this consequence was Henry David Thoreau, who remarked in *Walden* that "We are in great haste to construct a magnetic tele-graph from Maine to Texas; but Maine and Texas, it may be, have nothing important to communicate.... We are eager to tunnel under the Atlantic and bring the old world some weeks nearer to the new; but perchance the first news that will leak through into the broad flapping American ear will be that Princess Adelaide has the whooping cough."

Thoreau, as it turned out, was precisely correct. He grasped that the tele-graph would create its own definition of discourse; that it would not only permit but insist upon a conversation between Maine and Texas; and that it would require the content of that conversation to be different from what Ty-pographic Man was accustomed to.

The telegraph made a three-pronged attack on typography's definition of discourse, introducing on a large scale irrelevance, impotence, and incoher-ence. These demons of discourse were aroused by the fact that telegraphy gave a form of legitimacy to the idea of context-free information; that is, to the idea that the value of information need not be tied to any function it might serve in social and political decision-making and action, but may attach merely to its novelty, interest, and curiosity. The telegraph made information into a commodity, a "thing" that could be bought and sold irrespective of its uses or meaning.

But it did not do so alone. The potential of the telegraph to transform in-formation into a commodity might never have been realized, except for the partnership between the telegraph and the press. The penny newspaper, emerging slightly before telegraphy, in the 1830's, had already begun the pro-cess of elevating irrelevance to the status of news. Such papers as Benjamin Day's *New York Sun* and James Bennett's *New York Herald* turned away from the tradition of news as reasoned (if biased) political opinion and urgent com-mercial information and filled their pages with accounts of sensational events, mostly concerning crime and sex. While such "human interest news" played little role in shaping the decisions and actions of readers, it was at least local—about places and people within their experience—and it was not always tied to the moment. The human-interest stories of the penny newspapers had a timeless quality; their power to engage lay not so much in their currency as in their transcendence. Nor did all newspapers occupy themselves with such content. For the most part, the information they provided was not only local but largely functional—tied to the problems and decisions readers had to ad-dress in order to manage their personal and community affairs.

The telegraph changed all that, and with astonishing speed. Within months of Morse's first public demonstration, the local and the timeless had lost their central position in newspapers, eclipsed by the dazzle of distance and speed. In fact, the first known use of the telegraph by a newspaper occurred *one day* after Morse gave his historic demonstration of telegraphy's workability. Using the same Washington-to-Baltimore line Morse had constructed, the *Baltimore Patriot* gave its readers information about action taken by the House of Representatives on the Oregon issue. The paper concluded its report by noting: "...we are thus enabled to give our readers information from Washington up to two o'clock. This is indeed the annihilation of space."

For a brief time, practical problems (mostly involving the scarcity of telegraph lines) preserved something of the old definition of news as functional information. But the foresighted among the nation's publishers were quick to see where the future lay, and committed their full resources to the wiring of the continent. William Swain, the owner of the *Philadelphia Public Ledger*, not only invested heavily in the Magnetic Telegraph Company, the first commercial telegraph corporation, but became its president in 1850.

It was not long until the fortunes of newspapers came to depend not on the quality or utility of the news they provided, but on how much, from what distances, and at what speed. James Bennett of the *New York Herald* boasted that in the first week of 1848, his paper contained 79,000 words of telegraphic content—of what relevance to his readers, he didn't say. Only four years after Morse opened the nation's first telegraph line on May 24, 1844, the Associated Press was founded, and news from nowhere, addressed to no one in particular, began to crisscross the nation. Wars, crimes, crashes, fires, floods—much of it the social and political equivalent of Adelaide's whooping cough—became the content of what people called "the news of the day."

As Thoreau implied, telegraphy made relevance irrelevant. The abundant flow of information had very little or nothing to do with those to whom it was addressed; that is, with any social or intellectual context in which their lives were embedded. Coleridge's famous line about water everywhere without a drop to drink may serve as a metaphor of a decontextualized information environment: In a sea of information, there was very little of it to use. A man in Maine and a man in Texas could converse, but not about anything either of them knew or cared very much about. The telegraph may have made the country into "one neighborhood," but it was a peculiar one, populated by strangers who knew nothing but the most superficial facts about each other.

Since we live today in just such a neighborhood (now sometimes called a "global village"), you may get a sense of what is meant by context-free information by asking yourself the following question: How often does it occur that information provided you on morning radio or television, or in the morning newspaper, causes you to alter your plans for the day, or to take some action you would not otherwise have taken, or provides insight into some problem you are required to solve? For most of us, news of the weather will sometimes have such consequences; for investors, news of the stock market; perhaps an occasional story about a crime will do it, if by chance the crime occurred near where you live or involved someone you know. But most of our daily news is inert, consisting of information that gives us something to talk about but cannot lead to any meaningful action. This fact is the principal legacy of the telegraph: By generating an abundance of irrelevant information, it dramatically altered what may be called the "information-action ratio."

In both oral and typographic cultures, information derives its importance from the possibilities of action. Of course, in any communication environment, input (what one is informed about) always exceeds output (the possibilities of action based on information). But the situation created by telegraphy, and then exacerbated by later technologies, made the relationship between information and action both abstract and remote. For the first time in human history, people were faced with the problem of information glut, which means that simultaneously they were faced with the problem of a diminished social and political potency.

You may get a sense of what this means by asking yourself another series of questions: What steps do you plan to take to reduce the conflict in the Middle East? Or the rates of inflation, crime, and unemployment? What are your plans for preserving the environment or reducing the risk of nuclear war? What do you plan to do about NATO, OPEC, the CIA, affirmative action, and the monstrous treatment of the Baha'is in Iran? I shall take the liberty of answering for you: You plan to do nothing about them. You may, of course, cast a ballot for someone who claims to have some plans, as well as the power to act. But this you can do only once every two or four years by giving one hour of your time, hardly a satisfying means of expressing the broad range of opinions you hold. Voting, we might even say, is the next to last refuge of the politically impotent. The last refuge is, of course, giving your opinion to a pollster, who will get a version of it through a desiccated question, and then will submerge it in a Niagara of similar opinions, and convert them into—what else?—another piece of news. Thus, we have here a great loop of impotence: The news elicits from you a variety of opinions about which you can do nothing except to offer them as more news, about which you can do nothing.

Prior to the age of telegraphy, the information-action ratio was sufficiently close so that most people had a sense of being able to control some of the contingencies in their lives. What people knew about had action-value. In the information world created by telegraphy, this sense of potency was lost, precisely because the whole world became the context for news. Everything became everyone's business. For the first time, we were sent information which answered no question we had asked, and which, in any case, did not permit the right of reply.

We may say then that the contribution of the telegraph to public discourse was to dignify irrelevance and amplify impotence. But this was not all: Telegraphy also made public discourse essentially incoherent. It brought into being a world of broken time and broken attention, to use Lewis Mumford's phrase. The principal strength of the telegraph was its capacity to move information, not collect it, explain it or analyze it. In this respect, telegraphy was the exact opposite of typography. Books, for example, are an excellent container for the accumulation, quiet scrutiny and organized analysis of information and ideas. It takes time to write a book, and to read one; time to discuss its contents and to make judgments about their merit, including the form of their presentation. A book is an attempt to make thought permanent and to contribute to the great conversation conducted by authors of the past. Therefore, civilized people everywhere consider the burning of a book a vile form of anti-intellectualism. *But the telegraph demands that we burn its contents.* The value of telegraphy is undermined by applying the tests of permanence, continuity or coherence. The telegraph is suited only to the flashing of messages, each to be quickly replaced by a more up-to-date message. Facts push other facts into and then out of consciousness at speeds that neither permit nor require evaluation.

The telegraph introduced a kind of public conversation whose form had startling characteristics: Its language was the language of headlines—sensational, fragmented, impersonal. News took the form of slogans, to be noted with excitement, to be forgotten with dispatch. Its language was also entirely discontinuous. One message had no connection to that which preceded or followed it. Each "headline" stood alone as its own context. The receiver of the news had to provide a meaning if he could. The sender was under no obligation to do so. And because of all this, the world as depicted by the telegraph began to appear unmanageable, even undecipherable. The line-by-line, sequential, continuous form of the printed page slowly began to lose its resonance as a metaphor of how knowledge was to be acquired and how the world was to be understood. "Knowing" the facts took on a new meaning, for it did not imply that one understood implications, background, or connections. Telegraphic discourse permitted no time for historical perspectives and gave no priority to the qualitative. To the telegraph, intelligence meant knowing *of* lots of things, not knowing *about* them.

Thus, to the reverent question posed by Morse—What hath God wrought?—a disturbing answer came back: a neighborhood of strangers and pointless quantity; a world of fragments and discontinuities. God, of course, had nothing to do with it. And yet, for all of the power of the telegraph, had it stood alone as a new metaphor for discourse, it is likely that print culture would have withstood its assault; would, at least have held its ground. As it happened, at almost exactly the same time Morse was reconceiving the meaning of information, Louis Daguerre was reconceiving the meaning of nature; one might even say, of reality itself. As Daguerre remarked in 1838 in a notice designed to attract investors, "The daguerreotype is not merely an instrument which serves to draw nature...[it] gives her the power to reproduce herself."

Of course both the need and the power to draw nature have always implied reproducing nature, refashioning it to make it comprehensible and manageable. The earliest cave paintings were quite possibly visual projections of a hunt that had not yet taken place, wish fulfillments of an anticipated subjection of nature. Reproducing nature, in other words, is a very old idea. But Daguerre did not have this meaning of "reproduce" in mind. He meant to announce that the photograph would invest everyone with the power to duplicate nature as often and wherever one liked. He meant to say he had invented the world's first "cloning" device, that the photograph was to visual experience what the printing press was to the written word.

In point of fact, the daguerreotype was not quite capable of achieving such an equation. It was not until William Henry Fox Talbot, an English mathematician and linguist, invented the process of preparing a negative from which any number of positives could be made that the mass printing and publication of photographs became possible. The name "photography" was given to this process by the famous astronomer Sir John F. W. Herschel. It is an odd name since it literally means "writing with light." Perhaps Herschel meant the name to be taken ironically, since it must have been clear from the beginning that photography and writing (in fact, language in any form) do not inhabit the same universe of discourse.

Nonetheless, ever since the process was named it has been the custom to speak of photography as a "language." The metaphor is risky because it tends to obscure the fundamental differences between the two modes of conversation. To begin with, photography is a language that speaks only in particularities. Its vocabulary of images is limited to concrete representation. Unlike words and sentences, the photograph does not present to us an idea or concept about the world, except as we use language itself to convert the image to idea. By itself, a photograph cannot deal with the unseen, the remote, the internal, the abstract. It does not speak of "man," only of *a* man; not of "tree," only of *a* tree. You cannot produce a photograph of "nature," any more than a photograph of "the sea." You can only photograph a particular fragment of the here-and-now—a cliff of a certain terrain, in a certain condition of light; a wave at a moment in time, from a particular point of view. And just as "nature" and "the sea" cannot be photographed, such larger abstractions as truth, honor, love, falsehood cannot be talked about in the lexicon of pictures. For "showing of" and "talking about" are two very different kinds of processes. "Pictures," Gavriel Salomon has written, "need to be recognized, words need to be understood." By this he means that the photograph presents the world as object; language, the world as idea. For even the simplest act of naming a thing is an act of thinking—of comparing one thing with others, selecting certain features in common, ignoring what is different, and making an imaginary category. There is no such thing in nature as "man" or "tree." The universe offers no such categories or simplifications; only flux and infinite variety. The photograph documents and celebrates the particularities of this infinite variety. Language makes them comprehensible.

The photograph also lacks a syntax, which deprives it of a capacity to argue with the world. As an "objective" slice of space-time, the photograph testifies that someone was there or something happened. Its testimony is powerful but it offers no opinions—no "should-have-beens" or "might-have-beens." Photography is preeminently a world of fact, not of dispute about facts or of conclusions to be drawn from them. But this is not to say photography lacks an epistemological bias. As Susan Sontag has observed, a photograph implies "that we know about the world if we accept it as the camera records it." But, as she further observes, all understanding begins with our *not* accepting the world as it appears. Language, of course, is the medium we use to challenge, dispute, and cross-examine what comes into view, what is on the surface. The words *"true"* and *"false"* come from the universe of language, and no other. When applied to a photograph, the question, Is it true? means only, Is this a reproduction of a real slice of space-time? If the answer is "Yes," there are no grounds for argument, for it makes no sense to disagree with an unfaked photograph. The photograph itself makes no arguable propositions, makes no extended and unambiguous commentary. It offers no assertions to refute, so it is not refutable.

The way in which the photograph records experience is also different from the way of language. Language makes sense only when it is presented as a sequence of propositions. Meaning is distorted when a word or sentence is, as we say, taken out of context; when a reader or listener is deprived of what was said before, and after. But there is no such thing as a photograph taken out of context, for a photograph does not require one. In fact, the point of photography is to isolate images from context, so as to make them visible in a different way. In a world of photographic images, Ms. Sontag writes, "all borders ...seem arbitrary. Anything can be separated, can be made discontinuous, from anything else: All that is necessary is to frame the subject differently." She is remarking on the capacity of photographs to perform a peculiar kind of dismembering of reality, a wrenching of moments out of their contexts, and a juxtaposing of events and things that have no logical or historical connection with each other. Like telegraphy, photography recreates the world as a series of idiosyncratic events. There is no beginning, middle, or end in a world of photographs, as there is none implied by telegraphy. The world is atomized. There is only a present and it need not be part of any story that can be told.

That the image and the word have different functions, work at different levels of abstraction, and require different modes of response will not come as a new idea to anyone. Painting is at least three times as old as writing, and the place of imagery in the repertoire of communication instruments was quite well understood in the nineteenth century. What was new in the mid-nineteenth century was the sudden and massive intrusion of the photograph and other iconographs into the symbolic environment. This event is what Daniel Boorstin in his pioneering book *The Image* calls "the graphic revolution." By this phrase, Boorstin means to call attention to the fierce assault on language made by forms of mechanically reproduced imagery that spread unchecked throughout American culture—photographs, prints, posters, drawings, advertisements. I choose the word "assault" deliberately here, to amplify the point implied in Boorstin's "graphic *revolution.*" The new imagery, with photography at its forefront, did not merely function as a supplement to language, but bid to replace it as our dominant means for construing, understanding, and testing reality. What Boorstin implies about the graphic revolution, I wish to make explicit here: The new focus on the image undermined traditional definitions of information, of news, and, to a large extent, of reality itself. First in billboards, posters, and advertisements, and later in such "news" magazines and papers as *Life, Look,* the New York *Daily Mirror* and *Daily News,* the picture forced exposition into the background, and in some instances obliterated it altogether. By the end of the nineteenth century, advertisers and newspapermen had discovered that a picture was not only worth a thousand words, but, where sales were concerned, was better. For countless Americans, seeing, not reading, became the basis for believing.

In a peculiar way, the photograph was the perfect complement to the flood of telegraphic news-from-nowhere that threatened to submerge readers in a sea of facts from unknown places about strangers with unknown faces. For the photograph gave a concrete reality to the strange-sounding datelines, and attached faces to the unknown names. Thus it provided the illusion, at least, that "the news" had a connection to something within one's sensory experience. It created an apparent context for the "news of the day." And the "news of the day" created a context for the photograph.

But the sense of context created by the partnership of photograph and headline was, of course, entirely illusory. You may get a better sense of what I mean here if you imagine a stranger's informing you that the illyx is a subspecies of vermiform plant with articulated leaves that flowers biannually on the island of Aldononjes. And if you wonder aloud, "Yes, but what has that to do with anything?" imagine that your informant replies, "But here is a photograph I want you to see," and hands you a picture labeled *Illyx on Aldononjes.* "Ah, yes," you might murmur, "now I see." It is true enough that the photograph provides a context for the sentence you have been given, and that the sentence provides a context of sorts for the photograph, and you may even believe for a day or so that you have learned something. But if the event is entirely self-contained, devoid of any relationship to your past knowledge or future plans, if that is the beginning and end of your encounter with the stranger, then the appearance of context provided by the conjunction of sentence and image is illusory, and so is the impression of meaning attached to it. You will, in fact, have "learned" nothing (except perhaps to avoid strangers with photographs), and the illyx will fade from your mental landscape as though it had never been. At best you are left with an amusing bit of trivia, good for trading in cocktail party chatter or solving a crossword puzzle, but nothing more.

It may be of some interest to note, in this connection, that the crossword puzzle became a popular form of diversion in America at just that point when the telegraph and the photograph had achieved the transformation of news from functional information to decontextualized fact. This coincidence suggests that the new technologies had turned the age-old problem of information on its head: Where people once sought information to manage the real contexts of their lives, now they had to invent contexts in which otherwise useless information might be put to some apparent use. The crossword puzzle is one such pseudo-context; the cocktail party is another; the radio quiz shows of the 1930's and 1940's and the modern television game show are still others; and the ultimate, perhaps, is the wildly successful "Trivial Pursuit." In one form or another, each of these supplies an answer to the question, "What am I to do with all these disconnected facts?" And in one form or another, the answer is the same: Why not use them for diversion? for entertainment? to amuse yourself, in a game? In *The Image*, Boorstin calls the major creation of the graphic revolution the "pseudo-event," by which he means an event specifically staged to be reported—like the press conference, say. I mean to suggest here that a more significant legacy of the telegraph and the photograph may be the pseudo-*context*. A pseudo-context is a structure invented to give fragmented and irrelevant information a seeming use. But the use the pseudo-context provides is not action, or problem-solving, or change. It is the only use left for information with no genuine connection to our lives. And that, of course, is to amuse. The pseudo-context is the last refuge, so to say, of a culture overwhelmed by irrelevance, incoherence, and impotence.

Of course, photography and telegraphy did not strike down at one blow the vast edifice that was typographic culture. The habits of exposition, as I have tried to show, had a long history, and they held powerful sway over the minds of turn-of-the-century Americans. In fact, the early decades of the twentieth century were marked by a great outpouring of brilliant language and literature. In the pages of magazines like the *American Mercury* and *The New Yorker*, in the novels and stories of Faulkner, Fitzgerald, Steinbeck, and Hemingway, and even in the columns of the newspaper giants—the *Herald Tribune*, the *Times*—prose thrilled with a vibrancy and intensity that delighted ear and eye. But this was exposition's nightingale song, most brilliant and sweet as the singer nears the moment of death. It told, for the Age of Exposition, not of new beginnings, but of an end. Beneath its dying melody, a new note had been sounded, and photography and telegraphy set the key. Theirs was a "language" that denied interconnectedness, proceeded without context, argued the irrelevance of history, explained nothing, and offered fascination in place of complexity and coherence. Theirs was a duet of image and instancy, and together they played the tune of a new kind of public discourse in America.

Each of the media that entered the electronic conversation in the late nineteenth and early twentieth centuries followed the lead of the telegraph and the photograph, and amplified their biases. Some, such as film, were by their nature inclined to do so. Others, whose bias was rather toward the amplification of rational speech—like radio—were overwhelmed by the thrust of the new epistemology and came in the end to support it. Together, this ensemble of electronic techniques called into being a new world—a peek-a-boo world, where now this event, now that, pops into view for a moment, then vanishes again. It is a world without much coherence or sense; a world that does not ask us, indeed, does not permit us to do anything; a world that is, like the child's game of peek-a-boo, entirely self-contained. But like peek-a-boo, it is also endlessly entertaining.

Of course, there is nothing wrong with playing peek-a-boo. And there is nothing wrong with entertainment. As some psychiatrist once put it, we all build castles in the air. The problems come when we try to *live* in them. The communications media of the late nineteenth and early twentieth centuries, with telegraphy and photography at their center, called the peek-a-boo world into existence, but we did not come to live there until television. Television gave the epistemological biases of the telegraph and the photograph their most potent expression, raising the interplay of image and instancy to an exquisite and dangerous perfection. And it brought them into the home. We are by now well into a second generation of children for whom television has been their first and most accessible teacher and, for many, their most reliable companion and friend. To put it plainly, television is the command center of the new epistemology. There is no audience so young that it is barred from television. There is no poverty so abject that it must forgo television. There is no education so exalted that it is not modified by television. And most important of all, there is no subject of public interest—politics, news, education, religion, science, sports—that does not find its way to television. Which means that all public understanding of these subjects is shaped by the biases of television.

Television is the command center in subtler ways as well. Our use of other media, for example, is largely orchestrated by television. Through it we learn what telephone system to use, what movies to see, what books, records and magazines to buy, what radio programs to listen to. Television arranges our communications environment for us in ways that no other medium has the power to do.

As a small, ironic example of this point, consider this: In the past few years, we have been learning that the computer is the technology of the future. We are told that our children will fail in school and be left behind in life if they are not "computer literate." We are told that we cannot run our businesses, or compile our shopping lists, or keep our checkbooks tidy unless we own a computer. Perhaps some of this is true. But the most important fact about computers and what they mean to our lives is that we learn about all of this from television. Television has achieved the status of "metamedium"—an instrument that directs not only our knowledge of the world but our knowledge of *ways of knowing* as well.

At the same time, television has achieved the status of "myth," as Roland Barthes uses the word. He means by *myth* a way of understanding the world that is not problematic, that we are not fully conscious of, that seems, in a word, natural. A myth is a way of thinking so deeply embedded in our consciousness that it is invisible. This is now the way of television. We are no longer fascinated or perplexed by its machinery. We do not tell stories of its wonders. We do not confine our television sets to special rooms. We do not doubt the reality of what we see on television, are largely unaware of the special angle of vision it affords. Even the question of how television affects us has receded into the background. The question itself may strike some of us as strange, as if one were to ask how having ears and eyes affects us. Twenty years ago, the question, Does television shape culture or merely reflect it? held considerable interest for many scholars and social critics. The question has largely disappeared as television has gradually *become* our culture. This means, among other things, that we rarely talk about television, only about what is *on* television—that is, about its content. Its ecology, which includes not only its physical characteristics and symbolic code but the conditions in which we normally attend to it, is taken for granted, accepted as natural.

Television has become, so to speak, the background radiation of the social and intellectual universe, the all-but-imperceptible residue of the electronic big bang of a century past, so familiar and so thoroughly integrated with American culture that we no longer hear its faint hissing in the background or see the flickering gray light. This, in turn, means that its epistemology goes largely unnoticed. And the peek-a-boo world it has constructed around us no longer seems even strange.

There is no more disturbing consequence of the electronic and graphic revolution than this: that the world as given to us through television seems natural, not bizarre. For the loss of the sense of the strange is a sign of adjustment, and the extent to which we have adjusted is a measure of the extent to which we have been changed. Our culture's adjustment to the epistemology of television is by now all but complete; we have so thoroughly accepted its definitions of truth, knowledge, and reality that irrelevance seems to us to be filled with import, and incoherence seems eminently sane. And if some of our institutions seem not to fit the template of the times, why it is they, and not the template, that seem to us disordered and strange.

COMING TO TERMS

In "The Peek-a-Boo World," Neil Postman discusses the difficulty of feeling informed in an age of electronic media. Rather than empowering us, he says, information received through the news media now gives us a feeling of impotence or the sense that we are being overwhelmed, at times, with useless information. First, we'd like you to go back through Postman's text and mark those places in the text where he is making major claims about the way we get and use information in an electronic age. Then, from these passages, we'd like you to summarize Postman's critical argument in your journal and to explain some key terms in Postman's critical vocabulary: "information-action ratio," "context-free," and "mythic." This language seems to be a way which Postman offers us to sort out and make sense of the information the news media present us with. It probably would be difficult to challenge Postman's reading of the historical developments in telecommunications, that the volume of news information now produced has outstripped any human capacity to absorb, digest, and find some kind of "relevance" in it. But we're interested in your take on the way Postman has chosen to evaluate news information in terms of relevance, how it affects our daily lives and whether or not it has a significant context. In your journal, respond to the following questions: Is Postman's emphasis on relevance a helpful way to understand the cultural significance of the news? Do you see problems with the ways in which Postman defines this context of relevance? Do you watch the news or listen to the news for its practical value alone? What are other reasons you might want to (or need to) know the news?

READING AS A WRITER

Postman links his argument about the changing shape and status of information to a series of communication technologies: the book, the telegraph, photography, television. The introductions of these technologies in society are certainly historical events. We like Postman's sense of complexity here, however, in that he imagines these events in other ways as well: as metaphors and cultural symbols, as a record of the ways in which our society and culture have been changed through these technological advancements. Each event in Postman's version of history represents a change in technology, but it's also seen as representing other profound changes: the changing ways of how people see themselves as part of a community, the changing ways people determine what's important or relevant to their lives and understand and represent reality, truth, and knowledge. There is the sense of finality or the fateful in Postman's reading of history, we think, in that with each telecommunication change the world is remarkably altered: nothing will ever be the "same." In your journal, we'd like you to explain Postman's sense of what has changed with the development of the book, the telegraph, the photograph, and the television set. As you list these technological developments and the changes they have wrought, how would you represent these changes as part of a pattern? We've noticed that Postman did not include telecommunication advancements like the telephone, for example, or film or the Internet (maybe you can think of others). What do you think Postman might have to say about these developments as well? From your point of view, would they continue some kind of pattern of significant change? Or would they disrupt the pattern of history, challenging what Postman believes to be the consequences of telecommunication advancements?

WRITING CRITICISM

We'd like you to conduct a test of the value of Postman's way of reading the news. You might see the value of your critical use of Postman in terms of its usefulness or as a way of raising additional questions or issues (or both) concerning the television news media in our electronic or information age. You will need to watch and tape a half-hour segment of a local news broadcast. (We think, in order to be fair to Postman's argument, that you should use a local news program with area, national, and international news segments.) If you can tape several such segments for the purposes of this essay, all the better. If you don't have access to a VCR, you might want to practice the work of taking notes on a televised newscast a few times before selecting the episodes you'll actually use.

You'll need to take notes on what you see: the kinds of information presented, the types and variety of news stories, how these stories are introduced or how the newscasters comment on different stories, the ratio to each other of the different subjects of the news—crime, politics, economy, fashion, sports, weather, health, human interest, advice, investigative reports, breaking news, etc. You might also try to classify stories into the categories of local, national, or international interest. As part of your viewing research, you might consider how certain film images seem to dominate the narration of a story, sometimes to the extent that we really are too transfixed or disturbed to hear what a reporter is saying. However you choose to read the televised news, remember that you need to take detailed notes. As part of your project here, and as a way of gaining a slightly different take on Postman's argument, you might also ask two or three people you know (not members of your class, however) to watch the same local newscast. You might tell them that you need their reactions to the program as part of the research you're conducting. Tell them that you'll ask some questions after they have watched the newscast. But tell them no more. The questions you'd ask might be: What stories, images, or personalities do you remember from the broadcast? What would you say were the most important stories you recall? Why do you think you remember the stories that you did? Collect the responses to the questions.

Then return to the notes you've taken on the program you watched (and to the responses to the questions you posed to other viewers) as well as the journal entries you've already taken on Postman. Your work here is to apply Postman's ideas to your experience and to others' experience of viewing the local news. Although you might apply other ideas or concepts in your work, we want you to test Postman's concepts of "information-action ratio," "context-free," and "mythic"—specifically and on two levels. First, you'll need to explain these concepts and then, analyze and place the segments of the local news program into these categories, explaining your choices, your reasons for each determination. What has Postman's approach allowed you to see in the news? In what ways does this system have limitations or problems? Can you use this system of analysis in explaining why the other viewers in your study chose this or that story as important or memorable? Do the other viewers' self-explanations of why a story is important to them confirm or challenge what Postman has to say about information presented by the television news media?

Reading the Romance

Women, Patriarchy and Popular Literature

Janice A. Radway

Readers and Romances

Surrounded by corn and hay fields, the midwestern community of Smithton, with its meticulously tended subdivisions of single-family homes, is nearly two thousand miles from the glass-and-steel office towers of New York City where most of the American publishing industry is housed. Despite the distance separating the two communities, many of the books readied for publication in New York by young women with master's degrees in literature are eagerly read in Smithton family rooms by women who find quiet moments to read in days devoted almost wholly to the care of others. Although Smithton's women are not pleased by every romance prepared for them by New York editors, with Dorothy Evan's help they have learned to circumvent the industry's still inexact understanding of what women want from romance fiction. They have managed to do so by learning to decode the iconography of romantic cover art and the jargon of back-cover blurbs and by repeatedly selecting works by authors who have pleased them in the past.

In fact, it is precisely because a fundamental lack of trust initially characterized the relationship between Smithton's romance readers and the New

York publishers that Dorothy Evans was able to amass this loyal following of customers. Her willingness to give advice so endeared her to women bewildered by the increasing romance output of the New York houses in the 1970s that they returned again and again to her checkout counter to consult her about the "best buys" of the month. When she began writing her review newsletter for bookstores and editors, she did so because she felt other readers might find her "expert" advice useful in trying to select romance fiction. She was so successful at developing a national reputation that New York editors began to send her galley proofs of their latest titles to guarantee their books a review in her newsletter. She now also obligingly reads manuscripts for several well-known authors who have begun to seek her advice and support. Although her status in the industry does not necessarily guarantee the representivity of her opinions or those of her customers, it does suggest that some writers and editors believe that she is not only closely attuned to the romance audience's desires and needs but is especially able to articulate them. It should not be surprising to note, therefore, that she proved a willing, careful, and consistently perceptive informant.

I first wrote to Dot in December 1979 to ask whether she would be willing to talk about romances and her evaluative criteria. I asked further if she thought some of her customers might discuss their reading with someone who was interested in what they liked and why. In an open and enthusiastic reply, she said she would be glad to host a series of interviews and meetings in her home during her summer vacation. At first taken aback by such generosity, I soon learned that Dot's unconscious magnanimity is a product of a genuine interest in people. When I could not secure a hotel room for the first night of my planned visit to Smithton, she insisted that I stay with her. I would be able to recognize her at the airport, she assured me, because she would be wearing a lavender pants suit.

The trepidation I felt upon embarking for Smithton slowly dissipated on the drive from the airport as Dot talked freely and fluently about the romances that were clearly an important part of her life. When she explained the schedule of discussions and interviews she had established for the next week, it seemed clear that my time in Smithton would prove enjoyable and busy as well as productive. My concern about whether I could persuade Dot's customers to elaborate honestly about their motives for reading was unwarranted, for after an initial period of mutually felt awkwardness, we conversed frankly and with enthusiasm. Dot helped immensely, for when she introduced me to her customers, she announced, "Jan is just people!" Although it became clear that the women were not accustomed to examining their activity in any detail, they conscientiously tried to put their perceptions and judgments into words to help me understand why they find romance fiction enjoyable and useful. . . .

The Act of Reading the Romance: Escape and Instruction

By the end of my first full day with Dorothy Evans and her customers, I had come to realize that although the Smithton women are not accustomed to thinking about what it is in the romance that gives them so much pleasure, they know perfectly well why they like to read. I understood this only when their remarkably consistent comments forced me to relinquish my inadvertent but continuing preoccupation with the text. Because the women always responded to my query about their reasons for reading with comments about the pleasures of the act itself rather than about their liking for the particulars of the romantic plot, I soon realized I would have to give up my obsession with textual features and narrative details if I wanted to understand their view of romance reading. Once I recognized this it became clear that romance reading was important to the Smithton women first because the simple event of picking up a book enabled them to deal with the particular pressures and tensions encountered in their daily round of activities. Although I learned later that certain aspects of the romance's story do help to make this event especially meaningful, the early interviews were interesting because they focused so resolutely on the significance of the *act of romance reading* rather than on the meaning of the romance.

The extent of the connection between romance reading and my informants' understanding of their roles as wives and mothers was impressed upon me first by Dot herself during our first two-hour interview which took place before I had seen her customers' responses to the pilot questionnaire. In posing the question, "What do romances do better than other novels today?", I expected her to concern herself in her answer with the characteristics of the plot and the manner in which the story evolved. To my surprise, Dot took my query about "doing" as a transitive question about the *effects* of romances on the people who read them. She responded to my question with a long and puzzling answer that I found difficult to interpret at this early stage of our discussions. It seems wise to let Dot speak for herself here because her response introduced a number of themes that appeared again and again in my subsequent talks with other readers. My question prompted the following careful meditation:

> *It's an innocuous thing. If it had to be...pills or drinks, this is harmful. They're very aware of this. Most of the women are mothers. And they're aware of that kind of thing. And reading is something they would like to generate in their children also. Seeing the parents reading is...just something that I feel they think the children should see them doing.... I've got a woman with teenage boys here who says "you've got books like...you've*

*just got oodles of da...da...da...[counting an imaginary stack of books]."
She says, "Now when you ask Mother to buy you something, you don't
stop and think how many things you have. So this is Mother's and it is my
money." Very, almost defensive. But I think they get that from their fathers.
I think they heard their fathers sometime or other saying, "Hey, you're
spending an awful lot of money on books aren't you?" You know for a long
time, my ladies hid 'em. They would hide their books; literally hide their
books. And they'd say, "Oh, if my husband [we have distinctive blue sacks],
if my husband sees this blue sack coming in the house...." And you know,
I'd say, "Well really, you're a big girl. Do you really feel like you have to
be very defensive?" A while ago, I would not have thought that way. I
would have thought, "Oh, Dan is going to hit the ceiling." For a while Dan
was not thrilled that I was reading a lot. Because I think men do feet threatened. They want their wife to be in the room with them. And I think my
body is in the room but the rest of me is not (when I am reading).*

Only when Dot arrived at her last observation about reading and its ability to transport her out of her living room did I begin to understand that the
real answer to my question, which she never mentioned and which was the
link between reading, pills, and drinks, was actually the single word, "escape," a word that would later appear on so many of the questionnaires. She
subsequently explained that romance novels provide escape just as Darvon
and alcohol do for other women. Whereas the latter are harmful to both
women and their families, Dot believes romance reading is "an innocuous
thing." As she commented to me in another interview, romance reading is a
habit that is not very different from an "addiction."

Although some of the other Smithton women expressed uneasiness
about the suitability of the addiction analogy, as did Dot in another interview, nearly all of the original sixteen who participated in lengthy conversations agreed that one of their principal goals in reading was their desire to
do something *different* from their daily routine. That claim was borne out by
their answers to the open-ended question about the functions of romance
reading. At this point, it seems worth quoting a few of those fourteen replies
that expressly volunteered the ideas of escape and release. The Smithton
readers explained the power of the romance in the following way:

They are light reading—escape literature—I can put down and pick up effortlessly.

Everyone is always under so much pressure. They like books that let them escape.

Escapism.

*I guess I feel there is enough "reality" in the world and reading is a means of
escape for me.*

Because it is an Escape [sic], and we can dream and pretend that it is our life.

I'm able to escape the harsh world for a few hours a day.

They always seem an escape and they usually turn out the way you wish life really was.

The response of the Smithton women is apparently not an unusual one. Indeed, the advertising campaigns of three of the houses that have conducted extensive market-research studies all emphasize the themes of relaxation and escape. Potential readers of Coventry Romances, for example, have been told in coupon ads that "month after month Coventry Romances offer you a beautiful new escape route into historical times when love and honor ruled the heart and mind." Similarly, the Silhouette television advertisements featuring Ricardo Montalban asserted that "the beautiful ending makes you feel so good" and that romances "soothe away the tensions of the day." Montalban also touted the value of "escaping" into faraway places and exotic locales. Harlequin once mounted a travel sweepstakes campaign offering as prizes "escape vacations" to romantic places. In addition, they included within the books themselves an advertising page that described Harlequins as "the books that let you escape into the wonderful world of romance! Trips to exotic places …interesting places…meeting memorable people…the excitement of love…. These are integral parts of Harlequin Romances—the heartwarming novels read by women everywhere." Fawcett, too, seems to have discovered the escape function of romance fiction, for Daisy Maryles has reported that the company found in in-depth interviewing that "romances were read for relaxation and to enable [women] to better cope with the routine aspects of life."

Reading to escape the present is neither a new behavior nor one peculiar to women who read romances. In fact, as Richard Hoggart demonstrated in 1957, English working-class people have long "regarded art as escape, as something enjoyed but not assumed to have much connection with the matter of daily life." Within this sort of aesthetic, he continues, art is conceived as "marginal, as 'fun,'" as something "for you to *use.*" In further elaborating on this notion of fictional escape, D. W. Harding has made the related observation that the word is most often used in criticism as a term of disparagement to refer to an activity that the evaluator believes has no merit in and of itself. "If its intrinsic appeal is high," he remarks, "in relation to its compensatory appeal or the mere relief it promises, then the term escape is not generally used." Harding argues, moreover, on the basis of studies conducted in the 1930s, that "the compensatory appeal predominates mainly in states of depression or irritation, whether they arise from work or other causes." It is interesting to note that the explanations employed by Dot and her women to interpret their romance reading for themselves are thus representative in a general way of a form of behavior common in an industrialized society where work is clearly distinguished from and more highly valued than leisure despite the fact that individual labor is often routinized, regimented, and minimally challenging.

It is equally essential to add, however, that although the women will use the word "escape" to explain their reading behavior, if given another comparable choice that does not carry the connotations of disparagement, they will choose the more favorable sounding explanation. To understand why, it will be helpful to follow Dot's comments more closely.

In returning to her definition of the appeal of romance fiction—a definition that is a highly condensed version of a commonly experienced process of explanation, doubt, and defensive justification—it becomes clear that romance novels perform this compensatory function for women because they use them to diversify the pace and character of their habitual existence. Dot makes it clear, however, that the women are also troubled about the propriety of indulging in such an obviously pleasurable activity. Their doubts are often cultivated into a full-grown feeling of guilt by husbands and children who object to this activity because it draws the women's attention away from the immediate family circle. As Dot later noted, although some women can explain to their families that a desire for a new toy or gadget is no different from a desire to read a new romantic novel, a far greater number of them have found it necessary to hide the evidence of their self-indulgence. In an effort to combat both the resentment of others and their own feelings of shame about their "hedonist, behavior, the women have worked out a complex rationalization for romance reading that not only asserts their equal right to pleasure but also legitimates the books by linking them with values more widely approved within American culture. Before turning to the pattern, however, I want to elaborate on the concept of escape itself and the reasons for its ability to produce such resentment and guilt in the first place.

Both the escape response and the relaxation response on the second questionnaire immediately raise other questions. Relaxation implies a reduction in the state of tension produced by prior conditions, whereas escape obviously suggests flight from one state of being to another more desirable one. To understand the sense of the romance experience, then, as it is enjoyed by those who consider it a welcome change in their day-to-day existence, it becomes necessary to situate it within a larger temporal context and to specify precisely how the act of reading manages to create that feeling of change and differentiation so highly valued by these readers.

In attending to the women's comments about the worth of romance reading, I was particularly struck by the fact that they tended to use the word escape in two distinct ways. On the one hand, they used the term literally to describe the act of denying the present, which they believe they accomplish each time they begin to read a book and are drawn into its story. On the other hand, they used the word in a more figurative fashion to give substance to the somewhat vague but nonetheless intense sense of relief they experience by identifying with a heroine whose life does not resemble their own in certain crucial aspects. I think it important to reproduce this subtle distinction as accurately as possible because it indicates that romance reading releases women from their present pressing concerns in two different but related ways.

Dot, for example, went on to elaborate more fully in the conversation quoted above about why so many husbands seem to feel threatened by their wives' reading activities. After declaring with delight that when she reads her body is in the room but she herself is not, she said, "I think this is the case with the other women." She continued, "I think men cannot do that unless they themselves are readers. I don't think men are *ever* a part of anything even if it's television." "They are never really out of their body either," she added. "I don't care if it's a football game; I think they are always consciously aware of where they are." Her triumphant conclusion, "but I think a woman in a book isn't," indicates that Dot is aware that reading not only demands a high level of attention but also draws the individual *into* the book because it requires her participation. Although she is not sure what it is about the book that prompts this absorption, she is quite sure that television viewing and film watching are different. In adding immediately that "for some reason, a lot of men feel threatened by this, very, very much threatened," Dot suggested that the men's resentment has little to do with the kinds of books their wives are reading and more to do with the simple fact of the activity itself and its capacity to absorb the participants' entire attention.

These tentative observations were later corroborated in the conversations I had with other readers. Ellen, for instance, a former airline stewardess, now married and taking care of her home, indicated that she also reads for "entertainment and escape." However, she added, her husband sometimes objects to her reading because he wants her to watch the same television show he has selected. She "hates" this, she said, because she does not like the kinds of programs on television today. She is delighted when he gets a business call in the evening because her husband's preoccupation with his caller permits her to go back to her book.

Penny, another housewife in her middle thirties, also indicated that her husband "resents it" if she reads too much. "He feels shut out," she explained, "but there is nothing on TV I enjoy." Like Ellen's husband, Penny's spouse also wants her to watch television with him. Susan, a woman in her fifties, also "read[s] to escape" and related with almost no bitterness that her husband will not permit her to continue reading when he is ready to go to sleep. She seems to regret rather than resent this only because it limits the amount of time she can spend in an activity she finds enjoyable. Indeed, she went on in our conversation to explain that she occasionally gives herself "a very special treat" when she is "tired of housework." "I take the whole day off," she said, "to read."

This theme of romance reading as a special gift a woman gives herself dominated most of the interviews. The Smithton women stressed the privacy of the act and the fact that it enables them to focus their attention on a single object that can provide pleasure for themselves alone. Interestingly enough, Robert Escarpit has noted in related fashion that reading is at once "social and asocial" because "it temporarily suppresses the individual's relations with his [*sic*] universe to construct new ones with the universe of the work." Unlike television viewing, which is a very social activity undertaken in the presence of others and which permits simultaneous conversation and personal interaction, silent trading requires the reader to block out the surrounding world and to give consideration to other people and to another time. It might be said, then, that the characters and events of romance fiction populate the woman's consciousness even as she withdraws from the familiar social scene of her daily ministrations.

I use the word ministrations deliberately here because the Smithton women explained to me that they are not trying to escape their husbands and children "per se" when they read. Rather, what reading takes them away from, they believe, is the psychologically demanding and emotionally draining task of attending to the physical and affective needs of their families, a task that is solely and peculiarly theirs. In other words, these women, who have been educated to believe that females are especially and naturally attuned to the emotional requirements of others and who are very proud of their abilities to communicate with and to serve the members of their families, value reading precisely because it is an intensely private act. Not only is the activity private, however, but it also enables them to suspend temporarily those familial relationships and to throw up a screen between themselves and the arena where they are required to do most of their relating to others.

It was Dot who first advised me about this phenomenon. Her lengthy commentary, transcribed below, enabled me to listen carefully to the other readers' discussions of escape and to hear the distinction nearly all of them made between escape from their families, which they believe they do *not* do, and escape from the heavy responsibilities and duties of the roles of wife and mother, which they admit they do out of emotional need and necessity. Dot explained their activity, for instance, by paraphrasing the thought process she believes goes on in her customers' minds. "Hey," they say, "this is what I want to do and I'm gonna do it. This is for me. I'm doin' for you all the time. Now leave me, just leave me alone. Let me have my time, my space. Let me do what I want to do. This isn't hurting you. I'm not poaching on you in any way." She then went on to elaborate about her own duties as a mother and wife:

> As a mother, I have run 'em to the orthodontist. I have run 'em to the swimming pool. I have run 'em to baton twirling lessons. I have run up to school because they forgot their lunch. You know, I mean, really! And you do it. And it isn't that you begrudge it. That isn't it. Then my husband would walk in the door and he'd say, "Well, what did you do today?" You know, it was like, "Well, tell me how you spent the last eight hours, because I've been out working." And I finally got to the point where I would say, "Well, I read four books, and I did all the wash and got the meal on the table and the beds are

all made, and the house is tidy." And I would get defensive like, "So what do you call all this? Why should I have to tell you because I certainly don't ask you what you did for eight hours, step by step."—But their husbands do do that. We've compared notes. They hit the house and it's like "Well all right, I've been out earning a living. Now what have you been doin' with your time?" And you begin to be feeling, "Now really, why is he questioning me?"

Romance reading, it would seem, at least for Dot and many of her customers, is a strategy with a double purpose. As an activity, it so engages their attention that it enables them to deny their physical presence in an environment associated with responsibilities that are acutely felt and occasionally experienced as too onerous to bear. Reading, in this sense, connotes a free space where they feel liberated from the need to perform duties that they otherwise willingly accept as their own. At the same time, by carefully choosing stories that make them feel particularly happy, they escape figuratively into a fairy tale where a heroine's similar needs are adequately met. As a result, they vicariously attend to their own requirements as independent individuals who require emotional sustenance and solicitude.

Angie's account of her favorite reading time graphically represents the significance of romance reading as a tool to help insure a woman's sense of emotional well-being. "I like it," she says, "when my husband—he's an insurance salesman—goes out in the evening on house calls. Because then I have two hours just to totally relax." She continued, "I love to settle in a hot bath with a good book. That's really great." We might conclude, then, that reading a romance is a regressive experience for these women in the sense that for the duration of the time devoted to it they feel gratified and content. This feeling of pleasure seems to derive from their identification with a heroine whom they believe is deeply appreciated and tenderly cared for by another. Somewhat paradoxically, however, they also seem to value the sense of self-sufficiency they experience as a consequence of the knowledge that they are capable of making themselves feel good.

Nancy Chodorow's observations about the social structure of the American family in the twentieth century help to illuminate the context that creates both the feminine need for emotional support and validation and the varied strategies that have evolved to meet it. As Chodorow points out, most recent studies of the family agree that women traditionally reproduce people, as she says, "physically in their housework and child care, psychologically in their emotional support of husbands and their maternal rotation to sons and daughters." This state of affairs occurs, these studies maintain, because women alone are held responsible for home maintenance and early child care. Ann Oakley's 1971 study of forty London housewives, for instance, led her to the following conclusion: "In the housekeeping role the servicing function is far more central than the productive or creative one. In the roles of wife and mother, also, the image of women as servicers of men's and children's needs is prominent: women 'service' the labour force by catering to the physical needs of men (workers) and by raising children (the next generation of workers) so that the men are free *from* child-socialization and free *to* work outside the home." This social fact, documented also by Mirra Komarovsky, Helena Lopata, and others, is reinforced ideologically by the widespread belief that females are *naturally* nurturant and generous, more selfless than men, and, therefore, cheerfully self-abnegating. A good wife and mother, it is assumed, will have no difficulty meeting the challenge of providing all of the labor necessary to maintain a family's physical existence including the cleaning of its quarters, the acquisition and preparation of its food, and the purchase, repair, and upkeep of its clothes, even while she masterfully discerns and supplies individual members' psychological needs. A woman's interests, this version of "the female mystique" maintains, are exactly congruent with those of her husband and children. In serving them, she also serves herself.

As Chodorow notes, not only are the women expected to perform this extraordinarily demanding task, but they are also supposed to be capable of executing it without being formally "reproduced" and supported themselves. "What is...often hidden, in generalizations about the family as an emotional refuge," she cautions, "is that in the family as it is currently constituted no one supports and reconstitutes women affectively and emotionally—either women working in the home or women working in the paid labor force." Although she admits, of course, that the accident of individual marriage occasionally provides a woman with an unusually nurturant and "domestic" husband, her principal argument is that as a social institution the contemporary family contains no role whose principal task is the reproduction and emotional support of the wife and mother. "There is a fundamental asymmetry in daily reproduction" Chodorow concludes, "men are socially and psychologically reproduced by women, but women are reproduced (or not) largely by themselves."

That this lack of emotional nurturance combined with the high costs of lavishing constant attention on others is the primary motivation behind the desire to lose the self in a book was made especially clear to me in a group conversation that occurred late in my stay in Smithton. The discussion involved Dot, one of her customers, Ann, who is married and in her thirties, and Dot's unmarried, twenty-three-year-old daughter, Kit. In response to my question, "Can you tell me what you escape from?," Dot and Ann together explained that reading keeps them from being overwhelmed by expectations and limitations. It seems advisable to include their entire conversation here, for it specifies rather precisely the source of those felt demands:

Dot: All right, there are pressures. Meeting your bills, meeting whatever standards or requirements your husband has for you or whatever your children have for you.

Ann: Or that you feel you should have. Like doing the housework just so.

Dot: And they do come to you with problems. Maybe they don't want you to—let's see—maybe they don't want you to solve it, but they certainly want to unload on you. You know. Or they say, "Hey, I've got this problem."

Ann: Those pressures build up.

Dot: Yeah, it's pressures.

Ann: You should be able to go to one of those good old—like the MGM musicals and just...

Dot: True.

Ann: Or one of those romantic stories and cry a little bit and relieve the pressure and—a legitimate excuse to cry and relieve some of the pressure build-up and not be laughed at.

Dot: That's true.

Ann: And you don't find that much anymore. I've had to go to books for it.

Dot: This is better than psychiatry.

Ann: Because I cry over books. I get wrapped up in them.

Dot: I do too. I sob in books! Oh yes. I think that's escape. Now I'm not gonna say I've got to escape my husband by reading. No.

Ann: No.

Dot: Or that I'm gonna escape my kids by getting my nose in a book. It isn't any one of those things. It's just—it's pressures that evolve from being what you are.

Kit: In this society.

Dot: And people do pressure you. Inadvertently, maybe.

Ann: Yes, it's being more and more restrictive. You can't do this and you can't do that.

This conversation revealed that these women believe romance reading enables them to relieve tensions, to diffuse resentment, and to indulge in a fantasy that provides them with good feelings that seem to endure after they return to their roles as wives and mothers. Romance fiction, as they experience it, is, therefore, *compensatory literature*. It supplies them with an important emotional release that is proscribed in daily life because the social role with which they identify themselves leaves little room for guiltless, self-interested pursuit of individual pleasure. Indeed, the search for emotional gratification was the one theme common to all the women's observations about the function of romance reading. Maureen, for instance, a young mother of two intellectually gifted children, volunteered, "I especially like to read when I'm depressed." When asked what usually caused her depression, she commented that it could be all kinds of things. Later she added that romances were comforting after her children had been especially demanding and she felt she needed time to herself.

In further discussing the lack of institutionalized emotional support suffered by contemporary American women, Chodorow has observed that in many preindustrial societies women formed their own social networks through which they supported and reconstituted one another. Many of these networks found secondary institutional support in the local church—while others simply operated as informal neighborhood societies. In either case, the networks provided individual women with the opportunity to abandon temporarily their stance as the family's self-sufficient emotional provider. They could then adopt a more passive role through which they received the attention, sympathy, and encouragement of other women. With the increasing suburbanization of women, however, and the concomitant secularization of the culture at large, these communities became exceedingly difficult to maintain. The principal effect was the even more resolute isolation of women within their domestic environment. Indeed, both Oakley in Great Britain and Lopata in the United States have discovered that one of the features housewives dislike most about their role is its isolation and resulting loneliness.

I introduce Chodorow's observations here in order to suggest that through romance reading the Smithton women are providing themselves with another kind of female community capable of rendering the so desperately needed affective support. This community seems not to operate on an immediate local level although there are signs, both in Smithton and nationally, that romance readers are learning the pleasures of regular discussions of books with other women. Nonetheless, during the early group discussions with Dot and her readers I was surprised to discover that very few of her customers knew each other. In fact, most of them had never been formally introduced although they recognized one another as customers of Dot. I soon learned that the women rarely, if ever, discussed romances with more than one or two individuals. Although many commented that they talked about the books with a sister, neighbor, or with their mothers, very few did so on a regular or extended basis. Indeed, the most striking feature of the interview sessions was the delight with which they discovered common experiences, preferences, and distastes. As one woman exclaimed in the middle of a discussion, "We were never stimulated before into thinking why we like [the novels]. Your asking makes us think why we do this. I had no idea other people had the same ideas I do."

The romance community, then, is not an actual group functioning at the local level. Rather, it is a huge, ill-defined network composed of readers on the one hand and authors on the other. Although it performs some of the same functions carried out by older neighborhood groups, this female community is mediated by the distances of modern mass publishing. Despite the distance, the Smithton women feel personally connected to their favorite authors because they are convinced that these writers know how to make them happy. Many volunteered information about favorite authors even before they would discuss specific books or heroines. All expressed admiration for their favorite writers and indicated that they were especially curious about their private lives. Three-fourths of the group of sixteen had made special trips to autographing sessions to see and express their gratitude to the women who had given them so much pleasure. The authors reciprocate this feeling of gratitude and seem genuinely interested in pleasing their readers. Many are themselves romance readers and, as a consequence, they, too, often have definite opinions about the particular writers who know how to make the reading experience truly enjoyable.

It seems highly probable that in repetitively reading and writing romances, these women are participating in a collectively elaborated female fantasy that unfailingly ends at the precise moment when the heroine is gathered into the arms of the hero who declares his intention to protect her forever because of his desperate love and need for her. These women are telling themselves a story whose central vision is one of total surrender where all danger has been expunged, thus permitting the heroine to relinquish self-control. Passivity *is* at the heart of the romance experience in the sense that the final goal of each narrative is the creation of that perfect union where the ideal mate, who is masculine and strong yet nurturant too, finally recognizes the intrinsic worth of the heroine. Thereafter, she is required to do nothing more than *exist* as the center of this paragon's attention. Romantic escape is, therefore, a temporary but literal denial of the demands women recognize as an integral part of their roles as nurturing wives and mothers. It is also a figurative journey to a utopian state of total receptiveness where the reader, as a result of her identification with the heroine, feels herself the *object* of someone else's attention and solicitude. Ultimately, the romance permits its reader the experience of feeling cared for and the sense of having been reconstituted effectively, even if both are lived only vicariously.

COMING TO TERMS

Radway reports that when she first asked her subjects about why they read romance novels their responses often had to do with some notion of *escape.* But she soon finds out that what these women mean by *escape* is often something different from and more complex than what she (and perhaps most other academic critics) might have at first suspected. In re-reading Radway, then, we'd like you to try to figure out just what her subjects might actually be doing when they describe themselves as escaping with a romance novel. See if you can briefly describe the uses and limits of this form of escape in your journal: What is it exactly that these women are trying to escape from? And where are they trying to get to? What might the romance novel be said to provide which might be absent from these readers' lives? How do romance novels help them to do so in ways that, say, gothic novels, autobiographies, television soaps, game shows, or talk shows don't? In what ways, according to Radway, do the readers of romances form a common audience? How would you define this "typical" reader? In what ways might reading the romance be seen as something more than just a "mindless" form of escape? That is, what positive value or use (here, you might think of a cultural explanation) does this kind of reading seem to hold for these women? Finally, what other authors have you read in *Media Journal* who represent escapism through media in serious ways?

READING AS A WRITER

At the beginning of her piece, Radway casts herself as a kind of visitor (or perhaps even an intruder) to Smithton—a scholar who flies into town to study the lives and reading habits of Dot and her clients. You might imagine Radway as a kind of anthropologist, setting out to investigate, take some notes on, and try to understand a "culture" that is foreign to her: a culture defined by what it reads. How would you describe this culture? How would you represent the reading culture that Radway herself inhabits? We would like you to develop this idea of two competing reading cultures in your journal by responding to the following: As readers we are thus set up to expect a series of contrasts or differences between Radway and her subjects. As you go back through her text, we'd like you to look for points where Radway makes note of such differences. What sorts of details does she remark on? What seems to surprise her? What does Radway fail to understand, at least at first? What might be the source of that misunderstanding and how would you explain it in terms of different, competing reading cultures? What hints are we given about how Dot and her clients respond to Radway? On the other hand, what does Radway seem to be able to take for granted? That is, what kinds of tastes, responses, or experiences does she share in common with her subjects? What does she represent as "foreign"? With whom do you identify more: Radway or the Smithton women? Why? What tone would you ascribe to Radway's study of the "romance culture"? Is it hostile? sympathetic? empathetic?

WRITING CRITICISM

Radway tries to account for the pleasure or meaning that other women take in reading romance novels. Her challenge as a writer is to describe and understand the media behavior of others, to enter an experience that she does not, at least initially, share, to make sense of an unfamiliar culture of experience. In the process she must come to know better not just the media experience itself, but the "others" to whom this experience appeals. We'd like you to take on a similar project here, to try to describe and understand the media experience of others. The first step is to find at least two people (three or four would be better) who share an enthusiasm for a particular media experience or ritual. They could be fans of a particular musical performer, followers of a favorite soap opera, regular viewers of this sitcom or that game show, readers of a certain genre of fiction, frequent visitors to a particular chat room on the Internet, sharers in the same fashion or style of dress, denizens of the same club, fans of the same sports team, collectors of the same comic books, dedicated readers of a fanzine—two or more people who share an enthusiasm that involves the media or popular culture in some way. This should be an enthusiasm you associate with others more than yourself. It need not be totally foreign to you—although it could be. But it should be a culture of media experience that you're curious about because others (at least two others) find it more meaningful, important, fun, or worthwhile than you do.

The second step is to enter the experience of these others, the subjects of your study. There are, we think, two basic ways of accomplishing this. The first is to do what they do—to watch the soap opera they watch, read the books they read, listen to the performers they listen to, watch the ball games they watch, and so on. The other method is to question them about their experience: What is the experience like for them? What do they have to do or know in order to enjoy it? What is meaningful, pleasurable, or important about the experience for the people involved? How did they "get into" and what do they "get out of" doing it? Why do they keep doing it? Are they aware of how others might view their enthusiasm? In either case, you will want to take notes in your journal about what your subjects tell you, as well as the texts you've come to share in common with them.

The third step is to think about what you have learned through your questions and observations. How would you explain the appeal of this experience for your subjects? How does your explanation compare to what your subjects have told you? What remains mysterious, elusive, or unexplained about the appeal of the experience? What do you not "get" and why (do you think) don't you get it? How does your "reading" of the texts central to this experience differ from theirs? What is the value of this experience for those who do not easily enter into it? Is this value something you can appreciate, or imagine yourself sharing? Or does it remain somehow foreign to you? If it does remain foreign, how might you explain your difficulty in terms of competing cultures, the one you inhabit and the one you are investigating?

Like Radway, you may want to center your writing around issues in gender. Or other issues—age, education, commercialism, violence, race, sexuality, social class—may seem to shape the tastes and behavior of your subjects more. That is something you will need to determine. But you will still want to re-read Radway in order to see how she approaches her subjects—the questions she asks of them, the stances she takes toward them, and the ways she interprets their statements and responses. For you will need to decide what you want to make use of in Radway's methods and what you want to revise.

Trend-Spotting
It's All the Rage
Edward Rothstein

This week, with the coming of the new year, there may be a spirit of rebirth at large in the land, a sense of something new about to begin. But if so, the feeling will be all too familiar. The displacement of the old and the celebration of the new has become a year-round habit. In fact, we are always resetting our clocks, recalibrating our sense of newness, ringing out and ringing in at ever faster rates. We have created a culture founded almost entirely on trends.

In newsrooms, board rooms and classrooms, we "Braille the culture," as one professional trend spotter, Faith Popcorn, has famously put it. We run our fingertips along trend-bumps as they speed past. Sales of snoring remedies are up. Sales of exotic fruit drinks are down. Current events are news in opera (in recent years, Stewart Wallace's "Harvey Milk," John Anthony Davis's "X: The Life and Times of Malcolm X"). Nineteenth-century novels are big on screen (last year, Jane Austen; this year, Henry James). Television sitcoms celebrate chattering friendships ("Seinfeld," "Friends" and various imitators). Things are moving so fast that some fashion gurus declared the "Evita" look dead even before the Madonna film opened last week.

We give everything a name—decades, styles, movements, generations—in attention-getting capital letters. The Me Decade was named by Tom Wolfe in 1976, after it was more than half over. The 80's were slurred as the Decade of Greed. The Beats begat the Baby Boomers who begat the Punks who begat Gen X who now await newly named successors.

In advanced intellectual life, trend-spotting is becoming just as frenetic, with ideas and arguments taking on many of the characteristics of fashion. Structuralism was superseded by Semiology, which joined forces with Lacanianism, which was displaced by Deconstruction, which was superseded by Cultural Studies. Now everything is engulfed by Po-Mo—post-modernism—which sometimes seems to be declaring that all trends are created equal.

Nowhere is the swirl more frenzied than in pop culture, which hasn't even got time for names or seasons. A generation in pop-culture terms seems to measure about two years, and a trend can come and go seemingly in a matter of weeks. Funk, hip-hop, house music and gangsta rap jostle for attention, with their variations competing for new, revised monikers. In television, the 50's and 60's are joined by the 70's and 80's as Nick at Nite and the cable industry create trends out of recycled nostalgia. It seems as if we are always racing to catch up with these changes in taste and style, learning the new names and constantly seeking to find newer ones. We want to ride the crest of these waves; we hope we never float helplessly while the action is elsewhere. We are trend addicts, seeking to be on top, ahead, beyond or on the brink. In fact, the trendiest trend in culture right now is trend-spotting.

Culture is almost haphazardly strewn about us, on screens, bill-boards, in concert halls and art galleries, filtered through millions of minds, executed with thousands of techniques. It's there in advertising and in serious music, in the latest Hollywood blockbuster and in the dumbest television sitcom, in university classrooms and in political rallies.

But culture is increasingly difficult to decipher, so we seek the supposed essences in the midst of chaos, trends that give a semblance of order and connection in a world we are partly constructing, partly being swept away in. There is an element of anxiety in this quest, but somewhere in this mess—and much of it is a mess—there seems to be a message, or at least a mirror, offering some explanation of ourselves we cannot find elsewhere.

And if we, the consumers, seek trends, how much keener are the producers of our entertainment, the marketers who bet millions of dollars on whether a particular star is "hot" or a particular book will "take off." There is money to be made out of our obsessions.

Faith Popcorn, for example, whose invented name promises the snap and crackle of instant satisfaction along with the reliability and confidence of homespun religion, is a professional trend reader. Among her credentials, she notes that she correctly predicted the demise of wine coolers and the rise of gourmet coffees and that major corporations like American Express and Pepsico have paid her to reveal trends still hidden to their competitors.

Increasingly, even criticism joins in the quest, practicing what might be called Zeitgeist journalism, in which everything becomes a sign of who we are and where we are going. The hunt might even be what the best-selling trend spotter John Naisbitt called a Megatrend—or, more properly, a Metatrend.

The problem is that this restless search for trends can never come up with the kinds of answers we hope to find. If we say buddy movies are "in" one season, or action thrillers the next, this is not a matter of progress, bringing us greater understanding; it is just a change in preferences. Every trend is just additional evidence of change rather than another step toward stability.

This was not always the case. Such extreme quests for trends were once unnecessary partly because of something now quaintly known as tradition. Tradition was once an imposing, if porous and amorphous, presence, a sense of past achievement that provided the context within which new artworks were created. The tradition—or traditions—invoked in an artwork partly provided its premises, partly its style, partly even its subject matter.

Listeners could comprehend Beethoven because they had come to know the music of Haydn, Mozart and lesser talents who had similar ideas about musical structure and drama. French Impressionism achieved its impact partly by rejecting a tradition of academic painting. There was a time when tradition forcibly affected how paint was applied to canvas, which images would be used in a novel, whether one musical line could be combined with another.

There was plenty of movement within a tradition—which is why there are so many distinctive works—though the tradition still provided a frame of reference. Each work was not only addressing an audience of viewers, listeners or readers but was also conversing with the many other works that preceded it.

The early development of opera, for example, was related to Renaissance conceptions of Greek drama and to the notion that a link existed between the meanings of words and their sounds. Stravinsky, the critic Richard Taruskin has shown us, worked within multiple folk traditions of his native Russia, which the composer wed to the manners of Parisian modernism. The development of Pop Art is inconceivable without reference to Dadaism and the rejection of the tradition of esthetic meaning.

It is impossible to consider any significant achievement in the arts during the last five centuries without invoking the word "tradition." A rebellious act is as beholden to the tradition it rejects as a conservative one is to the tradition it upholds: that is what the avant-garde has been about and why it can now seem so formulaic.

Tradition implies expanse and history; trend implies brevity and sensation. Tradition invokes age; trend speaks of youth. Tradition demands reference to the past; trend demands iconoclasm and newness. Tradition is based on resemblance—how this artwork or that aspect of culture invokes or relies on what has come before; trend is based on difference—how this artwork is distinct from what has come before.

Tradition also provides a context for culture, a home. Artists work within a tradition; they are nestled by it, challenged by it, oppressed by it, but they cannot fully discard it. Trends nestle nobody. Tradition can be cautious to a fault; trends can be reckless to a fault. Tradition demands an active creator who tries to mold it for new purposes; trends create passive participants. Tradition requires dedication: "If you want it," wrote T. S. Eliot, "you must obtain it by great labor." A trend is almost always a cliché, always something that is widely accepted, requiring no proof; it attracts followers rather than leaders, crowds rather than individuals. (Trends can grow into traditions, but this is a long process requiring commitment, interest and labor.)

For a large part of cultural history, there was a balance between these two attitudes to the past and present—a balance guided by what T. S. Eliot called the presence of the past. But now that balance has shifted and the past has become a burden. The very word "tradition" has taken on the suggestion of something rigid, stultifying, restrictive, mindless. During the last hundred years, many artists have even cultivated this attitude; now it has become widely accepted. The prophecy that tradition will kill art has become self-fulfilling.

Consider the situation in the world of classical music. Despite new interpreters of the mainstream repertory and new compositions appearing on the margins of music culture, there is a sense of finality in the concert world. Within the last 30 years, changes to the Western music tradition have been truly marginal. Failure to pass on the tradition to new audiences may mean that both composers and listeners will cease to treat it as a living organism; that is one reason conversations with many composers, performers and managers tend to become morbid.

There is such a sense of finality, that many composers have deliberately sought other traditions in which to ground themselves. Minimalism was influenced by African and Asian music. Eastern European mysticism tried to discard three centuries of narrative drama and leap back into the Renaissance. Avant-garde groups have tried to adapt pop instrumentation and manners.

These are all evidence that the music of the concert hall from about 1780 to 1950—the core of the Western tradition—has ceased to have a compelling hold on creators, and that there is no secure tradition within which classical music can develop, only a series of nascent alternatives. Something has come to an end.

This situation leaves the way wide open for the most fashionable trends to make their way into the concert scene—ideas about programming (much crossover) or orchestras (more pops). The high arts are not completely vulnerable because the aura of tradition still hovers, slowing the pressure of trends, moderating them. In most of the high arts, the traditions are so weighty with achievement that they provide frames of reference, even when trends loom large.

In mass entertainments like film and television, however, tradition has much less depth or weight. It once seemed to have a chance: silent film began invoking the operatic tradition—in the design of movie houses, the musical accompaniments and the heightened use of gesture to suggest meaning. Film began to develop a tradition of its own.

But to the extent that a movie's creators care more about the audience than about the tradition—and that is now a very great extent—movies tend to be subject to the traditionless forces of trend. Genre films—action, horror, buddy or otherwise—dominate, along with sequels, partly because they provide an illusion of tradition by repeating formulas. It may even be that the recent tendency to make crowd pleasers based on the 19th-century novel is an attempt to provide some sense of a tradition in a form that has almost entirely lost what it once had.

Popular music, some of which has a long tradition connecting it with black American musical styles, also makes little obeisance to the richness of that tradition. The emphasis, even when drawing on contemporary forms of black music, is primarily on the new, and the different. While there are undoubtedly influences and traditions in popular music, innovation, mass audience appeal and iconoclasm tend to be the defining forces. Television is no different. There are exceptions, but for the most part, the medium needs to respond energetically to the demands of passing trends.

As in entertainment, so too, in intellectual life. The controversy over the canon in the universities—questioning whether the great books of the West should still be required reading—is a controversy over whether this tradition (which is actually a multiplicity of traditions) will be a presence in the future. But critics of the canon tend to treat this body of knowledge as if it were similar to everything else in our trend-ridden era, the reflection of simple ideas—in this case, negative and narrow ones like racism or imperialism. A tradition is treated as just another trend ready for replacement.

In fact, a developing academic discipline known as "cultural studies" is partly a study of trends and their meanings. Few distinctions are made between a Braille-like bump in popular culture or an imposing achievement in a highly developed tradition; for some scholars, both contain equally important information. The appeal of shopping centers is as crucial as the image of the body in Western art, the nature of the sports fan as central as the education of a scientist.

There is something amiss in these efforts to treat a tradition with no more seriousness than the latest passing fashion. When the early-19th-century German philosopher Hegel used the word "Zeitgeist," he was trying to outline the course of world history as a series of systematic transformations in human consciousness.

At any given time any Zeit—types of government, the kinds of art being created, aspects of religious belief—would all be reflections of something Hegel called the Spirit, the Geist. Each stage along history's way was seen as an embodiment of Spirit, containing tensions and contradictions that would be resolved in a succeeding era that would in turn fall prey to tensions that would give birth to yet another age.

That was the origin of the notion of trend, but the Zeitgeist was a trend suffused with tradition. We cannot now believe in such a progression, particularly because the key to our Zeitgeist lies in increasingly brief periods of Zeit and little conception of Geist. There is no direction or goal, no past to give context to the present.

The shared experience of fashion, television shows, hit movies and best sellers provide immediate bonds in a world without traditional ones. But those are fragile bonds, having little authority or significance. Instead of Eliot's "presence" of the past, we are left with the lingering ache of absence. Every day becomes a new year, every tradition little more than a trend.

COMING TO TERMS

Early in "Trend-Spotting: It's All the Rage," Edward Rothstein speaks about what from his vantage point has become a defining (and disturbing) characteristic of how we chronicle change and determine what's important in our culture: "This displacement of the old and the celebration of the new has become a year-round habit. In fact, we are always resetting our clocks, recalibrating our sense of newness, ringing out and ringing in at ever faster rates. We have created a culture founded almost entirely on trends." Like other writers in *Media Journal,* part of Rothstein's critical argument is concerned with mapping out some major shifts or changes in society. In Rothstein's case, this shift points to the consequences of losing touch with traditions. And according to Rothstein, we have become a people consumed with what's new, what's in style, and how to name these "new" phenomena. As you re-read Rothstein's article, respond to the following questions in your journal. In what ways do you find yourself agreeing with Rothstein's reading of contemporary culture? What observations or conclusions would you challenge? How is "Trend-Spotting: It's All the Rage" a cautionary tale? In other words, according to Rothstein, what is lost when a culture is more concerned with recognizing trends than with honoring traditions? What does our "extreme search for trends" reveal about what's important or valuable in our society? What does our need to name and categorize the "new" tell us?

READING AS A WRITER

What is central to Rothstein's critical argument is the careful and extended set of differences he marks between a tradition and a trend. In fact, Rothstein continues this work of marking distinctions between these two key concepts throughout his article, setting up a clear sense of contrast or opposition between these two ways of explaining culture. His definitions here extend far beyond what we might find in a dictionary. In your journal, we'd like you to create two columns, one labeled *tradition* and the other *trend.* Then re-read "Trend-Spotting: It's All the Rage" and, in your own words, list the important characteristics Rothstein names for both ways of making sense out of a culture. Finally, respond to the following questions in your journal. According to Rothstein, what attitude does a tradition take towards the past? What attitude does a trend suggest? How does Rothstein explain the origin of both a trend and a tradition? Does Rothstein explain how a trend might become a tradition? How might his reluctance to do so affect your reading of his article? How does Rothstein apply these two key concepts, say, to music, film, and education? Finally, what's your take on how Rothstein defines *tradition* and *trend*? What do you agree with and what do you contest? How would you extend or amplify Rothstein's ways of reading these words?

WRITING CRITICISM

Edward Rothstein not only claims that "extreme searching for trends" has affected the artistic and social fabric of contemporary society, but that this preoccupation with trends, this emphasis on the new and the rejection of the traditional have begun to characterize discussions of higher education as well. Rothstein illustrates this claim by pointing to "the controversy over the canon in the universities—questioning whether the great books of the West should still be required reading." This question of the canonical or traditional—for example, what texts should be taught in literature, history, and writing courses—currently remains a controversial subject in U.S. education, often as a part of a larger and an equally contentious conversation about multiculturalism, historical revision, and political correctness (issues which Rothstein might call *trends*). For Rothstein, the current tension in education might be represented by a choice: to choose the traditional or to choose the trendy.

As a way of extending and responding to Rothstein's concerns about higher education, we'd like you to write an essay in which you discuss your educational experiences in using *Media Journal* in your college or university composition course. Your task here is to imagine your work with media or popular culture in this course in terms of trends and traditions. (You may use Rothstein's way of defining trends and traditions, or you may offer your own way of explaining these terms. In either case, you'll need to show how you're using these critical terms early in your discussion.)

A place to begin this project would be to think about other writing or English courses you have taken previously and the ways in which these courses might have shared commonplaces. For example, what kinds of texts did you read in these classes? Would you say that these texts were canonical? Part of a tradition of literature? Part of a collection of, say, great books or great authors? Why do you think these particular texts were chosen? What did they contribute to your education, your understanding? Did these texts exhibit a particular attitude towards the past? Then pose and answer a similar set of questions involving your work with media or popular culture in this course.

Finally, examine the consequences, the advantages and disadvantages, of using media or popular culture in a college or university writing course.

What is gained and what is lost in relationship to these other courses you have taken? As part of this project, we'd like you to explore the relationship between educational trends and traditions in a way in which Rothstein seems not to be very interested. In his discussion, Rothstein sets up traditions and trends as oppositional categories: it's either a trend, or it's a tradition. We're not convinced, however, that current practices in and new approaches to education can or should be so neatly categorized. And we think it's not particularly helpful to say: "Shakespeare is a tradition, the writers in *Media Journal* are a trend." Instead we'd like you to consider in what ways the texts you've read in other English courses might be seen not only as differing from but also as sharing something of importance with the writers in *Media Journal.*

Letter from Skywalker Ranch

Why Is the Force Still with Us?

John Seabrook

I—The Summit

The biannual Star Wars Summit Meeting is an opportunity for the licensees who make Darth Vader masks and thirty-six-inch sculpted Yoda collectibles to trade strategy and say "May the Force be with you" to the retailers from F.A.O. Schwarz and Target who sell the stuff, and for everyone in the far-flung Star Wars universe to get a better sense of "how deeply the brand has penetrated into the culture," in the words of one licensee. Almost six hundred people showed up at this year's summit, which took place in early November in the Marin County Civic Center, in San Rafael, California; Star Warsians came from as far away as Australia and Japan. Those arriving by car were ushered to parking places by attendants waving glowing and buzzing Luke Skywalker lightsabres, which were one of the cooler pieces of new product seen at the summit this year; other arrivals, who were staying in the Embassy Suites next door, strolled across the parking lot in the early-morning sunshine of another beautiful day in Northern California.

It was interesting to stand by the door of the auditorium and reflect that all these people, representing billions of dollars of wealth, depended for their existence on an idea that seemed utterly *un*commercial at the time George Lucas began trying to sell it to studios, in the mid-seventies, when his thirteen-page treatment of "The Star Wars" (as it was then called) was rejected by Universal and by United Artists; only Alan Ladd, who was then at Fox, would gamble on it, over the heated objections of the Fox board. When Lucas would get down on the carpet with his toy airplanes and his talk of the Empire and the dark and light sides of the Force and the anthropology of Wookiees, even Lucas's friends thought, "George has lost it," according to the screenwriter Gloria Katz. "He almost had this tunnel vision when it came to his projects, but it seemed this time he was really out there. 'What's this thing called a Wookiee? What's a Jedi, George? You want to make a space opera?'"

The first movie went on to earn three hundred and twenty-three million dollars (more than any previous film had ever earned), and so far the trilogy has brought in about a billion three hundred million dollars in worldwide box-office sales and more than three billion more in licensing fees. Before "Star Wars," merchandise was used only to promote movies; it had no value apart from the films. But Star Wars merchandise became a business unto itself, and it inaugurated modern merchandising as we know it—the Warner Bros. Store, Power Rangers, the seventeen *thousand* different "101 Dalmatians" products that Disney has licensed so far—although Star Wars remains "the holy grail of licensing," in the words of one analyst. Last year, Star Wars action figures were the best-selling toy for boys and the second over-all best-seller, after Barbie. A large percentage of Star Wars action figures are actually bought by adults—Star Wars merchandise dealers—who are hoarding them, to speculate on the price. As a result, it has been difficult to find Star Wars toys, and some stores have limited the number of action figures that one person can buy. (The "vinyl-caped Jawa" action figures, which sold for around three dollars in 1978, are now worth about fourteen hundred.) In addition, LucasArts, which makes Star Wars CD-ROMs, is among the top five producers of video games for computers, while Star Wars novels are, book for book, the single most valuable active franchise in publishing. (Most of the twenty-odd novels Bantam has published since 1991 have made the *Times* hardcover- or paperback-best-seller list.) All this in spite of the fact that there has been no new Star Wars movie in theatres since "The Return of the Jedi," in 1983—although that is about to change.

What *is* it that makes people crave the Star Wars brand in so many different flavors? Somewhere between the idea and the stuff, it seemed to me—between the image of Luke gazing at the two setting suns on the planet Tatooine while he contemplated his destiny as a fighter pilot for the Rebel Alliance, and the twelve-inch Luke collectibles sold by Kenner—an alchemic transformation was taking place: dreams were being spun into desire, and desire forged into product. Here at the summit, you could feel this process drawing energy from the twin rivers of marketing and branding on the one hand, and from people's need to make sense of things on the other. In a world where the stories and images and lessons provided by electronic media seem to be replacing the stories and images and lessons people used to get from religion, literature, and painting, the lessons of Star Wars—that good is stronger than evil, that human values can triumph over superior technology, that even the lowliest of us can be redeemed, and that all this is relatively free of moral ambiguity—is a very powerful force indeed.

As an alien at the summit, I wasn't invited inside the auditorium, but several participants debriefed me on the goings on. In the morning, Howard Roffman, the vice-president of Lucasfilm Licensing, talked about the coming "Star Wars Trilogy Special Edition," a digitally enhanced version of the original trilogy. The first movie will appear on nearly two thousand screens starting January 31st, followed three weeks later by "The Empire Strikes Back" and two weeks after that by "The Return of the Jedi." Then, in 1999, the first of three *new* Star Wars films will début, with the second tentatively scheduled to follow in 2001 and the third in 2003. Yes, it is true, Roffman told the audience, that the first of the new "prequel" movies, which, as everyone present already knew, will begin to tell the back story to the original trilogy—what happened *before* Luke's adventures—will be directed by George Lucas himself, who, as everyone also knew, has not directed a film since the first Star Wars movie.

Roffman warned the licensees not to flood the market with Star Wars merchandise this winter—maybe because he was concerned about damaging the prequel's allure. He admitted that no one knew how well the "Special Edition" would do, because nobody has ever given a trilogy of movies already seen on television and video such a wide theatrical rerelease. Still, he believed that a lot of people would go to see "Star Wars" again, because seeing it for the first time had been such an important event. "'Star Wars' has a timeless quality," Roffman had told me earlier. "For a lot of people, it was a defining moment in their lives. There is a whole generation that remembers where they were when they first saw 'Star Wars.' Now that original generation has aged, and they'll be looking at the films through different eyes—plus they'll want to take their kids." A movie that was designed to appeal to a feeling like nostalgia in the first place would be revisited by people seeking to feel nostalgic for *that* experience, in the pursuit of an ever-receding vision of a mythic past.

Roffman told the crowd that he still got chills when he saw the opening shot of the rebel Blockade Runner pursued by the Imperial Star Destroyer. Tom Sherak, a Fox executive vice-president, who followed Roffman to the podium, joked that he got chills a little earlier in the film, when he saw the Fox logo come up on the screen. When Lucas approached Fox with the idea of rereleasing the Star Wars trilogy, Sherak told me, it was "like Christmas." Fox readily agreed to Lucas's proposal that the studio pay for the use of digital technology to fix some things about the original movies that had always bothered him. Along with every other studio, Fox is hoping to win the rights to distribute the three Star Wars prequel movies, and was willing to do almost anything to make George happy. The studio eventually spent fifteen million dollars on the digital enhancements, and will spend perhaps another twenty million marketing the "Special Edition," which is expected to make around a hundred million.

Then John Talbot, the director of marketing for Pepsi-Cola, showed off sketches for its R2-D2 coolers, which will have Star Wars-related Pepsi cans inside their heads and will probably be displayed in convenience stores. He talked about all the different ways the company would promote the "Special Edition": not only through the Pepsi label but through Frito-Lay, Pizza Hut, and Taco Bell as well—all part of an unprecedented two-billion-dollar commitment from Pepsi which will help drive the Star Wars brand even deeper into the culture, and insure that even if the story were to fade from the surface of the Earth it will remain buried underground in the form of Luke Skywalker pizza boxes and Obi-Wan sixteen-ounce beverage cups.

At five o'clock, a squadron of Imperial Guards, accompanied by white-shelled Imperial Storm Troopers, took the stage, followed by Darth Vader himself, sporting a costume from Lucasfilm Archives. (F.A.O. Schwarz and Neiman Marcus are actually selling limited edition full-size Vader mannequins for five thousand dollars each—a Star Wars product that comes close to having the quasi-religious status of a "prop.") Speaking a recorded version of what sounded like James Earl Jones's voice, the Dark Lord of the Sith sternly upbraided the audience for not inviting him to the summit, but then said that it was actually a good thing he hadn't been invited, because he'd been able to spend the day at "the Ranch" with someone far more important than "you mere merchandisers." With that, George Lucas strode onto the stage.

Wherever Lucas appears, he receives a standing ovation, and he got two today—one on taking the stage and the other on leaving it. It isn't simply Lucas's success as a filmmaker that people are applauding—the fact that, having begun his career right here at the Marin County Civic Center (he had used it as a location for his first film, the dark, dystopian "THX 1138"), Lucas had gone on to create in Star Wars and Indiana Jones two of the most valuable movie properties ever, and, with John Milius, had also had the idea for "Apocalypse Now," which Francis Ford Coppola ended up making. Nor is it his success as a businessman: Industrial Light & Magic, which started as essentially Lucas's model shop—a place to do the effects for the original "Star Wars"—has grown into the premier digital-imaging studio in the world, and is responsible for most of the milestones in computer graphics, including the cyborg in "Terminator 2," the dinosaurs in "Jurassic Park," and the Kennedy cameo in "Forrest Gump." No, you applaud for Lucas because he *is* Star Wars. It's difficult for brains braised in Star Wars from early adulthood to conceive of Lucas in any other terms. The purpose of a myth, after all, is to give people a structure for making sense of the world, and it happens that Lucas's heroic myths are an almost irresistible way of making sense of *him*. "It's like George is Yoda," Tom Sherak told me. "He doesn't say a lot of words, but as you're listening the passion this man feels comes through. He's talking about his own creation. It's this whole thing that just comes over you."

Earlier, during a break, marketing reps from Hasbro had given all the members of the audience their own Luke Skywalker lightsabres, and at Lucas's appearance people turned them on and waved them around. ("It was like being at a Jethro Tull concert in 1978," according to one participant.) After they'd finished, Lucas said a few words about his reasons for wanting to re-release the trilogy, which were chiefly that it would allow a new generation of fans to see the movies in theatres. He later told me he had made a point of keeping his own son, who is now four, from watching "Star Wars" on video, so he could show it to him in a theatre first.

Lucas's aura may be almost palpable, but his *prana*—the Sanskrit word for life force—is oddly blurry behind the looming shadow of his myth. (Of course, his lack of presence is also part of the myth: it's what makes him Yoda-like.) He is slight, and has a small, round belly, a short beard, black nerd-style glasses, and a vulnerable-sounding voice. According to the summiteers I talked to, the only really memorable moment in his ten minutes on the stage came when Alan Hassenfeld, the head of Hasbro, gave him a twelve-inch sculpted Obi-Wan Kenobi body with a George Lucas head, and the crowd went wild. A George Lucas toy! As one member of the audience said later, "What a collectible that would make!" George held up his George toy, and the people all cheered and waved their lightsabres.

II—The Ranch

Skywalker Ranch, the headquarters of Lucas's enterprises, is that part of the Star Wars universe which juts up above the top layer of the myth, into the real world. Deep in the hills of West Marin, the Ranch is a three-thousand-acre detailed evocation of a nineteenth-century ranch that never was. When Lucas was beginning to conceive the place, in the early eighties, he wrote a short story about an imaginary nineteenth-century railroad tycoon who retired up here and built the homestead of his dreams. In this story, the tycoon's Victorian main house dates from 1869; a craftsman-style library was added in 1910; the stable was built in 1870 and the brook house in 1913. Lucas gave his story to the architects of the Ranch and told them to build it accordingly. (He calls this style "remodel" architecture.)

Here, just as in "Star Wars," Lucas created a new world and then layered it with successive coats of mythic anthropology to make it feel used. He has made the future feel like the past, which is what George Lucas does best. "Star Wars" takes place in a futuristic, sci-fi world, but Lucas tells you at the beginning that it existed "a long time ago, in a galaxy far, far away." The movie has all the really cool parts of the future (interplanetary travel, flashy effects, excellent machines), but it also has the friendships, the heroism, and other reassuring conventions of the cinema-processed past (outlaw saloons, dashing flyboys, sinister nobles, brave knights, and narrow escapes). It makes you feel a longing for the unnameable thing that is always being lost (a feeling similar to the one you get from Lucas's second film, "American Graffiti," which helped make nostalgia big business), but it's a longing sweetened by the promise that in the future we'll figure out some way of getting that unnameable thing back. This was the deliciously sad desire that was being forged into product at the summit down in San Rafael, but up here at Lucas's domain the desire seemed to exist in a purer form.

To get to the Ranch, you head west from San Rafael on Lucas Valley Road, which was named long ago, for a different, unrelated Lucas. It's just a coincidence, but as you wind through the plump, grassy hills of Marin County—America's Tuscany—you feel as though you were entering a sort of Jurassic Park of entrepreneurial dreams, in which there are no coincidences, only destiny. A few yards beyond the automatic wooden gates, you come to a kiosk with a guard whose arm patch says "Skywalker Fire Brigade," and then you drive past Skywalker Inn, where each of the guest rooms—the John Ford room, the Akira Kurosawa room—is decorated in the style of the eponymous director. (Tim Burton was staying in one of them at the time of my visit, mixing sound for "Mars Attacks!")

As in "Star Wars," so at the Ranch, you get the sense that some all-controlling intelligence has rubbed itself over every element for a long time. The Technical Building, which is filled with the latest in sound-engineering and editing technology, including a full THX theatre and a soundstage that can seat a hundred-piece orchestra, looks precisely like a nineteenth-century California winery. (According to Lucas's story, it was originally built in 1880, and the interior was remodelled in the Art Deco style in 1934.) Trellises of Pinot Noir, Chardonnay, and Merlot grapes are growing outside, and the grapes are shipped to Francis Coppola's real winery in Napa Valley. Then there are the stables and a baseball field and people on bicycles with Skywalker license plates and an old hayrack and mower that look as if they'd been sitting beside the road for a hundred years. Lucas designed the place so that you can see only one building at a time from any one spot on the property, and although two hundred people come to the Ranch each day, you see no cars—only bicycles, except for the occasional Skywalker Fire Brigade vehicle. The Ranch has three underground parking garages, which can accommodate two hundred cars.

Some of Lucas's friends told me that they thought the Ranch was George's attempt to recapture not only America's legendary Western past but his own past, especially his golden days at U.S.C. film school, in the sixties, when he was a protégé of Coppola's (they met when he was a student observer on Coppola's film "Finian's Rainbow"). He soon became friendly with Steven Spielberg, Martin Scorsese, and Brian De Palma. They were all young artists, who just wanted to make artistically worthy films and weren't yet worried about topping one another's blockbusters. ("E.T." and "Jurassic Park" both surpassed the "Star Wars" box-office record; Lucas is hoping to take the record back from Spielberg with the prequel.) But not many filmmakers come up to the Ranch to conceive and write films; they come to use the technology in postproduction. As Spielberg once wrote, "George Lucas has the best toys of anybody I have ever known, which is why it's so much fun playing over at George's house."

From the outside, the main house looks like the big house on the Ponderosa, but the inside is more like the Huntington Gallery in San Marino, which used to be the home of the nineteenth-century railroad magnate Henry Huntington. Covering the walls is first-growth redwood that was salvaged from old bridges near Newport Beach and has been milled into panelling, and hanging on the redwood are selections from Lucas's collection of paintings by Norman Rockwell, another unironic American image-maker, with whom Lucas feels an affinity. To the Lucas fan, the most exciting things in the house are to be found in the two glass cases in the front hall, where the holiest relics are stored. The "real" lightsabre that Luke uses in "Star Wars" is here, and so are Indy's bullwhip and the diary that leads Indy to the Holy Grail. The Holy Grail itself is stored in the Archives Building.

Two days a week, Lucas comes to the Ranch to conduct business, usually driving himself along Lucas Valley Road. He spends most of the rest of his time at home, writing and looking after his three kids. The oldest, a teen-age girl, he adopted with his ex-wife and collaborator Marcia Lucas (she won an Oscar for editing "Star Wars"); the younger ones—a girl and a boy—he adopted as a single parent, after he and Marcia split up. Because my first visit to the Ranch fell on a Wednesday, which is not one of Lucas's regular days, I had been told there would be no meeting with him. But as we were touring the two-story circular redwood library with the stained-glass roof (there was a cat on the windowsill, a fire burning in the fireplace, and a Maxfield Parrish painting hanging over the mantel), word came that Lucas was here today after all, and would see me now. I was led down the back stairway into the dim recesses of the basement, which was sort of like going backstage at a high-tech theatre, and into a small, windowless room filled with editing machines, lit chiefly by light coming from two screens—a television and a glowing Power-Book. There sat the great mythmaker himself, wearing his usual flannel shirt, jeans, sneakers, and Swatch watch. He emerged from a dark corner of the couch to shake hands, then retreated into the dimness again.

Lucas conceived the Ranch as a complete filmmaking operation—his version of the ideal studio, one that would respect human values and creativity, as opposed to Hollywood studios, which he saw as evil and greedy and encouraging of mediocrity. Lucas's famous disdain of Hollywood is partly a result of his father's influence—George Lucas, Sr., was a conservative small-town businessman, who viewed all lawyers and film executives as sharpies and referred to Hollywood as Sin City—and partly a result of his own bitter experience with his first two films, "THX 1138" and "American Graffiti." The dark lords at Universal thought "Graffiti" was so bad that they weren't going to release the film at all, and then they were going to release it just on TV; finally, when Coppola, who had just made "The Godfather," offered to buy the movie, Universal relented and brought it out in the theatres, although not before Ned Tanen, then the head of Universal, cut four and a half minutes from it—a move that caused Lucas terrific anguish. Made for seven hundred and seventy-five thousand dollars, "Graffiti" went on to earn nearly a hundred and twenty million.

"I've always had a basic dislike of authority figures, a fear and resentment of grown-ups," Lucas says in "Skywalking," the 1993 biography by Dale Pollock (to whom I'm indebted for some of the details of Lucas's early career). When the success of the first Star Wars film allowed Lucas to "control the means of production," as he likes to say, he financed the second and third films himself, and he built the Ranch. In the beginning, the films edited at the Ranch were Lucas's own: he was busy working on the Star Wars movies and overseeing the Indiana Jones series (which he conceived the story for and produced; Spielberg directed). But before long people at the Ranch were spending less time on Lucas's films—-"Willow," "Radioland Murders," and the "Young Indiana Jones Chronicles" television series have been his main projects in the last few years—and more time doing other people's. Today, Lucas offers a full-service digital studio, where directors can write, edit, and mix films, and Industrial Light & Magic, down in San Rafael, can do the special effects.

Lucas is the sole owner of his companies; Gordon Radley, the president of Lucasfilm, estimated in *Forbes* that Wall Street would value them at five billion dollars. (*Forbes* estimated that Lucas himself was worth two billion.) He is an old-fashioned, paternalistic chairman of the board, who gives each of his twelve hundred employees a turkey at Thanksgiving, and who sits every month at the head of the locally carpentered redwood boardroom table in the main house and listens to reports from the presidents of the various divisions of his enterprise.

"My father provided me with a lot of business principles—a small-town retail-business ethic, and I guess I learned it," Lucas told me in his frail-sounding voice. "It's sort of ironic, because I swore when I was a kid I'd never do what he did. At eighteen, we had this big break, when he wanted me to go into the business"—George, Sr., owned an office-supply store—"and I refused, and I told him, 'There are two things I know for sure. One is that I will end up doing something with cars, whether I'm a racer, a mechanic, or whatever, and, two, that I will never be president of a company.' I guess I got outwitted."

Lucas's most significant business decision—one that seemed laughable to the Fox executives at the time—was to forgo his option to receive an additional five-hundred-thousand-dollar fee from Fox for directing "Star Wars" and to take the merchandising and sequel rights instead. The sequels did almost as well as the first movie, and the value of the Star Wars brand, after going into a hiatus in the late eighties, reëmerged around 1991, when

Bantam published "Heir to the Empire," by Timothy Zahn, wherein Princess Leia and Han Solo have children. The book surprised the publishing world by going to No. 1 on the *Times* hardcover-fiction list, and marketers quickly discovered a new generation of kids who had never seen the movies in theatres but were nevertheless obsessed with Star Wars.

Lucas's business success as the owner of Star Wars, however, has had the ironic result of taking him away from the thing that touched his audience in the first place. He told me he'd stopped directing movies because "when you're directing, you can't see the whole picture." He explained, "You want to take a step back, be the over-all force behind it—like a television executive producer. Once I started doing that, I drifted further and further away. Then I had a family, and that changed things—it's very hard to direct a movie and be a single parent at the same time." Then his company became a big business. "The company started as a filmmaking operation. I needed a screening room, then a place to do post, then mixing, then special effects—because, remember, I was in San Francisco. You can't just go down the street and find this stuff—you have to build it. Everything in the company has come out of my interests, and for a long time it was a struggle. But six years ago the company started coming into its own. The CD-ROM market, which we had been sitting on for fifteen years, suddenly took off, and digital-filmmaking techniques took off—they are five times larger than six years ago—and suddenly I had a big company, and I had to pay attention to it." He estimated that he now spends thirty-five percent of his time on his family, thirty-five percent on movies and thirty percent on the company.

Lucas's day-to-day activities in the main house include the management of the Star Wars story, which is probably the most carefully tended secular story on Earth. Unlike Star Trek, which is a series of episodes connected by no central narrative, Star Wars is a single story—"a finite, expanding universe," in the words of Tom Dupree, who edits Bantam's Star Wars novels in New York. Everyone in the content-creating galaxy of Star Wars has a copy of "The Bible," a burgeoning canonical document (currently a hundred and seventy pages long) that is maintained by "continuity editors" Allan Kausch and Sue Rostoni. It is a chronology of all the events that have ever occurred in the Star Wars universe, in all the films, books, CD-ROMs, Nintendo games, comic books, and role-playing guides, and each medium is seamlessly coördinated with the others. For example, a new Lucas story called "Shadows of the Empire" is being told simultaneously in several media. A recent Bantam novel introduces a smuggling enterprise, led by the evil Prince Xizor, within the time line of the original trilogy. Three weeks before Christmas, he crossed over into the digital realm: he began to appear in the new sixty-four-bit Nintendo playstations. Meanwhile, Dark Horse comics is also featuring the Prince, along with related characters, like the bounty hunter Boba Fett. (Boba had only a few minutes of screen time in the movies, but has become one of the most popular characters among gen-X fans.)

New developments in even the remotest corners of the Star Wars universe are always approved by Lucas himself. The continuity editors send him checklists of potential events, and Lucas checks yes or no. "When Bantam wanted to do the back story on Yoda," Dupree said, "George said that was off limits, because he wanted him to remain a mysterious character. But George has made available some time between the start of Episode Four, when Han Solo is a young pilot on the planet Corellia, and the end of the prequel, so we're working with that now." Although Lucas had once imagined Star Wars as a nine-part saga, he hasn't yet decided whether he wants to make the final three movies, but he has allowed licensees to build into the narrative space on the future side of "Jedi"; the later years of Han, Luke, Leia, and the rest of the characters have been colonized by other media.

"George creates the stories, we create the places," Jack Sorensen, the president of LucasArts, told me in explaining how the interactive division fits into the Lucas multimedia empire. "Say you're watching the movie, and you see some planet that's just in the background of the movie, and you go, 'Hey, I wonder what that planet is like?' There's a lot going on on a planet, you know. So we'll make that planet the environment for a game. In a film, George would have to keep moving, but the games give you a chance to explore."

I asked Sorensen to explain the extraordinary appeal of Star Wars, and he said, "I'm as perplexed by this as anyone, and I'm right in the midst of it. I travel all the time for this job, and I meet people abroad, in Italy, or in France, who are totally obsessed with 'Star Wars.' I met this Frenchman recently who told me he watches it every week. I don't really think this is caused by some evil master plan of merchandisers and marketers. The demand is already out there, and we're just meeting it—it would exist without us. I don't know if I want to say this in print, but I feel like Star Wars is the mythology of a non-sectarian world. It describes how people want to live. People all view politics as corrupt, much more so in Europe than here, and yet people are not cynical underneath—they want to believe in something pure, noble. That's Star Wars."

"Star Wars" still seems pure in some ways, but in other ways it doesn't. The qualities that seemed uncommercial about it at the time Lucas was trying to make it—most strikingly, that the literary elements (the characters and the plot) were subordinate to the purely cinematic elements (motion)—are the very qualities that seem commercial about it now. This is one way of measuring how "Star Wars" has changed both movies and the people who go to movies.

"I'm a visual filmmaker," Lucas told me. "I do films that are kinetic, and I tend to focus on character as it is created through editing and light, not stories. I started out as a harsh enemy of story and character, in my film-school days, but then I fell under Francis's mentorship, and his challenge to me after 'THX 1138' was to make a more conventional movie. So I did 'American Graffiti,' but they said that was just a montage of sounds, and then I did 'Star Wars.' I was always coming from pure cinema—I was using the grammar of film to create content. I think graphically, not linearly."

The first Star Wars movie is like a two-hour-long image of raw speed. If you saw it when you were young, this tends to be what you remember—the feeling of going really fast. Lucas is a genius of speed. His first ambition was to be a race-car driver, and it was only after he was nearly killed in a terrible accident, when he was eighteen—he lived because his seat belt unaccountably broke and he was hurled free of the car—that his interest shifted to film. (His first moving pictures were of race cars.) Perhaps the most memorable single image in "Star Wars" is the shot of the Millennium Falcon going into hyper-space for the first time, when the stars blur past the cockpit. Like all the effects in the movie, this works not because it is a cool effect (it's actually pretty low tech—merely "motion blur" photography) but because it's a powerful graphic distillation of the feeling the whole movie gives you: an image of pure kinetic energy which has become a permanent part of the world's visual imagination. (The other day, I was out running, and as a couple of rollerbladers went whizzing by I heard a jogger in front of me say to his friend, "It's like that scene in 'Star Wars' when they go into hyperspace.") Insofar as a media-induced state of speed has become a condition of modern life, Lucas was anticipating the Zeitgeist in "Star Wars."

The problems arise when Lucas has to slow down. In "Radioland Murders," the characters have to carry the narrative, but Lucas couldn't make this work, so he had to speed up the pace and turn the movie into farce. Also, because Lucas has little rapport with actors, his films tend to have only passable acting in them, which forces him to rely unduly on pace and editing. Mark Hamill once said, "I have a sneaking suspicion that if there were a way to make movies without actors, George would do it." Lucas is well known to be impatient with actors' histrionics, and has little interest in becoming involved in their discussions about method. Harrison Ford told me that Lucas had only two directions for the actors in "Star Wars"; he replayed them for me over hue-vos rancheros on the terrace at the Bel Air Hotel one Saturday morning, using his George-as-director voice—nasal, high, kind of whiny. The two directions were "O.K., same thing, only better," and "Faster, more intense." Ford said, "That was it: 'O.K., same thing, only better.' 'Faster, more intense.'"

For serious Star Wars fans—the true believers—Lucas's story tends to mingle with Luke's, the real one becoming proof that the mythic one can come true, and the mythic one giving the real one a kind of larger-than-life significance. Just as Luke is a boy on the backwater planet of Tatooine—he is obedient to his uncle, who wants him to stay on the farm, but he dreams that there might be a place for him somewhere out there in the larger world of adventure—so Lucas was a boy in the backwater town of Modesto, California. He dreamed of being a great race-car driver, but his liberal-bashing, moralizing father wanted him to stay at home and take over the family business. (Pollack's book recounts how his father liked to humiliate him every summer by chopping off his hair.) Just as a benevolent father figure (Obi-Wan) helps Luke in his struggle past his dark father, the older Coppola took young George under his wing at film school, and helped him get his first feature film made. And, just as Luke at the end of the first Star Wars film realizes his destiny and becomes a sort of hero-knight, so Lucas became the successful filmmaker, fulfilling a prophecy he'd made to his father in 1962, two years before he left Modesto for U.S.C., which was that he would be "a millionaire before I'm thirty."

But both Luke and Lucas have had to reckon with their patrimony. There's the famous scene on the Cloud City catwalk in "Empire" when Darth reveals that he is Luke's father. He has cut off Luke's hand and tries to turn him to the dark side, saying, "Join me, and together we can rule the galaxy as father and son!": Luke responds by leaping off the catwalk into the abyss. Just as Luke has to contend with the qualities he may have inherited from Darth Vader, so Lucas, in his career after "Star Wars," has stopped directing films and has become the successful, fiscally conservative businessman that his father always wanted him to be.

The scripts for the prequel, which Lucas is finishing now, make it clear that Star Wars, taken as a whole story and viewed in chronological order, is not really the story of Luke at all but the story of Luke's father, Anakin Skywalker, and how he, a Jedi Knight, was corrupted by the dark side of the Force and became Darth Vader. When I asked Lucas what Star Wars was ultimately about, he said, "Redemption." He added, "The scripts to the three films that I'm finishing now are a lot darker than the second three, because they are about a fall from grace. The first movie is pretty innocent, but it goes downhill from there, because it's more of a tragic story—that's built into it."

I said that the innocence was what many people found so compelling about the first Star Wars movie, and I asked whether it was harder for him, now that he is twenty years less innocent, to go back to work on the material.

"Of course your perspective changes when you get older and as you get battered by life," he said.

"Have you been battered by life?"

"Anyone who lives is going to get battered. Nothing comes easy."

I believed him. He was Yoda, after all. He had lived for almost nine hundred years. He had known the sons who triumph over their dark fathers only to find themselves in the murkier situation of being fathers themselves, and that knowledge had made him wise but it had also worn him out. That was the note of loss in his voice—the thing that the Star Wars trilogy didn't allow very often onto the screen, but it was there in the background, like the remnants of blue screens you could see in "Empire," in the thin outlines surrounding the Imperial Walkers on the ice planet Hoth. (These flaws have been digitally erased in the "Special Edition.")

Did Lucas worry about being turned to the dark side himself—did fame, money, or power tempt him? Or maybe it happened more slowly and subtly, with the temptation to stop being a filmmaker and become a kind of master toymaker instead, which is fun until you wake up one day and realize you have become one of your own toys. You could see that starting to occur in "Jedi": the Ewoks, those lovable furry creatures, seemed destined for the toy store even before they helped Luke defeat the Empire.

"The world is all yours," I said to Lucas. "You could have anything you wanted."

"Like what?" he asked. "What do I want? What do you want? Basically I just like to make movies, and I like raising a family, and whatever money I've made I've plowed back into the company. I've only ever been tempted by making movies—I don't need yachts, I'm not a party animal, and holding and using power over people never interested me. The only dangerous side of having this money is that I will make movies that aren't commercial. But, of course, 'Star Wars' was not considered commercial when I did it."

III—The Story

Since "Star Wars" sprang into Lucas's mind first as pictures, not as a narrative, he needed a line on which to hang his images. He studied Joseph Campbell's books on mythology, among other sources, taking structural elements from many different myths and trying to combine them into one epic story. One can go through "Star Wars" and almost pick out chapter headings from Campbell's "The Hero with a Thousand Faces": the hero's call to adventure, the refusal of the call, the arrival of supernatural aid, the crossing of the first threshold, the belly of the whale, and the series of ordeals culminating in a showdown with the angry father, when, at last, as Campbell writes, "the hero...beholds the face of the father, understands—and the two are atoned"— which is precisely what happens at the end of "Jedi." Lucas worked this way partly out of a conscious desire to create a story that would touch people— it was a kind of get-'em-by-the-archetypes-and-their-hearts-and-minds-will-follow approach—but it also appears to have been an artistic necessity.

"When I was in college, for two years I studied anthropology—that was basically all I did," Lucas said as we sat in his basement den. The glow of the blank TV screen illuminated his face. "Myths, stories from other cultures. It seemed to me that there was no longer a lot of mythology in our society—the kind of stories we tell ourselves and our children, which is the way our heritage is passed down. Westerns used to provide that, but there weren't Westerns anymore. I wanted to find a new form. So I looked around, and tried to figure out where myth comes from. It comes from the borders of society, from out there, from places of mystery—the wide Sargasso Sea. And I thought, Space. Because back then space was a source of great mystery. So I thought, O.K., let's see what we can do with all those elements. I put them all into a bag, along with a little bit of 'Flash Gordon' and a few other things, and out fell 'Star Wars.'"

He said that his intention in writing "Star Wars" was explicitly didactic: he wanted it to be a good lesson as well as a good movie. "I wanted it to be a traditional moral study, to have some sort of palpable precepts in it that children could understand. There is always a lesson to be learned. Where do these lessons come from? Traditionally, we get them from church, the family, art, and in the modern world we get them from media—from movies." He added, "Everyone teaches in every work of art. In almost everything you do, you teach, whether you are aware of it or not. Some people aren't aware of what they are teaching. They should be wiser. Everybody teaches all the time."

Lucas's first attempt at writing the story lasted from February, 1972, to May, 1973, during which he produced the thirteen-page plot summary that his friend Laddie (Alan Ladd) paid him fifteen thousand dollars to develop into a script. Like "American Graffiti," which had been a montage of radio sounds, this treatment was like a montage of narrative fragments. It took Lucas another year to write a first draft; he is an agonizingly slow writer. He composed "Star Wars" with No. 2 pencils, in tiny, compulsively neat-looking script, on green-and-blue-lined paper, with atrocious spelling and grammar. Realizing that he had too much stuff for one movie, he cut the original screenplay in half, and the first half was the germ for the three scripts he is finishing now; the second half became the original trilogy. (Lucas's way of making the future into the past is thus part of the stricture of Star Wars. Many people don't remember this, but the first movie was called, in full, "Star Wars, Episode IV: A New Hope.")

He showed his first draft to a few people, but neither Laddie nor George's friends could make much sense of it. "It only made sense when George would put it in context with past movies, and he'd say, 'It's like that scene in "Dam Busters,"'" Ladd told me, in his office on the Paramount lot. The only characters that Lucas seemed to have a natural affinity for were the droids, C-3PO and R2-D2. In early drafts, the first thirty pages focussed almost entirely on the droids; the feeling was closer to the Orwellian mood of "THC 1138." Based on his friends' reactions, Lucas realized he would need to put the humans on the screen earlier, in order to get the audience involved. But his dialogue was wooden and labored. "When I left you, I was but the learner," Darth says to Obi-Wan, before dispatching the old gentleman with his sabre. "I am the master now!" Harrison Ford's well-known remark about Lucas's dialogue, which he repeated for me, was "George, you can type this shit, but you sure can't say it."

Lucas also borrowed heavily from the film-school canon. The lightsabres and Jedi Knights were inspired by Kurosawa's "Hidden Fortress," C-3PO's look by Fritz Lang's "Metropolis," the ceremony at the end by Leni Riefenstahl's "Triumph of the Will." Alec Guinness does a sort of optimistic reprise of his Prince Feisal character from "Lawrence of Arabia," while Harrison Ford plays Butch Cassidy. Critics of "Star Wars" point to the film's many borrowings as evidence of Lucas's failure as a filmmaker. (It is said of Lucas, as it was of Henry Ford, that he didn't actually invent anything.) But "Star Wars" fans see this as a brilliant postmodernist commentary on the history of popular film. Lucas uses these references with a childlike lack of irony that Scorsese, De Palma, or Spielberg would probably be incapable of, because they grew up with the movies. The critic and screenwriter Jay Cocks, a longtime friend of Lucas's, told me, "I don't think George went to a lot of movies as a kid, like Marty and Brian. Before arriving at film school, he hadn't seen 'Alexander Nevsky' or 'Potemkin'; I don't think he even saw 'Citizen Kane.' I remember when I went up to this little cabin he and Marcia had in Northern California, and I saw he had some Flash Gordon comic strips, drawn by Alex Raymond, on his desk, and a picture of Eisenstein on his wall, and the combination of those two were really the basis for George's aesthetic in 'Star Wars.'"

But Lucas added many small personal details to the story as well, which is part of what gives his creation its sensuous feeling of warmth. According to Pollock, the little robot is called R2-D2 because that was how Walter Murch, the sound editor on "American Graffiti," asked for the Reel 2, Dialogue 2 tape when they were in the editing room one day, and for some reason "R2-D2" stuck in George's mind. The Wookiee was inspired by Marcia's female Alaskan malamute, Indiana (who also lent her name to Lucas's other great invention); the image of Chewie came to George in a flash one day as the dog was sitting next to Marcia in the front seat of the car. Many of Ben Burtt's sounds for the movie also had personal significance. He told me that the buzzing sound of the lightsabre was the drone that an old U.S.C. film-school projector made. The pinging sound of the blasters was the sound made one day when Burtt was out hiking and caught the top of his backpack on some guy wires.

Shooting of the interiors began in March, 1976, in Elstree Studios, outside London, and lasted four months. The English crew were awful to Lucas. "They just thought it was all very unsophisticated," Harrison Ford told me. "I mean, here's this seven-foot-tall man in a big hairy suit. They referred to Chewie as 'the dog.' You know, 'Bring the dog in.' And George isn't the best at dealing with those human situations—to say the least." Meanwhile, Lucas was struggling with Fox to get more than the nine million dollars that Fox wanted to spend (the movie ended up costing about ten million to make), and he was paranoid that the studio was going to take the movie away from him, or, treat it the way that Universal had treated "Graffiti."

Lawrence Kasdan, who co-wrote the scripts for "Empire" and "Jedi" and also wrote the script for the first Indiana Jones movie, thinks Lucas's antagonistic relationship with Hollywood is what "Star Wars" is really about. "The Jedi Knight is the filmmaker, who can come in and use the Force to impose his will on the studio," he told me. "You know that scene when Alec Guinness uses the Force to say to the Storm Trooper outside the Mos Eisley cantina, 'These aren't the droids you're looking for,' and the guy answers, 'These aren't the droids we're looking for'? Well, when the studio executive says, 'You won't make this movie,' you say, 'I *will* make this movie,' and then the exec agrees, 'You will make this movie.' I once asked him, 'George, where did you get the confidence to argue with Ned Tanen?' He said, 'Your power comes from the fact that you are the creator.' And that had an enormous influence on me."

When a rough cut of "Star Wars" had been edited together, George arranged a screening of the film at his house. The party travelling up from Los Angeles included Spielberg, Ladd, De Palma, a few other friends of George's, and some Fox executives. "Marty was supposed to come, too," the screenwriter Willard Huyck told me, referring to Martin Scorsese, "but the weather was bad and the plane was delayed and finally Marty just went home."

"Marty had an anxiety attack is what happened," Gloria Katz, Huyyck's wife and writing partner, put in. Huyck and Katz are old friends of Lucas's, who co-wrote "Graffiti" with him, and are the beneficiaries of George's generous gift of two points of "Star Wars" for their help on the script.

"Marty's very competitive, and so is George," Huyck went on. "All these guys are. And at that time Marty was working on 'New York, New York,' which George's wife, Marcia, was editing, and everybody was saying that it was going to be the big picture of the year—but here we were all talking about 'Star Wars.'"

Here, at this first screening of "Star Wars," a group of writers, directors, and executives, all with ambitions to make more or less artistically accomplished Hollywood films, were confronting the template of the future—the film that would in one way or another determine everyone's career. Not surprisingly, almost every one of them hated it. Polite applause in the screening room, no cheers—a "real sweaty-palm time," Jay Cocks, who was also there, said. It's possible that, just this once, before the tsunami of marketing and megatude closed over "Star Wars" forever, these people were seeing the movie for what it really was—a film with comic-book characters, an unbelievable story, no political or social commentary, lousy acting, preposterous dialogue, and a ridiculously simplistic morality. In other words, a bad movie.

"So we watch the movie," Huyck said, "and the crawl went on forever, there was tons of back story, and then we're in this spaceship, and then here's Darth Vader. Part of the problem was that almost none of the effects were finished, and in their place George had inserted World War Two dogfight footage, so one second you're with the Wookiee in the spaceship and the next you're in "The Bridges at Toko-Ri." It was like, *George, what-is-going-on?*"

"When the film ended, people were aghast," Katz said. "Marcia was really upset—she was saying, 'Oh my God, it's "At Long Last Love,"' which was the Bogdanovich picture that was such a disaster the year before. We said, 'Marcia, fake it, fake it. *Laddie is watching.*'"

Huyck: "Then we all got into these cars to go someplace for lunch, and in our car everyone is saying, 'My God, what a disaster!' All except Steven, who said, 'No, that movie is going to make a hundred million dollars, and I'll tell you why—it has a marvellous innocence and naïveté in it, which is

George, and people will love it.' And that impressed us, that he would think that, but of course no one believed him."

Katz: "We sat down to eat, and Brian started making fun of the movie. He was very acerbic and funny."

"Which is the way Brian *is*," Huyck said. "You know, 'Hey George, what were those Danish rolls doing in the princess's ears?' We all sat there very nervously while Brian let George have it, and George just sort of sank deeper into his chair. Brian was pretty rough. I don't know if Marcia ever forgave Brian for that."

When the movie opened, Huyck and Katz went with George and Marcia to hide out in Hawaii. On their second day there, Huyck told me, "Laddie called on the phone and said, 'Turn on the evening news.' Why? 'Just watch the evening news,' he said. So we turned it on, and Walter Cronkite was on saying, 'There's something extraordinary happening out there, and it's all the result of a new movie called "Star Wars."' Cut to lines around the block in Manhattan. It was absolutely amazing. We were all stunned. Now, of course, the studios can tell you what a movie is going to do—they have it down to a science, and they can literally predict how a movie is going to open. But in those days they didn't know."

Dean Devlin, one of the creators of "Independence Day," was fourteen years old when "Star Wars" came out. He remembers being sixth in line on the day the movie opened at Mann's Chinese Theatre, in Hollywood, on May 25, 1977. "I don't even know why I wanted to see it," he told me when I went to talk with him in his office, just off the commissary at Paramount. "I just knew I had to be there." The movie changed his life. "To me, 'Star Wars' was like the 'Sgt. Pepper's Lonely Hearts Club Band' of movies. You know how people like David Bowie say that that was the album that made them see what was possible in pop music? I think 'Star Wars' did the same thing for popcorn movies. It made you see what was possible." He added, "'Star Wars' was the movie that made me say, I want to do something like that." "Stargate" and "Independence Day" are what he has done so far, and the day after we talked he and his partner, Roland Emmerich, were leaving for Mexico, to hole up and work on their next film, "Godzilla."

"'Star Wars' was a serious breakthrough, a shift in the culture, which was possible only because George was this weird character," Lawrence Kasdan said. "After you saw it, you thought, My God, anything is possible. He opened up people's minds. I mean, the amazing thing about cinema technology is that it hasn't changed since it started. It's mind-boggling. With all these changes in technology, and the computer, we're still pulling this little piece of plastic through a machine and shining light through it. Very few directors change things. Welles did things differently. But everything is different after 'Star Wars.' All other pictures reflect its influence, some by ignoring it or rebelling against it." Every time a studio executive tells a writer that his piercing and true story needs an "action beat" every ten minutes, the writer has George Lucas to thank. "Ransom," the recent movie by Lucas's protégé Ron Howard, is two hours' worth of such action beats and little else.

Huyck suggested that one of the legacies of the trilogy's success is that a movie as fresh and unknowing as "Star Wars" wouldn't get made today. "Truffaut had the idea that filmmaking entered a period of decadence with the James Bond movies, and I'm not sure you couldn't say the same thing about 'Star Wars,' though you can't blame George for it," he told me. "'Star Wars' made movies big business, which got the studio executives involved in every step of the development, to the point of being on the set and criticizing what the director is doing. Now they all take these screenwriting courses in film school and they think they know about movies—their notes are always the same. Both 'American Graffiti' and 'Star Wars' would have a very difficult time getting through the current system. 'Star Wars' would get pounded today. Some executive would get to the point where Darth Vader is revealed to be Luke's father and he would say, 'Give me a break.'"

These days, Kasdan said, "the *word* 'character' might come up at a development meeting with the executives, but if you asked them they couldn't tell you what character is." He went on, "Narrative structure doesn't exist—all that matters is what's going to happen in the next ten minutes to keep the audience interested. There's no faith in the audience. They can't have the story happen fast enough."

Twenty years ago, Pauline Kael suggested that in sacrificing character and complexity for non-stop action "Star Wars" threatened to turn movies into comic books, and today, in a week when two of the three top movies are "101 Dalmatians" and "Space Jam," it is easy to believe she was right. (I know which movies are on top because I read last week's grosses, along with the N.F.L. standings, in Monday's *Times:* another aspect of the blockbuster mentality that "Star Wars" helped to launch.)

When I reminded Lucas of Kael's remark, he sighed and said, "Pauline Kael never liked my movies. It's like comparing novels and sonnets, and saying a sonnet's no good because it doesn't have the heft of a novel. It's not a valid criticism. After I did 'Graffiti,' my friends said, 'George, you should make more of an artistic statement,' but I feel 'Star Wars' did make a statement—in a more visual, less literary way. People said I should have made 'Apocalypse Now' after 'Graffiti,' and not 'Star Wars.' They said I should be doing movies like 'Taxi Driver.' I said, 'Well, "Star Wars" is a kids' movie, but I think it's just as valid an art film as "Taxi Driver."' Besides, I couldn't ever do 'Taxi Driver.' I don't have it in me. I could do 'Koyaanisqatsi,' but not 'Taxi Driver.' But of course if the movie doesn't fit what they think movies should be, it shouldn't be allowed to exist. I think that's narrow-minded. I've been trying to rethink the art of movies—it's not a play, not a book, not music or dance. People were aware of that in the silent era, but when the talkies started they lost track of it. Film basically became a recording medium."

IV—The Model Shop

When you're up at Skywalker Ranch, it's possible to imagine that the future is somehow going to end up saving the past after all, and that we will find on a new frontier what we lost a long time ago in a galaxy far, far away. But in San Rafael, at Industrial Light & Magic, this is harder to believe. Down here, where legions of young geeks work in high-tech, dark satanic mills and stare into computer screens as they build "polys" (the polygonal structures that are the basic elements of computer graphics), the future doesn't seem likely to save the past. The future destroys the past.

Take, for example, the I.L.M. model shop—a high-ceilinged space, full of creatures, ghosts, and crazy vehicles that were used in movies like "Ghostbusters" and "Back to the Future." It's like a museum of non-virtual reality. There are books on the torque capacity of different kinds of wrenches, cylinders of compressed gas, tools of all descriptions, and a feathery layer of construction dust covering everything. All the monsters, ships, weapons, and other effects in the original "Star Wars" were made in the model shop—by hand, more or less—by the team of carpenter-engineer-tinkerer-inventors assembled by Lucas. They used clay, rubber, foam latex, wire, and seven-foot-two hair-Wookiee suits to create their illusions. (The only computers used in making the original "Star Wars" were used to control the motion of the cameras.) The modelmakers were so good at what they did that they unleashed a desire for better and better special effects in movies, which ended up putting a number of the people in the model shop out of work. "Either you learn the computer or you might as well work as a carpenter," one former member of the model shop said.

Today, the model shop survives as a somewhat vestigial operation, producing prototypes of the creatures and machines that will then be rendered in computer graphics for the movie screen. (While I was visiting I.L.M., I saw two carpenters working on a C-3PO, whose head, arms, and legs were spread all over the workbenches—he looked the way he did after the Sand People got done with him—and I asked, naïvely, if I would see him in one of the new movies. No, he was just another prototype.) One model-shop old-timer is the artist Paul Huston, who is still around because he has learned how to use the computer. Huston did oil-on-canvas matte paintings for the original "Star Wars"; he did digital matte paintings for the "Special Edition." (No one does any offscreen matte painting at I.L.M. anymore.) He showed me a tape of some of his digital additions to the first film, which occur before and after the famous Mos Eisley cantina scene, where a "C.G." (computer graphic) Jabba the Hutt will now confront Han Solo in the docking bay where the Millennium Falcon is waiting. (Ford did his half of the scene with a human actor inside a Jabba suit in 1976, but the footage was never used, because it didn't look good enough.) In order to create a crashed spaceship that is part of the background at the Mos Eisley spaceport, Huston actually built a real model, out of various bits scavenged from airplane and helicopter kits—this gave it a kind of ad-hoc, gnarled thingeyness—and he took it outside and photographed it in natural light. Then he scanned that image into the computer and worked on it with Photoshop. A more digitally inclined artist would probably have built the crashed spaceship from scratch right on the computer.

"To me, it comes down to whether you look to the computer for your reality or whether you look out there in the world," Huston said, pointing outside the darkened C.G. room where we were standing to the bright sunlight and the distant Marin hills. "I still think it's not in the computer, it's out there. It's much more interesting out there." He turned back to the spaceport image on the screen. "Look at the way the light bounces off the top of this tower, and how these shadows actually deepen here, when you'd expect them to get lighter—you would never think to do that on a computer."

Huston seemed like kind of a melancholy guy, with sad eyes behind small wire-rimmed glasses. He said he had accepted the fact that creations like the C.G. T. Rex in "Jurassic Park" were the way of the future, and that in many cases C.G. makes effects better. The model shop could never get lava to glow properly, for example, but with C.G. it's easy. Also, C.G. is more efficient—you can get more work done. "The skills that you needed ten years ago to do modelling and matte painting were hard to attain," he said. "Making models was a process of trial and error. There were no books to tell you how to do it. It was catch as catch can. Today, with the computer, you can get modelling software, which has a manual, and follow the instructions."

He gestured around at the monitors, scanners, and other machinery all over the place. "Our company has taken the position that the computer is the future, and we have moved aggressively toward that," he said. "I can see that's what everyone wants. In model-making there aren't many shortcuts. And this is a business of turnover and scheduling and getting as much work through the pipeline as you can. If there is a shortcut to doing something on a computer, and it's going to cost a hundred and fifty thousand dollars, the client will say, 'Oh sure, let's do it.' But if you offer them a model-building project for a third of that there are no takers. Because that's not what clients are interested in. So we're trying to do everything synthetically, with computers, to see how far we can go with that. Part of the interest and excitement is in the surprise—you don't know what is going to happen."

C.G. is supposed to make films cheaper to produce, and its champions continue to promise the nirvana of a movie made by one person on a P.C. for next to nothing. Dean Devlin, a large consumer of C.G., told me about "this British kid who brought my partner Roland and me a full-length film he had shot for ten thousand dollars in a garage." He explained, "It was an excellent film with unbelievably good effects, and the way he shot it you couldn't tell it had mostly been done on a computer, and we looked at each other and said, 'It's coming.' A hundred-percent-C.G. live-action film, with no models." But so far C.G. has actually made movies more expensive than ever. When I asked Jim Morris, the president of Lucas Digital, I.L.M.'s parent company, why this was so, he said, "If we did the same shots now that we did in 'T2,' those shots would be cheaper today. The fly in the ointment is that very few directors are trying to do what has already been done. The line we hear ninety-nine percent of the time is 'I want something that no one has ever seen before.' No one comes in and says, 'O.K., I'll take three twisters.'"

A potential milestone in C.G. toward which I.L.M. is racing its competitors—Jim Cameron's Digital Domain and Sony's Image Works—is the creation of the first "synthespian": a C.G. human actor in a live-action movie. "Toy Story," made by Pixar, a company that grew out of I.L.M. and which Lucas sold to Steve Jobs ten years ago, was an all-C.G. movie, but it starred toys; no one has yet made a believable C.G. human. In "Forrest Gump" audiences saw a little bit of Kennedy: What about a whole movie starring Kennedy? Or a new James Dean movie, in which Dean is a synthespian made up of bits digitally lifted from old images of him?

Recalling Mark Hamill's remark about Lucas—"I have a sneaking suspicion that if there were a way to make movies without actors, George would do it"—I asked Morris if he thought synthetics would ever be the stars of movies. Morris acknowledged that that was "a holy grail for some digital artists" but added that he didn't think a synthetic would be much of an improvement over an actor. "You get a lot for your money out of an actor," he said. "And even if you did replace him with a synthetic you're still likely to need an actor's voice, and you'd have the animator to deal with, so what's the advantage? That said, it does make sense when you're talking about safety. You don't have to have human beings putting themselves at risk doing stunts when you can have synthetics doing them."

One of the lessons "Star Wars" teaches is that friends who stick together and act courageously can overcome superior weapons, machines, and any other kind of technology. At the end of the first movie, Luke, down to his last chance to put a missile in the Death Star's weak spot (a maneuver that will blow up the reactor and save the Rebel Alliance from destruction), hears Obi-Wan telling him to turn off the targeting computer in his X-Wing fighter and rely on the Force instead: "Use the Force, Luke." Clearly, the lesson is that the Force is superior to machines. My visit to I.L.M. made me wonder if that lesson was true. At the very least, the situation seemed to be a lot more complex than "Star Wars" makes it out to be. The gleeful embrace of the latest thing, of the coolest effect, is a force that is at least equal to most other human qualities. The technology to create never-before-seen images also creates the desire to see even more amazing images, which makes the technology more and more powerful, and the people who use it correspondingly less important. And the power of business is not always in the service of human values or of making good films. Lucas has benevolently followed his business instincts into the new world of dazzling, digitally produced effects, and has built in I.L.M. a company that currently earns tens of millions of dollars a year and will probably earn much more in the future—the future! But his success is a sad blow to the fantasy that men are more powerful than machines.

"The possibilities of synthetic characters have grown enormously, and George will rely much more on those in the new 'Star Wars' movies," Morris went on. "People in rubber suits worked for back then, and added a certain charm, but there is only so much you can do with that in terms of motion. In C.G., there really aren't physical limitations. There are only believability issues, and George's sensitivity grounds that—he decides what works on the screen. Yoda was a Frank Oz puppet in 'Empire,' and that worked for an old, wizened Yoda who could hardly move. But in the prequels we're going to have a sixty-years-younger Yoda, and we'll probably see a synthetic Yoda instead."

COMING TO TERMS

Seabrook organizes his discussion of the *Star Wars* trilogy into four individually subtitled sections or segments: The Summit, The Ranch, The Story, and The Model Shop. As you read each of these sections of "Why Is the Force Still with Us?", we want you to think about how these sections tell their own story but at the same time are part of a larger story. In your journal, we would like you to respond to the following questions: How are these four sections connected or how do they relate to each other? And why do they appear in the order they do? As you re-read Seabrook for a second time mark what you consider to be important places in the text of each section and record these words or passages in your journal under the appropriate subtitle. What kinds of commonplaces do you see which might connect the sections? Are words, ideas, or issues repeated? Are the sections arranged in chronological order? What would happen to the larger story if these sections were moved around?

READING AS A WRITER

Seabrook reads or interprets the *Star Wars* trilogy from multiple viewpoints. Reading from more than one perspective is a way of acknowledging the complexity of a text. Two critical strategies that Seabrook incorporates to make sense of George Lucas' films are biographical or psychoanalytical criticism and archetypal or mythic criticism. In psychoanalytical interpretation, a reader attempts to explain an author's work in terms of events or relationships in the author's life. And in archetypal criticism, a reader examines how an author makes use of cultural myths, important stories which are told and retold as a way of answering crucial human questions or imparting values and lessons important to a culture.

We would like you to reread Seabrook and mark those places in his discussion where he offers either psychoanalytical or archetypal readings of the *Star Wars* trilogy. Record these passages in your journal and then respond to the following questions: As a way of making sense of a text, what's your take on these ways of interpretation? How are they useful? What might be some problems in relying upon these methods of reading?

WRITING CRITICISM

We think one of the intriguing aspects of Seabrook's work is his attempt to understand the longevity and popularity of the *Star Wars* films. The appeal of these films, however, remains a complex issue. Seabrook acknowledges that the *Star Wars* trilogy is set in the future (at least as far as special effects and technology would suggest); yet, at the same time, Seabrook points out that *Star Wars* recalls the past. The story takes place "a long time ago, in a galaxy far, far away." One way to begin to understand the appeal of films like *Star Wars* might be to examine how the past is presented. People in contemporary times often look to the past with a sense of nostalgia, not so much with any desire to return to or live in the past, but to regain something which has been lost because of the pace, rush, or complexity of modern life. Maybe you can recall special family gatherings where older members of your family spoke about their pasts, when things were somehow different, somehow better, when people themselves were somehow different. One might say that we pay a price for technological advances or progress: something is gained, something is lost. With this sense of loss in mind, Seabrook writes that films like *Star Wars* are ways of "getting that unnameable thing back."

For this assignment we would like you to explore the ways in which some popular culture medium helps you get that unnameable thing back, how some popular culture medium set in a specific and particular time past returns something that has been lost in contemporary times and how this medium creates nostalgic interest in a past. Or as an alternative approach, how would you explain the continued appeal of some popular culture medium that recalls a specific past? For example, using Seabrook's discussion as a point of entry, how would you explain the popularity of a television program like *Happy Days* or a film like *American Graffiti*? *Why* might a person collect records, fashions, posters, or other memorabilia from the 1960s? World War II ended over 50 years ago, yet film depictions of that era are still quite popular. Westerns were once a staple of American television programming. Popular historical novels and romance novels are often set in the 19th century. In magazines we see collectibles—toys, plates, dolls—being advertised as being from a specific time past. Why might there be a nostalgic interest in these times past? What is it about life today that might make this past or that past seem so appealing?

Tales from the Cutting-Room Floor

The Reality of "Reality-Based" Television

Debra Seagal

May 6, 1992

Yesterday I applied for a job as a "story analyst" at *American Detective,* a prime-time "reality-based" cop show on ABC that I've never seen. The interview took place in Malibu at the program's production office, in a plain building next door to a bodybuilding gym. I walked past rows of bronzed people working out on Nautilus equipment and into a dingy array of padded dark rooms crowded with people peering into television screens. Busy people ran up and down the halls. I was greeted by the "story department" manager, who explained that every day the show has camera crews in four different cities trailing detectives as they break into every type of home and location to search, confiscate, interrogate, and arrest. (The crews have the right to do this, he told me, because they have been "deputized" by the local police department. What exactly this means I was not told.) They shoot huge amounts of videotape and it arrives every day, rushed to Malibu by Federal Express. Assistants tag and time-code each video before turning it over to the story department.

After talking about the job, the story-department manager sat me in front of a monitor and gave me two hours to "analyze" a video. I watched the camera pan through a dilapidated trailer while a detective searched for incriminating evidence. He found money in a small yellow suitcase, discovered a knife under a sofa, and plucked a tiny, twisted marijuana butt from a swan-shaped ashtray. I typed each act into a computer. It took me forty-five minutes to make what seemed a meaningless record. When I got home this afternoon there was a message on my phone machine from the story-department manager congratulating me on a job well done and welcoming me to *American Detective*. I am pleased.

May 18, 1992

Although we're officially called story analysts, in-house we're referred to as "the loggers." Each of us has a computer/VCR/print monitor/TV screen/headphone console looming in front of us like a colossal dashboard. Settling into my chair is like squeezing into a small cockpit. The camera crews seem to go everywhere: Detroit, New York, Miami, Las Vegas, Pittsburgh, Phoenix, Portland, Santa Cruz, Indianapolis, San Jose. They join up with local police teams and apparently get access to everything the cops do. They even wear blue jackets with POLICE in yellow letters on the back. The loggers scrutinize each hourlong tape second by second, and make a running log of every visual and auditory element that can be used to "create" a story. On an average day the other three loggers and I look at twenty to forty tapes, and in any given week we analyze from 6,000 to 12,000 minutes—or up to 720,000 seconds of film.

The footage comes from handheld "main" and secondary" cameras as well as tiny, wirelike "lock-down" cameras taped to anything that might provide a view of the scene: car doors, window visors, and even on one occasion—in order to record drug deals inside an undercover vehicle—a gear-shift handle. Once a videotape is viewed, the logger creates a highlight reel—a fifteen-minute distillation of the overall "bust" or "case." The tapes and scripts are then handed over to the supervising producer, who in turn works with technical editors to create an episode of the show, each of which begins with this message on the screen: "What you are about to see is real. There are no re-creations. Everything was filmed while it actually happened."

There are, I've learned, quite a few of these reality and "fact-based" shows now, with names like *Cops, Top Cops,* and *FBI: The Untold Stories.* Why the national obsession with this sort of voyeuristic entertainment? Perhaps we want to believe the cops are still in control. The preponderance of these shows is also related to the bottom line: they are extremely inexpensive to produce. After all, why create an elaborate car-chase sequence costing tens of thousands of dollars a minute when a crew with a couple of video cameras can ride around with the cops and get the "real" thing? Why engage a group of talented writers and producers to make intelligent and exciting TV when it's more profitable to dip into the endless pool of human grief?

I've just participated in my first "story meeting" with the supervising producer. He occupies a dark little room filled with prerecorded sounds of police banter, queer voice-over loops, segments of the *American Detective* theme song, and sound bites of angry drug-busting screams ("Stop! Police! Put your hands up, you motherfucker!"). A perpetual cold wind blows from a faulty air duct above his desk. He is tall, lanky, in his fifties; his ambition once was to be a serious actor. His job is to determine what images will be resurrected as prime-time, Monday-night entertainment. He doesn't look miserable but I suspect he is.

There are six of us in the story meeting, the producer, four loggers, and the story-department manager. Each logger plays highlight reels and pitches stories, most of which are rejected by the producer for being "not hot enough," "not sexy." Occasionally, I learned today, a highlight reel is made of a case that is still in progress, such as a stakeout. Our cameramen then call us on-site from their cellular phones during our story meeting and update us on what has been filmed that day, sometimes that very hour. The footage arrives the next morning and then is built into the evolving story. This process continues in a flurry of calls and Federal Express deliveries while the real drama unfolds elsewhere—Pittsburgh or San Jose or wherever. We are to hope for a naturally dramatic climax. But if it doesn't happen, I understand, we'll "work one out."

May 26, 1992

I'm learning the job. Among other tasks, we're responsible for compiling stock-footage books—volumes of miscellaneous images containing every conceivable example of guns, drugs, money, scenics, street signs, appliances, and interior house shots. This compendium is used to embellish stories when certain images or sounds have not been picked up by a main or secondary camera: a close-up of a suspect's tightly cuffed wrists missed in a rush, a scream muffled by background traffic noise. Or, most frequently, the shouts of the cops on a raid ("POLICE! Open the door! Now!") in an otherwise unexciting ramrod affair. Evidently the "reality" of a given episode is subject to enhancement.

Today the story-department manager gave me several videotapes from secondary and lock-down cameras at an undercover mission in Indianapolis. I've never been to Indianapolis, and I figured that, if nothing else, I'd get to see the city.

I was wrong. What I saw and heard was a procession of close-up crotch shots, nose-picking, and farting in surveillance vans where a few detectives waited, perspiring under the weight of nylon-mesh raid gear and semiautomatic rifles. Searching for the scraps of usable footage was like combing a beach for a lost contact lens. The actual bust—a sad affair that featured an accountant getting arrested for buying pot in an empty shoe-store parking lot—was perhaps 1 percent of everything I looked at. In the logic of the story department, we are to deplore these small-time drug busts not because we are concerned that the big drugs are still on the street but because a small bust means an uninteresting show. A dud.

Just before going home today, I noticed a little list that someone tacked up on our bulletin boards to remind us what we are looking for:

DEATH
STAB
SHOOT
STRANGULATION
CLUB
SUICIDE

June 3, 1992

Today was the first day I got to log Lieutenant Bunnell, which is considered a great honor in the office. Lieutenant Bunnell is the show's mascot, the venerated spokesperson. Only two years ago he was an ordinary narcotics detective in Oregon. Today he has a six-figure income, an agent, fans all over the country, and the best voice coach in Hollywood. He's so famous now that he's even stalked by his fans, such as the strange woman who walked into our office a few days ago wearing hole-pocked spandex tights, worn-down spike-heeled backless pumps, and a see-through purse. She'd been on his trail from Florida to California and wanted his home phone number. She was quietly escorted out the door to her dilapidated pickup truck.

At the beginning of each episode, Lieutenant Bunnell sets the scene for the viewer (much like Jack Webb on *Dragnet*), painting a picture of the crime at hand and describing the challenges the detectives face. He also participates in many of these raids, since he is, after all, still a police lieutenant. The standard fare: Act I, Bunnell's suspenseful introduction; Act II, Bunnell leads his team on a raid; Act III, Bunnell captures the bad suspect and throws him in the squad car, etc. The format of each drama must fit into an eleven-minute segment. So it is that although *American Detective* and its competitors seem a long way from *Dragnet, The Mod Squad, The Rookies*, et al.—all the famous old cop shows—they follow the same formula, the same dramatic arc, because this is what the viewers and advertisers have come to expect.

June 10, 1992

The producers are pleased with my work and have assigned me my own beat to log—Santa Cruz in northern California. Having spent several summers there as a teenager, I remember its forests, its eucalyptus and apple orchards. But today, two decades later, I strap on earphones, flip on the equipment, and meet three detectives on the Santa Cruz County Narcotic Enforcement Team.

Dressed in full SWAT-team regalia, they are Brooks, an overweight commander; Gravitt, his shark-faced colleague; and Cooper, a detective underling. The first image is an intersection in Santa Cruz's commercial district. While an undercover pal negotiates with a drug dealer across the street, the three detectives survey an unsuspecting woman from behind their van's tinted windows. It begins like this:

[*Interior of van. Mid-range shot of Commander Brooks, Special Agent Gravitt, and Detective Cooper*]

Cooper: Check out those volumptuous [*sic*] breasts and that volumptuous [*sic*] ass.

Brooks: Think she takes it in the butt?

Cooper: Yep. It sticks out just enough so you can pull the cheeks apart and really plummet it. [*Long pause*] I believe that she's not beyond fellatio either.

[*Zoom to close-up of Cooper*]

Cooper: You don't have true domination over a woman until you spit on 'em and they don't say nothing.

[*Zoom to close-up of Gravitt*]

Gravitt: I know a hooker who will let you spit on her for twenty bucks... [*Direct appeal to camera*] Can one of you guys edit this thing and make a big lump in my pants for me?

[*Zoom to close-up of Gravitt's crotch, walkie-talkie between his legs*]

June 15, 1992

I'm developing a perverse fascination with the magic exercised in our TV production sweatshop. Once our supervising producer has picked the cases that might work for the show, the "stories" are turned over to an editor. Within a few weeks the finished videos emerge from the editing room with "problems" fixed, chronologies reshuffled, and, when necessary, images and sound bites clipped and replaced by old filler footage from unrelated cases.

By the time our 9 million viewers flip on their tubes, we've reduced fifty or sixty hours of mundane and compromising video into short, action-packed segments of tantalizing, crack-filled, dope-dealing, junkie-busting cop culture. How easily we downplay the pathos of the suspect; how cleverly we breeze past the complexities that cast doubt on the very system that has produced the criminal activity in the first place. How effortlessly we smooth out the indiscretions of the lumpen detectives and casually make them appear as pistol-flailing heroes rushing across the screen. Watching a finished episode of *American Detective*, one easily forgets that the detectives are, for the most part, men whose lives are overburdened with formalities and paper-work. They ambush one downtrodden suspect after another in search of mar-ijuana, and then, after a long Sisyphean day, retire into red-vinyl bars where they guzzle down beers among a clientele that, to no small degree, resembles the very people they have just ambushed.

June 23, 1992

The executive producer is a tiny man with excessively coiffed, shoulder-length blond hair. He is given to wearing stone-washed jeans, a buttoned-to-the-collar shirt, and enormous cowboy boots; he also frequently wears a po-lice badge on his belt loop. As I log away, I see his face on the screen flashing in the background like a subliminal advertisement for a new line of L.A.P.D. fashion coordinates. He sits in on interrogations, preens the detectives' hair, prompts them to "say something pithy for the camera." He gets phone calls in surveillance vans and in detective briefing rooms. With a cellular phone flat against his ear, he even has conversations with his L.A. entourage—Lorimar executives, ABC executives, other producers—while he runs in his police jacket behind the cops through ghettos and barrios.

I am beginning to wonder how he has gained access to hundreds of cop cars from California to New Jersey. Clearly the cops don't fear they will be compromised; I see the bonding that takes place between them and the exec-utive producer, who, after a successful raid, presents them with *American Detective* plaques that feature their own faces. Their camaraderie is picked up continuously by the cameras. One of my colleagues has a photograph of our executive producer and Lieutenant Bunnell with their arms around a topless go-go dancer somewhere in Las Vegas; underneath it is a handwritten caption that reads, "The Unbearable Lightness of Being a Cop."

June 25, 1992

Today I logged in several hours of one detective sitting behind a steering wheel doing absolutely nothing. How a man could remain practically immo-bile for so long is beyond my comprehension. He sat and stared out the win-dow, forgetting that the tiny lock-down camera under his window visor was rolling. After an hour, it seemed as though *I* had become the surveillance cam-era, receiving his every twitch and breath through the intravenous-like cir-cuitry that connects me to my machine and my machine to his image. There was, finally, a moment when he shifted and looked directly at the camera. For a second our eyes met, and, flustered, I averted my gaze.

June 26, 1992

Today would have been inconsequential had not the supervising producer emerged from his air-conditioned nightmare and leaned over my desk. "We'll have a crew covering Detroit over the weekend," he said. "Maybe we'll get a good homicide for you to work on." I was speechless. I've never seen a homicide, and I have no interest in seeing one. But I'm working in a place where a grisly homicide is actually welcomed. I am supposed to look forward to this. After work, I prayed for benevolence, goodwill, and peace in Detroit.

June 29, 1992

My prayers must have worked—no Detroit homicide case came in today. That doesn't mean, however, that I'm any less complicit in what is clearly a sordid enterprise. This afternoon I analyzed a tape that features detectives busting a motley assortment of small-time pot dealers and getting them to "flip" on their connections. The freshly cuffed "crook" then becomes a C.I. (confidential informant). Rigged with hidden wires and cameras, the C.I. works for the detectives by setting up his friends in drug busts that lead up the ladder. In exchange for this, the C.I. is promised a more lenient sentence when his day comes up in court. Some of the C.I.s have been busted so many times before that they are essentially professional informants. Ironically, some have actually learned how the game is played by watching reality-based cop shows. This is the case with a nervous teenage first-time pot seller who gets set up and busted in a bar for selling half an ounce of pot. When the undercover cop flashes his badge and whips out his cuffs, a look of thrilled recognition brightens the suspect's face. "Hey, I know you!" he gasps. "You're what's-his-name on *American Detective,* aren't you? I watch your show every week! I know exactly what you want me to do!"

The cops are flattered by the recognition, even if it comes from a teenage crook caught selling pot. They seem to become pals with the C.I.s. Sometimes, however, they have to muscle the guy. The tape I saw today involves a soft-spoken, thirtysomething white male named Michael who gets busted for selling pot out of his ramshackle abode in the Santa Cruz mountains. He's been set up by a friend who himself was originally resistant to cooperating with the detectives. Michael has never been arrested and doesn't understand the mechanics of becoming a C.I. He has only one request: to see a lawyer. By law, after such a request the detectives are required to stop any form of interrogation immediately and make a lawyer available. In this case, however, Commander Brooks knows that if he can get Michael to flip, they'll be able to keep busting up the ladder and, of course, we'll be able to crank out a good show.

So what happens? Hunched in front of my equipment in the office in Malibu, this is what I see, in minute after minute of raw footage:

[*Michael is pulled out of bed after midnight. Two of our cameras are rolling and a group of cops surround him. He is entirely confused when Brooks explains how to work with them and become a confidential informant.*]

Michael: Can I have a lawyer?... I don't know what's going on. I'd really rather talk to a lawyer. This is not my expertise at all, as it is yours. I feel way outnumbered. I don't know what's going on....

Brooks: Here's where we're at. You've got a lot of marijuana. Marijuana's still a felony in the state of California, despite whatever you may think about it.

Michael: I understand.

Brooks: The amount of marijuana you have here is gonna send you to state prison.... That's our job, to try to put you in state prison, quite frankly, unless you do something to help yourself. Unless you do something to assist us....

Michael: I'm innocent until proven guilty, correct?

Brooks: I'm telling you the way it is in the real world.... What we're asking you to do is cooperate...to act as our agent and help us buy larger amounts of marijuana. Tell us where you get your marijuana....

Michael: I don't understand. You know, you guys could have me do something and I could get in even more trouble.

Brooks: Obviously, if you're acting as our agent, you can't get in trouble....

Michael: I'm taking your word for that?...

Brooks: Here's what I'm telling you. If you don't want to cooperate, you're going to prison.

Michael: Sir, I do want to cooperate—

Brooks: Now, I'm saying if you don't cooperate right today, now, here, this minute, you're going to prison. We're gonna asset-seize your property. We're gonna asset-seize your vehicles. We're gonna asset-seize your money. We're gonna send your girlfriend to prison and we're gonna send your kid to the Child Protective Services. That's what I'm saying.

Michael: If I get a lawyer, all that stuff happens to me?

Brooks: If you get a lawyer, we're not in a position to wanna cooperate with you tomorrow. We're in a position to cooperate with you right now. Today. Right now. Today....

Michael: I'm under too much stress to make a decision like that. I want to talk to a lawyer. I really do. That's the bottom line.

[*Commander Brooks continues to push Michael but doesn't get far.*]

Michael: I'm just getting more confused. I've got ten guys standing around me....

Brooks: We're not holding a gun to you.

Michael: Every one of you guys has a gun.

Brooks: How old is your child?

Michael: She'll be three on Tuesday.

Brooks: Well, children need a father at home. You can't be much of a father when you're in jail.

Michael: Sir!

Brooks: That's not a scare tactic, that's a reality.

Michael: That is a scare tactic.

Brooks: No, it isn't. That's reality.... And the reality is, I'm sending you to prison unless you do something to help yourself out....

Michael: Well, ain't I also innocent until proven guilty in a court of law?... You know what, guys? I really just want to talk to a lawyer. That's really all I want to do.

Brooks: How much money did you put down on this property?... Do you own that truck over there?

Michael: Buddy, does all this need to be done to get arrested?...

Brooks: Yeah. I'm curious—do you own that truck there?

Michael: You guys know all that.

Brooks: I hope so, 'cause I'd look good in that truck.

Michael: Is this Mexico?

Brooks: No. I'll just take it. Asset-seizure. And you know what? The county would look good taking the equity out of this house.

Michael: Lots of luck.

[*Commander Brooks continues to work on Michael for several minutes.*]

Michael: I feel like you're poking at me.

Brooks: I *am* poking at you.

Michael: So now I really want to talk to a lawyer now.

Brooks: That's fine. We're done.

[*Brooks huffs off, mission unaccomplished. He walks over to his pals and shakes his head.*]

Brooks: That's the first white guy I ever felt like beating the fucking shit out of.

If Michael's case becomes an episode of the show, Michael will be made a part of a criminal element that stalks backyards and threatens children. Commander Brooks will become a gentle, persuasive cop who's keeping our streets safe at night.

July 1, 1992

Today I got a video to analyze that involves a car chase. It includes the three Santa Cruz cops and a few other officers following two Hispanic suspects at top speed through a brussels-sprout field in the Central Valley. Our camera-men, wearing police jackets, are in one of their undercover vans during the pursuit. (One of them has his camera in one hand and a pistol held high in the other. The police don't seem to care about his blurred role.) When the suspects stop their car and emerge with their arms held high, the detectives bound out of their vans screaming in a shrill chorus ("Get on the ground, cocksucker!" "I'll blow your motherfucking head off."). I watch. Within sec-onds, the suspects are pinned to the ground and held immobile while cops kick them in the stomach and the face. Cooper is particularly angry because his van has bounced into a ditch during the pursuit. He looks down at one of the suspects. "You bashed my car," he complains. "I just got it painted, you motherfucker." With that he kicks the suspect in the head. Our main cam-eraman focuses on the detectives ambling around their fallen prey like hunters after a wild-game safari; a lot of vainglorious, congratulatory back-slapping ensues. Our secondary cameraman holds a long, extreme close-up of a sus-pect while his mouth bleeds into the dirt. "I feel like I'm dying," he wheezes, and turns his head away from the camera. I watch.

This afternoon, in the office, the video drew a crowd. One producer shook his head at the violence. 'Too bad," he said. "Too bad we can't use that foot-age." This was clearly a case of too much reality for reality-based TV. I couldn't help but wonder what the producers would do if these two suspects were beaten so badly that they later died. Would they have jeopardized their own livelihoods by turning over the video to the "authorities"?

September 21, 1992

I'm losing interest in the footage of detectives; now it is the "little people" who interest me, the people whose stories never make it past a highlight reel. I am strangely devoted to them. There is "the steak-knife lady" who waves her rusty weapon in front of a housing project in Detroit. I replay her over and over again. There is something about her: her hysteria, her insistence on her right to privacy, and her flagrant indignation at the cameras ("Get those cameras outta my face, you assholes!"); the way she flails her broken knife in self-defense at a drunk neighbor while her gigantic curlers unravel; the way she consoles her children, who watch with gaping mouths. This woman is *pissed.* She is *real.* Little does she know I'm going to be watching her in Malibu, California, while I sip my morning cappuccino, manipulating her image for my highlight reel. I feel like I'm in the old Sixties movie *Jason and the Argonauts,* in which Zeus and Hera survey the little humans below them through a heavenly pool of water that looks, oddly enough, like a TV screen.

And there is a skinny, mentally disturbed redhead who took in a boyfriend because she was lonely and friendless. Unknown to her, he is selling heroin out of her apartment. But in the eyes of the law she is considered an accomplice. When the cops interrogate her, all she can say about her boyfriend is, "I love him. I took him in because I love him. He's a little bit retarded or something. I took him in." Later she breaks down sobbing. She is terrified that her father will throw her into a mental institution. "I need love. Can't you understand that?" she cries to the policeman who is trying to explain to her why they are arresting her boyfriend. "I need love. That's all I need, sir."

There are, too, the hapless Hispanic families living in poverty, stashing marijuana behind tapestries of the Virgin Mary and selling it to some of the same white middle-class couch potatoes who watch reality-based cop shows. There are the emotionally disturbed, unemployed Vietnam veterans selling liquid morphine because their SSI checks aren't enough to cover the rent. And there are AIDS patients who get busted, their dwellings ransacked, for smoking small quantities of pot to alleviate the side effects of their medication.

In our office the stories of people like these collect dust on shelves stacked with *Hollywood Reporters*, cast aside because they are too dark, too much like real life. I feel overwhelmed by my ability to freeze-frame their images in time-coded close-ups. I can peer into their private lives with the precision of a lab technician, replaying painful and sordid moments. I am troubled that something of their humanity is stored indefinitely in our supervising producer's refrigerated video asylum. Some of their faces have even entered my dreamworld. This afternoon when I suggested that such unfortunates might be the real stars of our show, my boss snapped, "You empathize with the wrong people."

September 28, 1992

This morning I realized that watching hour after hour of vice has begun to affect me. After a raid, when the detectives begin to search for drugs, money, and weapons while our cameras keep rolling, I find myself watching with the intensity of a child foraging through a grassy backyard for an exquisitely luminous Easter egg. The camera moves through rooms of the unknown suspect as the detectives poke through bedrooms with overturned mattresses and rumpled, stained sheets, through underwear drawers and soiled hampers; into the dewy, tiled grottoes of bathrooms, past soap-streaked shower doors and odd hairs stuck to bathtub walls, clattering through rows of bottles, creams, tubes, and toothbrushes, their bristles splayed with wear. The exploration continues in kitchens, past half-eaten meals, where forks were dropped in surprise moments earlier, past grime-laden refrigerators and grease-pitted ovens, past cats hunched frozen in shock, and onward, sometimes past the body of a dog that has recently been shot by the police, now stiffening in the first moments of rigor mortis.

In the midst of this disarray the police sometimes find what they are so frantically looking for: abundant stacks of $100 bills stuffed in boots, behind secret panels and trap doors; heroin vials sealed in jars of cornmeal stashed in the dank corners of ant-infested cupboards; white powders in plastic Baggies concealed behind moldy bookshelves; discarded hypodermic needles in empty, economy-size laundry-detergent boxes; and thin, spindly marijuana plants blooming in tomato gardens and poppy fields. And, finally, on a lucky day, the guns: the magnums, automatics, shotguns, machine guns, and, in one case, assault rifles leaned against walls, their barrels pointed upward.

I feel as though my brain is lined with a stratum of images of human debris. Sitting at home in my small bungalow, I have begun to wonder what lurks behind the goodwill of my neighbors' gestures, what they are doing behind their porches and patios.

September 30, 1992

Today was stock-footage day. I spent ten hours finding, cutting, and filing still-shots of semiautomatic rifles and hypodermic needles. I am starting to notice signs that I am dispirited and restless. I spend long moments mulling over camera shots of unknown faces. Today I took my lunch break on the Malibu pier, where I sat transfixed by the glassy swells, the kelp beds, and minnows under the jetty. I know I can't go on much longer, but I need to pay the rent.

October 1, 1992

I've just worked through a series of videos of the Las Vegas vice squad as they go on a prostitute rampage with our cameramen and producers. Pulling down all-nighters in cheesy motel rooms, the detectives go undercover as our camera crew, our producers, and some of the detectives sit in an adjacent room, watching the live action through a hidden camera. It is, essentially, a voyeur's paradise, and definitely X-rated. The undercover cops' trick is to get the call girls into a position where they are clearly about to accept money for sexual acts. The scam goes something like this: "Hi, I'm John. Me and my buddies here are passing through town. Thought you gals might be able to show us a good time..."

"What did you have in mind?" they ask. The detectives respond with the usual requests for blow jobs. Maybe the undercover cops ask the girls to do a little dancing before getting down to real business. They sit back and enjoy the show. Sometimes they even strip, get into the motel's vibrating, king-size bed, and wait for just the right incriminating moment before the closet door bursts open and the unsuspecting woman is overwhelmed by a swarm of detectives and cameramen.

"He's my boyfriend!" many insist as they hysterically scramble for their clothes.

"What's his name?" the cops respond while they snap on the cuffs.

"Bill. Bob. Uh, John..."

It doesn't matter. The police get their suspect. The camera crew gets its footage. The cameras keep on rolling. And what I see, what the viewer will never see, is the women—disheveled, shocked, their clothes still scattered on musty hotel carpets—telling their stories to the amused officers and producers. Some of them sob uncontrollably. Three kids at home. An ex who hasn't paid child support in five years. Welfare. Food stamps. Some are so entrenched in the world of poverty and pimps that they are completely numb, fearing only the retribution they'll suffer if their pimps get busted as a result of their cooperation with the cops. Others work a nine-to-five job during the day that barely pays the rent and then become prostitutes at night to put food on the table. Though their faces are fatigued, they still manage a certain dignity. They look, in fact, very much like the girl next door.

I can't help but see how each piece of the drama fits neatly into the other: one woman's misery is another man's pleasure; one man's pleasure is another man's crime; one man's crime is another man's beat; one man's beat is another man's TV show. And all of these pieces of the drama become one big paycheck for the executive producer.

October 5, 1992

Today the executive producer—in the flesh, not on tape—walked into the office and smiled at me. I smiled back. But I was thinking: one false move and I'll blow your head off.

October 9, 1992

It would seem that there could not be any further strangeness to everything that I've seen, but, in fact, there is: almost all of the suspects we film, including the prostitutes, sign releases permitting us to put them on TV. Why would they actually want to be on TV even when they've been, literally, caught with their pants down? Could it be because of TV's ability to seemingly give a nobody a certain fleeting, cheap celebrity? Or is it that only by participating in the non-reality of TV can these people feel *more* real, more alive? I asked around to understand how the release process happens.

Usually a production coordinator—an aspiring TV producer fresh out of college—is assigned the task of pushing the legal release into the faces of overwhelmed and tightly cuffed suspects who are often at such peak stress levels that some can't recognize their own faces on their driver's licenses. "We'll show your side of the story," the production coordinator might say. Sometimes it is the police themselves who ask people to sign, suggesting that the cameras are part of a training film and that signing the form is the least of their present concerns. And to anyone in such a situation this seems plausible, since the entire camera crew is outfitted with police jackets, including the executive producer, who, with his "belt badge," could easily be mistaken for a cop in civilian attire. And, clearly, many of those arrested feel that signing anything will help them in court. In the rare event that a suspect is reluctant to sign the release, especially when his or her case might make for a good show, the *American Detective* officials offer money; but more frequently, it seems, the suspect signing the release form simply doesn't adequately read or speak English. Whatever the underlying motive, almost all of the arrested "criminals" willingly sign their releases, and thus are poised—consciously or not—-to participate in their own degradation before the American viewing public.

October 16, 1992

Today I saw something that convinced me I may be lost in this netherworld of videotape; I did, finally, get a homicide. The victim lived in Oregon and planned to save up to attend Reed College. She was a stripper who dabbled in prostitution to make ends meet. On the tape the cops find her on her bed clutching a stuffed animal, her skull bludgeoned open with a baseball bat. A stream of blood stains the wall in a red arc, marking her descent just three hours earlier.

The guy who killed her was a neighbor—blond, blue-eyed, wore a baseball cap, the kind of guy you'd imagine as the head of a Little League team, or a swim coach. He has that particularly American blend of affability, eagerness, and naiveté. When the cops ask him why he bludgeoned her repeatedly after clubbing her unconscious with the first stroke, he replies, "I don't know. I don't really know."

She was Asian, but you would never have known it from what was left of her. What one sees on the tape is that bloody red stain on the wall. We never know why he killed her. We never really know who she was. But it doesn't really matter. She is "just another prostitute." And she will be very good for the show's ratings.

October 19, 1992

This morning I explained my feelings to my boss. I said I "didn't feel good" about the work and had decided to quit. He understood, he said, for he'd once had certain ideals but had eventually resigned himself to the job.

Before departing, I asked a colleague if he was affected by the grief and vice on our monitors. "They're only characters to me," he replied. I noted this quietly to myself, and, with barely a good-bye to my other co-conspirators, I slipped out of the *American Detective* offices into the noon blaze of the California sun, hoping to recover what it is I've lost.

Editors' Note: American Detective *was canceled last summer by ABC, despite good ratings.*

COMING TO TERMS

As you read "Tales from the Cutting-Room Floor," something you may have overlooked is that Debra Seagal's article takes the form of a series of individually dated diary entries. We'd like you to re-read these diary entries and respond to the following questions in your journal. How do these entries tell a story of Seagal's changing attitudes toward her new job? The television professionals with whom she works? The criminals she logs in on videotape? Or reality-based television itself?

One way to address this question of change is to select one or two diary entries which you read as offering Seagal's first impressions on these subjects. How would you characterize these impressions? Then select entries which indicate some change in attitude or perspective? What has changed? Is the change dramatic, occurring in one entry in particular? Or is the change gradual? What do you make of Seagal's concluding line: "I slipped out of the *American Detective* offices into the blaze of the California sun, hoping to recover what it is I've lost." In what ways might these changes be read as a narrative of loss? What is it that she has lost?

Finally, as another way of coming to terms with her article, we'd like you to consider the significance of Seagal's choice of a diary format here to explore the ways in which television, especially shows like *American Detective,* collect, edit, and package "reality" for the viewing audience. Considering Seagal's subject here, what is gained by using the diary format? Instead of some other approach? In the traditional sense, a diary is a private record—a place of secrets, insights, dreams, experiences, and confessions. One might say that a diary has its own sense of "reality," and that's a private reality. Diaries are often sold with locks and keys: to read another's diary is a breach of trust. What are the consequences of making this private text public? publishing it? In what ways might writing (rewriting and editing) for an audience alter the reality of a diary, specifically Debra Seagal's diary?

READING AS A WRITER

Maybe it is true that *the camera doesn't lie,* but Debra Seagal would have us give serious consideration to a qualifying proposition: *The camera doesn't necessarily tell the truth either*—or at least, a complete version of it. As such, we think an important aspect of Seagal's project here is to expose and make us fully aware of just how much work—setting up multiple camera shots, logging and editing videotapes, splicing in stock sound recordings for authentic effects, pitching story lines to producers, for example—occurs behind the scenes of an episode of *American Detective* to produce "reality." Media critics sometimes speak of the belief that the television camera captures objective reality or opens a clear, unadulterated window on the real as the "transparency fallacy." Because of the camera angle, when it's turned off or on, or how the film is finally edited, the camera doesn't really offer a transparent view of anything.

We think that Seagal's discussion of how reality-based television might be manufactured, manipulated, or controlled lends itself in significant ways to a related set of issues in writing. First, writers can also be seen as promoting a specific version of reality by the language they choose to represent it. And secondly, to see a writer's use of language as "transparent," as in the case of the television camera, might be another illustration of the "transparency fallacy." Discussions of the degree of objectivity and subjectivity to be found in a writer's use of language have often focused on the *denotation* (the primary, dictionary definition of a word) and *connotation* (some kind of secondary meaning, often positive or negative) of words. Certainly some words are "loaded" or "colored" with additional meaning beyond their denotation. For example, to refer to a group of politicians as a *regime* means something quite different from calling the same group an *administration.* We think, however, that language use is not quite that simple or obvious. And the stakes are too high. Concluding that a piece of writing is objective or subjective by counting denotations and connotations misses a critical point: even a careful denotative choice of language is still a choice, among other possible choices, to represent something, and this selection encompasses interpretation.

We'd like you to put aside for a while a way of reading which either sees a writer's use of language as either objective or subjective, loaded or neutral. Rather we'd like you to reread Seagal's text and mark places in her text where Seagal represents aspects of her investigation, such as her descriptions of her co-workers at *American Detective,* the crime scenes she logs or tapes, the studio offices at work, the police she interacts with or views on screen, or even herself. Select one of these aspects of Seagal's work and record the ways in which she describes this aspect from passages throughout her diary entries. Underline words or phrases that strike you as important. Then answer the following questions in your journal. How does Seagal's descriptions of this one aspect suggest a consistency in her point of view or attitude? What words or phrases would you point to as evidence for your interpretation? Why? How would you characterize this point of view? Or do Seagal's descriptions reveal a shifting perspective or a changing point of view? If so, where does this shift of change begin and what words or phrases would you point to as indicating this shift? Finally, how would you describe this initial point of view and its revised or changed versions?

WRITING CRITICISM

What you are about to see is real. There are no recreations. Everything was filmed while it actually happened. —Preamble to each episode of *American Detective*

> DEATH
> STAB
> SHOOT
> STRANGULATION
> CLUB
> SUICIDE
> —Story analysts' bulletin board, *American Detective*

The two headnotes above can be read as media presenting strikingly different versions of "reality" or as Debra Seagal would have it, "the reality of reality-based television." We think Seagal's insider investigation of this popular genre of television cop shows raises a number of crucial issues concerned with media realism. She challenges us to rethink phrases commonly and casually used (but passionately debated) in discussions of media—realism, reality-based, realistic, true-to-life, etc.—and the degree to which reality can be manipulated. To one degree or another, one version of the history of media and popular culture can be read as a continuing narrative of how either artist and audience "work" together to create an acceptable version of reality (for example, a mutually clear understanding of what lines not to cross, what's too graphic, what subjects are just taboo) or how the audience and artist clash over what should or should not be represented in film, radio, television, magazines, songs, online, or even in comic books.

What to represent and to what degree to represent it realistically are questions that continue to play a dominant role in the critical conversation about media and popular culture. Debates on media realism—its value, problems, and dangers—are contentious and they seem to be never-ending. Critical moments of condemnation in which media and popular culture texts are assailed as sensationalized, exploitative, too graphic, degrading, immoral, pornographic, or prurient seem to co-exist or share history with moments of critical praise in which the same texts are proclaimed as cutting-edge, socially and artistically responsible, heralding new attitudes towards once forbidden subjects, or ushering in a new realism. This cultural ambiguity is constantly fueled by often oppositional stances based in religion, morality, artistic freedom, and constitutional rights of freedom of speech and the press. Under attack, the producers of controversial media and popular culture texts—whether they be journalists, graphic artists, film makers, talk show hosts, singers, television directors, etc.—seeing themselves as cultural messengers, will often invoke the expression *don't kill the messenger* just because the message is graphic and disturbing.

We'd like you to write an essay investigating the issues of realism in media or popular culture by showing how some text or medium succeeds at achieving an acceptable level of realism for its audience. To say a song, a magazine, talk show, sitcom or drama, film, comic book, online chat room, or romance novel, for example, is realistic, true-to-life, or reality-based (and leave it at that) would certainly be problematic. These texts or media, if read within the context of "Tales from the Cutting-Room Floor," might offer some rather varied approaches to realism. Seagal's work offers us an important point of entry into the critical conversation of media realism and provides a framework from which to explore how the appearance of reality in media and popular culture might be manipulated to achieve a certain effect and insure acceptance by an audience.

To return for a moment to the two headnotes we've cited from Seagal's article, it is the sense of the second headnote we'd like you to explore in this writing project: that is, your recognition and textual proof that this medium or that text is somehow more than a transparent window on reality. What would you argue is the reality formula for the text or medium you have selected? What is its code of self-regulation as far as presenting reality to an audience? In what ways has this text or medium censored itself? What might be its list of *de facto* "do's," "don't's," or "maybe's" as far as realism is concerned?

You will need to select a media or popular culture text that will allow you to describe its approach to realism in relationship to an audience. As part of this process, you'll need to explore this "consensual" level of realism in terms of the values, beliefs, and interests of this audience. We think it's important to remember that your reading of a cultural text here is not necessarily an endorsement of its approach to realism. And it's not necessary that you select a controversial medium or text. In discussions of media realism, more often than not, representations of sex and violence have garnered the most attention. We think, however, that your investigation of media realism might include other representations as well, such as representations of family, work, school, the professions, marriage, language, race or gender, religion, the justice system, and friendship. Finally, as part of your project, you might make use of other writers in *Media Journal*—like Alexander Nehamas, Benjamin DeMott, Eric Foner and John Sayles, M. Kadi, or Alan Light, among others—who comment on the uses of realism in media and popular culture.

Street Corners in Cyberspace

Andrew L. Shapiro

You probably didn't notice, but the Internet was sold a few months ago. Well, sort of: The federal government has been gradually transferring the backbone of the U.S. portion of the global computer network to companies such as I.B.M. and M.C.I. as part of a larger plan to privatize cyberspace. But the crucial step was taken on April 30, when the National Science Foundation shut down its part of the Internet, which began in the 1970s as a Defense Department communications tool. That left the corporate giants in charge.

Remarkably, this buyout of cyberspace has garnered almost no protest or media attention, in contrast to every other development in cyberspace—particularly Senator James Exon's proposed Communications Decency Act, which would criminalize "obscene, lewd, lascivious, filthy, or indecent" speech on computer networks. Yet issues of ownership and free speech are inextricably linked. Both raise the vexing question of what role—if any—government should play in cyberspace and, consequently, of what this new frontier will become.

The chorus of opposition to Exon's misguided proposal is right on; the bill deserves to be scrapped. But as cyberspace becomes privatized and commercialized, we should also be skeptical of the laissez-faire utopianism of many of Exon's critics. They say that cyberspace is a bastion of free expression and that users can regulate themselves. "No government interference!" is their rallying cry. In the context of censorship, this sounds right. But in the context of ownership, it is wrong. Speech in cyberspace will not be free if we

"Street Corners in Cyberspace," by Andrew L. Shapiro. Reprinted with permission from the July 3, 1995 issue of *The Nation*.

allow big business to control every square inch of the Net. The public needs a place of its own.

This seems simple enough. And yet the issue of who owns cyberspace has been overlooked because it's difficult in the abstract to see what's at stake here. So let's get concrete: Consider two models of cyberspace that represent what total privatization deprives us of and what it leaves us with.

In the first model—this is what we're being deprived of—you use a computer and modem to go on-line and enter a virtual world called Cyberkeley. As you meander down the sidewalk, you find a post office, libraries and museums, shopping malls full of stores, and private clubs that service a limitless variety of clientele, from those who want spiritual guidance, tips on gardening or legal advice to those with a penchant for live sex or racist hatemongering. You also encounter vibrant public spaces—some large like a park or public square, others smaller like a town hall or street corner. In these public forums some people are talking idly, others are heatedly debating social issues. A few folks are picketing outside a store where hard-core pornography is sold, others are protesting the post office's recently increased mail rates and one lone activist outside the Aryan Militia's hangout hands out leaflets urging racial unity. Most people are just passing through, though you and they can't help but take notice of the debaters, the demonstrators, even the leafleter.

In the second cyberspace model—which is what we're getting—you enter an on-line world called Cyberbia. It's identical to Cyberkeley, with one exception: There are no spaces dedicated to public discourse. No virtual sidewalks or parks, no heated debate or demonstrators catching your attention, no street-corner activist trying to get you to read one of her leaflets. In fact, you can shape your route so that you interact only with people of your choosing and with information tailored to your desires. Don't like antiabortion activists, homeless people, news reports about murders? No problem—you need never encounter them.

Cyberbia is tempting. People can organize themselves in exclusive virtual communities and be free of any obligation to a larger public. It is, perhaps, the natural result of a desire for absolute free choice; customization and contract. But, at least for now, cyberspace must be treated as an extension of, rather than an alternative to, the space we live in. Consequently, it should be clear that Cyberbia—like suburbia—simply allows inhabitants to ignore the problems that surround them off-line. In Cyberkeley, by contrast, people may be inconvenienced by views they don't want to hear. But at least there are places where bothersome, in-your-face expression flourishes and is heard. These public forums are essential to an informed citizenry and to pluralistic, deliberative democracy itself.

Unfortunately, cyberspace is shaping up to be more like Cyberbia than Cyberkeley. That's because the consensus among on-line boosters—from the terminally wired hackers to the cyberwonks at the Electronic Frontier Foundation to Newt Gingrich, Al Gore and other Third Wave lawmakers—is that all cyberspace should be privately owned and operated. While their fears of government abuse or inefficient centralization may be legitimate, presenting the choice as one between totalitarian control and a total absence of publicly owned space is misleading. These extreme alternatives prevent us from moving toward something like Cyberkeley—a mode of cyberspace that is mostly private, but which preserves part of this new domain as a public trust, a common space dedicated to citizens' speech. Without this hybrid vision, it is unlikely that we will realize the democratic possibilities of this new technology.

All this becomes readily apparent if one steps back from the cyberspace framework and contemplates how our anemic public discourse came to be. By the 1960s, crucial victories for freedom of expression had been won in the courts against government attempts to suppress labor activists, Communists, civil rights activists and war protesters. At the heart of these victories was the public forum doctrine, which minimizes government control of speech in areas such as sidewalks and parks, disallowing restrictions based on content or viewpoint. As the Supreme Court said in 1939, these special locations were "held in trust for the use of the public...for purposes of assembly, communicating thoughts between citizens, and discussing public questions."

From the 1960s to the present, as Owen Fiss of Yale Law School has argued, speech on the street corner has become increasingly silent as a result of two concurrent trends: the rise of electronic media and the privatization of spaces that had previously been public—a prime example being the center of commerce, now the shopping mall. For a brief time, with the F.C.C's fairness doctrine for broadcasters and a 1968 Supreme Court decision that granted labor picketers access to a privately owned mall, there was reason to believe that speech might remain diverse and unfettered in the new locales. But the Supreme Court slowly and surely chipped away at what Justice William Brennan called "robust public debate." It limited citizen access to radio and television, and it allowed speech to be restricted not only in private shopping malls but also in publicly owned spaces, such as airports and post offices [see David Cole, "In Your Space," March 14, 1994]. Property owners and even municipalities acting in "private" capacities were able to use the First Amendment to exclude dissenting voices by arguing that they should not have to associate with speech with which they disagreed.

Given this history, it should be clear how important public forums in cyberspace could be—as a way of keeping debate on-line robust *and* as a direct remedy for the dwindling number of such spaces in our physical environment. A remarkable incident shows further why this is so.

In 1990 the on-line service Prodigy started something of a revolt among some of its members when it decided to raise rates for those sending large volumes of e-mail. Protesters posted messages claiming they were being penalized for speaking frequently; they sent e-mail to Prodigy's on-line advertisers threatening a boycott. In response, Prodigy not only read and censored their messages, it summarily dismissed the dissenting members from the service. A spokesman for the company, which is a joint venture of Sears and I.B.M., wrote unapologetically in a *New York Times* opinion piece that the company would continue to restrict speech as it saw fit, including speech criticizing the company.

This example, extreme as it is, demonstrates the difference between Cyberkeley, which has public forums, and Cyberbia, which does not. In the latter, there is no way for the aggrieved Prodigy members to picket the on-line service in an area that its patrons will see (nor is there any high authority to which they can appeal); thus the company has no incentive to retrain from suppressing dissent. Sure, the dissenters can open their own shop, but they'll probably be lost in some distant corner of cyberspace. By contrast, in Cyberkeley, the protesters would have access to the virtual sidewalk outside Prodigy; they would be able to protest the company's rate hike—and the fact that they were kicked out simply for airing a contrarian view—in a way that Prodigy members and nonmembers alike would hear.

Undoubtedly, some will argue that the Prodigy incident was an anomaly and that cyberspeech is already as free as it needs to be. But like the protests of the Prodigy dissidents, all speech in cyberspace is, in three fundamental ways, less free than speech in a traditional public forum.

First, cyberspeech is expensive, both in terms of initial outlay for hardware and recurring on-line charges. For millions of Americans, this is no small obstacle, especially when one considers the additional cost of minimal computer literacy. While it is true that speech becomes cheaper once you've gained access to cyberspace—sending a message to 1,000 people costs little more than sending it to one—the initial threshold cost of this entry remains prohibitively high; the specter of information haves and have-nots is already upon us. This problem might be alleviated somewhat through subsidized access to cyberspace via public computer terminals (which have been established with success in Santa Monica and other cities), lower connection costs for low-income users (as with telephone service) and even something as idiosyncratic as Speaker Gingrich's tax breaks for laptops.

Second, speech on the Net is subject to the whim of private censors who are not accountable to the First Amendment. More and more, travelers in cyberspace are using commercial on-line services such as America Online and Compuserve, which, like Prodigy, have their own codes of decency and monitors who enforce them. Even those who prefer the more anarchic Usenet discussion groups are subject to regulation by self-appointed system operators and moderators. Since these censors are private agents, not state actors, disgruntled users have no First Amendment claim. Fortunately, most monitors, even at commercial on-line services, are not heavy-handed. But this is not always the case and, more important, there is no legal way to insure that it will be.

Third, speech in cyberspace can be shut out by unwilling listeners too easily. With high-tech filters, Net users can exclude all material from a specific person or about a certain topic. This feature may protect children from inappropriate speech, but adults should not be able to steer clear of "objectionable" views, particularly marginal political views, so easily.

Together, these three points demonstrate why on-line speech is less free than it seems and, again, why Cyberkeley is preferable to Cyberbia. They also show why the cyberwonks' solution to Net regulation—a laissez-faire print model that treats all users as writers rather than, say, speakers—is inadequate: We're either paying to publish in mass-circulation periodicals where editors are free to censor us or we're writing pamphlets no one knows about because there's no public space in which to distribute them.

So how do we reverse the onslaught of commercialization and move from Cyberbia to Cyberkeley? Congress could recognize the important public function of private on-line services and thus require them to save a place for dissenting speech, just as the Supreme Court required a privately owned "company town" to do in 1946. Eli Noam of Columbia University suggests as much when he says that computer networks "become political entities" because of their quasi-governmental role—taxing members, establishing rules, resolving conflicts. However, this remedy would probably be as unpopular today among courts and legislators as newspaper right-of-reply statutes and the F.C.C. fairness doctrine.

A more appropriate solution might be for Congress and state and local governments to establish forums in cyberspace dedicated explicitly to public discourse. This entails more than just setting up a White House address on the Internet or even starting the virtual equivalent of PBS or NPR. These public forums must be visible, accessible and at least occasionally unavoidable—they must be street corners in cyberspace.

Through regulation or financial incentives, Congress might be able to get users of commercial and academic on-line services to pass through a public gateway before descending into their private virtual worlds. This gateway could provide either a comprehensive list or a representative sample (depending on technological capacities) of issues that are being discussed in public cyberforums, which the user can—if she chooses—enter and exit with a simple click of the mouse. The entry point might also register how many citizens are speaking in each forum and how long the discussion there has been going on.

Though Net users might initially see the public gateway as an imposition, it really isn't different from the burden of exposure that they accept in their everyday lives. To get to a store or private club one must, at least momentarily, traverse the square or travel the street. As on a real public sidewalk, an actual pedestrian can try to ignore what's there and pass right by. Most probably will. But some will be enticed to listen and even to argue. More important, all will have at least the opportunity to hear truly free speech outside the control of private interests—and, like expression in any public forum, also from government censorship.

Even if Cyberkeley is technically feasible, skeptics of regulation from the left and right will question whether it is legal or even good policy. And this is precisely where the free choice between libertarianism and authoritarianism should be abandoned. In a liberal democracy, the people have a right to demand that government promote speech in a content neutral way, particularly speech that is drowned out by the voices of the moneyed and powerful.

Admittedly, this state obligation to make speech available that otherwise would be unheard may seem unfamiliar to First Amendment absolutists who take it as an article of faith that, in all matters pertaining to freedom of expression, government should simply stay out of the way. But in fact, just as the state protects citizens from unfair market conditions, it also has a role to play when the marketplace of ideas fails and there is outright domination of some views. This is the same reason our government gives—and should give more, as European nations do—grants to marginal artists, postal subsidies to small magazines of opinion and free use of cable channels to community organizations. As the Supreme Court said last year in *Turner Broadcasting v. FCC*, "Assuring that the public has access to a multiplicity of information sources is a governmental purpose of the highest order, for it promotes values central to the First Amendment."

Cyberkeley might also solve other Net snafus. For example, controversial political discussions on-line are sometimes destroyed by a phenomenon called "spamming," in which an individual endlessly replicates a message in order to silence his opponents. Open discourse is effectively shut down. In a public forum, government cannot censor speech, but it can impose minimal "time, place and manner" restrictions, particularly in order to maintain the fairness of public debate. Thus, just as the state might prohibit use of a 100-decibel loudspeaker at 2 A.M. in Times Square, it might also turn down the volume on those who drown out other speech by spamming.

More mundane problems might also be ameliorated. For example, as political campaigns buy time on commercial on-line services, public cyberforums could take on an important role for candidates with little money. While Bob Dole is chatting with America Online subscribers, his grass-roots opponents should be able to set up a soapbox on the virtual sidewalk outside.

If cyberspace is deprived of public forums, we'll get a lot of what we're already used to: endless home shopping, mindless entertainment and dissent-free chat. If people can avoid the unpalatable issues that might arise in these forums, going on-line will become just another way for elites to escape the very nonvirtual realities of injustice in our world. As the wired life grows exponentially in the coming years, we'll all be better off if we can find a street corner in cyberspace.

COMING TO TERMS

Shapiro proposes in his essay two different ways of imagining what "cyberspace" might look, feel, and work like: Cyberkeley and Cyberbia. Draw a line down the middle of a page in your journal. On one side, list the various terms, images, and ideas Shapiro associates with Cyberkeley; on the other, list those he associates with Cyberbia. In trying to extend Shapiro's sense of differences between these two visions of cyberspace, what other terms, thoughts, or experiences concerning the Internet (that you've personally experienced, heard others speaking about, or read about) would you include in the cyber region of Cyberkeley? Cyberbia? Finally, in drawing upon the two lists you've created, write a few paragraphs in which you first point out the major differences between these two ways of imagining cyberspace. We think that Shapiro is fashioning an opportunity for his readers to make a choice between these two ways of imagining cyberspace. The choice is not a simple one, however. Review the two lists you've created to gain a sense of the advantages and disadvantages of "living" in each cyber community. What is at stake in choosing one view of cyberspace over the other one? In other words, what is gained or lost in selecting either way of imagining cyberspace?

READING AS A WRITER

Shapiro organizes his piece around two extended and contrasted metaphors: one that pictures cyberspace as an urban neighborhood and another that views it as a suburb. In doing this work with metaphor, Shapiro is drawing upon another technique used by experienced writers: fashioning a critical argument through *analogy.* An analogy builds upon the similarity between like features of two things and works on more than one level. The analogy here might be stated as follows: The differences (the quality of cyberlife) in how we might use the Internet and Web are similar to the differences between the quality of life in living in a city and living in the suburbs. As such, Shapiro is making use of a well-established, long-standing comparison in U.S. culture. We think that Shapiro makes a deft move here as a writer: by creating this particular analogy, Shapiro takes an abstract concept like *cyberspace* and makes it concrete and knowable by giving it physical dimensions. But *urban* and *suburban* are more than locations or geographic markers; they are metaphors and they are cultural symbols, which have undergone various interpretations and shifts of meaning as U.S. culture and society have also undergone demographic changes. To live in a city or to live in the suburbs means different things at different times in history. To make the analogy work, Shapiro must "control" or stabilize the meaning of *urban* and *suburban.* We'd like you to consider Shapiro's success here by answering a set of questions in your journal. First, what do you associate with living in a city? Living in the suburbs? What aspects of city life does Shapiro emphasize in creating his image of Cyberkeley? Why does Shapiro give this name to his conception of a cybercity? Why, for example, doesn't Shapiro invite us to "live" in Cybutte (Butte, Montana) or Cyburbank (Burbank, California) or even Cyboston? What might be the cultural and historical significance in the name Berkeley, California? Or the University of California at Berkeley? And how does this significance become important to Shapiro's critical argument? Conversely, what images of the suburbs does Shapiro employ in building his vision of Cyberbia, and which ones does he gloss over? Are there ways of reworking Shapiro's metaphors that might make Cyberbia seem preferable to Cyberkeley? How does Shapiro guard against this possible revision?

WRITING CRITICISM

We would like you to undertake a project in which you begin to assess some of the strengths and limits, the possibilities and perils, of electronic discourse on the Internet and the Web. What we have in mind here, however, is an investigation that would not take the form of a traditional academic essay, but rather take the form of a written dialogue or transcript from an electronic discourse involving three principal *online* participants: Andrew Shapiro, M. Kadi, and you.

First, you will need to read (or re-read) M. Kadi's article "The Internet Is Four Inches Tall" found elsewhere in *Media Journal.* You'll need to take some notes from Kadi's discussion which you feel represent important aspects of Kadi's critical argument. (As a place to begin, you might respond to some of the suggested questions in the "Coming to Terms" and "Reading as a Writer" sections.) Like Shapiro, Kadi examines the Internet and the Web from a position of complexity. We think neither writer is intellectually satisfied with a reductive *pro* and *con* analysis. And, also like Shapiro, Kadi works towards giving her discussion of the abstract concept of cyberspace some kind of physical or concrete grounding: she introduces a typical online user as "J. Individual" and reigns in the incredible vastness of the Internet and Web in her title, "The Internet Is Four Inches Tall."

From your reading of both Shapiro and Kadi, you'll need to choose an issue or subject concerned with electronic discourse which strikes you as interesting and significant or controversial and one to which Kadi and Shapiro have something to contribute in their discussions. These contributions might involve similar or different perspectives or points of view on the subject or issue. Issues or subjects might include, but are not limited to, education and the Internet, writing and the Internet, democracy and the Internet, cultural diversity and the Internet, or political discussions and the Internet. Once you've selected an issue (we think it might be a good idea to limit your investigation to one or maybe two), you'll need to take some notes on places in Kadi's and Shapiro's texts where you think these authors are speaking about the same subject or issue. After this work, you should have a pretty good sense of how Kadi and Shapiro might "talk" to each other on an issue.

Next, you'll need to do some work or research to prepare yourself to enter this critical conversation from the position of actual *experience* on the Internet, observations on an actual online forum. So you'll need to locate a discussion group or network on a subject that not only interests you but also can be said to connect in some way to one of the issues or subjects Kadi and Shapiro have discussed. You'll need to monitor this online conversation closely and to take some notes in response to the following questions: Who participates in this discussion? What do they tell or reveal about themselves in their postings? How? How would you characterize the "discursive styles" of the various participants? In what ways do their postings resemble speech? Writing? Both? Or neither? Whose postings tend to get the most responses? Whose get ignored? Can you speculate why? Are there things that electronic discourse helps its users do that they are unable (or less able) to do in speech or writing? Would you describe their online conversation as a "robust political debate" (Shapiro) or would you be more inclined to describe this conversation as an illustration of the "wide, wide world of online existence only insofar as our already existing interests and prejudices dictate" (Kadi)? Or would you choose another way to describe the quality of the conversation? Would you say that the online user's discourse takes place in anywhere like "Cyberkeley" or "Cyberbia"? If not, what metaphor would you use instead? Do any of the online users recall Kadi's representation of "J. Individual"? Does your monitoring of this online conversation lead you to the conclusion that the Internet really is only four inches tall? What has struck you as new or different about electronic discourse? What have you found to be appealing or troubling (or both) about it?

Finally, we'd like you to pull all of this work together and create a format in which you appear to be having a conversation with Andrew Shapiro and M. Kadi and in which you contribute to the discussion based on your experiences and observations with electronic discourse on the Internet, by asking questions of, exchanging ideas with, and contesting or supporting what Kadi and Shapiro have to say. Your written response to this assignment might take the form of a facsimile or hard copy version of an online forum, with postings attributed to Kadi, Shapiro, and yourself. Or you might create a script from a dialogue between these participants. (You might take a look at "A Conversation Between Eric Foner and John Sayles" elsewhere in *Media Journal* to become familiar with this kind of format.) To give a sense of organization to this written text, you might focus your initial efforts on drafting some responses between Kadi and Shapiro on the specific electronic discourse subject or issue you've chosen and think about, based on your monitoring of the Internet, places where you might interrupt and add something. You don't have to quote the authors, but you can, of course. But you probably will find it easier to paraphrase or summarize or extend Kadi's and Shapiro's arguments. You should see your intellectual responsibility here, in any case, as fairly representing these authors. Another approach here would put you in the position of the moderator of an online forum, where you ask a similar set of questions of both authors.

Sex, Lies
and Advertising

Gloria Steinem

About three years ago, as *glasnost* was beginning and *Ms.* seemed to be ending, I was invited to a press lunch for a Soviet official. He entertained us with anecdotes about new problems of democracy in his country. Local Communist leaders were being criticized in their media for the first time, he explained and they were angry.

"So I'll have to ask my American friends," he finished pointedly, "how more *subtly* to control the press." In the silence that followed, I said, "Advertising."

The reporters laughed, but later, one of them took me aside: How *dare* I suggest that freedom of the press was limited? How dare I imply his newsweekly could be influenced by ads?

I explained that I was thinking of advertising's mediawide influence on most of what we read. Even newsmagazines use "soft" cover stories to sell ads, confuse readers with "advertorials," and occasionally self-censor on subjects known to be a problem with big advertisers.

But, I also explained, I was thinking especially of women's magazines. There, it isn't just a little content that's devoted to attracting ads, it's almost all of it. That's why advertisers—not readers—have always been a problem for *Ms.* As the only women's magazine that didn't supply what the ad world euphemistically describes as "supportive editorial atmosphere" or "complementary copy" (for instance, articles that praise food/fashion/beauty subjects to "support" and "complement" food/fashion/beauty ads), *Ms.* could never attract enough advertising to break even.

"Oh, *women's* magazines," the journalist said with contempt. "Everybody knows they're just catalogs—but who cares? They have nothing to do with journalism."

I can't tell you how many times I've had this argument in 25 years of working for many kinds of publications. Except as moneymaking machines—"cash cows" as they are so elegantly called in the trade—women's magazines are rarely taken seriously. Though changes being made by women have been called more far-reaching than the industrial revolution—and though many editors try hard to reflect some of them in the few pages left to them after all the ad-related subjects have been covered—the magazines serving the female half of this country are still far below the journalistic and ethical standards of news and general interest publications. Most depressing of all, this doesn't even rate an exposé.

If *Time* and *Newsweek* had to lavish praise on cars in general and credit General Motors in particular to get GM ads, there would be a scandal—maybe a criminal investigation. When women's magazines from *Seventeen* to *Lear's* praise beauty products in general and credit Revlon in particular to get ads, it's just business as usual.

I.

When *Ms.* began, we didn't consider *not* taking ads. The most important reason was keeping the price of a feminist magazine low enough for most women to afford. But the second and almost equal reason was providing a forum where women and advertisers could talk to each other and improve advertising itself. After all, it was (and still is) as potent a source of information in this country as news or TV and movie dramas.

We decided to proceed in two stages. First, we would convince makers of "people products" used by both men and women but advertised mostly to men—cars, credit cards, insurance, sound equipment, financial services, and the like—that their ads should be placed in a women's magazine. Since they were accustomed to the division between editorial and advertising in news and general interest magazines, this would allow our editorial content to be free and diverse. Second, we would add the best ads for whatever traditional "women's products" (clothes, shampoo, fragrance, food, and so on) that surveys showed *Ms.* readers used. But we would ask them to come in *without* the usual quid pro quo of "complementary copy."

We knew the second step might be harder. Food advertisers have always demanded that women's magazines publish recipes and articles on entertaining (preferably ones that name their products) in return for their ads; clothing advertisers expect to be surrounded by fashion spreads (especially ones that credit their designers); and shampoo, fragrance, and beauty products in general usually insist on positive editorial coverage of beauty subjects, plus photo credits besides. That's why women's magazines look the way they do. But if we could break this link between ads and editorial content, then we wanted good ads for "women's products," too.

By playing their part in this unprecedented mix of *all* the things our readers need and use, advertisers also would be rewarded: ads for products like cars and mutual funds would find a new growth market; the best ads for women's products would no longer be lost in oceans of ads for the same category; and both would have access to a laboratory of smart and caring readers whose response would help create effective ads for other media as well.

I thought then that our main problem would be the imagery in ads themselves. Carmakers were still draping blondes in evening gowns over the hoods like ornaments. Authority figures were almost always male, even in ads for products that only women used. Sadistic, he-man campaigns even won industry praise. (For instance, *Advertising Age* had hailed the infamous Silva Thin cigarette theme, "How to Get a Woman's Attention: Ignore Her," as "brilliant.") Even in medical journals, tranquilizer ads showed depressed housewives standing beside piles of dirty dishes and promised to get them back to work.

Obviously, *Ms.* would have to avoid such ads and seek out the best ones—but this didn't seem impossible. *The New Yorker* had been selecting ads for aesthetic reasons for years, a practice that only seemed to make advertisers more eager to be in its pages. *Ebony* and *Essence* were asking for ads with positive black images, and though their struggle was hard, they weren't being called unreasonable.

Clearly, what *Ms.* needed was a very special publisher and ad sales staff. I could think of only one woman with experience on the business side of magazines—Patricia Carbine, who recently had become a vice president of *McCall's* as well as its editor in chief—and the reason I knew her name was a good omen. She had been managing editor at *Look* (really *the* editor, but its owner refused to put a female name at the top of his masthead) when I was writing a column there. After I did an early interview with Cesar Chavez, then just emerging as a leader of migrant labor, and the publisher turned it down because he was worried about ads from Sunkist, Pat was the one who intervened. As I learned later, she told the publisher she would resign if the interview wasn't published. Mainly because *Look* couldn't afford to lose Pat, it *was* published (and the ads from Sunkist never arrived).

Though I barely knew this woman, she had done two things I always remembered: put her job on the line in a way that editors often talk about but rarely do, and been so loyal to her colleagues that she never told me or anyone outside *Look* that she had done so.

Fortunately, Pat did agree to leave *McCall's* and take a huge cut in salary to become publisher of *Ms.* She became responsible for training and inspiring generations of young women who joined the *Ms.* ad sales force, many of whom went on to become "firsts" at the top of publishing. When *Ms.* first started, however, there were so few women with experience selling space that Pat and I made the rounds of ad agencies ourselves. Later the fact that *Ms.* was asking companies to do business in a different way meant our saleswomen had to make many times the usual number of calls—first to convince agencies and then client companies besides—and to present endless amounts of research. I was often asked to do a final ad presentation, or see some higher decision-maker, or speak to women employees so executives could see the interest of women they worked with. That's why I spent more time persuading advertisers than editing or writing for *Ms.* and why I ended up with an unsentimental education in the seamy underside of publishing that few writers see (and even fewer magazines can publish).

Let me take you with use through some experiences, just as they happened:

• Cheered on by early support from Volkswagen and one or two other car companies, we scrape together time and money to put on a major reception in Detroit. We know U.S. carmakers firmly believe that women choose the upholstery, not the car, but we are armed with statistics and reader mail to prove the contrary: a car is an important purchase for women, one that symbolizes mobility and freedom.

But almost nobody comes. We are left with many pounds of shrimp on the table, and quite a lot of egg on our face. We blame ourselves for not guessing that there would be a baseball pennant play-off on the same day, but executives go out of their way to explain they wouldn't have come anyway. Thus begins ten years of knocking on hostile doors, presenting endless documentation, and hiring a full-time saleswoman in Detroit; all necessary before *Ms.* gets any real results.

This long saga has a semihappy ending: foreign and, later, domestic carmakers eventually provided *Ms.* with enough advertising to make cars one of our top sources of ad revenue. Slowly, Detroit began to take the women's market seriously enough to put car ads in other women's magazines, too, thus freeing a few pages from the hothouse of fashion-beauty-food ads.

But long after figures showed a third, even a half, of many car models being bought by women, U.S. makers continued to be uncomfortable addressing women. Unlike foreign carmakers, Detroit never quite learned the secret of creating intelligent ads that exclude no one, and then placing them in women's magazines to overcome past exclusion. (*Ms.* readers were so grateful for a routine Honda ad featuring rack and pinion steering, for instance, that they sent fan mail.) Even now, Detroit continues to ask, "Should we make special ads for women?" Perhaps that's why some foreign cars still have a disproportionate share of the U.S. women's market.

• In the *Ms.* Gazette, we do a brief report on a congressional hearing into chemicals used in hair dyes that are absorbed through the skin and may be carcinogenic. Newspapers report this too, but Clairol, a Bristol-Myers subsidiary that makes dozens of products—a few of which have just begun to advertise in *Ms.*—is outraged. Not at newspapers or newsmagazines, just at us. It's bad enough that *Ms.* is the only women's magazine refusing to provide the usual "complementary" articles and beauty photos, but to criticize one of their categories—*that* is going too far.

We offer to publish a letter from Clairol telling its side of the story. In an excess of solicitousness, we even put this letter in the Gazette, not in Letters to the Editors where it belongs. Nonetheless—and in spite of surveys that show *Ms.* readers are active women who use more of almost everything Clairol makes than do the readers of any other women's magazine—*Ms.* gets almost none of these ads for the rest of its natural life.

Meanwhile, Clairol changes its hair coloring formula, apparently in response to the hearings we reported.

• Our saleswomen set out early to attract ads for consumer electronics: sound equipment, calculators, computers, VCRs, and the like. We know that our readers are determined to be included in the technological revolution. We know from reader surveys that *Ms.* readers are buying this stuff in numbers as high as those of magazines like *Playboy;* or "men 18 to 34," the prime targets of the consumer electronics industry. Moreover, unlike traditional women's products that our readers buy but don't need to read articles about, these are subjects they want covered in our pages. There actually *is* a supportive editorial atmosphere.

"But women don't understand technology," say executives at the end of ad presentations. "Maybe not," we respond, "but neither do men—and we all buy it."

"If women *do* buy it," say the decision-makers, "they're asking their husbands and boyfriends what to buy first." We produce letters from *Ms.* readers saying how turned off they are when salesmen say things like "Let me know when your husband can come in."

After several years of this, we get a few ads for compact sound systems. Some of them come from JVC, whose vice president, Harry Elias, is trying to convince his Japanese bosses that there is something called a women's market. At his invitation, I find myself speaking at huge trade shows in Chicago and Las Vegas, trying to persuade JVC dealers that showrooms don't have to be locker rooms where women are made to feel unwelcome. But as it turns out, the shows themselves are part of the problem. In Las Vegas, the only women around the technology displays are seminude models serving champagne. In Chicago, the big attraction is Marilyn Chambers, who followed Linda Lovelace of *Deep Throat* fame as Chuck Traynor's captive and/or employee. VCRs are being demonstrated with her porn videos.

In the end, we get ads for a car stereo now and then, but no VCRs; some IBM personal computers, but no Apple or Japanese ones. We notice that office magazines like *Working Woman* and *Savvy* don't benefit as much as they should from office equipment ads either. In the electronics world, women and technology seem mutually exclusive. It remains a decade behind even Detroit.

• Because we get letters from little girls who love toy trains, and who ask our help in changing ads and box-top photos that feature little boys only, we try to get toy-train ads from Lionel. It turns out that Lionel executives *have* been concerned about little girls. They made a pink train, and were surprised when it didn't sell.

Lionel bows to consumer pressure with a photograph of a boy *and* a girl—but only on some of their boxes. They fear that, if trains are associated with girls, they will be devalued in the minds of boys. Needless to say, *Ms.* gets no train ads, and little girls remain a mostly unexplored market. By 1986, Lionel is put up for sale.

But for different reasons, we haven't had much luck with other kinds of toys either. In spite of many articles on child-rearing; an annual listing of nonsexist, multiracial toys by Letty Cottin Pogrebin; Stories for Free Children, a regular feature also edited by Letty; and other prizewinning features for or about children, we get virtually no toy ads. Generations of *Ms.* saleswomen explain to toy manufacturers that a larger proportion of *Ms.* readers have preschool children than do the readers of other women's magazines, but this industry can't believe feminists have or care about children.

• When *Ms.* begins, the staff decides not to accept ads for feminine hygiene sprays or cigarettes; they are damaging and carry no appropriate health warnings. Though we don't think we should tell our readers what to do, we do think we should provide facts so they can decide for themselves. Since the antismoking lobby had been pressing for healthy warnings on cigarette ads, we decide to take them only as they comply.

Philip Morris is among the first to do so. One of its brands, Virginia Slims, is also sponsoring women's tennis and the first national polls of women's opinions. On the other hand, the Virginia Slims theme, "You've come a long way, baby," had more than a "baby" problem. It makes smoking a symbol of progress for women.

We explain to Philip Morris that this slogan won't do well in our pages, but they are convinced its success with some women means it will work with *all* women. Finally, we agree to publish an ad for a Virginia Slims calendar as a test. The letters from readers are critical—and smart. For instance: Would you show a black man picking cotton, the same man in a Cardin suit, and symbolize the antislavery and civil rights movements by smoking? Of course not. But instead of honoring the tests results, the Philip Morris people seem angry to be proven wrong. They take away ads for *all* their many brands.

This costs *Ms.* about $250,000 the first year. After five years, we can no longer keep track. Occasionally, a new set of executives listen to *Ms.* saleswomen, but because we won't take Virginia Slims, not one Philip Morris product returns to our pages for the next 16 years.

Gradually, we also realize our naiveté in thinking we *could* decide against taking cigarette ads. They became a disproportionate support of magazines the moment they were banned on television, and few magazines could compete and survive without them; certainly not *Ms.*, which lacks so many other categories. By the time statistics in the 1980s showed that women's rate of lung cancer was approaching men's, the necessity of taking cigarette ads has become a kind of prison.

• General Mills, Pillsbury, Carnation, DelMonte, Dole, Kraft, Stouffer, Hormel, Nabisco: you name the food giant, we try it. But no matter how desirable the *Ms.* readership, our lack of recipes is lethal.

We explain to them that placing food ads *only* next to recipes associates food with work. For many women, it is a negative that works *against* the ads. Why not place food ads in diverse media without recipes (thus reaching more men, who are now a third of the shoppers in supermarkets anyway), and leave the recipes to specialty magazines like *Gourmet* (a third of whose readers are also men)?

These arguments elicit interest, but except for an occasional ad for a convenience food, instant coffee, diet drinks, yogurt, or such extras as avocados and almonds, this mainstay of the publishing industry stays closed to us. Period.

• Traditionally, wines and liquors didn't advertise to women: men were thought to make the brand decisions, even if women did the buying. But after endless presentations, we begin to make a dent in this category. Thanks to the unconventional Michel Roux of Carillon Importers (distributors of Grand Marnier, Absolut Vodka, and others), who assumes that food and drink have no gender, some ads are leaving their men's club.

Beermakers are still selling masculinity. It takes *Ms.* fully eight years to get its first beer ad (Michelob). In general, however, liquor ads are less stereotyped in their imagery—and far less controlling of the editorial content around them—than are women's products. But given the underrepresentation of other categories, these very facts tend to create a disproportionate number of alcohol ads in the pages of *Ms.* This in turn dismays readers worried about women and alcoholism.

• We hear in 1980 that women in the Soviet Union have been producing feminist *samizdat* (underground, self-published books) and circulating them throughout the country. As punishment, four of the leaders have been exiled. Though we are operating on our usual shoestring, we solicit individual contributions to send Robin Morgan to interview these women in Vienna.

The result is an exclusive cover story that includes the first news of a populist peace movement against the Afghanistan occupation, a prediction of *glasnost* to come, and a grass-roots, intimate view of Soviet women's lives. From the popular press to women's studies courses, the response is great. The story wins a Front Page award.

Nonetheless, this journalistic coup undoes years of efforts to get an ad schedule from Revlon. Why? Because the Soviet women on our cover *are not wearing makeup.*

• Four years of research and presentations go into convincing airlines that women now make travel choices and business trips. United, the first airline to advertise in *Ms.,* is so impressed with the response from our readers that one of its executives appears in a film for our ad presentations. As usual, good ads get good results.

But we have problems unrelated to such results. For instance: because American Airlines flight attendants include among their labor demands the stipulation they could choose to have their last names preceded by "Ms." on their name tags—in a long-delayed revolt against the standard, "I am your pilot, Captain Rothgart, and this is your flight attendant, Cindy Sue"— American officials seem to hold the magazine responsible. We get no ads.

There is still a different problem at Eastern. A vice president cancels a subscription for thousands of copies on Eastern flights. Why? Because he is offended by ads for lesbian poetry journals in the *Ms.* Classifieds. A "family airline," as he explains to me coldly on the phone, has to "draw the line somewhere."

It's obvious that *Ms.* can't exclude lesbians and serve women. We've been trying to make that point since our first issue included an article by and about lesbians, and both Suzanne Levine, our managing editor, and I were lectured by such heavy hitters as Ed Kosner, then editor of *Newsweek* (and now of *New York Magazine*), who insisted that *Ms.* should "position" itself *against* lesbians. But our advertisers have paid to reach a guaranteed number of readers, and soliciting new subscriptions to compensate for Eastern would cost $150,000, plus rebating money in the meantime.

Like almost everything ad-related, this presents an elaborate organizing problem. After days of searching for sympathetic members of the Eastern board, Frank Thomas, president of the Ford Foundation, kindly offers to call Roswell Gilpatrick, a director of Eastern. I talk with Mr. Gilpatrick, who calls Frank Borman, then the president of Eastern. Frank Borman calls me to say that his airline is not in the business of censoring magazines: *Ms.* will be returned to Eastern flights.

• Women's access to insurance and credit is vital, but with the exception of Equitable and a few other ad pioneers, such financial services address men. For almost a decade after the Equal Credit Opportunity Act passes in 1974, we try to convince American Express that women are a growth market—but nothing works.

Finally, a former professor of Russian named Jerry Welsh becomes head of marketing. He assumes that women should be cardholders, and persuades his colleagues to feature women in a campaign. Thanks to this 1980s series, the growth rate for female cardholders surpasses that for men.

For this article, I asked Jerry Welsh if he would explain why American Express waited so long. "Sure," he said, "they were afraid of having a 'pink' card."

• Women of color read *Ms.* in disproportionate numbers. This is a source of pride to *Ms.* staffers, who are also more racially representative than the editors of other women's magazines. But this reality is obscured by ads filled with enough white women to make a reader snowblind.

Pat Carbine remembers mostly "astonishment" when she requested African American, Hispanic, Asian and other diverse images. Marcia Ann Gillespie, a *Ms.* editor who was previously the editor in chief of *Essence*, witnesses ad bias a second time: having tried for *Essence* to get white advertisers to use black images (Revlon did so eventually, but L'Oréal, Lauder, Chanel, and other companies never did), she sees similar problems getting integrated ads for an integrated magazine. Indeed, the ad world often creates black and Hispanic ads only for black and Hispanic media. In an exact parallel of the fear that marketing a product to women will endanger its appeal to men, the response is usually, "But your [white] readers won't identify."

In fact, those we are able to get—for instance, a Max Factor ad made for *Essence* that Linda Wachner gives us after she becomes president—are praised by white readers, too. But there are pathetically few such images.

• By the end of 1986, production and mailing costs have risen astronomically, ad income is flat, and competition for ads is stiffer than ever. The 60/40 preponderance of edit over ads that we promised to readers becomes 50/50; children's stories, most poetry, and some fiction are casualties of less space; in order to get variety into limited pages, the length (and sometimes the depth) of articles suffers; and, though we do refuse most of the ads that would look like a parody in our pages, we get so worn down that some slip through. Still, readers perform miracles. Though we haven't been able to afford a subscription mailing in two years, they maintain our guaranteed circulation of 450,000.

Nonetheless, media reports on *Ms.* often insist that our unprofitability must be due to reader disinterest. The myth that advertisers simply follow readers is very strong. Not one reporter notes that other comparable magazines our size (say, *Vanity Fair* or *The Atlantic*) have been losing more money in one year than *Ms.* has lost in 16 years. No matter how much never-to-be-recovered cost is poured into starting a magazine or keeping one going, appearances seem to be all that matter. (Which is why we haven't been able to explain our fragile state in public. Nothing causes ad-flight like the smell of nonsuccess.)

My healthy response is anger. My not-so-healthy response is constant worry. Also an obsession with finding one more rescue. There is hardly a night when I don't wake up with sweaty palms and pounding heart, scared that we won't be able to pay the printer or the post office; scared most of all that closing our doors will hurt the women's movement.

Out of chutzpah and desperation, I arrange a lunch with Leonard Lauder, president of Estée Lauder. With the exception of Clinique (the brainchild of Carol Phillips), none of Lauder's hundreds of products has been advertised in *Ms.* A year's schedule of ads for just three or four of them could save us. Indeed, as the scion of a family-owned company whose ad practices are followed by the beauty industry, he is one of the few men who could liberate many pages in all women's magazines just by changing his mind about "complementary copy."

Over a lunch that costs more than we can pay for some articles, I explain the need for his leadership. I also lay out the record of *Ms.* more literary and journalistic prizes won, more new issues introduced into the mainstream, new writers discovered, and impact on society than any other magazine; more articles that became books, stories that became movies, ideas that became television series, and newly advertised products that became profitable; and, most important for him, a place for his ads to reach women who aren't reachable through any other women's magazine. Indeed, if there is one constant characteristic of the ever-changing *Ms.* readership, it is their impact as leaders. Whether it's waiting until later to have first babies, or pioneering PABA as sun protection in cosmetics, *whatever* they are doing today, a third to a half of American women will be doing three to five years from now. It's never failed.

But, he says, *Ms.* readers are not *our* women. They're not interested in things like fragrance and blush-on. If they were, *Ms.* would write articles about them.

On the contrary, I explain, surveys show they are more likely to buy such things than the readers of, say, *Cosmopolitan* or *Vogue.* They're good customers because they're out in the world enough to need several sets of everything: home, work, purse, travel, gym, and so on. They just don't need to read articles about these things. Would he ask a men's magazine to publish monthly columns on how to shave before he advertised Aramis products (his line for men)?

He concedes that beauty features are often concocted more for advertisers than readers. But *Ms.* isn't appropriate for his ads anyway, he explains. Why? Because Estée Lauder is selling "a kept-woman mentality."

I can't quite believe this. Sixty percent of the users of his products are salaried, and generally resemble *Ms.* readers. Besides, his company had the appeal of having been started by a creative and hardworking woman, his mother, Estée Lauder.

That doesn't matter, he says. He knows his customers, and they would *like* to be kept women. That's why he will never advertise in *Ms.*

In November 1987, by vote of the Ms. Foundation for Education and Communication (*Ms.*'s owner and publisher, the media subsidiary of the Ms. Foundation for Women), *Ms.* was sold to a company whose officers, Australian feminists Sandra Yates and Anne Summers, raised the investment money in their country that *Ms.* couldn't find in its own. They also started *Sassy* for teenage women.

In their two-year tenure, circulation was raised to 550,000 by investment in circulation mailings, and, to the dismay of some readers, editorial features on clothes and new products made a more traditional bid for ads. Nonetheless, ad pages fell below previous levels. In addition, *Sassy*, whose fresh voice and sexual frankness were an unprecedented success with young readers, was targeted by two mothers from Indiana who began, as one of them put it, "calling every Christian organization I could think of." In response to this controversy, several crucial advertisers pulled out.

Such links between ads and editorial content was a problem in Australia, too, but to a lesser degree. "Our readers pay two times more for their magazines," Anne explained, "so advertisers have less power to threaten a magazine's viability."

"I was shocked," said Sandra Yates with characteristic directness. "In Australia, we think you have freedom of the press—but you don't."

Since Anne and Sandra had not met their budget's projections for ad revenue, their investors forced a sale. In October 1989, *Ms.* and *Sassy* were bought by Dale Lange, owner of *Working Mother, Working Woman,* and one of the few independent publishing companies left among the conglomerates. In response to a request from the original *Ms.* staff—as well as to reader letters urging that *Ms.* continue, plus his own belief that *Ms.* would benefit his other magazines by blazing a trail—he agreed to try the ad-free, reader-supported *Ms.* you hold now and to give us complete editorial control.

II.

Do you think, as I once did, that advertisers make decisions based on solid research? Well, think again. "Broadly speaking," says Joseph Smith of Oxtoby-Smith, Inc., a consumer research firm, "there is no persuasive evidence that the editorial context of an ad matters."

Advertisers who demand such "complementary copy," even in the absence of respectable studies, clearly are operating under a double standard. The same food companies place ads in *People* with no recipes. Cosmetic companies support *The New Yorker* with no regular beauty columns. So where does this habit of controlling the content of women's magazines come from?

Tradition. Ever since *Ladies Magazine* debuted in Boston in 1828, editorial copy directed to women has been informed by something other than its readers' wishes. There were no ads then, but in an age when married women were legal minors with no right to their own money, there was another revenue source to be kept in mind: husbands. "Husbands may rest assured," wrote editor Sarah Josepha Hale, "that nothing found in these pages shall cause her [his wife] to be less assiduous in preparing for his reception or encourage her to 'usurp station' or encroach upon prerogatives of men."

Hale went on to become the editor of *Godey's Lady's Book,* a magazine featuring "fashion plates": engravings of dresses for readers to take to their seamstresses or copy themselves. Hale added "how to" articles, which set the tone for women's service magazines for years to come: how to write politely, avoid sunburn, and—in no fewer than 1,200 words—how to maintain a goose quill pen. She advocated education for women but avoided controversy. Just as most women's magazines now avoid politics, poll their readers on issues like abortion but rarely take a stand, and praise socially approved lifestyles, Hale saw to it that *Godey's* avoided the hot topics of its day: slavery, abolition, and women's suffrage.

What definitively turned women's magazines into catalogs, however, were two events: Ellen Butterick's invention of the clothing pattern in 1863 and the mass manufacture of patent medicines containing everything from colored water to cocaine. For the first time, readers could purchase what magazines encouraged them to want. As such magazines became more profitable, they also began to attract men as editors. (Most women's magazines continued to have men as top editors until the feminist 1970s.) Edward Bok, who became editor of *The Ladies' Home Journal* in 1889, discovered the power of advertisers when he rejected ads for patent medicines and found that other advertisers canceled in retribution. In the early 20th century, *Good Housekeeping* started its Institute to "test and approve" products. Its Seal of Approval became the grandfather of current "value added" programs that offer advertisers such bonuses as product sampling and department store promotions.

By the time suffragists finally won the vote in 1920, women's magazines had become too entrenched as catalogs to help women learn how to use it. The main function was to create a desire for products, teach how to use products, and make products a crucial part of gaining social approval, pleasing a husband, and performing as a homemaker. Some unrelated articles and short stories were included to persuade women to pay for these catalogs. But articles were neither consumerist nor rebellious. Even fiction was usually subject to formula: if a woman had any sexual life outside marriage, she was supposed to come to a bad end.

In 1956, Helen Gurley Brown began to change part of that formula by bringing "the sexual revolution" to women's magazines—but in an ad-oriented way. Attracting multiple men required even more consumerism, as the Cosmo Girl made clear, than finding one husband.

In response to the workplace revolution of the 1970s, traditional women's magazines—that is, "trade books" for women working at home—were joined by *Savvy, Working Woman,* and other trade books for women working in offices.

But by keeping the fashion/beauty/entertaining articles necessary to get traditional ads and then adding career articles besides, they inadvertently produced the antifeminist stereotype of Super Woman. The male-imitative, dress-for-success woman carrying a briefcase became the media image of a woman worker, even though a blue-collar woman's salary was often higher than her glorified secretarial sister's, and though women at a real briefcase level are statistically rare. Needless to say, these dress-for-success women were also thin, white, and beautiful.

In recent years, advertisers' control over the editorial content of women's magazines has become so institutionalized that it is written into "insertion orders" or dictated to ad salespeople as an official policy. The following are recent typical orders to women's magazines:

- Dow's Cleaning Products stipulates that ads for its Vivid and Spray 'n Wash products should be adjacent to "children or fashion editorial"; ads for Bathroom Cleaner should be next to "home furnishing/family" features; and so on for other brands. "If a magazine fails for 1/2 the brands or more," the Dow order warns, "it will be omitted from further consideration."
- Bristol-Myers, the parent of Clairol, Windex, Drano, Bufferin, and much more, stipulates that ads be placed next to "a full page of compatible editorial."
- S.C. Johnson & Son, makers of Johnson Wax, lawn and laundry products, insect sprays, hair sprays, and so on, orders that its ads *should not be opposite extremely controversial features or material antithetical to the nature/copy of the advertised product.*" (Italics theirs.)
- Maidenform, manufacturer of bras and other apparel, leaves a blank for the particular product and states: "The creative concept of the _____ campaign, and the very nature of the product itself appeal to the positive emotions of the reader/consumer. Therefore, it is imperative that all editorial adjacencies reflect that same positive tone. The editorial must not be negative in content or lend itself contrary to the _____ product imagery/message (e.g., *editorial relating to illness, disillusionment, large size fashions, etc.*)" (Italics mine.)
- The De Beers diamond company, a big seller of engagement rings, prohibits magazines from placing its ads with "adjacencies to hard news or anti/love-romance themed editorial."
- Procter & Gamble, one of this country's most powerful and diversified advertisers, stands out in the memory of Anne Summers and Sandra Yates (no mean feat in this context): its products were not to be placed in *any* issue that included *any* material on gun control, abortion, the occult, cults, or the disparagement of religion. Caution was also demanded in any issue covering sex or drugs, even for educational purposes.

Those are the most obvious chains around women's magazines. There are also rules so clear they needn't be written down: for instance, an overall "look" compatible with beauty and fashion ads. Even "real" nonmodel women photographed for a women's magazine are usually made up, dressed in credited clothes, and retouched out of all reality. When editors do include articles on less-than-cheerful subjects (for instance, domestic violence), they tend to keep them short and unillustrated. The point is to be "upbeat." Just as women in the street are asked, "Why don't you smile, honey?" women's magazines acquire an institutional smile.

Within the text itself, praise for advertisers' products has become so ritualized that fields like "beauty writing" have been invented. One of its frequent practitioners explained seriously that "It's a difficult art. How many new adjectives can you find? How much greater can you make a lipstick sound? The FDA restricts what companies can say on labels, but we create illusion. And ad agencies are on the phone all the time pushing you to get their product in. A lot of them keep the business based on how many editorial clippings they produce every month. The worst are products," like Lauder's as the writer confirmed, "with their own name involved. It's all ego."

Often, editorial becomes one giant ad. Last November, for instance, *Lear's* featured an elegant woman executive on the cover. On the contents page, we learned she was wearing Guerlain makeup and Samsara, a new fragrance by Guerlain. Inside were full-page ads for Samsara and Guerlain antiwrinkle cream. In the cover profile, we learned that this executive was responsible for launching Samsara and is Guerlain's director of public relations. When the *Columbia Journalism Review* did one of the few articles to include women's magazines in coverage of the influence of ads, editor Frances Lear was quoted as defending her magazine because "this kind of thing is done all the time."

Often, advertisers also plunge odd-shaped ads into the text, no matter what the cost to the readers. At *Women's Day,* a magazine originally founded by a supermarket chain, editor in chief Ellen Levine said, "The day the copy had to rag around a chicken leg was not a happy one."

Advertisers are also adamant about where in a magazine their ads appear. When Revlon was not placed as the first beauty ad in one Hearst magazine, for instance, Revlon pulled its ads from *all* Hearst magazines. Ruth Whitney, editor in chief of *Glamour,* attributes some of these demands to "ad agencies wanting to prove to a client that they've squeezed the last drop of blood out of a magazine." She also is, she says, "sick and tired if hearing that women's magazines are controlled by cigarette ads." Relatively speaking, she's right. To be as censoring as are many advertisers for women's products, tobacco companies would have to demand articles in praise of smoking and expect glamorous photos of beautiful women smoking their brands.

I don't mean to imply that the editors I quote here share my objections to ads: most assume that women's magazines have to be the way they are. But it's also true that only former editors can be completely honest. "Most of the pressure came in the form of direct product mentions," explains Sey Chassler, who was editor in chief of *Redbook* from the sixties to the eighties. "We got threats from the big guys, the Revlons, blackmail threats. They wouldn't run ads unless we credited them."

"But it's not fair to single out the beauty advertisers because these pressures came from everybody. Advertisers want to know two things: What are you going to charge me? What *else* are you going to do for me? It's a holdup. For instance, management felt that fiction took up too much space. They couldn't put any advertising in that. For the last ten years, the number of fiction entries into the National Magazine Awards has declined.

"And pressures are getting worse. More magazines are more bottom-line oriented because they have been taken over by companies with no interest in publishing.

"I also think advertisers do this to women's magazines especially," he concluded, "because of the general disrespect they have for women."

Even media experts who don't give a damn about women's magazines are alarmed by the spread of this ad-edit linkage. In a climate *The Wall Street Journal* describes as an unacknowledged Depression for media, women's products are increasingly able to take their low standards wherever they go. For instance: newsweeklies publish uncritical stories on fashion and fitness. *The New York Times Magazine* recently ran an article on "firming creams," complete with mentions of advertisers. *Vanity Fair* published a profile of one major advertiser, Ralph Lauren, illustrated by the same photographer who does his ads, and turned the lifestyle of another, Calvin Klein, into a cover story. Even the outrageous *Spy* has toned down since it began to go after fashion ads.

And just to make us really worry, films and books, the last media that go directly to the public without having to attract ads first, are in danger, too. Producers are beginning to depend on payments for displaying products in movies, and books are now being commissioned by companies like Federal Express.

But the truth is that women's products—like women's magazines—have never been the subjects of much serious reporting anyway. News and general interest publications, including the "style" or "living" sections of newspapers, write about food and clothing as cooking and fashion, and almost never evaluate such products by brand name. Though chemical additives, pesticides, and animal fats are major health risks in the United States, and clothes, shoddy or not, absorb more consumer dollars than cars, this lack of information is serious. So is ignoring the contents of beauty products that are absorbed into our bodies through our skins, and that have profit margins so big they would make a loan shark blush.

III.

What could women's magazines be like if they were as free as books? as realistic as newspapers? as creative as films? as diverse as women's lives? We don't know.

But we'll only find out if we take women's magazines seriously. If readers were to act in a concerted way to change traditional practices of *all* women's magazines and the marketing of *all* women's products, we could do it. After all, they are operating on our consumer dollars; money that we now control. You and I could:

- write to editors and publishers (with copies to advertisers) that we're willing to pay *more* for magazines with editorial independence, but will *not* continue to pay for those that are just editorial extensions of ads;
- write to advertisers (with copies to editors and publishers) that we want fiction, political reporting, consumer reporting—whatever is, or is not, supported by their ads;
- put as much energy into breaking advertising's control over content as into changing the images in ads, or protesting ads for harmful products like cigarettes;
- support only those women's magazines and products that take *us* seriously as readers and consumers.

Those of us in the magazine world can also use the carrot-and-stick technique. For instance: pointing out that, if magazines were a regulated medium like television, the demands of advertisers would be against FCC rules. Payola and extortion could be punished. As it is, there are probably illegalities. A magazine's postal rates are determined by the ratio of ad to edit pages, and the former costs more than the latter. So much for the stick.

The carrot means appealing to enlightened self-interest. For instance: there are many studies showing that the greatest factor in determining an ad's effectiveness is the credibility of its surroundings. The "higher the rating of editorial believability," concluded a 1987 survey by the *Journal of Advertising Research,* "the higher the rating of the advertising." Thus, an impenetrable wall between edit and ads would also be in the best interest of advertisers.

Unfortunately, few agencies or clients hear such arguments. Editors often maintain the false purity of refusing to talk to them at all. Instead, they see ad salespeople who know little about editorial, are trained in business as usual, and are usually paid by commission. Editors might also band together to take on controversy. That happened once when all the major women's magazines did articles in the same month on the Equal Rights Amendment. It could happen again.

It's almost three years away from life between the grindstones of advertising pressures and readers' needs. I'm just beginning to realize how edges got smoothed down—in spite of all our resistance.

I remember feeling put upon when I changed "Porsche" to "car" in a piece about Nazi imagery in German pornography by Andrea Dworkin—

feeling sure Andrea would understand that Volkswagen, the distributor of Porsche and one of our few supportive advertisers, asked only to be far away from Nazi subjects. It's taken me all this time to realize that Andrea was the one with the right to feel put upon.

Even as I write this, I get a call from a writer for *Elle,* who is doing a whole article on where women part their hair. Why, she wants to know, do I part mine in the middle?

It's all so familiar. A writer trying to make something of a nothing assignment; an editor laboring to think of new ways to attract ads; readers assuming that other women must want this ridiculous stuff; more women suffering for lack of information, insight, creativity, and laughter that could be on these same pages.

I ask you: Can't we do better than this?

COMING TO TERMS

Gloria Steinem describes here the difficulty of getting advertisers to advertise in *Ms.* a pioneering feminist magazine she helped found in 1972. Throughout her piece she recalls a variety of arguments she made to advertisers about why they should place ads in *Ms.* We'd like you to attend to these arguments in re-reading her piece, take some notes on them in your journal, and respond to the following questions: What are the major points Steinem makes in arguing her case with advertisers? What does she try to tell advertisers about *Ms* readers? What does she tell them about women? What does she try to tell advertisers about their own interest—that is, about the effective use of advertising? How would you characterize these arguments? As practical? Philosophical? Economic? Political? Once you have considered Steinem's arguments to advertisers, consider their replies to her: What do they say back when she tries to convince them to advertise? What kind of reasoning do they use? More importantly, what do they tend to do? Are they sympathetic to her point of view? Hostile? In these exchanges between Steinem and the advertisers, what's being understood and what's being misunderstood? Finally, in telling this story of her struggles with the existing corporate mentality, what sort of argument is Steinem trying to make to us as her readers? What conclusions might Steinem want us to draw from her story?

READING AS A WRITER

"I ask you: Can't we do better than this?" So Steinem ends her piece. But who are the "we" in her question? Who might be included? There is a rather obvious answer to this question, one which is a starting point in responding to our question, but a place we want you to move beyond as you re-read "Sex, Lies and Advertising." You already know that this piece appeared originally in a 1990 issue of *Ms.* magazine. So one audience of this article are the readers of the magazine, who one might assume are interested and sympathetic to what Steinem has to say. We think, however, that Steinem's article attempts to do much more than enlist the aid of an already supportive group of readers. Successful writers, like Steinem, who take a position of advocacy, who ask readers to rethink their viewpoints or change their minds on a subject, often anticipate an audience's potential objections or resistance to a point of view. Another way to think of this extended sense of audience is: How does a writer speak to readers who already may agree with her? And then how does this same writer speak to readers who might be ambivalent or even need a considerable amount of convincing? With this in mind, we'd like you to re-read "Sex, Lies and Advertising" to investigate the ways in which Steinem might be addressing both the potentially sympathetic as well as the potentially hostile reader. In doing this work, we'd like you to play both roles: the sympathetic reader and the unconvinced, ambivalent, or even hostile reader. We'd like you to go back through Steinem's text and mark places in her article where you initially support or agree with what she is saying. Then, in your journal, respond to the following questions: In these places, what has Steinem assumed to be some of the likely interests, values, and concerns you both might already share? Likewise, what does she seem to expect you to already know about women's magazines and women's issues? Next, what are some places in her article where you see Steinem attempting to challenge your beliefs? Or working to convince you of something? In what ways has Steinem anticipated your objections or resistance to the points she is making? What does she tell you that you do not already know about women's magazines and women's issues? How successful has Steinem been in convincing you? In what ways has she failed to convince you?

WRITING CRITICISM

Nobody reads a newspaper left-to-right, top-to-bottom. Instead we all learn to follow a story as it leaps from column to column, page to page, all the while ignoring—at least for the moment—the ads, images, and other stories that surround it. Similarly, we learn to "page" through a magazine, often in no particular order at all, following those articles that catch our interest, skipping over those that don't, perhaps glancing from time to time at the ads and photos that lie in between. In "Sex, Lies, and Advertising," Gloria Steinem resists this sort of training as a reader, and instead examines, as it were, the "whole page"—both ads and editorial copy—as if they formed a single, coherent text. And indeed her argument is that they do—that most certainly in the case of women's magazines, and increasingly in the case of most other publications, advertisers are able to push for what they call a "supportive editorial atmosphere" for their products, which means, in effect, that they can subtly influence much of what gets printed as "news" or "features." But Steinem not only makes a strong argument about the shaping of women's magazines, she also illustrates a powerful move in media criticism—to look at not just the "editorial" part of a text, but also the promotional materials that surround and encase it—ads, covers, layouts, images, packaging, and the like—when trying to assess the messages it offers to its readers or audience.

We'd like you to take on a similar project here. Begin by locating a media or popular culture text that strikes you as having an interesting mix of promotional and editorial copy. Newspapers and magazines are, of course, convenient and striking examples; but you might look to other media as well: The "morning drive" show on a radio station, for instance, with its mix of patter, news, skits, weather, traffic, commercials, and even, every now and then—music. Or an afternoon broadcast of a football game—as it shifts from action to replay to commentary to interview to preview to NFL public interest message to commercial—and back again. Or you might also look at how sitcoms or local news programs are shaped and scripted around commercial breaks (so that they often seem odd and jerky if viewed without them). You may have noticed a trend in local news programming in which the lines between *news* and *station promotion* begin to get blurred: What newscasters do as a promotion for the station becomes part of the *news*. And even many books now are enveloped in one level after another of promotional material—as you pick them from their lavish cardboard displays in bookstores or supermarkets, handle their shiny and intricate covers, and finally turn through pages of critical blurbs to reach the main text inside. A few publishers of paperback books have now broken a once "sacred" covenant: While magazines might advertise, *serious* books do not. It is now not unusual, however, to see advertisements in paperback books. To return for a moment to Steinem's invitation to learn an alternative way of reading media, both editorial and ads, as a single text, you might also consider applying her approach to film. While it is true that films are not interrupted by commercials (unless shown on network television), we think that Steinem's approach might be useful in reading the *whole* film experience. You might rent a film and pay attention to the "trailers," ads, and highlights for other films, as well as other kinds of commercials on the videotape. In any case, and as usual, you will want to get hold of a text that you can examine closely and quote from precisely—which may mean making use of a tape recorder or VCR.

Once you have your text, you will want to ask much the same questions of it that both editors and advertisers must also ask: What sort of person is likely to read (or look at or listen to) this? What are their probable interests, concerns, hopes, anxieties? Try to get past simple demographics here—*Playboy* is mostly read by men, *Town and Country* appeals to the upper-class, that sort of thing—and focus instead on the attitudes and values implied by your text. Recall the story Steinem tells about the president of Estée Lauder worrying that *Ms.* failed to reflect the "kept woman mentality" of his cosmetics. This is the level of nuance that you want to become attuned to. That is, you want to look at your text as trying to sell not just subscriptions or products, but beliefs, values, and attitudes about the world as well. In doing so, you will want to look closely, as Steinem does, at the relations set up between advertising and editorial copy. What happens when you look at both the features and the ads in a magazine as parts of the same continuous text? Or when you consider a sitcom and the commercials that are woven throughout it as part of the same broadcast? Do both seem to imagine the same sort of person as their reader (or watcher or listener)? Or are there ways in which they send out conflicting messages or signals? You may want to use this assignment as an opportunity to test out Steinem's claim that by long tradition "editorial copy directed to women has been informed by something other than its readers' wishes." Does this still hold true for women's magazines? Does it hold true across media? (For TV shows watched mainly by women, say?) Might much the same be said for texts directed at men, or for those aimed at a more mixed or "general" audience? In analyzing your text, how would you represent its readers', viewers', or listeners' wishes, and that other *something*? Where are the discrepancies located? Is there an agenda? Does it seek to control or alter opinion, open up new ways of thinking or protect the status quo? You might think of the differences between an audience's wishes and that other *something* in terms other than gender. Your text might be seen as sending out similar or conflicting messages about work, for instance, or social class, or age, or race, or religion. Whatever you do in the way of making use of what Steinem has to say, you will want, as always, to support what you have to say with specific examples from the texts you are studying.

The Machine in the Kitchen

John Thorne

Arriving a little late at America's culinary renaissance, I encountered the food processor and the microwave oven together sometime in the mid-1970s. As often happens at such a joint encounter, the two machines became firmly linked in my mind. At the time, however, I thought there was a clear distinction to be made between them. Although each was designed to make kitchen work faster and easier, the food processor seemed to be opening whole new culinary horizons to the home cook, while the microwave was nothing more than the boon tool of the I-hate-to-cook crowd, a hyped-up gadget that did things incredibly fast at the cost of doing them well.

If Carl Sontheimer, the entrepreneur who adapted and marketed the food processor in America, was a genius, it was as much for knowing not only that the time was ripe for this machine but exactly which cooks it was ripe *for*—those of us who thought we wanted to take cooking seriously. Other manufacturers copied the machine but missed the mark: They colored it pink, softened its formidable lines, and advertised it as another kitchen work-saver, a kind of turbo-blender.

If you had no interest in that sort of thing, a food processor was a mistaken purchase. The Dad who bought one as a surprise for Mom to help her out with the kitchen chores was in for a rude surprise: If Mom didn't own at least one French cookbook, she most likely put the machine away under the counter after the first exploratory spin.

In truth, the food processor was a work-*maker*. The sorts of things you were drawn to do with it you wouldn't ever think of trying without it: shredding your own rillettes, sieving your own quenelles, hand-mounting your own mayonnaise.

But to the fledgling serious cook, the Cuisinart, dressed in its spotless kitchen whites, presented itself not merely as a tool of professional chefs (which it was) but as a professionalizing one, the sole essential shortcut to chefdom. We had only to follow instructions to jump straight from *commis* to *gros bonnet*—at least in our own kitchen. Or so we thought.

Before we can begin to understand what machines have done to our cooking, we first need to take account of the impact of an earlier kitchen technology: the cookbook. My generation (by which I mean those who come of age under the lingering aura of the Kennedy presidency) did, I still think, bring a breath of fresh air into the American kitchen. College education had made our minds hungry for new experiences; now our mouths were catching up. We were open to a new kind of culinary adventuring: We wanted to be exposed to the connoisseurship of food as well as the eating of it.

But our strength was also our weakness. We had learned to think in the classroom, where the printed page brought what was, for most of us, our first real interactions with culture. Hence, it was also via the printed page that we expected to master French cooking—just as we had done French literature.

Unfortunately, as much as cookbooks attracted us—and seemed attractive in themselves—they were simply not up to the intellectual caliber of the books from which we received what was best about our education. Our understanding of English literature, for example, would have been quite different if it had never gotten beyond anthologies of "best" poems.

This, however, is the way food writing was—and still almost entirely is—done. Its great shameful secret is its utter intellectual poverty: It may sometimes tell you things you never knew, but nowhere does it make you think. Food writers collect recipes, which, like folk songs or stuffed birds, are considered ends enough in themselves. The recipe collection—the cookbook—is the original kitchen machine. If it did not exist, there could be no food processor, no microwave oven—probably no cooking at all as we now know it. Recipes collapse the fullness of lived experience into a mechanical succession of steps that, once parsed, can be followed by anyone. But the result—the made dish—is only a copy, a simulacrum whose true meaning lies somewhere else. This does not much matter in a cuisine whose coherence still lies in a complex amalgam of tradition, prejudice, shared skills, and that ultimate common denominator: available ingredients.

The danger arises when the use of recipes becomes so prevalent that this coherence is lost—because recipe cooking cannot bring it back. Even if a cook internalizes enough familiar recipes so as not to need to consult cookbooks often, this explains nothing about a cuisine in which all recipes are essentially beside the point. A food writer from a recipe-based cuisine likewise has no choice but to reduce all culinary experience into recipes. Even our best culinary writers present French cuisine as simply a standard repertoire of recipes, provincial and classical, that all French cooks prepare—some better and some worse than others. Recipes do play their role in French cooking, without a doubt, but that cuisine is much better explained as a complex, interacting network of artisanal skills. Because the language of recipe writing cannot capture the fragile ecology of such an artisanal culture, there is an inevitable rupture between the experience the French have of their cooking and the way that cookbook writers attempt to capture it.

This problem is not easily discernible in any individual recipe, and it was on individual recipes that my generation tended to concentrate. We treated each one the way we had been taught to read a poem: intensely, producing an ingenious personal reading, a kind of edible explication. How we got from the one to the other was of no great moment as long as the results showed brilliance. We were willing to learn ad hoc the necessary skills but even more willing to see them, as opportunity arose, relegated to a machine.

Most machines assist the cook without disrupting or denying that hard-earned repository of manual experience whose density gives depth and meaning to the dishes it produces. Just as a motor can help turn a steering wheel without robbing a driver of all feel for the road, so can another crank an eggbeater without denying the tactile mastery gained from working a whisk in batter.

The food processor is different. It is not an electric knife (which had already been invented, only to prove itself nothing much) but a cutting *machine*. It pushes the cook's hand aside, for it works far too quickly for the body to control it directly. Our responses are simply not fast enough; we learn to count seconds instead. Set the steel blade in place; fill the container with basil, pine nuts, garlic, and chunks of Parmesan; and switch the motor on. The machine begins to sauce the ingredients so immediately that they seem to flow together; for its operator, there is a genuine sensual delight in the way, so quickly, so smoothly, it pulps the basil, grinds the cheese, and, as the olive oil starts dribbling down the feeding tube, plumps it all into a thick and unctuous cream.

Sensual it truly is, but it is the sensuality of an observer, not a participant. The food processor does not enhance the cook's experience. Instead, that work is divided between the mind, which directs it, and the machine, which performs it. The body's part is reduced to getting out the tools and (mostly) cleaning up afterward.

Historically, the cost of this usurpation was not only the loss of kitchen work by which the body had formerly refreshed itself, exercising genuinely demanding skills and shaping work to the tempo of personal rhythms. There were two other unexpected consequences. The first was that, by strength of example, the food processor began to corrode the meaning of *all* kitchen work. What could not be done effortlessly, cleanly, perfectly, now became by comparison drudgery, all the more susceptible to replacement by some other, cleverer machine. The second consequence was one to which we *nouveaux cuisinistes* were especially vulnerable. Members of a wealthy, acquisitive culture who could pick the ingredients of our meals almost at will, we had now been given a machine that allowed us to prepare an almost unlimited number of complex dishes without any kind of physical restraint.

None of this, of course, was the food processor's fault. But just as the automobile changed the landscape of America in ways that no one had expected or prepared for, the food processor shifted the nature of culinary reality—more slowly for the culture at large but almost immediately in the microculture that was beginning to take food seriously. As it proliferated, it began rising the stakes for all, even for those who did not yet possess one. And it is this effect—in an equally unexpected way—that we have just begun to experience with the microwave.

Unlike the food processor, the microwave oven has no French ancestry or even, until relatively recently, any influential food-world friends. It lacks the assertively simple high-tech styling and the associations with haute cuisine that might have provided it with a Cuisinart-like cachet. In fact, with its buttons, buzzers, and revolving carousels, the microwave, from the very start, has seemed irredeemably prole—right down to the reason for owning one. For, no matter the culinary arguments voiced in its favor, the only truly compelling reason to own a microwave is still the first: It is the best medium yet devised for the most instant reheating of cooked food.

Even so, what made the microwave seem irresistible to its purchasers was not all that different from what made the food processor seem so desirable. Both fed a fantasy of participating in a cuisine whose rationale had already evaporated, no matter how delicious its dishes. For the microwave owner, however, that cuisine was the familiar American supper. With its assistance, a family could dispense with a regular cook and even a common dinnertime—and still sit down to the familiar trinity of meat, starch, and vegetable, served piping hot on a premium plastic plate.

What the proponents of this new work-free microwave cuisine do not understand—and what we seem to be in no position to teach them—is that the impoverishment cooking has undergone in our hands has left them even less able than we to imagine a cuisine not based on the promiscuous use of recipes or the necessary convenience of machines. We, too, are still convinced that, by following recipes and using devices such as food processors and microwave ovens, we are "saving" time. But by doing this, we have externalized the experience of cooking into a series of unwanted chores. We are being made more efficient—at the cost of a whole realm of experience.

To understand what has really happened to us, imagine attempting to reverse the process. Imagine *wanting* to take a whole afternoon to prepare supper leisurely—without food processor, microwave oven, or cookbook. To live, after all, is to experience things, and every time we mince an onion, lower the flame under a simmering pot, shape the idea and substance of a meal, we actually gain rather than lose lived time. Anyone who has seriously attempted to do such cooking knows that it requires a kind of commitment that we are no longer able to give our quotidian lives, for this experience has become too dear for casual expenditure.

The devices making more and more demands on our actual lived time do not themselves provide experience to compensate for what they have taken away. Instead, they provide a compressed, fictive experience that does not take place in genuinely lived time. As cooking has become more and more like watching television—the mind kept busy, the body left essentially uninvolved—it has forced onto our eating the same kind of pretend-life. A frozen microwave entrée like an Armóur Dinner Classic or a Swanson's Le Menu dinner might in this sense be viewed as a cookbook brought to its ultimate consummation: a food book that can be opened and its contents devoured. Like cookbooks, these food packages arouse appetite through an experience transmitted by glossy photographs, enticing prose, and a sense—conveyed through design and pricing—that we are being given the metaphorical equivalent of a good home-cooked meal.

Invert this insight and the cookbook reveals itself to us as a kind of pre-technological TV dinner. It, too, stimulates appetite by end-running actual sensual experience and appealing directly to the mind. It, too, wraps endless sameness in metaphors that the brain hungers for, steadily blurring the distinction between actual and fictional experience. "Look," it says to the tongue, "stop complaining. Tonight we're going to France to have moussaka Provençale with Mireille Johnston." But the tongue doesn't reply; it has long since stopped paying attention.

All recipes are built on the belief that somewhere at the beginning of the chain there is a cook who does not use them. This is the great nostalgia of our cuisine, ever invoking an absent mother-cook who once laid her hands on the body of the world for us and worked it into food. The promise of every cookbook is that it offers a way back onto her lap.

She's long gone, that lady. But without the fantasy of her, none of this would be bearable. Our cuisine becomes a Borges fable or an Escher print, a universe crammed with cooks all passing the same recipes around and around, each time cooking them faster and faster...until everything fuses together into a little black box. Put in a piece of food on a clean plate; a minute later, take out a cooked piece of food on a dirty one—and call it progress.

COMING TO TERMS

Thorne's piece is centered on an analysis of three technologies: the cookbook, the food processor, and the microwave oven. Respond to the following questions in your journal. What have these machines done to cooking, according to Thorne? What does he object to in considering their effects on the preparation of food? Or, to put it another way, what does he feel has been *lost* as a result of these technologies coming into the kitchen? Why does Thorne regret that loss? Which passages in his article describe this loss particularly well? What values or beliefs about food do you think Thorne is using to fashion his critique? In what ways does Thorne expand the context of this critique? In other words, besides food and cooking, what other subjects does Thorne address in "The Machine in the Kitchen"? Have you read other writers in *Media Journal* who consider similar discussions of technological progress?

READING AS A WRITER

As part of his argument about what the food processor and the microwave oven have done to the process and experience of cooking, Thorne refers to what he calls "an earlier kitchen technology: the cookbook." This may seem a somewhat odd name to give to a book. What does Thorne mean by calling the cookbook, and its list of recipes, a "technology"? Thorne has taken a rather common term (but certainly a culturally important one) and made its meaning more inclusive. In doing so, what connections does Thorne make between the cookbook as a technology and the more obvious technologies of the food processor and microwave oven? How does considering the cookbook a technology (rather than simply as a text) change how we might understand its meaning and function? You might want to think not so much about the *dictionary* definition of technology here, but rather about the *cultural* importance or meaning of this word. In what ways does our culture value technology? Are there negative effects when Thorne includes the cookbook as an innovation of technology?

WRITING CRITICISM

John Thorne tries to show us how technologies support certain ideas about food and the cooking of food. For him, the food processor is more than simply a device for chopping things, just as the microwave is more than a machine for heating things up. Both are also ways of reimagining food, of redefining the process of cooking. We'd like you to write an essay in which you try your hand at this sort of criticism. Locate some machine or technology that exists in your personal environment and that is not ordinarily an object of study or criticism. Have it at hand so you can study it as closely as may be needed. Then draft a piece of writing in which you show why this object deserves to be thought about in more depth, and, perhaps, to be criticized. It's not necessary to be negative in your evaluation. That is, you don't have to argue that the device is dangerous in its effects or objectionable in some way (although you certainly can argue this if you choose to). Rather, the point is to look at the device as more interesting than you (and your readers) may have previously thought, as connected to more things than it might seem at first glance, as expressing certain ideas that are worth considering—in short, as meaningful in some way that it may not have seemed before you sat down to write about it.

In choosing an object to write about, you might want to look around your home or dorm room at things you use but rarely think about—things that plug in, switch on, hide in drawers, rest on tabletops, await daily use. The more ordinary the device, the better. Some possibilities: the blender, coffee grinder, alarm clock, hair dryer, lawnmower, remote control, electronic pager, power drill, answering machine, make-up mirror, bathroom scale, garage-door opener, radar detector, car phone, intercom, exercise machine. Many offices offer another set of devices to think about: computers, copiers, phone systems, fax machines, and the like. Obviously these are only suggestions. Many other devices would work well for this assignment. The real challenge is to find something you want to think about.

Your writing will probably take on one version or the other of this argument: While this device *appears* to have a simple and obvious function—to cut the grass, grind the coffee, dry your hair, wake you up—when you really start to think about it, it can actually be seen to mean something else as well. (You might refer to Judith Williamson's article "Urban Spaceman" found elsewhere in *Media Journal* for another writer who is able to see a familiar object in new, unfamiliar ways. You might also look for some print advertisements for this object to see how this product has been advertised.) In any case, your task is to define what that something else is, to show that the device you selected is more interesting than it once seemed. To do so you will need somehow to get beyond the obvious uses, the common sense understandings of what your device is for. You will want to show that embedded in the machine are certain *ideas,* and your writing will be an attempt to discover what those ideas are— and to criticize them if necessary.

A Weight that Women Carry

The Compulsion to Diet in a Starved Culture

Sallie Tisdale

I don't know how much I weigh these days, though I can make a good guess. For years I'd known that number, sometimes within a quarter pound, known how it changed from day to day and hour to hour. I want to weigh myself now; I lean toward the scale in the next room, imagine standing there, lining up the balance. But I don't do it. Going this long, starting to break the scale's spell—it's like waking up suddenly sober.

By the time I was sixteen years old I had reached my adult height of five feet six inches and weighed 164 pounds. I weighed 164 pounds before and after a healthy pregnancy. I assume I weigh about the same now; nothing significant seems to have happened to my body, this same old body I've had all these years. I usually wear a size 14, a common clothing size for American women. On bad days I think my body looks lumpy and misshapen. On my good days, which are more frequent lately, I think I look plush and strong; I think I look like a lot of women whose bodies and lives I admire.

I'm not sure when the word "fat" first sounded pejorative to me, or when I first applied it to myself. My grandmother was a petite woman, the only one in my family. She stole food from other people's plates, and hid the debris of her own meals so that no one would know how much she ate. My mother was a size 14, like me, all her adult life; we shared clothes. She fretted

463

endlessly over food scales, calorie counters, and diet books. She didn't want to quit smoking because she was afraid she would gain weight, and she worried about her weight until she died of cancer five years ago. Dieting was always in my mother's way, always there in the conversations above my head, the dialogue of stocky women. But I was strong and healthy and didn't pay too much attention to my weight until I was grown.

It probably wouldn't have been possible for me to escape forever. It doesn't matter that whole human epochs have celebrated big men and women, because the brief period in which I live does not; since I was born, even the voluptuous calendar girl has gone. Today's models, the women whose pictures I see constantly, unavoidably, grow more minimal by the day. When I berate myself for not looking like whomever I think I should look like that day, I don't really care that no one looks like that. I don't care that Michelle Pfeiffer doesn't look like the photographs I see of Michelle Pfeiffer. I want to look—think I should look—like the photographs. I want her little miracles: the makeup artists, photographers, and computer imagers who can add a mole, remove a scar, lift the breasts, widen the eyes, narrow the hips, flatten the curves. The final product is what I see, have seen my whole adult life. And I've seen this: even when big people become celebrities, their weight is constantly remarked upon and scrutinized; their successes seem always to be *in spite* of their weight. I thought my successes must be, too.

I feel my self expand and diminish from day to day, sometimes from hour to hour. If I tell someone my weight, I change in their eyes: I become bigger or smaller, better or worse, depending on what that number, my weight, means to them. I know many men and women, young and old, gay and straight, who look fine, whom I love to see and whose faces and forms I cherish, who despise themselves for their weight. For their ordinary, human bodies. They and I are simply bigger than we think we should be. We always talk about weight in terms of gains and losses, and don't wonder at the strangeness of the words. In trying to always lose weight, we've lost hope of simply being seen for ourselves.

My weight has never actually affected anything—it's never seemed to mean anything one way or the other to how I lived. Yet for the last ten years I've felt quite bad about it. After a time, the number on the scale became my totem, more important than my experience—it was layered, metaphorical, *metaphysical,* and it had bewitching power. I thought if I could change that number I could change my life.

In my mid-twenties I started secretly taking diet pills. They made me feel strange, half-crazed, vaguely nauseated. I lost about twenty-five pounds, dropped two sizes, and bought new clothes. I developed rituals and taboos around food, ate very little, and continued to lose weight. For a long time afterward I thought it only coincidental that every passing week I also grew more depressed and irritable.

I could recite the details, but they're remarkable only for being so common. I lost more weight until I was rather thin, and then I gained it all back. It came back slowly, pound by pound, in spite of erratic and melancholy and sometimes frantic dieting, dieting I clung to even though being thin had changed nothing, had meant nothing to my life except that I was thin. Looking back, I remember blinding moments of shame and lightning-bright moments of clearheadedness, which inevitably gave way to rage at the time I'd wasted—rage that eventually would become, one again, self-disgust and the urge to lose weight. So it went, until I weighed exactly what I'd weighed when I began.

I used to be attracted to the sharp angles of the chronic dieter—the caffeine-wild, chain-smoking, skinny women I see sometimes. I considered them a pinnacle not of beauty but of will. Even after I gained back my weight, I wanted to be like that, controlled and persevering, live that underfed life so unlike my own rather sensual and disorderly existence. I felt I should always be dieting, for the dieting of it; dieting had become a rule, a given, a constant. Every ordinary value is distorted in this lens. I felt guilty for not being completely absorbed in my diet, for getting distracted, for not caring enough all the time. The fat person's character flaw is a lack of narcissism. She's let herself go.

So I would begin again—and at first it would all seem so…easy. Simple arithmetic. After all, 3,500 calories equal one pound of fat—so the books and articles by the thousands say. I would calculate how long it would take to achieve the magic number on the scale, to succeed, to win. All past failures were suppressed. If 3,500 calories equal one pound, all I needed to do was cut 3,500 calories out of my intake every week. The first few days of a new diet would be colored with a sense of control—organization and planning, power over the self. Then the basic futile misery took over.

I would weigh myself with foreboding, and my weight would determine how went the rest of my day, my week, my life. When 3,500 calories didn't equal one pound lost after all, I figured it was my body that was flawed, not the theory. One friend, who had tried for years to lose weight following prescribed diets, made what she called "an amazing discovery." The real secret to a diet, she said, was that you had to be willing to be hungry *all the time.* You had to eat even less than the diet allowed.

I believed that being thin would make me happy. Such a pernicious, enduring belief. I lost weight and wasn't happy and saw that elusive happiness disappear in a vanishing point, requiring more—more self-disgust, more of the misery of dieting. Knowing all that I know now about the biology and anthropology of weight, knowing that diets are bad for me and won't make me thin—sometimes none of this matters. I look in the mirror and think: Who am I kidding? *I've got to do something about myself.* Only then will this vague discontent disappear. Then I'll be loved.

For ages humans believed that the body helped create the personality, from the humors of Galen to W. H. Sheldon's somatotypes. Sheldon distinguished between three templates—endomorph, mesomorph, and ectomorph—and combined them into hundreds of variations with physical, emotional, and psychological characteristics. When I read about weight now, I see the potent shift in the last few decades: the modern culture of dieting based on the idea that the personality creates the body. Our size must be in some way voluntary, or else it wouldn't be subject to change. A lot of my misery over my weight wasn't about how I looked at all. I was miserable because I believed *I* was bad, not my body. I felt truly reduced then, reduced to being just a body and nothing more.

Fat is perceived as an *act* rather than a thing. It is antisocial, and curable through the application of social controls. Even the feminist revisions of dieting, so powerful in themselves, pick up the theme: the hungry, empty heart; the woman seeking release from sexual assault, or the man from the loss of the mother, through food and fat. Fat is now a symbol not of the personality but of the soul—the cluttered, neurotic, immature soul.

Fat people eat for "mere gratification," I read, as though no one else does. Their weight is *intentioned*, they simply eat "too much," their flesh is lazy flesh. Whenever I went on a diet, eating became cheating. One pretzel was cheating. Two apples instead of one was cheating—a large potato instead of a small, carrots instead of broccoli. It didn't matter which diet I was on; diets have failure built in, failure is in the definition. Every substitution—even carrots for broccoli—was a triumph of desire over will. When I dieted, I didn't feel pious just for sticking to the rules. I felt condemned for the act of eating itself, as though my hunger were never normal. My penance was to not eat at all.

My attitude toward food became quite corrupt. I came, in fact, to subconsciously believe food itself was corrupt. Diet books often distinguish between "real" and "unreal" hunger, so that *correct* eating is hollowed out, unemotional. A friend of mine who thinks of herself as a compulsive eater says she feels bad only when she eats for pleasure. "Why?" I ask, and she says, "Because I'm eating food I don't need." A few years ago I might have admired that. Now I try to imagine a world where we eat only food we need, and it seems inhuman. I imagine a world devoid of holidays and wedding feasts, wakes and reunions, a unique shared joy. "What's wrong with eating a cookie because you like cookies?" I ask her, and she hasn't got an answer. These aren't rational beliefs, any more than the unnecessary pleasure of ice cream is rational. Dieting presumes pleasure to be an insignificant, or at least malleable, human motive.

I felt no joy in being thin—it was just work, something I had to do. But when I began to gain back the weight, I felt despair. I started reading about the "recidivism" of dieting. I wondered if I had myself to blame not only for needing to diet in the first place but for dieting itself, the weight inevitably regained. I joined organized weight-loss programs, spent a lot of money, listened to lectures I didn't believe on quack nutrition, ate awful, processed diet foods. I sat in groups and applauded people who lost a half pound, feeling smug because I'd lost a pound and a half. I felt ill much of the time, found exercise increasingly difficult, cried often. And I though that if I could only lose a little weight, everything would be all right.

When I say to someone, "I'm fat," I hear, "Oh, no! You're not *fat*! You're just—" What? Plump? Big-boned? Rubenesque? I'm just *not thin*. That's crime enough. I began this story by stating my weight. I said it all at once, trying to forget it and take away its power; I said it to be done being scared. Doing so, saying it out loud like that, felt like confessing a mortal sin. I have to bite my tongue not to seek reassurance, not to defend myself, not to plead. I see an old friend for the first time in years, and she comments on how much my fourteen-year-old son looks like me—"except, of course, he's not chubby." "Look who's talking," I reply, through clenched teeth. This pettiness is never far away; concern with my weight evokes the smallest, meanest parts of me. I look at another woman passing on the street and think, "At least I'm not *that* fat."

Recently I was talking with a friend who is naturally slender about a mutual acquaintance who is quite large. To my surprise my friend reproached this woman because she had seen her eating a cookie at lunchtime. "How is she going to lose weight that way?" my friend wondered. When you are as fat as our acquaintance is, you are primarily, fundamentally, seen as fat. It is your essential characteristic. There are so many presumptions in my friend's casual, cruel remark. She assumes that this woman should diet all the time—and that she *can*. She pronounces whole categories of food to be denied her. She sees her unwillingness to behave in this externally prescribed way, even for a moment, as an act of rebellion. In his story "A Hunger Artist," Kafka writes that the guards of the fasting man were "usually butchers, strangely enough." Not so strange, I think.

I know that the world, even if it views me as overweight (and I'm not sure it really does), clearly makes a distinction between me and this very big woman. I would rather stand with her and not against her, see her for all she is besides fat. But I know our experiences aren't the same. My thin friend assumes my fat friend is unhappy because she is fat: therefore, if she loses weight she will be happy. My fat friend has a happy marriage and family and a good career, but insofar as her weight is a source of misery, I think she would be much happier if she could eat her cookie in peace, if people would shut up and leave her weight alone. But the world never lets up when you are her size; she cannot walk to the bank without risking insult. Her fat is seen as perverse bad manners. I have no doubt she would be rid of the fat if she could be. If my left-handedness invited the criticism her weight does, I would want to cut that hand off.

In these last several years I seem to have had an infinite number of conversations about dieting. They are really all the same conversation—weight is lost, then weight is gained back. This repetition finally began to sink in. Why did everyone sooner or later have the same experience? (My friend who had learned to be hungry all the time gained back all the weight she had lost and more, just like the rest of us.) Was it really our bodies that were flawed? I began reading the biology of weight more carefully, reading the fine print in the endless studies. There is, in fact, a preponderance of evidence disputing our commonly held assumptions about weight.

The predominant biological myth of weight is that thin people live longer than fat people. The truth is far more complicated. (Some deaths of fat people attributed to heart disease seem actually to have been the result of radical dieting.) If health were our real concern, it would be dieting we questioned, not weight. The current ideal of thinness has never been held before, except as a religious ideal; the underfed body is the martyr's body. Even if people can lose weight, maintaining an artificially low weight for any period of time requires a kind of starvation. Lots of people are naturally thin, but for those who are not, dieting is an unnatural act; biology rebels. The metabolism of the hungry body can change inalterably, making it ever harder and harder to stay thin. I think chronic dieting made me gain weight—not only pounds, but fat. This equation seemed so strange at first that I couldn't believe it. But the weight I put back on after losing was much more stubborn than the original weight. I had lost it by taking diet pills and not eating much of anything at all for quite a long time. I haven't touched the pills again, but not eating much of anything no longer works.

When Oprah Winfrey first revealed her lost weight, I didn't envy her. I thought, She's in trouble now. I knew, I was certain, she would gain it back; I believed she was biologically destined to do so. The tabloid headlines blamed it on a cheeseburger or mashed potatoes; they screamed OPRAH PASSES 200 POUNDS, and I cringed at her misery and how the world wouldn't let up, wouldn't leave her alone, wouldn't let her be anything else. How dare the world do this to anyone? I thought, and then realized I did it to myself.

The "Ideal Weight" charts my mother used were at their lowest acceptable-weight ranges in the 1950s, when I was a child. They were based on sketchy and often inaccurate actuarial evidence, using, for the most part, data on northern Europeans and allowing for the most minimal differences in size for a population of less than half a billion people. I never fit those weight charts, I was always just outside the pale. As an adult, when I would join an organized diet program, I accepted their version of my Weight Goal as gospel, knowing it would be virtually impossible to reach. But reach I tried; that's what one does with gospel. Only in the last few years have the weight tables begun to climb back into the world of the average human. The newest ones distinguish by gender, frame, and age. And suddenly I'm not off the charts anymore. I have a place.

A man who is attracted to fat women says, "I actually have less specific physical criteria than most men. I'm attracted to women who weigh 170 or 270 or 370. Most men are only attracted to women who weigh between 100 and 135. So who's got more of a fetish?" We look at fat as a problem of the fat person. Rarely do the tables get turned, rarely do we imagine that it might be the viewer, not the viewed, who is limited. What the hell is wrong with *them*, anyway? Do they believe everything they see on television?

My friend Phil, who is chronically and most painfully thin, admitted that in his search for a partner he finds himself prejudiced against fat women. He seemed genuinely bewildered by this. I didn't jump to reassure him that such prejudice is hard to resist. What I did was bite my tongue at my urge to be reassured by him, to be told that I, at least, wasn't fat. That over the centuries humans have been inclined to prefer extra flesh rather than the other way around seems unimportant. All we see now tells us otherwise. Why does my kindhearted friend criticize another woman for eating a cookie when she would never dream of commenting in such a way on another person's race or sexual orientation or disability? Deprivation is the dystopian ideal.

My mother called her endless diets "reducing plans." Reduction, the diminution of women, is the opposite of feminism, as Kim Chernin points out in *The Obsession*. Smallness is what feminism strives against, the smallness that women confront everywhere. All of women's spaces are smaller than those of men, often inadequate, without privacy. Furniture designers distinguish between a man's and a woman's chair, because women don't spread out like men. (A sprawling woman means only one thing.) Even our voices are kept down. By embracing dieting I was rejecting a lot I held dear, and the emotional dissonance that created just seemed like one more necessary evil.

A fashion magazine recently celebrated the return of the "well-fed" body; a particular model was said to be "the archetype of the new womanly woman...stately, powerful." She is a size 8. The images of women presented to us, images claiming so maliciously to be the images of women's whole lives are not merely social fictions. They are absolute fictions; they can't exist. How would it feel, I began to wonder, to cultivate my own real womanliness rather than despise it? Because it was my fleshy curves I wanted to be rid of, after all. I dreamed of having a boy's body, smooth, hipless, lean. A body rapt with possibility, a receptive body suspended before the storms of maturity. A dear friend of mine, nursing her second child, weeps at her newly voluptuous body. She loves her children and hates her own motherliness, wanting to be unripened again, to be a bud and not a flower.

Recently, I've started shopping occasionally at stores for "large women," where the smallest size is a 14. In department stores the size 12 and 14 and 16 clothes are kept in a ghetto called the Women's Department. (And who would want that, to be the size of a woman? We all dream of being "juniors" instead.) In the specialty stores the clerks are usually big women and the customers are big, too, big like a lot of women in my life—friends, my sister, my mother and aunts. Not long ago I bought a pair of jeans at Lane Bryant and then walked through the mall to the Gap, with its shelves of generic clothing. I flicked through the clearance rack and suddenly remembered the Lane

Bryant shopping bag in my hand and its enormous weight, the sheer heaviness of that brand name shouting to the world. The shout is that I've let myself go. I still feel like crying out sometimes: Can't I feel *satisfied*? But I am not supposed to be satisfied, not allowed to be satisfied. My discontent fuels the market; I need to be afraid in order to fully participate.

American culture, which has produced our dieting mania, does more than reward privation and acquisition at the same time: it actually associates them with each other. Read the ads: the virtuous runner's reward is a new pair of $180 running shoes. The fat person is thought to be impulsive, indulgent, but insufficiently or incorrectly greedy, greedy for the wrong thing. The fat person lacks ambition. The young executive is complimented for being "hungry"; he is "starved for success." We are teased with what we will *have* if we are willing to *have not* for a time. A dieting friend, avoiding the food on my table, says, "I'm just dying for a bite of that."

Dieters are the perfect consumers: they never get enough. The dieter wistfully imagines food without substance, food that is not food, that begs the definition of food, because food is the problem. Even the ways we *don't eat* are based in class. The middle class don't eat in support groups. The poor can't afford not to eat at all. The rich hire someone to not eat with them in private. Dieting is an emblem of capitalism. It has a venal heart.

The possibility of living another way, living without dieting, began to take root in my mind a few years ago, and finally my second trip through Weight Watchers ended dieting for me. This last time I just couldn't stand the details, the same kind of details I'd seen and despised in other programs, on other diets: the scent of resignation, the weighing-in by the quarter pound, the before and after photographs of group leaders prominently displayed. Jean Nidetch, the founder of Weight Watchers, says, "Most fat people need to be hurt badly before they do something about themselves." She mocks every aspect of our need for food, of a person's sense of entitlement to food, of daring to *eat what we want*. Weight Watchers refuses to release its own weight charts except to say they make no distinction for frame size; neither has the organization ever released statistics on how many people who lose weight on the program eventually gain it back. I hated the endlessness of it, the turning of food into portions and exchanges, everything measured out, permitted, denied. I hated the very idea of "maintenance." Finally I realized I didn't just hate the diet. I was sick of the way I acted on a diet, the way I whined, my niggardly, penny-pinching behavior. What I like in myself seemed to shrivel and disappear when I dieted. Slowly, slowly I saw these things. I saw that my pain was cut from whole cloth, imaginary, my own invention. I saw how much time I'd spent on something ephemeral, something that simply wasn't important, didn't matter. I saw that the real point of dieting is dieting—to not be done with it, ever.

I looked in the mirror and saw a woman, with flesh, curves, muscles, a few stretch marks, the beginnings of wrinkles, with strength and softness in equal measure. My body is the one part of me that is always, undeniably, here. To like myself means to be, literally, shameless, to be wanton in the pleasures of being inside a body. I feel *loose* this way, a little abandoned, a little dangerous. That first feeling of liking my body—not being resigned to it or despairing of change, but actually *liking* it—was tentative and guilty and frightening. It was alarming, because it was the way I'd felt as a child, before the world had interfered. Because surely I was wrong; I knew, I'd known for so long, that my body wasn't all right this way. I was afraid even to act as though I were all right: I was afraid that by doing so I'd be acting a fool.

For a time I was thin. I remember—and what I remember is nothing special—strain, a kind of hollowness, the same troubles and fears, and no magic. So I imagine losing weight again. If the world applauded, would this comfort me? Or would it only compromise whatever approval the world gives me now? What else will be required of me besides thinness? What will happen to me if I get sick, or lose the use of a limb, or, God forbid, grow old?

By fussing endlessly over my body, I've ceased to inhabit it. I'm trying to reverse this equation now, to trust my body and enter it again with a whole heart. I know more now than I used to about what constitutes "happy" and "unhappy," what the depths and textures of contentment are like. By letting go of dieting, I free up mental and emotional room. I have more space, I can move. The pursuit of another, elusive body, the body someone else says I should have, is a terrible distraction, a sidetracking that might have lasted my whole life long. By letting myself go, I go places.

Each of us in this culture, this twisted, inchoate culture, has to choose between battles: one battle is against the cultural idea, and the other is against ourselves. I've chosen to stop fighting myself. Maybe I'm tilting at windmills; the cultural ideal is ever-changing, out of my control. It's not a cerebral journey, except insofar as I have to remind myself to stop counting, to stop thinking in terms of numbers. I know, even now that I've quit dieting and eat what I want, how many calories I take in every day. If I eat as I please, I eat a lot of one day and very little the next; I skip meals and snack at odd times. My nourishment is good—as far as nutrition is concerned, I'm in much better shape than when I was dieting. I know that the small losses and gains in my weight over a period of time aren't simply related to the number of calories I eat. Someone asked me not long ago how I could possibly know my calorie intake if I'm not dieting (the implication being, perhaps, that I'm dieting secretly). I know because calorie counts and grams of fat and fiber are embedded in me. I have to work to *not* think of them, and I have to learn not to think of them in order to really live without fear.

When I look, *really* look, at the people I see every day on the street, I see a jungle of bodies, a community of women and men growing every which way like lush plants, growing tall and short and slender and round, hairy and hairless, dark and pale and soft and hard and glorious. Do I look around at the multitudes and think all these people—all these people who are like me and not like me, who are various and different—are not loved or lovable? Lately, everyone's body interests me, every body is desirable in some way. I see how muscles and skin shift with movement; I sense a cornucopia of flesh in the world. In the midst of it I am a little capacious and unruly.

I repeat with Walt Whitman, "I dote on myself...there is that lot of me, and all so luscious." I'm eating better, exercising more, feeling fine—and then I catch myself thinking, *Maybe I'll lose some weight.* But my mood changes or my attention is caught by something else, something deeper, more lingering. Then I can catch a glimpse of myself by accident and think only: That's me. My face, my hips, my hands. Myself.

COMING TO TERMS

Sallie Tisdale argues how we look physically tends to be viewed as an outward sign of who we are personally or spiritually—and thus that our current preoccupation with thinness can be read as an index to a whole set of other cultural fears and values. We'd like you re-read "A Weight that Women Carry" now in order to list and define what she says these cultural concerns are. What does fat "mean" in our culture, according to Tisdale? What worries or fears do we attach to it? What character attributes or flaws in a person do we associate with being fat? Conversely, what do we associate with being "thin," what values do we associate with it? If thinness has not always been valued in history, then why is it so prized today? In what ways does Tisdale explain her cultural reading of thin today? What is the cultural context for her critical argument here? Mark any passages in Tisdale's text that help you answer these questions, and then jot down some notes in your journal on the present cultural meanings of "fat" and "thin," as she defines them in her essay. Finally, we'd like you to begin to test Tisdale's way of reading body types within a cultural context. Do your own personal experiences confirm Tisdale's argument? What are these experiences? Does the portrayal of thin and fat characters on television or film confirm or challenge her argument? In either case, what programs and characters would you point to?

READING AS A WRITER

Locate a recent book or magazine article on dieting or weight control. Read it carefully, taking some notes in your journal as you do, not for the information it may have to offer about a specific diet plan or menus or calories or exercise, but for the *tone* it takes in addressing its readers. What attitude might it be said to take toward the audience? Does the article or book address its readers in a humorous or serious manner, helpful and sympathetic way, or a threatening or foreboding way? Does it focus more on the consequences of being overweight? Or does it concentrate on the success of losing weight? Does it speak to an audience as a friend, or does it speak from a position of detached authority, from a distance? Does it specify or suggest a particular age, gender, or lifestyle? What does it assume to be the likely urges, temptations, problems, or weaknesses of its readers? How does it promise to help them? Does the text speak to and picture its readers in any of the ways that Tisdale describes? If there are differences, what are they? As you respond to these questions in your journal—remember: books and articles on dieting are all asking people to make changes in their life. There is, thus, a similarity in critical argument. But what we're interested in here is your insight on the *attitude* of that argument, and the ways in which this attitude or tone helps you to understand how an author perceives an audience.

WRITING CRITICISM

Tisdale writes here about the body as a "text" that is consciously shaped and interpreted, and particularly about the pressures put on people (and especially women) to mold their physical selves in conformity with a certain ideal physical type. We'd like you to continue this inquiry into the "fashion of the body" and the ways in which this fashion might be interpreted within a cultural context. And, as a point of entry into this investigation, we'd like you to think about the ways media in particular urge people to think about their bodies, and to present them to (or hide them from) the gaze of others.

There are two broad forms that we imagine such an investigation might take. In the first, your subject would be yourself, your body. Here, you might want to add or respond to the lines of thinking begun by Tisdale, to reflect on how questions about body weight and size have impinged on your own life and experiences, choices and decisions. But you could consider other aspects of the "fashion of the body" as well: hair (big, trimmed, kinky, straight, dyed, balding, clean-shaven); skin (dark, light, tan, pale, smooth, wrinkled, pierced, tattooed); eyes (glasses, contacts, tinted contacts); muscles, makeup, nails, breasts, hips, legs, height, and so on. Obviously, any or all of these could seem the most trivial of subjects, if handled in a way that allows them to become so. It will be your responsibility as a writer to make sure that this is not the case, to show instead how a particular issue concerning your body and physical appearance is somehow linked to larger issues in the culture, how your reading of a body text might reveal cultural attitudes toward, for example, the values of success, social acceptance, sexual attractiveness, conformity/non-conformity.

The other approach would be to look at some aspect of how bodies are presented to us through the media. Again you would want to take advantage of the work Tisdale has already done, to make use of and expand upon her critical argument. We think the primary question here would be: What sort of body seems currently "in"? There could be more than one body type in fashion at the same time, so you might want to see your investigation as exploring a competitive field of several body types. You would also need to define carefully and limit the "in" of your study. "In" could be defined in a number of ways, by location (school or work, eastern United States or western United States) or population (age, race, gender, single or married, or actively dating). (You may be tempted to begin your work here with young, thin, and beautiful or tall, dark, and handsome, but you will want to move past the "princes and princesses" of never-ending fairy tales as quickly as possible, to say more than the obvious.) What values are associated with this or that particular body look? Who has it? Does it have a genealogy? In other words, where does this body look come from? What's its history? What are people now being encouraged to do with or to their bodies? Why? If a certain body type is "in," what is this body to be used for? How does mainstream culture respond to this body look? Has there been some kind of public reaction to this fashion of the body? If you take this approach, you will want to support what you have to say with the evidence of particular texts—images and phrasings from magazines, books, movies, TV, videos, department store displays, and the like.

One way of thinking about the differences between these two approaches is this: While the first encourages you to look at some of the personal effects of the cultural pressures placed on our bodies, the second asks you to examine the mechanisms through which those pressures are exerted.

"This Is a Naked Lady"

Gerard Van Der Leun

Back in the dawn of online when a service called The Source was still in flower, a woman I once knew used to log on as "This is a naked lady." She wasn't naked of course, except in the minds of hundreds of young and not-so-young males who also logged on to The Source. Night after night, they sent her unremitting text streams of detailed wet dreams, hoping to engage her in online exchanges known as "hot chat"—a way of engaging in a mutual fantasy typically found only through 1-900 telephone services. In return, "The Naked Lady" egged on her digital admirers with leading questions larded with copious amounts of double entendre.

When I first asked her about this, she initially put it down to "just fooling around on the wires."

"It's just a hobby," she said. "Maybe I'll get some dates out of it. Some of these guys have very creative and interesting fantasy lives."

At the start, The Naked Lady was a rather mousy person—the type who favored gray clothing of a conservative cut—and was the paragon of shy and retiring womanhood. Seeing her on the street, you'd never think that her online persona was one that excited the libidos of dozens of men every night.

But as her months of online flirtations progressed, a strange transformation came over her: She became (through the dint of her blazing typing speed) the kind of person that could keep a dozen or more online sessions of hot chat going at a time. She got a trendy haircut. Her clothing tastes went from Peck and Peck to tight skirts slit up the thigh. She began regaling me with descriptions of her expanding lingerie collection. Her speech became bawdier, her jokes naughtier. In short, she was becoming her online personality—lewd, bawdy, sexy, a man-eater.

The last I saw of her, The Naked Lady was using her online conversations to cajole dates and favors from those men foolish enough to fall into her clutches.

The bait she used was an old sort—sex without strings attached, sex without love, sex as a fantasy pure and simple. It's an ancient profession whose costs always exceed expectations and whose pleasures invariably disappoint. However, the "fishing tackle" was new: online telecommunications.

In the eight years that have passed since The Naked Lady first appeared, a number of new wrinkles have been added to the text-based fantasy machine. Groups have formed to represent all sexual persuasions. For a while, there was a group on the Internet called, in the technobabble that identifies areas on the net, alt.sex.bondage.golden.showers.sheep. Most people thought it was a joke, and maybe it was.

Online sex stories and erotic conversations consume an unknown and unknowable portion of the global telecommunications bandwidth. Even more is swallowed by graphics. Now, digitized sounds are traveling the nets, and digital deviants are even "netcasting" short movie clips. All are harbingers of things to come.

It is as if all the incredible advances in computing and networking technology over the past decades boil down to the ability to ship images of turgid members and sweating bodies everywhere and anywhere at anytime. Looking at this, it is little wonder that whenever this is discovered (and someone, somewhere, makes the discovery about twice a month), a vast hue and cry resounds over the nets to root-out the offending material and burn those who promulgated it. High tech is being perverted to low ends, they cry. But it was always so.

There is absolutely nothing new about the prurient relationship between technology and sexuality.

Sex, as we know, is a heat-seeking missile that forever seeks out the newest medium for its transmission. William Burroughs, a man who understands the dark side of sexuality better than most, sees it as a virus that is always on the hunt for a new host—a virus that almost always infects new technology first. Different genders and psyches have different tastes, but the overall desire seems about as persistent over the centuries as the lust for bread and salvation.

We could go back to Neolithic times when sculpture and cave painting were young. We could pick up the prehistoric sculptures of females with pendulous breasts and very wide hips—a theme found today in pornographic magazines that specialize in women of generous endowment. We could then run our flashlight over cave paintings of males whose members seem to exceed the length of their legs. We could travel forward in time to naughty frescos in Pompeii, or across continents to where large stones resembling humongous erections have for centuries been major destinations to pilgrims in India, or to the vine-choked couples of the Black Pagoda at Ankor Wat where a Mardi Gras of erotic activity carved in stone has been on display for centuries. We could proceed to eras closer to our time and culture, and remind people that movable type not only made the Gutenberg Bible possible, but that it also made cheap broadsheets of what can only be called "real smut in Elizabethan English" available to the masses for the very first time. You see, printing not only made it possible to extend the word of God to the educated classes, it also extended the monsters of the id to them as well.

Printing also allowed for the cheap reproduction and broad distribution of erotic images. Soon, along came photography; a new medium, and one that until recently did more to advance the democratic nature of erotic images than all previous media combined. When photography joined with photolithography, the two together created a brand new medium that many could use. It suddenly became economically feasible and inherently possible for lots of people to enact and record their sexual fantasies and then reproduce them for sale to many others. Without putting too fine a point on it, the Stroke Book was born.

Implicit within these early black-and-white tomes (which featured a lot of naked people with Lone Ranger masks demonstrating the varied ways humans can entwine their limbs and conceal large members at the same time) were the vast nascent publishing empires of *Playboy, Penthouse,* and *Swedish Erotica.*

The point here is that all media, when they are either new enough or become relatively affordable, are used by outlaws to broadcast unpopular images or ideas. When a medium is created, the first order of business seems to be the use of it in advancing religious, political, or sexual notions and desires. Indeed, all media, if they are to get a jump-start in the market and become successful, must address themselves to mass drives—those things we hold in common as basic human needs.

But of all these: food, shelter, sex and money; sex is the one drive that can elicit immediate consumer response. It is also why so many people obsessed with the idea of eliminating pornography from the earth have recently fallen back on the saying "I can't define what pornography is, but I know it when I see it."

They're right. You can't define it; you feel it. Alas, since everyone feels it in a slightly different way and still can't define it, it becomes very dangerous to a free society to start proscribing it.

And now we have come to the "digital age" where all information and images can be digitized; where all bits are equal, but some are hotter than others. We are now in a land where late-night cable can make your average sailor blush. We live in an age of monadic seclusion, where dialing 1-900 and seven other digits can put you in intimate contact with pre-op transsexuals in wet suits who will talk to you as long as the credit limit on your MasterCard stays in the black.

If all this pales, the "adult" channels on the online service CompuServe can fill your nights at $12.00 an hour with more fantasies behind the green screen than ever lurked behind the green door. And that's just the beginning. There are hundreds of adult bulletin board systems offering God Knows What to God Knows Who, and making tidy profits for plenty of folks.

Sex has come rocketing out of the closet and into the terminals of anyone smart enough to boot up FreeTerm. As a communications industry, sex has transmogrified itself from the province of a few large companies and individuals into a massive cottage industry.

It used to be, at the very least, that you had to drive to the local (or not-so-local) video shop or "adult" bookstore to refresh your collection of sexual fantasies. Now, you don't even have to leave home. What's more, you can create it yourself, if that's your pleasure, and transmit it to others.

It is a distinct harbinger of things to come that "Needless to say..." letters now appearing online are better than those published in Penthouse Forum, or that sexual images in binary form make up one of the heaviest data streams on the Internet, and that "amateur" erotic home videos are the hottest new category in the porn shops.

Since digital sex depends on basic stimuli that is widely known and understood, erotica is the easiest kind of material to produce. Quality isn't the primary criteria. Quality isn't even the point. Arousal is the point, pure and simple. Everything else is just wrapping paper. If you can pick up a Polaroid, run a Camcorder, write a reasonably intelligible sentence on a word processor or set up a bulletin board system, you can be in the erotica business. Talent has very, very little to do with it.

The other irritating thing about sex is that like hunger, it is never permanently satisfied. It recurs in the human psyche with stubborn regularity. In addition, it is one of the drives most commonly stimulated by the approved above-ground media (Is that woman in the Calvin Klein ads coming up from a stint of oral sex, or is she just surfacing from a swimming pool?) Mature, mainstream corporate media can only tease. New, outlaw media delivers. Newcomers can't get by on production values, because they have none.

Author Howard Rheingold has made some waves recently with his vision of a network that will actually hook some sort of tactile feedback devices onto our bodies so that the fantasies don't have to be so damned cerebral. He calls this vision "dildonics," and he has been dining out on the concept for years. With it, you'll have virtual reality coupled with the ability to construct your own erotic consort for work, play, or simple experimentation.

Progress marches on. In time, robotics will deliver household servants and sex slaves.

I saw The Naked Lady about three months ago. I asked her if she was still up to the same old games of online sex. "Are you kidding?" she told me. "I'm a consultant for computer security these days. Besides, I have a kid now. I don't want that kind of material in my home."

COMING TO TERMS

Gerard Van Der Leun makes several claims about an ongoing relationship between technology and sexuality. He writes, for example, "There is absolutely nothing new about the prurient relationship between technology and sexuality." And at another point Van Der Leun says, "Sex is a virus that infects new technology." To support these claims, he turns to history and offers a number of illustrations which (according to Van Der Leun) show how advancements in telecommunications technology engender new opportunities to express, explore, or exploit sexuality. Locate and re-read those passages in Van Der Leun's article where he offers these historical proofs, and make a list of them in your journal. Then answer the following questions. What other advancements might be added to this list (explain your reasons for including them)? Are there advancements which run contrary to Van Der Leun's claim about technology and sexuality? What do you think is the point of this historical evidence? Is Van Der Leun warning us or is he assuring us not to worry about the dangers of sexual license on the Internet?

READING AS A WRITER

In "This Is a Naked Lady," Van Der Leun frames his discussion of sex, technology, and media "outlaws" with an *anecdote,* opening and closing his article with stories of The Naked Lady, an online persona he first met eight years ago. Anecdotes might be read in any number of ways, as short, entertaining stories; but sometimes anecdotes can be read as reflective of something larger, as an alternative way of presenting a critical argument, for instance. In framing his article this way, Van Der Leun chooses an alternative way of introducing and providing a sense of closure to his discussion. (As readers, we tend to expect a rather explicit thesis or statement of an author's critical argument.) The intersection of narrative and expository writing found in "This Is a Naked Lady" offers us the opportunity to explore the critical relationships between two kinds of discourse that have been seen traditionally as being somehow oppositional, at least in the sense of what each discourse might accomplish intellectually. As a way of exploring this critical relationship between narrative and exposition, between story and critical argument, respond to the following questions in your journal. First, in your own words, how would you explain Van Der Leun's critical argument? Next, how does the anecdote of The Naked Lady illustrate, complicate, or extend this argument? Does the beginning anecdote remind you of a specific kind of narrative? In particular, what does the first line recall? A story with a moral? A parable? A fairy tale? Do the anecdotes of The Naked Lady confuse the issues of Van Der Leun's critical argument? Do they contribute to his argument? If the anecdotes were eliminated, how would this alter your reading of his article?

WRITING CRITICISM

While he seems most concerned with sexuality and its "prurient relationship" with technology, Gerard Van Der Leun can also be seen as using this issue as an illustration of an even more inclusive critical argument about media: "The point here is that all media, when they are either new enough or become relatively affordable, are used by outlaws to broadcast unpopular images or ideas." We were intrigued by Van Der Leun's linking of "unpopular images or ideas" with "outlaws," creating in a sense the *media outlaw. Outlaw* is a particularly rich word and is invested with multiple and, at times, contradictory cultural meanings in U.S. history. The legends of the American West, the bank robbers of the 1920s and 1930s, the heroes of many popular and folk ballads: all have been called outlaws. Those who inhabit a counterculture might be said to be cultural outlaws. And today, those who push the limits of legal behavior in areas of the environment, animal rights, or other political causes, might also earn the title of outlaw. Our relationship to the outlaw is complex. On one hand, an outlaw is a criminal or a fugitive from justice, but maybe the law broken is perceived to be immoral or beneficial only to a select minority; on the other hand, an outlaw might be an individual who rebels against a conventional way of thinking or doing things. As such, the American concept of outlaw can easily inspire a kind of cultural ambiguity, suggesting both an individual to be feared as well as romanticized and celebrated.

We'd like you to write an essay in which you explore this concept of *media outlaw* through a reading of some popular or media culture text. For this assignment you might choose a character from a television series or a film. Or you might consider a television or radio personality. The Internet offers several possibilities here: a chat room might be a place where cyber-outlaws meet and exchange ideas, for example. You might also select a columnist, journalist, or musical performer as well.

Your choice of media personality, celebrity, or artist here needs to illustrate Van Der Leun's cultural reading of "outlaw": someone who works in the currency of "unpopular images or ideas" through popular or media culture. As you do the work of this project, you'll need to address a set of questions: How does your choice of subject here show a media outlaw at work? In what ways does this media outlaw challenge, question, or violate the rules, conventions, or traditions of that medium? It's important here to establish—in Van Der Leun's terms—what the "popular" is. In other words, in order to show how someone might be "breaking the laws" of the traditional or conventional, you first need to define in some detail what the traditional or conventional is in terms of an artist or performer or the media or popular culture text itself. What are the rules, the conventions, the traditions, the usual expectations? And, as an extension of Van Der Leun's project, how might this performer, artist, personality, or text also be a *cultural outlaw* as well, violating cultural values or beliefs? Finally, Van Der Leun doesn't appear to be all that surprised that the Internet would develop a "prurient" relationship to sexuality. For the author, it seems almost predictable, business as usual, a given. How might your choice of media outlaw also be considered predictable? How is it that the untraditional becomes a tradition, that the unpopular becomes popular, that the outlaw becomes a stock figure or stereotype, a predictable response to culture and history?

Denim Downsize

Jay Weiser

I sat at my desk the other afternoon, dressed in business casual, and wistfully fingered my most recent pay stub. Since Labor Day—and a subsequent 45 percent downsizing—we've been allowed to dress more informally to boost staff morale. And I thought as I've thought before: perhaps it's no coincidence that business casual is sweeping the country just as white-collar job security evaporates.

In the beginning, for men who were managers or professionals in big organizations, there was the suit. In *Sex and Suits*, Anne Hollander traces its origin to the 1780s, but the modern ensemble of jacket and pants in a single wool fabric, dress shirt and vertical tie dates from the era of the first great industrial bureaucracies about a century ago. In Edward Steichen's 1908 photographs, Teddy Roosevelt and William H. Taft look just like suited folk today, minor differences in lapel size and collars aside. John Molloy's classic *Dress for Success* (1975) called the suit "the central power garment—the garment that establishes our position...in any in-person business situation," and laid out in absurd detail the appropriate fabrics, patterns and colors for suits and their associated shirts and ties. In the 1980s, acceptable professional wear for men became louder—blue shirts with white collars, bright suspenders, wild ties—but the suit remained invulnerable until the advent of casual Fridays around 1991.

By 1992, a year of lingering slow-motion Depression, a survey by Dockers manufacturer Levi Strauss showed that 26 percent of companies had a dress-down day. Now nearly three-quarters of the largest companies, including General Motors, Ford, Mobil, Chrysler, General Electric and even the conservative hold-out IBM, encourage casual dress for the office at least some of the time.

"Denim Downsize," by Jay Weiser, *New Republic*, 2/26/96, pp. 10–11. Reprinted with permission of the publisher.

But, unfortunately, dressing down seemed to go hand-in-hand with downsizing. Since 1989, about 3 million people have been laid off from their jobs. There has been some improvement—downsizings have apparently been decreasing since 1993, and in November 1995 unemployment was a relatively low 5.6 percent—but insecurity is rampant. Forty-three percent of white-collar workers who lost their jobs in the 1991-92 recession have had to settle for lower pay, and nearly 40 percent of the major firms that downsized in 1994 boosted hours for the survivors. Lifetime employment in one company is now dead and portable skills your only shelter. In this anxious environment, business casual seems, as Deloitte & Touche Atlanta managing partner David Passman put it, "a no-cost benefit." Too bad that real-dollar wages and benefits—the kind you live on—are down 5.5 percent since 1987. But in your Nautica ensemble, the new business-think goes, you'll feel blissful anyway.

Of course, business casual isn't merely a scheme to turn employers trembling beneath the budget axe into smiley worker bees. For one thing, many employees like it. And besides, a whole array of social forces—from the upstart entrepreneurship of the '80s to the influx of women into management—helped ensure the rise of business casual. In *Dress for Success*, John Molloy had extolled the IBM uniform of dark suits and white shirts, arguing that the right look imbued companies with *esprit de corps*. But within a few years, IBM was rocked by innovative Silicon Valley companies, where computer geeks ran the show and dress codes were unheard of. Meanwhile, the new generation of professional women, lacking a business uniform as standard as men's, veered between dressing as frilly non-players and power-suited yuppies-from-hell. It was an unappealing choice, but male co-workers still envied women's relative freedom and fantasized about going tie-less.

Within the business world, informality has spread well beyond business casual, and its influence has sometimes been deceptive. Nowadays, for example, convention would expect middle managers in many big companies to call the CEO by his first name. But are they really any more equal than they would have been thirty years ago, when they would have addressed him as "Mr."? In fact, the new informality masks an increasingly hierarchical power structure within companies, just as Gingrichian rhetoric about the "Opportunity Society" masks increasing economic inequality. In his 1995 book, *Company Man*, Anthony Sampson calls today's modern big corporations "monarchies" compared to the committee-run firms of twenty years ago. Monarchy is reflected in salary: today's typical American CEO makes 190 times the compensation of the typical worker; twenty years ago the multiple was only 40.

And the truth is business casual isn't quite the social-leveler it appears to be. It doesn't, after all, mean dressing as you would at home, with comfort or self-expression the paramount concerns. It has its own rules, even if they aren't always explicit. That's why it can actually be harder to get right than the old uniform was. (In recognition of the challenges, the helpful Levi Strauss has issued "A Guide to Casual Businesswear," while Neiman Marcus offers a free video, and Marshall Field sells a book on the subject.)

In the emerging casual consensus, t-shirts and torn jeans are taboo; as are the hip-hop fashions of African-American teenagers and the lawn-mowing outfits of the suburban gentry. Preppie, however, is in, along with the natural fabrics and understated designer stuff. The idea, as a *Cleveland Plain Dealer* service article for women puts it, is to "dress one level above," but not to be so fancy or high-style that you're conspicuous. I got the message; on my emergency business-casual shopping spree, I acquired two Ralph Lauren Polo sweaters. (I rationalize that Ralph's fake-English clothes really *are* designed for my lifestyle, since we're both from New York's outer boroughs.) My new Gap khakis probably aren't too déclassé, though I worry whether the cotton twill fabric is as lustrous as Tommy Hilfiger's.

Despite these subtle gradations, dress distinctions between professionals and the support staff are now blurred, except on the rare occasions when outsiders are visiting and the rules demand the old uniform. My $25 Land's End mesh polo shirt is affordable for a much larger market than the Izod Lacoste variety, but the only easily visible difference is the alligator monogram. Looking around my office, I've noticed that virtually all the men, from the mailroom guys to the department heads, are dressed similarly.

As companies have gotten more monarchical, all their employees have become *sans-culottes.* The functional and symbolic reasons for distinctive dress have diminished. The newly industrialized society that gave birth to the suit has metamorphosed into a mass service economy, with a relatively small proportion of the population laboring in dirty jobs that require plain, easily washable clothes. College-educated workers—the former suit-wearing classes—are now nearly as likely to lose their jobs as blue-collar workers. The image that comes to mind is from Busby Berkeley's *Gold Diggers of 1933*—those endless ranks of jobless World War I veterans trudging through the Piranesian production number "My Forgotten Man." Nowadays, they'd wear khakis and pinpoint oxford shirts instead of uniforms and helmets. Progress—of a sort.

COMING TO TERMS

Weiser here tries to decode the meanings both of a particular term—"business casual"—and the "look" that it represents. We'd like you to answer the following questions in your journal. What are the "rules" for casual dress in the workplace, according to Weiser? (Why are rules even needed, if the look is supposedly "informal"?) What does this particular look suggest to employees about their companies, their status, their future? In what ways does Weiser see the meanings of "business casual" as positive? In what ways does he view its implications as troubling?

READING AS A WRITER

Locate a newspaper or magazine article or advertisement that discusses "casual dressing" for the workplace. (You might make use of *The Readers' Guide to Periodical Literature* or the Internet here to find appropriate articles; Sunday supplements to newspapers often have articles concerned with the workplace; and if you choose an ad, make sure it has written copy or text as well as images.) List in your journal the terms your text uses to define or describe "business casual." In what ways does the language of your text reinforce or contradict what Weiser has to say? What common concerns does Weiser share with your text?

WRITING CRITICISM

Weiser links styles of dress to issues of status and power. One way of asserting authority through dress, obviously, is to wear a *uniform*—as do people in the military, the police, the clergy, some medical and technical workers, and the like. But, of course, uniforms can also be used to demonstrate a lack of authority—as is the case, for instance, with prisoners, medical patients, students, many restaurant and service workers, and so on. What is perhaps a more interesting and complex issue, though, is how people dress to advertise (or sometimes downplay) some form of personal authority when they are *not* required to wear a uniform. This is very often a concern of people who hold "professional" jobs: lawyers, doctors, managers, administrators, academics, teachers, social workers, and the like. For instance, almost any young person beginning to teach college has probably, at one point or the other, thought about how to dress in order to distinguish themselves from their students (or may have deliberately decided not to do so). Similarly, women in business, as Weiser notes, often have to negotiate between a desire to dress fashionably and a need to wear clothes that will suggest that they are to be taken seriously as men. One result of such pressures is that professionals in many fields often seem to end up wearing a kind of improvised "uniform" after all, or at least to follow a common fashion "code" that encourages a certain kind of "look" and discourages others.

We'd like you to write an essay which describes and interprets such a code of dress—an unofficial but nonetheless noticeable "uniform." Pick a group of people—workers in a particular office, college professors, salespeople in a mall, the members of a certain student clique or organization—who seem to you to share a common way of dressing, and whom you can observe and take some notes on. This does not mean that everyone in your group needs to dress exactly alike; indeed, you may learn much from observing people who are both part of the crowd and who, for some reason, stand out from it. But you should be able to identify various patterns and themes of dress, to describe a general "look" that most individuals in the group conform more or less closely to, or "pull off" more or less successfully. You will want to offer several examples of this fashion code or look, to suggest what variations it allows and what traits or features it requires (or at least seems strongly to encourage). If in doing so, it seems useful to refer to drawings or photographs, then make them part of your writing.

Once you have described the look you are studying, you will next need to explain and interpret it. For instance, if almost everyone in your college English department really does wear tweed jackets with patches of the elbows, then you will want to try to come up with a reason why they do so—to figure out, that is, what those patches might be meant to signify about their wearers. Ask yourself: What does this look tell *outsiders* about the members of this group? What does it tell the members of the group about themselves? How are these messages sent, through what details and variations of dress? To answer such questions well, you will need to be as precise in your observations and notetaking as you can, to describe the "texts" (that is, the clothes your subjects are wearing) you are writing about as fully and clearly as possible.

Michael Jordan Leaps the Great Divide

John Edgar Wideman

> *This old woman told me she went to visit this old retired bull-fighter who raised bulls for the ring. She had told him about this record that had been made by a black American musician, and he didn't believe that a foreigner, an American—especially a black American—could make such a record. He sat there and listened to it. After it was finished, he rose from his chair and put on his bullfighting equipment and outfit, went out and fought one of his bulls for the first time since he had retired, and killed the bull. When she asked him why he had done it, he said he had been so moved by the music that he just had to fight the bull.*
> —Miles, The Autobiography

When it's played the way it's spozed to be played, basketball happens in the air, the pure air; flying, floating, elevated above the floor, levitating the way oppressed peoples of this earth imagine themselves in their dreams, as I do in my lifelong fantasies of escape and power, finally, at last, once and for all, free. For glimpses of this ideal future game we should thank, among others, Elgin Baylor, Connie Hawkins, David Thompson, Helicopter Knowings, and of course, Julius Erving, Dr. J. Some venerate Larry Bird for reminding us how close a man can come to a perfect gravity-free game and still keep his head, his feet firmly planted on terra firma. Or love Magic Johnson for confounding boundaries, conjuring new space, passing lanes, fast-break and break-down lanes neither above the court nor exactly on it, but somehow whittling and expanding simultaneously the territory in which the game is enacted. But really, as we envision soaring and swooping, extending, refining the combat zone of basketball into a fourth, outer, other dimension, the dreamy ozone of flight without wings, of going up and not coming down till we're good and ready, then it's Michael Jordan we must recognize as the truest prophet of what might be possible.

A great artist transforms our world, removes scales from our eyes, plugs from our ears, gloves from our fingertips, teaches us to perceive reality differently. Proust said of his countryman and contemporary, the late-nineteenth-century Impressionist Auguste Renoir: "Before Renoir painted there were no Renoir women in Paris, now you can see them everywhere." Tex Winters, a veteran Chicago Bulls coach, a traditionalist who came up preaching the conventional wisdom that a lay-up is the highest-percentage shot, enjoys Michael Jordan's dunks, but, says MJ, "Every time I make one, he says, 'So whatever happened to the simple lay-up?' 'I don't know, Tex, this is how I've been playing my whole career.' You know, this stuff here and this stuff here [the hands are rocking, cradling, stuffing an imaginary ball] is like a lay-up to me. You know I've been doing that and that's the creativity of the game now. But it drives him nuts...and he says, 'Well, why don't you draw the foul?' I say I never have. The defense alters many of my shots, so I create. I've always been able to create in those situations, and I guess that's the Afro-American game I have, that's just natural to me. And even though it may not be the traditional game that Americans have been taught, it works for me. Why not?"

The lady is gaudy as Carnival. Magenta, sky-blue, lime, scarlet, orange swirl in the dress that balloons between her sashed waist and bare knees. Somebody's grandmother, gift-wrapped and wobbly on Madison Street, to-readoring through four lanes of traffic converging on Chicago's Stadium. Out for a party. Taxi driver says this is where they stand at night. Whore women, he calls them, a disgusted judgmental swipe in his voice, which until now has been a mellow tour-guide patter, pointing out the Sears Tower, Ditka's, asking me how tall is Michael Jordan. The tallest in basketball? Laughing at his memory of a photo of Manute Bol beside Muggsy Bogues. Claiming to have seen Michael Jordan at Shelter, a West Side club late on Wednesday, the night of Game Four after the Bulls beat Detroit last spring to even the best-of-seven NBA championship semifinal at two games apiece. Yes, with two other fellas. Tall like him. Lots of people asking him to sign his name. Autographs, you know. A slightly chopped, guttural, Middle-Eastern-flavored Chicagoese, patched together in the two and a half years he's resided in the States. "I came here as student. My family sent me three thousand dinars a year, and I could have apartment, pay my bills, drive a car.

Then hard times at home. Dinar worth much less in dollars. Four people, all of them, must work a month to earn one thousand dinars. No school now. I must work now. American wife and new baby, man."

The driver's from Jordan, but the joke doesn't strike me until I'm mumbling out the cab in front of Chicago Stadium. JORDAN, the country. Appearing in the same column, just above JORDAN, MICHAEL, in *Readers' Guide to Periodical Literature*, where I researched Michael Jordan's career. Usually more entries in each volume under JORDAN, MICHAEL, than any other JORDAN.

The other passenger sharing a cab from O'Hare to downtown Chicago is a young German from Hamburg, in the city with about one hundred thousand fellow conventioneers for the Consumer Electronics Show. It's while he's calculating exchange rates to answer the driver's question about the cost of a Mercedes in Germany that the lady stumbles backward from the curb into the street, blocking strings of cars pulled up at a light. She pirouettes. Curtsies. The puffy dress of many colors glows brighter, wilder against grays and browns of ravaged cityscape. Partially demolished or burnt-out or abandoned warehouses and storefronts line both sides of Madison. Interspersed between buildings are jerry-rigged parking lots where you'd leave your car overnight only if you had a serious grudge against it. I think of a mouth rotten with decay, gaps where teeth have fallen out. Competing for the rush of ballgame traffic, squads of shills and barkers hip-hop into the stalled traffic, shucking and jiving with anyone who'll pay attention. One looms at the window of our cab, sandwiched in, a hand-lettered sign tapping the windshield, begging us to park in his oasis, until the woman impeding our progress decides to attempt the curb again and mounts it this time, Minnie Mouse high heels firmly planted as she gives the honking cars a flounce of Technicolor behind and a high-fived middle finger.

The woman's black, and so are most of the faces on Madison as we cruise toward the stadium in a tide of cars carrying white faces. Closer, still plenty of black faces mix into the crowd—vendors, scalpers, guys in sneakers and silky sweat suits doing whatever they're out there doing, but when the cab stops and deposits me into a thin crust of dark people who aren't going in, I cut through quickly to join the mob of whites who are.

I forget I am supposed to stop at the press trailer for my credentials but slip inside the building without showing a pass or a ticket because mass confusion reigns at the gate. Then I discover why people are buzzing and shoving, why the gate crew is overwhelmed and defeated: Jesse Jackson. Even if it belongs to Michael Jordan this evening, Chicago's still Jackson's town too. And everyone wants to touch or be touched by this man who is instantly recognized, not only here in Chicago but all over the world. Casually dressed tonight, black slacks, matching black short-sleeved shirt that displays his powerful shoulders and arms. He could be a ballplayer. A running back, a tight end. But the eyes, the bearing are a quarterback's. Head high, he scans the whole field, checks out many things at once, smiles, and presses the flesh of the one he's greeting but stays alert to the bigger picture. When a hassled ticket-taker stares suspiciously at me, I nod toward Jesse, as if to say, I'm with him, he's the reason I'm here, and that's enough to chase the red-faced gate-keeper's scrutiny to easier prey. This minute exchange, insignificant as it may be, raises my spirits. Not because Jesse Jackson's presence enabled me to get away with anything—after all, I'm legit, certified, qualified to enter the arena—but because the respect, the recognizability he's earned reflects on me, empowers me, subtly alters others' perceptions of who I am, what I can do. When I hug the Reverend Jesse Jackson I try to impart a little of my apprecia-tion to the broad shoulders I grip. By just being out there, being heard and seen, by standing for something—for instance, an African-American's right, duty, and ability to aim for the stars—he's saved us all a lot of grief, bought us, black men, white, the entire rainbow of sexes and colors, more time to get our sorry act together. *Thank you* is what I always feel the need to say when I encounter the deep light of his smile.

Your town, man.

Brother Wideman, what are you doing here?

Writing about Michael Jordan.

We don't get any further. Somebody else needs a piece of him, a word, a touch from our Blarney stone, our Somebody.

In Michael's house the PA system is cranked to a sirenlike, earsplitting pitch, many decibels higher than a humane health code should permit. The Luvabulls, Chicago's aerobicized version of the Dallas Cowboy Cheerleaders, shake that fine, sculpted booty to pump up the fans. Very basic here. Primal-scream time. The incredible uproar enters your pores, your blood, your brain. Your nervous system becomes an extension of the overwhelming assault upon it. In simplest terms, you're ready for total war, transformed into a weapon poised to be unleashed upon the enemy.

From my third-row-end-zone folding chair, depth perception is nil. The game is played on a flat, two-dimensional screen. Under the opposite basket the players appear as they would crowded into the wrong end of a telescope.

Then, as the ball moves toward the near goal, action explodes, a zoom lens hurtles bodies at you larger, more intense than life. Middle ground doesn't ex-ist. You're surprised when a ref holds up both arms to indicate a three-point goal scored from behind the twenty-three-foot line. But your inability to gauge the distance of jump shots or measure the swift, subtle penetration, step by step, yard by yard as guards dribble between their legs, behind the back, spin-ning, dipping, shouldering, teasing their way upcourt, is compensated for by your power to watch the glacial increments achieved by big men muscling each other for position under your basket, the intimacy of those instants when the ball is in the air at your end of the court and just about everybody on both teams seems driven to converge into a space not larger than two telephone booths. Then it's grapple, grunt, and groan only forty feet away. You can read the effort, the fear, the focus in a player's eyes. For a few seconds you're on the court, sweating, absorbing the impact, the crash of big bodies into one another, wood buckling underfoot, someone's elbow in your ribs, shouts in your ear, the wheezes, sighs, curses, hearing a language spoken here and nowhere else except when people are fighting or making love.

MJ: What do I like about basketball? *Hmmm.* That's a good question. I started when I was twelve. And I enjoyed it to the point that I started to do things other people couldn't do. And that intrigued me more. Now I still enjoy it because of the excitement I get from fans, from the people, and still having the same ability to do things that other people can't do but want to do and they can do only through you. They watch you do it, then they think that they can do it. Or maybe they know it's something they can't do and ironically, that's why they feel good watching me. That drives me. I'm able to do something that no one else can do.

And I love competition. I've earned respect thanks to basketball. And I'm not here just to hand it to the next person. Day in and day out I see people take on that challenge, to take what I have earned. Joe Dumars, for one—I mean, I respect him, don't get me wrong. It's his job. I've got something that people want. The ability to gain respect for my basketball skills. And I don't ever want to give it away. Whenever the time comes when I'm not able to do that, then I'll just back away from the game.

JW: We've always been given credit for our athletic skill, our bodies. You've been blessed with exceptional physical gifts, and all your mastery of the game gets lost in the rush. But I believe your mind, the way you conceive the game, plays as large a role as your physical abilities. As much as any other player I've seen, you seem to play the game with your mind.

MJ: The mental aspect of the game came when I got into college. After winning the national championship at North Carolina in 1982, I knew I had the ability to play on that level, but there were a lot of players who had that ability. What distinguishes certain players from others is the mental aspect. You've got to approach the game strong, in a mental sense. So from my sophomore year on, I took it as a challenge to try and outthink the defense, outthink the next player. He might have similar skills, but if I can be very strong mentally, I can rise above most opponents. As you know, I went through college ball with Coach Dean Smith, and he's a very good psychological-type coach. He doesn't yell at you. He says one line and you think within yourself and know that you've done something wrong.... When I face a challenger, I've got to watch him, watch what he loves to do, watch things that I've done that haven't worked.... How can I come up with some weapon, some other surprises to overpower them?

JW: You don't just use someone on your team to work two-on-two. Your plan seems to involve all ten players on the court. A chesslike plan for all four quarters.

MJ: I think I have a good habit of evaluating situations on the floor, offensively, defensively, teammate or opponent. And somehow filling in the right puzzle pieces to click. To get myself in a certain mode or mood to open a game or get a roll going. For instance, the last game we lost to Detroit in the playoffs last spring: We're down twenty, eighteen, twenty-two points at the half. Came back to eleven or ten down. I became a point guard. Somehow I sensed it, sensed no one else wanted to do it or no one else was going to do it until I did it. You could see once I started pushing, started doing these things, everybody else seemed like they got a little bit higher, the game started to go higher, and that pushes my game a little higher, higher, higher. I kept pulling them up, trying to get them to a level where we could win.

Then, you know, I got tired, I had to sit out and rest. Let Detroit come all the way back. It hurts a little bit, but then again I feel good about the fact that I mean so much to those guys, in a sense that if I don't play, if I don't do certain things, then they're not going to play well. It's like when people say it's a one-man gang in Chicago. I take it as a compliment, but then it's unfair that I would have to do all that.

I can dictate what I want to do in the course of the game. I can say to my friends, Well, I'll score twelve points in the first quarter...then I can relax in the second quarter and score maybe six, eight. Not take as many shots, but in the second half I can go fifteen, sixteen, quick. That's how much confidence I have in my ability to dictate how many points I can score and be effective and give the team an opportunity to win.

I don't mind taking a beating or scoring just a few points in the first half, because I feel the second half I'm going to have the mental advantage. My man is going to relax. He feels he's got an advantage, he's got me controlled, that means he's going to let down his guard just a little bit. If I can get past that guard one time I feel that I've got the confidence to break him down.

On the same night in July that Michael Jordan slipped on a damp outdoor court at his basketball camp on the Elmhurst College campus, jamming his wrist and elbow, on a court in a park called Mill River in North Amherst, Massachusetts, my elbow cracked against something hard that was moving fast, so when I talked with Michael Jordan the next morning in Elmhurst, Illinois, on the outskirts of Chicago, my elbow was sore and puffy, his wrapped and packed in ice.

We'd won two straight in our pickup, playground run, pretty ordinary, local, tacky hoop that's fine if you're inside the game, but nothing to merit a spectator's attention. Ten on the court, nine or ten on the sidelines handing out or waiting to take on the winners, a small band of witnesses, then, for something extraordinary that happened next. On a breakaway dribble Sekou beats everybody to the hoop except two opponents who hadn't bothered to run back to their offensive end. It was that kind of game, spurts of hustle, lunch breaks while everybody else did the work. Sekou solo, racing for the hoop, two defenders, converging to cut him off, slapping at the ball, bodying him into a vise to stop his momentum. What happens next is almost too quick to follow. Sekou picks up his dribble about eight feet from the basket, turns his back to the goal, to the two guys who are clamping him, as if, outnumbered he's looking downcourt for help. A quick feint, shoulders and head dipped one way, and then he brings the ball across his body, slamming it hard against the asphalt, *blam*, in the space his fake has cleared. Ball bounds higher than the basket and for an instant I think he's trying a trick shot, bouncing the ball into the basket, a jive shot that's missing badly as the ball zooms way over the rim toward the far side of the backboard. While I'm thinking this and thinking Sekou's getting outrageous, throwing up a silly, wasteful, selfish shot even for the playground, even in a game he can dominate because he's by far the best athlete, while I'm thinking this and feeling a little pissed off at his hotdogging and ball hogging, the ball's still in the air, and Sekou spins and rises, a pivot off his back foot so he's facing the hoop, one short step gathering himself, one long stride carrying him around the frozen defenders and then another step in the air, rising till he catches the rock in flight and crams it one-handed down, *down* through the iron.

Hoop, poles, and backboard shudder. A moment of stunned silence, then the joint erupts. Nobody can believe what they've seen. The two players guarding kind of slink off. But it wasn't about turning people into chumps or making anybody feel bad. It was Sekou's glory. Glory reflected instantly on all of us because he was one of us out there in the game and he'd suddenly lifted the game to a higher plane. We were all larger and better. Hell, none of us could rise like Sekou, but he carried us up there with him. He needed us now to amend and goddamn and high-five and time-out. Time out, stop this shit right here. Nate, the griot, style-point judge, and resident master of ceremonies, begins to perform his job of putting into words what everybody's thinking. *Time out.* We all wander onto the court, to the basket that's still vibrating, including the two guys Sekou had rocked. Sekou is hugged, patted, praised. Skin smacks skin, slaps skin. Did you see that? Did you see that? I ain't never seen nothing like that. Damn, Sekou. Where'd you learn that shit? Learned it sitting right here. Right here when I was coming up. My boy Patrick. Puerto Rican dude, you know Patrick. He used to do it. Hey man, Patrick could play but Patrick didn't have no serious rise like that, man. Right. He'd bounce it, go get it, shoot it off the board. Seen him do it more than once. Sitting right here on this bench I seen him do it. Yeah, well, cool, I can believe it. But man that shit you did. One hand and shit…damn…damn, Sekou.

I almost told Michael Jordan about Sekou's move. Asked Michael if he'd ever attempted it, seen it done. Maybe, I thought, someday when I'm watching the Bulls on the tube Michael will do a Sekou for a national audience and I won't exactly take credit, but deep inside I'll be saying, Uh-huh, uh-huh.

Because we all need it, the sense of connection, the feeling we can be better than we are, even if better only through someone else, an agent, a representative, Mother Teresa, Mandela, one of us ourselves taken to a higher power, altered for a moment, alive in another's body and mind. One reason we need games, sports, the heroes they produce. To rise. To fly.

I didn't have to tell Michael Jordan Sekou's story. MJ earns a living by performing nearly on a regular basis similar magical feats for an audience of millions across the globe. I told him instead about my elbow. Tuesday night had been a bad night for elbows all over North America. Commiserating, solicitous about his injury, but also hopelessly vain, proud of mine, as if our sore elbows matched us, blood brothers meeting at last, a whole life-time of news, gossip, and stuff to catch up on. Since we couldn't very well engage in that one-on-one game I'd been fantasizing, not with him handicapped by a fat elbow, we might as well get on with the interview I'd been seeking since the end of the NBA playoffs. Relax and get it on in this borrowed office at Elmhurst, my tape recorder on the desk between us, Michael Jordan settled back in a borrowed swivel chair, alert, accommodating. Mellow, remarkably fresh after a protracted autograph session, signing one item apiece for each of the 350 or so campers who'd been sitting transfixed in a circle around him earlier that morning as he shared some of his glory with them in a luminous exchange that masked itself as a simple lesson in basketball basics from Michael Jordan.

JW: Your style of play comes from the playground, comes from tradition, the African-American way of playing basketball.

MJ: Can't teach it.

JW: When I was coming up, if a coach yelled "playground move" at you it meant there was something wrong with it, which also meant in a funny way there was something wrong with the playground, and since the playground was a black world, there was something wrong with you, a black player out there doing something your way rather than their way.

MJ: I've been doing it my way. When you come out of high school, you have natural, raw ability. No one coaches it, I mean, maybe nowadays, but when I was coming out of high school, it was all natural ability. The jumping, quickness. When I went to North Carolina, it was a different phase of my life. Knowledge of basketball from Naismith on…rebounds, defense, free-throw shooting, techniques. Then, when I got to the pros, what people saw was the raw talent I'd worked on myself for eleven, twelve years *and* the knowledge I'd learned at the University of North Carolina. Unity of both. That's what makes up Michael Jordan's all-around basketball skills.

JW: It seems to me we have to keep asserting the factors that make us unique. We can't let coaches or myths about body types take credit for achievements that are a synthesis of our intelligence, physical gifts, our tradition of playing the game a special way.

MJ: We were born to play like we do.

JW: Players like you and Magic have transformed the game. Made it more of a player's game, returned it closer to its African-American roots on the playground.

MJ: You know, when you think about it, passing like Magic's is as natural, as freewheeling, as creative as you can be. You can call it playground if you want, but the guy is great. And certainly he's transcended the old idea of point guard. You never saw a six-eight, six-nine point guard before he came around. No coach would ever put a six-eight guy back there.

JW: If you were big, you were told to go rebound, especially if you were big and black.

MJ: Rebound. Go do a jump lay-up, be a center, a forward. A man six-eight started playing, dribbling in his backyard. Said, I can do these things. Now look. Everybody's trying to get a six-nine point guard.

JW: For me, the real creativity of the game begins with the playground. Like last night, watching young guys play, playing with them, that's where the new stuff is coming from. Then the basketball establishment names it and claims it.

MJ: They claim it. But they can't. The game today is going away from the big guy, the old idea everybody's got to have a Jabbar, Chamberlain. Game today is going toward a versatile game. Players who rebound, steal, block, run the court, score, the versatile player who can play more than one position. Which Magic started. Or maybe he didn't start it, but he made it famous. This is where the game's going now.

JW: Other people name it and claim it. That kind of appropriation's been a problem for African-American culture from the beginning. Music's an obvious example. What kind of music do you like?

MJ: I love jazz. I love mellow music. I love David Sanborn. Love Grover Washington Jr. Rap...it's okay for some people. But huh-uh. Not in my house.

JW: Do you listen to Miles Davis?

MJ: Yeah.

JW: He talks about his art in a new biography he wrote with Quincy Troupe. When Miles relates jazz to boxing, I also hear him talking about writing, my art, and basketball, yours.

MJ: I know what you're saying.

JW: Right. There's a core of improvisation, spontaneity in all African-American arts.

MJ: I'm always working to put surprises, something new in my game. Improvisation, spontaneity, all that stuff.

What's in a name: *Michael*—archangel, conqueror of Satan. "Now war arose in heaven. Michael and his angels fighting against the dragon; and the dragon and his angels fought, but they were defeated" (Revelation 12:7); *Jordan*—the foremost river in Palestine, runs from the Lebanons to the Dead Sea, 125 miles, though its meanderings double that length. "Jordan water's bitter and cold …chills the body, don't hurt the soul…" (African-American traditional spiritual). *Michael Jordan*—a name worth many millions per annum. What's in that name that makes it so incredibly valuable to the people who have millions to spend for advertising what they sell, who compete for the privilege of owning, possessing Michael Jordan's name to adorn or endorse their products? In a country where Willie Horton's name and image helped win or lose a presidential election, a country in which one out of every four young black males is in prison, on parole or probation, a country where serious academics convene to consider whether the black male is an endangered species, how can we account for Michael Jordan's enormous popularity? Because MJ is an American of African descent, isn't he? Maybe we're more mixed up about race than we already know we are. Perhaps MJ is proof there are no rules about race, no limits to what a black man can accomplish in our society. Or maybe he's the exception that proves the rule, the absence of rules. The bedrock chaos and confusion that dogs us. At some level we must desire the ambiguity of our racial thinking. It must work for us, serve us. When one group wants something bad enough from the other, we reserve the right to ignore or insist upon the inherent similarities among all races, whichever side of the coin suits our purposes. One moment color-blind, the next proclaiming one group's whiteness, the other's blackness, to justify whatever mischief we're up to. It's this flip-flopping that defines and perpetuates our race problem. Our national schizophrenia and disgrace. It's also the door that allows MJ entry to superstardom, to become a national hero, our new DiMaggio, permits him to earn his small fraction of the billions we spend to escape rather than confront the liabilities of our society.

Sports Illustrated offers a free Michael Jordan video if you subscribe right away. Call today. In this ad, which saturates prime time on the national sports network, a gallery of young people, male and female, express their wonder, admiration, awe, and identification with Michael Jordan's supernatural basketball prowess. He can truly fly. The chorus is all white, good-looking, clean cut, casually but hiply dressed. An audience of consumers the ad both targets and embodies. A very calculated kind of wish fulfillment at work. A premeditated attempt to bond MJ with these middle-class white kids with highly disposable incomes.

In other ads, black kids wear fancy sneakers, play ball, compete to be a future Michael Jordan. There's a good chance lots of TV viewers who are white will enter the work force and become dutiful, conspicuous consumers, maybe even buy themselves some vicarious flight time by owning stuff MJ endorses. But what future is in store for those who intend to *be* the next MJ? Buy Jordan or be Jordan. Very different messages. Different futures, white and black. Who's zooming who?

In another national ad, why do we need the mediating figure of an old, distinguished-looking, white-haired, Caucasian gentleman in charge, giving instructions, leading MJ into a roomful of kids clamoring for MJ's magic power?

The Palace at Auburn Hills, the Detroit Pistons' home, contrasts starkly with Chicago Stadium. Chicago, the oldest NBA arena; the Palace one of the newest. Chicago Stadium is gritty, the Palace plush. In the Auburn Hills crowd an even greater absence of dark faces than in Chicago Stadium. But they share some of the same fight songs. *We will, we will, rock you.* And the Isley Brothers' classic "Shout." On a massive screen in the center of the Palace, cuts from the old John Belushi flick *Animal House* drive the Detroit fans wild. "Shout" is background music for an archetypal late-Fifties, early-Sixties frat party. White kids in togas twist and shout and knock themselves out. Pre-Vietnam American Empire PG-rated version of a Roman orgy. A riot of sloppy boozing, making out, sophomoric antics to the beat of a jackleg, black rhythm-and-blues band that features a frenzied, conk-haired singer, sweating, eyes rolling, gate-mouthed, screaming, "Hey-ay, hey-ay. Hey-ay, hey-ay, shout! C'mon now, shout." Minstrel auctioneer steering the action higher, higher. Musicians on the screen performing for their audience of hopped-up, pampered students fuse with the present excitement, black gladiators on the hardwood floor of the Palace revving up their whooping fans. Nothing is an accident. Or is it? Do race relations progress, or are we doomed to a series of reruns?

JW: What's the biggest misconception that is part of your public image? What's out there, supposedly a mirror, but doesn't reflect your features?

MJ: I'm fortunate that there are no big misunderstandings. My biggest concern is that people view me as being some kind of a god, but I'm not. I make mistakes, have faults. I'm moody, I've got many negative things about me. Everybody has negative things about them. But from the image that's been projected of me, I can't do any wrong. Which is scary. And it's probably one of the biggest fears I have. And I don't know how to open people's eyes. I mean, I'm not gong to go out and make a mistake so that people can see I make mistakes. Hey, you know, I try to live a positive life, love to live a positive life, but I do have negative things about me and I do make mistakes. And I'm so worried that if I make a mistake today, it can ruin the positive things I try to project. It's a day-in, day-out, nine-to-five job.

JW: A lot of pressure.

MJ: Pressure I didn't ask for, but it was given to me and I've been living, living with it.

JW: A kind of trap, isn't it, because you say, "I don't want everybody to think I'm a paragon of virtue. I'm a real person." But you also know in the back of your mind being a paragon is worth X number of dollars a day. So you don't like it but you profit from it.

MJ: Right. It has its advantages as well as disadvantages. Advantages financially. I'm asked to endorse corporations that are very prestigious as well as very wealthy. I have the respect of many, many kids as well as parents, their admiration. So it's not *just* the financial part, but the respect that I earn from the 350 kids in this camp and their parents, friends, equals the financial part of it. The respect I get from those people—that's the pressure.

JW: Not to let them down.

MJ: Not to let them down.

Postgame Chicago Stadium. Oldest arena in the league. Old-fashioned bandbox. Exterior built on monumental, muscle-flexing scale. Inside you have to duck your head to negotiate a landing that leads to steep steps descending to locker rooms. Red painted walls, rough, unfinished. Overhead a confusion of pipes, wires—the arteries of the beast exposed. Ranks of folding chairs set up for postgame interviews. MJ breaks a two-day silence with the media. Annoyed that the press misinterpreted some animated exchanges between teammates and himself.

   None of the usual question-and-answer interplay. MJ says his piece and splits, flanked by yellow-nylon-jacketed security men, three of them, polite but firm, benign sheep dogs guarding him, discouraging the wolves.

   In MJ's cubicle near the door hangs a greenish suit, a bright tropical-print shirt, yellows, black, beige, oranges, et cetera. He's played magnificently, admonished the press, clearly weary, but the effort of the game's still in his eyes, distracting, distancing him, part of him still in another gear; maybe remembering, maybe savoring, maybe just unable to cut it loose, the flow, the purity, the high of the game when every moment counts, registers, so as he undresses, the yellow jacket assigned to his corner has to remind him not to get naked, other people are in the room.

   A man chatting up the security person screening MJ's cubicle holds a basketball for MJ to sign. As MJ undresses he's invited to a reporter's wedding. The soon-to-be groom is jiving with another member of the press, saying something about maybe having a kid will help him settle down, improve his sense of responsibility. MJ, proud father of two-year-old Jeffrey Michael, interrupts. "You got to earn the respect of kids. Responsibility and respect, huh-uh. You don't just get it when you have a kid. Or a couple of kids."

   Word's sent in that Whitney Houston and her entourage are waiting next door for a photo session. Celebrity hugging and mugging. I say hi to a man hovering near MJ's corner. Thought he might be MJ's cut buddy or an older brother. Turns out to be MJ's father. Same dark, tight skin. Same compact, defined physique. "Yes, I think I'm Mister Jordan. Hold on a minute, let me check my social security card." A kidder like his son. Like his son, he gives you friendly pats. Or maybe it's son like father. In the father's face, a quality of youthfulness and age combined, not chronological age but the timeless serenity of a tribal elder, a man with position, authority, an earned place in a community. In the locker room of the Spectrum in Philly, I'd been struck by the same mix in MJ's face as he addressed reporters. Unperturbed by lights, cameras, mikes, the patent buzz and jostle of media rudeness swarming around him, he sits poised on a shelf in a dressing stall, his posture unmistakably that immemorial high-kneed, flex-thighed, legs-widespread, weight-on-the-buttocks squat of rural Africans. Long hands dangling between his knees, head erect, occasionally bowing as he retreats into a private, inner realm to consider a question, an answer before he spoke. Dignified, respectful. A disposition of body learned how many grandfathers ago, in what faraway place. Passed on, surviving in this strange land.

From the shower MJ calls for slip-ons. Emerges wrapped in a towel. A man, an African-American man in a suit of dark skin, tall, broad-shouldered, long-limbed, narrow-waisted, bony ankles and wrists, his body lean-muscled, sleek, race cars, cheetahs, a computer-designed body for someone intended to sprint and leap, loose and tautly strung simultaneously, but also, like your body and mine, a cage. MJ is trapped within boundaries he cannot cross, you cannot cross. No candid camera needs to strip him any further. Why should it? Whose interests would be served? Some of his business is not ours. Are there skeletons in his closet? Letters buried in a trunk, sealed correspondence? What are we seeking with our demands for outtakes, for what's X-rated in public figures' lives? In the crush of the locker room, one reporter (a female) whispers to another; we build these guys up so we can be around to sell the story when they fall.

Naked MJ. A price on his head. He pays it. We pay it. Hundreds of thousands of fans plunk down the price of a ticket to catch his act live. Millions of dollars are spent to connect products with the way this body performs on a basketball court. What about feedback? If body sells products, how do products affect body? Do they commodify it, place a price on it? Flesh and blood linked symbiotically with products whose value is their ability to create profit. If profit's what it's all about—body, game, product—does profit-making display ball playing? Are there inevitable moments when what's required of MJ the ballplayer is different from what's required of the PR Jordan created by corporate interests for media consumption?

The PR Michael Jordan doesn't need to win a ring. He's already everything he needs to be without a ring. Greatest pressure on him from this perspective is to maintain, replicate, duplicate whatever it is that works, sells. It's an unfunny joke when MJ says wryly, "Fans come to see me score fifty and the home team win." The PR Jordan doesn't really lose when he has a spectacular game and his teammates are mediocre. He's still MJ. But if he puts forth a kamikaze, all-or-nothing, individual effort and fails by forcing or missing shots, by coming up short, the blame falls squarely on his shoulders. If he plays that tune too often, will the public continue to buy it?

Can a player be bigger than the game? Is Jordan that good? Does he risk making a mockery of the game? (Recall John Lennon. His lament that the Beatles' early, extraordinary success pressured them to repeat themselves, blunted artistic innovation, eventually pushed them into self-parody, the songs they'd written taking over, consuming them.)

The young man from Mississippi, baby of eleven children, speaks with a slow, muddy, down-home drawl. We're both relaxing in the unseasonably warm sun, outside Hartford, the Bradley International Airport B Terminal. My designation: Chicago, Game Six of last spring's NBA Eastern Championship. He's headed for his sister's home in Indiana. A nice place, he hopes. Time to settle down after five, six years of roaming, but decent work's hard to find. "I sure hope Michael wins him a ring. Boy, I want to see that. Cause my man Michael's the best. Been the best awhile. He's getting up there a little bit in age, you know. Ain't old yet but you know, he can't do like he used to. Used to be he could go out anytime and bust sixty-three. He's still good, still the best now but he better go on and get him his ring."

JW: I'm jumping way back on you now. Laney High School, Wilmington, N.C., the late Seventies, early Eighties. How did it feel to sit on the bench?

MJ: I hated it...you can't help anybody sitting on the bench. I mean, it's great to cheer, but I'm not that type of person. I'm not a cheerleader.

JW: You couldn't make the team?

MJ: I was pissed. Because my best friend, he was about six six, he made the team. He wasn't good but he was six six and that's tall in high school. He made the team and I felt I was better.... They went into the playoffs and I was sitting at the end of the bench and I couldn't cheer them on because I felt I should have been on the team. This is the only time that I didn't actually cheer for them. I wanted them to lose. Ironically, I wanted them to lose to prove to them that I could help them. This is what I was thinking at the time: You made a mistake by not putting me on the team and you're going to see it because you're going to lose. Which isn't the way you want to raise your kids...but many kids now do think that way, only because of their desire to get out and show that they can help or they can give something.

I think to be successful, I think you have to be selfish, or else you never achieve. And once you get to your highest level, then you have to be unselfish. When I first came to the league, I was a very selfish person in the sense I thought for myself first, the team second...and I still think that way in a sense. But at the same time, individual accolades piled up for me and were very soothing to the selfishness that I'd had in myself. They taught me how, you know, to finally forget about the self and help out the team, which is where I am now. I always wanted the team to be successful but I felt, selfishly, I wanted to be the main cause.

> *Phi Gamma Delta was assailed Monday with charges of racism, becoming the second fraternity to face such accusations stemming from activities during Round-Up last weekend.*
>
> *At the root of the newest charges is a T-shirt sold and distributed by the fraternity—often known as the Fijis—during its annual "low-hoop" basketball tournament Saturday. On the T-shirt is the face of a "Sambo" caricature atop the body of professional basketball player Michael Jordan.*
>
> *Meanwhile, Delta Tau Delta continued its internal investigation into Friday's incident in which FUCK COONS and FUCK YOU NIGS DIE were spray-painted on a car destroyed with sledgehammers on fraternity property.*
>
> —The Daily Texan, University of Texas, Austin, 4/10/90

We're slightly lost. My wife and I in a rental car on the edge of Shreveport, Louisiana, leading two rented vans carrying the Central Massachusetts Cougars. We're looking for the Centenary College Gold Dome, where soon, quite soon, opening ceremonies of the AAU Junior Olympic Girls Basketball national tourney will commence. Suddenly we cross Jordan Street, and I know we'll find the Gold Dome, be on time for the festivities, that my daughter, Jamila, and her Cougar teammates will do just fine, whatever. Renoir women everywhere, Jordans everywhere.

Next day, while girls of every size, shape, color, creed, and ethnic background, from nearly every state in the Union, are playing hoop in local high school gyms, about three miles from the Cougars' motel, at the Bossier City Civic Center, Klansmen in hoods and robes distribute leaflets at a rally for former Klan leader David Duke, who's running for the U.S. Senate.

MJ: Well, I think real often in terms of role models, in terms of positive leaders in this world...Nelson Mandela, Bill Cosby.... We're trying to show there are outlets, there are guidelines, there are positive things you can look for and achieve. I mean, we're trying to give them an example to go for. I think that's the reason I try to maintain the position I have in the corporate world as well as in the community. That we can do this, prove that success is not limited to certain people, it's not limited to a certain color.... Grab a hand and pull someone up. Be a guide or a role model. Give some type of guidelines for other people to follow.

JW: What ways do you have of controlling this tidal wave of public attention? How do you remain separate from the Michael Jordan created "out there"?

MJ: Stay reachable. Stay in touch. Don't isolate. I don't try to isolate myself from anybody. I think if people can feel that they can touch or come up and talk to you, you're going to have a relationship and have some influence on them. My old friends, who I try to stay in touch with at all times, I think they always keep me close to earth.

About six years ago I was only forty-three, so I couldn't understand why the jumper was falling short every time, banging harmlessly off the front of the iron. After a summer-camp game my wife, Judy, had watched mainly because our sons, Dan and Jake, were also playing, I asked her if she'd noticed anything unusual about my jump shot. "Well, not exactly," she offered, "except when you jumped your feet never left the ground."

JW: When you're playing, what part of the audience is most important to you?

MJ: Kids. I can notice a kid enjoying himself. On the free-throw line, *bing,* that quick, I'll wink at him, smile, lick a tongue at him, and keep going and still maintain the concentration that I need for the game. That's my personality. I've always done it. I can catch eye-to-eye a mother, a father, anybody ...kid around with them as I'm playing in a serious and very intense game. That's the way I relax, that's the way I get my enjoyment from playing the game of basketball. Seeing people enjoying themselves and relating to them like I'm enjoying myself. Certainly letting them know they're part of the game.

The Game. I am being introduced to a group of young African-American men, fifteen or so high school city kids from Springfield, Massachusetts, who are attending the first-ever sessions of a summer camp at Hampshire College, an experiment intended to improve both their basketball skills and SAT scores. My host, Dennis Jackson (a coach, confidentially, at Five-Star hoop camp back when Michael Jordan, still a high schooler, made his debut into the national spotlight), is generously extolling my credentials: Author, professor, college ballplayer, Rhodes Scholar, but it isn't till Dennis Jackson says this man is working on an article about MJ that I know I have everybody's attention. I tell them that I haven't received an interview with MJ yet, he's cooling out after the long season, but they're eager to hear any detail of any MJ moment I've been privy to so far. An anecdote or two and then I ask them to free-associate. When I say MJ what comes to mind first? *Air, jam, slam-dunk, greatest, wow* are some of the words I catch, but the real action is in faces and bodies. No one can sit still. Suddenly it's Christmas morning and they're little boys humping down the stairs into a living room full of all they've been wishing for. Body English, *ohhs* and *ahhs.* Hey man, he can do anything, anything. DJ assures me that if Mike promised, I'll get my interview. *As good a player as MJ is, he's ten times the person.* We're all in a good mood, and I preach a little. Dreams. The importance of believing in yourself. My luck in having a family that supported me, instilled the notion I was special, that I could do anything, be anything. Dreams, goals. Treasuring, respecting the family that loves you and stands behind you.

What if you don't have no family?

The question stops me in my tracks. Silences me. I study each face pointed toward me, see the guys I grew up with, my sons, brothers, Sekou, MJ, myself, the faces of South African kids *toi-toing* in joy and defiance down a dirt street in Crossroads. I'd believed the young men were listening, and because I'd felt them open up, I'd been giving everything I had, trying to string together many years, many moments, my fear, anger, frustration, the hungers that had driven me, the drumbeat of a basketball on asphalt a rhythm under everything else, patterning the crazy-quilt chaos of images I was trying to make real, to connect with their lives. What if there's no family, no person, no community out there returning, substantiating the fragile dream a young man spins of himself? Silenced.

Michael said he didn't like politics. I'll always stand up for what I think is right but.... Politics, he said, was about making choices. And then you get into: This side's right and that side's wrong. Then you got a fight on your hands. Michael's friends say he naturally shies away from controversy. He's mellow. Middle-of-the-road. Even when the guys are just sitting around and start arguing about something, Michael doesn't like it. He won't take sides, tries to smooth things out. That's Michael.

Finally I respond to the young man at the circle's edge leaning back on his arms. He doesn't possess an NBA body. He may be very smart, but that can be trouble as easily as salvation. You have me, I want to say. I support you, love you, you are part of me, we're in this together. But I know damn well he's asking for more. Needs more. Really, he's asking for nothing while he's asking everything.

Find a friend. I bet you have a cut buddy already. Someone to hang with and depend on. One person who won't let you fall. And you won't let him fall. Lean on each other. It's hard, hard, but just one person can make a difference sometimes.

They look at the books I've written. Ask more questions. Mostly about Michael again. Some eyes are drifting back to the court. This rap's been nice, but it's also only a kind of intermission. The game. I should have brought my sneaks. Should be thirty years younger. But why, seeing the perils besetting these kids, would anyone want to be thirty years younger? And why, if you're Peter Pan playing a game you love and getting paid a fortune for it and adored and you can fly too, why would you ever want to grow a minute older?

The young men from Springfield say hello.

Nate, the historian from the court at Mill River, said: "Tell the Michael I say hello."

COMING TO TERMS

They say that it is the person who asks the questions who holds the power, who controls and directs what will be said—even if the focus of an interview would appear to be on its subject, the person answering the questions (in this case, Michael Jordan). Your first task as a reader of this piece, then, is to shift what might seem a natural focus on what Michael Jordan has to say, and instead to try to figure out what John Edgar Wideman wants to use the occasion of speaking to Jordan to say. What point is Wideman trying to make? Other than Jordan, what are the *subjects* of his interview/essay? Mark those passages and record some in your journal where you think Wideman's own focus and agenda become most clear. You could also include passages where you think Wideman has gotten off track. In order to determine what the *subjects* of this piece are we think you might become aware of three distinctive modes or uses of language in this piece: One where Wideman talks *to Jordan,* another where he talks *about Jordan,* and a third where he talks *about himself.* Record passages in your journal where you see Wideman twining these three strands together into a coherent piece. Do they contribute to form a single *subject*? What connections do you see Wideman making between Jordan as a person, Jordan as a celebrity or symbol, and Wideman's own experiences as a ballplayer, a fan, and a black man?

READING AS A WRITER

We would like you to take a closer look at John Wideman's title "Michael Jordan Leaps the Great Divide." Here, Wideman is using both allusion and metaphor. Specifically, the Great Divide is a geographical allusion to the Rocky Mountains in the United States. Maybe more interesting, however, is to think about the Great Divide as a metaphor and the ways it might connect to Wideman's writing. (A metaphor is sometimes defined as a word or phrase which is compared with an object, person, or idea it does literally represent.) One of the natural obstacles for early settlers in the 18th and 19th century was finding a way over or through the Rocky Mountains. If crossing the Great Divide can be understood as a metaphor for overcoming obstacles, then how might you explain its significance in terms of Michael Jordan's and John Wideman's lives? Jot down your ideas in your journal. Wideman also speaks about a number of American and international "celebrities" not usually associated with discussions of American basketball. In what ways might these people have "leaped the Great Divide"? Finally, it seems that John Wideman sees leaping the Great Divide in terms other than a sense of accomplishment or achievement. When Jordan, Wideman, or other people leap the Great Divide, what are the consequences, responsibilities, and dangers they encounter?

WRITING CRITICISM

John Wideman writes powerfully about what Michael Jordan means to him and other fans of basketball. He conducts, in other words, a "reading" of Jordan as an athlete who not only plays well but who also symbolizes something to people. We would like you to take on a similar project here, to write about an athlete—of either gender, at any level, in any sport—whom you know through the media and who has meant something to you. This could be someone who became a favorite, won your respect, made you curious, helped feed your interest in a sport, stuck in your mind, or somehow impressed you as interesting as a kind of public personality. You needn't be a sports fan to do this assignment. All you need to find is an athlete (famous or otherwise) who has left an impression—or, as we sometimes say, an image—in your mind.

In your writing you should try to distinguish between what this person does—his or her exploits as athlete—and what he or she "means" or "says" or "stands for." Your subject here is not only a particular athlete, but also the way in which athletes in general come to represent certain ideals, images, or aspirations in American culture or in another culture. As you write this essay, we would like you to attempt a cultural reading of the roles athletes play in the culture today. So, in a sense, you are not only speaking about your interest in this athlete, but are also attempting to explain the reasons why athletes are so important in contemporary society. You might respond to the following set of questions as a place to begin your cultural reading: Why are athletes seen as heroic? Why are they perceived as role models for children? Why do we pay attention when athletes endorse products or support certain political candidates for local or national office? What do athletes contribute to society? Remember: As you perform a cultural reading of sports in contemporary society, you are not necessarily endorsing this way of valuing sports or sports figures—rather you are reading a culture and you may or may not agree with how this culture reads sports.

In thinking about these matters, you will want to consider the role of media. In what ways does the media bring us closer to athletes? How does it distance us from them? What do we feel we "know" about famous athletes? What can we never know? How is our image of an athlete formed? How do media and culture contribute to this process of formation? What does this image do for us, or to us? In what ways does the media interfere with the pleasures we might take in the performance of good athletes? In what ways do the media feed that pleasure?

Urban Spaceman

Judith Williamson

A vodka advertisement in the London underground shows a cartoon man and woman with little headphones over their ears and little cassette-players over their shoulders. One of them holds up a card which asks, "Your place or mine?"—so incapable are they of communicating in any other way. The walkman has become a familiar image of modern urban life, creating troops of sleep-walking space-creatures, who seem to feel themselves invisible because they imagine that what they're listening to is inaudible. It rarely is: nothing is more irritating than the gnats' orchestra which so frequently assails the fellow-passenger of an oblivious walk-person—sounding, literally, like a flea in your ear. Although disconcertingly insubstantial, this phantom music has all the piercing insistency of a digital watch alarm; it is your request to the head-phoned one to turn it down that cannot be heard. The argument that the walkman protects the *public* from hearing one person's sounds, is back-to-front: it is the walk-person who is protected from the outside world, for whether or not their music is audible they are shut off as if by a spell.

The walkman is a vivid symbol of our time. It provides a concrete image of alienation, suggesting an implicit hostility to, and isolation from, the environment in which it is worn. Yet it also embodies the underlying values of precisely the society which produces that alienation—those principles which are the lynch-pin of Thatcherite Britain: individualism, privatization and "choice." The walkman is primarily a way of escaping from a *shared* experience or environment. It produces a privatized sound, in the public domain; a weapon of the individual against the communal. It attempts to negate *chance:* you never know what you are going to hear on a bus or in the streets, but the walk-person is buffered against the unexpected—an apparent triumph of individual control over social spontaneity. Of course, *what* the walk-person controls is very limited; they can only affect their *own* environment, and although this may make the individual *feel* active (or even rebellious) in social terms they are absolutely passive. The wearer of a walkman states that they expect to make no input into the social arena, no speech, no reaction, no intervention. Their own body is the extent of their domain. The turning of desire for control inward toward the body has been a much more general phenomenon of recent years; as if one's muscles or jogging record were all that one *could* improve in this world. But while everyone listens to whatever they want within their "private" domestic space, the peculiarity of the walkman is that it turns the inside of the head into a mobile home—rather like the building society image of the couple who, instead of an umbrella, carry a tiled roof over their heads (to protect them against hazards created by the same system that provides their mortgage).

This interpretation of the walkman may seem extreme, but only because first, we have become accustomed to the privatization of social space, and second, we have come to regard sound as secondary to sight—a sort of accompaniment to a life which appears as essentially visual. Imagine people walking round the streets with little TVs strapped in front of their eyes, because they would rather watch a favourite film or programme than see where they were going, and what was going on around them. (It could be argued that this would be too dangerous—but how about the thousands of suicidal cyclists who prefer taped music to their own safety?) This bizarre idea is no more extreme in principle than the walkman. In the visual media there has already been a move from the social setting of the cinema, to the privacy of the TV set in the living-room, and personalized mobile viewing would be the logical next step. In all media, the technology of this century has been directed toward a shift, first from the social to the private—from concert to record-player—and then of the private *into* the social—exemplified by the walkman, which, paradoxically, allows someone to listen to a recording of a public concert, in public, completely privately.

The contemporary antithesis to the walkman is perhaps the appropriately named ghetto-blaster. Music in the street or played too loud indoors *can* be extremely anti-social—although at least its perpetrators can hear you when you come and tell them to shut up. Yet in its current use, the ghetto-blaster stands for a shared experience, a communal event. Outdoors, ghetto-blasters are seldom used by only their individual owners, but rather act as the focal point for a group, something to gather around. In urban life "the streets" stand for shared existence, a common understanding, a place that is owned by no-one and used by everyone. The traditional custom of giving people the "freedom of the city" has a meaning which can be appropriated for ourselves today. There *is* a kind of freedom about *chance* encounters, which is why conversations and arguments in buses and bus-queues are often so much livelier than those of the wittiest dinner party. Help is also easy to come by on urban streets, whether with a burst shopping bag or a road accident.

It would be a great romanticization not to admit that all these social places can also hold danger, abuse, violence. But, in both its good and bad aspects, urban space is like the physical medium of society itself. The prevailing ideology sees society as simply a mathematical sum of its individual parts, a collection of private interests. Yet social life demonstrates the transformation of quantity into quality: it has something extra, over and above the characteristics of its members in isolation. That "something extra" is unpredictable, unfixed, and resides in interaction. It would be a victory for the same forces that have slashed public transport and privatized British Telecom, if the day were to come when everyone walked the street in headphones.

COMING TO TERMS

Judith Williamson looks here at the ways the walkman has reorganized how people interact with each other in city streets and other public places. As such, Williamson enters a critical conversation in which a number of writers in *Media Journal* have participated, one about the social and cultural changes witnessed as part of an ever-changing landscape of technological progress and advances in telecommunications. As you come to terms with Williamson's cultural and sociological reading of the walkman, you might want to return to other journal entries—written say for M. Kadi, Andrew Shapiro, Neil Postman, John Seabrook, Joshua Meyrowitz, or Travis Charbeneau—in which you already have written notes on how one or several of these authors have *come to terms* with technological progress. We think a multi-author context might be of value here in appreciating Williamson. While we certainly don't want to say that these other authors have somehow chosen only the most visible and dominant of technological advancements, we do think one of the reasons why we were drawn to "Urban Spaceman" is that Williamson writes against the grain of what might be said to be the most obvious choices in this continuing critical conversation. The walkman, as a cultural text, is easily overlooked, almost invisible against the larger cultural background. But Williamson sees the walkman as "a vivid symbol of our time." And she is concerned with the effects of this technology on not so much an individual as a *social* level. We'd like you to restate in your own words Williamson's critical argument in your journal, paying careful attention to the ways she uses the terms *private* and social in her analysis. Why is it for Williamson that the walkman is "a vivid symbol of our times"? You'll need to explain what she means by "our times." You might see this phrase in terms of the sociological effects of this technology. Or often times, when a writer uses the phrase "our times," it is being used as a textual cue for or a historical marker of some kind of change or the perception of difference, one period of history being different from another. Members of an older generation might say, "That wasn't done in our *times*." And members of a younger generation might respond, "But in *our times,* it's okay." What have been the changes in Williamson's reading of society and culture, symbolic in the walkman, that cause her to see the present distinctly as "our times"?

READING AS A WRITER

Williamson's short piece is in many ways a polemic *against* a certain, current mode of city life and a plea *for* another (maybe disappearing or even idealized), competing view of urban living. The walkman becomes a part of a cultural index to map out ways in which Williamson sees the social fabric of urban life changing as a result of this technology. As a polemic, Williamson's article presents a controversial argument, controversial in its subject, but also, we think, controversial in Williamson's representation of urban life before the impact of technology. At one point Williamson lauds the "freedom of the city." And at another point she remarks, "It would be a great romanticization not to admit that all these social spaces can also hold danger, abuse, violence." Williamson seems aware of the consequences of presenting a too idealized or romanticized version of the city, tempering what she has to say with maybe more realistic depictions of urban life. This attempt to find some kind of balance between these competing views leads to a *tension,* an unresolvable conflict in an author's text. We think if you even consider the title here, "Urban Spaceman," there is tension there as well, a bringing together of words whose meanings push against each other. We'd like you to explore the *tension* between these competing representations of urban or city life in your journal by responding to several questions: Where are the places in "Urban Spaceman" which you might point to as examples of both *realistic* and *idealized* representations of the city? Record the passages in your journal and explain your reading of each kind of representation. You might also consider the language Williamson uses to describe people who listen to walkmen and those who don't. What kind of person might be said to take advantage of Williamson's "freedom of the city"? Finally, how successful is Williamson in finding a balance in her article between representing places and in representing people?

WRITING CRITICISM

We'd like you to write an essay in which you test Williamson's views either against your own interpretation of similar cultural experiences or as part of a project examining how advertisements for telecommunication products might verify or dispute the claims Williamson makes about media technology in the 20th century.

If you own or use a walkman frequently, ask yourself if it shapes how you deal with others in the ways that Williamson suggests. If you think there are important differences, then see if you can begin to spell them out. If you don't listen to a walkman very much, then think about those times you are in the company of those who do—on the street, in the bus or subway, in the classroom, shops, or offices. Do you notice or respond to these people as Williamson does? In either case, see if you can describe in your writing the sort of impact the walkman has had on the social world of which you are part (your times), and be ready to explain how these effects are similar to or different from those described by Williamson.

Or, alternatively, you may want to see if you can extend her approach to some other technology. Identify another device that you think has in some way changed the ways people communicate or interact with each other. Some possible subjects might include: answering machines, cell phones, e-mail, voice mail, remote controls, VCRs, Xerox machines, personal computers, lap tops, security or surveillance cameras, and the like. Try to define both the kind of social interactions this technology encourages its users to enter into, and those that it might prevent or discourage them from taking part in. Do people use this machine or technology to get in touch with or to isolate themselves from others? (Or both? As when someone uses a phone answering machine to "screen" calls before picking up.) Or have you ever had a "conversation" with someone that consisted only of several exchanges of phone messages? In what ways does this machine regulate or limit the sorts of messages its users are able to send out or take in? For instance, knowing you may have only a minute or so before the tape shuts off on an answering machine puts an interesting sort of pressure on the kind of message you leave. On the other hand, some people write a "script" and rehearse before leaving a phone message. Conversely, knowing a text has been written on a computer—and thus can be easily altered and reprinted—may prompt a reader or teacher to offer more sweeping sorts of advice about how to revise it than they might otherwise suggest. E-mail grants us quick access to others, but are privacy and confidentiality casualties of this efficiency?

You might consider that car phones did begin as car phones, but now are cell phones which are used just about everywhere and found in places and situations where you might not expect people to be talking on a phone. Taking this as an example, you might also want to consider the kind of situations or events that your machine is normally found in or part of. Are there situations or places where its use would seem inappropriate? Is there a certain type of person you commonly picture as using this machine? What does using this machine "say" to other people who might be in the same space? What trends or impulses in our culture might you say the technology stands for? You may find it useful here to draw on the terms suggested by Williamson, or you may want to take this writing as an opportunity to challenge her views about public and private life.

We have one final approach you might want to consider for this writing project. In seeing the walkman as just one illustration of a dominant trend in telecommunications technology, Williamson writes: "In all media, the technology of this century has been directed toward a shift—from the social to the private—and then of the private *into* the social." We'd like you to test out this reading of history by selecting representative ads from several different decades to determine whether or not what Williamson has said rings true. You may certainly use the ads printed in *Media Journal* if you like. The ads you select need to be promoting some kind of telecommunications product or technology—telephone, typewriter, radio, television, stereo, copier, computer, and so on—and these ads need to originate from several different decades. Once you've selected your set of ads, you'll need to analyze both the written and visual messages they convey and interpret these messages in light of what Williamson has concluded about the media. For example, do the ads promise some kind of change in personal, family, or community relationships if *this* certain product is purchased? Do these products seem to promise to further communication between individuals or groups of people? Or do these products somehow promise to push, as Williamson says, "the private *into* the social"?

Projects in Criticism

The following are ideas for sustained critical projects—reading and writing tasks that might extend for a few weeks or even for the length of a term. Many draw on earlier reading and journal assignments in this book. Most ask you to do some kind of research—to find materials that may not be readily available, to talk with others about their experiences with the media, to work with many different texts, or to investigate the situations in which media texts are encountered and the uses to which they are put. None are conventional term paper assignments, but all are meant as ongoing and substantial projects—writings that will go through and build on a good deal of planning, research, drafting, and revision before they are done.

A Media Autobiography

The aim of this project is to have you take a sustained and critical look at the ways the media figure in your own life. To do so we would like you to compose a *media autobiography:* a piece (or series of pieces) in which you talk about the ways various media—music, movies, TV, books, newspapers, fashion, advertising, and the like—have entered into your life and the roles they have played in it. Think about the *effects* that the media have had on you, on the ways they have helped shape the sort of person you are, and the important *uses* you have made of the media.

For one example of the kind of writing we have in mind, see Dave Marsh's essay "Fortunate Son." There Marsh reflects on the meaning of a particular song at a particular moment during his high school days. This song, he says, changed his life. You need not find something so dramatic as a life-changing moment, but you should be able to reflect, as Marsh does, on the ways in which the media have entered into your experience and helped shape who you are.

In order to write your autobiography you will need to think not only about your media likes and dislikes but also about the various patterns and habits of using the media that you have developed over the years. This will take some time and careful thought. So you should begin serious work on your autobiography several weeks before it is due. We have tried to design the media journal assignments in this book to help you do so. We suggest that you go back through your journal writings as a way of gathering material for your autobiography, that in effect you think of them as sections or drafts of this longer project.

Your autobiography can take any form that you think best expresses what you want to say. You may want to write a fairly straightforward autobiographical narrative, the kind that begins with your childhood and moves to the present day ("The first time I remember watching television was..."). But you don't have to. You could center your writing instead on a single crucial event or text, on a particularly revealing moment or encounter with the media. Or you could write a diary, or a collage of moments. Or you may want to write a more analytic or critical piece—one that traces the effects of a certain issue or medium in your understanding of yourself. Or you may want to try to account for your fascination with a particular celebrity or performer, or to understand your attachment to a certain kind of music, or your interest in a certain writer or film director, or your absorption in a particular TV program. And since a sense of identity often stems not only from those decisions you have made to stand *with* someone or *for* something but also from those times you felt a need to stand *against* someone or something, you may want to write about those aspects of our culture that you choose to resist or protest or subvert. The only real requirement is that you write about some of the ways you have worked with or against the texts of our common culture in forging a sense of who you are or want to be.

Whichever approach you take to your autobiography, you should think of it as a project that draws from and builds on many of your other writings for this course. So give yourself enough time to write it with care and attention, and enough space to say what you want to say fully and well. (Most responses to this assignment run from six to twelve pages, typed, double-spaced.) Once again, in composing your autobiography you should feel free to draw on or incorporate the work you have done in your media journal in any way that seems useful to you (though you are not strictly required to do so). Imagine yourself as writing to the other members of your class—to a group of readers, that is, who have been thinking and talking about the ways people make use of the texts of popular culture in their everyday lives. Try to make what you have to say of real interest and use to that common project.

The Meanings of News

Traditionally we think of the news media as offering us information, providing us with facts we didn't have before. We follow the news in order to "stay in touch" with the world and remain "informed citizens." But several writers in this book are convinced that news reports, particularly on television, do more—or at least something different—than that. Joel Meyrowitz, for instance, says that TV news alters our sense of place and rearranges social relations. "We pay more attention to, and talk more about, fires in California, starvation in Africa and sensational trials in Rhode Island than the troubles of nearly everyone except perhaps a handful of close family, friends and colleagues," he writes. Neil Postman argues that the electronic media, starting with the telegraph, give us a feeling of "irrelevance, impotence, and incoherence" by flooding us with information. In his piece on Barbara Walters, Mark Crispin Miller views the celebrity interview as a ritual, in which Walters "acts out, and exploits, the viewers' ambivalence toward the superstars." Walters, he says, is "in the business, not of public information but of popular delusion."

For all these writers, the news is interesting precisely because it offers us something other than, or more than, mere "information." In certain respects, many of these writers assert, news may even interfere with the goal of remaining an "informed citizen." They all agree that whenever we encounter the news, more is happening than the simple transfer of facts from the news media to us. But they have different perspectives on what else is going on. In order to think about the news the way these authors do, first detach yourself from common sense understandings of what the news is about, and why we attend to it. Here, then, is your challenge in this assignment: to analyze a news text as "about" something other than the provision of information and the relaying of facts.

By a "news text" we mean anything that presents itself to us as news—a news magazine, newspaper, or news story in printed form, a news program or news-related show on television, or even a news broadcast on the radio. The text you pick for study may offer news about any number of topics: politics, sports, entertainment, local events, business, celebrities, fashion, etc. (In other words, you don't have to confine yourself to texts reporting "hard" news about world affairs and the like—though you may, of course, do so if you want.) But you should select a text that seems to offer its audience something beyond information; your task is to write about what this "something" is. Possible texts we can imagine using in this assignment include: the national newspaper *USA Today*, a supermarket tabloid like the *National Enquirer*, a local or community newspaper, a news magazine like *Time* or *Newsweek*, a popular prime-time news program like *60 Minutes* on CBS or *20-20* on ABC, one of the major network's evening newscasts, the news reports on MTV, a fanclub newsletter, a specialized news program like *Entertainment Tonight*, or just your local evening news at 5:00–6:00 P.M., or 10:00 P.M.–11:00 P.M. These are only suggestions; you can and should seek out your own texts. If possible, you should tape any TV or radio program you want to analyze. If that is not possible, you should take extensive notes on what you see and hear. If you choose to write about a print text, you should obtain a copy of the text in order to examine it closely.

The first thing to ask yourself is: In what way does this text present itself as "news"? That is, in what way does it suggest that we can become informed by attending to it? In what ways does it try to connect us with a world "out there"? How does it present itself as a "serious" and "responsible" medium? How do we know, in other words, that this text has something to do with news (or "newsiness") and the business of becoming informed? And what sort of world or cultural sphere does this text claim to inform us about? The second question—and the heart of the assignment—is: *What else* do you see going here? What else besides the provision of news and information? Here are some useful ways of asking that second question:

What is interesting, compelling, or fun about this text? What satisfactions or pleasures does it provide, other than the feeling of being informed? What else can you use it for, other than to inform yourself?

What role does the journalist (writer, reporter, correspondent, or anchor) play, beyond the role of information provider? What else does this person do or claim to do? What talents or qualities must this person possess in order to be good at what he does? Do these talents or qualities have much to do with providing information?

What kind of participation in the news does your text offer? What role are viewers, readers, or listeners asked to play? (Are they asked, for instance, to write or phone in their responses to the news?) What sort of reactions do they seem to be expected to have? What are they supposed to feel or believe or conclude after encountering this text?

How does the text look? (Look here especially at charts, pictures, photographs, diagram, design, editing—anything that contributes to the general appearance of the text.) What do the visual aspects of the text "say" that its accompanying written or spoken elements do not? What is the "tone" of the text? What does this tone "say" or accomplish? How do the visual aspects of the text contribute to this tone?

What parts of the text are *not* devoted to offering information? (Think, for instance, of the advertising in news magazines, the banter on TV news reports.) What is the relationship or connection between these other (nonnews) functions and the news or information function? Do the two work together? How so? Or do they work against each other? That is, do they pull the audience in different directions? Perhaps one function has fairly little to do with the other. In that case, you might ask yourself why they are present in the same text.

Your goal in writing is to describe what this particular text offers its audience *besides* information, and why. This may allow you to consider what functions the news in general may have in addition to informing people. Review the essays in this book by Meyrowitz, Postman, and Miller as you go about researching and writing your own piece. And you should not feel any need to deny that information is available from the news media, including the text you select. The question you've been asked to address is: what else is available? While that "something else" is the subject of your writing here, you may still want to acknowledge that your text does provide its users with real news and useful information.

Critics, Fans, and Writing

We want you to do some research here into the ways popular culture is written about by people who are not students, academics, or journalists—who are instead just "regular" fans. Begin by joining the fan club of a media figure, or by obtaining copies of a newsletter or fanzine about such a figure. (Most pop musicians and bands give fan club information on album covers. Some "alternative" bookstores and music outlets will have fanzines for sale. The reference section of your college library should also have a volume called the *Encyclopedia of Associations*. Look under "fan clubs" for a list of such clubs and their addresses.) After you obtain and read through your materials, write an essay in which you read them as evidence of one kind of "relationship" people strike up with media figures.

Why do you think people join this club (or read this magazine)? How might it add to their experiences as fans? What do they have to say about the star around whom the club is constructed? What sort of relationship with him or her does it promise? Are there interviews? Quotations from or information about his or her work? Is there anything that your materials seem to avoid mentioning or refuse to acknowledge?

What do your materials assume about the people to whom they are addressed? Are the people who read them also invited to become writers, to send in their own thoughts or photos or letters? What do the writers of these fan materials notice most or value about their subject? What are they trying to sell—other than the obvious (membership in the club, T-shirts, tapes, etc.)? In other words, what ideas or values are they trying to promote? And how are these related to the life and work of the star, as well as to the concerns of his or her fans?

In what ways do your materials support or argue against John Caughey's notions about how the "imaginary relationships" fans set up with media figures they admire? (See his essay "Imaginary Social Relationships" in *Media Journal*.) How might his analysis have been different if he considered such materials? What do they tell us about fame and fandom in our culture that he doesn't?

Plan ahead to do this assignment well. If you are joining a fan club, you need to write away at the beginning of the semester so that your materials arrive in time. Hand in copies of the materials you've studied along with your writing. And make sure you back up your analysis of these texts as you would your analysis of any others—with a close reading of their images and phrasings.

The Process of Viewing a Movie

The aim of these four writings is to have you reflect on how you form a response to a movie. What knowledge and expectations do you bring to your viewing of a film? Where do these come from? What roles might advertisements and other kinds of publicity play in forming them? Which words do you use to describe what you think about a movie? Where does this vocabulary come from? How do the remarks of other viewers—both in writing and in conversation—shape what you notice (or remember) as significant (or disappointing) about a movie? Your subject in these writings, then, should be yourself as you go about interpreting a film. Their goal is not simply for you to detail your responses to a particular movie but to clarify your own methods and inclinations as a moviegoer.

Part One: Pre-Viewing

Choose a movie that you're interested in seeing and try to make the knowledge and expectations you bring to your viewing of it as clear as possible. Begin with a close reading of a print advertisement for the movie. (Please include a copy of this ad with your writing.) What information does the ad give you about the film? How does it try to persuade you to see it? How does it single out this movie from the others advertised on the same pages? If you're familiar with TV or radio ads for the movie, or with previews shown in movie theaters or on entertainment programs, then discuss those, too. Again, your focus should be on what you're being told to look for in the movie, the hints or promises you're being given concerning what it will be about or be like. If you've talked about the movie with friends, or read or listened to what critics thought about it, then note what they had to say. If you're familiar with any of the actors in or makers of the film, tell how their past work influences what you expect from them this time. Similarly, think about the *genre* of the film: Is it a sequel? a remake? a take-off? Does it bring to mind any other recent movies? Try to be specific about where your information about the movie comes from. Think of your subject in this writing as everything you know about the movie—*without having seen it yet.*

Part Two: Viewing

Go see the movie. Note in your journal what surprised you about it, along with what met your expectations (for either good or bad). If while viewing the movie you realize there were certain things you had known about it but had failed to mention in Part One, note them down now, too. Then, in writing this part of the assignment, try to show how your present reading grew out of your "pre-viewing" of it, how your experience of the film was connected to (and perhaps in some ways reshaped) the expectations you brought to it as a viewer. Think of your task not as writing a "review" of the movie but as documenting your own evolving response to it.

Part Three: Re-Viewing

Do some research for this part of the assignment. Locate two or three "outside" texts you think will help you better understand the movie you've just seen. These texts may be either written or visual, critical or popular. That is, you may look for them in the library, at a newsstand, on television, or at a movie theater or video outlet. In addition to your own writings, you may also find it useful to review the movie you are studying at this time.) In any case, though, you should be careful to document precisely where you found them. Among other things, they might take the form of reviews, interviews, biographies, historical or critical studies, information about how the movie was made, or other movies or writings by people who worked on it. In your writing, explain how these texts add to or change your understanding of the movie you have seen. If possible, hand in copies of your "outside" texts with your writing.

Part Four: Reflections on the Process of Moviegoing

Write an essay on how you formed and changed your reading of the movie you have been working on. The relation of this piece to your previous three writings is up to you. You might choose, for instance, to rework some of the writing you've done in them as part of this final essay, or you may want to quote from and analyze that earlier work, or you may simply decide to start anew. (In addition to your own writing, you may also find it useful to review the movie you are studying at this time.) In any case, though, the aim of your writing here is to offer your reader some insight into how you go about interpreting a film, to use your experiences with this particular text as a way of talking about the knowledge, strategies, resources, and attitudes you characteristically bring to your viewing of movies. Address your writing to readers who are not familiar with the work you've done on the earlier parts of this assignment and who will thus need to learn how you have gotten to whatever point you are now at in your thinking and writing about both the film and your processes as a moviegoer.

Naming and Interpreting a Genre

The interpretation and analysis of popular genres is a common mode of media criticism. The "slasher" film, the romance novel, the family sitcom, the women's beauty magazine, the heavy metal album, the female stand-up comic—each of these can be considered genres. In this project, we would like you to locate, name, describe, and interpret a popular media genre (or subgenre) of your choosing. You might re-read "Movies and History" by Eric Foner and John Sayles if your interest in genre might make you want to investigate a specific kind of historical representation in film. Or you might look at William Kowinski's "The Malling of America" or Benjamin DeMott's "Visions of Black-White Friendships" to see the ways in which these authors narrow their analyses of film to a subgenre. Finally, we'd like you to return to John Seabrook's "Why Is the Force Still With Us?" before you choose a genre and begin to write. Seabrook's project considers the difficulty in neatly categorizing a film series like the *Star Wars* trilogy.

In selecting a genre, you have many choices available to you. A good way to get a handle on these choices is to visit some locations where they are displayed. A visit to the video store will evidence the many genres and subgenres within popular film—gangster movies, horror movies, boy-meets-girl movies, invaders-from-another-planet movies, and so on. A visit to a magazine stand will likewise show the wide variety of magazine genres—beauty magazines, fashion magazines, biker magazines, exercise magazines, and the like. A visit to a chain bookstore will display various genres of books— self-help books, get-rich-quick books, celebrity biographies, spy novels, and so forth. And a visit to a music store will reveal the dominant genres of popular music—heavy metal music, rap music, folk music, country music, and so on. There's no "television store," but you do have two handy alternatives: flip through *TV Guide* or a similar publication to learn the range of shows that are routinely available, or browse through the channels yourself.

Looking at genres tends to bring to mind subgenres. This is a wise strategy, and we recommend it to you. Picking out a smaller category within a genre will make your work both easier and more suggestive by narrowing the number of works you need to discuss in depth. There are many gangster films, for instance, but not so many black gangster films. Focusing on a subgenre also makes it easier to be original; you may be the first to locate and give a name to your genre, and this in itself is useful. The relationship between the genre and the subgenre you have selected also offers you a way to launch your discussion. In short, finding and naming a smaller genre within a big genre offer many advantages. This is the approach we suggest.

In choosing a genre or subgenre try to obtain the texts you want to write about in depth. It will be necessary to examine and re-examine these texts, and working from notes is difficult when there may be many things you failed to note or cannot remember. While this is not an absolute requirement, we recommend strongly that you be able to hold in your hand several examples of the genre you want to describe. (In the case of TV programs, you should try to videotape the programs you want to analyze.)

A word of warning about numbers: Two examples do not make a type. Three examples is still very shaky. You should be able to name several works that belong to the genre you're writing about—at least four or five, but perhaps as many as ten or twenty. These names should appear in your essay, although you won't need to talk about all of them. You should be able to discuss in your essay at least four or five examples of the genre, but the more cases you present the more convincing your discussion will be. While mentioning examples that add nothing to the analysis is pointless, inventing a genre that does not really exist aside from the two or three examples you may have found is also not worthwhile.

We suggest that you divide your writing into three basic tasks. The first should offer a name for your genre and a basic description of its "formula." This means identifying the typical elements or recurrent themes in the various texts that you have selected as good examples of the genre. The purpose of describing the basic formula is to convince your reader that your genre does in fact exist, that enough typical features are present to make up a type—or, to put it another way, that enough examples of the type exist to make it a useful category of analysis. Often a critic will compare the genre in question to related but slightly different genres. Also common is the strategy of singling out a classic or "pure" example of the formula—a text that exhibits almost all the common features of the type, or exhibits them most intensely. This can be a good way of illustrating what the genre is all about.

A second task, often accomplished in tandem with the first, is to describe the satisfactions of the genre—to explain why the formula "works," to account for its popularity, to indicate the pleasures it provides. This means entering into the experiences of the audience in an effort to discover what they find meaningful, enjoyable, comforting, or useful in the texts of the genre. You may yourself feel enthusiastic about the genre, or you may have certain reservations about it. Either way, it is important to account for the genre's success with the audience—to explain what the audience "gets" from the formula, to understand how and why the genre "works."

Your third and most difficult task involves the interpretation and criticism of your genre. Here your goal is to explain, not what the genre is, or why it works, but what it means, why we should care about it, why it matters that this genre has become popular. What can we learn from looking at this genre? Discuss this in terms of what is dangerous or objectionable about its allures—or what is liberating, uplifting, reliably pleasurable about them. Finally, speculate on where the genre is heading or how it is evolving.

Select a genre or subgenre that enables you to accomplish these tasks. As always, choose material that is somehow interesting to you, that you feel strongly about, that you find curious or compelling—something that you want to understand better. Understanding what makes a genre popular is, of course, the heart of this project.

The Media Generation(s)

Media experiences are one way in which we have come to define generations. In this assignment, we would like you to reflect on that process. Your task is to enter into the media experience of a different generation by interviewing a member of that generation about his or her uses of the media. Then, write about what you've learned—including what you've learned about your own generation and its media experience.

A generation can be defined in two ways. The traditional way is to consider yourself and your siblings one generation, your parents the "older" generation, your grandparents two generations older, and so on. By this definition, twenty-five to thirty-five years separate one generation from another. But modern culture, especially media culture, has given new meaning to the idea of a generation, shrinking the number of years required before a "new" generation is thought to arrive on the scene. Today, you may consider older siblings as members of an "older" generation. As a college student, you might look at kids who are twelve or thirteen and regard them as the "next" generation. In this newer sense, a generation is defined culturally—by common experiences, common tastes, common outlooks and attitudes—rather than biologically, through the cycles of birth and giving birth. But even the biological definition takes on cultural dimensions. We understand our parents' generation by their different experiences, tastes, and attitudes. Much of this involves the media and the different uses that generations make of media culture. At the same time, the media inform us about what different generations are all about. We build up an image of, say, the "sixties generation" through media images, commentaries, and associations. We feel we can identify our parents and their friends through the music they enjoy, the movies they like, and so on.

In this assignment, use *either* the cultural or biological definition of a "generation." The assignment begins when you select a subject—someone who belongs, in your mind, to a different generation than your own. This could be one of your parents or grandparents, a relative, a family friend, an older brother or sister, or just an acquaintance from an older generation. Or it could be a *younger* sibling, a teenager in your neighborhood, a kid you know, or will get to know. (If you employ the cultural definition of a generation, explain why you think your subject belongs to a different generation from your own. Indeed, this explanation should become part of your writing.) Tell your subject that you have an assignment to write about their experiences with the media and that you would like to ask her some questions. You might wish to audiotape her answers, but even if you do, also take notes on your interview with them in your journal.

You can start by getting some general information about where and when that person was born, where she was raised, her schooling, her work experiences, raising a family, and so on. Then shift to your subject's experiences with media. Your goal is to find out as much as you can about her media diet—both in the past and the present. You might want to start by asking: "Did you read much as a kid?" Or, "Did you watch a lot of television?" Or, "Did you listen to the radio a lot when you were growing up?" A more open-ended question may also work: "Looking back on it now, which of the major media—radio, TV, newspapers, books, movies—have been most important in your life?" For younger subjects your questions will be more about the present than the past.

You will also want to ask about the *significance* of different media in your subject's life. Try to find out which media, and which media experiences— that is, particular shows, movies, songs, features, performers, moments, events—stand out for him as important or worth remembering, and why. An easy way to get at this is simply to ask him about his "favorites"—of all kinds. Try to get a feel for what your subject seems to "get" from various media texts or performers—that is, the value or lasting effect of the experience. A good question might be something like, "What do you think you got from seeing so many movies?" Or, "What do you get from heavy metal that you can't get from other things in your life?"

In thinking about what your subject has to say, and in prompting him or her to say more, use your own experiences as a point of comparison. That is, if an interviewer was going to find out about your own media habits and the importance of the media in your life, what would he or she have to ask about to uncover the facts? You may thus want to use yourself as a comparison in some of your questions. For example, "I know that I could never live without my CD player. Is there anything you feel that way about?"

Your goal in asking such questions is not simply to generate a list of items and memories, but to gain an understanding of the "why" and "how" of your subject's experience—why she has developed her media habits, how the media have shaped her life, how her experience differs from yours, and what separates her generation from your own. So don't confine your questions to your subject's childhood; his current experiences are just as relevant to your aim here—which is to understand how the media have entered into his life and helped define his experiences, including his feelings about himself as belonging to a particular generation.

Here you might ask your subject to think in generational terms. Questions like these might be helpful: Do you ever think of yourself as belonging to a generation? Which one? What do members of your generation have in common? What do you think distinguishes your generation from others? Which events are crucial to understanding your generation? Which media products or performers are important to know about and understand? Which single album (or book, or movie, or media figure) holds the key to understanding your generation?

But your interview is just the first step in your research. Next, you investigate some of the media texts that, according to your subject, are important in understanding his or her experience, his or her generation. Your goal should be to grasp what these media texts means to those who are older (or younger) than yourself. Inevitably, you will find yourself comparing your own reactions to those of a different generation. This leads to the third stage in your research: asking yourself some of the same questions you asked your subject, or uncovering some of your own generation's media experience by asking your peers or classmates about the generation to which you belong.

As part of your research, you might benefit from a reading (or re-reading) of Louis Menand's "Life in the Stone Age." Or you might consider John Seabrook's work in "Why Is the Force Still With Us?" As his title suggests, Seabrook tries to understand why the appeal of the *Star Wars* trilogy seems to transcend any kind of defining generational interest. And as a caveat, you might weigh Edward Rothstein's objections to defining, naming things too quickly in a culture ("Trend-Spotting: It's All the Rage.")

When you have completed your interviews and research, you will be ready to draft a piece that reflects on another generation's media experience through (1) the analysis of your subject's comments and reflections; (2) your examination of media texts associated with your subject or your subject's generation; and (3) your thoughts and commentary on your own experience and that of your peers. Your writing may take whatever form you find most effective. You may want, for instance, to reproduce transcripts from your interviews, or simply to summarize or paraphrase what was said in them. But whichever form it takes, your writing should tell us not only what you learned about your subject but also about yourself and about the workings of the media.

Toward a Theory of Watching TV

Alexander Nehamas believes in "serious watching" and David Marc in the importance of watching television "actively," in "one's own terms," not accepting the ways producers might be packaging it. From a different and unusual vantage point, Stuart Fischoff analyzes the dynamics or interactions of host, guests, expert guests, and the studio audience of TV talk shows. And from an equally striking, behind-the-scenes perspective, Debra Seagal exposes the "reality" of a reality-based cop series.

The question is what to do with these competing views of television, how to relate your own thoughts on the medium to what these critics have said about it. In this assignment, then, we'd like you to imagine yourself as entering into a critical conversation about TV that already includes the voices of Marc, Nehamas, Fischoff, Seagal, among others in *Media Journal* who watch television in critical ways. Which of these writers best helps you account for your own experiences with television? Or do you prefer to stake out a position of your own that differs from what all of them have to say? Is there anything that they have all missed about TV that you want to add?

To make your position convincing, you need to do at least three things. First, show that you understand the arguments that these critics are making—that you can restate what each says in a fair and accurate way, and that you can point out the differences between them. Second, show what you think is significant about your own experiences with TV and why. You may have noticed, for instance, that while Marc and Nehamas are all talking about television, each seems to have a very different sense of what's "on" it—from sitcoms to dramas like *St. Elsewhere* to Pepsi commercials—and that their choice of examples shapes what they have to say in powerful ways. So think carefully about how you want to represent your experiences with TV, about which programs you want to cite as central to your dealings with it. Third, and finally, use your own experiences with TV to talk back to these critics, to define your own critical voice.